technical mathematics

technical mathematics

third edition

Harold S. Rice

Professor Emeritus of Mathematics
Wentworth Institute

Raymond M. Knight

Professor of Mathematics
Hudson Valley Community College

McGraw-Hill Book Company

New York	Kuala Lumpur	Panama
St. Louis	London	Rio de Janeiro
San Francisco	Mexico	Singapore
Düsseldorf	Montreal	Sydney
Johannesburg	New Delhi	Toronto

Library of Congress Cataloging in Publication Data

Rice, Harold S.
    Technical mathematics.

    Bibliography: p.
    1. Mathematics—1961   I. Knight, Raymond M.,
joint author. II. Title.
QA39.2.R52   1973   510   72-1365
ISBN 0-07-052200-6

Technical Mathematics

2 3 4 5 6 7 8 9 0   HDMM   7 9 8 7 6 5 4 3

This book was set in Medallion by York Graphic
Services, Inc. The editors were Bob Flowers, Cary
F. Baker, Jr., and J. W. Maisel; the designer was
Marsha Cohen; and the production supervisor was
Robert R. Laffler. The drawings were done by John
Cordes, J & R Technical Services, Inc.
The printer was Halliday Lithograph Corporation;
the binder, The Maple Press Company.

contents

preface

In this edition, as in previous ones, the emphasis is on problem solving within the context of engineering technologies.

It is expected that the student using this text has studied at least a year of elementary algebra and a year of geometry, however, review materials are included for students in need of a review.

Depending on the preparation of the student and the objectives of the course, it should be adequate for either a one- or two-semester course offered to beginning students in the engineering technologies.

Every effort has been made to present to the student problems which are meaningful within this context at a realistic level of sophistication in view of his preparation.

A consistent effort is made to avoid arbitrary statements which demand sheer memorization when an appeal to the student's reason and reference to familiar experiences would result in improved retention.

Underlying the entire treatment is an effort to develop student competence in problem solving. This is not limited to the routine "plug-in" aspects of problem solving. We have tried to assist him in understanding the basis of the mathematical models he uses, their relevance, and their limitations.

Not only have we tried to help him learn how to use mathematics, but of at least equal importance, we have tried to help him avoid misusing it.

Also, where practical, the language of the explanations has been altered to conform more closely to current usage in secondary education.

We have intentionally used simple numbers in the illustrative examples in order that the computational "dog work" shall not obscure the principle being explained.

The areas of duplication between this text and the authors' "Technical Mathematics with Calculus" have been reduced by omitting chapters on simultaneous quadratic equations; the binomial theorem; exponential functions; trigonometric formulas, identities, and equations; and complex numbers and position vectors. However, we have expanded the treatment of arithmetic, geometry, and graphical solutions.

Several changes in the order of topics have been made to improve the teachability of the text.

Exponents and Radicals (Chapter 4) is presented immediately after basic algebra. The chapter on Ratio, Variation, and Proportion has been presented earlier to follow directly after Linear Equations in One Variable. Geometry now appears as Chapter 7. In this order use can be made of the basic algebra; linear equations; and ratio, variation, and proportion. There is a wide range in the difficulty of the geometry problems. The demarcation between the basic and the more stimulating nonroutine problems has been clearly indicated. Some "word problems" in linear equations are now to be solved using two variables instead of one. This is in response to a student survey. An exercise in linear empirical equations has been added. There is an increased emphasis on graphical representation and calculation. However, Chapter 10 on Graphical Methods may be omitted without any loss of continuity. Much of the material normally presented in exponential functions is now treated in Chapter 11, Common Logarithms, using the compound interest law.

As a reflection of the "age of the astronaut," we have increased the number of problems related to earth science and space technology.

Thanks are due to the authors' colleagues for their continued interest and constructive criticism. Among these are Mrs. Ruby Painton, Dudley Smith, and Norman Swanson, all of Hudson Valley Community College in Troy, N.Y.

Miss Ella Hackett, also of Hudson Valley Community College, deserves our appreciation for her efforts in typing parts of the manuscript.

In particular, the authors wish to express their appreciation to Walter Granter of the Vermont Technical College, J. E. Lowry of Northeastern Oklahoma A & M College, Richard Wheeler of Wentworth Institute, David Hunter and J. B. Jones of Central Piedmont Community College, and Adelbert Goertz of the Pennsylvania State University for their valued suggestions.

Harold S. Rice

Raymond M. Knight

technical mathematics

introduction to arithmetic

The student is expected to acquire a reasonable skill in the use of elementary arithmetic. The daily, routine calculations should be just that—not a major undertaking. Although a desk calculator may be available, it will always be helpful to have the standby ability to make one's own calculations by hand. If the work can be done mentally without recourse to pencil and paper, all the better. A number of suggestions are offered in Appendix B to make the work less arduous without loss of accuracy.

Above all, there is no substitute for common sense in recognizing a reasonable answer.

The problems in the following exercise are intended to be representative of situations encountered in everyday living. It is assumed that the student can recall the basics of fractions, decimals, etc. Therefore no drill on these topics is included.

EXERCISE 1

The following problems depend in part on the material discussed in Appendix B, Computational Aids.

Apply the check of 9s to Probs. 1 to 6 (Sec. B.4).

1. Add: 43078
      6525
     30022
      1955
     20697

2. Add: 46824
       83021
        7155
      106052
        3846

3. Multiply: 82,456 × 3,641
5. Divide: 14,725 by 83

4. Multiply: 53,252 × 2,571
6. Divide: 10,184 by 47

In Probs. 7 to 17, compute to the appropriate number of significant digits by abbreviated multiplication or division (Secs. B.5 and B.7).

7. 460.52 × 2.8334
9. 0.36755 × 43.725843
11. 17,254 × 0.0038526
13. 75.431 ÷ 0.46843122
15. 7,263.8 ÷ 10.521
17. 65.81589 ÷ 0.75022

8. 17.603 × 8.482
10. 65.81589 × 0.75022
12. 562.47 ÷ 17.438
14. 0.05264 ÷ 7.51817
16. 1,020.4 ÷ 30.612

Find the squares of the following (Sec. B.12):

18. $8\frac{1}{2}$
21. 115
24. 51
27. 997

19. $12\frac{1}{2}$
22. 165
25. 103
28. 5.3

20. $6\frac{1}{4}$
23. 98
26. 196
29. 11.2

Find the products of the following (Sec. B.3):

30. 126 × 98
33. 52.4 × 9.99

31. 53 × 996
34. 823 × 998

32. 112 × 99.5

Find to five significant digits the square roots of the following, using the iterative process (Sec. B.17):

35. $\sqrt{630}$
38. $\sqrt{5250}$
41. $\sqrt{48}$

36. $\sqrt{85}$
39. $\sqrt{21.8}$
42. $\sqrt{120}$

37. $\sqrt{0.75}$
40. $\sqrt{750}$
43. $\sqrt{1248}$

Making use of the principles discussed in Appendix B, calculate the following at sight as far as possible:

44. 16 × 35

45. 24 × 25

46. 36×175 47. $\frac{1}{2} \times 6\frac{5}{8}$

48. $\frac{1}{4} \times 5\frac{1}{2}$ 49. $8\frac{1}{4} \times 6\frac{5}{16}$

50. $3\frac{5}{8} \times 4\frac{1}{2}$ 51. $\dfrac{1\frac{7}{8}}{2\frac{1}{2}}$

52. $\dfrac{2\frac{1}{4}}{7\frac{1}{2}}$ 53. $\dfrac{1\frac{3}{4}}{6\frac{1}{2}}$

54. $\dfrac{3\frac{1}{3}}{6\frac{1}{4}}$ 55. $8\frac{3}{4} \times 56$

56. $8\frac{1}{3} \times 48$ 57. $27\frac{1}{2} \times 36$

58. $63\frac{3}{4} \times 52\frac{1}{2}$

EXERCISE 2

1. A piece of property is valued at $28,500. What is the real estate tax at $75.20 per $1,000 evaluation?

2. A coil of wire weighs 55 lb. It is cut into sections, each 1 ft $7\frac{1}{2}$ in long and weighing $9\frac{1}{2}$ oz. How many sections can be obtained?

3. What decimal part of a mile is 25 ft 8 in? Find the answer to three significant digits.

4. An artificial earth satellite has a period of 1 h 47 min. If the length of its orbit is 30,650 mi, find its speed in miles per hour.

5. An alloy is $\frac{3}{8}$ copper, $\frac{5}{12}$ tin, and the balance lead. How much lead is there in 282 lb of alloy?

6. The center-to-center distance between two holes is $3\frac{9}{16}$ in. The diameters of the holes are $7\frac{1}{4}$ and $1\frac{5}{8}$ in, respectively. Find the width of the "neck" of metal between the edges of the holes.

7. An auto trip of 763.4 mi consumed 39.6 gal of gasoline. How many miles per gallon were obtained? Find the cost (to the nearest cent) of the gasoline used at 31.5 cents/gal.

8. Which is the better value, 1 lb 3 oz of canned fruit for 47 cents or 13 oz for 35 cents?

9. Find the cost of 1 lb 5 oz of hamburger at $1.23/lb.

10. How many sections $2\frac{5}{16}$ in long can be cut from a steel rod $56\frac{1}{2}$ in long? Allow $\frac{1}{16}$ in for each cut.

11. A runner can run 100 yd in $14\frac{3}{5}$ s. Find his speed in miles per hour.

12. The net earnings of a company for a given year were $17,403,615. There were 3,721,047 shares of stock outstanding. Find the net earning per share to the nearest cent.

13. What decimal part of a foot is $\frac{5}{32}$ in?

14. What decimal part of a pound is $4\frac{1}{2}$ oz?

15. A pile of paper is $3\frac{5}{8}$ in high. Each sheet is 0.0078 in thick. How many sheets are in the pile? Disregard air spaces.

16. Using a basically more accurate method, the pile of sheets in Prob. 15 was found

to weigh 5,200 g, while a single sheet weighed 12.7 g. What is the revised estimate of the number of sheets?

17. Eight centers are to be equally spaced on a straight line. The two end centers are $5\frac{3}{8}$ in apart. Find the distance between two adjacent centers to the nearest $\frac{1}{64}$ in.

18. A spray formula calls for 1 pt concentrate to 100 gal water. How many teaspoons would be added to 2 gal water (6 tsp = 1 fluid oz)?

19. Over a certain period the tax rate of a city increased from $46.20 to $73.50. Find the percentage increase to the nearest 1 percent. As a rule, unless otherwise indicated, the first number in chronological order is taken as the base. In this case the base is $46.20.

20. A repossessed TV is sold for $182, which represents a 35 percent loss on the cost. Find the cost.

21. A federal tax of $22\frac{1}{2}$ percent applies to a man's $6,000 salary. A surtax of 8 percent on the original tax is imposed.
    (a) Find the total tax on the $6,000.
    (b) What single tax (to the nearest 0.1 percent) would be equivalent to the schedule in part a?

22. In 1 year, $175.00 grows to $182.70 by addition of interest. Find the percent interest.

23. A man's take-home pay is $110.70 after deducting 18 percent withholding tax. What is his pay before the tax deduction?

24. A formula calls for 15 lb of pure salt. How many pounds of 60 percent salt solution should be used?

25. A dealer pays $4.95 wholesale for a transistor radio. For what selling price should he list the radio in order to make a 40 percent profit on the cost after giving a 10 percent discount on the list price?

26. Eleven shelves each 2 ft 10 in long are to be cut. Boards are available in 10-, 12-, 14-, and 16-ft lengths. How many boards and of what lengths should be chosen for minimum waste?

27. If spaceship Apollo 8 were to maintain an average speed of 5,000 mi/h, how long would it take to reach the planet Pluto if we take its distance to be 4,000,000,000 miles?

28. According to a recent U.S. Census Bureau survey, there is in the United States one birth every $7\frac{1}{2}$ s, one death every 17 s, one immigrant every $1\frac{1}{2}$ min, and one emigrant every 23 min. It then follows that there is a net gain of one person every t seconds. Find t.

29. Straight-line depreciation calls for a constant number of dollars annual loss of value. A machine is valued at $3,800 new, and at the end of 5 years it is worth $1,640. Find its value at the end of each year.

30. Compare the amounts obtained in Prob. 29 with those resulting from the *sum-of-years-digits* method. Here we add $1 + 2 + 3 + 4 + 5 = 15$. Then $\frac{5}{15}$ of the total depreciation will be written off during the first year, $\frac{4}{15}$ the second year, etc., $\frac{1}{15}$ being assigned to the fifth year.

31. A duplicating machine costs $195 new and has a turn-in value of $55 at the end of the sixth year. (*a*) Using the straight-line method, fill in the spaces in the table below. (*b*) Repeat, using the sum-of-years-digits method.

	Book value, straight line	Book value, sum-of-years digits
New	$195	$195
End of year 1	——	——
End of year 2	——	——
End of year 3	——	——
End of year 4	——	——
End of year 5	——	——
End of year 6	$55	$55

32. We often hear of very small concentrations in terms of parts per million (ppm). Verify the following approximations to 1 ppm.

> 1 oz of accelerator in $31\frac{1}{4}$ tons of concrete
> 1-g needle in 1 ton of hay
> 1 fluid oz of dye in 7,530 gal of water
> 1 min in 1.9 years
> $\frac{1}{16}$-in thickness in a pile 1 mi high

33. A man earns $171.20 for a 40-h week. A pair of shoes costs $27.50. How many hours of work would buy a pair of shoes? (These and similar figures are often a more accurate measure of "prosperity" than a wage scale alone or a cost-of-living index alone.)

34. On a scale drawing $\frac{3}{16}$ in represents 5 ft. What distance is represented by $5\frac{11}{32}$ in on the drawing?

35. A man works a 38-h week at $3.25/h with time-and-a-half for overtime. How much will he be paid for a 44-h work week?

EXERCISE 3

This exercise, which is to be done by inspection, will test your number sense. Select at sight the most nearly correct answer. Do not work out. It is suggested that the instructor announce a time limit for the exercise.

1. $18.96 + 1.37 + 2.98 + 4.87 = (25, 28, 32, 23)$
2. $12 \div \frac{1}{4} = (3, 16, 48, 11\frac{3}{4})$
3. $1\frac{1}{2}$ percent of $160 = (24, 8, 2.4, 240)$
4. $1,986 \div 51.214 = (40, 30, 500, 4)$
5. $\frac{41}{13} \times 29 = (60, 75, 6, 90)$
6. $33\frac{1}{3}$ percent of $58\frac{1}{2} = (15, 1,500, 20, 2,000)$

7. $\dfrac{x}{24} = \dfrac{162.3}{15.719}$ x = (25, 240, 2, 52)

8. $(2.59)^2 = (4, 10, 6, 5)$

9. $35.78 \div 0.05923 = (70, 6, \frac{1}{7}, 600, 7)$

10. $14.92 \times 0.003964 = (0.04, 0.06, 0.56, 5.0)$

11. $\frac{4}{15} - \frac{1}{6} = (\frac{1}{3}, \frac{1}{10}, \frac{8}{15}, \frac{7}{30})$

the slide rule and numerical tables

From time to time we shall review some of the methods whereby much of the drudgery of computation may be eliminated and the effectiveness of work increased. These methods include the use of the slide rule, tables, approximations, and graph paper, to name a few. In this chapter we shall obtain practice in the use of the slide rule and tables.

In this chapter and in Chap. 7 any commonly used handbook of tables may be referred to. For example, see the later editions of Burington's "Handbook of Mathematical Tables and Formulas," McGraw-Hill Book Company; Hudson's "Engineer's Manual," John Wiley & Sons, Inc.; "Standard Mathematical Tables," Chemical Rubber Publishing Co.

2.1 Scope of the Slide Rule

It is essential to have a clear idea of what can and what cannot be done on a slide rule. The slide rule is an instrument used for processes of multiplication, division, proportion, and calculation of simple powers and roots. Further discussion of the slide rule will be found in Secs. 11.31, 11.32 and 13.21 to 13.23; this chapter will deal only with the processes named above. Addition and subtraction cannot be performed on the slide rule. One of the most frequently asked questions is, "Is slide-rule computation

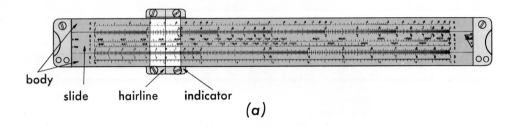

body   slide   hairline   indicator

(a)

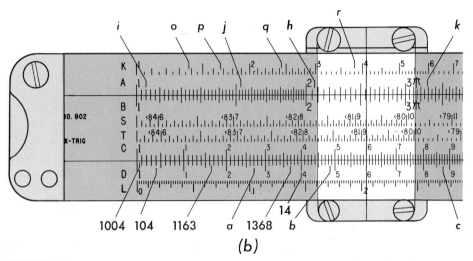

(b)

FIGURE 2.1

good enough?" The answer depends upon the nature of the problem. Many experimental data are accurate to no more than three significant figures. In such cases slide-rule computation is sufficiently accurate. Even in problems requiring more precise methods, the slide rule is accurate enough to detect gross errors and to estimate the order of magnitude of the answer.

The slide rule will not think for you. Like an automobile, it will take you to your destination if you provide the proper direction.

There are many types of slide rules, but our discussion will be limited to the general technique of operation common to most basic 10-in rules. (More versatile slide rules feature inverted and folded scales as well as log-log scales. Instruction in their use is given in the manual provided with the rule. Almost as many operations can be performed on the basic rule as on the rule having additional scales. The chief advantage of the latter lies in its greater ease of operation.) Additional practice in using the slide rule may be gained by solving many of the problems in Chap. 7.

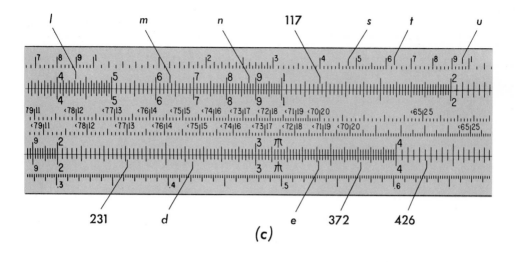

(c)

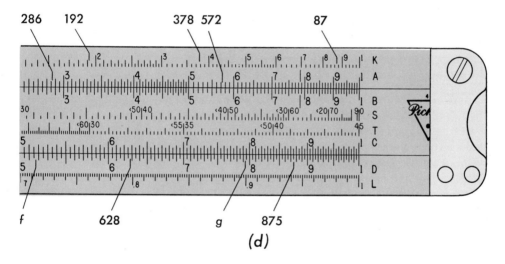

(d)

2.2 Description of the Slide Rule

The slide rule consists of three parts (Fig. 2.1): the *slide,* or central sliding part; the *body,* or the upper and lower bars between which the slide operates; and the *indicator,* which is the movable glass plate marked with a hairline.

The mark associated with the numeral 1 at an end of a scale is called the *index* of the scale. Two positions on two different scales are said to be *opposite* if the hairline can be made to cover both simultaneously.

To perform the operations mentioned in Sec. 2.1, we need use only the C, D, A, B, and K scales. The C and D scales are identical. The A and B scales are identical with each other, and the distance between two successive integers is one-half the distance between the same two integers on the C and D scales. The K scale is compressed even further, so that the distance between any two integers is one-third the distance between the same two integers on scales C and D.

The C and D scales are divided into nine principal divisions by *primary marks* bearing the large numbers 1, 2, 3, . . . , 8, 9, 1. The space between any two primary marks is divided into ten parts by *secondary marks*. These are not numbered, except between the primary marks 1 and 2, where they bear the small numerals 1 to 9. The space between two successive secondary marks is divided into two, five, or ten spaces by unnumbered *tertiary marks*.

The A and B scales each have two identical portions. Each portion is divided into nine principal divisions by numbered primary marks. Unnumbered secondary marks further divide each principal division, and still finer division is supplied by tertiary marks between the primary marks numbered 1 and 5.

The K scale is divided into three identical portions. They are divided much as the A and B scales are, but more coarsely.

2.3 Location of Numbers on the Scales; Accuracy of the Slide Rule

It must be remembered that the decimal point plays no part in locating a number on the C and D scales. The first significant digit is located by reference to the primary marks, the second by reference to the secondary marks, and the third by reference to the tertiary marks or to some point between tertiary marks, according to the portion of the scale being used. By estimating fractions of a graduation, a fourth digit may be located for numbers occurring between primary marks 1 and 2 on the C and D scales. This means that the maximum accuracy ordinarily available with a 10-in slide rule is 1 part in 1,000. Location of actual numbers will be best illustrated by referring to the examples in Fig. 2.1*b* to *d*.

EXERCISE 1

Read as closely as possible the points indicated in Fig. 2.1*b* to *d*.

D scale: *a b c d e f g*
A scale: *h i j k l m n*
K scale: *o p q r s t u*

2.4 Multiplication

Either the C and D scales or the A and B scales may be used. Ordinarily the C and D scales are preferable, since their larger-scale divisions make for greater accuracy. Their use is illustrated in Fig. 2.2 and in the following example.

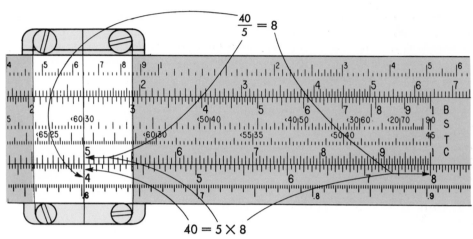

FIGURE 2.2

Example 1. Multiply 8 by 5.

Set the right index of C opposite 8 on the D scale. Move the indicator so the hairline covers 5 on the C scale. Directly below this 5 will be found the "slide-rule product" 4 on the D scale. The student must locate the decimal point himself, and common sense indicates that the answer is 40, not 4.

It will be appreciated that the same setting would be used for multiplying 80 by 50; 8,000 by 0.5; 0.08 by 500; etc.

If we were to try to use the above setting of the right index to multiply 8 by 12, a reading would be impossible, because the 12 on the C scale is located beyond the end of the D scale. In this case we would set the *left* index of the C scale over 8 on the D scale and locate on the D scale our answer 96, immediately under 12 on the C scale. It will be useful to remember that when the product of the first significant figures in multiplicand and multiplier is less than 10, the student should use the left index; when it is greater than 10, he should use the right index.

When multiplying more than two numbers, use the hairline to mark the position of the product of the first two. Without reading this product, use it as the multiplicand for the next multiplication.

If the magnitude of the product is not immediately obvious, a rough mental check will suffice.

Example 2. The slide-rule product of $1{,}440 \times 37.5 \times 0.08125$ is 439. It is evident that our product is approximately $1{,}500 \times 40 \times 0.08$, or 4,800. Hence the answer must be 4,390.

To multiply two numbers:

1. Set the proper index of C opposite either factor on the D scale.
2. Place the hairline over the other factor on the C scale, and read the significant digits of the product under the hairline on the D scale.
3. Determine the position of the decimal point by a rough mental estimate.
4. In the above steps, C and D may be replaced by B and A, respectively.

EXERCISE 2

Perform the following multiplications, retaining in the product as many significant figures as you think are justified:

1. 4.00 × 17.00	2. 7.50 × 1.20
3. 3.30 × 9.0	4. 64.0 × 0.375
5. 288 × 382	6. 321 × 1,069
7. 728 × 1,218	8. 617 × 1,645
9. 1,006 × 902	10. 1,258 × 1,562
11. 862 × 482	12. 66.2 × 10.3
13. 1.475 × 1,520	14. 0.981 × 0.693
15. 329 × 0.00352	16. 0.342 × 1.306
17. 8.14 × 0.0309	18. 2.46 × 330,000 × 3.14
19. 3.1 × 920 × 0.486 × 1,520	20. 0.1038 × 0.0063 × 28 × 9.82
21. 512 × 62.5 × 0.0027 × 87	22. 0.1047 × 0.00774 × 0.349 × 0.0562

2.5 Division

Since division is the inverse of multiplication, Fig. 2.2 may be used to illustrate division as well as multiplication. In this example we have the setting for $40 \div 5 = 8$.

To divide one number by another:

1. Bring the divisor on the C scale opposite the dividend on the D scale by means of the hairline.
2. Opposite the index of the C scale, read the significant figures of the quotient on the D scale. If desired, the indicator may be used to aid in this reading.
3. Determine the position of the decimal point by a rough mental estimate.
4. C and D may be replaced by B and A, respectively, as in multiplication.

EXERCISE 3

Perform the following divisions, retaining in the quotient as many significant figures as you think are justified:

1. 18.00 ÷ 50.0   2. 25.0 ÷ 3.00   3. 750 ÷ 5.50
4. 12.8 ÷ 72   5. 69.8 ÷ 4.78   6. 197.2 ÷ 858
7. 0.924 ÷ 21.0   8. 17.5 ÷ 1,646   9. 1 ÷ 37.5
10. 0.0752 ÷ 0.000718   11. 0.1804 ÷ 363   12. 1 ÷ 2.73
13. 0.1875 ÷ 0.078125   14. 0.005632 ÷ 18.432

2.6 Location of Decimal Point in Scientific Notation

Use of scientific notation makes work with very large or very small numbers easier and reduces the likelihood of error. (See Appendix A.)

Example 3. Multiply 538,000 by 0.00377.

In scientific notation, we have, roughly, $(5 \times 10^5) \times (4 \times 10^{-3})$. This product may be written $5 \times 4 \times 10^2 = 2 \times 10^3$. Since the product of 5.38 and 3.77 on the slide rule is 203 (before locating the decimal point), our answer must be 2.03×10^3, or 2,030.

Example 4. Divide 538,000 by 0.00377.

In scientific notation, we have, roughly, $(5 \times 10^5)/(4 \times 10^{-3}) = 1.2 \times 10^8$. Since the slide-rule reading is 1,427, our answer must be 1.427×10^8, or 142,700,000.

2.7 Combined Multiplication and Division

The easiest method of computing the quotient of two products is to alternate between division and multiplication. If we do all the multiplying first, and then all the dividing, more moves are required, with greater chance of error.

Example 5. Evaluate

$$\frac{825 \times 184}{227 \times 316}$$

Divide 825 by 227 in the usual way. The quotient will be found on the D scale under the C index. However, we do not read the value, since it serves merely as the multiplicand for 184. Without moving the slide, we move the indicator so that the hairline covers 184 on the C scale. The product will be found on the D scale under the hairline. Since this product is to be divided by 316, we do not read the value, nor do we move the hairline, but we move the slide so that 316 is under the hairline. Directly beneath the C index read approximately 2115. The answer is 2.115, with the last figure in doubt.

Sometimes the slide must be moved so that one C index is moved to the spot formerly occupied by the other C index (as marked with the hairline). This situation might have been avoided, with some sacrifice of accuracy, by using the A and B scales.

EXERCISE 4

Perform the following computations, expressing the answer in scientific notation. Retain in the answer as many significant figures as conditions justify.

1. $2.4 \times 6.5 \times 10.37$

2. $1{,}476 \times 37.8 \times 54.0$

3. $0.00842 \times 0.295 \times 6.1875$

4. $67.1 \times 0.000418 \times 3.0$

5. $32.00 \times 5.000 \times 1.900 \times 0.4000$

6. $\dfrac{1}{0.00532 \times 0.0612}$

7. $\dfrac{1.28 \times 3.56}{74.4}$

8. $\dfrac{15.8 \times 1.35}{0.031}$

9. $\dfrac{21.3 \times 0.054}{97.4 \times 3.80}$

10. $\dfrac{1{,}927}{412 \times 0.00592 \times 483}$

11. $\dfrac{24.6 \times 0.359}{296 \times 4.61 \times 98.7}$

12. $\dfrac{560{,}000 \times 0.0045 \times 12{,}500}{1{,}050{,}000 \times 0.072}$

2.8 Proportion

Proportions may usually be solved with only one setting of the slide. Observe in Fig. 2.3 that when 8 on the C scale is opposite 64 on the D scale, we find 5 opposite 4. This setting could therefore illustrate the solution of the proportion $5/x = 8/6.4$, in which we set 8 opposite 64 and, opposite 5, read $x = 4$. Note that all other pairs of numbers opposite each other have the same ratio (for example, $^{60}/_{48}$, $^{70}/_{56}$, $^{75}/_{60}$, $^{90}/_{72}$).

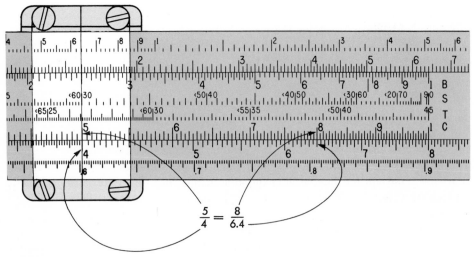

$$\frac{5}{4} = \frac{8}{6.4}$$

FIGURE 2.3

To solve a proportion, locate the numbers on the C and D scales in the same relative position as in the proportion $a/b = c/d$ (or $c/d = a/b$ if the setting for a falls to the right of the setting for c). When the C and D scales cannot accommodate the simultaneous settings of a over b and c over d, the A and B scales may be used.

Of course we can solve for x, obtaining $x = (5)(6.4)/8$ and, using the principle of Sec. 2.7, again obtain $x = 4$.

EXERCISE 5

In the following proportions, calculate x to three significant figures:

1. $\dfrac{x}{8.5} = \dfrac{32}{28.9}$ 2. $\dfrac{x}{21.5} = \dfrac{89}{79}$

3. $\dfrac{372}{x} = \dfrac{637}{9.31}$ 4. $\dfrac{8.2}{377} = \dfrac{0.323}{x}$

5. $\dfrac{18.3}{63.6} = \dfrac{x}{29}$ 6. $\dfrac{267}{8.75} = \dfrac{x}{192}$

7. $\dfrac{0.716}{x} = \dfrac{10.1}{168}$ 8. $\dfrac{795}{0.109} = \dfrac{42.3}{x}$

Note that in Prob. 5, for example, we may estimate the value of x by comparing either vertically or horizontally. Since 18.3 is about $(0.3)(63.6)$, x must be about $(0.3)(29)$, or 9. Again, since 29 is about $(\frac{1}{2})(63.6)$, x must be about $(\frac{1}{2})(18.3)$, or 9.

2.9 Squares and Square Roots

To find the square of a number, set the hairline of the indicator over that number on the D scale, and under the hairline read the square of that number on the A scale. We can also read from the C to the B scale in the same way. Figure 2.4 shows that $4^2 = 16$ (D to A scale), that $5^2 = 25$ (C to B scale), and that $8^2 = 64$ (right C index reading to right B index setting).

Figure 2.5 illustrates that $(1.428)^2 = 2.04$ and that $(2.53)^2 = 6.4$.

Example 6. Find 6^2.
Set the hairline over 6 on the D scale. Read 36 under the hairline on the A scale.

Example 7. Find $(8.62)^2$.
Set the hairline over 8.62 on the D scale. Read 74.3 under the hairline on the A scale.

Example 8. Find $(71,700)^2$.
On the A scale read 514 directly above the 717 on the D scale. Since $(71,700)^2 = (7.17 \times 10^4)^2$, the answer must be $51.4 \times 10^{4\times2} = 51.4 \times 10^8 = 5,140,000,000$.

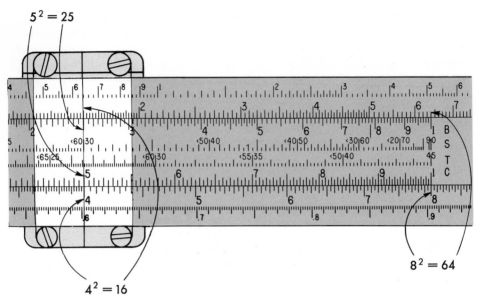

FIGURE 2.4

Example 9. Find $(0.00386)^2$.

On the A scale read 149 directly above the 386 on the D scale. Since $(0.00386)^2 = (3.86 \times 10^{-3})^2$, the answer must be $14.9 \times 10^{-3 \times 2} = 14.9 \times 10^{-6} = 0.0000149$.

The process of obtaining square roots is the reverse of that used in calculating squares. Therefore Figs. 2.4 and 2.5 may be used to show that $\sqrt{16} = 4$, $\sqrt{25} = 5$,

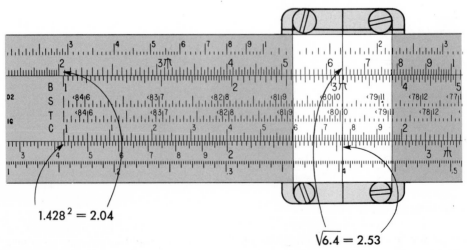

FIGURE 2.5

$\sqrt{64} = 8$, $\sqrt{2.04} = 1,428$, and $\sqrt{6.4} = 2.53$. We may indeed use Fig. 2.4 to read $\sqrt{64} = 8$; but if we had not had the operation of squaring 8 to guide us, we might have made an unfortunate choice in working from the 64 on the left half of the A scale; i.e., we might have read 253 directly below on the D scale. A quick check will show that 2.53 is the square root of 6.4. (This procedure is opposite to that of squaring 2.53, an operation originally illustrated in Fig. 2.5.)

It is evident that in reversing the procedure for squaring we must first determine which half of the A scale to choose as a starting point. There are a number of ways of making this choice. Perhaps the simplest is that used in the examples in Sec. 2.13. The procedure is as follows: Write down the number whose square root is to be found. Indicate the grouping of the digits as for longhand extraction of square root. Write in their proper positions the decimal point and the first significant digit in the square root. Select that half of the A scale which lies above the first significant figure (just determined) on the D scale.

Example 10. Find $\sqrt{6,870,000}$.

Indicating the grouping, the decimal point, and the first significant figure, we write

$$\overset{2}{\underline{\hspace{2.2cm}}}$$
$$\sqrt{6\ 87\ 00\ 00.}$$

It is apparent that we read from 687 on the *left* half of the A scale, since we find 2 on the D scale under that half. Accordingly, we read from 687 on the left half of the A scale directly below to 262 on the D scale. This indicates that the square root of 6,870,000 will be written

$$\overset{2\ \ 6\ \ 2\ \ 0.}{\underline{\hspace{2.2cm}}}$$
$$\sqrt{6\ 87\ 00\ 00.}$$

Example 11. Find $\sqrt{0.0000687}$.

A rough indication of the answer is found by proceeding as before and writing

$$\overset{0.\ \ 0\ \ 0\ \ 8}{\underline{\hspace{2.2cm}}}$$
$$\sqrt{0.\ 00\ 00\ 68\ 7}$$

In this case we evidently read down from the 687 on the *right* half of the A scale. We read 829 on the D scale, and the answer is clearly 0.00829.

2.10 Cubes and Cube Roots

To find the cube of a number, set the hairline over that number on the D scale and read its cube on the K scale under the hairline.

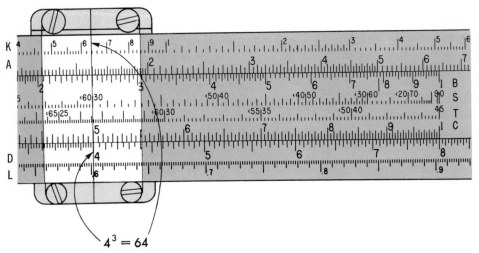

FIGURE 2.6

Example 12. Find $(6.35)^3$.

Set the hairline over 6.35 on the D scale. Read 256 under the hairline on the K scale.

Example 13. Find $(0.0439)^3$.

Read 846 on the K scale directly opposite 439 on the D scale. Since $(0.0439)^3 = (4.39 \times 10^{-2})^3$, the answer must be $84.6 \times 10^{-2 \times 3} = 84.6 \times 10^{-6} = 0.0000846$.

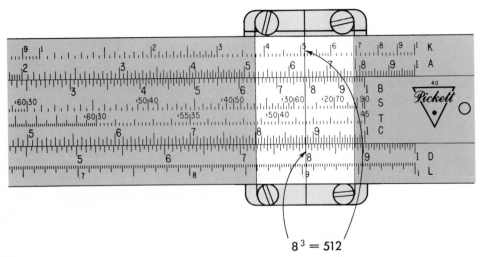

FIGURE 2.7

To find a cube root, we reverse the procedure for cubing and work from the K scale to the D scale. When working from the A scale to the D scale, we had to choose between two sections. On the K scale we must choose the proper section out of three. The process is exactly comparable with that used in finding a square root. Figures 2.6 to 2.8 show the settings for 4^3, 8^3, and 16^3, and therefore the settings for $\sqrt[3]{64}$, $\sqrt[3]{512}$, and $\sqrt[3]{4,100}$.

Example 14. Find $\sqrt[3]{0.0000048}$.

Indicating the grouping, the decimal point, and the first significant figure, we write

$$\begin{array}{c} 0.\ 0 \qquad 1 \\ \hline \sqrt[3]{0.000\ 004\ 8} \end{array}$$

Evidently we work from the *left* third of the K scale, since it lies opposite the 1+ on the D scale. Therefore we read from 48 on the left third of the K scale directly to 1,687 on the D scale. This indicates that the cube root of 0.0000048 will be written

$$\begin{array}{c} 0.\ 0 \qquad 1 \ \ 6\ 8\ 7 \\ \hline \sqrt[3]{0.000\ 004\ 8} \end{array}$$

Example 15. Find $\sqrt[3]{58,500,000}$.

Proceeding as before, we write

$$\begin{array}{c} 3 \\ \hline \sqrt[3]{58\ 500\ 000.} \end{array}$$

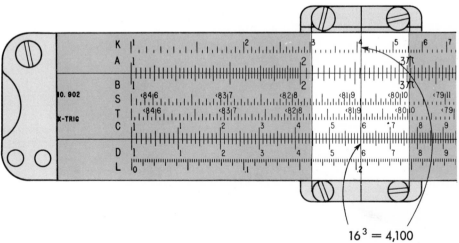

$16^3 = 4,100$

FIGURE 2.8

and find that the *middle* third of the K scale lies opposite the 3 on the D scale. Reading, therefore, from 585 in the middle section of the K scale, we find 388 on the D scale directly opposite. Hence the answer must be 388.

EXERCISE 6

Evaluate the following as accurately as the slide rule will allow:

1. $(18.00)^2$

2. $(2,730)^2$

3. $(6.05)^2$

4. $(34.5)^2$

5. $(167.8)^2$

6. $(0.854)^2$

7. $(0.1054)^2$

8. $(0.00782)^2$

9. $(517.23)^2$

10. $(0.030448)^2$

11. $(63 \times 426)^2$

12. $(9.1 \times 0.0119)^2$

13. $(0.027 \times 1.72 \times 7.95)^2$

14. $(5.1 \times 0.438 \times 14.12)^2$

15. $\left(\dfrac{1,647}{658}\right)^2$

16. $\left(\dfrac{69.8}{858}\right)^2$

17. $\left(\dfrac{852}{658}\right)^2$

18. $\left(\dfrac{61.4 \times 0.673}{2.16}\right)^2$

19. $\left(\dfrac{582 \times 1.104}{37.2 \times 8}\right)^2$

20. $\sqrt{49.2}$

21. $\sqrt{0.00702}$

22. $\sqrt{2,980}$

23. $\sqrt{0.837}$

24. $\sqrt{17,840}$

25. $\sqrt{0.0542}$

26. $\sqrt{0.000347}$

27. $\sqrt{45,897}$

28. $\sqrt{103.6}$

29. $\sqrt{7,080,000}$

30. $\sqrt{\dfrac{1.316}{0.016}}$

31. $\sqrt{\dfrac{1,127 \times 5.47}{21 \times 0.0025}}$

32. $\sqrt{\dfrac{11.26}{0.787 \times 24.8}}$

33. $(11.9)^3$

34. $(1.33)^3$

35. $(0.157)^3$

36. $(118.5)^3$

37. $(23.19)^3$

38. $(0.0342)^3$

39. $(478)^3$

40. $(6.375)^3$

41. $(783)^3$

42. $(0.0876)^3$

43. $(15.45 \times 0.132)^3$

44. $(8.75 \times 0.037)^3$

45. $\left(\dfrac{10.12}{17.58}\right)^3$

46. $\left(\dfrac{48.8}{4.21}\right)^3$

47. $\sqrt[3]{13.5}$

48. $\sqrt[3]{0.0474}$

49. $\sqrt[3]{76,215}$

50. $\sqrt[3]{2.85}$

51. $\sqrt[3]{9,384,260}$

52. $\sqrt[3]{0.00625}$

53. $\sqrt[3]{652 \times 0.725}$

54. $\sqrt[3]{1.447 \times 0.0298}$

55. $\sqrt[3]{\dfrac{8,747}{0.212}}$

56. $\sqrt[3]{\dfrac{158.5}{255}}$

2.11 Circumference and Area of a Circle

Because the problem of finding the circumference of a circle is simply one of finding the product of a given diameter and 3.14, we follow the regular procedure for multiplication.

To find the area of a circle when the radius is given, set the index of C opposite the radius on the D scale. Place the hairline over π on the B scale. The answer is found under the hairline on the A scale. Figure 2.9 shows that the area of a circle of radius 17.44 in is about 956 in².

To find the area of a circle when the diameter is given, set the index of C opposite the diameter on the D scale. Place the hairline over $\pi/4$, or 0.785 on the B scale. The area is found under the hairline on the A scale. Figure 2.10 shows that the area of a 9.03-in-diameter circle is 64.0 in².

EXERCISE 7

Make a copy of the following table of circles and fill in the blanks:

	Radius	Diameter	Circumference	Area
1.	6.3 mi	_____	_____	_____
2.	5¼ in	_____	_____	_____
3.	78 ft	_____	_____	_____
4.	9⁄16 in	_____	_____	_____
5.	_____	11.2 yd	_____	_____
6.	_____	31.6 ft	_____	_____
7.	_____	⅞ in	_____	_____
8.	_____	14½ in	_____	_____
9.	_____	_____	8.43 in	_____
10.	_____	_____	26.7 ft	_____
11.	_____	_____	_____	10.00 in²
12.	_____	_____	_____	63.8 ft²

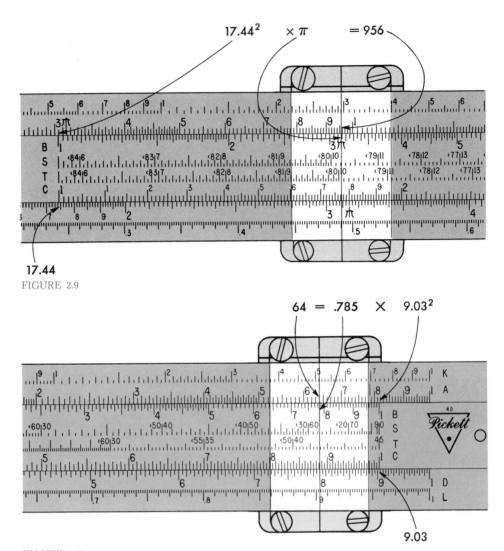

FIGURE 2.9

FIGURE 2.10

2.12 General Suggestions for Slide-rule Operation

As the student becomes skillful in the use of the slide rule, he will acquire various tricks of the trade which will make this tool even more effective. A few suggestions are outlined in the following paragraphs.

Before setting numbers on the slide rule, cancel or combine simple numbers to reduce the number of moves required.

Example 16. $\dfrac{2 \times 43 \times 3}{17 \times 61} = \dfrac{6 \times 43}{17 \times 61}$

A move is saved.

Example 17. $\dfrac{60 \times 105}{5 \times 2 \times 97} = \dfrac{6 \times 105}{97}$

Two moves are saved.

Increased accuracy in slide-rule division is possible through the calculation of the first few significant figures by long division.

Example 18. Divide 483 by 617, obtaining five significant figures in the quotient.

We may obtain the first two significant figures by long division and the next three by slide rule.

By slide rule 174 ÷ 617 yields the digits 282. Hence our answer is 0.78282.

$$
\begin{array}{r}
.78 \\
617\,\overline{)483.00} \\
431\ 9 \\
\hline
51\ 10 \\
49\ 36 \\
\hline
1\ 74
\end{array}
$$

In cases in which the dividend differs but little from the divisor, greatly improved accuracy may be obtained, as indicated by the following example.

Example 19. Divide 732.00 by 741.00.

$$\frac{732}{741} = \frac{741 - 9}{741} = 1 - \frac{9}{741}$$

By slide rule, 9 ÷ 741 = 0.01215. Hence

$$\frac{732.00}{741.00} = 1 - 0.01215 = 0.98785$$

A similar device is useful when multiplying by a number having a value near unity.

Example 20. Multiply 0.99756 by 610.95.

$$(0.99756)(610.95) = (1 - 0.00244)(610.95) = 610.95 - 1.49 = 609.46$$

Preliminary rearrangement of formulas is often helpful, as in logarithmic computation (Sec. 11.26).

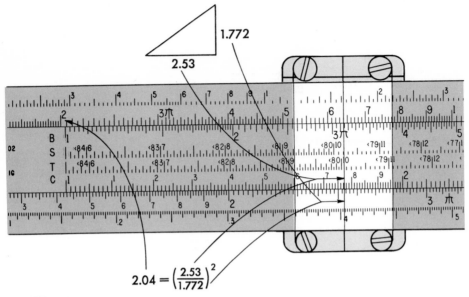

FIGURE 2.11

As an alternative to the factoring method, it is suggested that the familiar formulas of the Pythagorean theorem be modified before use on the slide rule, as indicated:

$$a = \sqrt{c^2 - b^2} \qquad \text{becomes} \qquad a = b\sqrt{(c/b)^2 - 1}$$
$$b = \sqrt{c^2 - a^2} \qquad \text{becomes} \qquad b = a\sqrt{(c/a)^2 - 1}$$
$$c = \sqrt{a^2 + b^2} \qquad \text{becomes} \qquad c = b\sqrt{(a/b)^2 + 1} \qquad (a > b)$$
$$c = a\sqrt{(b/a)^2 + 1} \qquad (b > a)$$

With the subtraction or addition of 1 done mentally, a minimum of moves is required.

Example 21. Find the hypotenuse of a right triangle whose sides are 2.53 and 1.772.

Letting 2.53, the larger side, be represented by a, and 1.772 by b, and referring to Fig. 2.11, we have the setting for obtaining $(2.53/1.772)^2$, which is seen to be 2.04. Adding 1 mentally to obtain 3.04, or $(a/b)^2 + 1$, we then refer to Fig. 2.12, showing the setting for $1.772\sqrt{3.04}$, which we read on the D scale as approximately 3.09.

EXERCISE 8

Compute the missing sides in the following right triangles in which c is the hypotenuse:

	a	b	c
1.	38	47	_____
2.	7.52	10.36	_____

3.	523	408	————
4.	$1\frac{7}{8}$ in	$2\frac{3}{4}$ in	————
5.	9.85	25.6	————
6.	————	0.0645	0.0892
7.	————	12.33	20.7
8.	————	$3\frac{1}{8}$ in	$4\frac{1}{4}$ in
9.	5,670	————	7,030
10.	0.1525	————	0.291
11.	$4\frac{7}{8}$ in	————	$5\frac{1}{2}$ in

In finding the hypotenuse of a right triangle containing a small acute angle, the approximation $c \approx a + (b^2/2a)$, where b is the smallest side, is useful. The smaller b becomes in relation to a, the closer the approximation. If b is as large as $0.2a$, then c, as calculated by the approximation, will be about 0.02 percent too large.

EXERCISE 9

Find c, evaluating $b^2/2a$ to three significant figures by slide rule.

	a	b
1.	52	3
2.	109	4
3.	15 in	$\frac{3}{8}$ in
4.	213	7
5.	1.75	0.08

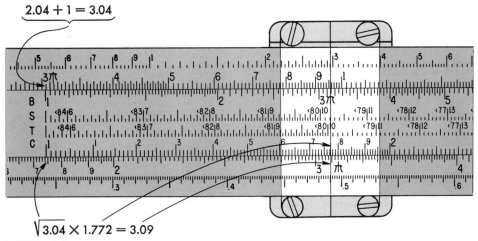

FIGURE 2.12

In the following sections we show how the usefulness of the tables can be extended by appropriate shifting of decimal points and by interpolation.

2.13 Extending the Range of the Tables

The table of squares may be extended beyond the given range by shifting the decimal point twice as many places in the square of the number as in the number itself, for $[x(10)^n]^2 = x^2(10)^{2n}$.

Example 22. Find $(7,650)^2$.

In most tables we shall have to use 765, since 7,650 is beyond the scope of the table. Looking up $(765)^2$, we find 585,225. Therefore $(7,650)^2 = 58,522,500$.

Example 23. Find $(1.38)^2$.

Looking up $(138)^2$, we find 19,044. Therefore $(1.38)^2 = 1.9044$.

Note: *The same method applies to finding the areas of circles, since the area varies as the square of the diameter.*

Example 24. Find the area of a $2\frac{3}{4}$-in circle.

Looking up the area of a circle of diameter 275, we find 59,396. Therefore the area of a 2.75-in circle is 5.9396 in².

The table of cubes may be extended by shifting the decimal point 3 times as many places in the cube of the number as in the number itself, for $[x(10)^n]^3 = x^3[10]^{3n}$.

Example 25. Find $(3.17)^3$.

Looking up $(317)^3$, we find 31,855,013. Therefore $(3.17)^3 = 31.855013$.

Example 26. Find $(0.431)^3$.

Looking up $(431)^3$, we find 80,062,991. Therefore $(0.431)^3 = 0.080062991$.

The use of the square-root table is governed by a rule similar to those just given. That is, we move the decimal point half as many places in the square root of the number as in the number itself. This is shown by the identity $\sqrt{x(10)^{2n}} = \sqrt{x}(10)^n$. For example,

$$\sqrt{5,090,000} = \sqrt{(509)(10,000)} = 100\sqrt{509}$$

The operation of this rule will be better understood by the following procedure:

Indicate the grouping of the digits of the number as for longhand extraction of square root. Write in their proper positions the decimal point and the first significant digit in the square root. Determine the subsequent digits as found in the square root of the tabulated number having the same grouping as the original number. Fill in these digits in proper sequence after the first significant digit already located.

Example 27. Find $\sqrt{5,090,000}$.

Indicating the grouping and the decimal point and first significant digit in the answer, we write

$$\frac{2}{\sqrt{5\ 09\ 00\ 00.}}$$

The first group, or "key group," is 5, whose square root is 2+. It is apparent that the tabulated number having the same digits and grouping is 509, of which the square root is 22.56103. This indicates that $\sqrt{5,090,000}$ will be written as follows:

$$\frac{2\ \ 2\ \ 5\ \ 6.\ 103}{\sqrt{5\ 09\ 00\ 00.}}$$

Example 28. Find $\sqrt{0.00509}$.

Our first indication of the answer is given by the grouping

$$\frac{0.\ 0\ \ 7}{\sqrt{0.00\ 50\ 90}}$$

In this example, the first group containing significant figures is 50, whose square root is 7+. Thus the grouping is seen to correspond to that of 5,090, of which the square root is 71.34424. Hence the square root of 0.00509 must be 0.07134424.

The procedure for extending the range of the cube-root table is analogous to that used with the square-root table. That is, we move the decimal point one-third as many places in the cube root of the number as in the number itself. This is shown by the identity $\sqrt[3]{x(10)^{3n}} = \sqrt[3]{x}(10)^n$.

Example 29. Find $\sqrt[3]{356,000}$.

Indicating the grouping, the decimal point, and the first significant digit in the answer, we write

$$\frac{7\ \ \ \ \ .}{\sqrt[3]{356\ \ 000.}}$$

The tabulated number having the same digits and grouping as 356,000 is 356, of which the cube root is 7.087341. Thus the cube root of 356,000 will be

$$\frac{7\ \ \ 0\ .87341}{\sqrt[3]{356\ \ 000.}}$$

Example 30. Find $\sqrt[3]{0.0356}$.
Grouping, we get

$$\begin{array}{c} 0.\ 3 \\ \hline \sqrt[3]{0.035\ \ 600} \end{array}$$

and the corresponding tabulated number is 35,600, of which the cube root is 32.89652. Hence we find

$$\begin{array}{c} 0.\ 3\quad \ 2\ \ 8\ 9\ 6\ 5\ 2 \\ \hline \sqrt[3]{0.035\ \ 6000} \end{array}$$

Example 31. Find $\sqrt[3]{0.00000356}$.
Grouping, we get

$$\begin{array}{c} 0.\ 0\quad \ 1 \\ \hline \sqrt[3]{0.000\ \ 003\ \ 560} \end{array}$$

Since the corresponding tabulated number is 3,560, of which the cube root is 15.26921, the desired cube root is

$$\begin{array}{c} 0.\ 0\quad \ 1\ \ 5\ 2\ 6\ 9\ 2\ 1 \\ \hline \sqrt[3]{0.000\ \ 003\ \ 560} \end{array}$$

which may be rounded off as desired.

2.14 Other Helpful Devices

Frequently interpolation may be avoided by judicious use of factoring; for example, $\sqrt{1,328} = \sqrt{4(332)} = 2\sqrt{332} = 2(18.2209) = 36.4418$.
The student should not overlook the advantages of rationalization in extracting roots of fractions. For example, $\sqrt{\tfrac{2}{3}}$ is much more conveniently found by evaluating $\tfrac{1}{3}\sqrt{6} = \tfrac{1}{3}(2.4495) = 0.8165$ than by finding $\sqrt{0.666667}$. These operations with radicals are explained in Chap. 4.
Square roots, for instance, may at times be found with increased accuracy by working backward from the table of squares. This is true if the table of square roots is given only to four or five significant figures or so.

Example 32. Find the square root of 174,126. From the square-root table:

	x	y $(= \sqrt{x})$
	174,000	417.13
1,000 $\{$ 126$\{$	174,126	? $\}$ 120
	175,000	418.33

Interpolation in the usual way gives us $\sqrt{174,126} = 417.13 + (126/1,000)(120) =$ 417.13 + 0.15 = 417.28 (precision to the nearest 0.01).

Using the table of squares:

Δx	x†	y(= x²)†	Δy
	⎛ 417.000	173,889 ⎞	
1	⎨ ?	174,126 ⎬ 237	835
	⎝ 418.000	174,724 ⎠	

†These values are exact.

(Δx and Δy, read "delta x" and "delta y," represent *changes* in x and y.) Interpolating, we obtain

$$417 + \frac{237}{835}\,1 = 417.284$$

Circumferences by Addition of Parts

Since $\pi(a + b) = \pi a + \pi b$, we may consider that the sum of the circumferences of two circles is equal to the circumference of a circle whose diameter is equal to the sum of the diameters of the first two circles.

Example 33. Find the circumference of a $5\frac{7}{16}$-in circle.

Divide 5.4375 into groups of not more than three significant figures each to avoid interpolation.

N	πN
5.4	16.96460
0.0375	0.11781
5.4375	17.08241

In most cases two or more superfluous digits will have to be rounded off to be consistent with the accuracy of the data of the problem, but interpolation will have been avoided.

It should be noted in passing that the method of "addition of parts" is applicable to any linear relation between two variables having a common zero, e.g., degrees to radians and inches to centimeters.

The use of the table of multiples of π is by no means limited to circumferences. For example, the area of an ellipse whose semiaxes are 6 and 8 is found by looking up the circumference of a circle of diameter 48. The circumference table is also useful in the calculation of surface speeds of drills, grinding wheels, pulleys, etc. In electrical engineering, multiples of π are involved in calculations of electromagnetic effects, capacitance, and inductance.

The reciprocal table may be used to evaluate fractions.

Example 34. What decimal part of a foot is $\frac{7}{16}$ in?

$$\frac{7}{16} \div 12 = \frac{7}{16} \times \frac{1}{12} = \frac{7}{192} = 7(\frac{1}{192}) = 7(0.0052083) = 0.03646$$

review of elementary algebra

It is assumed that the student using this text has successfully completed a thorough course in elementary algebra. This chapter is therefore a review in which we emphasize the techniques of algebraic manipulation as a foundation for problem solving.

3.1 Constants and Variables

In both arithmetic and algebra we use written symbols to represent numbers. These symbols are called *numerals*. In arithmetic we are restricted to numerals such as 5, 6, 8, etc., each of which represents a specific number. In algebra we often use a letter of the alphabet to represent an unspecified number. When letters of the alphabet are used in this way, they are also called numerals.

For example, we might let the letter a represent a number expressing the height of a rectangle and let the letter b represent the number expressing the length of the base of the same rectangle. Then we can formulate a rule for finding the area of any rectangle as follows: *The area of a rectangle is found by multiplying a times b.*

This rule will apply to any rectangle, but the specific numerical value of a and b will vary from rectangle to rectangle, depending on its dimensions. Appropriately enough, in this context a and b are called *variables*.

A constant is a numeral which, during a particular problem or discussion, represents one specific number. Of course the Arabic numerals, such as 5, 6, and 8, are always constants.

3.2 Positive and Negative Numbers

The first numbers with which we became familiar were the numbers used in arithmetic to represent weights of physical objects, distance, time, money, and the like. These numbers constitute what we call the set of *positive numbers*.

3.3 The set of Negative Numbers

A slightly more advanced topic deals with the set of *negative numbers*. The word negative in this context implies that negative numbers have certain properties which, in a sense, are opposite to corresponding properties attributed to positive numbers.

For example, if temperature above zero is designated by *positive* numbers, then temperature below zero is designated by *negative* numbers. If the elevation of points on the earth's surface above an arbitrarily chosen bench mark is designated by *positive* numbers, then the elevation of points below this bench mark is designated by *negative* numbers. The work done by the gas in a piston is ordinarily designated by *positive* numbers; consequently, the work done on the gas by the piston is designated by *negative* numbers. If the velocity of an object moving directly away from the center of the earth is designated by *positive* numbers, then the velocity of an object moving directly toward the center of the earth is designated by *negative* numbers.

We distinguish between positive and negative numbers by the symbols + and −, respectively. Thus, if a lifting force of 140 lb were designated by the positive number +140 lb, then a dead weight of 140 lb would be designated by the negative number −140 lb.

Further, if a credit were designated by +$4, then a debit of $4 would be designated by −$4.

Numbers used in elementary algebra are called *real* numbers, in contrast with *imaginary* numbers, which we shall consider at a later stage. Real algebraic numbers may be positive, negative, or zero, whereas arithmetic numbers do not include negative values. A number written without a sign is understood to be positive; thus the number 5 is understood to be +5.

Real numbers may also be classified as *rational* or *irrational*. Rational numbers can be expressed as the ratio between two integers (whole numbers). Irrational numbers cannot be expressed as the ratio between two integers.

It has been proved that certain numbers, among them $\sqrt{2}$, $\sqrt[3]{6}$, and π, for example, are irrational numbers. The name rational is derived from the same source as the word *ratio*. When used in this context therefore, the word rational has nothing to do with the relative sanity of a certain kind of number.

3.4 Order

In everyday usage we think of 100 automobiles as being a greater number of automobiles than 70 automobiles. Thus it is common to think of the number 100 as being greater than the number 70.

Using the symbol $>$ to mean "greater than" and the symbol $<$ to mean "less than," we may write

$100 > 70$

and

$70 < 100$

As we read the temperature scale in Fig. 3.1 from top down toward 0, we find

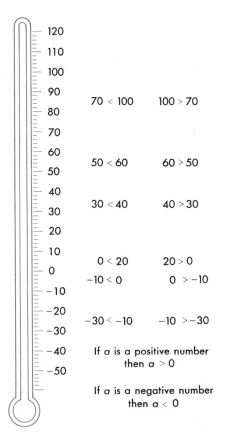

$70 < 100$ $100 > 70$

$50 < 60$ $60 > 50$

$30 < 40$ $40 > 30$

$0 < 20$ $20 > 0$

$-10 < 0$ $0 > -10$

$-30 < -10$ $-10 > -30$

If a is a positive number
then $a > 0$

If a is a negative number
then $a < 0$

FIGURE 3.1

that, in the sense we are discussing, each number we encounter is *smaller* than those above it. That is,

$70 < 100$
$30 < 40$
$0 < 15$

When we come to the number 0, we continue in the same pattern. For example,

$-10 < 0$
$-20 < -10$

and so on, even though $+20 > +10$.

Thus every positive number is greater than any negative number. If the number a is greater than 0 (written $a > 0$), then a must be a positive number. If the number a is less than 0 (written $a < 0$), then a must be a negative number.

3.5 Absolute Value

The velocity of a guided missile might be $+3,400$ ft/s on the upward trajectory, and a little later a similar velocity on the descent would be referred to as $-3,400$ ft/s. However, if we simply wish to designate the rate of motion without specifying sense or direction (that is, the *speed*), we write $|3,400|$ ft/s, which is called the *absolute value* of either $+3,400$ or $-3,400$ ft/s.

Accordingly we adopt the following definitions:

For any real number a,

$|a| = a$ If $a \geq 0$ (a is equal to or greater than 0)
$|a| = -a$ If $a < 0$ (a is less than 0)

Thus

The absolute value of a number is never negative.

For example,

$|+10| = +10$
$|-15| = +15$ (where $+15$ is the negative of -15)

It is evident that although $-15 < +10$ (-15 is less than $+10$), $|-15| > |+10|$ (the absolute value of -15 is greater than the absolute value of $+10$).

Example 1. Point A is 15 mi south of point B. Point C is 10 mi north of point B. We say that A is farther from B than C is, regardless of direction.

3.6 Some Properties of the Set of Real Numbers

Fundamental Laws of Addition and Multiplication

I. $a + b = b + a$ commutative law of addition

II. $ab = ba$ commutative law of multiplication

III. $a + (b + c) = (a + b) + c$ associative law of addition

IV. $a(bc) = (ab)c$ associative law of multiplication

V. $a(b + c) = ab + ac$ distributive law

Also, by the above laws, $a(b + c) = (b + c)a = ba + ca = a(c + b) = ac + ab$.

From these and other basic properties of the set of real numbers we could derive the rules for operating with them. However, we shall simply state these rules and illustrate them without formal justification.

3.7 Signs of Operation

The signs $+$, $-$, $\times$, $\div$, and $\sqrt{}$ have the same meaning as in arithmetic. However, the sign $\times$ is not commonly used in algebra. For example, multiplication of a by b is customarily written $a \cdot b$, or simply ab. Thus abc has the same meaning as $a \times b \times c$. If we wish to multiply the expression $a + b$ by the expression $x + y$, we write $(a + b)(x + y)$, the absence of any connecting sign indicating multiplication.

3.8 Addition

The operation of addition of signed numbers may well be illustrated by the balancing of credits and debits.

For example,

$(+7) + (+8) = +15$ (Adding credits of \$7 and \$8 results in a total credit of \$15.)

$(-3) + (-1) = -4$ (Adding debits of \$3 and \$1 gives us a total debit of \$4.)

$(-4) + (+7) = +3$ (Adding a debit of \$4 and a credit of \$7 gives us a net credit of \$3.)

$(-9) + (+2) = -7$ (Adding a debit of \$9 and a credit of \$2 results in a net debit of \$7.)

From the above it will be seen that the following rules are valid:

To add two numbers of the same sign, add their absolute values and prefix the sum by their common sign.

Example 2. Add the following:

$$(+7) + (+8) = +(|+7| + |+8|) = +(7 + 8) = +15$$
$$(-3) + (-1) = -(|-3| + |-1|) = -(3 + 1) = -4$$

To add two numbers of opposite sign, find the difference of their absolute values and prefix the result by the sign of the number having the greater absolute value.

Example 3. Add the following:

$$(-4) + (+7) = +(|+7| - |-4|) = +(7 - 4) = +3$$
$$(-9) + (+2) = -(|-9| - |+2|) = -(9 - 2) = -7$$

To find the sum, or net value, of several numbers of both positive and negative sign, add all the positive numbers and all the negative numbers and find the algebraic sum according to the preceding rule.

Example 4. Add 5, -3, -6, $+2$, -1, $+7$, and -8.
    Applying the commutative law of addition, we find that we are adding 5, $+2$, $+7$ and -3, -6, -1, -8, or $+14$ and -18, which equal -4.

3.9 Subtraction

Subtraction of two numbers is performed by reversing the sign of the subtrahend and proceeding as with the addition of signed numbers.

That is, if a and b are real numbers, the subtraction indicated by $a - b$ is defined such that

$$a - b = a + (-b)$$

Now the rules for addition stated in the preceding section will hold.
    Examples of this rule follow:

$$(+8) - (+5) = (+8) + (-5) = +3$$
$$(+4) - (+9) = (+4) + (-9) = -5$$
$$(-7) - (-3) = (-7) + (+3) = -4$$
$$(-2) - (-9) = (-2) + (+9) = +7$$
$$(-5) - (+6) = (-5) + (-6) = -11$$
$$(+7) - (-3) = (+7) + (+3) = +10$$

Note that in the second example it is possible to subtract a larger number from a smaller. Consider the embarrassing situation of the bank depositor who, although he has only a $4 balance, draws a check of $9 and is therefore $5 overdrawn!

In the last example a man may figure that he is $7 to the good after taking account of all his debits and credits. If, after taking stock, an outstanding debt of $3 is canceled (subtract $-$$3) or $3 in cash is received (add $+$$3), the result is the same in either case—his net worth is $10. Note that the minus sign functions in two capacities: (1) to denote the operation of subtraction, (2) to designate a negative quantity.

Example 5. $8 - 5 = 3$ could designate either (1) or (2) in the preceding sentence, but $8 + (-5) = 3$ or $-5 + 8 = 3$ definitely refers to the quantity -5.

3.10 Multiplication

The multiplication of signed numbers is governed by the following rules:

1. *The product of two numbers of like sign is positive.*

   In other words, the product of two positive real numbers or the product of two negative real numbers is the product of their absolute values.

   $$(+2)(+7) = |+2| \times |+7| = 2 \times 7 = +14$$
   $$(-2)(-7) = |-2| \times |-7| = 2 \times 7 = +14$$

2. *The product of two numbers of opposite sign is negative.*

   That is, the product of a positive real number and a negative real number is the negative of the product of their absolute values.

   $$(+2)(-7) = -(|+2| \times |-7|) = -(2 \times 7) = -14$$
   $$(-2)(+7) = -(|-2| \times |+7|) = -(2 \times 7) = -14$$

3. *The continued product of three or more numbers is positive if there is an even number of negative numbers, and negative if there is an odd number of negative numbers.*

   To this rule we may add a corollary:

   Even powers of both positive and negative numbers are positive. Odd powers of positive numbers are positive, and odd powers of negative numbers are negative. (See Sec. 4.1.)

Example 6. (a) The product of $+2$ and $+3$ is $+6$.
  (b) The product of $+2$ and -3 is -6.
  (c) The product of -2 and $+3$ is -6.
  (d) The product of -2 and -3 is $+6$.

These examples may be illustrated by the following commonplace situations:

(a) The temperature is rising at the rate of $2°$ an hour. At this rate it will be $6°$ warmer in 3 h.
(b) At the same rate it was $6°$ colder 3 h ago.
(c) The temperature is falling at the rate of $2°$ an hour. At this rate it will be $6°$ colder 3 h from now.
(d) At the rate in (c), it was $6°$ warmer 3 h ago.

4. *Changing the sign of a quantity is equivalent to multiplying that quantity by* -1.

3.11 Factors

When two or more numbers are multiplied together to form a product, each of the numbers multiplied is called a *factor* of the product.

It is important here to note that if one or more of these factors are zero, then the product is zero.

Any number which contains only itself and 1 as factors is called a *prime* number.

3.12 Division

The rules of sign governing division resemble those for multiplication, since dividing by a number x is equivalent to multiplying by the fraction 1/x (the reciprocal of x).

1. *The quotient of two real numbers of like sign is a positive real number.*

That is, the quotient of two positive real numbers or the quotient of two negative real numbers is a positive real number. Expressed more formally, the quotient of two real numbers of like sign is the quotient of their absolute values.

$$\frac{+8}{+2} = \frac{|+8|}{|+2|} = \frac{8}{2} = +4$$

$$\frac{-8}{-2} = \frac{|-8|}{|-2|} = \frac{8}{2} = +4$$

2. *The quotient of two real numbers of opposite sign is a negative real number.*

Expressed more formally, the quotient of a positive real number and a negative real number is the negative of the quotient of their absolute values.

$$\frac{-8}{+2} = -\frac{|-8|}{|+2|} = -\frac{8}{2} = -4$$

$$\frac{+8}{-2} = -\frac{|+8|}{|-2|} = -\frac{8}{2} = -4$$

In addition to the foregoing, the following principles apply to division:

1. *A number cannot be divided by zero. Such a quotient is undefined.*

For example, assume $5 \div 0 = n$. Then $5 = n \times 0$. But this is impossible since $n \times 0 = 0$.

2. *The value of a fraction is unchanged if both numerator and denominator are multiplied, or divided, by the same nonzero number.*
3. *Changing the sign of either the numerator or denominator of a fraction reverses the sign of the fraction. Reversing the sign of both numerator and denominator does not change the sign of the fraction.*
4. *Division of zero by zero is indeterminate.*

For example, assume $0 \div 0 = n$. Then $0 = 0 \times n$. But $0 \times n = 0$ whatever the value of n; hence n is indeterminate.

EXERCISE 1

1. Find the sum of each of the following pairs of numbers:
    (a) $+8, +5$ (b) $+7, -2$ (c) $+4, +10$ (d) $+6, -9$
    (e) $-3, -11$ (f) $-12, -8$ (g) $-9, +13$ (h) $-11, +4$
2. Subtract the second number from the first in each pair of numbers in Prob. 1.
3. Apply the commutative law to add:
    (a) $8 + 2 - 3 - 4 + 6 - 1 - 9 + 5$
    (b) $-3 + 7 + 2 - 6 - 5 + 1 - 8$
    (c) $4 + 10 - 3 + 7 - 6 - 9 + 2$
4. Find each of the following products:
    (a) $(8)(5)$ (b) $(-6)(7)$ (c) $(-12)(-4)$
    (d) $(9)(-6)$ (e) $(2)(-3)(4)$ (f) $(-5)(3)(-4)$
    (g) $(-3)(-6)(2)(-1)$

5. Perform the indicated operations:

(a) $20 \div 5$ (b) $24 \div (-8)$ (c) $(-30) \div 6$

(d) $(-35) \div (-7)$ (e) $\dfrac{-28}{(-7)(3)}$ (f) $\dfrac{(36)(15)}{-81}$

(g) $\dfrac{(42)(-35)}{(-49)(-18)}$

3.13 Algebraic Expression

An *algebraic expression* is a collection of number symbols combined through one or more of the operations of addition, subtraction, multiplication, division, and extraction of roots.

In an algebraic expression, a *term* is an indicated product of a finite number of factors. Thus, in the expression $ab - ac + bcxy$, the terms are ab, $-ac$, and $bcxy$.

An algebraic expression containing only one term is called a *monomial,* for example, bcm or $-ax$. An expression containing two terms is called a *binomial,* for example, $mn + xyz$; one containing three terms is called a *trinomial,* for example, $k - hn + w$. In general, expressions containing more than one term are called *poly-nomials*, or *multinomials*. A polynomial may also be defined as an algebraic expression in which the operations are limited to addition, subtraction, and multiplication. A *coefficient* is the multiplier of a term. When two or more quantities are multiplied together to form a product, each of the quantities is called a *factor* of the product. In the term $5xyz$, 5, x, y, and z are called factors of the expression. Usually the numerical part of the term is regarded as the coefficient, although any one or more of the factors could be taken as the coefficient of the remaining factors. Thus 5 would usually be considered the coefficient, but by the associative law, $5y$ could be considered the coefficient of xz. When the expression contains no numerical coefficient, a coefficient of 1 is understood. Thus bcm is the same as $1bcm$.

An *exponent* is a figure or letter placed to the right of and somewhat above a quantity called the *base* to indicate how many times the base is to be taken as a factor. Thus a^4 means $a \cdot a \cdot a \cdot a$ and is read "a fourth" or "a to the fourth *power.*"

It is important to maintain a clear distinction between coefficient and exponent: $3x$ means $x + x + x$, whereas x^3 means $x \cdot x \cdot x$.

When a quantity is written without an exponent, the exponent 1 is understood. Thus y means y^1.

Terms having a common literal factor are called *similar* terms. Similar terms can be added or subtracted by adding or subtracting their numerical coefficients and affixing the common literal parts. If the terms are not similar terms, these operations can only be indicated.

Example 7. Express the sum of $5cd^2$, cd^2, $-2cd^2$.

By the distributive law $5cd^2 + cd^2 - 2cd^2 = (5 + 1 - 2)cd^2 = 4cd^2$.

Example 8. $7xy$

$$5x^2$$
$$-2y^2$$
$$\overline{}$$
$$7xy + 5x^2 - 2y^2$$

Example 9. Add the following:

$$3x + 5y - 8$$
$$2x - 6y + 5$$
$$-4x - y + 8$$
$$\overline{+ x - 2y + 5}$$

Example 10. Subtract $3x^2 - 5xy + 2y^2$ from $5x^2 + 2xy - y^2$.

$$5x^2 + 2xy - y^2$$
$$3x^2 - 5xy + 2y^2$$
$$\overline{2x^2 + 7xy - 3y^2}$$

EXERCISE 2

Write the expressions in Probs. 1 to 8 in the most concise form possible by using coefficients and exponents. Use the commutative and associative laws where applicable.

1. $a + a + a - a + a$
2. $x - x - x + x - x - x$
3. $y \cdot y \cdot y \cdot y$
4. $a \cdot a \cdot b \cdot b \cdot b$
5. $4x \cdot y \cdot y \cdot z \cdot z \cdot z$
6. $(a + c)(a + c)(a + c)$
7. $w(xw)w(xw)$
8. $a(ab)(abc)(bc)c$

Find the numerical values in Probs. 9 to 30 if $a = 3$, $b = 4$, $c = -5$, and $x = 1$.

9. $a + b$
10. $2c - a$
11. $7(2a - c)$
12. $3a^2bc$
13. $4a^2 - b^2$
14. $(a + c)^2$
15. $b^2 - x^2$
16. $(3a - b)^3$
17. $(a + b)(a + b)$
18. $a + b(a + b)$
19. $(a + b)a + b$
20. $a + ba + b$
21. $(2a - b + c)^2$
22. $(c + b)^3x$
23. $\sqrt{a^2 + b^2}$
24. $\sqrt{(3a)^2 - b^2 - x^2}$
25. $\dfrac{4a + b}{c + 3x}$
26. $\dfrac{4a}{c} + \dfrac{b}{3x}$
27. $(4b + 3c - x)^4$
28. $\left(\dfrac{4a + 2b}{3x - c}\right)^2$
29. $3abc^2$
30. $(3abc)^2$

Perform the indicated additions in Probs. 31 to 42.

31. $\quad 7x$
$\quad\quad 5x$
$\quad\underline{-\ x}$

32. $\ -11ab$
$\quad\quad 4ab$
$\quad\quad\ ab$
$\quad\underline{-\ 2ab}$
$\quad\quad 8ab$

33. $\quad 9mn$
$\quad -2mn$
$\quad\underline{-\ mn}$
$\quad\quad 7mn$

34. $\quad 6cd$
$\quad -4cd$
$\quad\quad 7cd$
$\quad\underline{-8cd}$

35. $5x + 4y$
$\ \underline{x + 5y}$

36. $-3a + 2c$
$\ \underline{4a - 3c}$

37. $6m + 10p$
$\ \underline{6m -\ 2p}$

38. $4x^2 - 2xy +\ y^2$
$\ \underline{5x^2 + 3xy - 2y^2}$

39. $3k^2 - 2km +\ m^2$
$\quad \underline{k^2 \qquad\ + 4m^2}$

40. $2p^2 \qquad\ - 3q^2$
$\ \underline{5p^2 + 4pq - 2q^2}$

41. $4x - 2y$
$\ \underline{5x \qquad - 3z}$

42. $10c^2 + cm - 3m^2$
$\ \underline{3c^2 + cm - 4m^2}$

43–50. In Probs. 35 to 42 subtract the second expression from the first.

51. From the sum of $a^2 - 2a + 3$ and $2a^2 - a + 5$ subtract $4a^2 - 4a + 7$.

52. From $8c^2 - 2c + 11$ subtract the sum of $c^2 - 4c + 6$ and $5c^2 + 3c + 5$.

53. Subtract the sum of $9x^2 - xy$ and $3x^2 + xy - 2y^2$ from $6x^2 - 3xy - y^2$.

3.14 Symbols of Grouping

Parentheses, ( ), brackets, [ ], or braces, { }, indicate that the expression which they enclose is to be treated as a unit. The less frequently used vinculum, $\overline{\quad}$, written over an expression, has the same meaning.

Example 11. $5 + (2 + 7) = 5 + 9 = 14$

Example 12. $8 - (3 + 10) = 8 - 13 = -5$

Example 13. $6 - \overline{7 - 2} = 6 - 5 = 1$

Example 14. $4(5 + 2) = 4(7) = 28$. Note that we might write (by the distributive law)

$4(5 + 2) = 4(5) + 4(2) = 20 + 8 = 28$

The following remarks concerning parentheses will be understood to apply equally well to brackets and braces.

Parentheses are useful in enclosing expressions to be multiplied:

Example 15. $5(-4)$ means 5 times -4, or -20.

Example 16. To multiply the expression $a + 3$ by the expression $a - 5$, we write $(a + 3)(a - 5)$ (read "the quantity $a + 3$ times the quantity $a - 5$"). If $a = 9$, then $(9 + 3)(9 - 5) = (12)(4) = 48$.

Parentheses can be used to avoid ambiguity regarding the order of application of the fundamental operations.

Example 17. Does the expression $14 - 6 \div 2$ mean that 8 is to be divided by 2 or that 3 is to be subtracted from 14?

There is a rule stating that multiplication and division take precedence over addition and subtraction. However, such sloppiness is indefensible when the intent is easily clarified by the use of parentheses. If written $(14 - 6) \div 2 = 8 \div 2 = 4$ in one case or $14 - (6 \div 2) = 14 - 3 = 11$ in the other, the meaning is no longer questionable.

To insert or remove parentheses preceded by a plus sign, rewrite the affected quantities with their respective signs unchanged.

Example 18. $8 + (3 - 4) = 8 + 3 - 4 = 7$

To insert or remove parentheses preceded by a minus sign, rewrite the affected quantities with their respective signs reversed.

Example 19. $11 - (7 - 2) = 11 - 7 + 2 = 6$

Example 19 may be illustrated by the following common situation: A man with $11 in his possession buys a $7 article and returns for credit a $2 article.

If the purchase and refund were made at the same counter, the $2 refund would probably be applied at the time of purchase, so that $11 - (7 - 2) = 11 - 5 = 6$; that is, only $5 would change hands. If, on the other hand, the refund were made in another department after the $7 purchase, we would write $11 - 7 + 2 = 6$. (In either case the customer still has $6 in cash.)

Note that if we started with the expression $11 - 7 + 2$ and wished to enclose the last two quantities with a pair of parentheses preceded by a minus sign, we should have to write $11 - 7 + 2 = 11 - (7 - 2)$, since we have already seen that $11 - (7 - 2) = 11 - 7 + 2$, and the operation must be reversible.

When one set of grouping symbols is contained within another, the innermost set is removed first.

Example 20. Perform the indicated operations and apply the commutative and distributive laws in the following expression: $3x + 2 - [2y + 5 - (6y - 2x + 3)]$.

Beginning with the inner group and performing the indicated multiplications with due attention to signs, we obtain

$$3x + 2 - [2y + 5 - 6y + 2x - 3] = 3x + 2 - 2y - 5 + 6y - 2x + 3 = x + 4y$$

EXERCISE 3

In Probs. 1 to 10, simplify by removing the symbols of grouping as in Example 20.

1. $3a + 4b + (2a - b)$
2. $5a - 3c - (2a - 4c)$
3. $4x - 7y - (3x - 4y) + x + 3y$
4. $6u + 5 - (3v + 4w - 2) - 2u + 4v$
5. $8m - [4n - (-3m + 3n) + m]$
6. $a - \{2x + [3a - 5x - (4a + x)] - 2a\}$
7. $-7b + \{6c - [2b - c + (8c - 4b) + b]\}$
8. $\{8c + 3k - [5c + 2k]\} - \{3c + k - [7c - (c - 4k)]\}$
9. $11ab - 3a^2b - [5ab^2 - (6ab + 2a^2b) + ab] + 3ab^2$
10. $\{4x^2 - xy + [y^2 - 2xy] - x^2\} - \{3x^2 - [y^2 - (2xy + x^2) + xy]\}$

In Probs. 11 to 16, enclose the last three terms in parentheses preceded by (a) a minus sign and (b) a plus sign.

11. $5a + 2b - 4c + m - x$
12. $2a^2 + 3ab - 7b^2 + 4b$
13. $7mn + 3m^2 - 4n^2 - 8m + 5n + 2mn$
14. $6c^2 + 5cd + d^2 - 2m^2$
15. $9w^2 + 5y^3 - 2w^2y - 3wy^2 + w^3$
16. $-7x^2 - 5xy - 2y^2 + 4yz - z^2$

3.15 Some Fundamental Rules of Exponents

The following laws relate to positive integral exponents, but in Chap. 4 it will be shown that they are valid for fractional and negative exponents as well.

I. *Products* $a^m \cdot a^n = a^{m+n}$

Example 21. $c^2 \cdot c^3 = c^{2+3} = c^5$ because

$$c^2 \cdot c^3 = (c \cdot c)(c \cdot c \cdot c) = c \cdot c \cdot c \cdot c \cdot c = c^5$$

Example 22. $a \cdot a^2 \cdot a^5 = a^1 \cdot a^2 \cdot a^5 = a^{1+2+5} = a^8$

IIa. Powers $(a^m)^n = a^{mn}$

Example 23. $(y^2)^3 = y^{2\times3} = y^6$ because

$(y^2)^3 = y^2 \cdot y^2 \cdot y^2 = (y \cdot y)(y \cdot y)(y \cdot y) = y \cdot y \cdot y \cdot y \cdot y \cdot y = y^6$

IIb. $(ab)^n = a^n b^n$

Example 24. $(ab)^5 = (ab)(ab)(ab)(ab)(ab)$
$= (a \cdot a \cdot a \cdot a \cdot a)(b \cdot b \cdot b \cdot b \cdot b) = a^5 b^5$

IIc. $\left(\dfrac{a}{b}\right)^n = \dfrac{a^n}{b^n}$

Example 25. $\left(\dfrac{a}{b}\right)^4 = \left(\dfrac{a}{b}\right)\left(\dfrac{a}{b}\right)\left(\dfrac{a}{b}\right)\left(\dfrac{a}{b}\right) = \dfrac{a \cdot a \cdot a \cdot a}{b \cdot b \cdot b \cdot b} = \dfrac{a^4}{b^4}$

III. Quotients $a^m/a^n = a^{m-n}$ $(m > n, a \neq 0)$ (Read "m is greater than n, a is not equal to zero.")

Example 26. $x^5 \div x^3 = x^{5-3} = x^2$ because

$x^5 \div x^3 = \dfrac{\cancel{x} \cdot \cancel{x} \cdot \cancel{x} \cdot x \cdot x}{\cancel{x} \cdot \cancel{x} \cdot \cancel{x}} = x \cdot x = x^2$

A further definition will be given [Eq. (3), Sec. 4.1] to cover cases in which the power of the numerator is less than that of the denominator.

EXERCISE 4

Perform the indicated operations:

1. $x^2 \cdot x^4$
2. $c^3 \cdot c^5$
3. $y^6 \cdot y$
4. $k^3 \cdot k^4 \cdot k^2$
5. $a \cdot a^2 \cdot a^3 \cdot a^4$
6. $(b^2)^4$
7. $(m^5)^3$
8. $(3a)^2$
9. $(5x^3)^2$
10. $(-4c^2)^3$
11. $(-2b^4)^6$
12. $-(-5w^4)^3$
13. $x^7 \div x^4$
14. $h^{12} \div h^3$
15. $y^6 \div y$
16. $(-a)^5 \div a^2$
17. $(-b)^6 \div (-b)^5$
18. $(-b)^6 \div (-b^4)$

3.16 Multiplying by a Monomial

To multiply a monomial by a monomial, multiply the product of the numerical co-efficients by the product of the literal factors, making use of the law of exponents where it applies. Determine and apply the proper sign.

Example 27. Find the indicated product:

$$(-3axy^2)(5a^3x^2) \hspace{6cm} \text{ANS.: } -15a^4x^3y^2$$

To multiply a polynomial by a monomial, multiply each term of the polynomial by the monomial and write the sum of the products.

Example 28. $4x^2y(5a^2x - 7b^2y) = 20a^2x^3y - 28b^2x^2y^2$ (by the distributive law)

EXERCISE 5

Perform the following multiplications:

1. $(8mn)(4mx)$
2. $(-3h^2x)(6hx^2)$
3. $(9abc)(-4bcd)$
4. $(-2ck^2x)(-7c^2x^5)$
5. $(a^2m)(m^2x)(ax^2)$
6. $(3ac^2d)(-4cd^3)(-2a^4cd^2)$
7. $(-9a^2b^3c^4)(-4a^3b^4m^5)(-2b^4c^5m^6)$
8. $5(4h - 6k)$
9. $-3y(6m - 5t)$
10. $4ax(5ay + 9mx)$
11. $6a^2b(5a^2 - 7ab - 9b^2)$
12. $-2xy^2(1 - 2x + 3x^2 - 4x^3)$
13. $(-3hk)(-2hk^3)(4a^3h^2 - 11bk)$
14. $\tfrac{2}{3}ab^2c^3(12ax - 21by)$

3.17 Multiplying One Polynomial by Another Polynomial

To multiply one polynomial by another, arrange the terms in descending powers of one of the letters involved. Multiply the first polynomial by the successive terms of the second, arranging like terms of the product in columns, and add.

Example 29. Multiply $10x - 2x^3 + x^4 - 1$ by $x^2 + 2 + 6x^3$.

Arranging in descending powers of x,

$$
\begin{array}{l}
x^4 - \quad 2x^3 + 10x \quad - 1 \\
6x^3 + \quad x^2 + \quad 2 \\
\hline
6x^7 - 12x^6 \qquad\quad + 60x^4 - \quad 6x^3 \qquad\qquad\qquad \text{(multiplying by } 6x^3) \\
\quad + \quad x^6 - \quad 2x^5 \qquad\quad + 10x^3 - x^2 \qquad\qquad \text{(multiplying by } x^2) \\
\qquad\qquad\qquad + \quad 2x^4 - \quad 4x^3 \qquad\quad + 20x - 2 \quad \text{(multiplying by 2)} \\
\hline
6x^7 - 11x^6 - \quad 2x^5 + 62x^4 \qquad\qquad - x^2 + 20x - 2
\end{array}
$$

The answer may be checked with reasonable certainty by substituting some simple number for x. Substitution of the number 1 should be avoided since any power of 1 is also 1.

Let us check by letting $x = 2$; then

$$x^4 - 2x^3 + 10x - 1 = (2)^4 - 2(2)^3 + 10(2) - 1$$
$$= 16 - 16 + 20 - 1 = 19$$

and

$$6x^3 + x^2 + 2 = 6(2)^3 + (2)^2 + 2 = 48 + 4 + 2 = 54$$
$$(19)(54) = 1{,}026$$

$$6x^7 - 11x^6 - 2x^5 + 62x^4 - x^2 + 20x - 2$$
$$= 6(2)^7 - 11(2)^6 - 2(2)^5 + 62(2)^4 - (2)^2 + 20(2) - 2$$
$$= 768 - 704 - 64 + 992 - 4 + 40 - 2 = 1{,}026$$

EXERCISE 6

Perform the following multiplications:

1. $(a + b)(c + d)$
2. $(m + x)(m - y)$
3. $(h + k)(h - k)$
4. $(y^4 - y^2)(y^5 + y^3)$
5. $(2m + 5w)(3m - 7w)$
6. $(4x - 6y)(6x + 9y)$
7. $(4x - 2y - 13)(6x + 3y)$
8. $(3p^2 + 4pq + 5q^2)(2p - 3q)$
9. $(b - x - y)(b + x + y)$
10. $(c + d - 5)(c - d + 4)$
11. $(n^2 + 2n + 4)(n - 2)$
12. $(a^2 - 5ab + b^2)(a^2 + 5ab - b^2)$
13. $(x^4 - x^3 + 4x^2 - x + 1)(x + 1)$
14. $(c^3 + 2c^2 - 9c - 18)(c - 2)$

3.18 Removing Parentheses Used to Indicate Multiplication

In removing parentheses used to indicate multiplication, the usual rules of multiplication apply.

Example 30. Simplify the expression

$$3 + 2\{5 - 4x[1 - 6(x + 2)]\}$$

[Note that 1 and $-6(x + 2)$ do not make -5 of any one thing.]
    Performing the indicated multiplications, beginning with the innermost group, we obtain

$$3 + 2[5 - 4x(1 - 6x - 12)] = 3 + 2(5 - 4x + 24x^2 + 48x)$$
$$= 3 + 10 - 8x + 48x^2 + 96x$$
$$= 13 + 88x + 48x^2$$

EXERCISE 7

Simplify the following expressions:

1. $1 - 2\{1 + 3[1 - 4(1 - 5x)]\}$
2. $a - \{[b - c] - [a + b - c - 2(a - b + c)]\}$
3. $5\{4[3(2 + x)]\} - 5\{-4[-3(2 - x)]\}$
4. $y^2 - y\{y + z[x(y - z) + y(z - x) + z(x - y)]\}$

EXERCISE 8

In the following problems, in cases where the result can be expressed in more than one way, the use of the letter representing the smaller dimension is preferred.

1. Express in the simplest form the perimeter in Fig. 3.2.
2. Same as Prob. 1 for Fig. 3.3.
3. Same as Prob. 1 for Fig. 3.4.
4. Same as Prob. 1 for Fig. 3.5.

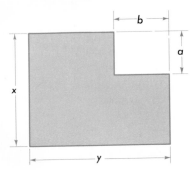

FIGURE 3.2

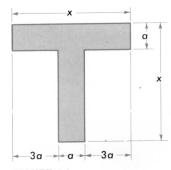

FIGURE 3.3

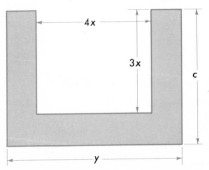

FIGURE 3.4

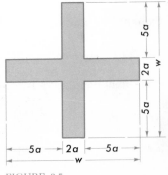

FIGURE 3.5

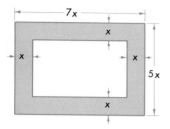

FIGURE 3.6

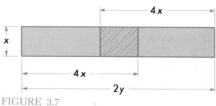

FIGURE 3.7

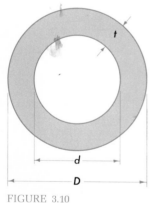

FIGURE 3.8

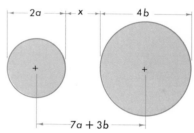

FIGURE 3.9

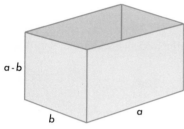

FIGURE 3.10

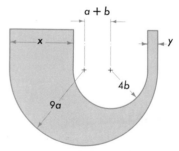

FIGURE 3.11

5–8. Express in the simplest possible form the area of each of Probs. 1 to 4.

9. Find the area of the form in Fig. 3.6.

10. Find the area of overlap in Fig. 3.7.

11. Find the outer surface and volume of the open-top box shown in Fig. 3.8.

12. Find in terms a and b the distance x in Fig. 3.9.

13. Find in terms of D and d the wall thickness t in Fig. 3.10.

14. Find in Fig. 3.11 the values of x and y in terms of a and b.

3.19 Dividing by a Monomial

To divide one monomial by another monomial, divide the coefficient of the dividend by the coefficient of the divisor, apply the rule of exponents to the literal factors, and prefix the proper sign.

Example 31. Divide $-20x^3yz^2$ by $-5xyz$.

By dividing like factors out of numerator and denominator, we write

$$\frac{-\overset{4}{\cancel{20}x} \cdot x \cdot x \cdot \cancel{y} \cdot z \cdot \cancel{z}}{-\cancel{5}\cancel{x} \qquad \cancel{y} \qquad \cancel{z}} = 4x^2z$$

Example 32. Divide $-12a^2bc^3$ by $16a^3cx^2$.

Again, by dividing like factors out of numerator and denominator, we obtain

$$\frac{-\overset{3}{\cancel{12}}\cancel{a} \cdot a \cdot b \cdot \cancel{c} \cdot c \cdot c}{\underset{4}{\cancel{16}}\cancel{a} \cdot a \cdot a \cdot \cancel{c} \cdot x \cdot x} = -\frac{3bc^2}{4ax^2}$$

To divide a polynomial by a monomial, divide each term of the polynomial by the monomial and write the sum of the quotients.

Example 33. $(24c^4x - 21c^2x^2 + 18cx^2) \div (-6c^2x) = -4c^2 + \dfrac{7x}{2} - \dfrac{3x}{c}$

EXERCISE 9

Obtain the following quotients:

1. $15c^3 \div 5c^2$
2. $-24h^3 \div 8h$
3. $35c^3d^2 \div (-7cd^2)$
4. $-42ab^2c^3 \div (-3ac^2)$
5. $12x^{12} \div 3x^3$
6. $-12ab^2m^3 \div 18a^3bm^2$
7. $40x^2y^3z \div 5wxz^3$
8. $-28a^4h^2k^3 \div (-20a^3h^3n^3)$
9. $(6ab - 8ac) \div 2a$
10. $(12c^2y + 20cw) \div 4c$
11. $(16km^2 - 24k^2m) \div (-8km)$
12. $(6xy - 30x^2y - 6xy^2) \div 6xy$
13. $(\pi r^2h + 2\pi rh) \div \pi rh$
14. $(35xy^4 - 20x^2y^3) \div 10xy^4$
15. $(20hk^3r^4 + 30h^2k^2r^2 - 21h^3k^2r^2) \div 12h^2k^2r^3$
16. $(10a^3bc - 12ab^2c^2 + 16ab^4) \div 8a^2b^2c^2$

3.20 Dividing by a Polynomial

Division of a polynomial by a polynomial is governed by essentially the same rules as long division in arithmetic.

Example 34. Divide $3x^2 + 8x + 9$ by $x + 2$.

$$
\begin{array}{r}
3x + 2 \\
x + 2 \overline{)3x^2 + 8x + 9} \\
\underline{3x^2 + 6x} \\
2x + 9 \\
\underline{2x + 4} \\
5
\end{array}
$$

ANS.: $3x + 2 + \dfrac{5}{x + 2}$

As in Example 29, the answer may be checked with reasonable certainty by substituting some simple number for x. In this example, let $x = 10$. Then

Dividend: $(3x^2 + 8x + 9) = 3(10)^2 + 8(10) + 9$
$$= 300 + 80 + 9 = 389$$

Divisor: $(x + 2) = 10 + 2 = 12$

Quotient: $\left(3x + 2 + \dfrac{5}{x + 2}\right) = 3(10) + 2 + \dfrac{5}{10 + 2}$

$$= 30 + 2 + \dfrac{5}{12} = 32\tfrac{5}{12}$$

Now by performing the corresponding numerical division, when $x = 10$, we obtain

$$
\begin{array}{r}
32 \\
12 \overline{)389} \\
\underline{36} \\
29 \\
\underline{24} \\
5
\end{array}
$$

ANS.: $32\tfrac{5}{12}$

Observe that the numerical quotients agree. This constitutes a reasonably good check on the algebraic division. Note that

$$\text{Quotient} = \frac{\text{dividend}}{\text{divisor}} = \text{integral portion} + \frac{\text{remainder}}{\text{divisor}}$$

Therefore we proceed to divide one polynomial by another as follows:

1. *Arrange each polynomial in either descending or ascending powers of some letter common to each expression.*
2. *Divide the first term of the dividend by the first term of the divisor to find the first term of the quotient.*

3. *Multiply the entire divisor by the first term of the quotient and subtract the result from the dividend.*

4. *Consider the remainder obtained in step 3 as a new dividend and repeat steps 2 and 3, continuing the process until there is either no remainder or a remainder whose first term cannot be evenly divided by the first term of the divisor.*

Example 35. Divide $8x - 13x^2 + 6x^3 - 12$ by $2x - 3$.

Rearranging, we obtain

$$
\begin{array}{r}
3x^2 - 2x + 1 \\
2x - 3 \,\overline{\smash{\big)}\, 6x^3 - 13x^2 + 8x - 12} \\
\underline{6x^3 - 9x^2} \\
-4x^2 + 8x \\
\underline{-4x^2 + 6x} \\
2x - 12 \\
\underline{2x - 3} \\
-9
\end{array}
$$

ANS.: $3x^2 - 2x + 1 - \dfrac{9}{2x - 3}$

Note that while remainders in arithmetical division are always positive, they may be either positive or negative in algebraic division.

To check, (integral portion)(divisor) + remainder = dividend. In this illustration,

$$(3x^2 - 2x + 1)(2x - 3) - 9 = 6x^3 - 13x^2 + 8x - 12$$

A check may also be obtained which is usually valid as in Sec. 3.17 by numerical substitution (avoiding the use of 0 and 1, or any value of x which makes the divisor equal zero).

Occasionally the dividend will be lacking one or more intermediate powers of a letter. In this case space should be left where such gaps occur.

Example 36. Divide $36x^4 - 25x^2 + 4$ by $3x - 2$.

It will be noted that there are no x^3 and x terms in the dividend. Hence we arrange as follows:

$$
\begin{array}{r}
12x^3 + 8x^2 - 3x - 2 \\
3x - 2 \,\overline{\smash{\big)}\, 36x^4 -25x^2 + 4} \\
\underline{36x^4 - 24x^3} \\
+ 24x^3 - 25x^2 \\
\underline{24x^3 - 16x^2} \\
-9x^2 \\
\underline{-9x^2 + 6x} \\
-6x + 4 \\
-6x + 4
\end{array}
$$

Example 37. Divide x^3 by $2x + 1$. Continue the division until a remainder containing no x term is obtained.

$$
2x + 1 \overline{\smash{\big)}\, x^3} \quad \dfrac{x^2}{2} - \dfrac{x}{4} + \dfrac{1}{8} - \dfrac{1}{8(2x + 1)}
$$

$$
\underline{x^3 + \dfrac{x^2}{2}}
$$

$$
- \dfrac{x^2}{2}
$$

$$
\underline{- \dfrac{x^2}{2} - \dfrac{x}{4}}
$$

$$
\dfrac{x}{4} + \dfrac{1}{8}
$$

$$
\underline{- \dfrac{1}{8}}
$$

EXERCISE 10

Divide:

1. $a^2 + 11a + 24$ by $a + 3$
2. $c^3 - 2c^2 - 2c + 1$ by $1 + c$
3. $8m^2 - 22mw + 15w^2$ by $2m - 3w$
4. $16z^4 - 1$ by $2z - 1$
5. $6k^6 - 48$ by $3k^2 - 6$
6. $8q^2 - q - 19q^3 + 15q^4 - 1$ by $5q^2 - 3q - 1$
7. $h^6 - 2h^3 + 1$ by $h^2 + h + 1$
8. $y^5 - 2y^4 - 13y^2 + 17y + 5$ by $3y + 1 + y^3 - 4y^2$
9. $64x^6 - 1$ by $2x - 1$
10. 1 by $1 + x$ (to four terms)
11. 1 by $x + 1$ (to four terms)

3.21 Special Products or Expansions

The following special products should be verified by the student and memorized both as literal formulas and as equivalent descriptive sentences.

$a(x + y + z) = ax + ay + az$ (see distributive law)   (1)

For example: $3ab^2(5ax - 8b) = 15a^2b^2x - 24ab^3$

$(x + y)(x - y) = x^2 - y^2$ (2)

For example: $(3cm^2 + 4x)(3cm^2 - 4x) = 9c^2m^4 - 16x^2$

$(x + y)^2 = x^2 + 2xy + y^2$ $\qquad$ (3)

For example: $(5ab + 4w)^2 = 25a^2b^2 + 40abw + 16w^2$

$(x - y)^2 = x^2 - 2xy + y^2$ $\qquad$ (4)

For example: $(2ax - 7x^2y)^2 = 4a^2x^2 - 28ax^3y + 49x^4y^2$

$(x + a)(x + b) = x^2 + (a + b)x + ab$ $\qquad$ (5)

For example: $(x + 5)(x - 2) = x^2 + (5 - 2)x - 10$
$$= x^2 + 3x - 10$$

$(ax + b)(cx + d) = acx^2 + (ad + bc)x + bd$ $\qquad$ (6)

For example: $(3x + 8)(2x - 7) = 6x^2 + (-21 + 16)x - 56$
$$= 6x^2 - 5x - 56$$

$(a + b)(c + d) = ac + ad + bc + bd$ $\qquad$ (7)

For example: $(3e - 2k)(2m + 5n) = 3e(2m + 5n) - 2k(2m + 5n)$
$$= 6em + 15en - 4km - 10kn$$

$(x - y)(x^2 + xy + y^2) = x^3 - y^3$ $\qquad$ (8)

For example: $(a - 3b)(a^2 + 3ab + 9b^2) = a^3 - 27b^3$

$(x + y)(x^2 - xy + y^2) = x^3 + y^3$ $\qquad$ (9)

For example: $(2c + m)(4c^2 - 2cm + m^2) = 8c^3 + m^3$

$(a + b + c)^2 = a^2 + b^2 + c^2 + 2ab + 2ac + 2bc$ $\qquad$ (10)

For example: $(2x - 3y + 5z)^2 = 4x^2 + 9y^2 + 25z^2 - 12xy + 20xz - 30yz$

There are additional special products which have not been listed above, because they arise too infrequently to be memorized. Some of these are as follows:

$(x - y)(x^{n-1} + x^{n-2}y + \cdots + xy^{n-2} + y^{n-1}) = x^n - y^n$ $\qquad$ (11)

$(x + y)(x^{n-1} - x^{n-2}y + \cdots + xy^{n-2} - y^{n-1}) = x^n - y^n$ $\qquad$ (12)

(for even values of n only)

$(x + y)(x^{n-1} - x^{n-2}y + \cdots - xy^{n-2} + y^{n-1}) = x^n + y^n$ $\qquad$ (13)

(for odd values of n only)

$(x + y)^3 = x^3 + 3x^2y + 3xy^2 + y^3$ $\qquad$ (14)

$(x - y)^3 = x^3 - 3x^2y + 3xy^2 - y^3$ $\qquad$ (15)

$(x^2 + xy + y^2)(x^2 - xy + y^2) = x^4 + x^2y^2 + y^4$ $\qquad$ (16)

EXERCISE 11

Obtain the following special products as far as possible as an oral exercise:

1. $4xy(5a - 7b)$
2. $-5m^2x(3bm - 8cx)$
3. $2ab^2c(3ax - 4by + 7cz)$
4. $-8hk(-2ah^2 - 9b^2k)$
5. $(x - 3)(x + 3)$
6. $(7 - c)(7 + c)$
7. $(8k - 3m)(8k + 3m)$
8. $(5c^2 - 4n)(5c^2 + 4n)$
9. $(9pq - 7r)(9pq + 7r)$
10. $(c + 7)^2$
11. $(8 + n)^2$
12. $(3a + 1)^2$
13. $(4b + 3)^2$
14. $(6c + 5k)^2$
15. $(12am + 7y)^2$
16. $(m - 6)^2$
17. $(6 - m)^2$
18. $(h - 8)^2$
19. $(5x - 1)^2$
20. $(1 - 4c)^2$
21. $(7t - 9)^2$
22. $(11w - 4)^2$
23. $(5h - 12n)^2$
24. $(d + 3)(d + 4)$
25. $(m + 7)(m - 3)$
26. $(y + 2)(y - 9)$
27. $(a - 4)(a - 11)$
28. $(x + 1)(x + 6)$
29. $(z - 4)(z - 12)$
30. $(2b + 3)(3b + 1)$
31. $(4c - 1)(2c + 5)$
32. $(5m + 3)(4m - 7)$
33. $(6x - 1)(5x - 2)$

3.22 Factoring

Factoring is the inverse of multiplication such as is illustrated by the problems in Exercise 11. We shall now start with a given product and determine what numbers or factors were multiplied to give this product. We are interested in determining any monomial factors and any prime polynomial factors, i.e., polynomial factors which are divisible by no rational integral expression except themselves or 1.

By reversing each of the expansion rules in Sec. 3.21 we obtain rules for factoring. Each example, upon interchange of right and left members of the identity, becomes an example of factoring.

In the factoring out of a common monomial factor of a polynomial, each literal factor of the monomial is taken the least number of times that it occurs in any one of the terms of the original polynomial.

Of course, factors may be checked by multiplication to give the original expression.

Example 38. Factor $18ab^4x^3 - 30a^2b^2c$.

We obtain (by the distributive law)

$6ab^2(3b^2x^3 - 5ac)$

since 6 is the highest common factor of 18 and 30, and the lowest power of a in any term is 1 and of b is 2, with c and x missing entirely from one term each.

Hint on Factoring. *If an expression appears to be of the type illustrated in Eq. (6), Sec. 3.21, after removing any monomial factor, test out the various factor combinations of the coefficients. We might go further and state that the factoring of any polynomial is simplified by first removing any monomial factors.*

Example 39. Factor $12x^2 + x - 6$.

     Writing the factors in a general form, we have

$$(ax + b)(cx - d)$$

We note that ac must equal 12, $(b)(-d)$ must equal -6, and $bc - ad$ must equal 1. Two possible factors might be $(6x - 2)$ and $(2x + 3)$. Instead of actually multiplying, all we need to do is to write $\dfrac{6 - 2}{2 + 3}$, representing the coefficients and constants arranged for multiplication. A glance at the cross products $(6)(3)$ and $(2)(-2)$ whose sum is $18 - 4$, or 14, shows that this is not the answer. A more careful inspection would have eliminated this choice as a possibility, since $6x - 2$ contains a factor of 2, whereas $12x^2 + x - 6$ does not. It will be seen from the arrangement $\dfrac{3 - 2}{4 + 3}$ that $(3x - 2)$ and $(4x + 3)$ are the required factors.

EXERCISE 12

Factor out the common monomial factor from each of the following:

1. $15x - 18y$
2. $16x + 40y$
3. $30ab - 42km$
4. $ax + ay$
5. $a^2 - ac$
6. $12xy - 20x^2$
7. $abc + bcm + cmx$
8. $a^2b^2 + a^2b + ab^2$
9. $42ax^2y - 24bx^3y^2 + 18cx^4y$
10. $15ab^2c^2 - 12a^2bc^2 + 8a^2b^2c$
11. $10a^2x^2 - 15abxy + 20b^2y^2$
12. $30ax^2y + 48bxy^2 - 36cxyz$
13. The expression $ab - ac$ may be written $a(b - c)$. Make use of this identity to state whether or not 63,549 and 63,574 have any common factors.

EXERCISE 13

Factor completely:

1. $x^2 - 4$
2. $c^2 - 25$
3. $36 - y^2$
4. $16z^2 - 1$
5. $49m^2 - 1$
6. $4a^2 - 9b^2$
7. $25d^2 - 64m^2$
8. $36 - 121a^2b^2$
9. $100y^2z^2 - 49c^2d^2$
10. $121k^4 - 1$
11. $16n^8 - 1$
12. $16 - 81x^{12}$
13. $(25)^2 - (24)^2$
14. $(10\frac{1}{2})^2 - (8\frac{1}{2})^2$

15. Show that $(97)^2$ may be calculated at sight by writing

$$(97)^2 = (97 + 3)(97 - 3) + 3^2 = 9,409$$

(See Sec. B.12.)

16. Show that the number 391 may be expressed as the difference between two squares and, as such, is factorable. Find the factors.

17. If n is any positive integer greater than unity, show that $n^3 - n$ is always divisible by 6.

18. Show that the difference between the squares of two consecutive odd integers must be divisible by 8.

19. Show that the sum of the squares of two odd numbers cannot be a perfect square.

EXERCISE 14

Find the missing terms needed to make the following expressions take the form of the trinomial squares $x^2 + 2xy + y^2$ or $x^2 - 2xy + y^2$. Insert only positive quantities within the parentheses.

1. $x^2 + (\ \ ) + 25$
2. $y^2 + (\ \ ) + 81$
3. $z^2 - (\ \ ) + 49$
4. $m^2 - (\ \ ) + 144$
5. $4a^2 + (\ \ ) + 1$
6. $36c^2 - (\ \ ) + 1$
7. $9n^2 + (\ \ ) + 25$
8. $16 - (\ \ ) + 121k^2$
9. $64b^2 - (\ \ ) + 25h^2$
10. $d^2 + 8d + (\ \ )$
11. $p^2 - 10p + (\ \ )$
12. $a^2 + 6ab + (\ \ )$
13. $c^2 - 14cq + (\ \ )$
14. $4w^2 - 28w + (\ \ )$
15. $9m^2 + 30mx + (\ \ )$
16. $25t^2 - 50tr + (\ \ )$

EXERCISE 15

Factor each of the following trinomial squares:

1. $x^2 - 4x + 4$
2. $b^2 + 10b + 25$
3. $36 - 12m + m^2$
4. $4y^2 - 20y + 25$
5. $9 + 42z + 49z^2$
6. $16c^2 + 24cd + 9d^2$
7. $64a^2 - 80an + 25n^2$
8. $49a^2b^2 - 112abc + 64c^2$
9. $81m^4 - 72m^2p + 16p^2$
10. $121x^2 + 154xyz^2 + 49y^2z^4$

EXERCISE 16

Factor each of the following expressions:

1. $y^2 + 7y + 12$
2. $a^2 + 9a + 20$
3. $m^2 - 12m + 35$
4. $c^2 - 15c + 36$
5. $x^2 + 19x + 48$
6. $b^2 - 2b - 24$

7. $h^2 + h - 30$ 8. $w^2 - 11w - 42$

9. $k^4 + k^2 - 72$ 10. $z^6 - 3z^3 - 108$

11. $x^2 + 21xy + 54y^2$ 12. $y^2 - 25yz + 66z^2$

13. $2a^2 + 3a + 1$ 14. $3c^2 + 10c + 3$

15. $4m^2 - 11m + 6$ 16. $7k^2 - 16k + 4$

17. $8b^2 + 18b + 9$ 18. $17n^2 + 6n - 11$

19. $54h^2 - 15h - 50$ 20. $48x^2 + 14x - 45$

Example 40. Factor

$$3a^2x^4 - 15a^2x^2 + 12a^2 = 3a^2(x^4 - 5x^2 + 4)$$
$$= 3a^2(x^2 - 4)(x^2 - 1)$$
$$= 3a^2(x + 2)(x - 2)(x + 1)(x - 1)$$

EXERCISE 17

In the following miscellaneous expressions first determine whether any factors exist. If an expression is factorable, it is suggested that any monomial factor be removed first. Factor completely into prime factors.

1. $9ac + 15bc$ 2. $2c^3 - 32c^2 + 128c$

3. $4a^2 - 100$ 4. $4m^2 - 20m + 100$

5. $6b^2 - 18b + 12$ 6. $4y^3 - 8y^2 - 32y$

7. $16k^2 + 36$ 8. $9am^2 - 30am + 9a$

9. $9h^2 - 34h + 25$ 10. $100z^2 - 60z + 36$

11. $81x^4 - 16$ 12. $35a^2b^2 - 80a^2b + 20a^2$

13. $4y^2 - 30y + 36$ 14. $24h^2 + 12h - 36$

15. $36t^2 + 120t + 100$ 16. $16rx^2 - 30rx + 9r$

17. $9m^2 - 12m + 16$ 18. $6c^2 - c + 4$

19. $a^3y^3 + 8$ 20. $a^2 + 2ab + 3ac + 6bc$

21. $x^2 + 4y^2 + z^2 + 4xy + 2xz + 4yz$ 22. $x^3 - \frac{1}{8}$

23. $x^3 - x + x^2 - 1$ 24. $27 - m^3$

25. $3ax + 3ay - bx - by$ 26. $a^2 + 4b^2 + 9c^2 - 4ab - 6ac + 12bc$

27. $bm^3 - bn^3$ 28. $y^3 - y + 4y^2 - 4$

29. $25c^2 + 9d^2 + f^2 + 30cd - 10cf - 6df$ 30. $x^6 + 7x^3 - 8$

31. So-called Pythagorean triangles are right triangles having integral sides. In order for the triangle to be primitive there must be no factor common to the sides. A formula for generating such triangles is

$$(m^2 - n^2)^2 + (2mn)^2 = (m^2 + n^2)^2$$

Any integers may be substituted for m and n. If the triangle is to be primitive, m and n must be prime to each other and one must be odd, the other even. Why?

Can you show that the factors 3, 4, and 5 must occur in the sides of a primitive triangle? (These factors may be distributed in various ways among the sides.)

There are an infinite number of such primitive triangles. A few examples are:

m	n	$m^2 - n^2$	$2mn$	$m^2 + n^2$
2	1	3	4	5
3	2	5	12	13
4	1	15	8	17
6	5	11	60	61

3.23 Properties of Fractions

The basic properties of algebraic fractions resemble those of arithmetical fractions. We list some of them below without formal justification.

1. *If both numerator and denominator of a fraction are multiplied or divided by the same number (other than zero), the value of the fraction remains unchanged.*
2. *Multiplying the numerator or dividing the denominator of a fraction by a given number multiplies the fraction by that number.*
3. *Dividing the numerator or multiplying the denominator of a fraction by a given number divides the fraction by that number.*
4. *If $\frac{m}{n}$ $(n \neq 0)$ is a rational number, then*

$$-\left(\frac{m}{n}\right) = \frac{-m}{n} = \frac{m}{-n} \tag{17}$$

Example 41. Simplify $\dfrac{a - b}{b - a}$.

From Eq. (17) we may write

$$\frac{a - b}{b - a} = -\frac{-(a - b)}{b - a} = -\frac{b - a}{b - a} = -1$$

5. *If an even number of factors in a fraction are multiplied by (-1), the sign of the fraction remains unchanged. If an odd number of factors are multiplied by (-1), the sign before the fraction must be reversed (Sec. 3.10).*

Example 42. The rule just expressed states in effect that we must introduce an even number of (-1)'s as factors. Given the expression

$$\frac{(a - 3)(a + 2)}{a - 8}$$

the following modifications are all equivalent:

$$-\frac{(3 - a)(a + 2)}{a - 8} \qquad -\frac{(a - 3)(-a - 2)}{a - 8} \qquad -\frac{(a - 3)(a + 2)}{8 - a}$$

$$\frac{(3 - a)(-a - 2)}{a - 8} \qquad \frac{(3 - a)(a + 2)}{8 - a} \qquad \frac{(a - 3)(-a - 2)}{8 - a}$$

$$-\frac{(3 - a)(-a - 2)}{8 - a}$$

In several of the modifications the binomial $a - 3$ in the original expression was multiplied by (-1) to give us $3 - a$, which is called the *negative* of $a - 3$ (and vice versa).

The original expression with its minimum of minus signs is usually preferred because of its simplicity.

To reduce a fraction to its lowest terms, express the numerator and denominator as products of their prime factors. Then divide both numerator and denominator by all prime factors common to both.

Example 43. $\dfrac{3a + 6b}{a^2 + 3ab + 2b^2} = \dfrac{3\cancel{(a + 2b)}}{(a + b)\cancel{(a + 2b)}} = \dfrac{3}{a + b}$

Example 44. $\dfrac{a^2 + a - 12}{9 - a^2} = \dfrac{(a - 3)(a + 4)}{(3 - a)(3 + a)}$

Multiplying by -1 in two places, we obtain

$$-\frac{\cancel{(a - 3)}(a + 4)}{\cancel{(a - 3)}(3 + a)} = -\frac{a + 4}{a + 3}$$

We may check by substituting some number such as 2 for a in the above equation. Then

$$\frac{a^2 + a - 12}{9 - a^2}$$

becomes

$$\frac{4 + 2 - 12}{9 - 4} = -\frac{6}{5}$$

and

$$-\frac{a + 4}{a + 3}$$

becomes

$$-\frac{2 + 4}{2 + 3} = -\frac{6}{5}$$

Note that any number divided out must be a factor of the entire numerator and denominator.

Example 45. $3bx/(a + b)$ does *not* equal $3x/a$, for b is not a factor of $a + b$; here we have *subtracted* b instead of *dividing*.

A numerical check will reveal this fallacy. If $a = 2$, $b = 3$, and $x = 5$ are substituted in the original fraction, we obtain

$$\frac{3bx}{a + b} = \frac{(3)(3)(5)}{2 + 3} = \frac{45}{5} = 9$$

whereas

$$\frac{3x}{a} = \frac{(3)(5)}{2} = \frac{15}{2} = 7.5$$

Example 46. $\dfrac{3bx}{ab + bc} = \dfrac{3\cancel{b}x}{\cancel{b}(a + c)} = \dfrac{3x}{a + c}$

EXERCISE 18

Express each of the following negative fractions as an equivalent fraction containing a minimum of negative signs:

1. $-\dfrac{y}{5}$ 　　2. $-\dfrac{3}{-a}$ 　　3. $-\dfrac{-x}{-2}$ 　　4. $-\dfrac{1}{1 - x}$

5. $-\dfrac{c - 3}{4}$ 　　6. $-\dfrac{m + 8}{7}$ 　　7. $-\dfrac{a - 1}{a + 1}$ 　　8. $-\dfrac{1 + k}{k - 1}$

9. $-\dfrac{-b-c}{a-c}$ 10. $-\dfrac{1}{(a-b)(a-c)}$ 11. $-\dfrac{x-y}{(x-z)(y+z)}$

12. $-\dfrac{(k-m)^2}{h-k}$ 13. $-\dfrac{(c-d)^3}{p+r}$ 14. $-\dfrac{(a+b)(b-c)}{(b+c)(c-d)}$

EXERCISE 19

Reduce the following fractions to lowest terms where possible (some may already be in lowest terms) using a minimum of negative signs in your answer.

1. $\dfrac{xy}{wx+xz}$ 2. $\dfrac{a}{a^2+a}$ 3. $\dfrac{am+an}{ab-ac}$

4. $\dfrac{8a-12b}{16a+24b}$ 5. $\dfrac{a+b}{a^2+2ab+b^2}$ 6. $\dfrac{k-m}{k^2-m^2}$

7. $\dfrac{2a+4b}{a^2-4b^2}$ 8. $\dfrac{x^2-4x+4}{x^2-4}$ 9. $\dfrac{6a+9b}{4a^2-9b^2}$

10. $\dfrac{2m^2-4m}{m^2+4m-12}$ 11. $\dfrac{9-a^2}{3a^2-9a}$ 12. $\dfrac{a^2-3a+2}{a^2+2a-3}$

13. $\dfrac{x^2-8x+16}{x^2-4}$ 14. $\dfrac{5cm^3-20c^3m}{10cm^2-10c^2m-20c^3}$ 15. $\dfrac{8a^2-16ab+8b^2}{4b^2-4a^2}$

16. $\dfrac{6x^2-12xy+18xz}{9mx+27mz-18my}$ 17. $\dfrac{18x^2-24x+32}{9x^3-16x}$ 18. $\dfrac{4c^2-17c+4}{12-3c}$

19. $\dfrac{(b-2c)^3}{2c^2+3bc-2b^2}$ 20. $\dfrac{3x^2-10xy+3y^2}{6y^2-18xy}$ 21. $\dfrac{x^2-5x+6}{x^2+5x-6}$

22. $\dfrac{a^2-b^2}{(a-b)^2}$ 23. $\dfrac{a^2+b^2}{(a+b)^2}$ 24. $\dfrac{x-y}{y-x}$

25. $\dfrac{(a-m)^2}{(m-a)^2}$

3.24 Multiplying Two or More Fractions

To multiply two or more fractions involving polynomials, write the indicated product of all the factors of the numerators over the indicated product of all the factors of the denominators. Reduce the product to lowest terms if it is not already in lowest terms.

Example 47. $\dfrac{24a^2-54b^2}{35x^2y} \cdot \dfrac{28xy^2}{30a^2+15ab-90b^2}$

$$= \dfrac{6(2a-3b)(2a+3b)}{35x^2y} \cdot \dfrac{28xy^2}{15(2a-3b)(a+2b)}$$

$$= \dfrac{8y(2a+3b)}{25x(a+2b)}$$

3.25 Dividing One Fraction by Another

To divide one fraction by another, multiply the first fraction by the reciprocal of the second and proceed as in Sec. 3.24.

In general, if the product of two numbers is 1, then each is said to be the *reciprocal* of the other. Thus the reciprocal of a/b is b/a since

$$\frac{a}{b} \cdot \frac{b}{a} = 1$$

In Example 48 below, the reciprocal of

$$\frac{k^2 + 6km + 9m^2}{km^2 - 4m^3}$$

is

$$\frac{km^2 - 4m^3}{k^2 + 6km + 9m^2}$$

since

$$\frac{k^2 + 6km + 9m^2}{km^2 - 4m^3} \cdot \frac{km^2 - 4m^3}{k^2 + 6km + 9m^2} = 1$$

Example 48. $\dfrac{k^2 + 9km + 18m^2}{k^2 - 9km + 20m^2} \div \dfrac{k^2 + 6km + 9m^2}{km^2 - 4m^3}$

$$= \frac{k^2 + 9km + 18m^2}{k^2 - 9km + 20m^2} \cdot \frac{km^2 - 4m^3}{k^2 + 6km + 9m^2}$$

$$= \frac{(k + 6m)\cancel{(k + 3m)}}{\cancel{(k - 4m)}(k - 5m)} \cdot \frac{m^2\cancel{(k - 4m)}}{\cancel{(k + 3m)}(k + 3m)}$$

$$= \frac{m^2(k + 6m)}{(k - 5m)(k + 3m)}$$

EXERCISE 20

Carry out the indicated operations:

1. $\dfrac{a^2 - b^2}{16} \cdot \dfrac{12}{a - b}$

2. $\dfrac{x^2 + 2x + 1}{18cw^3} \cdot \dfrac{12c^3w}{x^2 - 1}$

3. $\dfrac{x^2 - x + 12}{abc} \cdot \dfrac{bcd}{x^2 + x - 6}$

4. $\dfrac{6x + 12y}{10a + 5} \cdot \dfrac{100a^2 - 25}{9x^2 - 81y^2}$

5. $\dfrac{a^2 + 2ab + b^2}{a^2 - b^2} \cdot \dfrac{a^2 - 2ab + b^2}{a^2 - b^2}$

6. $\dfrac{x + 3y}{2a + 1} \cdot \dfrac{1 - 4a^2}{x^2 - 9y^2}$

7. $\dfrac{m^2 + 2mn + n^2}{4m^2 + 4mn} \cdot \dfrac{6m^2}{n^2 - m^2}$

8. $\dfrac{x^2 + x - 2}{7b^2x^2 - 14b^2x + 7b^2} \cdot \dfrac{14abx - 28ab}{1 - 2x + x^2}$

9. $\dfrac{a + 2}{15} \div \dfrac{a^2 - 4}{10a}$

10. $\dfrac{a - b}{x + y} \div \dfrac{x - y}{a + b}$

11. $\dfrac{4a - 6x}{9ay} \div \dfrac{9x - 6a}{12y^2}$

12. $\dfrac{15ax^2}{x^2 - 9} \div \dfrac{25a^2bx}{x^2 + x - 12}$

13. $\dfrac{14xy^2z}{4x^2 - 9} \div \dfrac{10x^2y}{6x + 9}$

14. $\dfrac{3c^2 - 12cm + 12m^2}{2c^2 + 2c - 12} \div \dfrac{6c^2 - 24m^2}{8ac + 24a}$

15. $\dfrac{a^2 - 25}{4a - 8} \cdot \dfrac{8 - 2a^2}{5a^2 + 5a - 10} \div \dfrac{5 + a}{15a^2 - 15}$

16. $\dfrac{4x^2 - 20x + 9}{4x^2 - 15x + 9} \cdot \dfrac{5x - 15}{4x - 18} \div (12x^2 - 3)$

3.26 Least Common Multiple

The *least common multiple* (LCM) of several algebraic expressions is the product of all the factors of the various expressions, each factor being taken the greatest number of times that it occurs in any one of the given expressions. In other words, the least common multiple is the smallest quantity which is divisible by all the several algebraic expressions.

Example 49. Find the LCM of $6ax - 9x^2$, $8a^4 - 18a^2x^2$, and $16a^3 + 48a^2x + 36ax^2$.
    The factors of each are $3x(2a - 3x)$, $2a^2(2a - 3x)(2a + 3x)$, and $4a(2a + 3x)^2$. The LCM of 3, 2, and 4 is 12; hence the required LCM is $12a^2x(2a - 3x)(2a + 3x)^2$, each factor appearing to the highest power to which it occurs in any one expression.

3.27 Addition and Subtraction of Fractions

These operations in algebra are entirely comparable with the corresponding arithmetical procedures.

To add or subtract fractions:

1. Reduce to lowest terms any fractions which may be so reduced.
2. Find the least common denominator, which is the LCM of all the denominators.
3. For each fraction, divide the least common denominator (LCD) by its own denominator and multiply both numerator and denominator by this quotient.
4. Write the new numerators obtained in step 3 over a single denominator (LCD). Each polynomial numerator is placed in parentheses preceded by the sign of the fraction from which it originated.
5. Remove the grouping signs, collect terms, and reduce the resulting fraction to lowest terms.

Example 50. Simplify the expression

$$\frac{a}{b} + \frac{c}{b}$$

By the distributive law

$$\frac{a}{b} + \frac{c}{b} = \frac{1}{b}(a) + \frac{1}{b}(c) = \frac{1}{b}(a + c) = \frac{a + c}{b}$$

Example 51. $\dfrac{a}{b} + \dfrac{c}{d} = \dfrac{ad}{bd} + \dfrac{bc}{bd} = \dfrac{1}{bd}(ad + bc) = \dfrac{ad + bc}{bd}$

Note. *The answer cannot be ad + bc, nor can it be* $\dfrac{a + c}{b + d}$. *If we let a = 2, b = 3, c = 4, and d = 5,*

$$\frac{a}{b} + \frac{c}{d} = \frac{2}{3} + \frac{4}{5} = \frac{22}{15}$$

whereas

$$ad + bc = (2)(5) + (3)(4) = 22$$

and

$$\frac{a + c}{b + d} = \frac{2 + 4}{3 + 5} = \frac{6}{8} = \frac{3}{4}$$

both obviously wrong.

Example 52. Simplify

$$\frac{2x^2 - 1}{6x^2} - \frac{2x - 5}{10x} - \frac{2}{15}$$

The LCM of the denominators is $30x^2$. Hence by step 3 we obtain

$$\frac{5(2x^2 - 1)}{30x^2} - \frac{3x(2x - 5)}{30x^2} - \frac{4x^2}{30x^2}$$

and by step 4

$$\frac{5(2x^2 - 1) - 3x(2x - 5) - 4x^2}{30x^2} = \frac{10x^2 - 5 - 6x^2 + 15x - 4x^2}{30x^2}$$

$$= \frac{15x - 5}{30x^2}$$

$$= \frac{3x - 1}{6x^2}$$

Example 53. Add

$$\frac{5x}{2x + 4} + \frac{3x - 1}{x^2 - 4}$$

The factors of the denominators are $2(x + 2)$ and $(x - 2)(x + 2)$. Hence the LCM (and LCD) is $2(x - 2)(x + 2)$. By step 3 we obtain

$$\frac{5x(x - 2)}{2(x - 2)(x + 2)} + \frac{2(3x - 1)}{2(x - 2)(x + 2)}$$

and by step 4

$$\frac{5x(x - 2) + 2(3x - 1)}{2(x - 2)(x + 2)} = \frac{5x^2 - 10x + 6x - 2}{2(x - 2)(x + 2)} = \frac{5x^2 - 4x - 2}{2(x - 2)(x + 2)}$$

Example 54. Combine

$$\frac{3x^2 - 4}{x^2 - 10x + 25} + \frac{7x - 9}{20 - 4x} + 2$$

Factoring denominators,

$$\frac{3x^2 - 4}{(x - 5)^2} + \frac{7x - 9}{4(5 - x)} + \frac{2}{1}$$

(Note that we may write integral expressions such as 2 as $\frac{2}{1}$.) If we write the fraction

$$+ \frac{7x - 9}{4(5 - x)} \qquad \text{as} \qquad - \frac{7x - 9}{4(x - 5)}$$

we obtain a common factor $x - 5$. Therefore the LCD is $4(x - 5)^2$, and we obtain

$$\frac{4(3x^2 - 4)}{4(x - 5)^2} - \frac{(x - 5)(7x - 9)}{4(x - 5)^2} + \frac{8(x - 5)^2}{4(x - 5)^2}$$

$$= \frac{4(3x^2 - 4) - (x - 5)(7x - 9) + 8(x - 5)^2}{4(x - 5)^2}$$

$$= \frac{12x^2 - 16 - (7x^2 - 44x + 45) + 8(x^2 - 10x + 25)}{4(x - 5)^2}$$

$$= \frac{12x^2 - 16 - 7x^2 + 44x - 45 + 8x^2 - 80x + 200}{4(x - 5)^2}$$

$$= \frac{13x^2 - 36x + 139}{4(x - 5)^2}$$

Example 55. Occasionally, two fractions that have the same denominator or one denominator the negative of the other may first be combined to advantage.

Simplify

$$1 + \frac{2a}{2a-1} + \frac{8a^2-4}{1-4a^2} - \frac{2}{4a^2-1} = 1 + \frac{2a}{2a-1} - \frac{8a^2-4}{4a^2-1} - \frac{2}{4a^2-1}$$

$$= 1 + \frac{2a}{2a-1} - \frac{8a^2-2}{4a^2-1}$$

$$= 1 + \frac{2a}{2a-1} - 2$$

$$= \frac{2a}{2a-1} - \frac{2a-1}{2a-1} = \frac{1}{2a-1}$$

EXERCISE 21

Simplify each of the following:

1. $\dfrac{7x}{10} - \dfrac{x}{6}$

2. $\dfrac{5w}{8y} - \dfrac{7w}{12y}$

3. $\dfrac{x}{yz} + \dfrac{y}{xz} + \dfrac{z}{xy}$

4. $\dfrac{3x-2y}{6} - \dfrac{5x+y}{10}$

5. $\dfrac{7a+b}{8} + \dfrac{5a-4b}{12} - \dfrac{-2a+3b}{9}$

6. $\dfrac{7x-3a^2y}{6a^2b} - \dfrac{3x-4by}{8b^2}$

7. $\dfrac{2a+3b-c}{4ab} - \dfrac{a-2b+3c}{6bc} + \dfrac{-3a-b+2c}{8ac}$

8. $\dfrac{1}{a+b} - \dfrac{1}{a+c}$

9. $\dfrac{m+5}{5} - \dfrac{m+6}{6}$

10. $\dfrac{x+y}{x-y} - \dfrac{x-y}{x+y}$

11. $a - \dfrac{c^2}{a}$

12. $\dfrac{4}{5a} - 3$

13. $b + 2 + \dfrac{1}{b}$

14. $\dfrac{6}{k+3} + k - 2$

15. $\dfrac{a-2b}{a^2-b^2} + \dfrac{2}{a-b}$

16. $\dfrac{k+2m}{k^2-9m^2} + \dfrac{4}{3m-k}$

17. $\dfrac{x-5}{x-6} + \dfrac{2x-8}{x^2-10x+24}$

18. $y^2 + y + 1 + \dfrac{1}{y-1}$

19. $\dfrac{2-c}{c^2+c-6} - \dfrac{5}{9-c^2} - \dfrac{4-c}{c^2-7c+12}$

20. $\dfrac{a+b}{(b-c)(c-a)} - \dfrac{b+c}{(a-c)(a-b)} + \dfrac{a+c}{(a-b)(b-c)}$

21. $\dfrac{3}{c-d} + \dfrac{4d}{(c-d)^2} - \dfrac{5d^2}{(c-d)^3}$

22. $\dfrac{1}{a-b} - \dfrac{2b}{a^2-ab} + \dfrac{b^2}{a^3-a^2b}$

23. $\dfrac{x^2}{2} - \dfrac{x}{4} + \dfrac{1}{8} - \dfrac{1}{8(2x+1)}$ (see Example 37)

24. $\dfrac{1}{b}(a+bx)^2 - \dfrac{a}{b}(a+bx)$ 25. $\dfrac{1}{b^2}\left[\dfrac{b}{(a+bx)^2} - \dfrac{ab}{(a+bx)^3}\right]$

26. $\dfrac{1}{ad-bc}\left(\dfrac{d}{c+dx} - \dfrac{b}{a+bx}\right)$ 27. $\dfrac{b}{a(a+bx)^2} + \dfrac{1}{a^2}\left(\dfrac{b}{a+bx} - \dfrac{1}{x}\right)$

28. $\dfrac{1}{b^3}\left[\dfrac{b}{(a+bx)^2} - \dfrac{2ab}{(a+bx)^3} + \dfrac{a^2b}{(a+bx)^4}\right]$

3.28 Complex Fractions

A complex fraction is a fraction whose numerator or denominator or both contain fractions.

Example 56. $\dfrac{\dfrac{a^2}{b^2} - 1}{\dfrac{a}{b} + 1}$

is a complex fraction. For convenience we shall refer to $a^2/b^2 - 1$ as the primary numerator and $a/b + 1$ as the primary denominator, while b^2 and b will be known as *secondary denominators.*

To simplify a complex fraction reduce the primary numerator and the primary denominator each to a single fraction and divide the primary numerator by the primary denominator.

Example 57. Simplify

$$\dfrac{4 - \dfrac{4}{a} + \dfrac{1}{a^2}}{1 - \dfrac{1}{4a^2}}$$

Simplifying the primary numerator, we obtain

$$4 - \dfrac{4}{a} + \dfrac{1}{a^2} = \dfrac{4a^2}{a^2} - \dfrac{4a}{a^2} + \dfrac{1}{a^2} = \dfrac{4a^2 - 4a + 1}{a^2} = \dfrac{(2a-1)^2}{a^2}$$

Simplifying the primary denominator, we have

$$1 - \frac{1}{4a^2} = \frac{4a^2}{4a^2} - \frac{1}{4a^2} = \frac{4a^2 - 1}{4a^2} = \frac{(2a + 1)(2a - 1)}{4a^2}$$

Then

$$\frac{(2a - 1)^2}{a^2} \div \frac{(2a + 1)(2a - 1)}{4a^2} = \frac{(2a - 1)(2a - 1)}{a^2} \cdot \frac{4a^2}{(2a + 1)(2a - 1)} = \frac{4(2a - 1)}{2a + 1}$$

This operation can often be performed more easily and quickly by finding the LCM of the *secondary* denominators. Multiply the *primary* numerator and denominator by the LCM just found and reduce to lowest terms.

Using the same complex fraction

$$\frac{4 - \dfrac{4}{a} + \dfrac{1}{a^2}}{1 - \dfrac{1}{4a^2}}$$

as in the preceding illustration, we note the LCM of the secondary denominators is $4a^2$. We then write

$$\frac{4a^2 \left(4 - \dfrac{4}{a} + \dfrac{1}{a^2} \right)}{4a^2 \left(1 - \dfrac{1}{4a^2} \right)} = \frac{16a^2 - 16a + 4}{4a^2 - 1} = \frac{4(4a^2 - 4a + 1)}{(2a + 1)(2a - 1)}$$

$$= \frac{4(2a - 1)(2a - 1)}{(2a + 1)(2a - 1)} = \frac{4(2a - 1)}{2a + 1}$$

The short method often makes it possible to simplify less involved complex fractions at sight; e.g.,

$$\frac{x - \dfrac{1}{x}}{1 + \dfrac{1}{x}} = \frac{x \left(x - \dfrac{1}{x} \right)}{x \left(1 + \dfrac{1}{x} \right)} = \frac{x^2 - 1}{x + 1} = x - 1$$

A special type of complex fraction is the continued fraction, which is typified by the expression

$$\cfrac{1}{1 + \cfrac{2}{1 + \cfrac{1}{1 + \cfrac{3}{x}}}}$$

Simplification is best effected by starting with the end fraction.

$$\cfrac{1}{1+\cfrac{2}{1+\cfrac{1}{1+\cfrac{3}{x}}}} = \cfrac{1}{1+\cfrac{2}{1+\cfrac{x}{x+3}}} = \cfrac{1}{1+\cfrac{2}{\cfrac{2x+3}{x+3}}} = \cfrac{1}{1+\cfrac{2x+6}{2x+3}}$$

$$= \cfrac{1}{\cfrac{4x+9}{2x+3}} = \cfrac{2x+3}{4x+9}$$

EXERCISE 22

Simplify the following complex fractions:

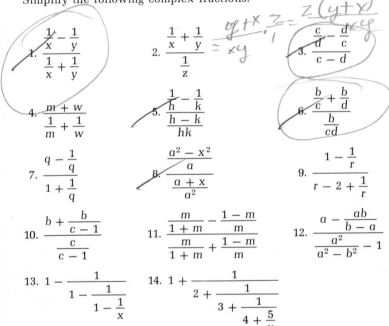

1. $\dfrac{\dfrac{1}{x}-\dfrac{1}{y}}{\dfrac{1}{x}+\dfrac{1}{y}}$

2. $\dfrac{\dfrac{1}{x}+\dfrac{1}{y}}{\dfrac{1}{z}}$

3. $\dfrac{\dfrac{c}{d}-\dfrac{d}{c}}{c-d}$

4. $\dfrac{m+w}{\dfrac{1}{m}+\dfrac{1}{w}}$

5. $\dfrac{\dfrac{1}{h}-\dfrac{1}{k}}{\dfrac{h-k}{hk}}$

6. $\dfrac{\dfrac{b}{c}+\dfrac{b}{d}}{\dfrac{b}{cd}}$

7. $\dfrac{q-\dfrac{1}{q}}{1+\dfrac{1}{q}}$

8. $\dfrac{\dfrac{a^2-x^2}{a}}{\dfrac{a+x}{a^2}}$

9. $\dfrac{1-\dfrac{1}{r}}{r-2+\dfrac{1}{r}}$

10. $\dfrac{b+\dfrac{b}{c-1}}{\dfrac{c}{c-1}}$

11. $\dfrac{\dfrac{m}{1+m}-\dfrac{1-m}{m}}{\dfrac{m}{1+m}+\dfrac{1-m}{m}}$

12. $\dfrac{a-\dfrac{ab}{b-a}}{\dfrac{a^2}{a^2-b^2}-1}$

13. $1-\dfrac{1}{1-\dfrac{1}{1-\dfrac{1}{x}}}$

14. $1+\dfrac{1}{2+\dfrac{1}{3+\dfrac{1}{4+\dfrac{5}{x}}}}$

exponents and radicals

In Chap. 3 we briefly discussed positive integral exponents. In this chapter we shall extend our study to include negative, zero, and fractional exponents and the equivalent radical notation.

4.1 Laws of Positive Integral Exponents

We recall from Sec. 3.15 that a positive integral exponent of a number indicates how many times that number is to be taken as a factor. Thus we define the symbol a^n when a is any number and n is a positive integer.

Definition 1. If n is a positive integer and a is any number, then

$$a^n = a \cdot a \cdot a \cdot a \cdot \cdots \cdot a \qquad \text{(to } n \text{ factors of } a\text{)}$$

Observe that a is one of n equal factors of a^n. The symbol a^n is read "the nth power of a," or "a to the nth power." The symbol a is called the *base*, and n is called the *exponent*.

By applying Definition 1, we may write

$a^4 = a \cdot a \cdot a \cdot a$
$a^3 = a \cdot a \cdot a$
$a^2 = a \cdot a$
$a^1 = a$

The symbol a^2 is called the "second power" of a, or "a squared." The symbol a^3 is called the "third power" of a, or "a cubed."

The following observations can be made regarding the signs of the powers of numbers.

An even power of a positive number is a positive number.

$$3 \cdot 3 \cdot 3 \cdot 3 = 3^4 = +81$$

An even power of a negative number is a positive number.

$$(-3)(-3)(-3)(-3) = (-3)^4 = +81$$

An odd power of a negative number is a negative number.

$$(-3)(-3)(-3) = (-3)^3 = -27$$

An odd power of a positive number is a positive number.

$$(+3)(+3)(+3) = (+3)^3 = +27$$

Equations (1) to (5) below result as a direct consequence of Definition 1.

$$3^2 \cdot 3^5 = (3 \cdot 3)(3 \cdot 3 \cdot 3 \cdot 3 \cdot 3) = 3^7$$

or

$$3^2 \cdot 3^5 = 3^{(2+5)} = 3^7$$

In general,

$$a^m \cdot a^n = a^{(m+n)} \tag{1}$$

For example,

$$x^7 \cdot x^5 = x^{12} \qquad \text{and} \qquad 2^3 \cdot 2^4 = 2^7 = 128$$

The power of a power can be illustrated as

$$(3^2)^5 = (3 \cdot 3)(3 \cdot 3)(3 \cdot 3)(3 \cdot 3)(3 \cdot 3)$$

$$3 \cdot 3 \cdot 3 \cdot 3 \cdot 3 \cdot 3 \cdot 3 \cdot 3 \cdot 3 \cdot 3 = 3^{2 \cdot 5} = 3^{10}$$

or

$$(3^5)^2 = (3 \cdot 3 \cdot 3 \cdot 3 \cdot 3)(3 \cdot 3 \cdot 3 \cdot 3 \cdot 3)$$
$$= 3 \cdot 3 \cdot 3 \cdot 3 \cdot 3 \cdot 3 \cdot 3 \cdot 3 \cdot 3 \cdot 3 = 3^{5 \cdot 2} = 3^{10}$$

In general,

$$(a^m)^n = (a^n)^m = a^{mn} \tag{2}$$

The quotient of two powers of the same base can be illustrated as

$$\frac{3^7}{3^2} = \frac{3 \cdot 3 \cdot 3 \cdot 3 \cdot 3 \cdot 3 \cdot 3}{3 \cdot 3} = 3 \cdot 3 \cdot 3 \cdot 3 \cdot 3 = 3^5$$

or

$$\frac{3^7}{3^2} = 3^{(7-2)} = 3^5$$

As a slightly different example,

$$\frac{3^2}{3^7} = \frac{3 \cdot 3}{3 \cdot 3 \cdot 3 \cdot 3 \cdot 3 \cdot 3 \cdot 3} = \frac{1}{3 \cdot 3 \cdot 3 \cdot 3 \cdot 3} = \frac{1}{3^5}$$

or

$$\frac{3^2}{3^7} = \frac{1}{3^{(7-2)}} = \frac{1}{3^5}$$

In general,

$$\frac{a^m}{a^n} = \begin{cases} a^{(m-n)} & \text{if } m > n \text{ and } a \neq 0 \\ \dfrac{1}{a^{(n-m)}} & \text{if } m < n \text{ and } a \neq 0 \end{cases} \tag{3}$$

The restrictions on Eq. (3) are necessary at this point since we are considering positive integral exponents only, in addition to the obvious requirement that $a \neq 0$

since division by zero is undefined. For example,

$$\frac{x^8}{x^5} = x^3 \qquad \frac{x^5}{x^8} = \frac{1}{x^3} \qquad \frac{5^7}{5^4} = 5^3 = 125$$

$$\frac{5^6}{5^8} = \frac{1}{5^2} = \frac{1}{25}$$

The power of a product can be illustrated as

$$(3 \cdot 2)^5 = (3 \cdot 2)(3 \cdot 2)(3 \cdot 2)(3 \cdot 2)(3 \cdot 2)$$
$$= 3 \cdot 3 \cdot 3 \cdot 3 \cdot 3 \cdot 2 \cdot 2 \cdot 2 \cdot 2 \cdot 2 = 3^5 \cdot 2^5$$

As a direct consequence of Definition 1, if a and b are any numbers, then

$$(a \cdot b)^n = a^n b^n \tag{4}$$

For example,

$$(3xy^3)^4 = 3^4 x^4 (y^3)^4 = 81x^4 y^{12}$$

and

$$(2x^2)^3 = (2)^3 (x^2)^3 = 8x^6$$

The quotient of two like powers can be illustrated as

$$\frac{3^5}{2^5} = \frac{3 \cdot 3 \cdot 3 \cdot 3 \cdot 3}{2 \cdot 2 \cdot 2 \cdot 2 \cdot 2} = \frac{3}{2} \cdot \frac{3}{2} \cdot \frac{3}{2} \cdot \frac{3}{2} \cdot \frac{3}{2} = \left(\frac{3}{2}\right)^5$$

In general,

$$\frac{a^n}{b^n} = \left(\frac{a}{b}\right)^n \qquad b \neq 0 \tag{5}$$

For example,

$$\left(\frac{3}{5}\right)^4 = \frac{3^4}{5^4} = \frac{81}{625} \qquad \text{and} \qquad \left(\frac{4x^2}{3y^3}\right)^3 = \frac{4^3(x^2)^3}{3^3(y^3)^3} = \frac{64x^6}{27y^9}$$

In the special case where $a = 1$, we may write

$$\frac{1}{3} \cdot \frac{1}{3} \cdot \frac{1}{3} \cdot \frac{1}{3} = \left(\frac{1}{3}\right)^4 = \frac{1 \cdot 1 \cdot 1 \cdot 1}{3 \cdot 3 \cdot 3 \cdot 3} = \frac{1^4}{3^4} = \frac{1}{3^4}$$

In general,

$$\left(\frac{1}{b}\right)^n = \frac{1}{b^n} \qquad b \neq 0 \tag{6}$$

EXERCISE 1

Carry out the indicated operations according to the laws of exponents.

1. $x^4 \cdot x^2$ 2. $y^j \cdot y^6$ 3. $b^x \cdot b^3$ 4. $c^4 \cdot c^n$

5. $3^2 \cdot 3^3$ 6. $a^x \cdot a^{1-x}$ 7. $a^{2y} \cdot a^{3y}$ 8. $(-2)^4$

9. $(-1)(-4)^2$ 10. $-(-5)^2$ 11. $(-2x)^3$ 12. $-(2x)^3$

13. $-(-2x)^3$ 14. $(a^2)^5$ 15. $(y^3)^4$ 16. $(-x^3)^2$

17. $(a^n)^2$ 18. $(a^2)^n$ 19. $(a^n)^n$ 20. $(-5x^4)^2$

21. $(\frac{5}{3})^2$ 22. $(-\frac{3}{4})^3$ 23. $(\frac{2}{3}x)^5$ 24. $(-\frac{3}{2}x)^3$

25. $(0.2x)^3$ 26. $(-0.3x^2)^4$ 27. $(a^n b^2)^3$ 28. $(a^2 b^3)^n$

29. $(4x^2 y^n)^n$ 30. $(-2x^3)(-2x)^3$ 31. $xy^4(xy)^4$ 32. $\dfrac{x^7}{x^4}$

33. $\dfrac{a^5}{a^3}$ 34. $\dfrac{b^8}{b}$ 35. $\dfrac{y^2}{y^6}$ 36. $\dfrac{x^3}{x^8}$

37. $\dfrac{a}{a^5}$ 38. $\dfrac{a^2 b^3}{b^5}$ 39. $\dfrac{x^3 y^3}{x^2 y^4}$ 40. $\dfrac{x^5}{x^7 y^3}$

41. $\dfrac{(6x^3)^2}{(4x^2)^3}$ 42. $\dfrac{(8a^2 b^3)^2}{(6ab^2)^3}$ 43. $\dfrac{(xy^2)^n}{(x^n y)^2}$ 44. $\dfrac{a^n b^{n+3}}{a^{n+1} b^{n+2}}$

45. $\dfrac{10ab^3}{15(ab)^3}$ 46. $\dfrac{0.6ab^2}{(0.2ab)^2}$ 47. $\dfrac{(ab^2 c^3)^n}{(a^n b^n c^n)^2}$ 48. $\dfrac{(3)^{4m}}{(9)^{2m}}$

49. $\dfrac{4^4 + 4^4 + 4^4 + 4^4}{2^4}$ 50. $2^n + 2^n + 2^n + 2^n$

4.2 Roots and Radicals

If $a^n = b$, where n is a positive integer, not only is b equal to the nth power of a, but by definition a is said to be an nth root of b. If $n = 2$, a is called the "second root," or the "square root" of b. When $n = 3$, a is called the "third root," or the "cube root" of b. When $n = 5$, a is called the "fifth root" of b, and so on.

Example 1. Since $2^3 = 8$, 2 is a cube root of 8.

Example 2. Since $(-2)^3 = -8$, -2 is a cube root of -8.

Example 3. Since $(-3)^2 = 9$, and $3^2 = 9$, both -3 and 3 are square roots of 9.

Example 4. Since $2^5 = 32$, 2 is a fifth root of 32.

From the above examples it is apparent that a number may have more than one nth root. On the other hand, we have, at this point, no method at all for finding an *even* root of a *negative* number. That is, we have no method for finding a square root, fourth, sixth, eighth, etc., root of any negative number. For example, we can find the square root of $+25$, and we know that there are two such roots: $+5$ and -5. However, we have, at the moment, no method for finding two *equal* factors which will multiply and give -25. The fact that $(-5)(+5) = -25$ has no bearing on the problem, because -5 and $+5$ are not equal factors.

We shall deal with the problem of finding even roots of negative numbers in Sec. 4.14.

For the present we shall exclude from the discussion any consideration of even roots of negative numbers.

This brings us to the definition of principal root.

4.3 Principal Roots

The nth root of the quantity b is written $\sqrt[n]{b}$, except that when $n = 2$, we simply write $\sqrt{b}$.

In this notation $\sqrt[n]{b}$ is called a *radical,* the symbol $\sqrt{}$ is called a *radical sign,* n is the *index* of the root, and b is called the *radicand.*

Without further definition the symbol $\sqrt[n]{b}$ would be somewhat ambiguous. For example, does $\sqrt{9}$ means $+3$ or -3? The answer is found in the concept of *principal root.*

Excluding even roots of negative numbers, which we are not yet ready to discuss, the principal root of a number has the same sign as that number itself. Thus $\sqrt{+16} = +4$, and $+4$ is the principal square root of $+16$; $\sqrt[3]{-8} = -2$, and the principal cube root of -8 is -2; $\sqrt[3]{+8} = +2$, and the principal cube root of $+8$ is $+2$.

Example 5. $\sqrt{9} = 3$. Note that $\sqrt{9}$ is not equal to -3 even though -3 is one of the square roots of 9. When we wish to designate the negative square root of 9, we write $-\sqrt{9}$, so that $-\sqrt{9} = -3$ and $\sqrt{9} = 3$.

Example 6. $\sqrt{36} = 6$, since $6^2 = 36$.

Example 7. $\sqrt[3]{-125} = -5$, since $(-5)^3 = -125$.

Example 8. $\sqrt[4]{16} = 2$, since $2^4 = 16$.

Example 9. $\sqrt[5]{243} = 3$, since $3^5 = 243$.

Example 10. $\sqrt{\frac{9}{25}} = \frac{3}{5}$, since $(\frac{3}{5})^2 = \frac{9}{25}$.

Example 11. $\sqrt[3]{8x^6y^9} = 2x^2y^3$, since $(2x^2y^3)^3 = 8x^6y^9$.

The foregoing examples suggest a more formal definition of principal root which applies only with respect to *real numbers.*

Definition 2. Let a, b, and n be real numbers, where n is a positive integer. The symbol $\sqrt[n]{a}$ is used to designate the *principal* nth root of a.

1. If a is a positive number and n is either an even or an odd number,

$$\sqrt[n]{a} = \text{positive } n\text{th root of } a \quad \text{(see Examples 6, 8, 9, 10, and 11)} \qquad (7)$$

2. If a is a negative number and n is an odd number,

$$\sqrt[n]{a} = \text{only real } n\text{th root of } a, \text{ this root being a negative number}$$
$$\text{(see Example 7)} \qquad (8)$$

3. If a is a negative number and n is an even number,

$$\sqrt[n]{a} = \text{an imaginary number (to be discussed in Sec. 4.14)} \qquad (9)$$

EXERCISE 2

In Probs. 1 to 8, write in radical form the principal roots of the numbers indicated and find their values.

1. The cube root of -8
2. The cube root of 8
3. The fourth root of 81
4. The fifth root of -32
5. The square root of $\frac{1}{25}$
6. The sixth root of $\frac{1}{64}$
7. The square root of 49
8. The cube root of $\frac{27}{125}$

Find the principal value of each of the following:

9. $\sqrt{25}$ 10. $\sqrt{121}$ 11. $\sqrt{64}$ 12. $\sqrt{225}$

13. $\sqrt[3]{27}$ 14. $\sqrt[3]{64}$ 15. $\sqrt{144}$ 16. $\sqrt[3]{-27}$

17. $\sqrt[4]{16}$ 18. $\sqrt[3]{1}$ 19. $\sqrt[3]{-1}$ 20. $\sqrt[5]{32}$

21. $\sqrt{0.04}$ 22. $\sqrt[3]{0.125}$ 23. $\sqrt{1600}$ 24. $\sqrt{\frac{9}{49}}$

25. $\sqrt[3]{-\frac{27}{125}}$ 26. $\sqrt[3]{\frac{1}{8}}$ 27. $\sqrt{4a^2b^4c^6}$ 28. $\sqrt[3]{8m^6n^9}$

29. $\sqrt[4]{\dfrac{a^4 x^8}{c^{12}}}$ 	30. $\sqrt[5]{\dfrac{-x^{10} y^{20}}{32}}$

31. $\sqrt{0.0169x^{16}}$ 	32. $\sqrt[3]{0.008a^{27}}$

4.4 Rational and Irrational Numbers

In Sec. 3.3 the distinction between rational and irrational numbers was brought out. A real *rational* number can be expressed as the *ratio* of one integer to another. *Irrational* numbers are numbers that cannot be so expressed.

An example of an irrational number is $\sqrt{2}$. There exists a real number, approximately 1.414, which, when squared, equals 2. But this number ($\sqrt{2}$), though real, cannot be expressed by an integer or the ratio of two integers. Other examples of real but irrational numbers are $\sqrt[3]{5}$, $\sqrt{6}$, $\sqrt[5]{-7}$, and π.

4.5 Fractional Exponents

So far we have been considering only positive integral values of n. We shall now attempt to preserve the laws already established for positive integral exponents (with necessary added restrictions) and apply them to negative, fractional, and zero exponents. Then we shall see what significance we must, as a consequence, attribute to negative, fractional, and zero exponents.

By Definition 1 the expression 9^2 means that 9 is taken twice as a factor, but it would be meaningless to say that $9^{1/2}$ indicates that 9 is taken one-half a time as a factor.

However, if in Eq. (1) we let $m = n = \frac{1}{2}$, we have

$$9^{1/2} \cdot 9^{1/2} = 9^{1/2+1/2} = 9^1 = 9$$

Evidently, then, $9^{1/2}$ is one of two equal factors of 9. Hence we are prompted to write

$$9^{1/2} = \sqrt{9} = 3$$

Similarly, $5^{1/3}$ assumes meaning if we write

$$5^{1/3} \cdot 5^{1/3} \cdot 5^{1/3} = 5^{(1/3+1/3+1/3)} = 5^1 = 5$$

Since $5^{1/3}$ is in this case one of three equal factors of 5,

$$5^{1/3} = \sqrt[3]{5}$$

In general, if n is a positive integer,

$$a^{1/n} \cdot a^{1/n} \cdot a^{1/n} \cdots \text{(to } n \text{ factors)} = a^{n/n} = a^1 = a$$

where $a^{1/n}$ is one of n equal factors of a. Thus we are led to make the following definition.

Definition 3. If a is a real number and n is a positive integer, then

$$a^{1/n} = \sqrt[n]{a} \tag{10}$$

where $\sqrt[n]{a}$ designates the *principal* nth root of a. We recall, however, that $\sqrt[n]{a}$ does not exist in the set of real numbers if a is a negative number while n is an even number. However, even roots of negative numbers do have meaning, and are known as *imaginary numbers*. We shall discuss these in Sec. 4.14.

Let us see what significance we may attach to the expression $5^{1/4} \cdot 5^{1/4} \cdot 5^{1/4}$. Following Definition 1, we may write

$$5^{1/4} \cdot 5^{1/4} \cdot 5^{1/4} = 5^{3/4}$$

Since $5^{1/4}$ is evidently one of three equal factors of $5^{3/4}$, it seems reasonable to write

$$5^{3/4} = (5^{1/4})^3 = (\sqrt[4]{5})^3 \tag{11}$$

Now let us see what significance we can attach to the expression $5^{3/4} \cdot 5^{3/4} \cdot 5^{3/4} \cdot 5^{3/4}$. Following Definition 1, we may write

$$5^{3/4} \cdot 5^{3/4} \cdot 5^{3/4} \cdot 5^{3/4} = 5^{12/4} = 5^3 \tag{12}$$

However, $5^{3/4}$ is evidently one of four equal factors of 5^3. Thus we may write

$$5^{3/4} = \sqrt[4]{5^3}$$

Consequently, from Eqs. (11) and (12) it appears that

$$5^{3/4} = (\sqrt[4]{5})^3 = \sqrt[4]{5^3}$$

and we are prompted to make the following definition.

Definition 4. If m and n are both integers with n a positive number, and if $\sqrt[n]{a}$ exists as a real number, then

$$a^{m/n} = (\sqrt[n]{a})^m = \sqrt[n]{a^m} \tag{13}$$

The symbol $\sqrt{a^2}$ deserves special attention. From the laws already established we find that

$$\sqrt{(+3)^2} = \sqrt{+9} = +3$$

and

$$\sqrt{(-3)^2} = \sqrt{+9} = +3$$

Accordingly, we define $\sqrt{a^2}$ to be

$$\sqrt{a^2} = |a| \tag{13a}$$

not simply a.

4.6 Zero exponent

If a is any number except zero, we can define the symbol a^0 by writing

$$a^0 = 1 \qquad a \neq 0 \tag{14}$$

This interpretation is entirely consistent, since by Sec. 3.15

$$\frac{a^m}{a^m} = 1$$

But by Eq. (3), Sec. 4.1,

$$\frac{a^m}{a^m} = a^{m-m} = a^0$$

Hence the two results must be equivalent.

4.7 Negative Exponents

We may define the symbol a^{-n} as follows:

$$a^{-n} = \frac{1}{a^n} \quad \cdot \quad \text{and} \qquad a^n = \frac{1}{a^{-n}} \tag{15}$$

This conclusion follows from Eq. (3), Sec. 4.1, if we let $m = 0$.

$$\frac{a^m}{a^n} = a^{m-n}$$

And if $m = 0$,

$$\frac{1}{a^n} = \frac{a^0}{a^n} = a^{0-n} = a^{-n}$$

Note that from Eq. (15) any *factor* (not any term) of the numerator or denominator of a fraction may be transferred to the denominator or numerator, as the case may be, by reversing the sign of its exponent.

Example 12. $16^{3/4} = \sqrt[4]{16^3}$ or more conveniently $(\sqrt[4]{16})^3 = 2^3 = 8$

Example 13. $27^{2/3} = \sqrt[3]{27^2}$ or more conveniently $(\sqrt[3]{27})^2 = 3^2 = 9$

Example 14. $12^0 = 1$

Example 15. $x^0 = 1 \qquad x \neq 0$

Example 16. $(4x - 7)^0 = 1 \qquad x \neq \frac{7}{4}$

Example 17. $3x^{-2} = \frac{3}{x^2} \qquad \left(\text{note that we do not obtain } \frac{1}{9x^2}\right)$

Example 18. $(3x)^{-2} = \frac{1}{(3x)^2} = \frac{1}{9x^2}$

Example 19. $8^{-2/3} = \frac{1}{8^{2/3}} = \frac{1}{(\sqrt[3]{8})^2} = \frac{1}{2^2} = \frac{1}{4}$

Example 20. $\frac{5x^3y^2}{z^4} = 5x^3y^2z^{-4}$

Example 21. $\frac{4a^3c}{by^4} = 4a^3b^{-1}cy^{-4}$

Example 22. $\frac{1}{x^{-1} - y^{-1}} = \frac{1}{1/x - 1/y} = \frac{1}{(y-x)/xy} = \frac{xy}{y-x}$

(A simple solution at sight depends upon multiplying numerator and denominator by xy. See Sec. 3.28.)

EXERCISE 3

Find the value of each expression.

1. $16^{1/4}$ 2. $36^{1/2}$ 3. 5^{-2} 4. 7^{-1}

5. $(-3)^{-3}$ 6. $(49)^{-1/2}$ 7. $(\tfrac{2}{3})^{-3}$ 8. $(\tfrac{1}{32})^{1/5}$

9. $(-\tfrac{1}{32})^{-1/5}$ 10. $(0.09)^{1/2}$ 11. $(0.16)^{-1/2}$ 12. $\dfrac{3a^0 - b^0}{a^0 + (3b)^0}$

13. $(-3)^4$ 14. $-(-3)^4$ 15. $(-3)^{-4}$ 16. $-(3)^{-4}$

17. 10^{-3} 18. $(-\tfrac{1}{6})^{-1}$ 19. $4^{-1} + 2^{-2}$ 20. $(0.001)^{-1/3}$

21. $(-8)^{5/3}$ 22. $(-27)^{4/3}$ 23. $(8^{-1} - 4^{-1})^{1/3}$ 24. $(-\tfrac{27}{64})^{-2/3}$

Write each expression without negative or zero exponents and simplify.

25. $x^{2/3} \cdot x^{1/2}$ 26. $y^5 \cdot y^{-5}$ 27. $(16x^{16})^{1/2}$

28. $(3x^{4/3})^3$ 29. $(-5^{1/2}x^{3/4}y)^4$ 30. $(2^{1/3}x^{1/4}y^{1/6})^{12}$

31. $-(-3a^{0.4}b^{0.6})^5$ 32. $4^{-2}ab^{-3}c$ 33. $\dfrac{3x^{-3}y^2}{6^{-1}z^{-2}}$

34. $\dfrac{3^{-1}x^6}{8y^{-3}}$ 35. $\dfrac{x^0y^{-1}}{x^2w^{-4}}$ 36. $\left(\dfrac{10^{-3}a^6}{90}\right)^{1/2}$

37. $(x^{1/2} - x^{-1/2})^2$ 38. $(m^{1/2} + n^{1/2})(m^{1/2} - n^{1/2})$

39. $(a^{1/2} - 3a^{-1/2})(a^{1/2} + 3a^{-1/2})$ 40. $(x^{3/2} + x^{1/2})^4$

Evaluate, simplify, or solve as indicated:

41. $\dfrac{2^{n+4} + 2^n + 2^n}{2(2^{n+3} - 2^{n+1})}$ 42. $(a^{-1} - b^{-1})(a - b)^{-1}$ 43. $(2^{1/2} - 2^{-1/2})^{-2}$

44. If $4^{n+3} - 120 = 4^{n+1}$, solve for n.   45. $2^n + 2^n$ 46. If $9^x + 9^{x-1} = 30$, solve for x.

Write the following expressions without denominators, using negative exponents where necessary.

47. $\dfrac{a}{b^3}$ 48. $\dfrac{1}{xy}$ 49. $\dfrac{z^2}{4^2}$ 50. $\dfrac{3k}{ab^2}$

51. $\dfrac{x^{1/2}y^2}{x^{1/3}y^{5/2}}$ 52. $\left(\dfrac{a^{1/2}b^{1/3}}{x^2}\right)^{-6}$ 53. $\left(\dfrac{a^3x^{-1}}{2by^{-2}}\right)^{-3}$ 54. $\left(\dfrac{64x^{1/3}}{y^{2/3}z^{1/2}}\right)^{1/6}$

Write the following expressions in radical notation.

55. $x^{1/2}$ 56. $y^{3/2}$ 57. $a^{-1/2}$ 58. $8x^{2/3}$

59. $x^{3/4}y^{1/4}$ 60. $(4xy^2)^{1/3}$ 61. $(5x^2)^{-1/3}$ 62. $(a^3b^2)^{1/6}$

Using fractional exponents, write the following expressions without radicals.

63. $\sqrt{a^3}$ 64. $\sqrt[3]{x^2y}$ 65. $\sqrt[4]{x+y}$ 66. $\sqrt[6]{a^3b^2}$

67. $\sqrt[4]{a^2b^4c^6}$ 68. $\sqrt{9ab^2c^3}$ 69. $\sqrt[3]{\dfrac{a^2}{8x^6}}$ 70. $\sqrt{16x^2y^3z^4}$

4.8 Laws of Radicals

The following laws of radicals follow directly from the laws of exponents, where we assume (1) that m and n are positive integers and (2) that the radical represents a real number.

$$\sqrt[n]{ab} = \sqrt[n]{a} \cdot \sqrt[n]{b} \tag{16}$$

$$\sqrt[n]{\frac{a}{b}} = \frac{\sqrt[n]{a}}{\sqrt[n]{b}} \tag{17}$$

Also, see Eq. (13).

Example 23. $(\sqrt[3]{x^2})^3 = x^2$

To remove factors from the radicand, apply Eq. (16) and remove from the radicand any factor which is a perfect nth power.

Example 24. $\sqrt{45} = \sqrt{(3)(3)(5)} = \sqrt{(3)^2(5)} = \sqrt{3^2} \cdot \sqrt{5} = 3\sqrt{5}$

Example 25. $\sqrt{4a^2x^2 - 36a^2y^2} = \sqrt{4a^2(x^2 - 9y^2)} = \sqrt{4a^2} \cdot \sqrt{x^2 - 9y^2}$
$$= 2a\sqrt{x^2 - 9y^2}$$

Example 26. $\sqrt[3]{48x^4y^6z^8} = \sqrt[3]{(2)(2)(2)(2)(3)x^4y^6x^8}$
$$= \sqrt[3]{(2)^3x^3y^6z^6(2)(3)xz^2}$$
$$= \sqrt[3]{(2)^3x^3y^6z^6} \cdot \sqrt[3]{(2)(3)xz^2} = 2xy^2z^2\sqrt[3]{6xz^2}$$

Note that this example is worked very conveniently by shifting temporarily to exponential notation.

$$\sqrt[3]{48x^4y^6z^8} = [(2)^4(3)x^4y^6z^8]^{1/3} = (2)^{4/3}(3)^{1/3}x^{4/3}y^{6/3}z^{8/3}$$

In any case in which an exponent is an improper fraction, divide the expression involved into two factors, one raised to an integral power, the other raised to a power expressed by an exponent in the form of a proper fraction. Thus, from the above example, we obtain

$$2^1 \cdot 2^{1/3} \cdot 3^{1/3} x^1 \cdot x^{1/3} \cdot y^2 \cdot z^2 \cdot z^{2/3}$$

By the commutative property,

$$(2xy^2z^2)(2^{1/3} \cdot 3^{1/3}x^{1/3}z^{2/3}) = 2xy^2z^2(2 \cdot 3xz^2)^{1/3} = 2xy^2z^2 \sqrt[3]{6xz^2}$$

Exponential notation is often useful in simplifying certain types of radicals involving literal powers and roots.

Example 27. Rationalize $\sqrt[n-3]{a^{2n-1}}$.
In exponential notation this becomes $a^{(2n-1)/(n-3)}$.
But

$$\frac{2n-1}{n-3} = 2 + \frac{5}{n-3}$$

Therefore

$$a^{(2n-1)/(n-3)} = a^{2+5/(n-3)} = a^2 \cdot a^{5/(n-3)} = a^2 \sqrt[n-3]{a^5}$$

EXERCISE 4

Simplify the following expressions by removing factors from the radicands.

1. $\sqrt{8}$	2. $\sqrt{12}$	3. $\sqrt{40}$	4. $\sqrt{54}$
5. $5\sqrt{18}$	6. $2\sqrt{27}$	7. $3\sqrt{32}$	8. $4\sqrt{63}$
9. $\sqrt[3]{24}$	10. $\sqrt[3]{-32}$	11. $5\sqrt[3]{54}$	12. $7\sqrt[3]{192}$
13. $\sqrt[3]{-250}$	14. $4\sqrt[3]{135}$	15. $5\sqrt[3]{96}$	16. $3\sqrt[5]{96}$
17. $\sqrt{x^7}$	18. $a^2\sqrt{c^9}$	19. $\sqrt{x^3y^3}$	20. $a\sqrt{a^2b}$
21. $\sqrt{12x^3y^5}$	22. $\sqrt{50a^3b^6c^9}$	23. $\sqrt{ax^2+bx^2}$	24. $\sqrt{a^2b^2+a^2c^2}$
25. $\sqrt{4m^2-16n^2}$	26. $\sqrt{18x^3-27x^2y}$	27. $\sqrt[n]{a^{n+1}}$	28. $\sqrt[n-1]{a^n}$
29. $\sqrt[n]{a^{n+3}}$	30. $\sqrt[n-1]{a^{2n}}$	31. $\sqrt[n+1]{a^{2n+3}}$	32. $\sqrt[n-2]{a^{2n-3}}$
33. $\sqrt[3]{x^6y^{6+a}}$	34. $\sqrt[4]{x^9y^{10+a}}$	35. $\sqrt[3n]{x^{8n}y^{6n}}$	36. $\sqrt[3k-1]{x^{6k}}$

In Probs. 37 to 42, simplify the radicals as before; then compute their values by using a table of square roots and cube roots.

37. $\sqrt{5300}$ 38. $\sqrt{0.000105}$ 39. $\sqrt{1375}$

40. $\sqrt{1648}$　　　41. $\sqrt[3]{6032}$　　　42. $\sqrt[3]{1875}$

Any quantity multiplying a radical may be introduced under the radical sign by raising to a power corresponding to the index of the radical.

Example 28. $5\sqrt{2} = \sqrt{5^2} \cdot \sqrt{2} = \sqrt{(25)(2)} = \sqrt{50}$

Example 29. $3ab^2 \sqrt[3]{2a^2b} = \sqrt[3]{(3ab^2)^3} \cdot \sqrt[3]{2a^2b} = \sqrt[3]{(27a^3b^6)(2a^2b)}$
$$= \sqrt[3]{54a^5b^7}$$

4.9 Rationalizing the Denominator in a Radical

This term refers to the elimination of fractions within a radical. The procedure is desirable in that it usually simplifies numerical computations.

To rationalize the denominator of a radical having an index n, multiply numerator and denominator of the fraction by the smallest quantity which will make the denominator a perfect nth power. Extract the nth root of the denominator [see Eq. (17)].

Example 30. $\sqrt{\dfrac{3}{7}} = \sqrt{\dfrac{3}{7} \cdot \dfrac{7}{7}} = \sqrt{\dfrac{21}{49}} = \dfrac{\sqrt{21}}{\sqrt{49}} = \dfrac{\sqrt{21}}{\sqrt{7^2}} = \dfrac{\sqrt{21}}{7}$

Example 31. $\sqrt{\dfrac{c^3}{x^3}} = \sqrt{\dfrac{c^3 \cdot x}{x^3 \cdot x}} = \dfrac{\sqrt{c^3x}}{\sqrt{x^4}} = \dfrac{c\sqrt{cx}}{x^2}$

Example 32. $\sqrt[3]{\dfrac{5m^4}{16a^2x^4}} = \sqrt[3]{\dfrac{5m^4 \cdot 4ax^2}{16a^2x^4 \cdot 4ax^2}} = \dfrac{\sqrt[3]{20am^4x^2}}{\sqrt[3]{64a^3x^6}} = \dfrac{m\sqrt[3]{20amx^2}}{4ax^2}$

Example 33. $\sqrt{\dfrac{xy}{x+y}} = \sqrt{\dfrac{xy}{x+y} \cdot \dfrac{x+y}{x+y}} = \dfrac{\sqrt{xy(x+y)}}{\sqrt{(x+y)^2}} = \dfrac{\sqrt{xy(x+y)}}{x+y}$

Example 34. $\sqrt{\dfrac{a}{b} - \dfrac{b}{a}} = \sqrt{\dfrac{a^2 - b^2}{ab}} = \sqrt{\dfrac{ab(a^2 - b^2)}{a^2b^2}} = \dfrac{\sqrt{ab(a^2 - b^2)}}{\sqrt{a^2b^2}}$
$$= \dfrac{\sqrt{ab(a^2 - b^2)}}{ab}$$

Note that, without rationalizing the denominator, we should have to evaluate an expression such as $\sqrt{3/7}$ by writing

$$\sqrt{3/7} = \sqrt{0.42857} = 0.6547$$

or

$$\frac{\sqrt{3}}{\sqrt{7}} = \frac{1.732}{2.646} = 0.6547$$

It can readily be appreciated how much easier and faster it is to rationalize and evaluate $\sqrt{21}/7 = 4.5826/7 = 0.6547$. Tables are assumed to be used in all cases.

EXERCISE 5

In the following expressions, rationalize the denominators and remove any perfect powers from the radicands. Where possible, use tables to compute the decimal values of Probs. 1 to 17.

1. $\sqrt{1/2}$

2. $\sqrt{2/3}$

3. $6\sqrt{1/8}$

4. $9\sqrt{1/12}$

5. $6\sqrt{1/15}$

6. $15\sqrt{5/18}$

7. $\sqrt[3]{1/2}$

8. $\sqrt[3]{1/4}$

9. $8\sqrt[3]{1/16}$

10. $10\sqrt[3]{1/25}$

11. $6\sqrt[4]{1/27}$

12. $6\sqrt[4]{1/8}$

13. $\frac{1}{2}\sqrt{8/3}$

14. $\sqrt{25/32}$

15. $4\sqrt{27/80}$

16. $\sqrt[3]{16/9}$

17. $\frac{1}{3}\sqrt[3]{36/25}$

18. $\sqrt{\frac{1}{a}}$

19. $\sqrt{\frac{1}{x^3}}$

20. $x\sqrt{\frac{1}{x^5}}$

21. $\sqrt[3]{\frac{1}{a^2}}$

22. $ab\sqrt[3]{\frac{1}{a^4}}$

23. $\sqrt[3]{\frac{m}{mx^2}}$

24. $\sqrt{\frac{3x}{2y}}$

25. $\sqrt{\frac{8a^3b}{27c^3}}$

26. $\frac{a}{b}\sqrt[3]{\frac{1}{ab}}$

27. $\sqrt{\frac{a-b}{a+b}}$

28. $\sqrt{x-2+\frac{1}{x}}$

29. $\sqrt[n]{\frac{1}{x}}$

30. $\sqrt[n+1]{\frac{1}{c^n}}$

31. $\sqrt{\frac{1}{a^{2n+1}}}$

32. $\sqrt[3]{\frac{1}{x^{3n-1}}}$

33. $\sqrt[8]{\frac{3^7 + 3^7 + 3^7}{4^7 + 4^7 + 4^7 + 4^7}}$

It should be remembered that rationalization is not always desirable. In evaluating Prob. 27 above by a desk calculator, given numerical values for a and b, there would be no advantage in using the "simplified" form $\frac{\sqrt{a^2 - b^2}}{a + b}$.

4.10 Reduction of the Order of a Radical

The order (index) of a radical can be reduced if the radicand can be expressed as a perfect power whose exponent contains a factor which is also contained in the index of the radical.

To reduce the order of a radical:

1. *Express the radicand as a perfect power of some rational number.*
2. *Transform the radical to equivalent fractional-exponent notation.*
3. *Reduce the fractional exponent to lowest terms.*
4. *Convert back to radical form.*

Example 35. Reduce the order of $\sqrt[6]{8a^3b^3}$.

$$\sqrt[6]{8a^3b^3} = \sqrt[6]{(2ab)^3} = (2ab)^{3/6} = (2ab)^{1/2} = \sqrt{2ab}$$

In general, a radical may be said to be simplified if the procedures of the preceding sections have been carried out where applicable. That is, it contains no negative or fractional exponents, the denominator of the radicand has been rationalized, all possible factors have been removed from the radicand, and the order is as small as possible.

EXERCISE 6

Simplify each of the following:

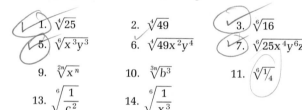

1. $\sqrt[4]{25}$ 2. $\sqrt[4]{49}$ 3. $\sqrt[6]{16}$ 4. $\sqrt[6]{36}$

5. $\sqrt[6]{x^3y^3}$ 6. $\sqrt[4]{49x^2y^4}$ 7. $\sqrt[6]{25x^4y^6z^8}$ 8. $\sqrt[10]{x^6y^8}$

9. $\sqrt[2n]{x^n}$ 10. $\sqrt[3n]{b^3}$ 11. $\sqrt[6]{1/4}$ 12. $\sqrt[4]{1/25}$

13. $\sqrt[6]{\dfrac{1}{c^2}}$ 14. $\sqrt[6]{\dfrac{1}{x^3}}$

4.11 Addition and Subtraction of Radicals

Two radicals are said to be the *same* if they have the same radicand and the same index. Multiples of the same radical can be added or subtracted by combining their coefficients. To add or subtract expressions containing radicals, simplify each radical and collect and combine all multiples of the same radical. Addition of radicals which remain unlike after simplification can only be indicated. The decimal value of numerical problems can be computed approximately by the use of tables.

Example 36. Add $\sqrt{96} - 3\sqrt{2/3} + \frac{1}{2}\sqrt{150} - 2\sqrt{3/8}$.
     Simplifying each radical,

$$4\sqrt{6} - \sqrt{6} + \frac{5}{2}\sqrt{6} - \frac{1}{2}\sqrt{6}$$

By the distributive law,

$$(4 - 1 + \tfrac{5}{2} - \tfrac{1}{2})\sqrt{6} = 5\sqrt{6}$$

Example 37. Add $\sqrt{5x} - \sqrt{5/x} + 2\sqrt{x/5} - \sqrt[6]{25x^2}$.
Simplifying each radical,

$$\sqrt{5x} - \frac{1}{x}\sqrt{5x} + \frac{2}{5}\sqrt{5x} - \sqrt[3]{5x}$$

Adding coefficients of like radicals,

$$\left(1 - \frac{1}{x} + \frac{2}{5}\right)\sqrt{5x} - \sqrt[3]{5x} = \left(\frac{7x - 5}{5x}\right)\sqrt{5x} - \sqrt[3]{5x}$$

EXERCISE 7

Simplify and combine where possible.

1. $\sqrt{2} + 6\sqrt{2}$ 2. $3\sqrt{2} - \sqrt{18}$

3. $3\sqrt{75} - 2\sqrt{27} + \sqrt{48}$ 4. $\sqrt{40} - 2\sqrt{90} + \sqrt{\tfrac{2}{5}}$

5. $\sqrt{98} + \tfrac{1}{3}\sqrt{72} - 2\sqrt{\tfrac{1}{8}}$ 6. $3\sqrt{20} - \sqrt[4]{25} - \tfrac{2}{3}\sqrt{180}$

7. $3\sqrt{28} + 4\sqrt{\tfrac{1}{7}} + \sqrt{112}$ 8. $\sqrt{60} - \sqrt[3]{\tfrac{3}{5}} - \sqrt[5]{\tfrac{5}{27}}$

9. $\tfrac{1}{2}\sqrt{24} + 3\sqrt{\tfrac{2}{3}} - 2\sqrt{\tfrac{3}{2}}$ 10. $3\sqrt[3]{6} + 2\sqrt[3]{48} - 4\sqrt[6]{36}$

11. $2\sqrt{28} - 4\sqrt{63} + 14\sqrt{\tfrac{1}{7}}$ 12. $5\sqrt[3]{108} - 2\sqrt[3]{32} - 4\sqrt[3]{\tfrac{1}{2}}$

13. $\sqrt{x} + \sqrt{x^3} + \sqrt{x^5}$ 14. $\sqrt{12x} - \sqrt{75x^3} + \sqrt{\dfrac{3}{x}}$

15. $\sqrt{\dfrac{x + y}{x - y}} - \sqrt{\dfrac{x - y}{x + y}}$ 16. $\sqrt{\dfrac{x + 2}{x - 1}} - \sqrt{\dfrac{x - 1}{x + 2}}$

17. $\sqrt{(a + b)^3} - \sqrt{a + b}$ 18. $\sqrt{x^5} - 4\sqrt{x^3} + 4\sqrt{x}$

19. $\sqrt{18} - \sqrt{\tfrac{1}{3}} + \sqrt{12} - \sqrt{\tfrac{1}{8}}$

20. $\sqrt{12x^4 + 48x^3 + 48x^2} - \sqrt{12x^4 + 12x^3 + 3x^2}$

21. $\sqrt{\dfrac{a}{b} - 3} + 2\sqrt{1 - \dfrac{3b}{a}} - \dfrac{3}{a}\sqrt{a^3b^2 - 3a^2b^3}$

4.12 Multiplication and Division of Radicals

A radical may ordinarily be multiplied or divided by another radical of the same order by using the formulas

$$\sqrt[n]{a} \cdot \sqrt[n]{b} = \sqrt[n]{ab} \tag{18}$$

and

$$\frac{\sqrt[n]{a}}{\sqrt[n]{b}} = \sqrt[n]{\frac{a}{b}} \tag{19}$$

When radicals of different orders are multiplied or divided, they must first be reduced to radicals of the same order. (The common index is usually the LCM of the original indices.)

Example 38. Multiply $\sqrt[3]{4x^2y^2}$ by $\sqrt[3]{12a^2x^2}$.

$$\sqrt[3]{4x^2y^2} \cdot \sqrt[3]{12a^2x^2} = \sqrt[3]{48a^2x^4y^2} = \sqrt[3]{(8x^3)(6a^2xy^2)} = 2x \sqrt[3]{6a^2xy^2}$$

Example 39. Divide $\sqrt[4]{8a^3c^2}$ by $\sqrt[4]{54ac^3}$.

$$\frac{\sqrt[4]{8a^3c^2}}{\sqrt[4]{54ac^3}} = \sqrt[4]{\frac{8a^3c^2}{54ac^3}} = \sqrt[4]{\frac{4a^2}{27c}} = \sqrt[4]{\frac{4a^2}{27c} \cdot \frac{3c^3}{3c^3}} = \frac{1}{3c} \sqrt[4]{12a^2c^3}$$

Example 40. Multiply $2\sqrt{6} - \sqrt{3}$ by $3\sqrt{3} - \sqrt{2}$.

$$2\sqrt{6} - \sqrt{3}$$
$$3\sqrt{3} - \sqrt{2}$$
$$\overline{6\sqrt{18} - 3(3) - 2\sqrt{12} + \sqrt{6}} = 18\sqrt{2} - 9 - 4\sqrt{3} + \sqrt{6}$$

Example 41. Multiply $\sqrt{2}$ by $\sqrt[3]{4}$.

$$\sqrt{2} \cdot \sqrt[3]{4} = \sqrt[6]{8} \cdot \sqrt[6]{16} = \sqrt[6]{128} = 2\sqrt[6]{2}$$

Alternative solution:

$$\sqrt{2} \cdot \sqrt[3]{4} = \sqrt{2} \cdot \sqrt[3]{2^2}$$
$$= 2^{1/2} \cdot 2^{2/3}$$
$$= 2^{7/6}$$
$$= 2^1 \cdot 2^{1/6}$$
$$= 2\sqrt[6]{2}$$

Example 42. Divide $\sqrt[3]{9}$ by $\sqrt{3}$.

$$\sqrt[3]{9} \div \sqrt{3} = \sqrt[6]{81} \div \sqrt[6]{27} = \sqrt[6]{81/27} = \sqrt[6]{3}$$

Alternative solution:

$$\sqrt[3]{9} \div \sqrt{3} = \sqrt[3]{3^2} \div \sqrt{3}$$
$$= 3^{2/3} \div 3^{1/2}$$
$$= 3^{2/3-1/2}$$
$$= 3^{1/6} = \sqrt[6]{3}$$

4.13 Division by an Irrational Binomial

In one important case of division of radicals, the divisor is a binomial in which one or both terms contain a second-degree radical. The division is effected by rationalizing the binomial divisor. This is done by multiplying the numerator and denominator of the fraction by the *conjugate* of the denominator, i.e., the denominator with the sign between its terms reversed. Simplify the new numerator and denominator. Similar methods can be applied to other special forms. This method is extremely important in Sec. 4.16, dealing with complex numbers.

Example 43. Rationalize the denominator of

$$\frac{12}{5 - \sqrt{7}}$$

Multiplying numerator and denominator by the conjugate of $5 - \sqrt{7}$, which is $5 + \sqrt{7}$, we obtain

$$\frac{12}{5 - \sqrt{7}} = \frac{12(5 + \sqrt{7})}{(5 - \sqrt{7})(5 + \sqrt{7})}$$
$$\frac{12(5 + \sqrt{7})}{5^2 - (\sqrt{7})^2} = \frac{12(5 + \sqrt{7})}{25 - 7} = \frac{12(5 + \sqrt{7})}{18} = \frac{2(5 + \sqrt{7})}{3}$$

Example 44. Rationalize the denominator of

$$\frac{\sqrt{6} + 3\sqrt{2}}{3\sqrt{6} + 2\sqrt{3}}$$

Multiplying numerator and denominator by $3\sqrt{6} - 2\sqrt{3}$,

$$\frac{\sqrt{6} + 3\sqrt{2}}{3\sqrt{6} + 2\sqrt{3}} = \frac{(\sqrt{6} + 3\sqrt{2})(3\sqrt{6} - 2\sqrt{3})}{(3\sqrt{6} + 2\sqrt{3})(3\sqrt{6} - 2\sqrt{3})}$$

$$= \frac{(3)(6) - 2\sqrt{18} + 9\sqrt{12} - 6\sqrt{6}}{(3\sqrt{6})^2 - (2\sqrt{3})^2}$$

$$= \frac{18 - 6\sqrt{2} + 18\sqrt{3} - 6\sqrt{6}}{54 - 12}$$

$$= \frac{6(3 - \sqrt{2} + 3\sqrt{3} - \sqrt{6})}{42}$$

$$= \frac{3 - \sqrt{2} + 3\sqrt{3} - \sqrt{6}}{7}$$

Example 45. Rationalize the denominator of

$$\frac{\sqrt{a+1} - \sqrt{a-1}}{\sqrt{a+1} + \sqrt{a-1}}$$

Multiplying numerator and denominator by $\sqrt{a+1} - \sqrt{a-1}$,

$$\frac{(\sqrt{a+1} - \sqrt{a-1})(\sqrt{a+1} - \sqrt{a-1})}{(\sqrt{a+1} + \sqrt{a-1})(\sqrt{a+1} - \sqrt{a-1})} = \frac{a+1 - 2\sqrt{a^2-1} + a - 1}{(a+1) - (a-1)}$$

$$= \frac{2a - 2\sqrt{a^2-1}}{2} = a - \sqrt{a^2-1}$$

EXERCISE 8

Perform the indicated operations and simplify the results. Where possible, use tables to compute the decimal values of answers containing no literal expressions.

1. $\sqrt{2} \cdot \sqrt{3}$ 2. $\sqrt{3} \cdot \sqrt{6}$ 3. $\sqrt{7} \cdot \sqrt{14}$

4. $3\sqrt{2} \cdot \sqrt{10}$ 5. $2\sqrt{11} \cdot 3\sqrt{11}$ 6. $4\sqrt{15} \cdot \sqrt{21}$

7. $\sqrt{6a} \cdot \sqrt{2ab}$ 8. $2\sqrt{xy} \cdot \sqrt{yz}$ 9. $\sqrt{abc} \cdot \sqrt{bcd}$

10. $\sqrt[3]{4} \cdot \sqrt[3]{6}$ 11. $\sqrt[3]{18} \cdot \sqrt[3]{15}$ 12. $\sqrt[3]{xy^2z} \cdot \sqrt[3]{x^2yz}$

13. $\sqrt[3]{4} \cdot \sqrt{6}$ 14. $\sqrt[3]{9} \cdot \sqrt{3}$ 15. $\sqrt[4]{8} \cdot \sqrt{2}$

16. $\sqrt[4]{27} \cdot \sqrt{3}$ 17. $(\sqrt{10} + \sqrt{3})(\sqrt{10} - \sqrt{3})$

18. $(7 + \sqrt{5})(7 - \sqrt{5})$ 19. $(5 - 3\sqrt{2})(5 + 3\sqrt{2})$

20. $(5\sqrt{6} - 2\sqrt{10})(5\sqrt{6} + 2\sqrt{10})$ 21. $(\sqrt{6} - \sqrt{3})^2$

22. $(7 - 2\sqrt{5})^2$ 23. $(\sqrt{3} + \sqrt{2})(3\sqrt{3} - 2\sqrt{2})$

24. $(2\sqrt{6} - \sqrt{3})(\sqrt{6} + 3\sqrt{2})$ 25. $(\sqrt{6} + 2\sqrt{10})(2\sqrt{15} - \sqrt{6})$

26. $\dfrac{\sqrt{6}}{\sqrt{2}}$ 27. $\dfrac{\sqrt{35}}{\sqrt{15}}$ 28. $\dfrac{5\sqrt{22}}{\sqrt{11}}$

29. $\dfrac{6\sqrt{3}}{\sqrt{15}}$

30. $\dfrac{\sqrt{ab}}{\sqrt{ac}}$

31. $\dfrac{\sqrt{21xy}}{\sqrt{14yz}}$

32. $\dfrac{\sqrt[3]{4}}{\sqrt{2}}$

33. $\dfrac{\sqrt{6}}{\sqrt[3]{3}}$

34. $\dfrac{\sqrt[4]{a^3}}{\sqrt[4]{a}}$

35. $\dfrac{\sqrt[4]{27}}{\sqrt{3}}$

36. $\dfrac{\sqrt{2}}{\sqrt[4]{8}}$

37. $\dfrac{6}{\sqrt{7}-2}$

38. $\dfrac{10}{\sqrt{13}-3}$

39. $\dfrac{24}{10-\sqrt{28}}$

40. $\dfrac{15}{\sqrt{17}-\sqrt{7}}$

41. $\dfrac{21}{2\sqrt{5}-\sqrt{6}}$

42. $\dfrac{\sqrt{6}}{2\sqrt{3}-\sqrt{2}}$

43. $\dfrac{\sqrt{10}}{3\sqrt{5}+2\sqrt{2}}$

44. $\dfrac{3\sqrt{2}-\sqrt{3}}{2\sqrt{3}+\sqrt{2}}$

45. $\dfrac{4\sqrt{5}+3\sqrt{3}}{3\sqrt{5}-2\sqrt{3}}$

46. $\dfrac{\sqrt{a-b}}{\sqrt{a}-\sqrt{b}}$

47. $\dfrac{\sqrt{xy}+\sqrt{yz}}{\sqrt{xyz}}$

48. $\dfrac{a-x}{\sqrt{a}+\sqrt{x}}$

49. $\dfrac{x}{\sqrt{x^2-a^2}} + \dfrac{a^2x}{\sqrt{(x^2-a^2)^3}}$

   Hint: Do not rationalize the denominator.

50. $\dfrac{1}{15}\left[(3x^2-2a^2)(3x)\sqrt{a^2+x^2}+6x\sqrt{(a^2+x^2)^3}\right]$

51. $\dfrac{1}{x+\sqrt{x^2+a^2}}\left(\dfrac{x}{\sqrt{x^2+a^2}}+1\right)$

52. $-\dfrac{2}{3b^2}\left[(2a-bx)\dfrac{b}{2\sqrt{a+bx}}-b\sqrt{a+bx}\right]$

53. $\dfrac{2}{3a^2}\left[a(ax+b)^{1/2}+(ax-2b)\dfrac{1}{2}(ax+b)^{-1/2}(a)\right]$

54. $\dfrac{2}{\sqrt{b}}\dfrac{1}{1+(ax-b)/b}\dfrac{1}{2}\left(\dfrac{ax-b}{b}\right)^{-1/2}\left(\dfrac{a}{b}\right)$

55. Show that when x approaches 0, the value of

$$\dfrac{\sqrt{3+x}-\sqrt{3}}{x}$$

   approaches $1/(2\sqrt{3})$. *Hint:* Rationalize the numerator.

56. Simplify

$$\dfrac{(x-a)^{1/2}+a(x-a)^{-1/2}}{x-a}$$

   Hint: Rationalize the numerator.

4.14 Imaginary Numbers

Thus far we have been unable to give meaning to an even root of a negative number. For example, up to this point such symbols as $\sqrt{-4}$, $\sqrt[6]{-64}$, etc., have been quite meaningless.

However, by enlarging our concepts of numbers, we are able to give meaning to even roots of negative numbers. We shall begin by giving particular attention to the symbol $\sqrt{-1}$, which we shall call the *imaginary unit*.

It is customary to assign the same meaning to the letters i or j as to the symbol $\sqrt{-1}$. The letter i is standard nomenclature in strictly mathematical work, while j is more common in electrical technology. We shall follow the technology convention and let

$$j = \sqrt{-1} \tag{20}$$

or

$$j^2 = -1 \tag{21}$$

More generally, an imaginary number is represented by the symbol

$$\sqrt[n]{-P}$$

where n is an even number and P is any positive real number. However, we shall limit our discussion of imaginary numbers to cases where n is 2.

From Eq. (4), in which a and b are any numbers, imaginary or otherwise, we may write

$$(j\sqrt{P})^2 = j^2(\sqrt{P})^2$$

By applying Eq. (4),

$$(j\sqrt{P})^2 = j^2 P$$

or

$$(j\sqrt{P})^2 = (-1) \cdot P = -P$$

Thus $j\sqrt{P}$ is seen to be one of two equal factors of $-P$ and is therefore a *square root* of $-P$.

Even though it can be shown that $-j\sqrt{P}$ is another square root of $-P$, the symbol $j\sqrt{P}$ is defined by common agreement as

$$j\sqrt{P} = \sqrt{-P} \tag{22}$$

In Eq. (22) P is any positive real number and $\sqrt{P}$ designates the *principal* square root of P.

For example,

$$\sqrt{-25} = +j5$$

Powers of j

By definition,

$$j^2 = -1 \qquad\qquad (23)$$

Then it follows from the basic rules of algebra that

$$j^0 = 1$$
$$j^1 = j$$
$$j^2 = -1$$
$$j^3 = j^2 \times j = -j$$
$$j^4 = j^2 \times j^2 = 1$$
$$j^5 = j^4 \times j = j$$
$$j^6 = j^4 \times j^2 = -1$$

In general, where n is any integer (positive, negative, or zero),

$$j^{4n} = +1 \qquad\qquad (24)$$
$$j^{4n+1} = +j \qquad\qquad (25)$$
$$j^{4n+2} = -1 \qquad\qquad (26)$$
$$j^{4n+3} = -j \qquad\qquad (27)$$

By use of Eqs. (24) to (27), any power of j may be determined at a glance. Observe that any real, integral power of j must be either unity or j or their negatives.

EXERCISE 9

Find the value of

1. j^{10} 2. j^6 3. j^{15} 4. j^{985} 5. j^{-25}

4.15 Operations with Imaginary Numbers

Addition and Subtraction of Imaginary Numbers

Imaginary numbers may be added and subtracted according to the rules of algebra already discussed.

Example 46. Add $\sqrt{-9}$ and $\sqrt{-49}$.

From Eq. (22) we may write

$$\sqrt{-9} = j\sqrt{9} = j3$$
$$\sqrt{-49} = j\sqrt{49} = j7$$
$$\sqrt{-9} + \sqrt{-49} = j3 + j7 = j10$$

Note: An alternative expression is $3j + 7j = 10j$.

Example 47. Subtract $\sqrt{-25}$ from $\sqrt{-121}$.

From Eq. (22) we may write

$$\sqrt{-25} = j\sqrt{25} = j5$$
$$\sqrt{-121} = j\sqrt{121} = j11$$
$$\sqrt{-121} - \sqrt{-25} = j11 - j5 = j6$$

Multiplication of Imaginary Numbers

Multiplication of imaginary numbers can be accomplished as illustrated below.

Example 48. Multiply $\sqrt{-5}$ by $\sqrt{-7}$.

Both $\sqrt{-5}$ and $\sqrt{-7}$ are imaginary numbers and consequently must be changed to the form $j\sqrt{b}$ before multiplying.

$$\sqrt{-5} = j\sqrt{5}$$
$$\sqrt{-7} = j\sqrt{7}$$

Then

$$\sqrt{-5} \times \sqrt{-7} = j\sqrt{5} \cdot j\sqrt{7}$$

or

$$\sqrt{-5} \times \sqrt{-7} = j^2 \cdot \sqrt{5 \cdot 7} = j^2\sqrt{35} = -\sqrt{35} = -5.916$$

Observe that Eq. (16) does not apply to even roots of negative numbers and

$$\sqrt{-5} \times \sqrt{-7} \neq \sqrt{(-5)(-7)}$$
$$\sqrt{-5} \times \sqrt{-7} \neq \sqrt{+35}$$

Example 49. Evaluate $(\sqrt{-5})^2$.

From Eq. (22)

$$\sqrt{-5} = j\sqrt{5}$$

Then

$$(\sqrt{-5})^2 = (j\sqrt{5})^2 = j^2(\sqrt{5})^2$$

From Eqs. (13a) and (21), therefore,

$$-(\sqrt{5})^2 = -|5| = -5$$

Also

$$(\sqrt{-5})^2 = j^2(\sqrt{5})^2 = (j)^2(5) = -5$$

Example 50. Evaluate $\sqrt{(-5)^2}$.
    From Definition 1, we may write

$$(-5)(-5) = (-5)^2 = +25$$

Then from Sec. 4.3,

$$\sqrt{(-5)^2} = \sqrt{25} = +5$$

Compare with Example 49.

Example 51. Multiply $\sqrt{-5}$, $\sqrt{-7}$, and $\sqrt{-15}$.

$$\sqrt{-5} = j\sqrt{5}$$
$$\sqrt{-7} = j\sqrt{7}$$
$$\sqrt{-15} = j\sqrt{15}$$
$$\sqrt{-5} \times \sqrt{-7} \times \sqrt{-15} = j\sqrt{5} \times j\sqrt{7} \times j\sqrt{15}$$
$$= j^3\sqrt{5} \times \sqrt{7} \times \sqrt{15} = j^3\sqrt{525} = j^35\sqrt{21} = -j5\sqrt{21}$$

Example 52. Multiply $3\sqrt{-6} + 2\sqrt{-5}$ by $3\sqrt{-3} - 7\sqrt{-11}$.

$$3\sqrt{-6} = j3\sqrt{6}$$
$$2\sqrt{-5} = j2\sqrt{5}$$
$$3\sqrt{-3} = j3\sqrt{3}$$
$$7\sqrt{-11} = j7\sqrt{11}$$
$$j3\sqrt{6} + j2\sqrt{5}$$
$$\underline{j3\sqrt{3} - j7\sqrt{11}}$$
$$j^29\sqrt{18} + j^26\sqrt{15} - j^221\sqrt{66} - j^214\sqrt{55}$$

Substituting for j^2 its numerical value, we obtain

$$-9\sqrt{18} - 6\sqrt{15} + 21\sqrt{66} + 14\sqrt{55}$$

Evaluating the radicals, we obtain

$$(-9 \times 4.243) - (6 \times 3.873) + (21 \times 8.124) + (14 \times 7.416)$$
$$= -38.187 - 23.238 + 170.604 + 103.824 = 213.00$$

Division of Imaginary Numbers

Example 53. Divide $\sqrt{-30}$ by $\sqrt{-15}$.

$$\frac{\sqrt{-30}}{\sqrt{-15}} = \frac{j\sqrt{30}}{j\sqrt{15}} = \frac{\sqrt{30}}{\sqrt{15}} = \sqrt{\frac{30}{15}} = \sqrt{2}$$

EXERCISE 10

Simplify:

1. $\sqrt{-27} + \sqrt{-147} - \sqrt{12}$

2. $\sqrt{-150} - \sqrt{-54} + \sqrt{216}$

3. $\sqrt{-18} + \sqrt{-72} - \sqrt{128} + \sqrt{32}$

4. $\sqrt{-63} - \sqrt{-567} + \sqrt{112}$

5. $\sqrt{-80} + \sqrt{-20} + \sqrt{-320} - \sqrt{-245}$

6. $\sqrt{-128} + \sqrt{-72} - \sqrt{-800} + \sqrt{-392}$

7. $\sqrt{-196} - \sqrt{-64} + \sqrt{-80} + \sqrt{80}$

8. $\sqrt{-192} + \sqrt{-12} - \sqrt{-48} + \sqrt{-256}$

9. $\sqrt{-32} - \sqrt{-392} + \sqrt{-128}$

10. $\sqrt{-3} \times \sqrt{-3}$

11. $\sqrt{-25y^4} \times \sqrt{-9x^2}$

12. $(-4\sqrt{-4})(-3\sqrt{-9})$

13. $-\sqrt{-27a^2} \times \sqrt{-75b^2}$

14. $j\sqrt{3} \times j\sqrt{5}$

15. $-j\sqrt{7} \times j\sqrt{15}$

16. $-j\sqrt{2} \times (-j\sqrt{5})$

17. $j5 \times j8$

18. $j5 \times (-j8)$

19. $(-j2) \times (-j3)$

20. $(2\sqrt{-5} - 3\sqrt{-2})(2\sqrt{-5} + 3\sqrt{-2})$

21. $(\sqrt{-2} + \sqrt{-3})^2$

22. $\sqrt{-5} \times \sqrt{-7} \times \sqrt{-2}$

23. $\sqrt{-3} \times \sqrt{-10} \times \sqrt{-5}$

24. $\sqrt{-15} \times \sqrt{-5} \times \sqrt{-3}$

25. $(\sqrt{-5} + \sqrt{-3})(\sqrt{-6} - \sqrt{-2})$

26. $(\sqrt{-2} + \sqrt{-5})(\sqrt{-7} + \sqrt{-6})(\sqrt{-4} - \sqrt{-9})$

27. $(\sqrt{-12} + \sqrt{-11})(\sqrt{-3} + \sqrt{-15})(\sqrt{-10} - \sqrt{-19})$

28. $\sqrt{-144} \div \sqrt{-100}$

29. $\sqrt{-294} \div \sqrt{-54}$

30. $j3 \div j^25$

31. $\sqrt{-160} \div \sqrt{8}$

32. $\sqrt{-320} \div \sqrt{-128}$

33. $\sqrt{-63} \div \sqrt{-18}$

34. $\sqrt{96} \div \sqrt{-32}$

35. $\sqrt{48} \div \sqrt{96}$

36. $(\sqrt{-392} \div \sqrt{-49}) \times \sqrt{196}$

37. $(\sqrt{-128} \div \sqrt{-80})(\sqrt{32} \div \sqrt{-16})$

4.16 Complex Numbers

A *complex number* is an expression in the form $a + jb$, where a and b are real numbers and $j = \sqrt{-1}$.

Complex numbers are used extensively in the solution of alternating-current-circuit problems.

For the present, we shall concentrate on the purely manipulative techniques involving complex numbers, leaving the applications until we have had a chance to cover more advanced algebra and trigonometry.

Notice that in the general form of a complex number $a + jb$, if $a = 0$, we have a pure imaginary number. If $b = 0$, we have a real number.

Addition of Complex Numbers

Example 54. Add the complex numbers $2 + j6$, $3 - j5$, and $4 + j2$.

$$
\begin{array}{l}
2 + j6 \\
3 - j5 \\
\underline{4 + j2} \\
9 + j3
\end{array}
$$

Subtraction of Complex Numbers

Example 55. Subtract $5 + j9$ from $8 + j4$.

$$
\begin{array}{l}
8 + j4 \\
\underline{5 + j9} \\
3 - j5
\end{array}
$$

Example 56. Subtract $8 - j5$ from $3 + j2$.

$$
\begin{array}{r}
3 + j2 \\
\underline{8 - j5} \\
-5 + j7
\end{array}
$$

Multiplication of Complex Numbers

Complex numbers are multiplied according to procedures already described in Sec. 4.15. After multiplication, any powers of j except the first are replaced by their equivalents (Sec. 4.14).

Example 57. Multiply $2 + j7$ by $3 - j6$.

2 + j7
3 − j6
‾‾‾‾‾‾‾‾‾‾‾‾‾‾‾‾‾
6 + j21 − j12 − j²42

Since $j^2 = -1$, the product can be written

6 + j21 − j12 + 42

or

48 + j9

Example 58. Find the product of the general complex numbers $(a + jb)$ and $(c + jd)$. The multiplication is performed exactly as is the multiplication of any two binomials.

$a + jb$
$c + jd$
‾‾‾‾‾‾‾‾‾‾‾‾‾‾‾‾‾‾‾‾‾‾
$ac + jbc + jad + j^2bd$ $\hspace{3cm}$ (28)

By definition, $j^2 = -1$; therefore expression (28) becomes

$ac + jbc + jad - bd$ $\hspace{3cm}$ (29)

Grouping real and imaginary numbers, expression (29) becomes

$(ac - bd) + j(bc + ad)$

Division of Complex Numbers

In division of complex numbers we use the principle of rationalization described in Sec. 4.13.

Example 59. Divide $5 + j8$ by $2 - j7$.
The problem can be arranged as

$$\frac{5 + j8}{2 - j7}$$

If we multiply the denominator by the conjugate of the denominator, i.e., by $2 + j7$ (Sec. 4.13), we obtain

$(2 - j7)(2 + j7) = 4 + 49 = 53$

Thus it is a basic principle of complex numbers that *the product of a complex number and its conjugate is a real number.*

In the present problem we can multiply both numerator and denominator by the conjugate of the denominator and obtain a fraction with a complex numerator and a real denominator.

$$\frac{(5 + j8)(2 + j7)}{(2 - j7)(2 + j7)} = \frac{(5 + j8)(2 + j7)}{53}$$

The numerator can now be multiplied in the usual way.

$$
\begin{array}{r}
5 + j8 \\
2 + j7 \\
\hline
10 + j16 + j35 + j^{2}56 \\
10 + j16 + j35 - 56 \\
\hline
-46 + j51
\end{array}
$$

The original fraction is now

$$\frac{-46 + j51}{53}$$

If it is desired to separate the number into its real and imaginary parts, it can be written

$$-\frac{46}{53} + j\frac{51}{53}$$

or

$$-0.868 + j0.962$$

Example 60. Divide the complex number $a + jb$ by the complex number $c + jd$. Rewriting the problem in the form of a fraction,

$$\frac{a + jb}{c + jd} \tag{30}$$

Multiplying numerator and denominator of expression (30) by $c - jd$, which is called the *complex conjugate* of $c + jd$ (Sec. 4.13),

$$\frac{a + jb}{c + jd} = \frac{(a + jb)(c - jd)}{(c + jd)(c - jd)} \tag{31}$$

Performing the indicated multiplication to obtain the numerator of Eq. (31),

$$\frac{\begin{array}{l} a + jb \\ c - jd \end{array}}{ac + jbc - jad - j^2bd}$$

Since by definition $j^2 = -1$, the above product becomes

$$ac + jbc - jad + bd$$

or after simplification, the numerator of Eq. (31) becomes

$$(ac + bd) + j(bc - ad)$$

Performing the indicated multiplication to obtain the denominator of Eq. (31),

$$\frac{\begin{array}{l} c + jd \\ c - jd \end{array}}{c^2 + jcd - jcd - j^2d^2} \tag{32}$$

Simplifying expression (32), we obtain

$$c^2 + d^2$$

Equation (31) can now be written

$$\frac{a + jb}{c + jd} = \frac{(ac + bd) + j(bc - ad)}{c^2 + d^2}$$

or

$$\frac{a + jb}{c + jd} = \frac{ac + bd}{c^2 + d^2} + j\frac{bc - ad}{c^2 + d^2}$$

EXERCISE 11

Multiply the following:

1. $(2 + j3)(7 + j5)$
2. $(-3 - j6)(2 + j8)$
3. $(2 + j7)(4 - j5)(3 + j10)$
4. $(3 + j5)(3 - j5)$
5. $-(2 + j7)(2 - j7)$
6. $(5 + \sqrt{-6})(3 - \sqrt{-8})(2 + \sqrt{-5})$
7. $(2 + j5)^5$
8. $(1.4 + j3.6)(0.7 + j9.1)$

9. $(-0.42 - j2.7)(0.63 + j4.2)$ 10. $(-8.6 + j3.6)(4.7 + j5.1)$

11. $\left(-\dfrac{1}{2} + j\,\dfrac{\sqrt{3}}{2}\right)^3$

Divide the following:

12. $\dfrac{1}{1 + j}$ 13. $\dfrac{3 + j6}{4 - j8}$

14. $\dfrac{2 - j9}{3 + j5}$ 15. $\dfrac{1 + j}{2j^5}$

16. $\sqrt{4^{-2}} \div \sqrt{0.625}$ 17. $(3 - \sqrt{-5}) \div (3 + \sqrt{-5})$

18. $\dfrac{1}{4 + 2\sqrt{-3} - \sqrt{-7}}$ 19. $\dfrac{1}{\sqrt{-2} - 1}$

20. $\dfrac{2\sqrt{-5} + 5\sqrt{-2}}{2\sqrt{-5} - 5\sqrt{-2}}$ 21. $\dfrac{\sqrt{5} - \sqrt{-5}}{\sqrt{2} + \sqrt{-2}}$

22. $\dfrac{3 + j2}{j}$ 23. $\dfrac{j}{5 - 3j}$

24. $\left(\dfrac{2 + 6j}{2 - 7j}\right)^2$

Simplify:

25. $\dfrac{(2 + 6j)(3 - 7j)}{(5 - 2j) + (6 - 4j)}$ 26. $\dfrac{(1 + 7j)(2 - 3j)}{(4 - 6j) - (3 - 5j)}$

27. $\dfrac{(1.8 - 3.2j)(0.8 + 0.6j)}{(1.8 - 3.2j) + (0.8 + 0.6j)}$ 28. $\dfrac{(2 + 3j)(6 - 4j)}{(2 + 3j) + (6 - 4j)}$

Linear equations in one variable

In this chapter we shall continue with the study of the "language" of algebra. We shall see how it is developed from our ordinary "language" and thus makes possible the solution of "word" problems.

5.1 Equations

An *equation* is a statement that two expressions represent the same number. The two expressions are called the *members,* or *sides,* of the equation. However, an equation containing one or more unknowns is not only a statement of fact, it also proposes a problem.

The equation $x + 5 = 13$ states the fact that 13 is 5 more than some number x, as yet unknown. The equation also proposes the problem of finding what number x represents. If 13 is 5 more than x, it follows that x is 5 less than 13, or $x = 13 - 5 = 8$, and the problem is solved.

To *solve* an equation, then, we find a numerical value for the variable (in this case x) which makes the equation a true sentence. Such values of the variable are said to "satisfy" the equation and are called *solutions,* or *roots,* of the equation.

To put the same idea somewhat more formally, we might say that the equation

$$x + 5 = 13 \tag{1}$$

has been converted to the equation

$$x = 13 - 5$$

or

$$x = 8 \tag{2}$$

by subtracting 5 from both members.

If we had subtracted, say, 2 from each member of Eq. (1), we would have obtained the equation

$$x + 5 - 2 = 13 - 2$$

or

$$x + 3 = 11 \tag{3}$$

However, in solving Eq. (1) we chose to subtract 5, instead of 2, from each number, because 5 and only 5, when so used, will make x appear alone on one side. The process of adding or subtracting the same quantity from both members of an equation is often known as *transposition*.

Consider now the equation

$$x - 2 = 6 \tag{4}$$

The statement of fact is that 6 is 2 less than some as yet unknown number x. The proposed problem is to find what number x is. If 6 is 2 less than x, it follows that x is 2 more than 6, or

$$x = 6 + 2 = 8$$

More formally, if we add 2 to both sides of the equation, we obtain

$$x - 2 + 2 = 6 + 2$$
$$x = 8 \tag{5}$$

The equation

$$\frac{x}{4} = 2 \tag{6}$$

states that 2 is one-fourth of x and implies the question, "What is x?" If one-fourth of a single x is 2, then one whole x must be 4 times 2, or 8. In other words, we can multiply both members of Eq. (6) by 4 and obtain

$$4\,\frac{x}{4} = (4)(2)$$

$$x = 8$$

Finally, let us consider the equation

$$3x = 24 \tag{7}$$

This equation states that 24 is equal to 3 times a certain number x whose value is at present unknown. It also implies the question, "What is x?" If 3 x's are equal to 24, then a single x must be equal to one-third of 24, or 8. We can obtain the same result by dividing both sides of the equation by 3.

$$3x \div 3 = 24 \div 3$$

5.2 Equivalent, Identical, and Conditional Equations

Two or more equations are *equivalent* if their respective members become equal when the same value (or values) of the unknown letter is used in each case.

Thus Eqs. (1), (3), (4), (6), and (7) are equivalent equations. Each may be obtained from any of the others by one or more of the processes of subtraction, addition, multiplication, and division illustrated above. In each case a true statement results when 8 is substituted for x.

An *identical* equivalent equation is a statement of equality holding true for all permissible values of the unknown letter. [A permissible value is one for which all the expressions appearing in the equation are defined. As an example, in the equation $(x^2 - 1)/(x - 1) = x + 1$, the value $x = 1$ is not permissible, the left-hand member being undefined, since this value leads to division by zero.] The equation $x^2 - 9 = (x + 3)(x - 3)$ is an identical equation. It is often written $x^2 - 9 \equiv (x + 3)(x - 3)$, the sign $\equiv$ being read "is identically equal to."

A *conditional* equation holds true for one or more but not all permissible values of the variable. Thus each of the equivalent equations (1) to (7) is also a conditional equation. Most of the equations with which we shall deal in this chapter are conditional equations.

5.3 Degree of an Equation

The *degree of a term* in an equation is obtained by adding the exponents of the variables in that term. The *degree of an equation* corresponds to the term of highest degree occurring in the equation. For example, if x and y are variables, all other letters representing constants, the following equations are of the first degree: $3x - 4 = b^2 + x$, $x + y = 7$. Second-degree equations may be exemplified by

$$x^2 - 3x + xy - y = 10 \qquad y^2 + 2y = x - 3$$

A typical third-degree equation is $x^2 - x^2y = y + 6$.

A first-degree equation is usually called a *linear equation*. We shall see why in Chap. 8. For the present we shall be concerned only with first-degree equations or equations which can be solved by the same methods as we use in solving first-degree equations.

5.4 Solving a Linear, or First-degree, Equation

The linear equations solved in Sec. 5.1 were particularly easy to solve. The methods used were informal and largely intuitive. To solve more difficult linear equations we shall need a more systematic strategy.

Usually this strategy consists in obtaining a sequence of equivalent equations, the last of which shows the variable alone as one member, the constants constituting the other member.

There are definite rules by which we can operate on a given equation in order to be sure of obtaining an equivalent equation. These are called *permissible* operations.

Permissible operations are as follows:

1. Adding the same number to both members of the given equation
2. Subtracting the same number from both members of the given equation
3. Multiplying both members of the given equation by the same *nonzero* number
4. Dividing both members of the given equation by the same *nonzero* number

Where appropriate, the following devices are useful:

If fractions occur in the equation, first make sure that they are in their lowest terms. Usually it will be best to clear the equation of fractions by multiplying both members by the LCD. Remove any parentheses.

By applying operations 1 and/or 2 above, transform the equation so that all terms containing the variable are on one side of the equation and the remaining terms are on the other side.

Apply the distributive law by collecting like terms and expressing in a factored form the collection of terms containing the variable. Divide each member of the equation by the coefficient of the variable. Finally, check the solution by substituting it for the variable in the original equation.

Example 1. Solve the equation

$$\frac{3x}{4} + \frac{1}{6} = 2x - \frac{7}{3}$$

Multiplying by the LCD, 12, we obtain

$$9x + 2 = 24x - 28 \tag{8}$$

By subtracting 9x from both members of Eq. (8) and adding 28 to both members of Eq. (8), we obtain

$$2 + 28 = 24x - 9x$$

Combining like terms,

$$30 = 15x$$

Dividing by 15,

$$^{30}\!/_{15} = 2 = x$$

To check this solution we substitute 2 for x in the given equation, obtaining

$$\frac{3(2)}{4} + \frac{1}{6} \overset{?}{=} 2(2) - \frac{7}{3}$$

$$\frac{3}{2} + \frac{1}{6} \overset{?}{=} 4 - \frac{7}{3}$$

$$\frac{9}{6} + \frac{1}{6} \overset{?}{=} \frac{12}{3} - \frac{7}{3}$$

$$\frac{10}{6} \overset{?}{=} \frac{5}{3}$$

$$\frac{5}{3} = \frac{5}{3}$$

Example 2. Solve the equation

$$\frac{8}{x-4} - \frac{6}{x-3} = \frac{2}{x-6}$$

Multiplying by the LCD, which is $(x - 4)(x - 3)(x - 6)$,

$$8(x - 3)(x - 6) - 6(x - 4)(x - 6) = 2(x - 4)(x - 3)$$

Removing parentheses,

$$8x^2 - 72x + 144 - 6x^2 + 60x - 144 = 2x^2 - 14x + 24 \tag{9}$$

By subtracting $2x^2$ from both members of Eq. (9) and adding $14x$ to both members of the same equation, we obtain

$$8x^2 - 6x^2 - 2x^2 - 72x + 60x + 14x = 24$$

Collecting like terms,

$$2x = 24$$

Dividing by 2,

$$x = 12$$

To check this solution we substitute 12 for x in the given equation; thus

$$\frac{8}{12 - 4} - \frac{6}{12 - 3} \stackrel{?}{=} \frac{2}{12 - 6}$$

or

$$\frac{8}{8} - \frac{6}{9} \stackrel{?}{=} \frac{2}{6}$$

or

$$1 - \frac{2}{3} \stackrel{?}{=} \frac{1}{3}$$

and we find the identity

$$\frac{1}{3} = \frac{1}{3}$$

We note that one of the permissible operations is the operation of multiplying both members of an equation by the same *nonzero* number.

If the occasion arises in which it seems expedient to multiply both members of an equation by an expression containing the variable, or unknown, then certain precautions are necessary. This will be illustrated in the following example.

Example 3. Solve for x if

$$\frac{x}{x-3} = 5 + \frac{3}{x-3} \tag{10}$$

By multiplying both members of Eq. (10) by the least common denominator, which in this case is $(x-3)$, we obtain

$$x = 5(x-3) + 3 \tag{11}$$

or

$$x = 5x - 15 + 3 \tag{12}$$

or

$$x = 5x - 12 \tag{13}$$

Subtracting 5x from both members of Eq. (13),

$$-4x = -12$$

or

$$4x = 12$$

and

$$x = 3 \tag{14}$$

As a check let us substitute $x = 3$ in Eq. (10).

$$\frac{3}{3-3} \stackrel{?}{=} 5 + \frac{3}{3-3}$$

or

$$\frac{3}{0} \stackrel{?}{=} 5 + \frac{3}{0} \tag{15}$$

Since division by zero is undefined, it follows that 3 is not a solution to Eq. (10). A value of the variable, such as $x = 3$, which satisfies a derived equation but does not satisfy the given equation is called an *extraneous solution,* or an *extraneous root.*

The possibility of obtaining an extraneous root is a strong argument for checking *all solutions.*

It is quite obvious from an inspection of Eq. (10) that 3 cannot be a root of that equation since this value of x will lead to a zero denominator.

As a matter of fact, Eq. (10) has no roots, as can be shown by subtracting $3/(x - 3)$ from both members to obtain

$$\frac{x}{x - 3} - \frac{3}{x - 3} = 5$$

or

$$\frac{x - 3}{x - 3} = 5 \tag{16}$$

If $x = 3$ in Eq. (16), then the left member becomes 0/0, which is indeterminate. If x is any other number, we obtain the absurdity

$$1 = 5$$

In general, we may conclude that if we multiply both members of a given equation by an expression containing the unknown, the new equation may have more solutions than the given equation.

Furthermore, if we divide both members of a given equation by an expression involving the unknown, a root may be lost.

The first of these situations may arise when we have in effect, perhaps inadvertently, multiplied both members of an equation by zero.

The second will be illustrated in the following example, where we may inadvertently divide both members of an equation by zero.

Example 4. Solve for x if

$$x^2 - 3x = 2x \tag{17}$$

By dividing both members of Eq. (17) by x, we obtain

$$x - 3 = 2 \tag{18}$$

or

$$x = 5$$

This value of x checks in Eq. (17). However, by inspection of Eq. (17), we see that x = 0 is also a solution, which was lost in the process of dividing both members of Eq. (17) by x to obtain Eq. (18).

Example 5. Solve for x the equation

$$3(3x - a) + 2a = a(ax - 3) + 6$$

(This is called a *literal equation,* since in addition to the variable x it contains a letter a, considered here to represent a constant.)

Removing parentheses,

$$9x - 3a + 2a = a^2x - 3a + 6$$

By subtracting 9x + 6 and adding 3a to both members, we obtain

$$-3a + 2a + 3a - 6 = a^2x - 9x$$

Collecting like terms,

$$2a - 6 = x(a^2 - 9)$$

Dividing by $a^2 - 9$,

$$\frac{2a - 6}{a^2 - 9} = \frac{2}{a + 3} = x$$

We may check by substituting $2/(a + 3)$ for x in the original equation, but such a check in a literal equation is apt to entail more work than the original solution. A reasonably certain check may be obtained by assigning to a some arbitrary simple numerical value. We should avoid using 1 or 0 or any number which will make any denominator equal to zero.

In this illustration, if we let $a = -2$, then x = 2. Substituting in the original equation,

$$3[(3)(2) - (-2)] + 2(-2) = (-2)[(-2)(2) - 3] + 6$$

or

$$3(6 + 2) - 4 = (-2)(-4 - 3) + 6$$

or

$$20 = 20$$

EXERCISE 1

Solve the following linear equations. When more than one letter appears, consider the last letter (in alphabetical order) as the unknown quantity and all other letters as known quantities.

1. $13x - 8 = 8x + 2$

2. $9x - 1 = 2x + 6$

3. $7x + 4 = x - 8$

4. $5 - 2x = x + 20$

5. $2 - 3x + 7 = 8x + 3 - x$

6. $11x + 3 - 4x = 16 - 2x + 2$

7. $5y - (3y - 2) = 10$

8. $7 - (8x + 1) = 18$

9. $6(w + 5) - 12 = 3(3w - 1) + 4w$

10. $26 - 5(3 - 2z) = z - 4(z + 9)$

11. $(r + 1)^2 = r^2 + 9$

12. $(2x - 3)^2 = 4x^2 - 15$

13. $(x - 2)^3 = x^2(x - 6)$

14. $(y + 1)(y - 2) = y^2 + 5$

15. $(z + 1)(z + 5) = (z + 2)(z + 3)$

16. $(2w + 1)(3w + 1) = (6w - 1)(w + 2)$

17. $\dfrac{x}{3} + \dfrac{x}{4} = \dfrac{7}{2}$

18. $\dfrac{5}{8}x - 1 = 1 + \dfrac{7}{10}x$

19. $\dfrac{x}{10} + \dfrac{x}{12} + \dfrac{x}{15} = x - 6$

20. $\dfrac{1}{6}(x - 2) = \dfrac{1}{8}(x + 1)$

21. $\dfrac{1}{8}(1 - y) - \dfrac{1}{10}(2 - y) - \dfrac{1}{12}(3 + y) = 0$

22. $0.2x = 46 - 0.03x$

23. $0.103 - 0.1x = 0.02x - 0.13x + 0.11$

24. $\dfrac{2}{x} + \dfrac{3}{x} = 10$

25. $\dfrac{4}{w} + 3 = 4 - \dfrac{3}{w}$

26. $\dfrac{1}{4y} - \dfrac{1}{6y} = \dfrac{1}{8}$

27. $\dfrac{3}{5x} - \dfrac{1}{2x} = \dfrac{1}{40}$

28. $\dfrac{8}{x + 4} = \dfrac{6}{x - 4}$

29. $\dfrac{5}{3y + 2} = \dfrac{7}{5y - 2}$

30. $\dfrac{4}{7z + 3} = \dfrac{3}{6z + 2}$

31. $\dfrac{4x - 3}{2x + 6} = \dfrac{6x - 2}{3x + 11}$

32. $\dfrac{6y - 3}{3y + 2} = \dfrac{2y + 1}{y + 2}$

33. $\dfrac{4}{5x + 5} - \dfrac{7}{10x + 10} = \dfrac{1}{20}$

34. $\dfrac{8}{x - 2} - \dfrac{5}{x - 11} = \dfrac{3}{x - 5}$

35. $\dfrac{7}{x + 1} - \dfrac{4}{x - 1} = \dfrac{3}{x + 5}$

36. $\dfrac{2x}{x^2 - 4} - \dfrac{4}{x^2 - 4} = \dfrac{2}{2x - 3}$

37. $\dfrac{2}{1 - 2w} + \dfrac{2}{7 - 2w} = 1 - \dfrac{4w^2 - 1}{4w^2 - 16w + 7}$

38. $ax + b = c$

39. $t - sx = r$

40. $6abx = 9a^3b^2c$

41. $\dfrac{x}{a} = \dfrac{a}{4}$

42. $\dfrac{b}{y} = \dfrac{c}{d}$

43. $\dfrac{m}{x} = k$

44. $\dfrac{c}{x} = a + 1$

45. $ax + bx = 3a + 3b$

46. $mx - h = hx - m$

47. $2mx + n^2 = 2nx + m^2$

48. $3ay = 5by + 2$

49. $ax + 2 = bx + 8$

50. $b(x + 1) = c$

51. $\dfrac{x}{c} - a = b$

52. $\dfrac{x}{a} + \dfrac{x}{b} = a^2 - b^2$

53. $\dfrac{a}{x} = b + c$

54. $\dfrac{m + n}{x} = m^2 + mn$

55. $(a - b)x - a^2 = (a + b)x$

56. $c(1 + w) + d(1 + w) = w(c + d + 1)$
57. $b(b - 2y) + c(c - 2y) + 2bc = 0$
58. $mn(z^2 - 1) = (m + nz)(n + mz)$
59. $(p - y)(y + q) - r(y + r) = (r - y)(y + r) + pq$

60. $\dfrac{ax}{b} + \dfrac{bx}{a} = 1$

61. $\dfrac{c + x}{m} = \dfrac{x}{c + m}$

62. $a^2 = \dfrac{a + c}{x} + c^2$

63. $\dfrac{w - c}{w - d} = \dfrac{c^2}{d^2}$

64. $\dfrac{by^2}{c - ay} + b + \dfrac{by}{a} = 0$

65. $\dfrac{m}{n - p} + \dfrac{n - p}{z} = \dfrac{m}{n + p} + \dfrac{n + p}{z}$

66. By inspection determine the number of roots in

    (a) the equation $y + \dfrac{3}{y - 5} = -5 + \dfrac{3}{y - 5}$

    (b) the equation $y + \dfrac{3}{y - 5} = 5 + \dfrac{3}{y - 5}$

5.5 Formulas

Perhaps the most common occurrence of the literal equation is the formula. In most mathematical or engineering handbooks will be found a number of formulas for determining length, area, volume, etc. Frequently it will be desirable to transform a formula or solve for one of the factors previously regarded as a known quantity. For example, the volume of a cone is expressed by the formula $V = \pi r^2 h/3$, in which it is assumed that the radius r and the height h are known and the volume V is to be found. Should there be repeated occasions when the volume and radius are known and the height is to be found, we should solve the formula for h. Following the usual procedure for solving literal linear equations, we obtain $h = 3V/\pi r^2$.

    In Exercise 2 below, the transformed version of the formula should be left in the most convenient form for computation. Few specific directions can be offered, but it is suggested that a number of constants related to each other by multiplication or division may well be combined into a single constant. Furthermore, a multiplying constant is usually more easily handled than a dividing constant. The student's most reliable guide is his imagination. If he will regard himself as the user of the formula he has produced, he will so arrange it that it requires a minimum of effort for actual computation.

If we consider the recently discussed formula $h = 3V/\pi r^2$, we see that possible variations are

(a) $h = \dfrac{3}{\pi} \cdot \dfrac{V}{r^2} = \dfrac{0.9549V}{r^2}$

(b) $h = \dfrac{V}{(\pi/3)r^2} = \dfrac{V}{1.047r^2}$

If a table of areas of circles is available, the original form would probably be best; otherwise Eq. (a) would be preferable. On the other hand, it is doubtful that Eq. (b) would be a good choice under any circumstances.

5.6 Mathematical Operations with Dimensional Units

In setting up equations derived from physical problems, it must be remembered that each side of the equation must reduce to like units or dimensions; i.e., the equation must be dimensionally correct. Operations with dimensional symbols are subject to the usual laws of algebra. Some common examples are:

Length $\times$ length = area $\qquad$ (ft)(ft) = ft^2

Length $\times$ length $\times$ length = volume $\qquad$ (ft)(ft)(ft) = ft^3

Area $\times$ length = volume $\qquad$ ft$^2 \times$ ft = ft^3

$\dfrac{\text{Weight}}{\text{Volume}}$ = density† $\qquad$ lb $\div$ ft$^3 = \dfrac{\text{lb}}{\text{ft}^3}$

$\dfrac{\text{Distance}}{\text{Time}}$ = average velocity $\qquad$ ft $\div$ s = $\dfrac{\text{ft}}{\text{s}}$

$\dfrac{\text{Change in velocity}}{\text{Time}}$ = average acceleration $\qquad$ $\dfrac{\text{ft}}{\text{s}} \div$ s = $\dfrac{\text{ft}}{\text{s}^2}$

$\dfrac{\text{Force}}{\text{Area}}$ = pressure† $\qquad$ lb $\div$ ft$^2 = \dfrac{\text{lb}}{\text{ft}^2}$

(Weight density) $\times$ height = pressure $\qquad$ $\dfrac{\text{lb}}{\text{ft}^3} \times$ ft = $\dfrac{\text{lb}}{\text{ft}^2}$

†Gravitational units (the language of the "man on the street"), rather than absolute units, are used.

Pressure × volume = work†

$$\frac{\text{lb}}{\text{ft}^2} \times \text{ft}^3 = (\text{ft})(\text{lb})$$

Distance × force = work†

$$\text{ft} \times \text{lb} = (\text{ft})(\text{lb})$$

Power × time = work†

$$\frac{(\text{ft})(\text{lb})}{\text{s}} \times \text{s} = (\text{ft})(\text{lb})$$

$$\frac{\text{Work}}{\text{Time}} = \text{power†}$$

$$(\text{ft})(\text{lb}) \div \text{s} = \frac{(\text{ft})(\text{lb})}{\text{s}}$$

5.7 Analysis of a Formula

Whenever a formula is developed or applied for the first time, it is instructive to subject it to a few screening tests. The most common of these are the test for dimensional soundness and the test for applicability to special conditions.

The application of these tests can be illustrated to advantage by referring to a formula such as that given in Fact 125, Chap. 7, for the volume of the frustum of a cone. The formula is

$$V = \tfrac{1}{3}\pi h(r_1^2 + r_1 r_2 + r_2^2)$$

If the indicated multiplication is carried out, the terms will be hr_1^2, $hr_1 r_2$, and hr_2^2—all third-degree terms in a length unit. Such terms are therefore expressions of volume (Sec. 5.6), and the formula is dimensionally sound.

If we assume $r_1 = r_2$, as would be the case in a cylinder, we obtain

$$V = \tfrac{1}{3}\pi h(r_1^2 + r_1 r_1 + r_1^2)$$

or

$$V = \pi r_1^2 h$$

the formula for the volume of a cylinder.

If we assume $r_1 = 0$, as in a cone, we obtain

$$V = \tfrac{1}{3}\pi h r_2^2$$

the formula for the volume of a cone.

If we take $h = 0$, the entire expression assumes the value zero, as is to be expected.

†Gravitational units (the language of the "man on the street"), rather than absolute units, are used.

5.8 Dimensional Units in Conversions

Dimensional units, properly handled, will not only check the dimensional soundness of a formula, they will also indicate what steps are necessary to make a required conversion.

Example 6. If costs \$30/h to run a boat when sailing at 10 mi/h. Find the operating cost in dollars per mile.

$$\frac{\$30}{1 \text{ h}} \div \frac{10 \text{ mi}}{1 \text{ h}} = \frac{\$30}{1 \text{ h}} \times \frac{\overset{3}{\cancel{1 \text{ h}}}}{\underset{1}{\cancel{10} \text{ mi}}} = \$3/\text{mi}$$

Example 7. The density of mercury is 13.6 g/cm³. Express the density in pounds per cubic inch.

$$\frac{13.6 \text{ g}}{\text{cm}^3} = \frac{13.6 \text{ g}}{1 \text{ cm}^3} \times \frac{1 \text{ lb}}{454 \text{ g}} \times \frac{(2.54)^3 \text{ cm}^3}{1 \text{ in}^3} = 0.492 \text{ lb/in}^3$$

EXERCISE 2

Solve the following formulas for the letter indicated. Leave your answer in the most convenient form for computation, and be prepared to justify your choice. The formulas are identified in the right-hand column.

	Solve for	Description of formula
1. $Q = \dfrac{WL}{T}$	T	Latent heat of vaporization
2. $X = \dfrac{1}{2\pi f C}$	C	Reactance of a capacitor
3. $I = \dfrac{E - e}{R}$	e	Current flowing through armature of generator
4. $V = \dfrac{V_t + V_0}{2}$	V_0	Average speed of uniformly accelerating body
5. $\dfrac{E}{e} = \dfrac{R + r}{r}$	r	Voltage drop
6. $T = \dfrac{1}{a} + t$ $\left(\text{first solve for } \dfrac{1}{a}\right)$	a	Temperature-conversion formula
7. $C = \dfrac{Kab}{b - a}$	a	
8. $S = \dfrac{rl - a}{r - 1}$	a, r	Geometric progression

9. $\rho = \dfrac{m}{d - L} - \dfrac{m}{d + L}$　　　　　m

10. $\dfrac{e}{x} = C(e - b) + \dfrac{b}{x}$　　　　　x

11. $Q = 0.000477EIT$　　　　　T　　Electrical equivalent of heat

12. $d = \frac{1}{2}at^2 - \frac{1}{2}a(t - 1)^2$　　　　　t　　Distance covered by falling body

13. $C = \frac{5}{9}(F - 32)$　　　　　F　　Celsius-Fahrenheit tempera-
　　　　　　　　　　　　　　　　　　ture conversion

14. $A = \dfrac{m}{t}(p + t)$　　　　　t　　Thickness of pipe

15. $H = \dfrac{0.4\pi NI}{L}$　　　　　I　　Magnetic intensity

16. $M = 10.5C + 35.2\left(W - \dfrac{C}{8}\right)$　　　　　C　　Theoretical amount of air re-
　　　　　　　　　　　　　　　　　　quired to burn solid fuel

17. $\dfrac{1}{x} + \dfrac{1}{nx} = \dfrac{1}{f}$　　　　　x　　Photographic enlargement

18. $S = \left(\dfrac{\pi d^2}{2} + \dfrac{\pi dl}{r}\right) \div \dfrac{\pi d^2 l}{4rc}$　　　　　r　　Exposed surface of cylinders

19. $S = T - \dfrac{1.299}{N}$　　　　　N　　Tap-size drill for U.S. standard
　　　　　　　　　　　　　　　　　　thread

20. $W = \dfrac{2PR}{R - r}$　　　　　R　　Differential pulley

21. $\dfrac{1}{R} = \dfrac{1}{r_1} + \dfrac{1}{r_2}$　　　　　R　　Parallel resistances

22. $I = \dfrac{E}{r + (R/n)}$　　　　　n　　Current produced by cells in
　　　　　　　　　　　　　　　　　　parallel

23. $T = T_1\left(1 - \dfrac{n - 1}{n} \cdot \dfrac{h}{h_0}\right)$　　　　　n

24. $x - y = xy$　　　　　y

25. $V = \dfrac{h}{6}(B + 4M + b)$　　　　　M　　Prismoidal formula

26. $wf = \left(\dfrac{w}{k} - 1\right)\dfrac{1}{k}$　　　　　w

27. $V_1 = V_0(1 + 0.00366t)$　　　　　t　　Expansion of gases

5.9 Suggestions for the Solution of Applied Problems Involving Linear Equations

There is no pat formula for the solution of "word" problems. One should not feel discouraged if he finds this section of the work difficult, for this is a common reaction.

This situation is due largely to the fact that the translation of a verbal or descriptive problem (commonly called a word problem) involves a minimum of dependence on mechanical rules. Such translation requires some exercise of ingenuity and of the skill that comes only with experience. If this seems a bleak outlook, it should be remembered that these remarks apply equally well to carpentry, machine design, music, painting, etc. The feeling of accomplishment in solving the more interesting applied problems is not experienced by those content merely to follow mechanical procedures.

While a rigid, ironclad procedure cannot be offered, a few general suggestions which have proved helpful can be outlined.

1. *Read the problem through once to get the general idea.*
2. *Carefully reread the problem, noting what is given and what is wanted.*
3. *A rough graphical representation will often suggest relationships leading to a solution (see Sec. 9.18).*
4. *Represent the unknown quantity by some appropriate symbol, such as v for velocity, t for time, w for weight.*

The symbol should be quantitative: it should refer to the number of units in the unknown; e.g., "Let t represent the time for the trip in hours"; "Let w represent the weight of copper in pounds." Be explicit.

All related quantities should be expressed in consistent units. For example, if it has been decided to express a velocity in miles per hour, all distances must be expressed in miles and all times must be expressed in hours.

If you consider your data to be inadequate, make an approximation by any reasonable means.

1. *Try to discover two expressions which are equal. Form them into an equation and check dimensionality.*

The word *dimension* as used here refers to the *kind* of a quantity with which we are dealing. For example, the weight w of an object may be measured in *units* of pounds, ounces, tons, grams, etc. However, regardless of the units used, w is said to have the dimension of weight.

The speed v of a moving automobile could be expressed in miles per hour, feet per second, or even furlongs per fortnight. In each case, however, v is said to have the dimension of speed, even though the units used are quite different.

Parenthetically, it is worth mentioning that the word "per" means "for each." That is, if an automobile is traveling at the rate of 20 miles *per* hour, this means that it is traveling at the rate of 20 miles *for each* hour.

2. *Solve the equation and check the solution against the original worded statement, not the equation you derived from it.*

Numerous laws of mathematics, natural sciences, etc., such as the following, may serve as the basis of equality:

The sum of the parts equals the whole.
Distance equals rate times time.
Weight equals volume times density.
Number of units times the unit cost equals the total cost.
Principal times the rate of interest times the time equals the interest.
Amount of material times the fraction of a particular ingredient equals the amount of ingredient.
The square of the hypotenuse of a right triangle equals the sum of the squares of the legs.

3. *Make a quick "commonsense" check to see if your answer seems reasonable.*

Example 8. A tourist having $4\frac{1}{2}$ h at his disposal rides out into the country on a bus at a rate of 19 mi/h. He plans to walk back at a rate of $3\frac{1}{2}$ mi/h. How far can he ride without arriving late at his starting point?

The one-way distance is required. The total time and the outgoing and return rates are given. Let $d =$ number of miles one way. We know that distance $=$ rate $\times$ time, or distance/rate $=$ time. We can set up an equation:

$$\text{Outbound time} + \text{return time} = \text{total time} \tag{19}$$

Equation (19) is dimensionally correct since it is expressed as

$$\text{A time} + \text{a time} = \text{a time}$$

Now, since

$$\text{Time} = \frac{\text{distance}}{\text{rate}}$$

we may write

$$\underset{\text{(outbound)}}{\frac{\text{Distance}}{\text{Rate}}} + \underset{\text{(return)}}{\frac{\text{distance}}{\text{rate}}} = \text{total time}$$

$$\frac{d}{19} + \frac{d}{3\frac{1}{2}} = 4\frac{1}{2} \tag{20}$$

Equation (20) is consistent with respect to the units used since both distances are in units of miles, both rates are in units of miles per hour, and time is expressed in hours.

Then

$$3\tfrac{1}{2}d + 19d = (19)(3\tfrac{1}{2})(4\tfrac{1}{2})$$
$$22\tfrac{1}{2}d = 299\tfrac{1}{4}$$
$$d = 13.3 \text{ mi}$$

Checking against the original statement,

$$\frac{13.3}{19} = 0.7 \text{ h outbound}$$

$$\frac{13.3}{3.5} = \frac{3.8 \text{ h return}}{4.5 \text{ h total}}$$

Alternatively,

Distance = distance
(outbound)   (return)

or

Rate × time = rate × time
  (outbound)   (return)

If t is the number of hours outbound, this equation becomes

$$(19)(t) = 3\tfrac{1}{2}(4\tfrac{1}{2} - t)$$

or

$$19t = 15.75 - 3.5t$$

and $t = 0.7$ h outbound and $(0.7)(19) = 13.3$ mi one-way distance.

Example 9. How much 80 percent by volume antifreeze (alcohol) must be added to 5 qt of 30 percent antifreeze to raise the strength to 65 percent?

We wish to know the amount of 80 percent antifreeze needed. We are given the concentrations of the initial solutions and the final mixture and the amount of the weaker solution. Let n = number of quarts of 80 percent solution required. Then $n + 5$ = number of quarts of mixture finally obtained. We can set up an equation saying that quarts of pure alcohol in the 5 qt of the 30 percent solution plus the quarts of pure alcohol in the n qt of 80 percent solution taken equals the quarts of pure alcohol in the $(n + 5)$ qt of 65 percent final mixture. Hence

$$(0.30)(5) + 0.80n = 0.65(n + 5)$$
$$1.5 + 0.8n = 0.65n + 3.25$$
$$0.15n = 1.75$$

$$n = \frac{1.75}{0.15} = 11\frac{2}{3} \text{ qt of 80 percent antifreeze required}$$

Check: $0.80 \times 11\frac{2}{3} = 9\frac{1}{3}$ qt pure alcohol in 80 percent solution
$0.30 \times 5 = 1\frac{1}{2}$ qt pure alcohol in 30 percent solution
$\text{Total} = 10\frac{5}{6}$ qt pure alcohol in mixture
$0.65 \times 16\frac{2}{3} = 10\frac{5}{6}$ qt pure alcohol in mixture (check)

It may be helpful to set up a table such as the following:

	Volume, qt	Active ingredient, qt
80% solution	n	$0.80n$
30% solution	5	$(0.30)(5) = 1.5$
Total	$n + 5$	$0.80n + 1.5$

But $0.80n + 1.5 = 0.65(n + 5)$, as above.

EXERCISE 3

Set up and solve linear equations in terms of one unknown† for the following problems. Check against the original statement.

1. The difference between two numbers is 7. Their sum is 53. Find the numbers. (See Prob. 1, Exercise 12, Chap. 9.)
2. One number is $1\frac{1}{2}$ times as large as another. The sum of the numbers is 35. Find the numbers.
3. The sum of two numbers is 45. The quotient obtained by dividing the larger by the smaller is 2, with a remainder of 3. Find the numbers.
4. Separate 22 into two parts such that one part is 3 times the other.
5. Find three consecutive numbers whose sum is 72.
6. Find four consecutive even numbers whose sum is 68.
7. Find five consecutive odd numbers whose sum is 85.
8. One side of a right triangle is 8 in, and the hypotenuse is 2 in longer than the other side. Find the sides of the triangle.
9. The difference between the squares of two consecutive numbers is 37. Find the numbers.

† A few of these problems are repeated in Chap. 9, where two unknowns are used. In many cases the choice between one and two unknowns is a matter of individual preference.

10. The denominator of a fraction exceeds the numerator by 8. The reduced value is $\frac{7}{9}$. What is the fraction?

11. The reduced value of a fraction is $\frac{2}{3}$. If 16 is added to both numerator and denominator, the reduced value of the new fraction is $\frac{10}{13}$. What was the original fraction?

12. The reduced value of a fraction is $\frac{4}{5}$. If 21 is deducted from both numerator and denominator, the reduced value of the new fraction is $\frac{1}{3}$. Find the original fraction.

   If we represent the original fraction as $\frac{4x}{5x}$, then we may write $\frac{4x - 21}{5x - 21} = \frac{1}{3}$.
   (Also see Prob. 12, Exercise 12, Chap. 9.)

13. The length of a rectangle exceeds the width by 14 ft. The perimeter is 160 ft. Find the dimensions.

14. The perimeter of an isosceles triangle is 50 in. The base is 11 in longer than one of the equal sides. Find the sides of the triangle.

15. Ten boys agreed to buy a canoe, dividing the expense equally. Two more boys joined the group, reducing the share of each by $1.50. How much did the canoe cost?

16. A boy worked 16 days for a certain amount of money. Had he received $0.90/day more, he could have earned his money in 1 day less time. Find his daily wages.

17. A man bought 30 acres of land for $1,280. Part of the land cost $40/acre; the rest cost $50/acre. How much land was sold at $40/acre? (See Prob. 13, Exercise 12, Chap. 9.)

18. A sum of money consists of dimes and quarters. There are 9 more dimes than quarters. If the total value is $10, find the number of dimes.

19. A man invested $7,000, part at $5\frac{1}{4}$ percent and the rest at 6 percent. If the interest for 1 year was $384, find the amount invested at 6 percent. (See Prob. 15, Exercise 12, Chap. 9.)

20. How much solder containing 50 percent tin and how much type metal containing 15 percent tin must be mixed to make 80 lb of solder containing 40 percent tin?

   If $t =$ lb of type metal, then $80 - t =$ lb of 50 percent solder. It follows that lb of tin in original solder + lb of tin in type metal = total tin or

$$0.50(80 - t) + 0.15t = (0.40)(80)$$

   (See also Prob. 16, Exercise 12, Chap. 9.)

21. A 12-qt cooling system is filled with 25 percent antifreeze. How many quarts must be drawn off and replaced with pure antifreeze to raise the strength to 45 percent? (See Prob. 17, Exercise 12, Chap. 9.)

22. Two trains leave opposite terminals of a 200-mi line at the same time. If the rates are 40 and 35 mi/h, respectively, when will they meet?

23. A group of bicyclists maintains an average rate of 12 mi/h. One hour and 45 min after they leave, a motorist sets out to overtake them. If his rate is 40 mi/h, how long will he take?

24. In a municipal election a total of 82,347 votes were cast for two nominees, A and B. If A received 6,051 votes more than B, find the number of votes each contestant received. (See Prob. 18, Exercise 12, Chap. 9.)

25. The difference between the squares of two consecutive odd numbers is 48. Find the numbers.

26. A pile of dimes and quarters has a value of $4.60. There are three more quarters than dimes. How many quarters are there?

We may set up a table as follows:

	Number of coins	Value of coins (cents)
Quarters	q	$25q$
Dimes	$q - 3$	$10(q - 3)$
Total		$25q + 10(q - 3) = 460$ cents

27. A grocer estimated that his supply of sugar would last 30 days. Because he sold 20 lb/day more than he expected, it lasted only 24 days. How many pounds did he have?

28. The reduced value of a certain fraction is $\frac{2}{3}$, and its denominator exceeds its numerator by 4. Find the fraction.

29. At a high-school game the price of admission was $0.25 for each adult and $0.10 for each child. If the turnstile showed 397 persons at the game and the gate receipts were $56.80, how many adults attended? (See Prob. 19, Exercise 12, Chap. 9.)

30. In the formula $N = 0.907(D/d - 0.94)^2 + 3.7$, N is the number of wires of diameter d that can be contained in a conduit of diameter D. What must be the ratio of conduit diameter to wire diameter in order that the conduit may hold 100 wires?

31. A bonus of $30,000 is to be distributed among 500 employees of a factory. There are 50 men with 20 years' service, 100 men with 10 years' service, and 350 men with 5 years' service. Each 20-year man is to receive twice as much as a 10-year man, and each 10-year man is to receive twice as much as a 5-year man. How much should each of the 5-year, 10-year, and 20-year men receive?

32. A farmer bought 100 acres of land for $3,880. Part cost $50/acre; the rest $18/acre. Find the number of acres bought at each price.

33. A man can clear a wood lot in 6 days, while his son could do it alone in 12 days. How long would the job take if they worked together?

34. A tank can be filled by two pipes in 6 min, while the first pipe alone would require 10 min. How long would it take the second pipe alone to fill the tank?

35. How much time would be required to fill the tank in Prob. 34 if the first pipe operated as an inlet and the second pipe as an outlet? The tank is being used for leaching purposes.

36. A job can be completed by A in $22\frac{1}{2}$ h, by B in $16\frac{2}{3}$ h, and by C in 18 h. Find the time required when all are working together.

37. How many gallons of water must be mixed with 500 gal of 96 percent (by volume) sulfuric acid to reduce the strength to 80 percent?

38. A uniform 20-in bar weighing 1 lb is in equilibrium when a 3-lb weight is suspended from one end and a 2-lb weight from the other. Find the position of the point of support.

39. A bar of metal contains 20 percent silver, and a second bar 12 percent. How many ounces of each must be taken to make a 40-oz bar containing $14\frac{1}{2}$ percent silver?

40. Eight thousand dollars is invested, some at $6\frac{1}{2}$ and the rest at $4\frac{1}{2}$ percent. How much must be invested at $6\frac{1}{2}$ percent to ensure a total annual income of $420?

41. How much high-speed tool steel containing 18 percent tungsten and how much steel containing 12 percent tungsten should be mixed to make 3,000 lb containing 14.6 percent tungsten? (See Prob. 20, Exercise 12, Chap. 9.)

42. What will be the final temperature when 42 lb of water at 135°F is mixed with 70 lb of water at 60°F? In problems of heat exchange involving no change of state, weight × specific heat × temperature change for the warm body = weight × specific heat × temperature change for the cooler body.

43. A bookrack is to be 44 in high overall. The stock is $\frac{7}{8}$ in thick. There are to be five shelf spaces, each one having $\frac{3}{4}$ in more vertical space than the one above it. Find the height of each shelf space. (There are to be six thicknesses of stock.)

44. If the equation $C = \frac{5}{9}(F - 32)$ represents the relationship between the Celsius and Fahrenheit readings for any temperature, find the temperature at which the two will be equal.

45. A machine having an initial value of $1,450 depreciates in value each year $50 less than during the preceding year. At the end of the sixth year it has a value of $190. What is its value at the end of the first year?

Let d = first year's depreciation, then $d - 50$ = second year's depreciation, etc., and we obtain the equation

$$1,450 - 190 = d + (d - 50) + (d - 100) + (d - 150) + (d - 200) + (d - 250)$$

46. A $39\frac{3}{4}$-in-diameter iron tire is to be shrunk onto a 40-in wheel. If the linear coefficient of expansion of iron is 0.000006 per degree Fahrenheit, to what temperature must the tire be heated from 70°F to fit on the wheel?

47. A poorly compensated watch, when carried vertically in the pocket, gains 11 s in 9 h and, when laid down horizontally, loses 28 s in 13 h. How many hours out of 24 in the horizontal position would result in no net gain or loss in a 24-h period?

48. A group of men pay $1.75 each toward the expenses of a luncheon. An excess of $1.25 results. If each had paid $1.50, there would have been $2.50 lacking. How many men were there?

49. A guy rope runs from the top of a derrick to a stake 7 ft from the base. If the rope is lengthened 15 ft, it will reach a stake 32 ft from the base. Find the height of the derrick.

50. Fifty pounds of high-grade solder spatters containing 50 percent lead and 50 percent tin are to be melted with some type metal containing 90 percent lead and 10 percent tin. The resulting alloy is to be a low-grade solder containing 75 percent lead and 25 percent tin. How many pounds of type metal will be needed?

A table based on tin would appear as follows:

	Lb metal	Lb tin
Type metal	w	$0.10w$
Solder	50	$(0.50)(50) = 25$
Total	$w + 50$	$0.10w + 25$

from which $0.10w + 25 = (0.25)(w + 50)$.

Or we could set up a table based on lead:

	Lb metal	Lb lead
Type metal	w	$0.90w$
Solder	50	$(0.50)(50) = 25$
Total	$w + 50$	$0.90w + 25$

whence $0.90w + 25 = (0.75)(w + 50)$.

51. A utility company petitioned to change its rate on electrical energy from a straight $5\frac{1}{2}$ cents/kWh to $4\frac{1}{4}$ cents/kWh plus 75 cents/month service charge. What monthly cost would be unaffected by the change in rate?

52. A man pays an income tax of $118.06 on an income which is taxed as follows: 1 percent on all income over $1,500, plus 3 percent on all income over $2,500, plus 5 percent on all income over $4,000. What is his income?

53. How many gallons of 35 percent antifreeze solution must be added to 3 gal of 80 percent antifreeze solution to reduce the strength to 60 percent?

54. A dairyman has 1,000 qt of milk containing 4.8 percent butterfat, but the city in which he sells his milk requires only 4 percent butterfat. How many quarts of cream testing 20 percent butterfat may be separated from the milk and still satisfy the legal minimum requirement?

55. The outbound trip of a bus was made at 24 mi/h, while the return trip was made at 30 mi/h. Find the one-way distance if the total running time was $4\frac{1}{2}$ h.

56. Position B is 5 mi due east of position A. An enemy battery is firing due north of A. The sound of the firing reaches A 2.5 s before it reaches B. Find the distance of the battery from A if sound travels at 1,100 ft/s.

57. The indicated airspeed of a Piper Cub plane is 120 mi/h. If a 40-mi/h west wind is blowing, how far west can the plane fly and still return to the airport $2\frac{1}{4}$ h after taking off?

58. A man has $3\frac{1}{4}$ h at his disposal. He can ride out on a bus at 16 mi/h and walk back at $3\frac{1}{2}$ mi/h. How far can he ride and still be able to return to his starting point within the allotted time?

59. At two stations A and B on a railroad line, the prices of soft coal are $25.50/ton and $27/ton, respectively. If the distance between A and B is 180 mi and coal can be shipped for $1\frac{1}{2}$ cents/(ton)(mi), find the location on the line between A and B at which it will be immaterial to a consumer whether he buys from A or B.

60. A plane has an airspeed of 300 mi/h. It flies directly into a head wind for 48 min and returns along the same route in 42 min. Find the wind speed.

61. The explosion of a floating mine was heard 8 s sooner through the water than through the air. If the speed of sound is 4,800 ft/s through the water and 1,125 ft/s through the air, how far away was the explosion?

62. To measure the flow of chlorine which would corrode an ordinary mechanical meter, pure oxygen is "bled" into the stream of chlorine gas at the rate of 5.3 ft³/min. If the chlorine, before dilution, contained 2.8 percent oxygen, and afterward contained 12.1 percent oxygen, find the original flow of chlorine gas in cubic feet per minute.

    Let v = original gas flow (ft³/min). Then, using an oxygen balance,

$$0.028v + 5.3 = 0.121(v + 5.3)$$

63. In the gambrel roof in Fig. 5.1 the upper set of rafters has a $\frac{1}{2}$ pitch and the lower set a pitch of 2. Find the lengths of AB and BC.

    Hint: Let $AD = h$. Then $DB = 2h$ (why?). EC will then equal $18 - 2h$ and BE will equal $2(18 - 2h)$ (why?).

64. A manufacturer wishes to make a competitive line of copper-coated iron wire which he can sell for $0.27/lb and make a gross profit of 50 percent above the cost of materials. If copper costs $0.32/lb and iron costs $0.12/lb, find the percentage of copper in the wire.

65. A man earning $180/week goes out on strike for 24 weeks. During the strike he receives $50/week unemployment benefit. Upon returning to work he receives

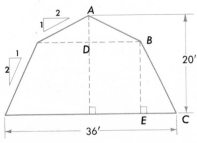

FIGURE 5.1

$210/week. How many weeks after returning to work will it take to recover the loss incurred by going out on strike?

66. In order to weigh a uniform bar of iron 8 ft long, a 5-lb weight was hung 8 in from one end, whereupon the bar was found to balance $45\frac{1}{2}$ in from that end. Find the weight of the bar.

67. A planer has a ratio of cutting speed to return speed of 1:2.75. Find the cutting speed in feet per second for a 5-ft stroke when the planer is making 15 cycles every 2 min.

68. A valve operates off a camshaft turning at a speed of 1,200 r/min. What is the travel of the valve if it opens at the rate of 2 ft/s and closes at the rate of 10 ft/s?

69. A man aged 50 has savings, etc., valued at $24,500. He considers that this amount will increase by $1,000 each year while he is employed. After retirement he will be obliged to reduce the principal by $1,800 each year. At what age can he retire so that the principal will not be exhausted before he reaches age 85?

70. A contractor must finish a job in 35 days. Only one power shovel can be used on the job at a time. The contractor owns a shovel that can do the job in 40 days and can rent, for $50/day, a second shovel that can do the job in 20 days. What is the smallest amount the contractor will have to pay for renting the shovel?

71. A range burner uses up a 54-gal drum of fuel oil in 31 days. The drum is filled the first of the month; after 10 days a water heater is also started, and the oil lasts 8 days longer. How many gallons did the heater use per day?

72. A contractor agrees to put in a concrete foundation within a certain time limit. He is to receive $40/day worked, plus $56 for each day by which completion of the job precedes the time limit. He finished the job in two-thirds of the time allowed and received $816. What was the time limit?

73. A man may retire at age 62 with social-security benefits of $150/month for the rest of his life, or he may retire at age 65 with benefits of $195/month. At what age will the total benefits from each plan be equal?

74. A number is composed of six digits, of which the first is 1. When the number is multiplied by 3, the order of digits remains the same except that the 1 is transposed to the units place. What is the number?

75. Using the formula for the length of a pulley belt in Prob. 75, page 154, find the missing diameters of the step-cone pulleys in Fig. 5.2 so that the same belt may be used in each of the three positions.

 The diameter of pulley $f = 15 \div 2\frac{1}{2} = 6$ in. Referring to Prob. 75, Chap. 7, it can be seen that, since l and a are constant, the sum $D + d$ must also be constant. As $e + f = 15 + 6 = 21$, this must be the constant total of $D + d$. Therefore $c + d = 21$, and $c = d = 10\frac{1}{2}$ in. $a + b = 21$, but $b = 3a$. Therefore $4a = 21$, or $a = 5\frac{1}{4}$ in and $b = 15\frac{3}{4}$ in.

76. A closed manometer tube is shown in Fig. 5.3. A length of air column AB is sealed off at atmospheric pressure with mercury. More mercury is then poured into the

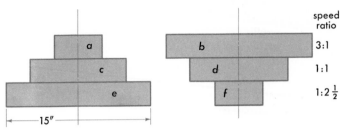

FIGURE 5.2

open arm at D, with new levels at C and D. Find, to the nearest 0.1 in of mercury, the original atmospheric pressure.

77. A man pays 33 cents/gal for gasoline. He gets 15 mi/gal from his car. He can have his engine overhauled for $150, after which he can expect to get 22 mi/gal of gasoline and save 50 cents per 100 mi on oil. After how many miles will he recover the cost of overhauling?

Let m equal the number of miles driven to recover the overhaul cost. Without the overhaul 33 cents/gal $\div$ 15 mi/gal = 33 cents/gal $\times$ $\frac{1}{15}$ gal/mi = $\frac{33}{15}$ cents/mi, or $\frac{33}{15}m$ cents for m miles.

With overhaul the gas cost will be $\frac{33}{22}m$ cents for m miles. Extra cost of oil without overhaul will be $0.5m$ cents for m miles. Since the cost of gas without overhaul + extra cost of oil equals the cost of gas after overhaul + $150 (or 15,000 cents) cost of overhaul, we may write

$$\tfrac{33}{15}m + 0.5m = \tfrac{33}{22}m + 15,000$$

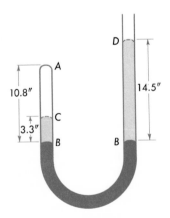

FIGURE 5.3

78. A family has 1 qt of milk delivered daily. The bill for the month of November was $9.66. It is known that the price advanced 2 cents/qt at some time before the middle of the month. What was the price per quart before the increase? When did the increase become effective? Prices are in integral cents.

79. A bus went up a hill at 25 mi/h and returned at 40 mi/h. What was the average speed for the round trip?

ratio, proportion, and variation

The comparison of quantities, often indicated by their ratio, is a common occurrence. In business it often takes the form of percentage (e.g., profit, loss, discount, interest rate). In engineering, physics, and other sciences, we find it in such characteristics as specific gravity, specific heat, and atomic weight.

6.1 Ratio

Quantities of the same kind may be compared (1) by stating the difference in magnitude of the two quantities or (2) by finding the quotient of the two quantities when expressed in the same units.

Thus, in a group consisting of 100 men and 20 women, we may say that (1) there are 80 more men than women or (2) there are five times as many men as women.

The second type of comparison is called the *ratio* of the quantities. It is expressed as a fraction, in this illustration $^{100}/_{20}$, which is then reduced to lowest terms, $^5/_1$. We may say that the number of men is to the number of women as 5 is to 1.

If the numbers are prime to each other, the division indicated by the fraction is often carried out; e.g., the rate of 56 to 39 is 1.436 to 1 (to four significant figures).

Note that the comparison must be made between like dimensions, e.g., between a length and a length—not between, say, a length and a weight. The dimensions must be expressed in the same units; e.g., the ratio between 2 lb and 12 oz is $^{32}/_{12}$, or $^8/_3$.

Occasionally, although incorrectly, we do sometimes apply the word "ratio" to comparisons between unlike quantities. For example, we sometimes speak of the economy ratio of an automobile expressed as the ratio of miles traveled to gallons of gasoline consumed. If the automobile will travel 100 miles on 5 gal of gasoline, the economy ratio is said to be 20 mi/gal.

Such comparisons might more properly be termed "rates" (e.g., miles per hour, dollars per cubic foot, etc.).

As another example, the scale ratio of a certain map may be indicated as 500 ft to 1.5 in. This means that if the distance between two points on the ground is 500 ft, then the corresponding points on the map are 1.5 in apart. Sometimes, with rather loose usage of the equality sign, this information is expressed as 500 ft = 1.5 in.

Since 500 ft is equivalent to $500 \times 12 = 6{,}000$ in, we can express the scale ratio as 6,000 in to 1.5 in. The scale ratio can then be expressed as

$$6{,}000/1.5 \quad \text{or} \quad 4{,}000/1$$

The latter form is much to be preferred.

It will be seen that percentage simply expresses the ratio between a given number of units and 100 units of the same kind.

It should be noted also that the ratio of two like quantities has no units of its own—it is said to be dimensionless. A good illustration of this is π, the ratio between the circumference and diameter of any circle.

EXERCISE 1

Express each of the following ratios in the simplest form:

1. 4 h to 40 min
2. 5 lb to 10 oz
3. 8 in to 2 yd
4. 330 ft to 1 mi
5. 480 in² to 1 ft²
6. 6 ft³ to 1 yd³
7. 13,200 ft² to 1 acre
8. 1°7′30″ to 360° (angular measurement)
9. Circumference of a 9-in circle to circumference of a 2-ft circle
10. Area of an 8-in square to area of a 12-in square
11. $6ax^2$ to $8a^2x$
12. $6xy - 4y^2$ to $9x^2 - 6xy$
13. $\frac{5}{8}$ to $\frac{15}{32}$
14. $3\frac{1}{3}$ to $6\frac{1}{4}$

6.2 Proportions

A statement of equality between two ratios is called a *proportion*. Thus $\$3/\$2 = 84$ mi/56 mi is a proportion and is read "$3 is to $2 as 84 mi is to 56 mi." The four quantities are called the *terms* of the proportion. The first and fourth terms are called the *extremes*, and the second and third terms are called the *means*, of a proportion.

Among the properties of proportions, perhaps the most important may be listed in terms of the proportion

$$\frac{a}{b} = \frac{c}{d}$$

1. *The product of the extremes is equal to the product of the means. Thus* $ad = bc$.
2. *The terms are in proportion by inversion. Thus* $b/a = d/c$.
3. *The terms are in proportion by alternation. Thus* $a/c = b/d$.

Each of these statements may be confirmed by the ordinary operations of equations.

If the means of a proportion are equal, as in the proportion $m/x = x/n$, it is called a mean proportion, and x is said to be the *mean proportional* between m and n. Strictly speaking, $x = \pm \sqrt{mn}$. However, we shall confine ourselves to the positive root.

A *continued proportion* between six or more quantities is illustrated by the expression $a:b:c = x:y:z$. This is a compact notation for expressing the simultaneous proportions

$$\frac{a}{b} = \frac{x}{y} \qquad \frac{a}{c} = \frac{x}{z} \qquad \frac{b}{c} = \frac{y}{z} \qquad \text{or} \qquad \frac{a}{x} = \frac{b}{y} = \frac{c}{z}$$

Example 1. The sum of \$4,628.20, representing the profits of a venture, is to be divided among three partners in the ratio of $2:3:7$. How much does each receive?

We may represent the amounts by $2x$, $3x$, and $7x$, since

$$2x:3x:7x = 2:3:7$$

Therefore

$$
\begin{aligned}
2x + 3x + 7x &= \$4{,}628.20 \\
12x &= 4{,}628.20 \\
x &= 385.683 \\
2x &= 771.37 \\
3x &= 1{,}157.05 \\
7x &= 2{,}699.78
\end{aligned}
$$

EXERCISE 2

In Probs. 1 to 10 solve the proportions for x.

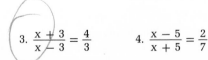

1. $\dfrac{3}{4} = \dfrac{x}{14 - x}$ 2. $\dfrac{5}{3} = \dfrac{20 - x}{x}$ 3. $\dfrac{x + 3}{x - 3} = \dfrac{4}{3}$ 4. $\dfrac{x - 5}{x + 5} = \dfrac{2}{7}$

5. $\dfrac{a}{b} = \dfrac{x}{c}$ 6. $\dfrac{m}{n} = \dfrac{k}{x}$ 7. $\dfrac{a}{b} = \dfrac{c-x}{x}$ 8. $\dfrac{h}{k} = \dfrac{x}{k-x}$

9. $\dfrac{x+a}{x+b} = \dfrac{x-c}{x-d}$ 10. $\dfrac{x-a}{x+b} = \dfrac{x+a}{x-k}$

In Probs. 11 to 20 find the mean proportional between the given quantities.

11. 6 and 24   12. 8 and 50   13. $2\frac{1}{2}$ and $\frac{5}{8}$

14. $8\frac{1}{3}$ and 12   15. $2a^2x$ and $8b^2x$ 16. x/y and y/x

17. $x^2 - y^2$ and $\dfrac{4(x+y)}{x-y}$ 18. $x^2 - 2xy + y^2$ and $x^2 + 2xy + y^2$

19. $\dfrac{18ab^3}{c}$ and $\dfrac{2ac}{b}$

20. $x + a$ and $x - a$ (Find the approximate value if a is very small relative to x.)

6.3 Variation

If two variables x and y are related so that $y = kx$, where k is a constant, we say that y *varies directly* as x, or simply y *varies* as x, or y is *proportional* to x. The constant k is called the *constant of proportionality*, or the constant of variation.

An equivalent notation is $y \propto x$, in which the symbol $\propto$ is read "varies as" or "is proportional to." This is *not* an equation but can be converted to the notation of an equation $y = kx$ by introducing the proportionality constant k.

For example, we say that the circumference of a circle varies as the diameter, or $C \propto D$, or $C = kD$. In this instance the proportionality constant k is π.

We note that the relation between y and x is a mutual one. That is, if y varies as x, then $y = kx$; but x is also proportional to y, or $x = (1/k)y$. Therefore, if y varies as x with the constant of proportionality equal to k, then x varies as y with the constant of proportionality equal to $1/k$.

Inverse Variation

If one variable varies directly as the reciprocal of a second, then the first is said to *vary inversely* as the second, or the first is said to be *inversely proportional* to the second. Thus $y = k \cdot 1/x$, or $y = k/x$. Since it is evident that $xy = k$, it follows that if the product of two variables is constant, either variable varies inversely as the other.

A common example of inverse variation is found in the inverse relationship between rate and time when a fixed distance is being covered. Here $rt = d$, or $t = d/r$. The constant of proportionality here is d.

The number of days needed to complete a job is inversely related to the number of men on the job. Here $mt = k$, or $t = k/m$. The constant of proportionality k is the size of the job in man-days. (m = number of men, t = number of days.)

Joint Variation

If a variable x varies directly as the product of y and z, that is, if $x = kyz$, we say that x varies jointly as y and z.

If $x = ky \cdot 1/z = ky/z$, it may be said that x varies directly as y and inversely as z.

Types of variation can be combined. For example, the electrical resistance of a wire varies directly as the length and inversely as the square of the diameter. Translated into algebraic notation, this may be written $R = kl/d^2$.

Solution of Variation Problems

Usually we know the type of variation involved and a set of corresponding values of the variables. From these data we can determine the proportionality constant k and any one missing value in a second set of values of the variables.

Example 2. We are given the relation that x is directly proportional to the square root of y and inversely proportional to z. Further, it is known that $x = 4$ when $y = 9$ and $z = 15$. Find the formula for x and find x when $y = 25$ and $z = 40$.

Expressing the relationship in algebraic notation, we have the equation of variation:

$$x = \frac{k\sqrt{y}}{z} \tag{1}$$

Substituting in Eq. (1) $x = 4$, $y = 9$, and $z = 15$, we have

$$4 = \frac{k\sqrt{9}}{15} = \frac{3k}{15}$$

or

$$k = 20 \tag{2}$$

From Eqs. (1) and (2) we write the formula

$$x = \frac{20\sqrt{y}}{z} \tag{3}$$

To solve the last part of our problem, we substitute $y = 25$ and $z = 40$ in Eq. (3), obtaining

$$x = \frac{20\sqrt{25}}{40} = \frac{(20)(5)}{40} = 2.5 \tag{4}$$

The illustration points up the following rules:

1. *Translate the statement of variation into an equation involving an unknown constant of proportionality.*
2. *Solve for the proportionality constant by substituting given data.*
3. *Substitute the value of the constant of proportionality in the equation of variation.*
4. *Use the formula or equation in step 3 to obtain the missing value of a variable when a second set of values of the other variables is given.*

There will be occasions when there are insufficient data to evaluate k or when we are not interested in determining k. To continue with the illustration just given, we represent the first set of values by x_1, y_1, and z_1, and the second set by x_2, y_2, and z_2. Hence we obtain

$$x_1 = \frac{k\sqrt{y_1}}{z_1} \tag{5}$$

and

$$x_2 = \frac{k\sqrt{y_2}}{z_2} \tag{6}$$

Dividing Eq. (5) by Eq. (6),

$$\frac{x_1}{x_2} = \frac{z_2\sqrt{y_1}}{z_1\sqrt{y_2}} \tag{7}$$

Substituting values in Eq. (7),

$$\frac{4}{x_2} = \frac{40\sqrt{9}}{15\sqrt{25}}$$

$$\frac{4}{x_2} = \frac{120}{75}$$

$$x_2 = 2.5$$

It should be noted that the computation in Eq. (7) will often be facilitated if we write it in the form

$$\frac{x_1}{x_2} = \frac{z_2}{z_1}\sqrt{\frac{y_1}{y_2}}$$

If, for example, $y_1 = 63$ and $y_2 = 28$, the preference is obvious when we write

$$\frac{\sqrt{63}}{\sqrt{28}} = \sqrt{\frac{63}{28}} = \sqrt{\frac{9}{4}} = \frac{3}{2}$$

Similarly, if the equation includes the expression $w_1{}^2/w_2{}^2$, the alternative form $(w_1/w_2)^2$ is often preferable. For example, suppose $w_1 = {}^{15}\!/_{32}$ and $w_2 = {}^9\!/_{16}$. Then

$$\frac{({}^{15}\!/_{32})^2}{({}^9\!/_{16})^2} = \left(\frac{{}^{15}\!/_{32}}{{}^9\!/_{16}}\right)^2 = ({}^{15}\!/_{32} \times {}^{16}\!/_9)^2 = ({}^5\!/_6)^2 = {}^{25}\!/_{36}$$

The process of eliminating the proportionality constant may be used to demonstrate that variation and proportion are equivalent ideas. In general, for repeated application of the same operation, it is convenient to evaluate k. However, for occasional application, the actual value of k is seldom required, and a simpler solution results from its elimination.

Given

$$y_1 = kx_1 \tag{8}$$

and

$$y_2 = kx_2 \tag{9}$$

(y varies directly as x.) Dividing Eq. (8) by Eq. (9),

$$\frac{y_1}{y_2} = \frac{x_1}{x_2} \tag{10}$$

(y is directly proportional to x.)

Given

$$y_1 = \frac{k}{x_1} \tag{11}$$

and

$$y_2 = \frac{k}{x_2} \tag{12}$$

(y varies inversely as x.) Dividing Eq. (11) by Eq. (12),

$$\frac{y_1}{y_2} = \frac{x_2}{x_1} \qquad \text{or} \qquad x_1 y_1 = x_2 y_2 \tag{13}$$

(y is inversely proportional to x.)

Common examples of this relation are:

1. Boyle's law. $P_1 V_1 = P_2 V_2$, temperature remaining constant.
2. Time required to complete a certain job (constant number of man-hours); e.g., if 10 men can do a given job in 6 h, 4 men can finish the same job in 15 h. That is, $m_1 h_1 = m_2 h_2$.

There is no substitute for common sense in solving a proportion. For example, it is required to reduce 950 ml of oxygen at 450°K and 840 mm to standard temperature and pressure (273°K and 760 mm). We know that cooling the gas from 450 to 273°K will cause it to contract and that lowering the pressure from 840 to 760 mm will make it expand; therefore we apply the fractions 273/450 for shrinkage and 840/760 for expansion. As a result we write

$$V_{stp} = 950\left(\frac{273}{450}\right)\left(\frac{840}{760}\right) = 637 \text{ ml}$$

6.4 Dimensionality of the Proportionality Constant

It is frequently instructive to determine the dimensionality of k or the units in which k is expressed.

Example 3. It is known that the weight of a rectangular block of wood varies jointly with the thickness, width, and length. If the weight is expressed in pounds and the dimensions in inches, determine in what units k is expressed.

The equation of variation is

$$W = ktwl$$

Solving for k,

$$\frac{W}{twl} = k$$

Substituting units,

$$\frac{lb}{(in)(in)(in)} = \frac{lb}{(in)^3}$$

Hence k is expressed in pounds per cubic inch—a density figure.

Example 4. The volume of a given quantity of a gas varies inversely as the pressure (temperature remaining constant). Determine the units of k if volume is expressed in cubic inches and pressure in pounds per square inch.

The equation of variation is

$$V = \frac{k}{P} \quad \text{or} \quad PV = k$$

Substituting units,

$$\frac{lb}{in^2} \, in^3 = in \cdot lb$$

Hence k is expressed in inch-pounds—a work unit.

EXERCISE 3

Express each of the relations in Probs. 1 to 5 as an equation containing an unknown constant of proportionality.

1. W varies jointly as x and y.
2. Q varies directly as x and inversely as y.
3. V varies directly as the cube of x and inversely as d.
4. M varies directly as b and inversely as the square root of c.
5. R varies directly as w and the square root of x and inversely as the cube of h.

In Probs. 6 to 10 write the formula for the first variable in terms of the other variables and the computed value of k.

6. H varies directly as x. $H = 8$ when $x = 20$.
7. N varies inversely as y. $N = 20$ when $y = 0.35$.
8. Q varies jointly as a, b, and c. $Q = 300$ when $a = 3$, $b = 7.5$, and $c = 8$.
9. V varies directly as m and inversely as the square of t. $V = 2$ when $m = 15$ and $t = 6$.
10. R varies directly as the fourth power of T and inversely as the square root of x. $R = \frac{1}{3}$ when $T = 2$ and $x = 36$.

EXERCISE 4

In Probs. 1 to 7 determine the numerical value and the units of k that refer to the units in which the first set of values of the variables is expressed. Solve for the unknown value of the variable in the second set of values.

1. P varies inversely as V. If $V = 30$ in^3 when $P = 84$ lb/in^2 find V when $P = 63$ lb/in^2.
2. R varies directly as l. If $R = 6.8$ Ω when $l = 23.5$ ft, find R when $l = 31.8$ ft.

3. v varies directly as t. If $v = 45$ ft/s when $t = 25$ s, find v when $t = 1$ min.
4. W varies directly as d^2. If $W = 12$ oz when $d = 8$ in, find W when $d = 1$ ft.
5. N varies inversely as d^2. If $N = 10,890$ plants/acre when set $d(= 2$ ft$)$ apart, find N when $d = 5\frac{1}{2}$ ft.
6. m varies inversely as d. If $m = 12$ men when $d = 10$ days, find m when $d = 8$ days.
7. v varies jointly as the square root of g and the square root of h. If $v = 3.8$ ft/s when $g = 32$ ft/s^2 and $h = 0.17$ ft, find v when $g = 30$ ft/s^2 and $h = 8$ in.

In Probs. 8 to 12 determine the unknown quantity without solving for k.

8. C varies directly as d^2. If $C = 80$ when $d = 12$, find C when $d = 15$.
9. v varies directly as $\sqrt{h}$. If $v = 28$ when $h = 3$, find v when $h = 12$.
10. R varies directly as l and inversely as d^2. If $R = 35$ when $l = 110$ and $d = 0.006$, find R when $l = 75$ and $d = 0.004$.
11. V varies directly as r^4 and p and inversely as l. If $V = 120$ when $r = 0.012$, $p = 20$, and $l = 30$, find V when $r = 0.016$, $p = 36$, and $l = 25$.
12. a varies directly as v^2 and inversely as r. If $a = 540$ when $v = 84$ and $r = 5$, find a when $v = 119$ and $r = 4$.

6.5 Applications

A few of the more common principles of mathematics and the physical sciences that may be expressed as variations are as follows:

Areas of similar figures vary as the squares of corresponding dimensions.

Volumes of similar solids vary as the cubes of corresponding dimensions. (Note that a 1-in square and a 2-in square, each cut from a $\frac{1}{4}$-in sheet of steel, are not similar solids, since not all three dimensions are doubled. Weights are as $1^2:2^2$, not $1^3:2^3$.)

Volumes of gases vary inversely as the absolute pressure and directly as the absolute temperature.

In any given chemical reaction between substances A and B, the reacting amount of A varies directly as the reacting amount of B.

The time required to finish a given job varies inversely as the number of men working on the job.

The rate of energy reception (heat, light, magnetism, etc.) varies inversely as the square of the distance from the source of energy (the inverse square law).

The revolutions per minute of two pulleys belted together vary inversely as their diameters.

The revolutions per minute of two gears in mesh vary inversely as the number of teeth.

Rate of heat conduction through a flat plate varies jointly as the area of one face of the plate and the difference between the temperatures of the opposite faces and inversely as the thickness of the plate.

Electrical resistance of a conductor varies directly as the length and inversely as the cross-section area.

In order to reduce drudgery it is suggested that the slide rule be used whenever accuracy of three significant figures is sufficient.

EXERCISE 5

1. Hydrogen used for inflation of balloons may be made by passing steam over red-hot scrap iron. If 8.5 lb of iron will make 78 ft^3 of hydrogen, how much iron would be needed to make 500 ft^3 of hydrogen?
2. Seven men take eighteen 8-h days to finish a job. How large is this job in man-hours? How many men will be needed to finish a like job in twelve 7½-h days?
3. A train usually makes its run in 2 h 25 min at an average speed of 42 mi/h. How long would it take if the speed were reduced to 34 mi/h?
4. The weight of 195 machine screws is 8½ oz. Find the number in 2 lb 7 oz.
5. If 75 iron washers weigh 9½ oz, how many washers are there in a batch weighing 4 lb?
6. The resistance of a spool of enameled magnet wire was 955 Ω. A piece 1 ft 9½ in long was cut off and found to have a resistance of 5.37 Ω. Find the length of wire originally on the spool.
7. The air-line distance between two points is 235 mi. They are 5⅛ in apart on the map. What is the air-line distance between two points 3¹³⁄₁₆ in apart on the same map?
8. A 16-in disk cut from a piece of sheet steel weighs 5.65 lb. What will be the diameter of a disk weighing 2.08 lb cut from the same piece of stock?
9. A 2½-in cast-iron sphere weighs 1.87 lb. How much will a 3½-in cast-iron sphere weigh?
10. Expenditures by the city of Newton in a certain year were approximately $39,770,000. The tax rate was $88.60 per $1,000 of assessed valuation. If an additional expenditure of $1,000,000 is proposed, to be financed by a higher tax rate, find the revised tax rate.
11. A formula calls for 15 ml of a 24 percent solution. How many milliliters of a 20 percent solution would be equivalent to this?
12. The profits of a partnership amount to $1,283.67. This is to be divided among four partners in proportion to their investments. Johnson, Miller, Spencer, and Weston invested $842, $1,363, $1,759, and $1,876, respectively. Find each man's share of the profits to the nearest cent.
13. A pilot flew a glider plane a distance of 87 mi in 50 min. During this time he descended from 12,800 to 8,000 ft. How much longer could he have remained aloft, and how far would he have glided?
14. Given two solids of the same volume, the weights are proportional to the density. What is the weight of a bronze casting formed by a pine pattern weighing 4¼ lb? Density of pine = 28 lb/ft^3; density of bronze = 550 lb/ft^3.

15. If 50 sheets of paper form a pile $\frac{9}{32}$ in thick, approximately, how many sheets are contained in a pile $2\frac{3}{8}$ in thick?

16. The analysis of a paint shows 46 percent vehicle and 54 percent pigment. The analysis of the pigment shows 15 percent zinc oxide, 60 percent titanium dioxide, and 25 percent lithopone. What is the percentage of each pigment in the ready-mixed paint?

17. The gravitational acceleration at the earth's surface may be taken as 32.2 ft/s². What will be the value of the gravitational constant for a guided missile 150 mi above the earth's surface? Gravitational force is inversely proportional to the square of the distance from the center of the earth. Assume the earth's radius is 3,960 mi.

18. Given the formula $x = m\sqrt{f}/[W(a + b^2/k)]$. Indicate, quantitatively where possible, the effect on x of doubling each of the other letters one at a time.

19. What is the area of a room scaling $5\frac{1}{2} \times 7\frac{1}{4}$ in on a floor plan if the scale is $\frac{3}{8}$ in = 1 ft?

20. From a 1:2:3 mixture of cement, sand, and gravel, 84 yd³ of concrete is to be made. How many cubic yards of each will be needed, allowing for a 20 percent shrinkage on mixing?

21. A "pie chart" is often used to show how the tax dollar is spent (Fig. 6.1). If $1,190,000 is allocated to the street department, $860,000 to the police and fire departments, and $1,870,000 to schools, compute to the nearest degree the size of each section of the "pie."

22. An inverted cone is 32 in high and 18 in in diameter at the top. If water flowing into the cone at a uniform rate reaches a depth of 12 in in 5 min, how much longer would it take to reach a depth of 15 in?

23. The outline of an estate is cut out from a map and found to weigh 42.78 g. A rectangular section 5 × 8 in was cut from the same sheet and found to weigh 5.31 g. If the scale of the map is 1 in to 150 ft, find the number of acres in the estate.

FIGURE 6.1

24. The gravity rate of flow of water from the bottom of a tank varies directly as the square root of the depth h of the water in the tank. The rate of flow v was 20 gal/min when the water was 9 ft deep.
    (a) Derive a formula for v in terms of h.
    (b) Find v when $h = 15$ ft.
    (c) Find h when $v = 12$ gal/min

25. If brand A of frozen strawberries is priced at $3\frac{1}{4}$ oz for 24 cents and brand B sells at $5\frac{1}{2}$ oz for 39 cents, which is the better buy?

26. If 42 vitamin tablets weigh 13 g, how many tablets are there in a lot weighing 68 g?

27. The odometer (mileage indicator) of a car correctly registers a 500-mi trip when the car is equipped with tires measuring 27 in diameter at the tread. What mileage would be indicated for the same trip when using 28-in-diameter tires?

28. In a fish-and-game survey 400 fish were taken from a pond, tagged by coloring with a harmless dye, and released. A few days later 5 tagged fish were counted among 62 that were taken from the pond. Approximately how many fish were in the pond originally?

29. If $3\frac{3}{4}$ in on a road map corresponds to 5 h traveling time, what traveling time corresponds to $4\frac{1}{2}$ in on the map?

30. A motorist gets 18 mi/gal of gasoline on his car. He spends \$90/year for gasoline. What would be his annual gasoline bill if he were to get 20 mi/gal?

31. The braking distance of a car is proportional to the square of the speed. This is illustrated in the following table:

$V_{mi/h}$	20	40	60
d_{ft}	100	400	900

    Find the braking, or stopping, distance from a speed of 30 mi/h.

32. If the "1940 dollar" will buy only 64 cents worth of goods today, what should a man earning \$5,600 in 1940 receive now merely to break even?

33. In parts a to h compute the missing quantities, where gear A drives gear B. Select appropriate gears from stock where an asterisk replaces the number of teeth.

    Assume that gears with the following numbers of teeth are available from stock: 96, 84, 72, 66, 60, 54, 48, 42, 36, 30, 27, 24, 21, and 18.

	No. teeth in gear A	No. teeth in gear B	r/min gear A	r/min gear B
(a)	42	96	56	—
(b)	60	—	150	125
(c)	48	84	—	112

(d)	—	36	45	105
(e)	*	*	360	1,200
(f)	*	*	350	550
(g)	*	*	200	640
(h)	*	*	180	840

34. Referring to Table 1.1, which is an excerpt from Appendix D, we wish to set up a scale model of the solar system in which a $\frac{1}{4}$-in-diameter pellet represents Earth. Confirm (a) the model distances of Earth, Venus, Mars, Jupiter, and α Centauri from the sun; (b) the model distance of the moon from Earth; (c) the model diameters of the moon, Venus, Mars, Jupiter, and the sun.

	Relative mass†	Actual diameter, mi	Scale diameter, in	Actual distance, mi	Scale distance
Earth	1	7,920	$\frac{1}{4}$	9.3×10^7	244 ft
Moon	0.0120	2,160	0.068	2.4×10^5‡	$7\frac{1}{2}$ in
Venus	0.81	7,700	0.24	6.73×10^7	$176\frac{1}{2}$ in
Mars	0.107	4,200	0.132	1.42×10^8	373 ft
Jupiter	318	88,700	2.80	4.84×10^8	1,272 ft
Sun	332,000	864,000	27.2		
α Centauri				4.31 light-years	12,580 mi

†Data for Probs. 35 and 37.
‡Distance from Earth.

35. The acceleration of gravity (g) is proportional to M/r^2, where M is the relative mass of the attracting body and r is the distance from its center. It is 32.2 ft/s² on the surface of the earth, whose relative mass = 1 and whose radius = 3,960 mi. Refer to the table of Prob. 34 to determine the value of g on the surface of (a) the moon, (b) Venus, (c) Mars, (d) Jupiter, (e) the sun. Confirm your answers as found in Appendix D.

36. Find the value of g 1,500 mi above the earth's surface.

37. Assume the earth-moon distance = 240,000 mi center to center. At a point on the earth-moon line 20,000 mi from the center of the moon calculate the net gravitational pull and its direction.

Most of the essential data needed in the solution of Probs. 38 to 40 will be found in Appendix D.

38. Neglecting air friction, the escape velocity $v_e = \sqrt{2gr}$, where r is the distance from the center of the attracting body, and g is the gravitational constant at that point. Using values of g listed in the table in Appendix D, and employing consistent units, verify the escape velocity as given in the same table for (a) Earth, (b) the moon, (c) Mars, (d) Venus.

39. Calculate the velocity of escape from a point 1,000 mi above the earth's surface.
40. The weight of an object at any point is proportional to the gravitational acceleration at that point. What would be the weight of a 150-lb astronaut on the surface of (a) the moon, (b) Mars, (c) Jupiter?
41. Water is running out of the bottom of a full tank 80 in high at the rate of 5 gal/min. How deep is the water when the flow is 3 gal/min? Rate of flow is proportional to $\sqrt{\text{depth}}$.
42. Water is flowing into a conical container 12 in in diameter at the top and 20 in tall. After 10 min of operation the water is 8 in deep. How much longer will it take to fill the tank?
43. A man 5 ft 4 in tall weighs 140 lb. Another man, of about the same build, is just 6 ft tall. How much would you expect the taller man to weigh?
44. The distance of the horizon at sea varies directly as the square root of the elevation of the observer above sea level. If the horizon is $4\frac{1}{2}$ mi distant at $13\frac{1}{2}$ ft elevation, find the distance at 380 ft elevation.
45. In Prob. 44 how high above sea level must a lighthouse be to be visible 15 mi out to sea?
46. A city of 50,000 people is supplied by a 25-in water main. If a future population of 120,000 is anticipated within 30 years, what size main (to the nearest inch) will provide for this population?
47. The horsepower required to drive a motorboat varies as the cube of the speed through the water. If 5 hp drives a boat at 10 mi/h, what size motor is needed to maintain a speed of 14 mi/h?
48. The braking distance of a car is proportional to the square of the velocity. If a car traveling 35 mi/h requires 115 ft to stop, what distance must be allowed when traveling 45 mi/h (assuming constant operating conditions)?
49. The tensile strength of a round bar is proportional to the square of the diameter. If a $\frac{3}{8}$-in-diameter rod supports 40,000 lb, how much will a $\frac{5}{16}$-in-diameter rod support?
50. A ball, starting from rest, rolls 8 ft down an incline during the first 4 s. How far will it roll during the fifth second? Distance is proportional to the square of the time.
51. A cone of slant height 12 in holds 3 qt of water. How far from the vertex should marks be placed on the slant height to indicate 1 qt? 2 qt?
52. A lot of soda ash containing 52 percent by weight water of crystallization is bought at $17\frac{1}{2}$ cents/lb. When the material is sold at retail, the moisture content is found to have dropped in storage to 37 percent. What should be the retail price per pound to realize a 40 percent profit based on the cost? [Note: Comparable purchase price after drying would be $17.5\left(\dfrac{63}{48}\right) = 23.0$ cents/lb. Why?]
53. A formula calls for $22\frac{1}{2}$ lb of soda ash as originally bought in Prob. 52. Find the proper amount of the material to use after partial drying in storage.

54. It has been stated that 90 percent of the capital in a certain country is held by 8 percent of the population. What is the ratio of the capital of an "advantaged" citizen to that of a "disadvantaged" one?

55. Ten men are working on a job which must be completed in 18 days. After 8 days the work is only one-third done. How many *extra* men must be hired if the time limit is to be met?

56. A formula calls for $7\frac{1}{2}$ gal of 25 percent solution. If only a 22 percent solution is available, how many gallons must be used?

57. The velocity of sound in air is independent of the density and pressure of the air and varies directly as the square root of the absolute temperature. An experimental unit carried on a rocket indicated a sound velocity of 977 ft/s at an altitude of 20 mi. What was the Fahrenheit temperature at that altitude if at 70°F the speed of sound in the air is 1,130 ft/s?

58. The lens opening (f stop) of a camera is inversely proportional to the square of the shutter speed if the same exposure is to be obtained. If 1/100 s is given at f/5.6: (a) What size opening would be used at 1/50 s? (b) How many seconds exposure would be needed when using f/32?

59. Directions for thinning varnish call for adding one part by weight of solvent to two parts of varnish. If the specific gravity of the solvent = 0.7 and that of the varnish = 0.95, what are the proportions by volume?

applications in geometry

A working knowledge of geometry is essential for solving many practical problems. When data are unavailable, the engineer, technician, or machinist is compelled to obtain his information indirectly. The word *geometry* means earth measurement—a science in which many dimensions are either inaccessible or too large for direct measurement. We assume that you have already learned some geometry; so in this chapter we are dealing chiefly with applications. A list of some of the more important facts in geometry appears on pages 168–177, and in the problems below, reference is made to them in brackets. For additional theorems or formulas see any standard geometry text or mathematical tables.

A glance at the problems in this chapter might lead to the impression that many different theorems and formulas are involved in their solution. However, in the majority of cases, one uses essentially only such familiar relationships as those dealing with similar figures, properties of the circle, and the Pythagorean theorem. It might be added that it is usually a good idea to draw a radius to any point of tangency on a circle. If two or more circles are involved, it is often helpful to draw their line of centers.

Many times it is a good idea to construct a scale drawing. The drawing will emphasize the relationships among the data and indicate the approximate answer. If the problem can be laid out to scale, it can be solved.

You should watch for situations involving parallel lines cut by a transversal.

You should also be alert for possible angle bisectors. Equal angles so formed may suggest the construction of either congruent or similar triangles.

It is suggested that you solve the problems in this chapter by slide rule as far as possible. You will gain skill in its use, and the drudgery of longhand computation will be reduced.

It is not expected that any particular class will attempt all or even most of the problems in this chapter. The problems are arranged roughly in order of difficulty, and selections may be made accordingly.

The material is divided into three sections:

I. Probs. 1 to 34. Basic.
II. Probs. 35 to 84. Somewhat less elementary.
III. Probs. 85 to 156. Many of these are challenging and nonroutine.

Asterisks (*) precede problems that are relatively more difficult compared with others in that section.

Examine each problem critically. Are the given data of the type considered reliable? Are any unreasonable or unwarranted assumptions involved? Answers should be calculated to a point deemed consistent with data given.

EXERCISE 1 (GROUP I)

1. Find angle x in Fig. 7.1.
2. Find angle a in Fig. 7.2.
3. Find angles a, b, c, and d in Fig. 7.3. [3, 8, 23]
4. Find x and y in Fig. 7.4. [83–85]
5. An exterior angle of a triangle is 73°. One opposite interior angle is 39°. Find the other opposite interior angle. [30]
6. Find angle a in Fig. 7.5.
7. Find angles a and b in Fig. 7.6. [28]
8. Two angles of a triangle are 41°30′ and 97°. The included side is 4.75. Find the shorter missing side. [14]
9. The altitude to the hypotenuse of a right triangle divides the hypotenuse into segments 3.70 and 8.43 in long. Find the altitude. [98]

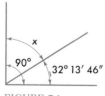

FIGURE 7.1

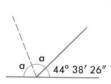

FIGURE 7.2

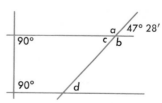

FIGURE 7.3

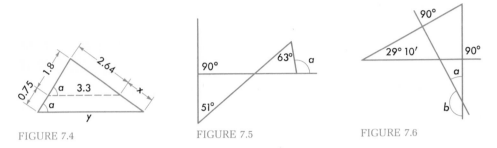

FIGURE 7.4   FIGURE 7.5   FIGURE 7.6

10. Find to the nearest 0.1 in² the area of a triangle whose sides are 19, 28, and 39 in. [102, 103]

11. If the angles of a triangle are in the ratio 1:2:3, find the angles and the ratio of the longest side to the shortest side. [29, 109]

12. Find the area of a square whose diagonal is 30 in. [102]

13. Two sides of a triangle are 13 and 15 in, and the altitude upon the third side is 12 in. Find the third side and the area. [99, 102]

14. (a) Find the area of a 51°24′ sector of a 15-in circle. [66]
    (b) Find the length of the arc of the sector. [67]

15. Find the length of one side and the area of a rhombus whose diagonals are 15 and 22 in. [99]

16. Find the tensile strength of a hexagonal brass bar $\frac{5}{8}$ in across flats. Assume 1 in² will support 30,000 lb.

17. A circle is inscribed in a triangle whose vertex angle is 48°14′. Find the angle formed at the center of the circle by lines drawn to the extremities of the base. [72]

18. What diameter hole should be drilled clear through a $\frac{3}{8}$-in plate of cast iron to remove 2 oz? The density of cast iron is 0.26 lb/in³. [127]

19. A circular coil of wire is approximately 20 in in average diameter and contains 63 turns. About how many feet of wire are there in the coil? [61]

20. A $\frac{1}{4}$-in wire is wound around a 1$\frac{1}{2}$-in cylinder with a lead of $\frac{5}{8}$ in. How much wire is required for 12 turns? (Compute the length of the centerline of the wire.) [99]

21. Find the area of the three-cornered section enclosed by three mutually tangent 15-in circles. [66, 108]

22. The base angles of a trapezoid are 60° and 45°. The upper base is 15, and the leg adjacent to the 60° angle is 10. Find the area of the trapezoid. [104]

23. Find the angle between the bisectors of the two angles of a trapezoid which are formed by the bases and one nonparallel side. [29]

24. Find the angle between the bisectors of the acute angles of a right triangle. [29]

25. Two consecutive angles of a quadrilateral are 83° and 105°. Find the angle between the bisectors of the other two angles. [45]

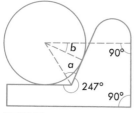

FIGURE 7.7

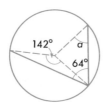

FIGURE 7.8

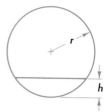

FIGURE 7.9

26. Find the angles a and b in Fig. 7.7. [70, 72]

27. Find the angle a in Fig. 7.8. [14]

28. The diagonals of a quadrilateral are 16 and 22 in and are perpendicular to each other. Find the area of the quadrilateral.

29. Two sides of a parallelogram are 10 and 12 in long, and the included angle is 135°. Find the length of the longer diagonal. [99, 107]

30. In an isosceles trapezoid the parallel sides are 11 and 19 in long, and the diagonals are each 17 in long. Find its area. [99, 104]

31. The lengths of the sides of a triangle are 3, 4, and 5. What is the ratio of the areas of the two smaller triangles into which it is divided by a perpendicular drawn to the longest side from the opposite vertex? [98]

32. An equilateral triangle and a regular hexagon are inscribed in the same circle. Find the ratio between the areas of hexagon and triangle.

33. A counterweight on a 60-in driving wheel is in the form of a segment subtending a central angle of 60°. Find the weight if it is 4 in thick and made of cast iron (density 0.26 lb/in³). [66, 108]

*34. Find the area of a segment $3\frac{1}{2}$ in high in a circle 18 in in diameter. The area A of a segment is given closely by $A = \frac{4}{3}h^2 \sqrt{2r/h - 0.608}$ (Fig. 7.9). The smaller h is in relation to r, the closer is the approximation.

(GROUP II)

*35. Two circles of radii 3 and 9 in are externally tangent to each other. Find the length of the common external tangent (i.e., of a line whose extremities are externally tangent to these circles at points other than the point of their mutual tangency). [70, 74]

36. A pyramid is 5 in square on the base and 7 in high. Find the volume, total area, and lateral edge.

37. Two 12-in circles have their centers 6 in apart. What is the area of the portion common to both? [66, 108]

38. Given the following figures, each having a perimeter of 30 in: (*a*) a right triangle having sides 5, 12, and 13 in; (*b*) an isosceles triangle whose sides are 11, 11, and 8 in; (*c*) a 10-in equilateral triangle; (*d*) a rectangle 4 × 11 in; (*e*) a $7\frac{1}{2}$-in square;

(f) a regular hexagon 5 in on a side; (g) a circle whose circumference is 30 in. Compare their areas.

39. Find the area of the end of the gable roof shown in Fig. 7.10. [102, 104]

40. Find the cost of painting the top of a hemispherical dome 42 ft 6 in in diameter at $1.80/yd². [128]

41. What is the total force acting on a 5½-in-diameter piston if the pressure is 120 lb/in²? [63]

42. Find the weight of an open-top sheet-iron cylinder 28 in high and 22 in in diameter. One square foot of sheet iron weighs 10½ oz. [63, 126]

43. A coil of ⅛-in copper wire weighs 140 lb. Find the length of wire in the coil if copper weighs 556 lb/ft³. [127]

44. Find the weight of a 12-ft length of cast-iron pipe 10 in inside diameter, ¾ in wall thickness, at 450 lb/ft³. [127]

*45. Find the lateral surface of the frustum of a right pyramid whose height is 12 in, lower base 28 in², and upper base 18 in². (Hint: Can you use the principle of Prob. 93?)

46. Find the weight of a hollow brass spherical shell 8 in outside diameter with ⅜ in wall thickness. One cubic foot of brass weighs 540 lb. [129]

*47. (a) Would the sphere in Prob. 46 float in freshwater? One cubic foot of freshwater weighs 62.4 lb.

(b) How would you estimate the wall thickness of an 8-in sphere that would barely float?

*48. (a) How many 2-in-diameter disks may be stamped out of the rectangular sheet shown in Fig. 7.11? (Circles are tangent.)

(b) What is the percent waste?

(c) As the dimensions of the rectangular sheet are increased indefinitely, the scrap metal approaches an irreducible minimum percentage. What is this percentage?

*49. Assume three mirrors to be arranged at right angles, each to the other two, like the inside corner of a cube. Show that an incident ray, directed to reflect from

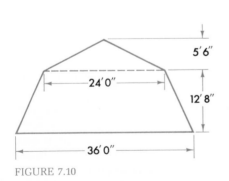

FIGURE 7.10

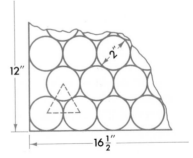

FIGURE 7.11

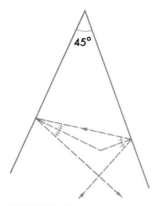

FIGURE 7.12

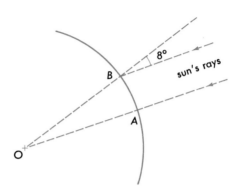

FIGURE 7.13

each of the three faces in succession, will ultimately be reflected back on a path parallel to itself. This would ensure that the reflected ray (from a satellite) would always be beamed back to the station without excessive scattering.

*50. In Fig. 7.12 two mirrors form an "optical square." Show that the incident and reflected rays intersect at right angles. (All rays lie in a plane perpendicular to the intersection of the two mirrors.) [29, 45]

51. In Fig. 7.13 the sun is directly overhead at point A on the earth. At the same time at a point B 550 mi away the sun is 8° from the vertical. Estimate the diameter of the earth. (The sun and points A, B, and O are in the same plane.) [22, 67]

52. Apollo 8 traveled at 3,625 mi/h in a circular orbit 70 mi above the moon's surface. If the diameter of the moon is 2,160 mi, what was the period of Apollo 8? (The period is the time required to make one revolution.)

53. In the rectangle shown in Fig. 7.14 the diagonal is braced by members a and b. Find the lengths of a and b. [98]

*54. In a right triangle whose legs are a and b and hypotenuse c show that the diameter of the inscribed circle $D = a + b - c$ (Fig. 7.15). [55, 72]

55. Find the total surface and volume of a cone whose height and base diameter are 17 and 20 in, respectively. See Prob. 93. [121, 123]

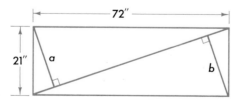

FIGURE 7.14

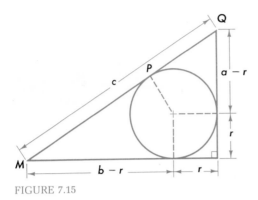

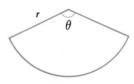

FIGURE 7.15

FIGURE 7.16

*56. The flat sector-shaped sheet in Fig. 7.16 when rolled up will form the cone described in Prob. 55. Find the radius and central angle of this "development." [67]

57. Find the volume and lateral surface of the frustum of a cone having base diameters of 13 and 22 in and a height of 20 in. [125]

58. The frustum of a cone is $5\frac{1}{4}$ in high. The diameters of its bases are $8\frac{3}{4}$ and $12\frac{1}{2}$ in. Find the height of the original cone. [118]

59. In Fig. 7.17 two cuts are made approximately 11 in from each corner of a 30-in square to make a regular duodecagon (12 sides). What distance should be measured when working with a 14-in square? How many degrees are there in angle a? [48, 97]

60. The scale ratio of a certain aerial photograph is 1:25,000. What is the area in acres of a town if the area in the photograph is 34.23 in²? [105]

61. What is the area of an orchard which appears in the photograph of Prob. 60 as a square 0.1 in on a side?

*62. In Fig. 7.18 find the length of a cross brace ab and the distance aw of the rivet from one end. [85]

63. A pumping station at O on a river is to deliver water to A and to B. When angle

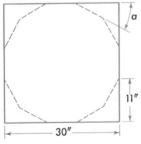

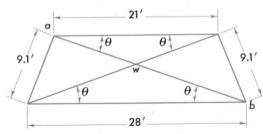

FIGURE 7.17

FIGURE 7.18

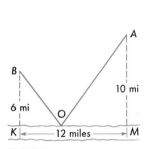

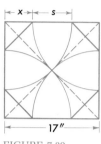

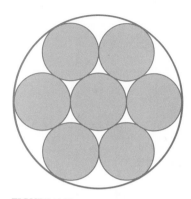

FIGURE 7.19   FIGURE 7.20   FIGURE 7.21

KOB equals angle AOM, the total length of piping is a minimum. Find this length in Fig. 7.19. [85]

*64. A regular octagon may be cut from a square by the method shown in Fig. 7.20. Find x and s. What percent of the material is wasted? [106]

65. Find the altitude to the longest side of a triangle whose sides are 8, 12, and 15 in. [102, 103]

66. Find the fraction of the large circle occupied by the seven small circles in Fig. 7.21. [65]

67. Two 10-in pulleys are connected by a belt which crosses itself at an angle of 60°. Find the length of the belt (Fig. 7.22). [60, 109]

68. Find R in Fig. 7.23. [74]

69. In Fig. 7.24 the centers of a 6-in and a 12-in circle are 15 in apart. How long is AB? [26, 60]

70. Find the angle α in Fig. 7.25. [76]

71. A circular pool 65 ft in diameter is surrounded by a walk having an area of 600 ft². Find the width of the walk to the nearest ½ in.

*72. A horizontal cylindrical tank 30 in in diameter contains 400 gal when full. How many gallons remain when the depth of liquid is 8 in? (See Prob. 34 above.)

73. Find the length of stock needed for the piece shown in Fig. 7.26. (Find the length of the centerline.) [67]

10″ diam   10″ diam

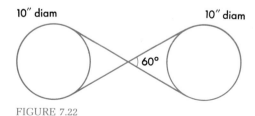

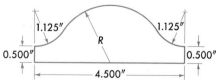

FIGURE 7.22   FIGURE 7.23

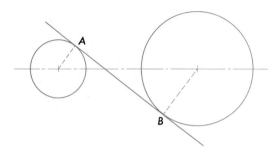

FIGURE 7.24

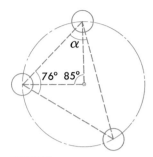

FIGURE 7.25

74. A 12,000-gal cylindrical tank is to be installed in a factory. Because of headroom the height is restricted to 15 ft. What should be the diameter of the tank to the nearest inch? [127]

75. Using the approximate formula $l = \pi(D + d)/2 + 2a$, where l is the length of an open belt connecting two pulleys whose diameters D and d differ only slightly and whose center-to-center distance is a, find the length of belting required for a 12- and a 16-in pulley 5 ft apart (a, l, D, and d are expressed in like units).

76. If water is flowing through an 8-in pipe with an average linear rate of 0.35 ft/s, find the rate of flow in gallons per minute.

77. What is the linear rate of flow in Prob. 76 when water is passing through a $\frac{3}{4}$-in-diameter metering orifice?

78. Find the diameter of the flywheel, a fragment of which is shown in Fig. 7.27 [100]

79. Given a regular hexagon of side s (Fig. 7.28), derive a formula for its area in terms of s. [108]

*80. Repeat Prob. 79 for a regular octagon of side s (Fig. 7.29). [107]

81. Find, in terms of s, the area of the regular dodecagon shown in Fig. 7.30. [108]

*82. Find the percentage error in taking the area of the square as equal to that of the circle in Fig. 7.31. Line segment AB is divided into five equal parts.

83. A 3-in-diameter piston ring should have an end clearance of 0.010 to 0.015 in.

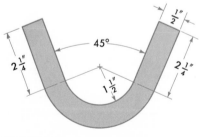

FIGURE 7.26

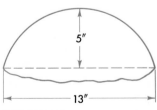

FIGURE 7.27

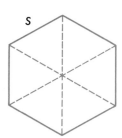

FIGURE 7.28

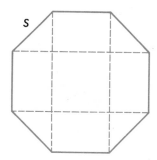

FIGURE 7.29

If the end clearance was 0.010 in when the ring was installed, how much could the diameter increase with wear before the clearance exceeded 0.015 in?

*84. A strip of metal 12 in wide is bent along the centerline to form a V trough. What angle between the sides would correspond to the maximum water capacity of the trough? [102]

(GROUP III)

*85. The following data were obtained in finding the diameter of the bore in a capillary glass tube: weight of empty tube, 14.603 g; weight of tube containing a thread of mercury $11\frac{1}{4}$ in long, 16.385 g. Find the diameter of the bore to the nearest 0.001 in if 1 g = 0.0353 oz and 1 in³ of mercury weighs 7.84 oz. [127]

86. Find the volume of a square pyramid with edges 7 in long. [123]

87. Find the altitude to one of the equal sides of an isosceles triangle whose sides are 5, 5, and 6. [102]

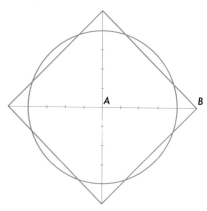

FIGURE 7.31

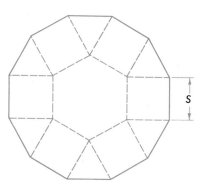

FIGURE 7.30

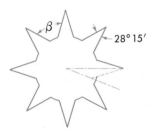

FIGURE 7.32

88. A regular hexagonal pyramid has a base edge 10 in long. If the lateral faces are inclined 60° to the base, find the altitude of the pyramid.

*89. An empty rectangular swimming pool is 2 ft deep at one end, and the bottom slopes uniformly to a depth of 10 ft at the other end. After cleaning, the pool is being refilled at a constant rate. If after $2\frac{1}{2}$ h the water is 2 ft deep at the deep end, how much longer will it take to fill the pool? [105]

90. Find angle β in Fig. 7.32. [30]

*91. A farmer traced the map of his farm on a sheet of No. 4 zinc and cut out the outline of his map. Then, to the same scale (1 in = 50 ft), he cut out a 4-in square. The 4-in square weighed $\frac{1}{2}$ oz, while the zinc map weighed 3 lb $10\frac{1}{2}$ oz. How many acres were there in the farm?

92. The U.S. Coast and Geodetic Survey maps have a scale of about 1 in to the mile. It is desired to measure areas on one of these maps by laying a grid made of celluloid over the area to be measured. The lines on the grid form squares. What should be the spacing of the rulings if each square represents 100 acres?

*93. Find the roof area, the top view of which is given in Fig. 7.33, if all parts of the roof have a $\frac{3}{4}$ pitch (3-in rise in every 4 in taken horizontally). Area of plane figure outlined by eaves = $\frac{4}{5}$ × (roof area). Why? Note that the principle involved here may be applied to the computation of the lateral surface of any cone, pyramid, or frustum having "uniform pitch."

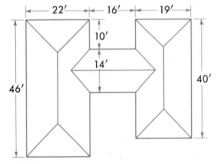

FIGURE 7.33

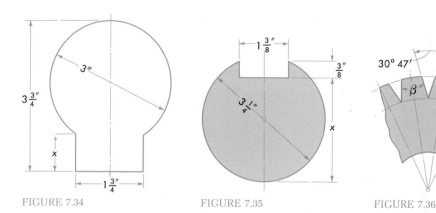

FIGURE 7.34   FIGURE 7.35   FIGURE 7.36

94. Find x in Fig. 7.34. [99]
95. Using 3,960 mi as the radius of the earth, find the length in miles of 1′ longitude (a) at 45°N, (b) at 60°N.
96. Find the distance x in Fig. 7.35. [99]
*97. Find the angle β in Fig. 7.36. (There are 21 teeth.) See Prob. 90.
98. Find the angles of a triangle if the first angle is $2\frac{1}{2}$ times the second and the third angle is $\frac{1}{4}$ of the second. [29]
99. The circumference of a circle exceeds its diameter by 13. Find the diameter.
100. The hypotenuse of a 45° right triangle exceeds one of the equal sides by 9. Find one of the equal sides. [107]
101. Find the height of a segment whose width is 24 in if the diameter of the circle is 40 in. [69, 99]
102. The side of an equilateral triangle exceeds the altitude by 5. Find a side of the triangle. [109]
103. A circle of what diameter has numerically the same diameter in inches as it has area in square inches? [63]
104. Find the length of one of the equal sides and the area of an isosceles right triangle having a perimeter of 20 ft. [107]
105. The area of a 4-ft walk surrounding a circular pond is 1,496 ft². Using $\pi = \frac{22}{7}$, find the diameter of the pond.
*106. A length of rope hangs from the top of a flagpole to the ground. Six feet of rope lies coiled up on the ground. It is found that the free end of the rope can be brought out 30 ft from the base of the pole and yet be touched to the ground. How high is the flagpole? [99]
107. An isosceles triangle has a perimeter of 32 in; the altitude to the base is 8 in. Find the length of one of the equal sides.
108. An oval running track has semicircular ends and straight sides. The overall length is three times the width. If one lap is 600 ft, find the length and width of the oval.

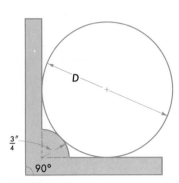

FIGURE 7.37

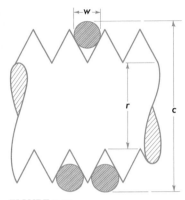

FIGURE 7.38

109. Find the diameter of the circle shown in Fig. 7.37. [107]
110. Three equal circles are externally tangent, each to each. A belt passed around these three circles is 60 in long. Find, in terms of π, the radius of a circle.
*111. Using the three-wire system for determining the root diameter r of a 60° sharp-V thread, develop a formula for r in terms of c and w (Fig. 7.38).
*112. Show that the annular area in Fig. 7.39 equals the length of the centerline times the width w.
*113. A 4-in circle is inscribed in a right triangle ABC. The circle is tangent to the hypotenuse AB at point D. If $AD = 8$ in, find BC. [72, 99]
*114. The profile AB of the nose of an experimental rocket in Figs. 7.40 and 7.41 is an arc of a circle whose center C is on the line BC. If AD is 24 ft and BD is $2\frac{1}{2}$ ft, find the radius r of the cross section at distance x from the nose tip, making x successively 6, 12, and 18 ft. Let R = radius of curvature (ft). Then, since

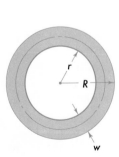

FIGURE 7.39

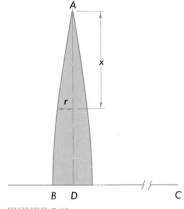

FIGURE 7.40

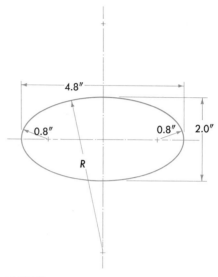

FIGURE 7.42

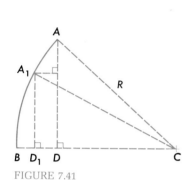

FIGURE 7.41

$AC = R$, $DC = R - 2.5$, and $AD = 24$, we have $R^2 = (R - 2.5)^2 + (24)^2$, or $R = 116.45$ ft, and $DC = 113.95$ ft. At the 6-ft station $A_1C = 116.45$, $A_1D_1 = 18$, and $D_1C = \sqrt{(116.45)^2 - (18)^2} = 115.05$. $D_1D = 115.05 - 113.95 = 1.10$ ft radius cross section at the 6-ft station.

      Repeat this procedure to find D_2C at the 12-ft station and D_3C at the 18-ft station.

*115. Find the radius R used in the construction of the approximate ellipse shown in Fig. 7.42. [99]

*116. In Fig. 7.43 show that $L = 2\sqrt{Rr}$. [70, 99]

*117. Given a square $ABCD$ in which $AB = 3\frac{1}{2}$ in; a circle is drawn passing through vertex C and tangent to AB and AD. Find the diameter of the circle. [107]

118. In Fig. 7.44 derive an expression for h in terms of the other dimensions.

119. In Fig. 7.45 derive an expression for w in terms of the other dimensions.

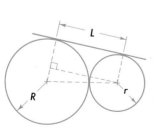

FIGURE 7.43

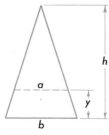

FIGURE 7.44

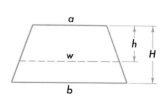

FIGURE 7.45

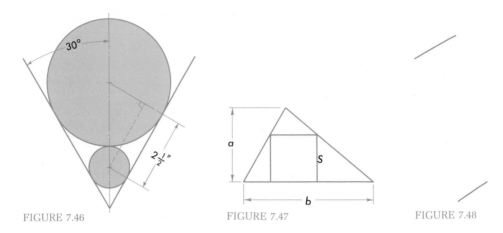

FIGURE 7.46   FIGURE 7.47   FIGURE 7.48

120. A circle of diameter D is inscribed in a 45° right triangle with leg S. Derive a formula for D in terms of S. (See Prob. 54.)

121. Find the diameters of the circles in Fig. 7.46. [109]

122. A square is inscribed in a triangle. Show that $s = ab/(a + b)$ (Fig. 7.47). [85]

*123. In Fig. 7.48 the stars photograph as traces (arcs) in a time exposure of the night sky. Estimate the length of exposure by scaling any measurements you think necessary. Polaris is beyond the lower right-hand corner of the photo. [67]

*124. A 6-ft offset in a pipeline consists of two equal reverse arcs. If the offset is accomplished in a distance of 10 ft (measured parallel to the straight pipe), find the radius of curvature of an arc (Fig. 7.49). [74, 85]

*125. The volume of a prismoid is given by $V = \frac{1}{6}h(b + 4M + B)$, where h = height and b, M, and B are the areas of the upper base, mid-section, and lower base, respectively. Applying this formula to the wedge in Fig. 7.50, develop a formula for V in terms of w, h, l, and L.

*126. A cone of base diameter d and slant height S is developed from a sector of central angle θ. Derive a formula for θ (in degrees) in terms of d and S.

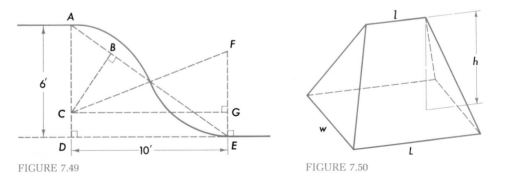

FIGURE 7.49   FIGURE 7.50

127. Show that if, from a point within an equilateral triangle, perpendiculars are drawn to the sides, the sum of these three perpendiculars equals the altitude of the triangle. (This relation is the basis of triangular coordinates.)

128. Figure 7.51 illustrates a nomogram for solving equations of the type $1/a + 1/b = 1/c$. To solve this equation for c where a and b are given, draw a 45° line from the origin.

Draw another line connecting A on one scale with B on the other. The location of the point of intersection of these two lines, which is C, may be referred to either axis.

With reference to Fig. 7.51 it can be shown that right triangles AMC and CPB are similar, and that therefore we may write the equation

$$\frac{a - c}{c} = \frac{c}{b - c}$$

Multiplying both members by $c(b - c)$, we obtain

$$(a - c)(b - c) = c^2 \qquad \text{or} \qquad ab - ac - bc + c^2 = c^2$$

We wish to show that $1/a + 1/b = 1/c$. Can you supply the missing steps?

If three or more reciprocals are to be added, as in the equation $1/a + 1/b + 1/d = 1/e$, the value of c as found above is connected with that of d on the opposite axis, and the value e obtained. This process can be continued for the addition of any number of reciprocals.

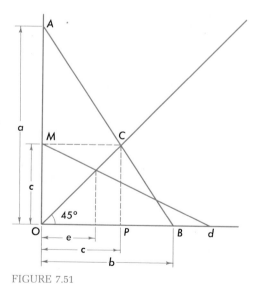

FIGURE 7.51

The problem $1/a = 1/c - 1/b$ can, of course, be solved by reversing the process. (Connect the points B and C to find A.)

If the student finds much occasion to use this method, he would do well to have the $45°$ line already drawn and to locate point c by laying a straightedge over points a and b, rather than drawing an actual line on the paper.

Applications of this nomogram† which immediately come to mind are work problems, parallel resistances, and focal length of lenses (see Sec. 10.3).

129. Figure 7.52 illustrates another nomogram for solving equations of the type $1/a + 1/b = 1/c$. Show that $1/am + 1/bm = 1/cm$. [23, 85]

130. Show that the "Z chart" in Fig. 7.53 may be used to solve the proportion

$$\frac{l_1}{l_2} = \frac{l_3}{l_4} \qquad [85]$$

131. Figure 7.54 illustrates a method of laying out an arc of a circle through points A, C, and B, where the center is inaccessible. From points A and B as centers lay out arcs of radius AB. Draw the radius from A through C to intersect the arc at D. Repeat B through C to E. Mark off a succession of convenient equal distances in both directions from D and E. Draw radii from points 1, 2, 3, etc., on arcs to A and to B. Show that intersection points 1, 2, 3, etc., lie on the required arc. [76]

*132. Find the short diameter b in the approximate ellipse in Fig. 7.55. [99]

133. Two 15-in pulleys connected by an open belt are 10 ft apart (center to center). If the belt splice passes a given point 125 times per minute, find the speed (revolutions per minute) of one of the pulleys.

*134. A roll of paper 4 ft 6 in in diameter is wound on a 6-in hub. Find the length of the paper in the roll if the sheet is 0.0080 in thick. What assumption is made here?

†The terms "nomogram," "nomograph," and "alignment chart" are synonymous.

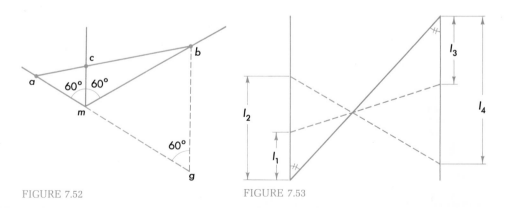

FIGURE 7.52   FIGURE 7.53

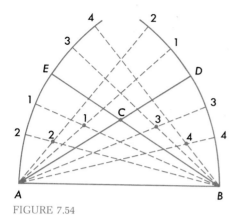

FIGURE 7.54

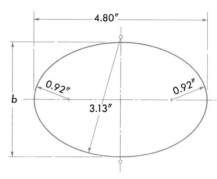

FIGURE 7.55

*135. In Fig. 7.56 the speed of the upper pulley is 450 r/min. Find the speed of the lower pulley when the idler belt is in the position shown. What is the extreme range of speeds possible in the lower pulley?†

*136. On one side of a river is a rectangular playground 100 ft wide measured back from the riverbank. An observer on the opposite bank notices that the fence posts of the sides parallel to the riverbank line up at *A* and *B*, as illustrated in Fig. 7.57. Assuming the fence posts to be uniformly spaced, find the width of the river. [85]

137. A semicircular piece of copper having a 6-in radius is rolled into a cone by bringing the two radii together. Find the capacity of the cone in cubic inches. [123]

*138. Find *m* in Fig. 7.58. [70, 85]

†The idler belt is the only point of contact between the pulleys.

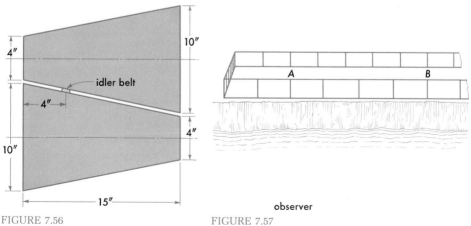

FIGURE 7.56

FIGURE 7.57

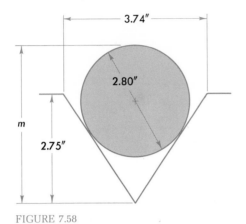

FIGURE 7.58

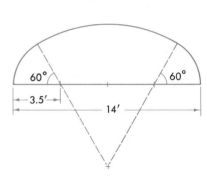

FIGURE 7.59

139. A slab of granite is being moved on logs 6 in in diameter. How far will the slab advance when a log revolves once?

140. Find the area of the arch in Fig. 7.59. [66]

141. Find the percentage error in taking $4PM$ as equal to the circumference of the circle in Fig. 7.60. Assume the radius is 1. Would the final answer be affected by assuming a different value for the radius?

*142. A hollow bronze sphere is 11 in in diameter and weighs 52 lb. The wall thickness is assumed to be uniform and is to be determined without mutilating the sphere. Assume the density of bronze is 0.30 lb/in³. How might we test for uniformity of wall thickness?

*143. A picture AB on a vertical wall (Fig. 7.61) is viewed from eye level CE. Find the distance CD corresponding to the maximum subtended angle ADB (best angle for viewing).

      Can you show that $CD = \sqrt{(AC)(BC)}$? [76]

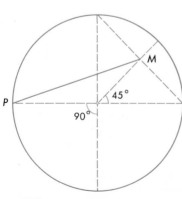

FIGURE 7.60

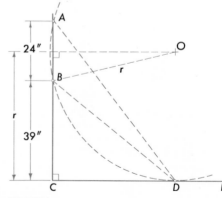

FIGURE 7.61

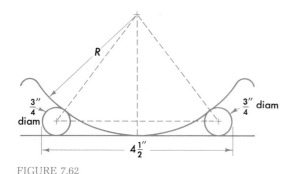

FIGURE 7.62

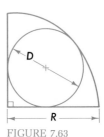

FIGURE 7.63

*144. As an auto traveled at constant speed, a point P on the tread of the tire traveled through space with varying speed, as shown by values taken at intervals during one revolution of the wheel.

Time, s	0	0.02	0.04	0.06	0.08	0.10	0.12	0.14	0.16
Speed, ft/s	0	38	69.5	91.5	99	91.5	69.5	38	0

Find the speed of the car and the diameter of the wheel.

145. Figure 7.62 illustrates a method of determining the radius of curvature of a cylindrical surface, the axis being inaccessible. Determine the radius R.

146. In Fig. 7.63 express R as a function of D. [107]

147. In Fig. 7.64 find the diameter of the smaller circle. [74]

148. An artificial earth satellite is 5,000 mi above the earth's surface. If the radius of the earth is 3,960 mi, what fraction of the earth's surface can the satellite "see"? (Area of curved surface of spherical segment or zone $= 2\pi Rh$, where $R =$ radius of sphere and $h =$ height of segment or zone.) (Fig. 7.65)
Show that $1/h = 1/h_1 + 1/R$.

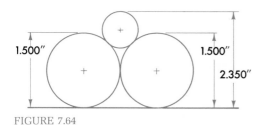

FIGURE 7.64

FIGURE 7.65

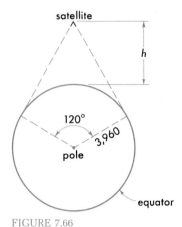

FIGURE 7.66

149. If three satellites having equatorial orbits are evenly spaced, what must be their altitude so that a satellite is visible from any part of the equator? (Fig. 7.66)

150. Given the period of one pass of satellite Faith 7 = 88 min and the diameter of the earth = 7,920 mi, find the distance l in Fig. 7.67 between two successive north-south equatorial crossings. (The displacement results from the rotation of the earth during this interval.)

*151. The shadow of a mountain on the moon at first quarter is observed at point P, located at about one-third of a radius from the center of the disk. The length of the shadow is approximately 0.002 times the radius of the moon. What is the height of the mountain if the moon's diameter is 2,160 mi? (Fig. 7.68)

152. Assuming the earth to be a perfect sphere exactly 7,920 mi in diameter, how many significant figures would be needed in the value of π to calculate the length of the equator to the nearest foot?

*153. Show algebraically that the volume of a spherical shell is approximately the product of the outer surface times the wall thickness. Volume of a sphere = $(\frac{4}{3})\pi$ (radius)³, R and r are the outer and inner radii, respectively, and $r \gg R - r$. ($\gg$ means "much greater than.")

154. If a right triangle is constructed with legs QR and PR equal to 1 and 2, respec-

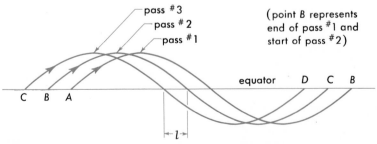

FIGURE 7.67

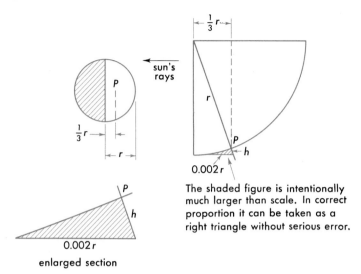

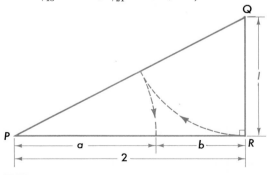

enlarged section

FIGURE 7.68

tively, and arcs drawn as indicated (Fig. 7.69), show that $a = \sqrt{5} - 1$ and $b = 3 - \sqrt{5}$. Show also that the ratio $a/b = (\sqrt{5} - 1)/(3 - \sqrt{5}) = (1 + \sqrt{5})/2 \approx 1.618$.

We say that the line segment PR has been divided into extreme and mean ratio, that is, $b/a = a/(a + b)$.

The constant 1.618 has been variously referred to as ϕ or τ. It has been called the "golden mean." The most pleasing shape of a rectangular picture is commonly considered to be about 1.618 : 1.

The intriguing Fibonacci sequence 1, 1, 2, 3, 5, 8, 13, 21, 34, etc., in which each term is formed from the sum of the two preceding terms, crops up not only in art but in plant life.

Furthermore, as we take the ratio between two successive terms, we obtain values ever closer to 1.618 (for example, $\frac{5}{3} = 1.667$, $\frac{8}{5} = 1.600$, $\frac{13}{8} = 1.625$, $\frac{21}{13} = 1.615$, $\frac{34}{21} = 1.619$, etc.).

FIGURE 7.69

Facts in Geometry

Angles, Straight Lines, Rectilinear Figures

1. Two angles are complementary when their sum is a right angle.
2. Complements of the same angle or of equal angles are equal.
3. Two angles are supplementary when their sum is a straight angle.
4. Supplements of the same angle or of equal angles are equal.
5. The sum of all the angles about a point in a plane is two straight angles.
6. If two adjacent angles have their exterior sides in a straight line, they are supplementary.
7. The sum of all successive adjacent angles around a point on one side of a straight line is a straight angle.
8. If one straight line intersects another straight line, the opposite or vertical angles are equal. $\angle a = \angle c$ (Fig. I, page 174)
9. Two triangles are congruent if two sides and the included angle of one are equal, respectively, to two sides and the included angle of the other.
10. Two triangles are congruent if two angles and the included side of one are equal, respectively, to two angles and the included side of the other.
11. Two triangles are congruent if the three sides of one are equal, respectively, to the three sides of the other.
12. Two right triangles are congruent if the hypotenuse and an acute angle of one are equal to the hypotenuse and an acute angle of the other.
13. Two right triangles are congruent if the hypotenuse and a leg of one are equal to the hypotenuse and a leg of the other.
14. In an isosceles triangle the angles opposite the equal sides are equal.
15. An equilateral triangle is equiangular, and conversely.
16. If one side of a triangle is greater than a second side, the angle opposite the first side is greater than the angle opposite the second side, and conversely.
17. The hypotenuse of a right triangle is greater than either leg.
18. The perpendicular bisector of a line segment is the locus of all points equidistant from the extremities of the segment.
19. A line passing through two points, each equidistant from the end points of a line segment, is the perpendicular bisector of the segment.
20. The perpendicular is the shortest line segment that can be drawn from a given point to a given line.
21. The bisector of an angle is the locus of all points equidistant from the sides of the angle.
22. If two parallel straight lines are cut by a transversal, the alternate interior angles are equal. $\angle a = \angle b$ (Fig. I)
23. If two parallel lines are cut by a transversal, the exterior-interior angles on the same side of the transversal are equal. $\angle b = \angle c$ (Fig. I)
24. If two parallels are cut by a transversal, the two interior angles on the same side of the transversal are supplementary, and conversely. $\angle b + \angle d = 180°$ (Fig. I)

25. Two lines parallel to a third line are parallel.
26. Two lines perpendicular to a third line are parallel.
27. If two angles have their sides respectively parallel, they are equal provided both pairs of parallels extend (a) in the same direction from their vertices or (b) in opposite directions. Otherwise they are supplementary. $\angle c = \angle e$, $\angle a = \angle e$, $\angle d + \angle e = 180°$ (Fig. I)
28. Two angles whose sides are perpendicular, each to each, are either equal or supplementary. $\angle a = \angle b$, $\angle c + \angle b = 180°$ (Fig. II)
29. The sum of the angles of a triangle is equal to 180°.
30. An exterior angle of a triangle is equal to the sum of the two opposite interior angles.
31. In a right triangle the sum of the two acute angles equals 90°. $\angle a + \angle d = 90°$ (Fig. II)
32. Two triangles are congruent if two angles and a side of one are equal to two angles and the corresponding side of the second.
33. Two right triangles are congruent if a leg and an acute angle of one are equal to the corresponding leg and acute angle of the other.
34. If two right triangles have the two legs of one equal, respectively, to the two legs of the other, the triangles are congruent.
35. The midpoint of the hypotenuse of a right triangle is equidistant from the vertices of the triangle.
36. The area of a parallelogram is equal to the product of its base and altitude.
37. The area of a rhombus is equal to one-half the product of its diagonals.
38. Parallelograms having equal bases and equal altitudes are equivalent, or equal in area.
39. A diagonal of a parallelogram divides it into two congruent triangles.
40. Any two consecutive angles of a parallelogram are supplementary.
41. The diagonals of a rhombus are perpendicular.
42. If two sides of a quadrilateral are equal and parallel, the figure is a parallelogram.
43. The opposite sides of a parallelogram are equal; the opposite angles are also equal.
44. The diagonals of a parallelogram bisect each other.
45. The sum of the interior angles of a polygon of n sides is equal to $(n-2)180°$.
46. Each interior angle of a regular polygon of n sides is equal to $[(n-2)/n]180°$.
47. The sum of the exterior angles of a polygon formed by producing its sides in succession at one extremity equals 360°.
48. Each exterior angle of a regular polygon of n sides is equal to $360°/n$.
49. The central angle of a regular polygon of n sides contains $360/n°$.
50. Segments of parallels included between parallels are equal. $x_1 = x_2$ and $y_1 = y_2$ (Fig. I)
51. A regular polygon is a polygon which is both equilateral and equiangular.
52. If three or more parallels intercept equal parts on one transversal, they intercept equal parts on every transversal. Given $a = b = c$, then $d = e = f$ (Fig. III)
53. The line which joins the midpoints of two sides of a triangle is parallel to the

third side and is equal to one-half the third side. Given $b = c$ and $e = f$, then $x_1 = \frac{1}{2}x_2$ (Fig. III)

54. The line which joins the midpoints of the legs of a trapezoid is parallel to the bases and equal to their arithmetic mean.
55. The intersection of the bisectors of the angles of a triangle determines the center of the inscribed circle. (Fig. IV)
56. The intersection of the perpendicular bisectors of the sides of a triangle determines the center of the circumscribed circle. $a_1 = a_2$, $b_1 = b_2$, $c_1 = c_2$ (Fig. V)
57. The medians of a triangle intersect at a point which cuts off two-thirds of each median from its vertex.

Circles

58. If a diameter of a circle bisects a chord that is not a diameter, it is perpendicular to the chord. (Fig. VI)
59. In a circle, or in equal circles, equal chords are equidistant from the center, and conversely. If $AB = AC$, then $PO = OQ$ (Fig. VI)
60. A tangent to a circle is perpendicular to the radius drawn to the point of contact. Tangent at $M \perp MO$ (Fig. VI)
61. The circumference of a circle is expressed by the formula $c = \pi d$, or $c = 2\pi r$. $\pi = c/d = 3.1416$, approximately.
62. The circumferences of two circles have the same ratio as their radii or as their diameters.
63. The area of a circle is given by $A = \pi r^2$, or $A = (\pi/4)d^2$.
64. The area of a circle is equal to one-half the product of its radius and its circumference.
65. The areas of two circles have the same ratio as the squares of their radii, or as the squares of their diameters, or as the squares of their circumferences.
66. The area of a sector of a circle is to the area of the circle as the angle of the sector is to $360°$.
67. The length of an arc of a circle is to the circumference of the circle as the central angle of the arc is to $360°$.
68. In the same circle, or in equal circles, equal central angles subtend equal arcs and equal chords.
69. A diameter perpendicular to a chord bisects the chord and the arc subtended by the chord. $d_1 = d_2$, $a_1 = a_2$ (Fig. VI)
70. A radius drawn to the point of contact is perpendicular to a tangent to a circle. (Fig. VI)
71. Through three points not in the same straight line, one circle, and only one, can be drawn.
72. The two tangents drawn to a circle from an outside point are equal and make equal angles with a line drawn from the point to the center of the circle. $\overline{AC} = \overline{BC}$, $\angle ACO = \angle BCO$ (Fig. VII)

73. If two circles intersect, their line of centers is perpendicular to their common chord at its midpoint.

74. If two circles are tangent to each other, the line of centers passes through the point of contact. (Fig. VII)

75. The number of degrees in a central angle equals the number of degrees in its intercepted arc.

76. An inscribed angle is measured by one-half its intercepted arc. (Fig. VIII)

77. An angle inscribed in a semicircle is a right angle. $\angle ABC = 90°$ (Fig. VIII)

78. An angle formed by two chords intersecting within a circle is measured by one-half the sum of the intercepted arcs.

79. An angle formed by a tangent and a chord drawn from the point of contact is measured by one-half the intercepted arc. $\angle ACE = \frac{1}{2}(a_1 + a_2)°$ (Fig. VI)

80. An angle formed by two secants, by two tangents, or by a secant and a tangent meeting outside the circle is measured by one-half the difference between the intercepted arcs. $\angle ACB = \frac{1}{2}(AQB - APB)°$ (Fig. VII)

81. Inscribed angles which intercept the same arc are equal. (Fig. VIII)

82. Parallel lines intercept equal arcs on a circle. $l_1 = l_2$ (Fig. VI)

Proportion, Similar Figures

83. A line parallel to one side of a triangle and meeting the other two sides divides these sides proportionally. $a_1/a_2 = b_1/b_2$ (Fig. IX)

84. If two lines are cut by a number of parallels, the corresponding segments are proportional. (Fig. IX)

85. If two triangles are similar, their corresponding sides are in proportion and their corresponding angles are equal, and conversely. (Fig. IX)

86. In any triangle the bisector of an angle divides the opposite side into segments which are proportional to the other two sides.

87. Equal or equivalent polygons have the same area.

88. Similar polygons are polygons whose corresponding angles are equal and whose corresponding sides are proportional.

89. Congruent polygons are both equivalent and similar.

90. Corresponding sides of congruent polygons are equal.

91. Corresponding angles of congruent polygons are equal.

92. The corresponding altitudes of two similar triangles have the same ratio as any two corresponding sides.

93. If two triangles are similar to a third triangle, they are similar to each other.

94. If two triangles have an angle of one equal to an angle of the other and the including sides proportional, the triangles are similar.

95. If two triangles have their sides respectively proportional, they are similar. (Fig. IX)

96. If three or more lines pass through the same point and intersect two parallel lines, they intercept proportional segments on the parallel lines. $a_1/a_2 = b_1/b_2$ (Fig. X)

97. In two similar polygons any two corresponding dimensions are to each other as any other two corresponding dimensions; also the perimeters are to each other as any two corresponding dimensions.

98. The altitude to the hypotenuse of a right triangle forms two right triangles which are similar to each other and to the whole right triangle and is a mean proportional between the segments of the hypotenuse. $x_1/h = h/x_2$ (Fig. II)

99. In a right triangle the square of the hypotenuse is equal to the sum of the squares of the legs.

100. If two chords in a circle intersect, the product of the segments of one chord is equal to the product of the segments of the other chord. $(AM)(MB) = (RM)(MC)$ (Fig. VIII)

101. The perpendicular from any point on a circle to a diameter of the circle is the mean proportional between the segments of the diameter. $KQ/d_1 = d_1/QR$ (Fig. VI)

102. The area of a triangle is equal to one-half the product of its base and its altitude.

103. The area of a triangle equals $\sqrt{s(s-a)(s-b)(s-c)}$, where a, b, and c are the sides and $s = \frac{1}{2}(a + b + c)$.

104. The area of a trapezoid is equal to the altitude multiplied by the average of the bases.

105. The areas of two similar polygons are to each other as the squares of any two corresponding dimensions.

Regular Polygons

106. The diagonal of a square is equal to one side multiplied by $\sqrt{2}$.

107. The hypotenuse of a 45° right triangle is equal to $\sqrt{2}$ times a leg. (Fig. XI)

108. The area of an equilateral triangle having a side S is given by the formula $A = (S^2/4)\sqrt{3}$. (Fig. XII)

109. In a 30°–60° right triangle the longer leg is $\sqrt{3}$ times the shorter leg and the hypotenuse is twice the shorter leg. (Fig. XII)

110. A circle may be circumscribed about, and a circle may be inscribed in, any regular polygon.

111. The area of a regular polygon is equal to one-half the product of its perimeter and its apothem (a). (Fig. XIII)

112. Of isoperimetric polygons having the same number of sides, the regular polygon has the greatest area.

113. Of all polygons equivalent in area and having the same number of sides, the regular polygon has the smallest perimeter.

114. If a circle is divided into three or more equal arcs, the chords of these arcs form a regular inscribed polygon and the tangents at the points of division form a regular circumscribed polygon.

115. If the midpoints of the arcs of a regular inscribed polygon are joined to the extremities of the respective sides, a regular inscribed polygon of double the number of sides is formed.

Solid Figures

116. The volumes of two similar solid figures are to each other as the cubes of corresponding dimensions.
117. The surfaces of two similar solid figures are to each other as the squares of corresponding dimensions.
118. Pyramidal or conic sections parallel to the base are figures similar to the base. Corresponding dimensions in two sections are proportional to the distances of the sections from the vertex, and corresponding areas are proportional to the squares of these distances.

Statements 119 to 127 refer to *right* prisms, cylinders, cones, and pyramids.

119. The lateral area of a prism is equal to the product of a lateral edge by the perimeter of the base of any right section.
120. The volume of a prism is equal to the product of the base by the altitude.
121. The lateral area of a regular pyramid or cone is equal to half the product of the slant height by the perimeter of the base. (Fig. XIV)
122. If a pyramid is cut by a plane parallel to the base:
    (a) The lateral edges and the altitudes are divided proportionally. $e_1/e_2 = h_1/h_2$ (Fig. XIV)
    (b) The section is a polygon similar to the base. (Fig. XIV)
    (c) The area of the section is to the area of the base as the square of the distance of the section from the vertex is to the square of the altitude. $A_1/A_2 = e_1^2/e_2^2$ (Fig. XIV)
123. The volume of a cone or pyramid is equal to one-third the product of the base by the altitude.
124. The volume of the frustum of a cone or pyramid is equal to $(\frac{1}{3})(h)(b + \sqrt{bB} + B) = (\frac{1}{6})(h)(b + 4M + B)$. (Fig. XV)
125. The volume of the frustum of a cone is equal to $(\frac{1}{3})(\pi h)(r^2 + rR + R^2)$. (Fig. XVI)
126. The lateral area of a right circular cylinder is equal to the product of the altitude by the circumference of the base.
127. The volume of a right circular cylinder is equal to the product of the base by the altitude.
128. The surface of a sphere is given by $A = 4\pi^2$.
129. The volume of a sphere is given by $V = \frac{4}{3}\pi r^3 = \frac{1}{6}\pi d^3$.
130. The intersection of a plane with the surface of a sphere is a circle. If the plane

passes through the center of the sphere, the intersection with the spherical surface is called a *great circle*.

131. The shortest distance between two points on a spherical surface is along the great circle connecting the points. (The arc must be $\leq 180°$.)

132. The area (curved surface) of a spherical segment or zone is equal to $2\pi Rh$, where R is the radius of the sphere and h is the height of the segment or zone.

133. Through the ends of a diameter of a sphere any number of great circles may be drawn.

134. Through two points on a spherical surface which are not ends of a diameter, only one great circle can be drawn.

135. Through three points on a spherical surface, one and only one circle can be drawn.

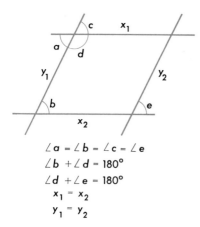

$\angle a = \angle b = \angle c = \angle e$
$\angle b + \angle d = 180°$
$\angle d + \angle e = 180°$
$x_1 = x_2$
$y_1 = y_2$

FIGURE I

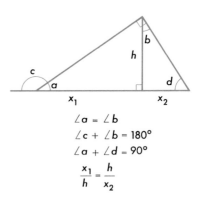

$\angle a = \angle b$
$\angle c + \angle b = 180°$
$\angle a + \angle d = 90°$
$\dfrac{x_1}{h} = \dfrac{h}{x_2}$

FIGURE II

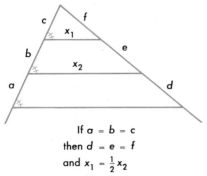

If $a = b = c$
then $d = e = f$
and $x_1 = \frac{1}{2}x_2$

FIGURE III

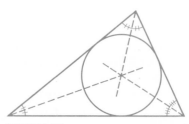

FIGURE IV

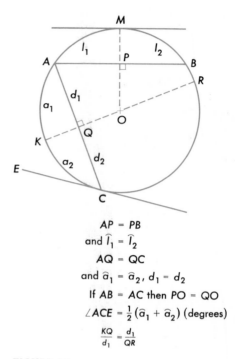

$$AP = PB$$
$$\text{and } \widehat{l}_1 = \widehat{l}_2$$
$$AQ = QC$$
$$\text{and } \widehat{a}_1 = \widehat{a}_2, d_1 = d_2$$
$$\text{If } AB = AC \text{ then } PO = QO$$
$$\angle ACE = \frac{1}{2}(\widehat{a}_1 + \widehat{a}_2) \text{ (degrees)}$$
$$\frac{KQ}{d_1} = \frac{d_1}{QR}$$

FIGURE VI

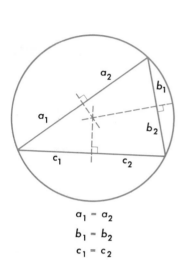

$$a_1 = a_2$$
$$b_1 = b_2$$
$$c_1 = c_2$$

FIGURE V

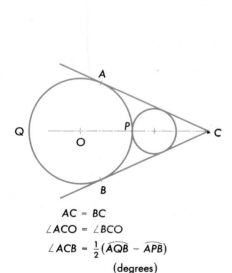

$$AC = BC$$
$$\angle ACO = \angle BCO$$
$$\angle ACB = \frac{1}{2}(\widehat{AQB} - \widehat{APB})$$
$$\text{(degrees)}$$

FIGURE VII

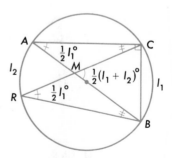

$$AB \text{ is a diameter}$$
$$\text{then } \angle ACB = 90°$$
$$(AM)(MB) = (RM)(MC)$$

FIGURE VIII

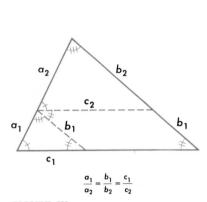

$$\frac{a_1}{a_2} = \frac{b_1}{b_2} = \frac{c_1}{c_2}$$

FIGURE IX

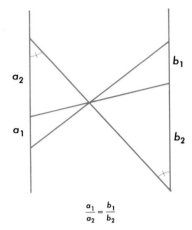

$$\frac{a_1}{a_2} = \frac{b_1}{b_2}$$

FIGURE X

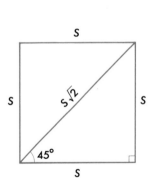

FIGURE XI

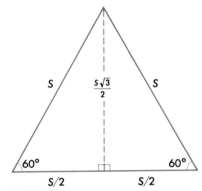

FIGURE XII

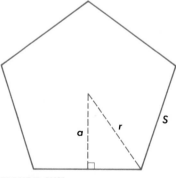

FIGURE XIII

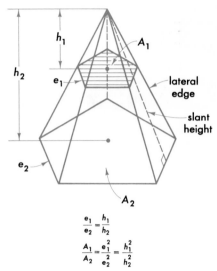

$$\frac{e_1}{e_2} = \frac{h_1}{h_2}$$

$$\frac{A_1}{A_2} = \frac{e_1^2}{e_2^2} = \frac{h_1^2}{h_2^2}$$

FIGURE XIV

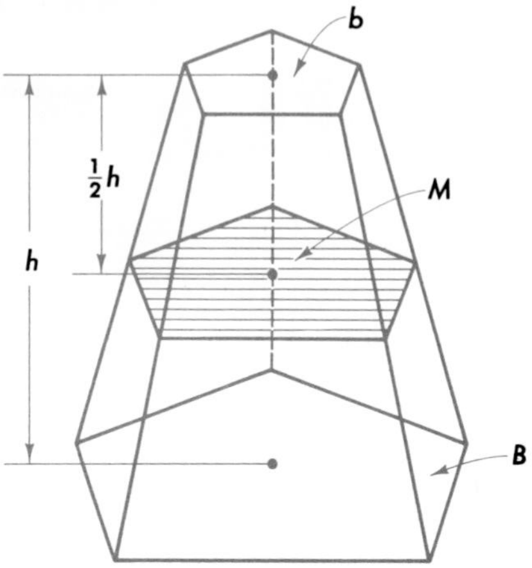

FIGURE XV

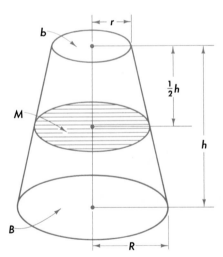

FIGURE XVI

linear functions

8

In this chapter we shall briefly introduce the basic idea of a function. Then we shall discuss the graphs of linear functions. This is commonly known as the analytic geometry of the straight line.

8.1 Ordered Pairs of Numbers

If we wish to convert the temperature of a room expressed in degrees Celsius to the corresponding temperature expressed in degrees Fahrenheit, the following equation applies.

$$F = 1.8C + 32 \tag{1}$$

where F = temperature in degrees Fahrenheit
C = temperature in degrees Celsius

By substituting a value of the variable C in Eq. (1), the corresponding value of the variable F may be calculated.

In this approach, the variable F is appropriately called the *dependent* variable and the variable C is called the *independent* variable.

Observe that for each value of the independent variable C there is a unique

TABLE 8.1

Independent variable	Dependent variable
-20	-4
-10	$+14$
0	$+32$
$+10$	$+50$
$+20$	$+68$

(only one) corresponding value for the dependent variable F. For example, if we substitute successive values for C, such as -20, -10, 0, $+10$, and $+20$, we find that corresponding values of F are -4, $+14$, $+32$, $+50$, and $+68$, respectively. These are tabulated in Table 8.1.

Here we have five pairs of numbers. Each pair consists of a value of the independent variable and the corresponding value of the dependent variable. By convention such pairs of numbers are often written in the form $(-20,-4)$, $(-10,+14)$, $(0,+32)$, $(+10,+50)$, and $(+20,+68)$. Observe that the value of the independent variable is written first, followed by the value of the dependent variable. Since the order in which these numbers are written is critical, they are called *ordered* pairs of numbers.

8.2 Functions

When we use the word "function" we shall refer to a set of ordered pairs of numbers, for example (x,y), such that for each value of the first variable x, there corresponds a unique value of the second variable y. Since the value of the variable y is considered to depend upon the value of the variable x, we refer to y as the dependent variable, and to x as the independent variable.

We also sometimes refer to y as the "value" of a function of x.

The *domain* of the function is the set of numbers from which specific values of the independent variable x may be chosen. The *range* of the function is the set of numbers among which are found the corresponding values of the dependent variable y.

Unless otherwise expressed or implied, both the range and the domain of the function involved will be the largest possible set of real numbers.

We observe that there are three important aspects of the discussion of functions so far.

1. The set of x values, or the *domain* of the function
2. The set of y values, or the *range* of the function
3. Some type of association between y and x such that a unique value of y can be determined if x is given

There are four ways by which the type of association between the independent and dependent variables can be described:

1. By a worded statement: The circumference of a circle is the product of π and the diameter.
2. By an equation: $y = 5x^2$.
3. By a table of values enumerating convenient, corresponding values for the variables involved in an equation.
4. By a graph of the function. The graph of a function is the set of points whose coordinates represent, respectively, the set of ordered pairs of numbers which constitutes the function. This is discussed in greater detail in Sec. 8.5.

In physical problems involving the function idea we often attempt to distinguish between *cause* and *effect* even though by doing so we may arrive at some quite arbitrary distinctions. Even so, it is often convenient to try to represent the cause by the independent variable and the effect by the dependent variable.

In each of the following illustrations it will be noted that the dependent variable is mentioned before the independent variable.

Rate of growth of vegetation depends upon the temperature.
Time of high tide depends upon the position of the moon.
Price of an article depends upon supply and demand.
Flow of electric current through a fixed resistance depends upon the voltage.
Postage to a given destination depends upon the weight of the package.

The primary meaning of the word "function" was given above. However, it is commonly used in a slightly different sense. Thus if

$$A = \pi r^2$$

we often say that A is a function of r. In this sense the word function is identified with the dependent variable rather than with an ordered pair of numbers.

8.3 Functional Notation

When, for any reason, the exact relation between variables is not to be expressed, a general form is used. Thus, instead of writing $A = \pi r^2$, we might use the more general form $A = f(r)$, which states that A is a function of r without giving the exact relation. It is read, "A is a function of r," or more simply, "A equals f of r."

If we let y represent the value of some function of x, we may write $y = f(x)$. Then $f(6)$ is the value of y when $x = 6$; likewise $f(-1)$ is the value y assumes when

$x = -1$; and so on. In general, $f(a)$ is the value of y which results when a is substituted for x in the expression for $f(x)$.

It should be remembered that $f(x)$ does not mean f times x; in fact, f does not represent a number. In the expression $y = f(x)$, f simply indicates that a functional relationship exists between y and x.

In the present context it is quite common to use y and $f(x)$ interchangeably. Thus we may let either y or $f(x)$ designate the dependent variable.

Example 1. Suppose that $y = x^2 - 3x - 4$. We may then write

$$y = f(x) = x^2 - 3x - 4$$

Then, when $x = 5$,

$$y = f(5) = (5)^2 - 3(5) - 4 = 6$$

Also, when $x = -6$,

$$y = f(-6) = (-6)^2 - 3(-6) - 4 = 50$$

EXERCISE 1

If $y = f(x) = 3x^2 - 2x - 5$, find

1. $f(0)$ 2. $f(-1)$ 3. $f(\frac{2}{3})$ 4. $f(a + 1)$ 5. $f(4) - f(-2)$

If $y = f(x) = 12x/(16 - x^2)$, find

6. $f(-2)$ 7. $f(5)$ 8. $f(\frac{1}{2})$ 9. $f(3a - 4)$ 10. $f(2a) - f(a)$

8.4 Rectangular Coordinates

To form a *rectangular system of coordinates* we draw a horizontal line X'X and a vertical line Y'Y. Their point of intersection O is called the *origin*. X'X is called the X axis, and Y'Y is called the Y axis.

The location of any point in the plane of the coordinate axes may be expressed by stating its distance and direction from each of the two axes. Distances above X'X and distances to the right of Y'Y are positive, whereas distances below X'X and distances to the left of Y'Y are negative.

The distance of a point measured to the right or left of the Y axis is called the *abscissa*, or X coordinate, of the point. The distance of a point from the X axis (upward or downward) is called the *ordinate*, or Y coordinate, of the point. In Fig. 8.1, point

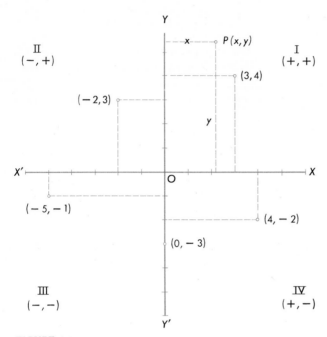

FIGURE 8.1

P is said to have the *coordinates* (x,y) and may be referred to as the point (x,y). In Fig. 8.1 the points $(3,4)$, $(4,-2)$, $(0,-3)$, $(-5,-1)$, and $(-2,3)$ are also designated. It should be noted that the coordinates are enclosed in parentheses and separated by a comma and that the abscissa is always written first (hence the term "ordered pair"). It will also be seen that the axes divide the plane into four quadrants, numbered as shown. The signs of the coordinates of points in each quadrant are also indicated.

The position of any point can be determined if its coordinates are known, and conversely, the coordinates of a point appearing on the coordinate plane can be determined by measurement.

EXERCISE 2

Locate the points having the coordinates given below.

1. $(2,5)$ 2. $(4, \sqrt{3})$ 3. $(-2,6)$ 4. $(-\sqrt{2},3)$
5. $(8,-3)$ 6. $(7,-4)$ 7. $(-3,-5)$ 8. $(-6,-2)$
9. $(0,5)$ 10. $(-\frac{5}{2},0)$ 11. $(0,0)$ 12. $(1\frac{1}{2},-\frac{2}{3})$

EXERCISE 3

Identify the closed figures formed by plotting the given points and joining in order with straight lines.

1. $(5,2)$, $(5,-4)$, $(-1,-4)$, $(-1,2)$
2. $(0,4)$, $(7,4)$, $(7,-1)$, $(0,-1)$
3. $(3,2)$, $(8,0)$, $(0,-9)$
4. $(2,4)$, $(5,-3)$, $(-1,-3)$
5. $(-1,6)$, $(5,2)$, $(-1,2)$
6. $(5,5)$, $(7,-3)$, $(-4,-3)$, $(-6,5)$
7. $(1,3)$, $(4,2)$, $(6,8)$
8. $(5,3)$, $(12,4)$, $(7,9)$, $(0,8)$

8.5 Graph of a Function

When we sketch the graph of a function, we say that we "graph the function."

Suppose that we are given some function $y = f(x)$. The first step in graphing the function is to make a table listing a number of different values of x and opposite them to write the corresponding values of y. We take these pairs of corresponding values of x and y as the coordinates of points which we plot. These points all lie on the graph of $y = f(x)$. Accordingly, we may define the *graph*, or *locus*, of a function as consisting of a system of points whose coordinates satisfy the relation $y = f(x)$.

A *linear function* of x is a first-degree polynomial in x having the form $mx + b$, where m and b are constants. The graph of such a function is always a straight line; hence the name linear function.

Example 2. Graph the function $y = f(x) = \frac{2}{3}x - 4$.

Form a table of arbitrarily chosen values of x together with the corresponding values of $f(x)$, that is, of y.

x	-6	-3	0	$1\frac{1}{2}$	3	5	8	12
y	-8	-6	-4	-3	-2	$-\frac{2}{3}$	$1\frac{1}{3}$	4

In Fig. 8.2 the points $(-6,-8)$, $(-3,-6)$, etc., are plotted, and a smooth line is drawn through these points.

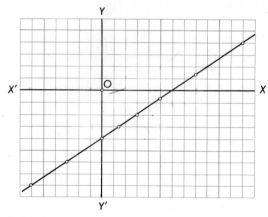

FIGURE 8.2

It is evident that we might plot any number of points whose coordinates satisfy the equation $y = \frac{2}{3}x - 4$. We infer, then, that we have a straight line which may be extended indefinitely in either direction. Theoretically, two points are sufficient to determine a straight line. However, in graphing any linear function, we should compute a minimum of three values of the function. For the sake of accuracy, these points should be located some distance apart. Any points not lying on the same straight line should be checked for errors.

For a given section of coordinate paper the position of the axes and the number of units assigned to a division on the paper should be chosen so as to make the desired portion of the graph as large as possible.

In general, if a function of x is defined by a formula, its graph is a smooth curve or, in some cases, two or more disconnected smooth curves. The term "smooth curve" is used in a broad sense to denote a line which may be straight or curved but which does not show any abrupt change in direction. For greater clarity it may be desirable to use different scales on the X and Y axes. Perhaps the most common example of an independent variable is time. Time units, with rare exceptions, are scaled horizontally.

It is good practice to see that each division on the axes corresponds to one, two, or five times an integral power of 10, as is the case with the scales of a slide rule.

Example 3. Graph the function $y = f(x) = -\frac{3}{2}x$.

Form a table listing arbitrary values of x and corresponding values of y:

x	8	5	2	0	−2	−5	−8
y	−12	−7½	−3	0	3	7½	12

In Fig. 8.3 the points $(8, -12)$, $(5, -7\frac{1}{2})$, etc., are plotted. They will be seen to lie in one straight line which is drawn through them.

EXERCISE 4

Plot the graphs of the following equations in which $y = f(x)$, indicating any intersections with the axes:

1. $y = x$
2. $y = 3x$
3. $y = \frac{1}{2}x$
4. $y = -x$
5. $y = -2x$
6. $y = -\frac{1}{3}x$
7. $y = 2$
8. $y = -3$
9. $x = 4$
10. $x = -2\frac{1}{2}$
11. $y = x + 3$
12. $y = 2x - 1$
13. $y = \frac{1}{2}x + 2$
14. $x = 0$
15. $y = 0$

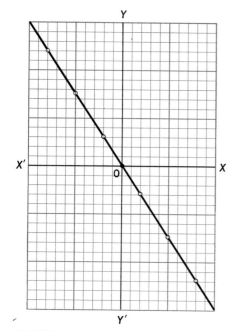

Y

X'

O

X

Y'

FIGURE 8.3

8.6 Slope of a Straight Line

Referring to Fig. 8.4, draw any straight line not parallel to the axes. Let P and Q be any two points on the line; denote the coordinates of P by (x_1, y_1) and the coordinates of Q by (x_2, y_2).

The steepness or slope of the line is then expressed by the equation

$$m = \frac{y_2 - y_1}{x_2 - x_1} \tag{2}$$

provided $x_2 - x_1 \neq 0$ or $x_2 \neq x_1$. Hence the *slope* of the straight line passing through two given points is equal to the difference of the ordinates of the points divided by the difference of their abscissas taken in the same order.

It is apparent from the rule just stated that the slope of a line could also be expressed by the fraction $(y_1 - y_2)/(x_1 - x_2)$. This is, of course, in agreement with the law of signs as applied to fractions.

Example 4. Find the slope of the line passing through $(-3,5)$ and $(9,11)$ (Fig. 8.5).

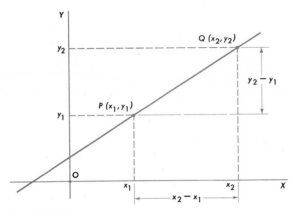

FIGURE 8.4

Referring to Eq. (2), we may consider $x_1 = -3$, $x_2 = 9$, $y_1 = 5$, $y_2 = 11$. The slope is therefore

$$\frac{11 - 5}{9 - (-3)} = \frac{6}{12} = \frac{1}{2}$$

Example 5. Find the slope of the line passing through $(4,4)$ and $(10,-5)$ (Fig. 8.5).

Taking $x_1 = 4$, $x_2 = 10$, $y_1 = 4$, and $y_2 = -5$, we have

$$\text{Slope} = \frac{-5 - 4}{10 - 4} = \frac{-9}{6} = -\frac{3}{2}$$

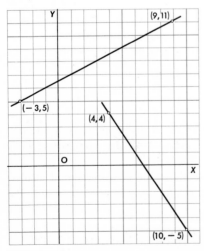

FIGURE 8.5

The following characteristics of slope may be cited:

1. *A line sloping upward to the right has a positive slope; a line sloping downward to the right has a negative slope.*
2. *The slope of a horizontal line is zero.*
3. *The slope of a vertical line is undefined* (since $x_2 - x_1 = 0$ and division by zero is undefined.)
4. *If we move a point to the right along a straight line, the slope corresponds to the progress made in a vertical direction when the point has advanced one unit horizontally.* Thus, if the point rises $1\frac{1}{2}$ units while at the same time moving one unit to the right, the slope of the line is $1\frac{1}{2}$; if the point falls $\frac{4}{5}$ of a unit while advancing one unit to the right, the slope is $-\frac{4}{5}$. Scale units are to be used in all cases.

EXERCISE 5

Find the slope of the straight line passing through the following points:

1. $(-2,1)$ and $(4,3)$ 2. $(1,-3)$ and $(4,3)$
3. $(0,4)$ and $(8,-2)$ 4. $(-2,5)$ and $(3,0)$
5. Origin and $(4,10)$ 6. $(-3,-1)$ and $(5,-1)$
7. What is the slope of (a) a 30° line, (b) a 45° line, (c) a 60° line, (d) a 120° line, (e) a 135° line, and (f) a 150° line? Assume equal X and Y scales. (The angle in question is measured counterclockwise from the x axis to the given line. See Fig. 8.15.)
8. An 8 × 8 square is cut up into four sections which are apparently reassembled to form a 5 × 13 rectangle. Where did the extra unit of area come from? (See Fig. 8.6.)

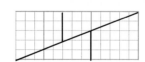

FIGURE 8.6

If three points $P(x_1,y_1)$, $Q(x_2,y_2)$, and $R(x_3,y_3)$ lie in a straight line (Fig. 8.7), the slope of PQ must be equal to that of QR. Hence, by Eq. (2), for three points lying in a straight line,

$$\frac{y_3 - y_2}{x_3 - x_2} = \frac{y_2 - y_1}{x_2 - x_1} \tag{3}$$

The student will recognize that this is the type of straight-line equation which forms the basis of interpolation.

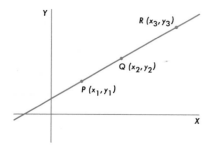

FIGURE 8.7

Example 6. The following data are thought to represent a straight line. Confirm this algebraically.

x	y	Δx	Δy	$\dfrac{\Delta y}{\Delta x}$
5	1			
		4	6	1.5
9	7			
		6	9	1.5
15	16			
		2	3	1.5
17	19			
		8	12	1.5
25	31			

In this table Δx represents the change in x from the preceding value. Thus the first value in the Δx column is found by writing $9 - 5 = 4$. Likewise the first value in the Δy column is found by writing $7 - 1 = 6$. The $\Delta y/\Delta x$ column (read "delta y over delta x") indicates the ratio of Δy to the corresponding Δx, and therefore represents the slopes of straight-line segments connecting consecutive pairs of points. Since these ratios are equal, all the points must be *collinear* (lie in the same straight line).

This test is routinely used to check against errors in data expected to be represented by a straight line.

8.7 Rate of Change of Linear Functions

It will be noted that Eq. (2) for slope expresses the *ratio* of the *change in y* to the *change in x*. We may then extend Eq. (2) to read

$$m = \frac{y_2 - y_1}{x_2 - x_1} = \frac{\Delta y}{\Delta x}$$

The delta concept of rates of change is of utmost importance in the study of calculus.

To express the concepts of Sec. 8.6 in functional notation, we shall find the rate of change in the value of a linear function by starting with a definite value of the independent variable, for example x, to which we shall assign the value x_1 and let x change by a certain amount Δx (Fig. 8.8a). Then the rate of change in the value of this function is

Rate of change in $f(x) = \dfrac{f(x_1 + \Delta x) - f(x_1)}{\Delta x}$

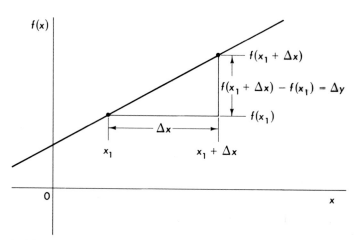

FIGURE 8.8a

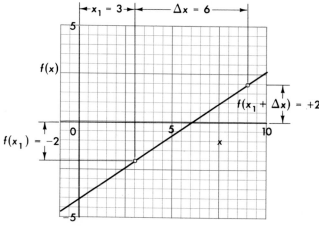

FIGURE 8.8b

We shall agree that $+\Delta x$ denotes an increase in the value of x and that $-\Delta x$ denotes a decrease in the value of x.

If we let

$$\Delta y = f(x_1 + \Delta x) - f(x_1)$$

then the rate of change in $f(x)$ with respect to x, or what amounts to the same thing, the rate of change in y with respect to x, is

$$\frac{\Delta y}{\Delta x} = \frac{f(x_1 + \Delta x) - f(x_1)}{\Delta x} \tag{4}$$

By comparing Eq. (4) above with Eq. (2) in Sec. 8.6 we note that the rate of change in the value of a linear function is numerically equal to the slope of the graph of that linear function.

Example 7. For the linear function represented in Fig. 8.8b find $f(x_1)$ and $f(x_1 + \Delta x)$ by direct reading of the graph when $x_1 = 3$ and $\Delta x = 6$. Also find the rate of change of this function.

By reading the graph, $f(x_1) = f(3) = -2$ and $f(x_1 + \Delta x) = f(3 + 6) = +2$.

$$\frac{f(x_1 + \Delta x) - f(x_1)}{\Delta x} = \frac{\Delta y}{\Delta x} = \frac{2 - (-2)}{6} = \frac{2}{3}$$

In this example we could, with equal logic, have reasoned that as x decreases by 6 units, then y decreases by 4 units and $\Delta y/\Delta x = -4/-6 = 2/3$. However, it is usually more convenient to deal consistently with increasing abscissas.

Example 8. If a straight line passes through the points $(-5, -16)$ and $(+10, +20)$, how much does y change when x increases by 0.005?

$$\text{Slope} = m = \frac{20 - (-16)}{10 - (-5)} = \frac{20 + 16}{10 + 5} = \frac{36}{15} = \frac{12}{5} = +2.4$$

In other words,

$$\frac{\Delta y}{\Delta x} = + \frac{12}{5} = +2.4$$

In this example Δx is given to be 0.005 and we are required to find Δy. Therefore

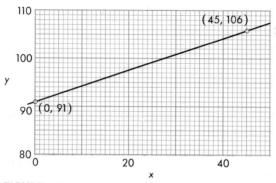

FIGURE 8.9

$$\frac{\Delta y}{0.005} = 2.4$$

$$\Delta y = 2.4 \times 0.005 = 0.012$$

EXERCISE 6

1. For the linear function displayed in Fig. 8.9 find $f(x_1 + \Delta x)$ by direct reading of the graph. Also calculate the rate of change of $f(x)$ per unit increase of x when (a) $x_1 = 9$, $\Delta x = 6$; (b) $x_1 = 21$, $\Delta x = 12$; (c) $x_1 = 27$, $\Delta x = 15$. (d) How much does y change when x increases by 0.012? (This graph shows the expansion at constant pressure of a gas occupying 91 ml at 0°C and warming to 45°C. Can you attribute any physical significance to the value of the slope?) [*Note*: $y = f(x)$.]

2. As in Prob. 1, find the rate of change of the function in Fig. 8.10. (a) $x_1 = 4$, $\Delta x = 3$;

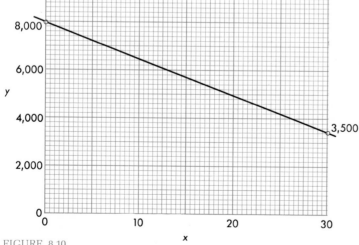

FIGURE 8.10

(b) $x_1 = 12$, $\Delta x = 5$; (c) $x_1 = 20$, $\Delta x = 8$. (d) How much does y change when x increases by 0.04? (Figure 8.10 represents the "straight-line" depreciation of a machine over a period of 30 years.) [*Note:* $y = f(x)$.]

8.8 Determination of a Straight Line

The equation of a particular straight line may be determined (1) if the slope of the line and the coordinates of a point on the line are known or (2) if the coordinates of two points on the line are known.

Observe that in Fig. 8.11 the straight-line graph intersects the Y axis at the point whose coordinates are $(0,b)$. The ordinate b is called the *y intercept*.

According to Sec. 8.6, the slope of the graph in Fig. 8.11 is given by

$$m = \frac{y - b}{x - 0} \tag{5}$$

By solving Eq. (5) for y, we obtain

$$mx = y - b$$

or

$$y = mx + b \tag{6}$$

where m = slope
b = y intercept

The y intercept is the value of y when $x = 0$. Similarly, the x *intercept* is the value of x when $y = 0$.

Equation (6) is called the *slope-intercept* form of a linear equation because both the slope and the y intercept are explicitly evident.

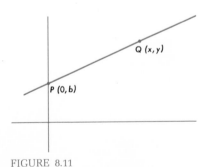

FIGURE 8.11

The slope m is the coefficient of the independent variable x, and the y intercept is the constant term b. Equation (6) is the most widely used of the various straight-line equations.

8.9 Graphical Evaluation of m and b

We can find the equation for the line segment $\overline{AB}$ in Fig. 8.12 directly from the graph.

By inspection we would select two arbitrary points on the line, such as $C(4,2)$ and $D(16,-1)$. It can be seen that

$$m = \frac{-1-2}{16-4} = -\frac{1}{4}$$

If the line is extended to cut the y axis, we read the y intercept $= 3$. Therefore the equation is $y = -\frac{1}{4}x + 3$. (Does the x intercept confirm this equation?)

EXERCISE 7

Determine the equation for each line in Fig. 8.13 in the form $y = mx + b$. Where possible, confirm your answer by determining the x intercept by inspection.

1. Line A 2. Line B 3. Line C 4. Line D
5. Line E 6. Line F 7. Line G
   Note: Try to devise your own method of solving Probs. 6 and 7.

8.10 Algebraic Evaluation of m and b

The slope-intercept form of a linear equation is probably the most generally useful form for our purpose. However, there are other forms, and we shall mention two of them briefly.

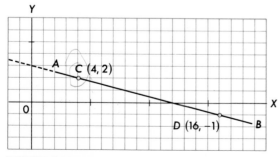

FIGURE 8.12

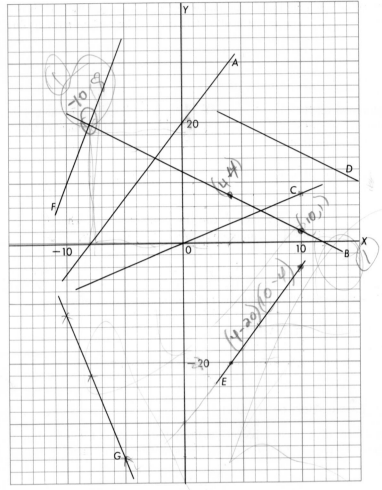

FIGURE 8.13

For example, consider the linear equation

$$y = -\tfrac{2}{3}x + 18 \tag{7}$$

By multiplying both members of Eq. (7) by 3 we obtain

$$3y = -2x + 54$$

Then we may write

$$2x + 3y = 54 \tag{8}$$

which is a form equivalent† to Eq. (7). By dividing Eq. (8) by 54 we obtain the *intercept* form

$$\frac{x}{27} + \frac{y}{18} = 1 \tag{9}$$

which is still another form equivalent to the slope-intercept form shown as Eq. (7).

When satisfactory graphical data are not available (as in Sec. 8.9), we apply algebraic methods of evaluating *m* and *b*.

Example 9. Find the equation of the straight line having a slope of $\frac{2}{3}$ and passing through point (4,5).

By substituting $\frac{2}{3}$, which is the given slope, in Eq. (6), we have

$$y = \tfrac{2}{3}x + b$$

We also know from the data given in the statement of the problem that when x = 4, y = 5. Thus we may write

$$5 = \tfrac{2}{3} \cdot 4 + b$$

Solving for *b*, we obtain

$$5 = \tfrac{8}{3} + b$$

or

$$b = 5 - \tfrac{8}{3} = \frac{15 - 8}{3} = \tfrac{7}{3}$$

and

$$y = \tfrac{2}{3}x + \tfrac{7}{3}$$

which is the equation of the straight line having a slope of $\frac{2}{3}$ and passing through the point (4,5).

Example 10. Find the slope and *y* intercept of the line whose equation is

$$3x + 2y = 8$$

†This is the simplest equivalent form having integral coefficients.

Solving the equation for y, we obtain

$$y = -\tfrac{3}{2}x + 4$$

Hence the line has a slope of $-\tfrac{3}{2}$ and a y intercept of 4.

A common example of the slope-intercept form occurs in the equation $l_1 = kW + l$, which states that in a spring balance the length of the relaxed spring is extended by k length units per W units of weight applied, producing a length l_1 under tension.

If the student has any doubt that the graph of $y = mx + b$ is a straight line, he may be reassured by noting the similarity to a flight of stairs. If we tabulate a series of values as in plotting a graph, we obtain the following table:

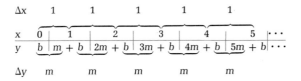

Δx	1		1		1		1		1	
x	0	1	2	3	4	5	···
y	b	m + b	2m + b	3m + b	4m + b	5m + b	···
Δy	m		m		m		m		m	

The change in x from point to point has been represented by Δx. The change in y from point to point has been represented by Δy.

It will be noted that when x increases uniformly in steps of 1, y also increases uniformly, but in steps of m. The ratio of vertical progress to horizontal progress is therefore uniform, and the slope is everywhere equal to m. Referring to Fig. 8.14, we may draw an analogy to a flight of steps, each tread being equal to 1, and each riser equal to m.

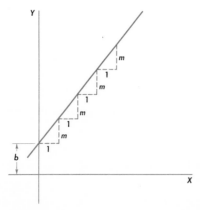

FIGURE 8.14

Example 11. Find the equation of the straight line passing through the points $(-2,3)$ and $(6,9)$.

From Eq. (2) the slope m is given by

$$m = \frac{9-3}{6-(-2)} = \frac{9-3}{6+2} = \frac{6}{8} = 0.75$$

From Eq. (6) we may now write

$$y = 0.75x + b$$

From the given conditions of the problem we know that when $x = 6$, then $y = 9$. Substituting these values in the above equation, we obtain

$$9 = 0.75 \times 6 + b$$

or

$$b = 9 - 0.75 \times 6 = 9 - 4.5 = 4.5$$

Then

$$y = 0.75x + 4.5$$

We could also have used the point $(-2,3)$. Then we would have written

$$3 = 0.75 \times (-2) + b = -1.5 + b$$

and

$$b = 3 + 1.5 = 4.5$$

Example 12. Find the equation of the line having an x intercept of -4 and a y intercept of 10.

The coordinates of the x intercept are $(-4,0)$. The coordinates of the y intercept are $(0,10)$. Therefore, from Eq. (2), the slope m is given by

$$m = \frac{10-0}{0-(-4)} = \frac{10}{4} = +2.5$$

From Eq. (6) we may write

$$y = 2.5 \cdot x + b$$

The y intercept b is given directly. Thus

$$y = 2.5x + 10$$

Example 13. Find the intercepts of the straight line whose equation is $3x - 4y = 12$.
   Since the y intercept is the value of y when $x = 0$ and the x intercept is the value of x when $y = 0$, we may solve for these intercepts by replacing x and y in turn by zero.

When $x = 0$: $-4y = 12$ or $y = -3$ (y intercept)
When $y = 0$: $3x = 12$ or $x = 4$ (x intercept)

8.11 Lines Parallel to the Axes

If a line is parallel to the X axis, its slope m is equal to zero, for by Eq. (1), $m = (y_2 - y_1)/(x_2 - x_1)$, or $m = 0/(x_2 - x_1) = 0$ (since $y_2 = y_1$). Substituting $m = 0$ in $y = mx + b$, we obtain $y = b$. This is reasonable, for it indicates that everywhere on the line the ordinate is b. Hence the line is parallel to the X axis and b units above it or below it, according to the sign of b.
   Since a line parallel to the Y axis is vertical, it has no finite slope, and the form $y = mx + b$ does not apply, for $x_2 = x_1$ and by Eq. (1), $m = (y_2 - y_1)/(x_2 - x_1)$, or $m = (y_2 - y_1)/0$; since division by zero is undefined, m is undefined. Moreover, it should be clear that a vertical line a units to the right or left of the Y axis must have the equation $x = a$, where a may be either positive or negative.
   It follows that every line constituting the grid of a sheet of graph paper has a unique equation (e.g., each horizontal line and each vertical line in Fig. 8.13 is represented by its own distinct equation).

8.12 Lines Passing through the Origin

If the line $y = mx + b$ passes through the origin, the y intercept b must equal zero, and the equation cannot have a constant term.

Example 14. Show that the line passing through the points (12,8) and (9,6) also passes through the origin.
   From Eq. (2) the slope m is given by

$$m = \frac{6 - 8}{9 - 12} = \frac{-2}{-3} = \frac{2}{3}$$

and the equation for this straight line can be written

$y = \frac{2}{3}x + b$

However, when $x = 12$, $y = 8$, and by substituting these values in the above equation, we find

$8 = \frac{2}{3}(12) + b$

or

$8 = 8 + b$

and $b = 0$.

Thus the y intercept is 0, and the graph therefore does pass through the origin, and the equation is

$y = \frac{2}{3}x$

EXERCISE 8

Write the equations for lines passing through the given point P and having the slope indicated. (See Fig. 8.15, where $\angle A$ is given in degrees. See also Sec. 8.6.)

In Probs. 5 to 8, assume equal x and y scale moduli.

1. $P(4,6)$; $m = \frac{1}{2}$
2. $P(-2,3)$; $m = 1$
3. $P(5,-1)$; $m = 2$
4. $P(-3,-2)$; $m = -\frac{2}{3}$
5. $P(3,7)$; $\angle A = 45°$
6. $P(-3,6)$; $\angle A = 30°$
7. $P(2,-5)$; $\angle A = 60°$
8. $P(0,0)$; $\angle A = 135°$
9. $P(2,7)$; horizontal
10. $P(4,-3)$; vertical

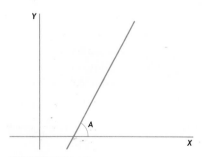

FIGURE 8.15

Convert each of the following equations to the slope-intercept form $y = mx + b$, finding the slope and y intercept in each case.

11. $x + y = 8$ 12. $2x - y = 6$ 13. $x - 3y = 12$

14. $2x + 3y = 6$ 15. $4y - x = 10$ 16. $3x - 4y - 6 = 0$

17. $7x - 5y + 14 = 0$ 18. $\frac{1}{2}x - \frac{1}{3}y = 1$ 19. $4y = 6$

20. $3x = 2y$ 21. $3y + 2x = 0$

Find the equation of the line passing through each of the following given pairs of points, leaving the answer in the form $Ax + By = C$, where A, B, and C are integers containing no common factor.

22. $(3,1), (5,3)$ 23. $(3,2), (6,3)$ 24. $(1,-1), (3,3)$

25. $(2,-4), (-6,2)$ 26. $(0,0), (-4,-6)$ 27. $(3,0), (-5,2)$

28. $(6,3), (-4,-2)$ 29. $(-2,5), (4,5)$

Find the equations of the lines having the following given intercepts:

30. x intercept: 3; y intercept: -6 31. x intercept: -4; y intercept: 2

32. x intercept: 10; y intercept: 4   33. x intercept: -9; y intercept: -6

34. x intercept: 6; y intercept: -10 35. x intercept: -8; y intercept: 6

36. x intercept: 8; y intercept: 14   37. x intercept: -15; y intercept: -10

38. Problem 51, page 125, is repeated below. Express the current and proposed schedules in the form $y = mx + b$.

A utility company petitioned to change its rate on electrical energy from a straight $5\frac{1}{2}$ cents/kWh to $4\frac{1}{4}$ cents/kWh plus 75 cents/month service charge. What monthly cost would be unaffected by the change in rate?

39. Plot the tax vs. taxable income (Fig. 8.16) according to the following schedule:

If income is:	Tax is:
Less than $4,000	20% of income
Between $4,000 and $8,000	$800 + 22% of excess over $4,000
Between $8,000 and $12,000	$1,680 + 26% of excess over $8,000
Between $12,000 and $16,000	$2,720 + 30% of excess over $12,000
Between $16,000 and $20,000	$3,920 + 34% of excess over $16,000
Between $20,000 and $24,000	$5,280 + 38% of excess over $20,000
Between $24,000 and $28,000	$6,800 + 43% of excess over $24,000

The entire "curve" will be seen to consist of seven distinct line segments, each representable by an equation over its length.

Express in the form $y = mx + b$ the equation for each line segment $\overline{OA}$, $\overline{AB}$, $\overline{BC}$, etc.

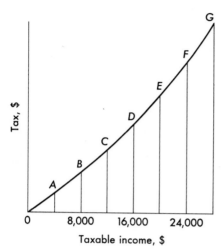

FIGURE 8.16

8.13 Parallel Lines

If two lines are parallel, their slopes are equal.

The comparison is most conveniently made by reducing both equations to the slope-intercept form, $y = mx + b$ [Eq. (6)].

8.14 Perpendicular Lines

In Fig. 8.17 let AC and CD be two perpendicular lines, where $C(b,c)$ is their common point and $A(a,0)$ and $D(d,0)$ are their x intercepts. Let $B(b,0)$ be the foot of the perpendicular CB drawn from C to AD.

Since the triangles ABC and BCD in Fig. 8.17 are similar (why?), we may write

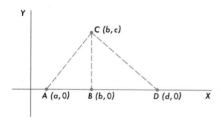

FIGURE 8.17

$$\frac{\overline{CB}}{\overline{BD}} = \frac{\overline{AB}}{\overline{CB}}$$

or

$$\overline{CB^2} = \overline{AB} \cdot \overline{BD}$$

Expressed in terms of the coordinates, this relationship becomes

$$c^2 = (b - a)(d - b) \tag{10}$$

The slope m_1 of AC is $(c - 0)/(b - a)$, or

$$m_1 = \frac{c}{b - a}$$

The slope m_2 of CD is $(c - 0)/(b - d)$, or

$$m_2 = \frac{c}{b - d}$$

Multiplying the values of the two slopes together, we obtain

$$m_1 m_2 = \frac{c^2}{(b - a)(b - d)} = -\frac{c^2}{(b - a)(d - b)}$$

Since from Eq. (10)

$$c^2 = (b - a)(d - b)$$

we may write

$$m_1 m_2 = -\frac{c^2}{c^2} = -1$$

or

$$m_1 = -\frac{1}{m_2} \tag{11}$$

Hence the slopes of perpendicular lines are negative reciprocals.

8.15 The Midpoint Formula

In Fig. 8.18 the point Q is midway between points P and R and lies on the line segment $\overline{PR}$. Thus

$$\overline{PQ} = \overline{QR}$$

Since the triangles PQa and QRb are congruent (Fact 12, Chap. 7), we may write

$$x_2 - x_1 = x_3 - x_2$$

Then

$$2x_2 = x_3 + x_1$$

or

$$\text{Abscissa of midpoint} = x_2 = (\tfrac{1}{2})(x_3 + x_1) \tag{12}$$

The student should verify the fact that

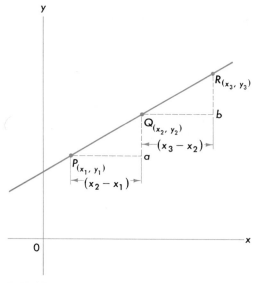

FIGURE 8.18

Ordinate of midpoint $= y_2 = (\tfrac{1}{2})(y_3 + y_1)$ **(13)**

It is intuitively evident that x_2 must be the arithmetic average of x_1 and x_3 and that y_2 must be the average of y_1 and y_3.

8.16 Distance between Two Points

Given the points $P_1(x_1, y_1)$ and $P_2(x_2, y_2)$ in Fig. 8.19, locate the point $Q(x_1, y_2)$ and draw lines $\overline{P_1 P_2}$, $\overline{P_1 Q}$, and $\overline{P_2 Q}$. Then in right triangle $P_1 P_2 Q$,

$$\overline{P_1 P_2}^2 = \overline{P_1 Q}^2 + \overline{P_2 Q}^2$$

In terms of the coordinates this becomes

$$\overline{P_1 P_2}^2 = (y_1 - y_2)^2 + (x_1 - x_2)^2$$

The distance between P_1 and P_2 is then

$$\overline{P_1 P_2} = \sqrt{(y_1 - y_2)^2 + (x_1 - x_2)^2}$$ **(14)**

This rule is valid regardless of the quadrants in which P_1 and P_2 are located, provided only that due regard is paid to signs when evaluating $y_1 - y_2$ and $x_1 - x_2$.
In short, we find the absolute values of $P_1 Q$ and $P_2 Q$ and determine the hypotenuse $P_2 P_1$ of the triangle in the usual way.
Obviously, in cases where $y_1 = y_2$ or $x_1 = x_2$, we have a horizontal line or a vertical line, respectively, and the answer follows by inspection.

EXERCISE 9

1. Given the equation $3x - 4y = 24$, corresponding to the general equation $Ax + By = C$, modify only one quantity at a time (A, B, or C) to make the line

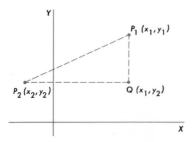

FIGURE 8.19

(a) pass through the origin; (b) vertical; (c) horizontal; (d) parallel to $6x - y = 11$; and (e) perpendicular to $8x - 12y = 13$.

2. Given the equation $5x + 8y = 20$, proceed as in Prob. 1 to make the line (a) pass through the origin; (b) vertical; (c) horizontal; (d) parallel to $15x - 10y = 23$; and (e) perpendicular to $4x - 15y = 17$.

Find the distance between P_1 and P_2 in Probs. 3 to 10.

3. $P_1(2,5)$; $P_2(2,11)$ 4. $P_1(3,-4)$; $P_2(3,3)$

5. $P_1(-4,7)$; $P_2(5,7)$ 6. $P_1(6,1)$; $P_2(0,9)$

7. $P_1(1,5)$; $P_2(4,9)$ 8. $P_1(-2,6)$; $P_2(3,9)$

9. $P_1(8,-4)$; $P_2(-1,-6)$ 10. $P_1(7,-8)$; $P_2(3,-5)$

11. Find the coordinates of the point midway between the given points.

(a) $(2,7)$ and $(8,3)$; (b) $(-6,5)$ and $(9,-5)$

12. Without plotting, determine what kind of figure is enclosed by the lines $x + y = -3$, $x - y = 3$, and $y = 3$.

13. What points on the line $2x + 3y = 6$ are equidistant from the axes?

14. Write the equation for the line passing through the origin and perpendicular to the line $y = \frac{3}{2}x - 4$.

15. Write the equation for the line parallel to the line $2x + 3y = 9$ and having its x intercept $= 7$.

16. Find the distance between the lines $y = 2x + 3$ and $y = 2x + 8$.

17. Given points $A(2,7)$ and $B(14,3)$, what is the equation for the perpendicular bisector of the line segment AB?

18. Find the distance from the point $(6,7)$ to the line $3x + 4y = 30$.

19. A tangent is drawn to the circle $x^2 + y^2 = 40$ at point $(6,2)$ (Fig. 8.20). The slope of the tangent is -3. (Why?) Show that the equation of the tangent is

$$y = -3x + 20$$

      Using the properties of symmetry, rather than developing independent solutions, determine the equations of lines tangent at the points indicated.

20. Locate the point on the Y axis which is equidistant from the point $(-4,3)$ and $(8,11)$.

8.17 Area of a Triangle

The area of the triangle PQR (Fig. 8.21) is evidently equal to the area of the triangle PQM plus the area of the trapezoid $MQRN$ minus the area of the triangle PRN.

      In terms of the coordinates we have

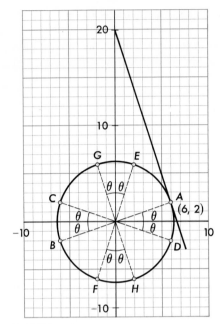

FIGURE 8.20

Area of $PQM = \frac{1}{2}(x_3 - x_1)(y_3 - y_1)$

Area of $MQRN = \frac{1}{2}(x_2 - x_3)[(y_2 - y_1) + (y_3 - y_1)]$

Area of $PRN = \frac{1}{2}(x_2 - x_1)(y_2 - y_1)$

(The order of points must be counterclockwise.) Therefore

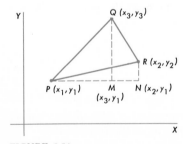

FIGURE 8.21

Area of PRQ = *PQM* + *MQRN* − *PRN*

$$= \tfrac{1}{2}(x_3 - x_1)(y_3 - y_1) + \tfrac{1}{2}(x_2 - x_3)[(y_2 - y_1) + (y_3 - y_1)]$$
$$\qquad\qquad\qquad\qquad\qquad - \tfrac{1}{2}(x_2 - x_1)(y_2 - y_1)$$
$$= \tfrac{1}{2}(x_3y_3 - x_1y_3 - x_3y_1 + x_1y_1 + x_2y_2 - x_3y_2 - 2x_2y_1$$
$$\qquad\qquad + 2x_3y_1 + x_2y_3 - x_3y_3 - x_2y_2 + x_1y_2 + x_2y_1 - x_1y_1)$$
$$= \tfrac{1}{2}(-x_1y_3 - x_3y_2 - x_2y_1 + x_3y_1 + x_2y_3 + x_1y_2)$$
$$= \tfrac{1}{2}[x_1(y_2 - y_3) + x_2(y_3 - y_1) + x_3(y_1 - y_2)] \qquad\qquad \textbf{(15)}$$

If any of the vertices lie outside the first quadrant, proper attention must be paid to signs. (For an application of this principle to surveying see Sec. 16.7.)

Example 15. Find the area of the triangle in Fig. 8.22.
From the figure we write

$$x_1 = -3 \qquad y_1 = -5 \qquad x_2 = 17 \qquad y_2 = 2 \qquad x_3 = 3 \qquad y_3 = 8$$

Substituting in Eq. (15),

$$\text{Area of } \Delta PQR = \tfrac{1}{2}\{(-3)(2 - 8) + 17[8 - (-5)] + 3(-5 - 2)\}$$
$$= \tfrac{1}{2}(18 + 221 - 21)$$
$$= 109$$

A simple method of computing the area of the triangle in Example 15 without memorizing a formula consists in enclosing the triangle in a rectangle and subtracting from the area of the rectangle the combined area of the three smaller right triangles that are formed. From Fig. 8.22 the dimensions may be shown as in Fig. 8.23.

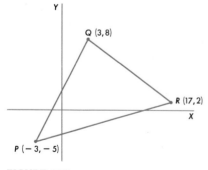

FIGURE 8.22

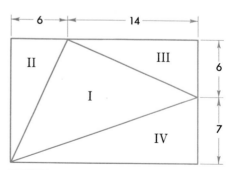

FIGURE 8.23

Area of triangle I = area of rectangle − (area of triangles II + III + IV)

$$= (13)(20) - \tfrac{1}{2}[(6)(13) + (14)(6) + (7)(20)]$$
$$= 260 - \tfrac{1}{2}(302)$$
$$= 109$$

The area of a polygon of n sides may be found in like manner by first dividing by diagonals into $n - 2$ triangles.

EXERCISE 10

Find the area of the triangles represented by the following vertices:

1. (0,0), (0,8), (12,0)   2. (−2,−5), (−4,4), (7,1)
3. (−5,−3), (2,3), (9,−7)   4. (−5,6), (3,8), (4,−4)
5. (−5,0), (0,7), (8,−4)   6. (1,4), (10,5), (3,−4)
7. Prove that the triangle in Prob. 5 is isosceles.

Find the areas of the polygons formed by joining the following points in the order given:

8. (−2,2), (2,6), (8,3), (5,−3)
9. (0,−4), (3,4), (12,0), (10,−3)
10. (−3,−1), (0,8), (12,4), (9,−5)
11. (−2,−4), (−3,3), (4,6), (7,2), (4,−3)
12. Prove that the figure in Prob. 10 is a rectangle.

simultaneous linear equations

In this chapter we shall develop methods for the solution of simultaneous linear equations. These principles will be applied to the solution of "word" problems. This entails a translation from the English language to the language of algebra, and is often quite frustrating to the student.

An effective aid in this process is the use of freehand, dimensioned graphs in the setting up of algebraic equations.

9.1 Graphical Solution of a System of Two Equations

If in a system of two or more equations, all the equations have a common solution, they are called *simultaneous equations*. A solution of two simultaneous equations in two unknowns, x and y, is a pair of corresponding values of x and y which simultaneously satisfies both equations.

Example 1. Solve graphically

$$x + y = 1 \tag{1}$$
$$x - 2y = 7 \tag{2}$$

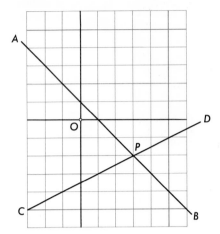

FIGURE 9.1

In Fig. 9.1, *AB* is the graph of Eq. (1) and *CD* is the graph of Eq. (2). All points of *AB* have coordinates satisfying Eq. (1), and all points on *CD* have coordinates satisfying Eq. (2); therefore the intersection point *P* has coordinates satisfying the two equations *simultaneously*. Since two straight lines can intersect in but one point, *P* is the only point having the property of a common solution. In this example the coordinates of *P* are $(3, -2)$; therefore $(x = 3, y = -2)$ is the only solution of this system of equations.

If the graphs of the equations are parallel lines, there is no solution and the equations are termed *inconsistent*.

Example 2. Given the system of equations

$$3x - 5y = 6 \tag{3}$$
$$6x - 10y = 10 \tag{4}$$

The graphs of these two equations are parallel lines (Fig. 9.2), each having a slope

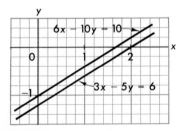

FIGURE 9.2

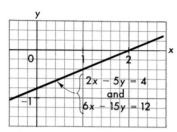

FIGURE 9.3

of $\frac{2}{5}$; since the lines cannot intersect, the equations have no common solution. This conclusion might have been reached without plotting since, on dividing Eq. (4) by 2, we have $3x - 5y = 5$. Since the expression $3x - 5y$ cannot, at the same time, be equal to 6 and to 5, there are no values of x and y satisfying both equations simultaneously.

If the graphs of the two equations are the same line, any solution of one equation is also a solution of the other, and the system has an unlimited number of solutions. Such equations are known as *dependent*, or equivalent, equations. (See Sec. 5.2.)

Example 3. Given the system

$$2x - 5y = 4 \tag{5}$$
$$6x - 15y = 12 \tag{6}$$

Solve for x and y (see Fig. 9.3).

Since Eq. (6) is reducible to Eq. (5) by division by 3, the equations are not independent but are equivalent.

It will be appreciated that graphical solutions are only approximate; when we have to estimate fractional scale divisions, the approximation is rougher.

EXERCISE 1

In the following problems determine, by inspection if possible, whether each pair of equations is inconsistent or dependent or possesses one common solution. In the last case solve graphically, estimating fractional answers to the nearest 0.1. It is suggested that the x and y intercepts and one or two other well-spaced points be used in the plotting.

1. $x + y = 8$
   $x - y = 2$

2. $x + y = -4$
   $x - y = 8$

3. $2x - y = 3$
   $2x - 3y = 11$

4. $3x - y = 4$
   $2y - 6x = -8$

5. $2x + 5y = 8$
   $3x - 2y = -7$

6. $4x - 6y = 8$
   $6x - 9y = 14$

7. $2x = 7y$
   $6x - 5y = 16$

8. $3x + 2y = 12$
   $3y - 2x = 4$

9. $4x - 3y = 10$
   $6y - 8x = -20$

10. $5x - 8y = 0$ 11. $2x - y = 5$ 12. $8x - 12y = 10$
     $7y - 4x = 3$ $x + y = 2$ $18y - 12x = 15$
13. $12x + 5y = 0$ 14. $2x - 7y + 9 = 0$
     $8x - 10y = 8$ $5x + 3y - 6 = 0$

9.2 Algebraic Methods of Solution

Algebraic methods of solving systems of simultaneous equations depend on eliminating one of the unknowns. There are four common methods of solution: addition or subtraction, substitution, comparison, and determinants. We shall confine our discussion to the first three of these.

9.3 Elimination by Addition or Subtraction

Example 4. Solve for x and y:

$$4x - 7y = 29 \tag{7}$$
$$6x + 5y = -3 \tag{8}$$

Multiplying Eq. (7) by 3,

$$12x - 21y = 87 \tag{9}$$

Multiplying Eq. (8) by 2,

$$12x + 10y = -6 \tag{10}$$

Subtracting Eq. (10) from Eq. (9),

$$-31y = 93 \tag{11}$$

Dividing Eq. (11) by -31,

$$y = -3 \tag{12}$$

Substituting -3 for y in Eq. (7),

$$4x + 21 = 29$$

or

$$x = 2 \tag{13}$$

Check by substituting $x = 2$ and $y = -3$ in Eq. (8):

$12 - 15 = -3$ (check)

The above procedure may be summarized as follows:

To solve a system of linear equations by elimination by addition or subtraction:

1. *Multiply both members of both equations, if necessary, by nonzero numbers, which will cause the coefficients of one of the unknowns to have the same absolute value in both equations.*
2. *To eliminate this unknown, the two equations obtained in step 1 are added if the matched coefficients have opposite signs. They are subtracted if they have like signs.*
3. *Solve the equation formed in step 2 for the unknown contained in it. Substitute the result in one of the original equations to obtain the other unknown.*
4. *Check by substituting the values of the unknowns in the original equation not used in step 3.*

Note: If both members of an equation contain a common factor, it may be advisable to divide through by that factor.

Clear of any fractions before eliminating an unknown (see exception, Sec. 9.7.)

It will usually be preferable to eliminate the unknown whose coefficients have the smaller LCM.

Example 5. Referring to Fig. 9.2, we see the graphical appearance of two inconsistent equations.

If we were to attempt an algebraic solution by the method of Example 4, we would arrive at an absurd result such as $0 = 2$. This would be the algebraic indication of inconsistency.

Example 6. Figure 9.3 shows the graphical appearance of two equivalent, or dependent, equations. An attempt to apply the algebraic method of Example 4 would lead to the unrewarding result $0 = 0$. This, then, would indicate that the two equations are equivalent.

9.4 Elimination by Substitution

Example 7. Solve for x and y:

$$7x - 3y = 10 \tag{14}$$
$$5x - 2y = 8 \tag{15}$$

Solve Eq. (15) for y:

$$5x - 8 = 2y \tag{16}$$

or

$$y = \frac{5x - 8}{2} \tag{17}$$

Substitute $(5x - 8)/2$ for y in Eq. (14):

$$7x - 3\left(\frac{5x - 8}{2}\right) = 10 \tag{18}$$

Solve Eq. (18) for x, first eliminating fractions:

$$14x - 3(5x - 8) = 20$$
$$14x - 15x + 24 = 20$$
$$x = 4$$

Substitute $x = 4$ in Eq. (17):

$$y = \frac{(5)(4) - 8}{2}$$

or

$$y = 6$$

Hence the solution of the system is $x = 4$, $y = 6$.
Check by substituting $x = 4$ and $y = 6$ in Eq. (14):

$$(7)(4) - (3)(6) = 10 \qquad 28 - 18 = 10$$

To solve a system of two linear equations by elimination by substitution:

1. *Solve one equation for one unknown (preferably the unknown having the simplest coefficient) in terms of the other.*
2. *The expression obtained in step 1 is to be substituted for the equivalent unknown in the other equation.*
3. *Solve the equation obtained in step 2 for the second unknown.*

4. *Substitute the value of the second unknown in the expression obtained in step 1, and solve for the first unknown.*
5. *Check by substituting values of both unknowns in the original equation not used in step 1.*

9.5 Elimination by Comparison

Example 8. Solve the following for x and y:

$$4x + 7y = 3 \tag{19}$$
$$6x - 5y = 20 \tag{20}$$

Solving Eq. (19) for x,

$$x = \frac{3 - 7y}{4} \tag{21}$$

Solving Eq. (20) for x,

$$x = \frac{5y + 20}{6} \tag{22}$$

From Eqs. (21) and (22) we may write

$$\frac{3 - 7y}{4} = \frac{5y + 20}{6}$$
$$3(3 - 7y) = 2(5y + 20)$$
$$9 - 21y = 10y + 40$$
$$-31y = 31$$
$$y = -1$$

Complete the solution as in elimination by addition or subtraction, obtaining $x = 2\frac{1}{2}$.

Note. *The method of addition or subtraction is the most generally useful, especially when the coefficients of one unknown have the same absolute value in both equations.*
The method of substitution is most convenient to apply when the unknown which is to be expressed in terms of the other has a coefficient of 1.

There is seldom a case of solution by comparison which could not have been done at least as easily by either of the first two methods.

EXERCISE 2

Solve the following systems of equations by the most convenient method:

1. $3x - 2y = 4$
   $x = y$

4. $7x - 5y = 161$
   $x = 4y$

7. $8x - 5y = 58$
   $x + y = 4$

10. $4x + 5y = 30$
    $6x + y = 19$

13. $11y - 4z = 41$
    $7y + 2z = 42$

16. $12x - 11z = -5$
    $9x + 4z = 33$

19. $11x - 3y + 34 = 0$
    $8x + 10y - 24 = 0$

22. $4(x + z) = 22$
    $6(x - z) = 15$

25. $\dfrac{x}{4} + 4y = 28$

$\quad 7x - \dfrac{y}{5} = 36$

28. $\dfrac{x - 2}{10} + 3y = 8$

$\quad \dfrac{x + 2}{6} - \dfrac{y - 7}{9} = 2$

2. $9x - 5y = 42$
   $x + y = 0$

5. $9x + 8y = 77$
   $x - y = 1$

8. $6x - 10y = 4$
   $x + y = 14$

11. $x + y = 13$
    $x - y = 5$

14. $5w + 6y = 17$
    $3w - 2y = 27$

17. $7x - 3y = 41$
    $4x + 5y = 10$

20. $13y + 5z - 33 = 0$
    $7y - 9z - 47 = 0$

23. $0.5x + 0.4w = 0.26$
    $0.7x + 0.3w = 0.26$

26. $\dfrac{x}{3} - 4y = -58$

$\quad 5x - \dfrac{y}{6} = 27\frac{1}{2}$

3. $8y - 5x = 18$
   $x - y = 0$

6. $3x - 2y = 10$
   $x - y = 1$

9. $2x + y = 5$
   $x + 2y = 19$

12. $5x + 2z = 61$
    $3x - 2z = 43$

15. $9x - 13w = -3$
    $6x - 7w = 3$

18. $9w + 8z = 15$
    $5w + 7z = 16$

21. $3(x + y) = 33$
    $5(x - y) = 25$

24. $0.3x + 0.4y = 14.5$
    $0.4x + 0.3y = 13.5$

27. $\dfrac{w + 3}{2} + 5z = 9$

$\quad \dfrac{z + 9}{10} - \dfrac{w - 2}{3} = 0$

Solve Probs. 29 and 30 by slide rule to whatever degree of accuracy is possible.

29. $0.103x + 0.950y = 10.47$
    $1.068x - 2.74y = 19.63$

30. $5.95x + 5.17y = 3.39$
    $4.62x - 12.3y = 5.56$

Note. In the slide-rule solution of Prob. 30, instead of finding the LCM of 5.95 and 4.62, we would usually prefer to eliminate an unknown, say x, by dividing the first equation through by 5.95 and the second by 4.62.

9.6 Literal Linear Equations

A system involving letters other than the unknowns is usually best solved by finding each unknown independently by elimination by addition or subtraction.

Example 9. Solve for x and y:

$$mx + ny = 2(m^2 - n^2) \tag{23}$$
$$x - y = m + n \tag{24}$$

Multiply Eq. (24) by n:

$$nx - ny = mn + n^2 \tag{25}$$

Add Eqs. (23) and (25):

$$
\begin{array}{l}
mx + ny = 2m^2 \qquad\;\; - 2n^2 \\
nx - ny = \qquad\;\; mn + \;\; n^2 \\
\hline
mx + nx = 2m^2 + mn - \;\; n^2
\end{array}
\tag{26}
$$

Factor Eq. (26):

$$x(m + n) = (m + n)(2m - n) \tag{27}$$

Divide Eq. (27) through by $m + n$:

$$x = 2m - n$$

Multiply Eq. (24) by m:

$$mx - my = m^2 + mn \tag{28}$$

Subtract Eq. (28) from Eq. (23):

$$
\begin{array}{l}
mx + \;\; ny = 2m^2 \qquad\;\; - 2n^2 \\
mx - \;\; my = \;\; m^2 + mn \\
\hline
ny + my = \;\; m^2 - mn - 2n^2
\end{array}
\tag{29}
$$

Factor Eq. (29):

$$y(n + m) = (m + n)(m - 2n) \tag{30}$$

Divide Eq. (30) through by $m + n$:

$$y = m - 2n$$

But note that y is more easily found here by substitution. Rewriting Eq. (24) as

$x - m - n = y$ and substituting $x = 2m - n$, we find at once that $2m - n - m - n = m - 2n = y$.

If answers are at all involved, checking may best be done by numerical substitution, as in Example 5 of Chap. 5.

9.7 Equations Linear in the Reciprocals of the Unknowns

Equations of the type $a/x + b/y = c$ are usually best solved without removing the unknowns from the denominators.

Example 10. Solve the system

$$\frac{9}{x} - \frac{15}{y} = 1 \tag{31}$$

$$\frac{14}{x} + \frac{20}{y} = 3 \tag{32}$$

Multiply Eq. (31) by 4:

$$\frac{36}{x} - \frac{60}{y} = 4 \tag{33}$$

Multiply Eq. (32) by 3:

$$\frac{42}{x} + \frac{60}{y} = 9 \tag{34}$$

Add Eqs. (33) and (34):

$$\frac{78}{x} = 13$$

$$78 = 13x$$

or

$$x = 6$$

Substitute $x = 6$ in Eq. (31):

$$\frac{9}{6} - \frac{15}{y} = 1$$

$$-\frac{15}{y} = -\frac{1}{2}$$

$$y = 30$$

The solution of the system is $x = 6$, $y = 30$.

Example 11. Solve the system

$$\frac{3}{8x} - \frac{1}{2y} = 2 \tag{35}$$

$$\frac{5}{6x} - \frac{5}{3y} = -8\frac{1}{3} \tag{36}$$

Multiply Eq. (35) by 10:

$$\frac{15}{4x} - \frac{5}{y} = 20 \tag{37}$$

Multiply Eq. (36) by 3:

$$\frac{5}{2x} - \frac{5}{y} = -25 \tag{38}$$

Subtract Eq. (38) from Eq. (37):

$$\frac{15}{4x} - \frac{5}{2x} = 45 \tag{39}$$

Multiply Eq. (39) by 4x:

$$15 - 10 = 180x$$
$$x = \tfrac{1}{36}$$

Substitute $x = \tfrac{1}{36}$ in Eq. (35):

$$\frac{3}{8(1/36)} - \frac{1}{2y} = 2$$

$$\frac{27}{2} - \frac{1}{2y} = 2$$

$$-\frac{1}{2y} = -\frac{23}{2}$$

$$y = \frac{1}{23}$$

Hence the solution of the system is $x = \tfrac{1}{36}$, $y = \tfrac{1}{23}$.

EXERCISE 3

Solve the following systems for x, y, z, or w, as the case may be.

1. $3x + y = 4c$
   $x - y = 4d$

2. $x + y = 5a$
   $2x - 3y = 5b$

3. $2x - 3y = a + 4b$
   $x + 2y = 4a - 5b$

4. $5x - 4y = 6c - 15d$
   $3x + y = 7c - 9d$

5. $ax + 2by = 4$
   $2ax - 6by = 3$

6. $12ax - 5by = -16$
   $6ax + 7by = 30$

7. $2ax + by = 10c$
   $ax - 3by = -9c$

8. $3ax - 4by = 18ab$
   $2ax + by = ab$

9. $bx + cy = 2bc$
   $cx + by = b^2 + c^2$

10. $ax - by = 2a^2 - 6ab + 2b^2$
    $bx + ay = 3a^2 - 3b^2$

11. $2nx + 2my = m^2 + n^2$

    $4nx - mx + my = mn + 2n^2$

12. $ax + 2cy = 5ac$

    $\dfrac{2x}{c} - \dfrac{y}{a} = \dfrac{-5}{ac}$

13. $\dfrac{1}{x} + \dfrac{1}{y} = 6$

    $\dfrac{1}{x} - \dfrac{1}{y} = 4$

14. $\dfrac{3}{x} - \dfrac{2}{y} = 10$

    $\dfrac{4}{x} + \dfrac{1}{y} = 28$

15. $\dfrac{10}{x} + \dfrac{6}{y} = 7$

    $\dfrac{14}{x} - \dfrac{9}{y} = 4$

16. $\dfrac{15}{x} + \dfrac{14}{w} = 5$

    $\dfrac{25}{x} - \dfrac{6}{w} = 1$

17. $\dfrac{1}{2x} + \dfrac{1}{3y} = 12$

    $\dfrac{1}{4x} + \dfrac{1}{9y} = 5$

18. $\dfrac{5}{6x} - \dfrac{3}{4z} = 6$

    $\dfrac{7}{9x} + \dfrac{1}{6z} = 7\tfrac{1}{3}$

9.8 Linear Equations in More Than Two Unknowns

Like a system in two unknowns, a system of linear equations in more than two unknowns may have a solution or may be dependent or inconsistent. Only the first case will be considered here. For a system to have a unique solution, it must have as many independent equations as unknowns.

Example 12. Solve for x, y, and z:

$$6x - 5y - 2z = 2 \tag{40}$$
$$4x + y + 3z = 10 \tag{41}$$
$$5x + 3y + 7z = 13 \tag{42}$$

Multiply Eq. (41) by 5:

$$20x + 5y + 15z = 50$$

Add Eq. (40): $\quad \underline{6x - 5y - 2z = 2}$

$$26x + 13z = 52 \tag{43}$$

Multiply Eq. (41) by 3:

$$12x + 3y + 9z = 30$$

Subtract Eq. (42): $\dfrac{5x + 3y + 7z = 13}{}$

$$7x \qquad + 2z = 17 \tag{44}$$

Note that Eq. (43) may be divided throughout by 13. Therefore we will obtain smaller coefficients in Eq. (45) by multiplying by $\frac{2}{13}$ rather than by 2.

$$4x + 2z = \quad 8 \tag{45}$$

Subtract Eq. (44): $\dfrac{7x + 2z = \quad 17}{}$

$$-3x \qquad = -9$$
$$x = 3$$

Substitute $x = 3$ in Eq. (44):

$$(7)(3) + 2z = 17$$
$$z = -2$$

Substitute $x = 3$ and $z = -2$ in Eq. (41):

$$(4)(3) + y + 3(-2) = 10$$
$$y = 4$$

Check the solution by substituting the values of x, y, and z in Eq. (40) or (42).

The procedure for solving a system of three linear equations in three unknowns may be summarized as follows:

1. *Select the unknown most easily eliminated and eliminate this unknown from one pair of equations.*
2. *Eliminate the same unknown from one of the equations just used and the third equation.*
3. *Solve the two equations so obtained as in Sec. 9.3.*
4. *Substitute the two values found in step 3 in the simplest of the original equations and solve for the third unknown.*
5. *Check by substituting the values of the unknowns in one of the two original equations not used in step 4.*

Simultaneous linear equations in more than three unknowns may be solved by an extension of the above method.

EXERCISE 4

Solve the following systems of equations:

1. $x + y = 7$
   $x + z = 8$
   $y + z = 9$

2. $x + y = 4$
   $x + z = 7$
   $y + z = 1$

3. $x + y + z = 8$
   $x - y + z = 2$
   $x + y - z = 12$

4. $x - y - z = 1$
   $x + y - z = 6$
   $x - y + z = -1$

Suggestion for Probs. 5 and 6: Add the equations and divide the sum through by the common factor.

5. $2x + y + z = 7$
   $x + 2y + z = 15$
   $x + y + 2z = 18$

6. $x + y + 3z = 9$
   $x + 3y + z = \frac{7}{3}$
   $3x + y + z = 7$

7. $2x + 3y - 4z = 11$
   $3x - 5y + z = -3$
   $6x + 2y - 7z = 18$

8. $7x + 4y - 3z = -15$
   $5x - 6y + 2z = 41$
   $x - 10y - 9z = -8$

9. $6x - 8y + 3z = -5$
   $9x + 20y - 4z = 25$
   $15x + 12y + 7z = 12$

10. $10x + 9y + 7z = 32$
    $11x - 6y - 8z = 20$
    $-9x + 12y + 10z = -22$

9.9 Matrices

Rectangular arrays of numbers such as

$$\begin{bmatrix} a_1 & b_1 & c_1 \\ a_2 & b_2 & c_2 \\ a_3 & b_3 & c_3 \end{bmatrix} \quad \begin{bmatrix} a_1 & b_1 & c_1 \\ a_2 & b_2 & c_2 \end{bmatrix} \quad \begin{bmatrix} a_1 \\ a_2 \\ a_3 \end{bmatrix}$$

are called *matrices*. Any one of them is called a *matrix*. The numbers a_1, a_2, a_3, b_1, b_2, b_3, and so on, are called *elements* of the matrix. In the present text, all the elements will be real numbers.

Horizontal lines of numbers such as a_1, b_1, c_1, etc., are called *rows*. Vertical lines of numbers such as a_1, a_2, a_3, etc., are called *columns*.

In this text we shall be interested only in *square* matrices. These are matrices having the same number of rows as columns.

A square matrix having two rows and two columns is said to be a 2 by 2 (sometimes written 2×2), or a *second-order*, matrix. A square matrix having six rows and six columns is said to be a 6 by 6, or a sixth-order, matrix, and so on.

9.10 Determinants

With each square matrix we may associate a unique number, to be defined presently, called the *determinant* of the matrix.

For example, the determinant of the square matrix

$$\begin{bmatrix} a_1 & b_1 \\ a_2 & b_2 \end{bmatrix} \quad \text{is symbolized by} \quad \begin{vmatrix} a_1 & b_1 \\ a_2 & b_2 \end{vmatrix}$$

The value of this number is defined to be

$$\begin{vmatrix} a_1 & b_1 \\ a_2 & b_2 \end{vmatrix} = a_1 b_2 - a_2 b_1$$

As a convenience we do often refer to the symbol

$$\begin{vmatrix} a_1 & b_1 \\ a_2 & b_2 \end{vmatrix}$$

for example, as a "determinant" in the sense that this symbol represents the number which is the determinant of the matrix

$$\begin{bmatrix} a_1 & b_1 \\ a_2 & b_2 \end{bmatrix}$$

Example 13. Find the value of $\begin{vmatrix} 4 & 5 \\ 3 & 7 \end{vmatrix}$.

$$\begin{vmatrix} 4 & 5 \\ 3 & 7 \end{vmatrix} = (4)(7) - (3)(5) = 28 - 15 = 13$$

Example 14. Evaluate $\begin{vmatrix} 6 & 3 \\ -2 & 5 \end{vmatrix}$.

$$\begin{vmatrix} 6 & 3 \\ -2 & 5 \end{vmatrix} = (6)(5) - (-2)(3) = 30 + 6 = 36$$

Example 15. Evaluate $\begin{vmatrix} 7 & x \\ 5 & -y \end{vmatrix}$.

$$\begin{vmatrix} 7 & x \\ 5 & -y \end{vmatrix} = (7)(-y) - (5)(x) = -7y - 5x$$

EXERCISE 5

Evaluate the following determinants:

1. $\begin{vmatrix} 3 & 5 \\ 5 & 8 \end{vmatrix}$ 　　2. $\begin{vmatrix} 3 & a \\ 4 & b \end{vmatrix}$ 　　3. $\begin{vmatrix} 4 & -6 \\ 2 & 5 \end{vmatrix}$

4. $\begin{vmatrix} 7 & -x \\ 6 & y \end{vmatrix}$ 　　5. $\begin{vmatrix} 9a & 5b \\ 4b & a \end{vmatrix}$ 　　6. $\begin{vmatrix} 6x & -7x \\ 5y & -6y \end{vmatrix}$

9.11 Solution of Simultaneous Linear Equations by Determinants

A system of linear equations is said to be in *standard form* if, as in the equations below [Eqs. (46)], the terms containing the unknowns are in the left-hand members and in the same order in all equations. The right-hand members contain only constants.

$$a_1x + b_1y = k_1$$
$$a_2x + b_2y = k_2 \tag{46}$$

Given such a system of equations in x and y, the solution can be shown to be

$$x = \frac{k_1b_2 - k_2b_1}{a_1b_2 - a_2b_1} \qquad y = \frac{a_1k_2 - a_2k_1}{a_1b_2 - a_2b_1} \tag{47}$$

provided $a_1b_2 - a_2b_1$ is not equal to zero.

In determinant notation we can write

$$k_1b_2 - k_2b_1 = \begin{vmatrix} k_1 & b_1 \\ k_2 & b_2 \end{vmatrix} \qquad a_1k_2 - a_2k_1 = \begin{vmatrix} a_1 & k_1 \\ a_2 & k_2 \end{vmatrix}$$

and

$$a_1b_2 - a_2b_1 = \begin{vmatrix} a_1 & b_1 \\ a_2 & b_2 \end{vmatrix}$$

We can therefore write the solution of Eq. (47) in the form

$$x = \frac{\begin{vmatrix} k_1 & b_1 \\ k_2 & b_2 \end{vmatrix}}{\begin{vmatrix} a_1 & b_1 \\ a_2 & b_2 \end{vmatrix}} \qquad y = \frac{\begin{vmatrix} a_1 & k_1 \\ a_2 & k_2 \end{vmatrix}}{\begin{vmatrix} a_1 & b_1 \\ a_2 & b_2 \end{vmatrix}} \qquad \text{if} \qquad \begin{vmatrix} a_1 & b_1 \\ a_2 & b_2 \end{vmatrix} \neq 0 \tag{48}$$

The solution [Eqs. (48)] is therefore a general formula applicable to any system of two simultaneous linear equations in two unknowns when arranged in standard form. It will be noted that both denominators of Eqs. (48) are the same. This determinant

$$\begin{vmatrix} a_1 & b_1 \\ a_2 & b_2 \end{vmatrix}$$

is called the *determinant of the system*. Its elements are the coefficients of the unknown quantities arranged in the same relative position as in the system of equations when the equations are arranged in standard form.

The numerator of the fraction which expresses the value of x differs from the denominator in that the constants k_1 and k_2 replace a_1 and a_2, the coefficients of x; and the numerator of the fraction which expresses the value of y differs from the denominator in that the constants k_1 and k_2 replace b_1 and b_2, the coefficients of y. This procedure may be summarized in *Cramer's rule*:

Arrange the given equations in standard form. In the solution, the value of an unknown is given by a fraction whose denominator is the determinant of the system and whose numerator is the same determinant except that the coefficients of the unknown have been replaced by the constants k_1 *and* k_2.

Example 16. Solve by determinants

$$4x + 7y = -19$$
$$5x - 3y = 35$$

We have

$$a_1 = 4 \qquad b_1 = 7 \qquad k_1 = -19$$
$$a_2 = 5 \qquad b_2 = -3 \qquad k_2 = 35$$

Hence

$$x = \frac{\begin{vmatrix} -19 & 7 \\ 35 & -3 \end{vmatrix}}{\begin{vmatrix} 4 & 7 \\ 5 & -3 \end{vmatrix}} = \frac{57 - 245}{-12 - 35} = \frac{-188}{-47} = 4$$

$$y = \frac{\begin{vmatrix} 4 & -19 \\ 5 & 35 \end{vmatrix}}{\begin{vmatrix} 4 & 7 \\ 5 & -3 \end{vmatrix}} = \frac{140 + 95}{-12 - 35} = \frac{235}{-47} = -5$$

Example 17. Solve by determinants

$$x - y = m - n$$
$$mx - ny = 2m^2 - 2n^2$$

Here

$$a_1 = 1 \qquad b_1 = -1 \qquad k_1 = m - n$$
$$a_2 = m \qquad b_2 = -n \qquad k_2 = 2m^2 - 2n^2$$

Hence

$$x = \frac{\begin{vmatrix} m - n & -1 \\ 2m^2 - 2n^2 & -n \end{vmatrix}}{\begin{vmatrix} 1 & -1 \\ m & -n \end{vmatrix}} = \frac{-mn + n^2 + 2m^2 - 2n^2}{-n + m}$$

$$= \frac{2m^2 - mn - n^2}{m - n} = 2m + n$$

$$y = \frac{\begin{vmatrix} 1 & m - n \\ m & 2m^2 - 2n^2 \end{vmatrix}}{\begin{vmatrix} 1 & -1 \\ m & -n \end{vmatrix}} = \frac{2m^2 - 2n^2 - m^2 + mn}{-n + m}$$

$$= \frac{m^2 + mn - 2n^2}{m - n} = m + 2n$$

If the equations $a_1x + b_1y = k_1$ and $a_2x + b_2y = k_2$ have no unique simultaneous solution, they are either inconsistent or dependent. Graphically, this means that the lines are either parallel or coincident. In either case the slopes are equal, and $-(a_1/b_1) = -(a_2/b_2)$, or $a_1b_2 - a_2b_1 = 0$. If the two lines coincide, their y intercepts are equal, or $k_1/b_1 = k_2/b_2$; that is, $k_1b_2 - k_2b_1 = 0$.

These facts may be summarized as follows:

If $a_1b_2 - a_2b_1 \neq 0$:   *The equations have a unique simultaneous solution.*

If $a_1b_2 - a_2b_1 = 0$
and $k_1b_2 - k_2b_1 \neq 0$:   *The equations are inconsistent and have no solution.*

If $a_1b_2 - a_2b_1 = 0$
and $k_1b_2 - k_2b_1 = 0$:   *The equations are dependent and have an infinite number of solutions.*

EXERCISE 6

In the following systems, use determinants to ascertain whether the equations are inconsistent, dependent, or have a unique common solution. Determine any common solutions.

1. $2x - 3y = 5$
   $4x - 6y = 8$

2. $3x - 7y = 1$
   $6x + 5y = 40$

3. $6x + 9y = 15$
   $8x + 12y = 20$

4. $3x - 4y - 15 = 0$
   $7x + 2y - 52 = 0$

5. $4x + 15y = 7a$
   $10x - 9y = 2a$

6. $\dfrac{2x - y}{3} = 5$

   $\dfrac{11x + 2y}{5} = 6$

7. $ax + by = (a - b)^2$
   $ax - by = a^2 - b^2$

8. $bx + ay = 2ab$
   $ax + by = a^2 + b^2$

9. $17x + 11y = 13$
   $7x - 9y = 73$

9.12 Determinants of the Third Order

A third-order determinant is the determinant of a third-order matrix and may be symbolized as

$$\begin{vmatrix} a_1 & b_1 & c_1 \\ a_2 & b_2 & c_2 \\ a_3 & b_3 & c_3 \end{vmatrix}$$

By definition its value may be written

$$\begin{vmatrix} a_1 & b_1 & c_1 \\ a_2 & b_2 & c_2 \\ a_3 & b_3 & c_3 \end{vmatrix} = a_1 b_2 c_3 + a_2 b_3 c_1 + a_3 b_1 c_2 - a_3 b_2 c_1 - a_2 b_1 c_3 - a_1 b_3 c_2$$

There are various ways of evaluating a determinant of the third order. If the student does not expect to use determinants of higher order than the third, the following is an easily remembered rule:

1. *Repeat the first and second columns of the determinant at the right of the determinant.*
2. *Form the products of the numbers in each full diagonal running downward to the right.*
3. *Form the products of the numbers in each full diagonal running upward to the right.*
4. *Subtract the algebraic sum of the products in step 3 from the algebraic sum of the products in step 2.*

This method is valid for a determinant of the third order, *but not for one of any higher order.*

The arrangement is illustrated as follows:

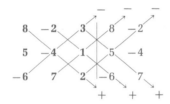

$$a_1 b_2 c_3 + a_2 b_3 c_1 + a_3 b_1 c_2 - a_3 b_2 c_1 - a_2 b_1 c_3 - a_1 b_3 c_2$$

Example 18. Evaluate the third-order determinant

$$\begin{vmatrix} 8 & -2 & 3 \\ 5 & -4 & 1 \\ -6 & 7 & 2 \end{vmatrix}$$

Arranging as described, we write

and obtain

$$(8)(-4)(2) + (-2)(1)(-6) + (3)(5)(7) - (-6)(-4)(3) - (7)(1)(8) - (2)(5)(-2)$$
$$= -64 + 12 + 105 - 72 - 56 + 20$$
$$= -55$$

9.13 Minors and Cofactors

The definitions given in this section are of great importance in preparation for the study of Sec. 9.14.

Figure 9.4 shows a third-order determinant in which the row and column containing a_1 have been deleted.

The second-order determinant

$$\begin{vmatrix} b_2 & c_2 \\ b_3 & c_3 \end{vmatrix}$$

FIGURE 9.4

remains. This second-order determinant is called the *minor* of the element a_1.

In general, if one row and one column are deleted from a determinant of order n (where n is an integer), the remaining array forms a determinant of order $(n - 1)$. This determinant is called the *minor* of the element common to the deleted row and column.

For example, Fig. 9.5 shows a fourth-order determinant for which, of course, $n = 4$. If we delete a column and a row, then a determinant of order

$$4 - 1 = 3$$

remains.

In particular, if we delete the second row and the third column as shown in Fig. 9.5, the third-order determinant

$$\begin{vmatrix} a_1 & b_1 & d_1 \\ a_3 & b_3 & d_3 \\ a_4 & b_4 & d_4 \end{vmatrix}.$$

remains. This determinant is said to be the *minor* of c_2.

For purposes of discussion and explanation, it is convenient to identify by number the row and column in which a given element lies (Fig. 9.6). Here, for example, the element d_2 is located in the second row and the fourth column. Its row number is therefore 2, and its column number is 4.

The element c_4 is located in the fourth row and the third column. Its row number is therefore 4, and its column number is 3.

An element is called an *even* element if the sum of its row number and its column

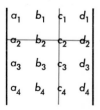

FIGURE 9.5

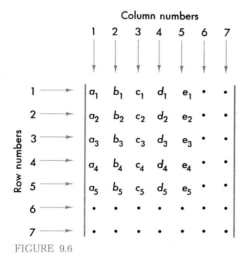

FIGURE 9.6

number is an even number. An element is called an *odd* element if the sum of its row number and its column number is an odd number.

The element d_2 in Fig. 9.5 is therefore an even element since the sum of its row and column number is $2 + 4 = 6$, which is an even number.

The element c_4 in Fig. 9.5 is an odd element since the sum of its row and column number is $4 + 3 = 7$, which is an odd number.

The *cofactor* of an *even* element is defined to be its own minor. The *cofactor* of an *odd* element is defined to be the *negative* of its own minor.

The cofactor of the element d_2 in Fig. 9.5 is therefore

$$\begin{vmatrix} a_1 & b_1 & c_1 \\ a_3 & b_3 & c_3 \\ a_4 & b_4 & c_4 \end{vmatrix}$$

The cofactor of the element c_4 in Fig. 9.5 is

$$-\begin{vmatrix} a_1 & b_1 & d_1 \\ a_2 & b_2 & d_2 \\ a_3 & b_3 & d_3 \end{vmatrix}$$

9.14 Evaluation of Determinants by Cofactors

The diagonal method described in Sec. 9.12 may be used to evaluate third-order determinants but none of higher order. The method of cofactors about to be discussed may be used for evaluating both third- and higher-order determinants.

The evaluation of third- and higher-order determinants may be accomplished by the application of the following theorem, which is stated without proof.

Theorem. *The value of any third- or higher-order determinant is equal to the algebraic sum of the products formed by multiplying each element of a row (or column) by its own cofactor.*

Example 19. Evaluate the third-order determinant

$$\begin{vmatrix} a_1 & b_1 & c_1 \\ a_2 & b_2 & c_2 \\ a_3 & b_3 & c_3 \end{vmatrix}$$

by the method of cofactors.

This evaluation is illustrated by the equation

$$\begin{vmatrix} a_1 & b_1 & c_1 \\ a_2 & b_2 & c_2 \\ a_3 & b_3 & c_3 \end{vmatrix} = a_1 \begin{vmatrix} b_2 & c_2 \\ b_3 & c_3 \end{vmatrix} - a_2 \begin{vmatrix} b_1 & c_1 \\ b_3 & c_3 \end{vmatrix} + a_3 \begin{vmatrix} b_1 & c_1 \\ b_2 & c_2 \end{vmatrix} \tag{49}$$

Here we chose to use the elements of column 1. These elements are a_1, a_2, and a_3. We observe that a_1 and a_3 are even elements. Hence the signs associated with the first and third terms are positive. The element a_2 is an odd element. Therefore the algebraic sign associated with the second term is negative.

The cofactor of a_1 is $\begin{vmatrix} b_2 & c_2 \\ b_3 & c_3 \end{vmatrix}$, the cofactor of a_2 is $-\begin{vmatrix} b_1 & c_1 \\ b_3 & c_3 \end{vmatrix}$, and the cofactor of a_3 is $\begin{vmatrix} b_1 & c_1 \\ b_2 & c_2 \end{vmatrix}$.

The following display may be helpful as a memory aid for determining the signs of the various terms involved in evaluating a given determinant.

In this diagram each $-$ sign occupies the same relative position as an odd element. Each $+$ sign occupies the same relative position as an even element.

$$\begin{vmatrix} + & - & + & - & \cdot & \cdot \\ - & + & - & + & \cdot & \cdot \\ + & - & + & - & \cdot & \cdot \\ - & + & - & + & \cdot & \cdot \\ \cdot & \cdot & \cdot & \cdot & \cdot & \cdot \\ \cdot & \cdot & \cdot & \cdot & \cdot & \cdot \end{vmatrix}$$

The general procedure of expansion by minors is applicable to determinants of fourth or higher order. Thus a fourth-order determinant will have as minors of its elements four determinants of the third order. Each of these third-order determinants will expand into three second-order determinants. In this way a fourth-order determinant expands into 12 second-order determinants, a fifth-order determinant into 60 of the second order, and so on.

If one of the elements of a determinant is zero, the work may be simplified by using minors of the elements in the same row or column (see Example 21).

Example 20. Evaluate the determinant

$$\begin{vmatrix} 8 & -2 & 3 \\ 5 & -4 & 1 \\ -6 & 7 & 2 \end{vmatrix}$$

Expanding by elements of the third column (we might have chosen any column or any row), we obtain

$$\begin{vmatrix} 8 & -2 & 3 \\ 5 & -4 & 1 \\ -6 & 7 & 2 \end{vmatrix} = 3\begin{vmatrix} 5 & -4 \\ -6 & 7 \end{vmatrix} - 1\begin{vmatrix} 8 & -2 \\ -6 & 7 \end{vmatrix} + 2\begin{vmatrix} 8 & -2 \\ 5 & -4 \end{vmatrix}$$

$$= 3[(5)(7) - (-6)(-4)] - 1[(8)(7) - (-6)(-2)] + 2[(8)(-4) - (5)(-2)]$$
$$= 3(35 - 24) - 1(56 - 12) + 2(-32 + 10) = 33 - 44 - 44 = -55$$

Example 21. Evaluate the determinant

$$\begin{vmatrix} 3 & 0 & -2 & 1 \\ 4 & -5 & 2 & -3 \\ -1 & 6 & 0 & -4 \\ 7 & 1 & -6 & 5 \end{vmatrix}$$

Expanding by elements of the first row (since it contains a zero),

$$3\begin{vmatrix} -5 & 2 & -3 \\ 6 & 0 & -4 \\ 1 & -6 & 5 \end{vmatrix} - 0\begin{vmatrix} 4 & 2 & -3 \\ -1 & 0 & -4 \\ 7 & -6 & 5 \end{vmatrix} + (-2)\begin{vmatrix} 4 & -5 & -3 \\ -1 & 6 & -4 \\ 7 & 1 & 5 \end{vmatrix} - 1\begin{vmatrix} 4 & -5 & 2 \\ -1 & 6 & 0 \\ 7 & 1 & -6 \end{vmatrix}$$

Taking elements of the second row of the first determinant, since it contains a zero, the first term equals

$$3\left\{-6\begin{vmatrix} 2 & -3 \\ -6 & 5 \end{vmatrix} + 0\begin{vmatrix} -5 & -3 \\ 1 & 5 \end{vmatrix} - (-4)\begin{vmatrix} -5 & 2 \\ 1 & -6 \end{vmatrix}\right\}$$

$$= 3[-6(10 - 18) + 0 + 4(30 - 2)] = 3(48 + 112) = 480$$

The value of the second term is zero.

Taking elements of the first row, the third term equals

$$-2\left\{4\begin{vmatrix} 6 & -4 \\ 1 & 5 \end{vmatrix} - (-5)\begin{vmatrix} -1 & -4 \\ 7 & 5 \end{vmatrix} + (-3)\begin{vmatrix} -1 & 6 \\ 7 & 1 \end{vmatrix}\right\}$$

$$= -2[4(30+4) + 5(-5+28) - 3(-1-42)]$$
$$= -2(136+115+129) = -760$$

Taking elements of the second row, the fourth term equals

$$-1\left\{-(-1)\begin{vmatrix} -5 & 2 \\ 1 & -6 \end{vmatrix} + 6\begin{vmatrix} 4 & 2 \\ 7 & -6 \end{vmatrix} - 0\begin{vmatrix} 4 & -5 \\ 7 & 1 \end{vmatrix}\right\}$$

$$= -1[30 - 2 + 6(-24-14) - 0] = -1(28-228) = 200$$

Hence the value of the fourth-order determinant is

$$480 - 0 - 760 + 200 = -80$$

9.15 Simplifying Determinants

The following theorem is of considerable importance in reducing the labor of evaluating determinants.

Theorem. *The value of a determinant is not changed if all the elements in any column or row are multiplied by the same number and either added to or subtracted from the corresponding elements in another column or row.*

The proof of this theorem may be found in texts on college algebra. We shall not prove it formally, but we shall illustrate its use.

By reference to Eq. (49), it is evident that if one of the elements a_1, a_2, or a_3 is zero, then the product of it and its associated cofactor will also be zero, and no particular arithmetical work is involved in evaluating the product.

Frequently we can, by the use of the above theorem, take a given determinant and from it write an equal determinant in which one or more elements in a row or column are zero. If we then choose our elements from this row or column, we know without further calculation that the products of these zero elements and their associated cofactors are zero.

By this method let us evaluate the third-order determinant

$$\begin{vmatrix} 8 & -2 & 3 \\ 5 & -4 & 1 \\ -6 & 7 & 2 \end{vmatrix}$$

given in Example 20.

Multiply the third column by 5 and subtract from the first column, obtaining

$$\begin{vmatrix} -7 & -2 & 3 \\ 0 & -4 & 1 \\ -16 & 7 & 2 \end{vmatrix}$$

Multiply the third column by 4 and add to the second column, obtaining

$$\begin{vmatrix} -7 & 10 & 3 \\ 0 & 0 & 1 \\ -16 & 15 & 2 \end{vmatrix}$$

This determinant may be expressed as

$$-1 \begin{vmatrix} -7 & 10 \\ -16 & 15 \end{vmatrix} = -1(-105 + 160) = -55$$

Note that in this problem we were fortunately able to retain the element 1, which reduced the arithmetical labor also.

As another example let us simplify the fourth-order determinant

$$\begin{vmatrix} 3 & 0 & -2 & 1 \\ 4 & -5 & 2 & -3 \\ -1 & 6 & 0 & -4 \\ 7 & 1 & -6 & 5 \end{vmatrix}$$

given in Example 21.

Multiply the fourth column by 3 and subtract from the first column, obtaining

$$\begin{vmatrix} 0 & 0 & -2 & 1 \\ 13 & -5 & 2 & -3 \\ 11 & 6 & 0 & -4 \\ -8 & 1 & -6 & 5 \end{vmatrix}$$

Multiply the fourth column by 2 and add to the third column, obtaining

$$\begin{vmatrix} 0 & 0 & 0 & 1 \\ 13 & -5 & -4 & -3 \\ 11 & 6 & -8 & -4 \\ -8 & 1 & 4 & 5 \end{vmatrix}$$

This reduces to

$$\begin{vmatrix} 13 & -5 & -4 \\ 11 & 6 & -8 \\ -8 & 1 & 4 \end{vmatrix}$$

with a -1 factor in front.

Multiply the second column of the above third-order determinant by 8 and add to the first column, obtaining

$$-1\begin{vmatrix} -27 & -5 & -4 \\ 59 & 6 & -8 \\ 0 & 1 & 4 \end{vmatrix}$$

Multiply the second column of the above third-order determinant by 4 and subtract from the third column, obtaining

$$-1\begin{vmatrix} -27 & -5 & 16 \\ 59 & 6 & -32 \\ 0 & 1 & 0 \end{vmatrix}$$

This reduces to

$$(-1)(-1)\begin{vmatrix} -27 & 16 \\ 59 & -32 \end{vmatrix} = (-27)(-32) - (59)(16) = -80$$

EXERCISE 7

Evaluate each of the following determinants as directed by the instructor:

1. $\begin{vmatrix} 2 & -1 & 3 \\ 4 & 5 & -2 \\ 1 & -3 & 1 \end{vmatrix}$
2. $\begin{vmatrix} 5 & 4 & 3 \\ 0 & 6 & 2 \\ 1 & 1 & 7 \end{vmatrix}$

3. $\begin{vmatrix} 4 & 0 & -2 \\ 7 & 8 & 3 \\ -5 & 0 & 1 \end{vmatrix}$
4. $\begin{vmatrix} 8 & 7 & -6 \\ 2 & 1 & 4 \\ 3 & -1 & 0 \end{vmatrix}$

5. $\begin{vmatrix} 6 & 2 & 1 \\ -3 & 4 & -5 \\ 3 & -1 & 2 \end{vmatrix}$
6. $\begin{vmatrix} 7 & 0 & -2 \\ -3 & 1 & 4 \\ 6 & 3 & 5 \end{vmatrix}$

7. $\begin{vmatrix} 4 & -4 & a \\ 3 & 5 & a \\ 2 & -1 & a \end{vmatrix}$
8. $\begin{vmatrix} 11 & 2 & -3 \\ -10 & 1 & 4 \\ 9 & 5 & 3 \end{vmatrix}$

9. $\begin{vmatrix} 2 & 3 & a+b \\ 1 & -4 & b+c \\ 5 & -1 & a+c \end{vmatrix}$ 10. $\begin{vmatrix} m & -2 & 3 \\ 4 & 2 & 5 \\ 1 & -1 & m \end{vmatrix}$

11. $\begin{vmatrix} 1 & 2 & 0 & 5 \\ 3 & 0 & 7 & 4 \\ 0 & -6 & 1 & -1 \\ 4 & 3 & 0 & -2 \end{vmatrix}$ 12. $\begin{vmatrix} 7 & 2 & 5 & -1 \\ 4 & 3 & 6 & 1 \\ 8 & 0 & 9 & -2 \\ -3 & 4 & 2 & 1 \end{vmatrix}$

13. $\begin{vmatrix} 5 & 2 & -3 & 8 \\ 4 & -1 & 0 & -2 \\ 0 & 3 & -4 & 1 \\ 6 & 0 & 3 & 5 \end{vmatrix}$ 14. $\begin{vmatrix} 6 & 4 & 0 & -1 \\ 1 & 3 & -5 & m \\ 2 & 0 & -4 & 0 \\ 7 & -3 & 1 & -2m \end{vmatrix}$

15. Show that the equivalent of Eq. (15), page 207, in determinant form is

$$A = \frac{1}{2}\begin{vmatrix} x_1 & y_1 & 1 \\ x_2 & y_2 & 1 \\ x_3 & y_3 & 1 \end{vmatrix}$$

9.16 Determinant Formulas for the Solution of Linear Equations in Three Unknowns

The application of Cramer's rule to the solution of a system of three linear equations in three unknowns is entirely comparable with that used in the solution of two simultaneous equations in two unknowns.

In solving the system

$$a_1 x + b_1 y + c_1 z = k_1$$
$$a_2 x + b_2 y + c_2 z = k_2$$
$$a_3 x + b_3 y + c_3 z = k_3$$

we obtain, by a method entirely analogous to that used in Sec. 9.11,

$$x = \frac{\begin{vmatrix} k_1 & b_1 & c_1 \\ k_2 & b_2 & c_2 \\ k_3 & b_3 & c_3 \end{vmatrix}}{\begin{vmatrix} a_1 & b_1 & c_1 \\ a_2 & b_2 & c_2 \\ a_3 & b_3 & c_3 \end{vmatrix}} \qquad y = \frac{\begin{vmatrix} a_1 & k_1 & c_1 \\ a_2 & k_2 & c_2 \\ a_3 & k_3 & c_3 \end{vmatrix}}{\begin{vmatrix} a_1 & b_1 & c_1 \\ a_2 & b_2 & c_2 \\ a_3 & b_3 & c_3 \end{vmatrix}} \qquad z = \frac{\begin{vmatrix} a_1 & b_1 & k_1 \\ a_2 & b_2 & k_2 \\ a_3 & b_3 & k_3 \end{vmatrix}}{\begin{vmatrix} a_1 & b_1 & c_1 \\ a_2 & b_2 & c_2 \\ a_3 & b_3 & c_3 \end{vmatrix}} \qquad (50)$$

As in second-order determinants, the common denominator is called the *determinant of the system.*

As in the case of two equations in two unknowns, if the determinant of the system equals zero, the equations are either inconsistent, with no solution, or dependent, with infinitely many solutions.

Example 22. Use determinants to solve the system

$$10x + 3y - 6z = -9$$
$$7x + 5y + 4z = 12$$
$$8x - 2y - 9z = -2$$

We note the following values:

$a_1 = 10$	$b_1 = 3$	$c_1 = -6$	$k_1 = -9$
$a_2 = 7$	$b_2 = 5$	$c_2 = 4$	$k_2 = 12$
$a_3 = 8$	$b_3 = -2$	$c_3 = -9$	$k_3 = -2$

Hence

$$x = \frac{\begin{vmatrix} -9 & 3 & -6 \\ 12 & 5 & 4 \\ -2 & -2 & -9 \end{vmatrix}}{\begin{vmatrix} 10 & 3 & -6 \\ 7 & 5 & 4 \\ 8 & -2 & -9 \end{vmatrix}}$$

Expanding each determinant by elements of the first row,

$$x = \frac{(-9)\begin{vmatrix} 5 & 4 \\ -2 & -9 \end{vmatrix} - (3)\begin{vmatrix} 12 & 4 \\ -2 & -9 \end{vmatrix} + (-6)\begin{vmatrix} 12 & 5 \\ -2 & -2 \end{vmatrix}}{(10)\begin{vmatrix} 5 & 4 \\ -2 & -9 \end{vmatrix} - (3)\begin{vmatrix} 7 & 4 \\ 8 & -9 \end{vmatrix} + (-6)\begin{vmatrix} 7 & 5 \\ 8 & -2 \end{vmatrix}}$$

$$= \frac{-9[(5)(-9) - (-2)(4)] - 3[(12)(-9) - (-2)(4)] - 6[(12)(-2) - (-2)(5)]}{10[(5)(-9) - (-2)(4)] - 3[(7)(-9) - (8)(4)] - 6[(7)(-2) - (8)(5)]}$$

$$= \frac{-9(-45 + 8) - 3(-108 + 8) - 6(-24 + 10)}{10(-45 + 8) - 3(-63 - 32) - 6(-14 - 40)}$$

$$= \frac{333 + 300 + 84}{-370 + 285 + 324} = \frac{717}{239} = 3$$

By a similar method y and z can be calculated, giving

$$y = \frac{\begin{vmatrix} 10 & -9 & -6 \\ 7 & 12 & 4 \\ 8 & -2 & -9 \end{vmatrix}}{\begin{vmatrix} 10 & 3 & -6 \\ 7 & 5 & 4 \\ 8 & -2 & -9 \end{vmatrix}} = -5 \qquad z = \frac{\begin{vmatrix} 10 & 3 & -9 \\ 7 & 5 & 12 \\ 8 & -2 & -2 \end{vmatrix}}{\begin{vmatrix} 10 & 3 & -6 \\ 7 & 5 & 4 \\ 8 & -2 & -9 \end{vmatrix}} = 4$$

EXERCISE 8

Solve the following systems of simultaneous linear equations by determinants:

1. $x + 2y + z = 20$
   $3x - y + 2z = 22$
   $2x + y - 4z = 7$

2. $4x + 7y + 2z = 21$
   $5x + 8y - 3z = 16$
   $-3x - 5y + 9z = 5$

3. $5x + 2y = -17$
   $3x + 7z = 23$
   $4y + 6z = 36$

4. $3x + 5y - 4z = 9a$
   $2x - 3y + 4z = 7b$
   $-6x + y + 3z = b - a$

Suggestion for Prob. 6: Let $a = 1/x$, $b = 1/y$, $c = 1/z$, and solve for a, b, and c.

5. $-7x + y + 4z = -16a$
   $x + y - 4z = -8b$
   $-x + 7y - 4z = 16c$

6. $\dfrac{1}{x} - \dfrac{1}{y} + \dfrac{2}{z} = 7$

   $\dfrac{2}{x} + \dfrac{2}{y} - \dfrac{3}{z} = -2$

   $-\dfrac{3}{x} + \dfrac{1}{y} + \dfrac{1}{z} = 1$

7. $x - 2y - 3z + 4w = -4$
   $3x - z + 2w = 16$
   $5y + 2z - 3w = 20$
   $4x + y - 7w = 10$

8. $x + 2y + 3z = 11$
   $-x + 2z + 3w = 9$
   $2x - 3y - w = -23$
   $y - z + 2w = -1$

Problems in Exercises 2, 3, 4, and 12 may be worked by determinants as directed by the instructor.

9.17 Application of Simultaneous Linear Equations to Kirchhoff's Law

Figure 9.7 is an idealized diagram of considerable importance if batteries are operated in parallel. Ordinarily, in such problems the given circuit parameters are the resistances R_1, R_2, and R_3 and the battery voltages E_1 and E_2. The problem is to solve for the

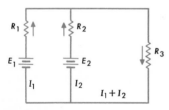

FIGURE 9.7

currents I_1 and I_2. The arrows next to the resistors indicate the direction in which the current is assumed to flow. Should either I_1 or I_2 come out negative, it simply means that the assumed direction of current flow was wrong, but if the solution is otherwise correct, the absolute value of the current will be correct.

The basic equations for the solution of this and similar problems are

$$E_1 = R_1 I_1 + R_3(I_1 + I_2) \tag{51}$$
$$E_2 = R_2 I_2 + R_3(I_1 + I_2) \tag{52}$$

EXERCISE 9

1. Find I_1 and I_2 when the circuit parameters are as follows:

	R_1	R_2	R_3	E_1	E_2
(a)	0.05	0.10	0.75	4.5	5.2
(b)	1.50	2.30	0.50	32.0	35.5
(c)	0.10	0.01	5.00	6.0	7.5

2. (a) Solve Eq. (51) for I_1 and find under what conditions I_1 must be negative.
(b) Solve Eq. (52) for I_2 and find under what conditions I_2 must be negative.
3. If $E_1 = 5$, $E_2 = 6$, and $R_3 = 10$, find the value of R_2 so that $I_1 = 0$.
4. Show that if

$$\frac{R_2}{R_3} + 1 = \frac{E_2}{E_1}$$

then $I_1 = 0$. [Refer to Eqs. (51) and (52).]

Figure 9.8 illustrates a "bridge" circuit which is widely used for electrical measurements as well as other applications.

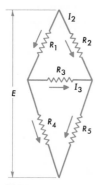

FIGURE 9.8

We shall assume that the known quantities are E, R_1, R_2, R_3, R_4, and R_5. Having been given these quantities, we shall solve for I_1, I_2, and I_3. As in the previous group of problems, the arrows next to the resistors indicate the assumed direction of current flow. Should either I_1, I_2, or I_3 come out to be negative, it means that the actual direction of current flow is exactly opposite to the assumed direction. This will not affect the absolute value of the current.

The basic equations involved are the following:

$$E = R_1I_1 + R_4(I_1 - I_3)$$
$$E = R_2I_2 + R_5(I_2 + I_3)$$
$$E = R_1I_1 + R_3I_3 + R_5(I_3 + I_2)$$

EXERCISE 10

1. Find I_1, I_2, and I_3 from the following data:

	E	R_1	R_2	R_3	R_4	R_5
(a)	10	2	3	4	5	6
(b)	5	1	5	3	8	4
(c)	1	2	6	9	5	3
(d)	2	5	9	5	3	8
(e)	25	6	9	2	5	8

2. If $I_3 = 0$, show that $R_1/R_4 = R_2/R_5$.

EXERCISE 11

Solve the following problems, using determinants.

1. Three bars of metal A, B, and C have the following weight composition: A contains 6 parts gold, 2 parts silver, and 1 part copper; B, 3 parts gold, 4 parts silver, and 2 parts copper; C, 1 part gold, 3 parts silver, and 5 parts copper. How many ounces of each must be taken to make 28 oz of alloy containing equal parts of gold, silver, and copper?
2. The following table gives the weight composition of three alloys A, B, and C.

	Copper, %	Tin, %	Zinc, %
A	80	10	10
B	20	40	40
C	50	...	50

How many pounds each of A, B, and C must be mixed to produce 600 lb of an alloy containing 60 percent copper, 10 percent tin, and 30 percent zinc?

3. A man wishes to apply to his garden 50 lb of a fertilizer containing 5 percent phosphorus and 12 percent nitrogen. He has available brand A containing 4 percent phosphorus and 10 percent nitrogen, brand B containing 7 percent phosphorus and 11 percent nitrogen, and brand C containing negligible phosphorus and 18 percent nitrogen. How much of each brand should be taken?

4. A waste mixed acid left over from nitrating is composed of 61.8 percent sulfuric acid, 20.1 percent nitric acid, and 18.1 percent water. It is required to make a mixture of 1,000 lb containing 60 percent sulfuric acid, 23 percent nitric acid, and 17 percent water. Solutions of sulfuric acid (98 percent) and nitric acid (90 percent) are available. How many pounds each of waste acid, sulfuric acid, and nitric acid must be taken if no additional water is used?

5. A beam AF weighing 400 lb/ft is supported at points A, B, C, D, E, and F. $AB = 5$ ft, $BC = 6$ ft, $CD = 7$ ft, $DE = 5$ ft, and $EF = 4$ ft. To determine the load carried by each support, the following equations must be solved for W_2, W_3, W_4, and W_5:

$$
\begin{aligned}
11W_2 + 3W_3 &= -34{,}100 \\
6W_2 + 26W_3 + 7W_4 &= -118{,}000 \\
7W_3 + 24W_4 + 5W_5 &= -93{,}600 \\
5W_4 + 16W_5 &= -37{,}800
\end{aligned}
$$

6. A girder AE weighing p lb/ft is supported at five equally spaced points A, B, C, D, and E. $AB = d$ ft. To determine how much of the load each support bears, solve the following equations for W_2, W_3, and W_4 in terms of p and d. ($W_1 = W_5 = 0$.)

$$
W_1 + 4W_2 + W_3 = -\frac{pd^2}{2}
$$

$$
W_2 + 4W_3 + W_4 = -\frac{pd^2}{2}
$$

$$
W_3 + 4W_4 + W_5 = -\frac{pd^2}{2}
$$

7. Fifty people wish to reach a place $27\frac{1}{2}$ mi away. The only available transportation is a bus having a capacity of 30 people and a speed of 35 mi/h. The party is divided into two roughly equal groups, which start at the same time. The first group starts on foot, walking at an average rate of 4 mi/h. The second group rides on the bus a certain distance and then walks the rest of the way at an average rate of 3 mi/h. The bus returns to meet the first group and to carry it the rest of the way. How far should each group walk in order that all may arrive at their destination at the same time? How many hours are required for the transfer?

9.18 The Graph As an Aid to Setting Up Algebraic Equations

Rough graphical representation is often a powerful way of setting up an equation for algebraic solution. Figure 9.9 represents a freehand sketch of the relationships involved in Example 1, page 260. By dimensioning the sketch, which need not be drawn to scale, we obtain the equation

$$295 = 40t† + 50(t + \tfrac{1}{2})$$

from which it readily follows that $t = 3$ h after the second truck leaves, or the meeting time is 4:30 P.M. The distance from A is $50(3 + \tfrac{1}{2}) = 175$ mi.

Perhaps the reader's first reaction is that we have here a means of scaling an answer from a graph drawn accurately to scale. Although this is true, the objective is the setting up of equations, based on dimensions and relations appearing in a rough freehand sketch.

The problems illustrated below have been selected from Exercise 3, Chap. 5, as typical examples.

Prob. 47, page 124 (Fig. 9.10). Draw a line OA with a slope of $^{11}/_9$ s/h. Draw from the 24-h point line BC with a slope of $-^{28}/_{13}$ s/h. Drop the perpendicular DE dividing the 24 h into $24 - h$ vertical hours and h horizontal hours.

Algebraically, since $\Delta y = m\,\Delta x$, the gain in $24 - h$ h in the vertical position is $DE = {}^{11}/_9(24 - h)$. The loss in h h in a horizontal position is $DE = -^{28}/_{13}h$. Since the net gain or loss is zero, we have $^{11}/_9(24 - h) + (-^{28}/_{13}h) = 0$.   ANS.: $h = 8.7$ h

---

†Using the relation $\Delta y = m\,\Delta x$, we have $\Delta y = (-40)(t) = -40t$ as measured from A to B, or the distance as measured from B to A is $40t$.

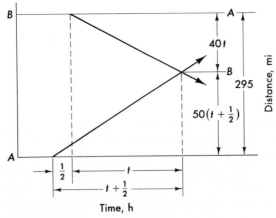

FIGURE 9.9

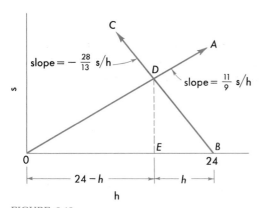

FIGURE 9.10

Prob. 50, page 125 (*Fig. 9.11*). Locate point A at (50,25). Draw line OA. From A draw line AB with a slope of 0.1 lb tin/lb metal. Draw line OC with a slope of 0.25 lb tin/lb metal. From the intersection point D of lines AB and OC drop the perpendicular DE. Since D is on the 25 percent line, it must represent an alloy containing 25 percent tin. Drop the perpendicular AG. This length represents 25 lb, which is the weight of tin in 50 lb of solder. DF is the weight of tin in w lb of type metal, or $DF = 0.1w$. (Since $\Delta y = m \Delta x$) total tin $= AG + DF = 25 + 0.1w$. Total tin also $= DE = 0.25(w + 50)$. Therefore $25 + 0.1w = 0.25(w + 50)$. ANS.: $w = 83\frac{1}{3}$ lb

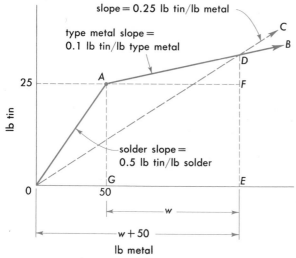

FIGURE 9.11

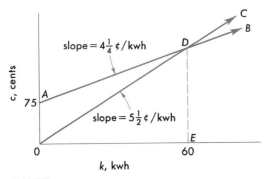

FIGURE 9.12

Prob. 51, page 125 (Fig. 9.12). From point $A = 75$ draw line AB with a slope of $4\frac{1}{4}$ cents/kWh. From the slope-intercept equation (7) the equation for this line must be $c = 4\frac{1}{4}k + 75$. Draw line OC with a slope of $5\frac{1}{2}$ cents/kWh. The equation for this line is evidently $c = 5\frac{1}{2}k$. At the intersection point D we have $c = 4\frac{1}{4}k + 75 = 5\frac{1}{2}k$.

ANS.: $k = 60$ kWh

Prob. 61, page 126 (Fig. 9.13). Draw line OA with a slope of 4,800 ft/s, representing the progress of sound through water. Similarly draw OB with a slope of 1,125 ft/s for passage through the air. Where the horizontal distance between OA and OB = 8, draw the horizontal line CD. Drop perpendiculars HK and GL. Then the time for passage through the air is $d/1,125$ (since $\Delta x = \Delta y/m$). Similarly the time through water is $d/4,800$. But $d/1,125 - d/4,800 = 8$.

ANS.: $d = 11,750$ ft

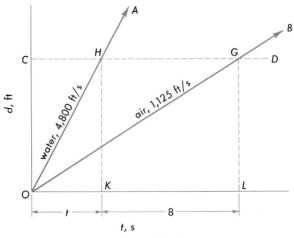

FIGURE 9.13

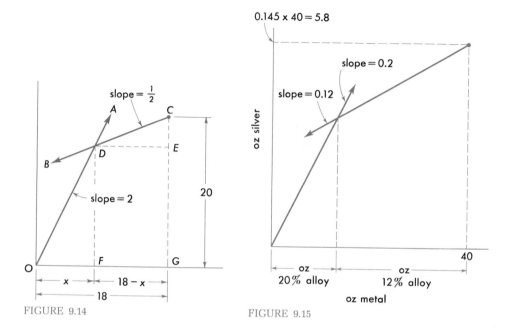

FIGURE 9.14

FIGURE 9.15

Prob. 63, page 126 (Fig. 9.14). Draw line OA with slope $= 2$. Locate point C at (18,20). From C draw CB with slope $= \frac{1}{2}$ intersecting OA at D. Draw horizontal line DE from D and vertical line DF from D. Then $DF = 2x$ ($\Delta y = m\,\Delta x$), and $CE = \frac{1}{2}(18 - x)$. But $DF + CE = CG$, or $2x + \frac{1}{2}(18 - x) = 20$, from which $x = 7\frac{1}{3}$ ft, and the rest of the solution follows readily. (Note that Fig. 9.14 is similar to Fig. 9.15 used in solving Prob. 39, page 124.)

9.19 Applications of Simultaneous Linear Equations

EXERCISE 12

Solve the following problems as directed. Problems marked with an asterisk (*) may be solved conveniently by graphical methods. A few of these problems are repeated from Chap. 5, where one unknown was used. In this exercise they are to be solved using two unknowns for comparison.

*1. The sum of two numbers is 53 and their difference is 7. Find the numbers.

2. A fraction reduces to $\frac{2}{3}$ if 2 is added to its numerator. It reduces to $\frac{1}{2}$ if 1 is added to its denominator. What is the fraction?

*3. A boat travels 60 mi upstream in 10 h, making the return trip downstream in 8 h. Find the rate of the current and of the boat in still water.

*4. Two planes are 60 mi apart. If they are flying toward each other, they will pass in 5 min. If they are headed in the same direction, the faster plane will overtake the slower in 45 min. What are their speeds?

5. If the width of a rectangle is increased by 2 ft and the length by 12 ft, the area will be increased by 480 ft². If the width is increased by 12 ft and the length by 2 ft, the area will be increased by 660 ft². Find the dimensions of the original rectangle.

*6. A plumber and his helper together receive $69.60, the plumber working 5 h and the helper 6 h. At another time the plumber works 8 h and the helper 7½ h, and they receive together $99.60. What are the hourly wages of each?

7. Three men and 6 boys can do in 2 days what 1 man and 8 boys can do in 3 days. Find the time required for 1 man alone and for 1 boy alone to do the work.

8. Six men and 5 boys take 2 days to do what 9 men and 15 boys can do in 1 day. If the labor cost for the job is $504, find the daily wage for a man and for a boy.

9. Three disks A, B, and C are externally tangent each to each. If the center-to-center distances AB, AC, and BC are 6¾, 6¼, and 8½ in, respectively, find the diameters.

10. Find the diameters of the disks in Fig. 9.16.

11. Mr. Warner and his son Phil agree to paint a house for $600. After working together for 10 days the job is ⅝ finished, and Mr. Warner leaves Phil to finish the job alone, which he does in 18 more days. How should the $600 be divided between father and son?

12. The reduced value of a fraction is ⅘. If 21 is deducted from both numerator and denominator, the reduced value of the new fraction is ⅓. Find the original fraction.
    If $n =$ the numerator and $d =$ the denominator, we have $n/d = ⅘$ and $(n - 21)/(d - 21) = ⅓$.

*13. A man bought 30 acres of land for $1,280. Part of the land cost $40/acre; the rest cost $50/acre. How much land was sold at $40/acre?

14. A sum of money consists of dimes and quarters. There are 9 more dimes than quarters. If the total value is $10, find the number of dimes.

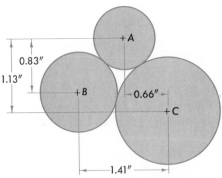

FIGURE 9.16

15. A man invested $7,000, part at $4\frac{1}{2}$ percent and the rest at 6 percent. If the interest for one year was $348, find the amount invested at 6 percent.

*16. How much solder containing 50 percent tin and how much type metal containing 15 percent tin must be mixed to make 80 lb of solder containing 40 percent tin?

We may set up a table as follows:

	Lb original material	Tin content, lb
Solder	s	$0.50s$
Type metal	t	$0.15t$
Total	$s + t = 80$	$0.50s + 0.15t = (0.40)(80) = 32$

The two equations may be solved simultaneously for s and t.

*17. A 12-qt cooling system is filled with 25 percent antifreeze. How many quarts must be drawn off and replaced with pure antifreeze to raise the strength to 45 percent?

18. In a municipal election a total of 82,347 votes were cast for two nominees A and B. If A received 6,051 votes more than B, find the number of votes each contestant received.

*19. At a high-school game the price of admission was $0.25 for each adult and $0.10 for each child. If the turnstile showed that 397 persons attended the game and the gate receipts were $56.80, how many adults attended?

*20. How much high-speed tool steel containing 18 percent tungsten and how much steel containing 12 percent tungsten should be mixed to make 3,000 lb containing 14.6 percent tungsten?

21. A stick of wood is to be cut into four equal sections for braces, as shown in Fig. 9.17. Find x and y, making no allowance for saw kerfs.

*22. The resistance R_t of a nickel wire at any Celsius temperature t is equal to $R_0(1 + at)$, where R_0 is the resistance at $0°C$ and a is the temperature coefficient of resistance. If the resistance at $20°C$ is $26.82\ \Omega$, and at $32°C$ is $28.10\ \Omega$, find a and R_0.

23. Express the area of the ring in Fig. 9.18 in terms of AB. Given a regular 6-in

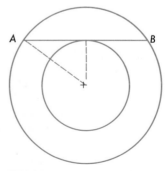

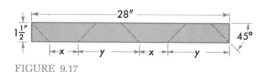

FIGURE 9.17   FIGURE 9.18

pentagon, find the difference between the areas of the circumscribed and inscribed circles. How does your answer compare with that obtained by taking a 6-in heptagon?

24. Two stations A and B on the same railroad line are on opposite sides of a mountain pass. A is 10 mi east of the divide, and B is 15 mi west of the divide. The grade is the same on both sides. If the running time from A to B is 44 min and the running time from B to A is 50 min, find the upgrade and downgrade speeds.

*25. A gas company charges $\$a$ service charge plus $\$b/1{,}000$ ft^3. Find a and b if 12,000 ft^3 of gas costs $\$6.40$ and 20,000 ft^3 costs $\$10.00$.

Example 23. A small firm has a taxable income of $\$390{,}000$. The state tax is 10 percent of that portion remaining after the federal tax is paid. The federal tax is 25 percent of that portion remaining after the state tax is paid. What are the state and federal taxes?

Let s = state tax (\$) and f = federal tax (\$); then

$$s = 0.10(390{,}000 - f)$$
$$f = 0.25(390{,}000 - s)$$

These equations can be rearranged as follows:

$$10s + f = 390{,}000$$
$$0.25s + f = 97{,}500$$

Subtraction yields

$$9.75s = 292{,}500 \qquad \text{or} \qquad s = \$30{,}000 \qquad \text{and} \qquad f = \$90{,}000$$

Checking,

$$\text{State tax} = 0.10(390{,}000 - 90{,}000) = \$30{,}000$$
$$\text{Federal tax} = 0.25(390{,}000 - 30{,}000) = \$90{,}000$$

26. A Penn-Central train running between New York City and Albany can make up 20 min time lost in starting by averaging 6 mi/h faster than usual. If on another occasion, by giving another train the right of way, it reduces its average speed by 8 mi/h and will finish its run 36 min late, find the distance and usual running time from New York to Albany.

27. A man ordered a number of pamphlets from a job printer. The printer made a fixed charge for each order plus a certain amount for each pamphlet. The total cost amounted to $\$42$. Several months later the customer placed a somewhat larger order for additional pamphlets. This order cost $\$54$. Had the customer been able to anticipate his needs and ordered all the pamphlets at one time, they would

have cost only $84. How much was the fixed charge? If the second order contained 80 more pamphlets than the first, find the number of pamphlets in each order and the unit cost.

28. A marksman fires at a target 480 yd away. He hears the bullet strike 2.4 s after he fires. An observer standing 420 yd from the target and 240 yd from the marksman hears the bullet strike 1.6 s after he hears the report of the rifle. Find the velocities of sound and of the bullet, assuming each to be uniform.

29. The difference between two quantities equals their product. Their sum less their product equals the smaller divided by the larger. What are the numbers?

30. The product of two quantities equals their difference. Their product plus their sum equals the larger divided by three times the smaller. Find the numbers.

31. A traveler has 400 lb of baggage, for which he must pay $4.50 transportation because of excess over the weight carried without charge. If a friend takes charge of part of the baggage, the traveler will pay $1.80 and the friend $1.20 for excess above the weight carried free. How much baggage is allowed to go free on one ticket?

32. Two laborers A and B were employed at different wages. Over a given pay period, A received $24 more than B. If A had been idle 2 days while B worked the entire period, they would have received equal amounts. If B had been idle 2 days with A working the entire period, A would have received $45 more than B. How long was a pay period and what was the daily wage of each man?

*33. In a race of 100 yd, A beats B by $\frac{1}{5}$ s. In the second trial, A gives B a start of 4 yd and B wins by $\frac{1}{5}$ s. Find the time required for A and B each to run 100 yd.

*34. Two cars are traveling in the same direction around a 1-mi circular racetrack. The faster car overtakes the slower car every 3 min. When the cars are traveling in opposite directions, they pass each other every 18 s. Find their speeds.

*35. Two runners A and B run around a 1-mi circular track at 18 and 22 ft/s, respectively. If they start simultaneously at the same point:

    (a) How often will B overtake A going in the same direction?

    (b) How often will B pass A going in the opposite direction? Compare with the satellite problems on pages 270–272.

Example 24. A time-honored problem is that of the motorcycle courier who starts from the rear of a troop column, rides to the head of the column, and immediately returns to the rear. How far did the courier ride if the column is 6 mi long and marching at 4 mi/h? The speed of the motorcycle is 30 mi/h. How long was the courier riding?

    According to Fig. 9.19, the courier traveled about 12.2 mi (6.9 mi to the head of the column and 5.3 mi to the rear). The total time was about 0.4 h, or 24 min.

    This problem is a good example of the opportunity offered by maintaining the proper relative values while changing other conditions.

    In this illustration let us consider rates of the cyclist relative to that of the troop column. It is evident that cyclist is riding at $30 - 4$, or 26, mi/h relative to the troop column while riding to the head of the column, and $30 + 4$, or 34, mi/h relative to

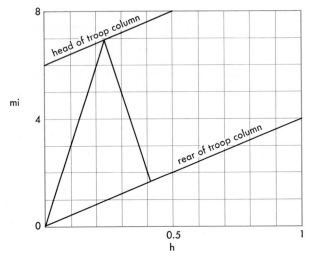

FIGURE 9.19

the column while riding to the rear. If the 6-mi column were stationary, then our problem would be to find the total time for the cyclist to advance 6 mi at 26 mi/h and return at 34 mi/h. The solution is obvious:

$$\frac{6}{34} + \frac{6}{26} = 0.1765 + 0.231, \text{ or } 0.408 \text{ h}$$

and confirms the answer already obtained by graphical means.

36. A bicyclist rides at a steady 20 mi/h parallel to a double-track rapid-transit line on which trains are running in both directions at regular intervals. Every 11 min 40 s a train overtakes him going in the same direction. Every 5 min a train passes him going in the opposite direction.
   (a) What is the speed of a rapid-transit train?
   (b) What is the headway between trains in miles? In minutes?
*37. A bus and a hiker start at the same time, one from A to B, the other from B to A. If they arrive at their destinations 5 min and 3 h, respectively, after passing one another, compare their rates of speed. What other information can be obtained from these data?
38. In balancing the equation for the combustion of methyl alcohol, the coefficients a, b, and c must be determined in the equation

$$a\text{CH}_3\text{OH} + b\text{O}_2 \longrightarrow c\text{H}_2\text{O} + a\text{CO}_2$$

According to a hydrogen balance,

$$4a = 2c$$

An oxygen balance requires that

$$a + 2b = c + 2a$$

Find the smallest integral values of a, b, and c satisfying the above equations.

39. The solution of copper in nitric acid proceeds according to the equation

$$aCu + 2bHNO_3 \longrightarrow aCu(NO_3)_2 + cNO + bH_2O$$

The nitrogen balance requires that

$$2b = 2a + c$$

The oxygen balance requires that

$$6b = 6a + c + b$$

Find the smallest integral values of a, b, and c satisfying the above equations.

40. If the sum of two numbers is represented by S and their product is represented by P, express the sum of the cubes of these numbers in terms of S and P.

*41. In how many ways may a bill of $1.37 be paid using only 5-cent and 8-cent stamps?

You will develop an equation containing two unknowns. Ordinarily there would be an infinite number of solutions. In this case roots must be integral (Diophantine equations). They must also be positive. Often there may be many solutions, sometimes none (for example, $2x + 4y = 15$). Plot your equation. Can you see any pattern in your answers?

9.20 Linear Empirical Equations

The underlying relationship between the variables in a set of experimental data is often more clearly revealed if expressed in the form of an equation. Because of the limitations of measured data, such an equation is called empirical. An *empirical formula* is one whose reliability is based upon a limited number of observations and is not necessarily supported by any established theory or law. It is based upon immediate experience rather than logical or mathematical conclusions. The term is not applied in a derogatory sense since many currently accepted scientific laws have been the outgrowth of empirical equations.

A vast amount of work in this area is done, not to discover or verify fundamental scientific laws but to discover and exhibit the performance characteristics of various equipment such as automobile engines, electronic gear, air conditioners, and the like.

It is common observation that experimentally obtained data rarely fall on a straight line when plotted, even though the general relationship appears to be linear. In such cases the constants of the most suitable linear equation may be derived by a variety of methods. One method consists in simply drawing a straight line through two representative points visually selected. This is done when the data are so rough that extreme refinements would be a waste of time and might even be misleading. The method of least squares is indicated when the data are sufficiently refined to warrant expenditure of considerable time. The method of averages represents a fair compromise between these two extremes. In this chapter we shall limit our discussion of derivation of constants to the method of averages.

Example 25. The following data were obtained on the amount of fuel oil (G gal/ 24 h) required to heat a house at various outdoor temperatures ($T° = 24$ h average). To avoid complicating factors, only clear, relatively calm days were chosen.

ΔT	$T°$	G	ΔG	$\Delta G/\Delta T$
	9	13.0		
9			−2.0	−0.22
	18	11.0		
7			−1.8	−0.26
	25	9.2		
10			−2.2	−0.22
	35	7.0		
11			−2.5	−0.23
	46	4.5		
13			−2.8	−0.22
	59	1.7		

To satisfy ourselves as to the degree of linearity, we may plot these data or determine the constancy of the ratio $\Delta G/\Delta T$. These ratios have been tabulated above and indicate a fairly close, but not exact, correspondence to linearity. Figure 9.20 indicates the same thing.

In view of the approximate nature of the relationship, we shall use the *method of averages*. This method requires that the data be divided into two groups as nearly equal as possible and the values averaged. Hence we obtain

$$T_1 = \text{average of first three } T\text{'s} = \frac{9 + 18 + 25}{3} = 17.3$$

$$T_2 = \text{average of last three } T\text{'s} = \frac{35 + 46 + 59}{3} = 46.7$$

$$G_1 = \text{average of first three } G\text{'s} = \frac{13.0 + 11.0 + 9.2}{3} = 11.1$$

$$G_2 = \text{average of last three } G\text{'s} = \frac{7.0 + 4.5 + 1.7}{3} = 4.4$$

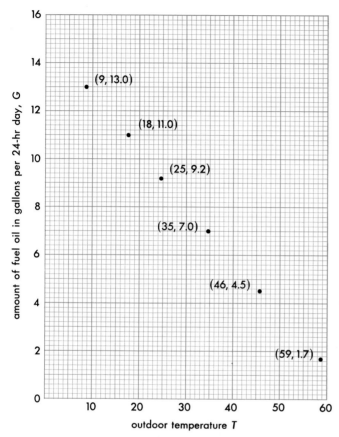

FIGURE 9.20

We shall now obtain the equation of the straight line through the points whose coordinates are (T_1, G_1) and (T_2, G_2).

This equation will be in the slope-intercept form

$$G = mT + b \tag{53}$$

Let the point P_1 (Fig. 9.21) be the point whose coordinates are (T_1, G_1), or (17.3,11.1). Let the point P_2 be the point whose coordinates are (T_2, G_2), or (46.7,4.4). The slope m of this line is

$$m = \frac{G_2 - G_1}{T_2 - T_1} = \frac{4.4 - 11.1}{46.7 - 17.3} = -\frac{6.7}{29.4} = -0.228$$

By substituting this value of slope in Eq. (54), we obtain

$$G = -0.228T + b \tag{55}$$

The coordinates of point P_1 in Fig. 9.21 are (17.3,11.1). (Observe that we could have chosen the point P_2 just as well.)

By substituting 17.3 for T and 11.1 for G in Eq. (55), we obtain

$$11.1 = -0.228 \times 17.3 + b$$

from which

$$11.1 = -3.94 + b$$

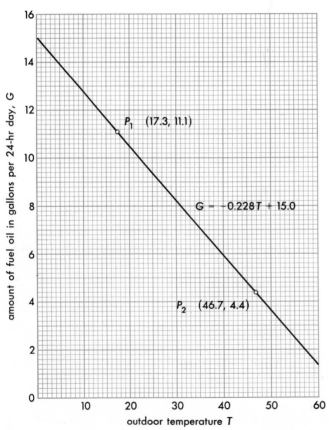

FIGURE 9.21

or

$b = 11.1 + 3.94 = 15.04$

or

$b = 15.0$ (to three significant digits)

The linear equation for G in terms of T in this instance is therefore

$$G = -0.228T + 15.0 \tag{56}$$

Some students might prefer to substitute the values (17.3,11.1) and (46.7,4.4) in the equation $G = mT + b$, obtaining the simultaneous equations

$11.1 = m(17.3) + b$
$4.4 = m(46.7) + b$

Subtracting, we have $6.7 = -29.4m$, or $m = -0.228$. Completing the solution as in Sec. 8.19, $b = 15.0$, or

$$G = -0.228T + 15.0$$

as above.

According to Eq. (56), when the outdoor temperature is $0°F$, the fuel consumption is 15.0 gal/24 h and decreases by 0.228 gal for each degree rise in temperature.

It will be noted in the above examples that we have solved for the dependent variable in terms of the independent variable. Although this is the usual practice, it is occasionally preferable to solve for the independent variable. Problem 9 in the following exercise illustrates this point. In this case the initial reading of the instrument is in terms of the dependent variable (temperature). Then we proceed to determine the corresponding independent variable (concentration of alcohol).

Other examples of this reversal of common procedure are conversion of voltage reading to temperature when checking a thermocouple; conversion of a manometer reading of pressure difference to rate of fluid flow; computation of wind speed from rpm of an anemometer; etc. (see Sec. 8.2).

Incidentally, wherever the time element is involved, it is almost without exception the independent variable.

EXERCISE 13

1. Show, without plotting, that the points in the following table fall exactly in a straight line. (See Example 6, page 188.)

x	20	28	34	44	48
y	50	38	29	14	8

(a) Derive the equation for y in terms of x.
(b) Find y when x = 100.
(c) Find x when y = −13.
(d) At what value is x = y?
(e) How could your answer in (d) be obtained graphically?

2. Show, without plotting, that the following points fall exactly in a straight line.

x	− 10	15	65	165	240
y	− 5	− 3	1	9	15

(a) Derive the equation for y in terms of x.
(b) Find the x intercept.
(c) Find y when x = 40.

3. Calculate, by the method of averages, the empirical linear equation best fitting the following data:

x	8	18	29	37	48	51	62	68
y	18	25	32	36	44	45	52	56

When x = 55 what is the probable value of y?

4. The overall cost of owning and operating a popular car is given in the following table:

A, mi/year	B, overall cost, cents/mi	C, total cost, $/year
5,000	25.0	———
10,000	14.5	———
15,000	11.0	———
20,000	9.25	———

Complete the column for values of C ($/year). Set up an equation of the type C = kA + b, where k and b are constants.

Plot A as abscissas versus C as ordinates. Do the points lie exactly in a straight line? What is the physical significance of the constants k and b?

5. Find the empirical equation, as in Prob. 3, for the following data:

x	− 11	1	9	12	19	25	34	45
y	− 3	5	10	12	17	22	28	35

Derive the empirical linear equations best representing the data in the tables below. Indicate whether the linear relationship is exact or approximate. If approximate, use the method of averages.

6. The following readings were taken in calibrating a spring where W is the load in pounds and d is the extension in inches. Derive an equation expressing W as a function of d.

W	3.0	5.0	7.0	8.5	11.0	12.0
d	0.20	0.45	0.80	1.0	1.3	1.5

7. The barometer reading in inches of mercury (P) varies with the altitude above sea level in feet (h) as follows:

h	200	500	1,200	1,800	2,500	3,200
P	29.72	29.36	28.58	27.93	27.20	26.48

What linear equation best describes h as a function of P within the range of the data? What is the sea-level pressure?

8. The following prices were quoted for printing a quantity of booklets, where N is the number of booklets and C is the cost (dollars) of the lot.

N	100	250	500	800	1,200
C	30	45	70	100	140

Derive a formula for C in terms of N. What is the significance of the constants?

9. The boiling point ($t°C$) of dilute alcohol solutions is related to the weight percent concentration of alcohol (p) as shown in the following table:

p	0	2	4	7	10
t	100	97.2	95.2	92.5	90.3

What formula will indicate approximately the alcoholic content when the boiling point is known? It is required that the values $p = 0$, $t = 100$ satisfy the equation. (Base your equation on two points, the first being $p = 0$, $t = 100$, the other point being the average of the other four.)

10. The boiling point of water ($t°C$) is related to the atmospheric pressure (P mm) as follows:

P	777	766	760	750	740.4	731.4	723.3
t	100.62	100.22	100.00	99.63	99.27	98.93	98.62

It is required that the equation shall be satisfied by the values $P = 760$, $t = 100$ and that it shall represent the best fit for the remaining data.

Derive an equation for P in terms of t.

Derive an equation for t in terms of P.

11. The speed of revolution of an anemometer (R r/min) was checked against the wind speed (V mi/h) to yield the following calibration data:

V	5.0	11.0	17.0	20.0	25.0	30.0	33.0	37.0
R	50	120	190	230	300	360	410	470

Derive a linear equation for V in terms of R. What is the significance of the constants?

12. The resistance R_t of a nickel wire at various Celsius temperatures (t) is shown in the following table:

t	10	25	45	70
R_t	50.4	55.2	61.6	69.6

Derive an equation of the form $R_t = R_{20}[1 + a(t - 20)]$, where R_{20} is the resistance at $20°C$.

What is the nature of a?

13. The velocity of sound in air (meters per second) at any Celsius temperature (t) is given by the formula

$$V = 331.7\sqrt{1 + \frac{t}{273}}$$

Evaluate V at $t = -30, -20, -10, 0, 10, 20, 30, 40$ and derive a linear formula for V to best fit these points.

graphical methods

In contrast to Chap. 9, we shall deal in this chapter with actual graphical solutions read from lines drawn accurately on a coordinate grid.

10.1 Graphical Solutions of Simultaneous Linear Equations

In situations where a certain type of problem has to be solved repeatedly to an accuracy of perhaps two or three significant digits, a graphical solution may be more efficient than a strictly algebraic one. Furthermore, a graphical solution often emphasizes important aspects of the problem which would otherwise be overlooked.

Graphical methods are often used as a first approximation to sift out masses of raw data. After the most promising of these data have been thus identified and isolated, they may be reduced by more refined means.

In Secs. 10.2 to 10.4 we shall illustrate a few of the more elementary graphical methods which apply to selected problems.

10.2 Related Rates

One of the more elementary classes of problems which are particularly adaptable to graphical solutions is that involving distance, time, and uniform speed. Consider the following example.

Example 1. The two points A and B are 295 mi apart on a certain throughway. A truck leaves point A at 1:00 P.M. and travels toward point B at an average speed of 50 mi/h. Another truck leaves point B at 1:30 P.M. and travels toward point A at an average speed of 40 mi/h. How far from point A will they meet and at what time?

At the end of each hour after 1:00 P.M. the truck entering the throughway at point A is 50 mi farther from point A. That is,

Clock time, P.M.	*Distance from point A, mi*
1:00	0
2:00	50
3:00	100
⋮	⋮

These points are plotted in Fig. 10.1a. Observe that the slope of the straight line passing through these points is numerically equal to the speed of the truck, that is, 50 mi/h.

At the end of each hour after 1:30 P.M. the truck entering the throughway at point B is 40 mi closer to point A. That is,

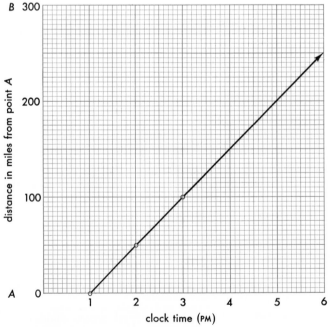

FIGURE 10.1a

Clock time, P.M.	Distance from point A, mi
1:30	295 − 0 = 295
2:30	295 − 40 = 255
3:30	285 − 80 = 215
⋮	⋮

These points are plotted in Fig. 10.1b.

Figure 10.2 shows both of these graphs plotted with reference to the same coordinate axes. By inspection of this figure we find that the trucks meet at 4:30 P.M. at a point 175 mi from point A.

Example 2. A truck travels at an average rate of 40 mi/h to its destination and immediately begins the return trip at an average rate of 60 mi/h. The round trip requires 5 h. Find the one-way distance.

Figure 10.3a shows the graph of distance vs. time for the outbound trip. The slope of the graph is numerically equal to 40 mi/h.

At this point we shall have occasion to introduce the idea of *directed speed* (or velocity). If then we designate the directed speed on the outbound trip as +40 mi/h, we designate the directed speed on the return trip as −60 mi/h.

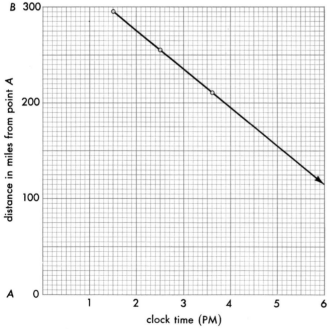

FIGURE 10.1b

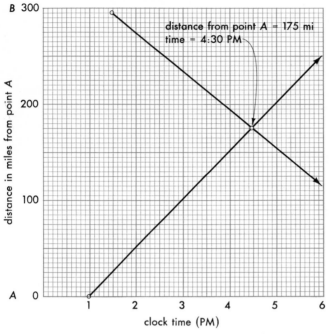

FIGURE 10.2

The graph of distance vs. time on the return trip will therefore have a slope of −60 mi/h. From the conditions of the problem we also know that the graph passes through the point (5,0). This straight-line graph is shown in Fig. 10.3*b*.

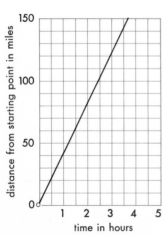

FIGURE 10.3*a*

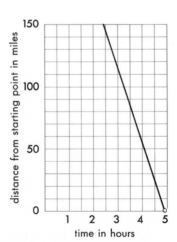

FIGURE 10.3*b*

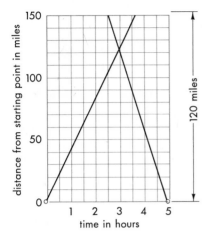

FIGURE 10.4

Both graphs are shown in Fig. 10.4, from which we observe that the one-way distance is 120 mi.

EXERCISE 1

It is recommended that the graph paper used in this exercise be ruled 10 lines/in with heavy tenth lines, using 8 in along the Y axis and 6 in along the X axis.

The recommended scales are as tabulated below.

Problem	Y axis	X axis
1	Distance, mi	Time, h
	10 mi/in	1 h/in
2	Distance, mi	Time, h
	50 mi/in	1 h/in
3	Distance, mi	Time, h
	10 mi/in	1 h/in
4	Distance, mi	Time, h
	2 mi/in	1 h/in
5	Distance, mi	Time, h
	50 mi/in	½ h/in
6	Seconds	Time, h
	5 s/in	5 h/in
7	Distance, ft	Time, s
	2,000 ft/in	2 s/in
8	Distance, mi	Wind speed, mi/h
	50 mi/in	10 (mi/h)/in

Solve the problems below by graphical methods.

1. A group of bicyclists maintains an average rate of 12 mi/h. One hour and 45 min after they leave a certain point, a motorist sets out from the same point to overtake them. If his rate is 40 mi/h, how long will it take?
2. Two trains leave opposite terminals of a 300-mi line at the same time. If the rates are 40 and 35 mi/h, respectively, when will they meet?
3. The outbound trip of a bus was made at 24 mi/h, while the return trip was made at 30 mi/h. Find the one-way distance if the total running time was $4\frac{1}{2}$ h.
4. A man has $3\frac{1}{4}$ h at his disposal. He can ride out on a bus at 16 mi/h and walk back at $3\frac{1}{2}$ mi/h. How far can he ride and still be able to return to his starting point in the allotted time?
5. The indicated airspeed of a small plane is 120 mi/h. If a 40 mi/h west wind is blowing, how far west can the plane fly and still return to the airport $2\frac{1}{4}$ h after taking off?
6. A poorly compensated watch when worn on the wrist gains 11 s in 9 h and when laid down horizontally loses 28 s in 13 h. How many hours out of the 24 in a horizontal position would result in no gain or loss in a 24-h period?
7. The explosion of a floating mine was heard 8 s sooner through the water than through the air. If the speed of sound is 4,800 ft/s through the water and 1,125 ft/s through the air, how far away was the explosion?
8. A plane has an airspeed of 300 mi/h. It flies directly into a head wind for 48 min and returns along the same route in 42 min. Find the wind speed.

10.3 Equations in the Form $1/a + 1/b = 1/c$

Figure 10.5 shows the graphs of two linear equations. These equations are

$$y = -\frac{b}{a}x + b \tag{1}$$

and

$$y = x \tag{2}$$

By algebraic methods let us find the value of y in terms of a and b which will satisfy both of these equations simultaneously.

Since by Eq. (2) $y = x$, we may replace x by y in Eq. (1), obtaining

$$y = -\frac{b}{a}y + b$$

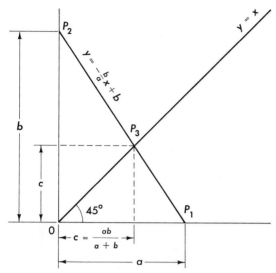

FIGURE 10.5

Multiplying both members by a,

$$ay = -by + ab$$

Adding by to both members and factoring,

$$y(a + b) = ab$$

and

$$y = \frac{ab}{a + b}$$

where the quantity $ab/(a + b)$ designates the ordinate (and abscissa) of point P_3 in Fig. 10.5. It will tend to avoid ambiguity in subsequent work if we represent this specific ordinate by some letter of the alphabet other than y. We therefore let

$$c = \frac{ab}{a + b} \tag{3}$$

Equation (3) can be transformed into an equivalent form which is simpler in appearance and actually more useful in many applied problems.

To develop this equivalent equation we invert (take the reciprocal) of both members of Eq. (3), obtaining

$$\frac{1}{c} = \frac{a+b}{ab}$$

which is equivalent to writing

$$\frac{1}{c} = \frac{1}{b} + \frac{1}{a} \tag{4}$$

This relationship is displayed in Fig. 10.5, in which the coordinates of P_1, P_2, and P_3 are $(a,0)$, $(0,b)$, and (c,c), respectively.

In Fig. 10.6 we have let $a = 36$ and $b = 45$. By direct reading of the nomograph we find that $c = 20$. This solution should be verified by substituting $a = 36$, $b = 45$, and $c = 20$ in Eq. (4).

Example 3. The formula

$$\frac{1}{R_1} + \frac{1}{R_2} = \frac{1}{R}$$

is encountered in electricity where R_1 and R_2 are two resistances connected in parallel.

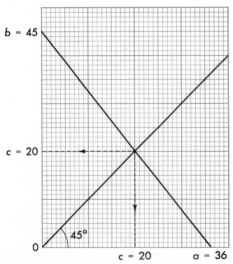

FIGURE 10.6

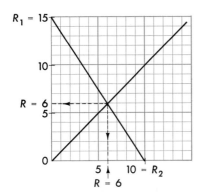

FIGURE 10.7

The resistance R is the equivalent resistance of the parallel combination. If $R_1 = 15\ \Omega$ and $R_2 = 10\ \Omega$, find graphically the value of the resistance R. See Fig. 10.7.

By direct reading we see that $R = 6\ \Omega$.

Example 4. Assume an additional resistance $R_3 = 7.5\ \Omega$ to be connected in parallel with R_1 and R_2. If we connect our previous answer, 6 Ω, with $R_3 = 7.5$, we obtain the joint resistance in parallel of R_1, R_2, and $R_3 = 3.3\ \Omega$. See Fig. 10.8.

Example 5. A routine job in a machine shop requires an operation on the milling machine and an operation on the drill press. The milling operation can produce 15 pieces/h, and the drill press operation can produce 10 pieces/h. How many finished parts are produced per hour?

Evidently it requires $\frac{1}{15}$ h on the drill press plus $\frac{1}{10}$ h on the milling machine to obtain a finished piece.

If we let $P = $ number of finished pieces per hour, then $1/P = $ total time required

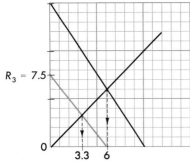

FIGURE 10.8

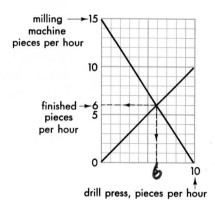

FIGURE 10.9

to finish 1 piece, and

$$\frac{1}{P} = \frac{1}{10} + \frac{1}{15}$$

Figure 10.9 is a graphical solution of the problem. By inspection of this graph, the number of finished pieces per hour is found to be 6 p.

EXERCISE 2

Note: A grid in the form shown in Fig. 10.10 is recommended for all problems in this exercise.

Solve the following problems by graphical methods.

1. A 12-Ω and a 24-Ω resistor are connected in parallel. Find the resistance of the combination.
2. Two resistors are connected in parallel. Their combined resistance is 5 Ω. The resistance of one of them is 11.25 Ω. Find the value of the other resistor.
3. A common lens formula is

$$\frac{1}{D_i} + \frac{1}{D_o} = \frac{1}{f}$$

where D_i = image distance
D_o = object distance
f = focal length

Find f if $D_i = 5$ and $D_o = 20$.

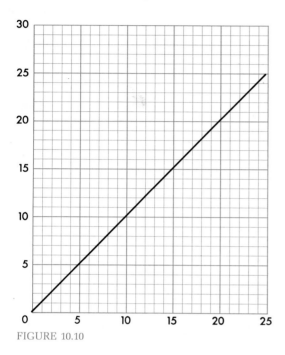

FIGURE 10.10

4. A man alone can clear a woodlot in 6 days, while his son alone can do it in 12 days. How long would the job take if they worked together?
5. A job can be done by A in 22½ h, by B in 16⅔ h, and by C in 18 h. Find the time required when all work together.
6. A tank can be filled by two pipes in 6 min, while the first pipe would require 10 min. How long would it take the second pipe to fill the tank alone?
7. How much time would it require to fill the tank in Prob. 6 if the first pipe operated as an inlet and the second pipe operated as an outlet?
8. A certain machine-tool job requires two operations. The first operation can handle 21 pieces/h. The second operation can handle 28 pieces/h. How many finished pieces are produced per hour?
9. A convex lens has a focal length of 5 in. Can you determine the image and object distances if the image distance is to be 2½ times the object distance? (This relation is required if we wish to enlarge a photo 2½ times.)

In the following artificial earth satellite (AES) problems we are considering only AES in equatorial orbit, i.e., following the earth's equator east to west or west to east.

Example 6. Consider an AES having a true period of 8 h and moving east to west. If A is the apparent, or observed, AES period in hours, the earth will make $A/24$ rotation in A h, and the satellite will make $A/8$ revolution in A h. Hence we arrive at the equation $A/8 + A/24 = 1$ (complete circumference). Dividing through by A, we obtain the equation $1/8 + 1/24 = 1/A$, or $A = 6$ h apparent period. In other words, the AES returns over the same spot on the equator every 6 h (Figs. 10.11 and 10.12a and b†).

Example 7. If the AES in Example 6 follows a W-E orbit, we have the equation $A/8 - A/24 = 1$, or $A = 12$ h apparent period (Figs. 10.13 and 10.14†).

It will be helpful to keep in mind that when the satellite is moving from west to east (in the same direction as the earth) its apparent period is longer than its true period, whereas for E-W orbit, opposing the earth's rotation, the reverse is true.

†Note that some of the nomographs have been "stretched out" horizontally for easier reading.

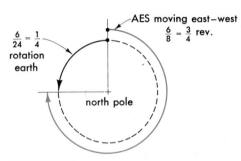

FIGURE 10.11

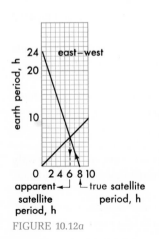

FIGURE 10.12a

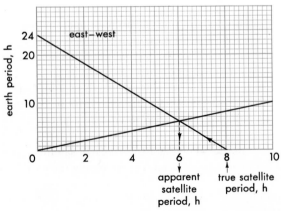

FIGURE 10.12b

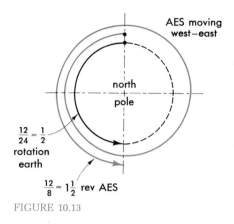

AES moving
west−east

north
pole

$\frac{12}{24} = \frac{1}{2}$
rotation
earth

$\frac{12}{8} = 1\frac{1}{2}$ rev AES

FIGURE 10.13

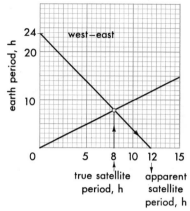

FIGURE 10.14

EXERCISE 3

Solve each of the following AES problems both algebraically and graphically.

1. (a) An AES is moving east in equatorial orbit. Its apparent period is 96 min. Each
   hour the observer on earth is carried east 1/24 rotation (while the satellite is
   also moving east). What is the true period? (See Prob. 35, page 249; also Fig. 10.15.)
   (b) An AES moving east has a true period of 144 min. What is its apparent period
   as measured by the observer?
2. (a) An AES moving west has a true period of 108 min. What is its apparent period?
   (b) The apparent period of an AES in E-W orbit is 138 min. Find its true period
   (Fig. 10.16).

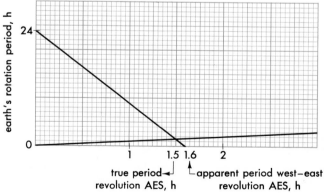

FIGURE 10.15

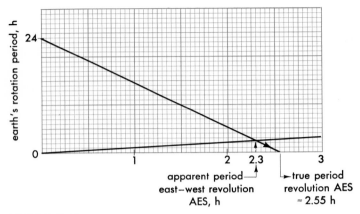

FIGURE 10.16

3. An AES in W-E orbit has an apparent period of 117 min as measured by the observer's watch. What is its true period?
4. Find the true period of an AES moving E-W if its apparent period is 169 min.

10.4 Two-component Mixtures

Example 8. A typical problem deals with a metallic alloy C which is to be made from a mixture of two alloys A and B.

Alloy A contains 42 percent silver, and alloy B contains 90 percent silver. How many ounces of alloy A and how many ounces of alloy B will be needed to make 40 oz of alloy C containing 60 percent of silver?

The most natural way to solve this problem graphically would be to represent ounces of alloy A as ordinates and ounces of alloy B as abscissas. We would then develop and plot two equations:

$$A + B = 40 \qquad \text{based on total weight of metal}$$

and

$$0.42A + 0.90B = (0.60)(40) \qquad \text{based on silver content}$$

Each equation is most conveniently plotted by locating and connecting its A and B intercepts (Fig. 10.17).

We shall deliberately abandon this approach for the time being in order to present another method. While possibly lacking the simplicity of the above conventional approach, it is more adaptable to routine repetitive problems of this kind.

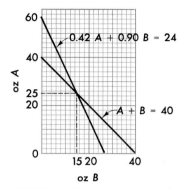

FIGURE 10.17

We can approach this problem from a geometric standpoint (Fig. 10.18). However, the properties of the straight line can also be applied to this problem and used to confirm the result obtained in Figs. 10.17 and 10.18.

To retain generality we shall represent the ounces of A, B, and C by W_a, W_b, and W_c, respectively, and the percent silver content of A, B, and C by P_a, P_b, and P_c, respectively (Fig. 10.19).

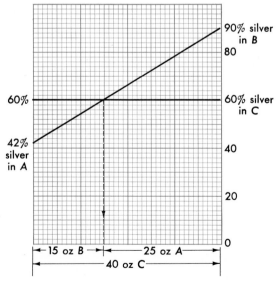

FIGURE 10.18

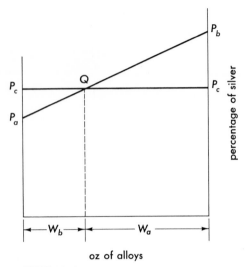

FIGURE 10.19

The slope of the diagonal line is

$$\frac{P_b - P_a}{W_b + W_a}$$

The y intercept is P_a.

Therefore its equation in slope-intercept form must be

$$y = \frac{P_b - P_a}{W_b + W_a} x + P_a \tag{5}$$

The coordinates of point Q are (W_b, P_c). Substituting in Eq. (5),

$$P_c = \frac{P_b - P_a}{W_b + W_a} W_b + P_a$$

or

$$(P_c - P_a)(W_b + W_a) = (P_b - P_a)W_b$$

Multiplying the indicated products and adding $P_a W_b + P_a W_a$ to both members, we obtain the equation

$$P_c W_b + P_c W_a = P_b W_b + P_a W_a \tag{6}$$

Equation (6) may be transformed to read $W_a(P_c - P_a) = W_b(P_b - P_c)$, or

$$\frac{W_b}{W_a} = \frac{P_c - P_a}{P_b - P_c} \tag{7}$$

Equation (7) agrees with the geometric statement that the sides of similar triangles (in Fig. 10.19) are in proportion.

By adding 1 to each member of Eq. (7), we obtain

$$\frac{W_a + W_b}{W_a} = \frac{P_b - P_a}{P_b - P_c}$$

But $W_a + W_b = W_c$; therefore

$$\frac{W_c}{W_a} = \frac{P_b - P_a}{P_b - P_c} \tag{8}$$

Inverting both members of Eq. (8) and multiplying the resulting equation by W_c, we have

$$W_a = \frac{P_b - P_c}{P_b - P_a} W_c \tag{9}$$

Checking against Example 8, we find

$$W = \left(\frac{90 - 60}{90 - 42}\right)(40) = 25 \text{ oz} \qquad \text{as expected}$$

An interesting sidelight on Eq. (6) appears if we rearrange it so as to read

$$P_c(W_a + W_b) = P_c W_c = P_a W_a + P_b W_b \tag{10}$$

This is a balance on silver stating that

Concentration $\times$ weight of C = concentration $\times$ weight of A +

concentration $\times$ weight of B

If we solve Eq. (7) for P_c, we have

$$P_c = \frac{P_a W_a + P_b W_b}{W_a + W_b} \tag{11}$$

This is a formula for a weighted average.

EXERCISE 4

The graph paper recommended for Exercise 1 will also be satisfactory for this exercise. Recommended scales are given below.

Problem	Y axis	X axis
1	Percent tin 20%/in	Weight, lb 50 lb/in
2	Percent antifreeze 20%/in	Gallons 2 gal/in
3	Percent tungsten 5%/in	Weight, lb 500 lb/in
4	Percent tin 20%/in	Weight, lb 20 lb/in
5	Percent silver 5%/in	Weight, oz 10 oz/in
6	Percent antifreeze 20%/in	Quarts 2 qt/in
7	Percent sulfuric acid 20%/in	Gallons 100 gal/in
8	Percent butterfat 5%/in	Quarts 200 qt/in

Solve the following problems (repeated from Exercise 3, Chap. 5) by graphical methods.

1. Fifty pounds of high-grade solder spatters containing 50 percent lead and 50 percent tin is melted with some type metal containing 90 percent lead and 10 percent tin. The resulting alloy is to be a low-grade solder containing 75 percent lead and 25 percent tin. How many pounds of type metal will be needed?
2. How many gallons of 35 percent by volume antifreeze solution must be added to 3 gal of 80 percent antifreeze solution to reduce the strength to 60 percent?
3. How much high-speed tool steel containing 18 percent tungsten and how much steel containing 12 percent tungsten should be mixed to make 3,000 lb of alloy containing 14.6 percent tungsten?
4. How much solder containing 50 percent tin and how much type metal containing 15 percent tin must be mixed to make 112 lb of solder containing 40 percent tin?
5. A bar of metal contains 20 percent silver, and a second bar contains 12 percent silver. How many ounces of each must be used to make a 40-oz bar containing $14\frac{1}{2}$ percent silver?
6. A 12-qt cooling system is filled with 25 percent antifreeze. How many quarts must be drawn off and replaced with pure antifreeze to raise the strength to 45 percent?

7. How many gallons of water must be mixed with 500 gal of 96 percent sulfuric acid to reduce the strength to 80 percent?

8. A dairyman has 1,000 qt of milk containing 4.8 percent butterfat, but the city in which he sells his milk requires only 4 percent butterfat. How many quarts of cream testing 20 percent butterfat may be separated from the milk and still satisfy the legal minimum requirement?

EXERCISE 5

The following problems are repeated from Exercise 12, Chap. 9, as an exercise in graphical solution.

Prob. 1. The difference between two numbers is 7. Their sum is 53. Find the numbers.

Prob. 3. A boat travels 60 mi upstream in 10 h, making the return trip downstream in 8 h. Find the rate of the current and of the boat in still water.

Prob. 6. A plumber and his helper together receive $69.60, the plumber working 5 h and the helper 6 h. At another time the plumber works 8 h and the helper $7\frac{1}{2}$ h, and they receive together $99.60. What are the hourly wages of each?

Prob. 14. A sum of money consists of dimes and quarters. There are 9 more dimes than quarters. If the total value is $10, find the number of dimes.

Prob. 16. How much solder containing 50 percent tin and how much type metal containing 15 percent tin must be mixed to make 80 lb of solder containing 40 percent tin?

Prob. 25. A gas company charges $a service charge plus $b/1,000 ft³. Find a and b if 12,000 ft³ of gas costs $6.40 and 20,000 ft³ costs $10.00.

Prob. 34. Two cars are traveling in the same direction around a 1-mi circular racetrack. The faster car overtakes the slower car every 3 min. When the cars are traveling in opposite directions, they pass each other every 18 s. Find their speeds.

Prob. 35. Two runners A and B run around a 1-mi circular track at 18 and 22 ft/s, respectively. If they start simultaneously at the same point:
(a) How often will B overtake A going in the same direction?
(b) How often will B pass A going in the opposite direction?

Logarithms

One of the most effective devices for saving time and effort in mathematical computations when a desk calculator is not available is the logarithm. By means of logarithms, processes of multiplication and division are replaced by addition and subtraction, and those of raising to a power and extracting a root are replaced by multiplication and division. In fact, some mathematical operations, previously difficult, admit of ready solution by logarithms.

11.1 The Concept of a Logarithm

Given the equation $x = b^y$, the following assertions may be made:

1. If b is any positive real number other than unity, then for every real value of y there exists one and only one positive number x, defined by the equation $x = b^y$.
2. Conversely, for every positive value of x there exists a unique real number y such that $x = b^y$.

If

$$x = b^y \qquad (b > 0 \text{ and } b \neq 1) \tag{1}$$

then y is called the logarithm of x to the base b and is written

$$y = \log_b x \tag{2}$$

For example,

$3^2 = 9$ is equivalent to $\log_3 9 = 2$
$5^3 = 125$ is equivalent to $\log_5 125 = 3$
$4^{-1} = \frac{1}{4}$ is equivalent to $\log_4 \frac{1}{4} = -1$
$5^{-2} = \frac{1}{25}$ is equivalent to $\log_5 \frac{1}{25} = -2$
$49^{1/2} = 7$ is equivalent to $\log_{49} 7 = \frac{1}{2}$
$15^0 = 1$ is equivalent to $\log_{15} 1 = 0$

Substituting $\log_b x$ for y in the equation $x = b^y$, we have $x = b^{\log_b x}$, from which it follows that

The logarithm of a number is the exponent indicating the power to which it is necessary to raise the base to equal the given number.

The above assertions imply that:

1. For any given logarithm y to a given base there exists a unique number x corresponding to that logarithm.
2. Every positive number x has a unique logarithm for any given positive base other than unity.

Since any number other than zero raised to the zero power equals 1, it follows that the logarithm of 1 to any base equals zero.

EXERCISE 1

In Probs. 1 to 9 write each exponential expression in logarithmic notation.

1. (a) $2^3 = 8$ (b) $2^{-6} = \frac{1}{64}$ (c) $7^{-2} = \frac{1}{49}$
2. (a) $3^5 = 243$ (b) $10^{-2} = 0.01$ (c) $(\frac{1}{6})^2 = \frac{1}{36}$
3. (a) $5^4 = 625$ (b) $3^{-3} = \frac{1}{27}$ (c) $(\frac{2}{3})^3 = \frac{8}{27}$
4. (a) $2^5 = 32$ (b) $7^0 = 1$ (c) $b^0 = 1$
5. (a) $3^{-4} = \frac{1}{81}$ (b) $6^2 = 36$ (c) $(\frac{1}{5})^{-2} = 25$
6. (a) $4^{1/2} = 2$ (b) $8^{1/3} = 2$ (c) $9^{3/2} = 27$
7. (a) $16^{3/4} = 8$ (b) $25^{1/2} = 5$ (c) $27^{2/3} = 9$
8. (a) $125^{2/3} = 25$ (b) $32^{3/5} = 8$ (c) $b^1 = b$
9. (a) $27^{-1/3} = \frac{1}{3}$ (b) $16^{-5/4} = \frac{1}{32}$ (c) $36^{-3/2} = \frac{1}{216}$

Let it be required to find the value of $\log_7 49$. Setting $\log_7 49 = y$, we paraphrase the question by asking ourselves what power of 7 equals 49. Since $7^2 = 49$, it is apparent that $\log_7 49 = 2$.

EXERCISE 2

Find the values of the following logarithms:

1. (a) $\log_5 25$ (b) $\log_2 8$ (c) $\log_3 81$
2. (a) $\log_4 8$ (b) $\log_8 16$ (c) $\log_{16} 8$
3. (a) $\log_9 27$ (b) $\log_{27} 81$ (c) $\log_{125} 25$
4. (a) $\log_7 \sqrt{7}$ (b) $\log_{12} 1$ (c) $\log_3 \frac{1}{3}$
5. (a) $\log_2 \frac{1}{8}$ (b) $\log_{10} 0.001$ (c) $\log_4 \frac{1}{32}$
6. (a) $\log_{125} 5$ (b) $\log_{16} 4$ (c) $\log_{16} \sqrt{4}$

Another version of the logarithmic relationship is exemplified by the equation $\log_5 x = 3$. Here we wish to know what number is equal to the third power of 5. Since $5^3 = 125$, we have $\log_5 125 = 3$.

It may be required to find the base of a logarithmic equation. For example, $\log_b 16 = 4$. This is read "log 16 to what base equals 4?" In other words, what number b raised to the fourth power equals 16? The answer is evidently 2. Hence $\log_2 16 = 4$.

EXERCISE 3

Find the value of x in Probs. 1 to 4.

1. (a) $\log_2 x = 3$ (b) $\log_5 x = 2$ (c) $\log_3 x = 4$
2. (a) $\log_5 x = 0$ (b) $\log_6 x = -1$ (c) $\log_8 x = \frac{4}{3}$
3. (a) $\log_{10} x = 2$ (b) $\log_3 x = -2$ (c) $\log_{27} x = \frac{2}{3}$
4. (a) $\log_{16} x = \frac{3}{4}$ (b) $\log_8 x = -\frac{2}{3}$ (c) $\log_{25} x = -\frac{1}{2}$

In Probs. 5 to 10 find the value of the base b.

5. (a) $\log_b 9 = 2$ (b) $\log_b 8 = 3$ (c) $\log_b 4 = \frac{2}{3}$
6. (a) $\log_b 1{,}000 = 3$ (b) $\log_b 25 = 2$ (c) $\log_b 3 = \frac{1}{2}$
7. (a) $\log_b 15 = 1$ (b) $\log_b 1 = 0$ (c) $\log_b \frac{1}{4} = -2$
8. (a) $\log_b \frac{1}{27} = -\frac{3}{2}$ (b) $\log_b \frac{1}{25} = -\frac{2}{3}$ (c) $\log_b 7 = -\frac{1}{2}$
9. (a) $\log_b 9 = -\frac{2}{3}$ (b) $\log_b 0.01 = 2$ (c) $\log_b \frac{1}{16} = -\frac{4}{3}$
10. (a) $\log_b 1{,}000 = \frac{3}{2}$ (b) $\log_b 6 = -\frac{1}{2}$ (c) $\log_b \frac{1}{64} = -6$

11.2 Properties of Logarithms

Since logarithms are exponents, the rules relating to the use of logarithms resemble the laws of exponents.

The most common relationships are the following:

$$\log_b PQ = \log_b P + \log_b Q \tag{3}$$

$$\log_b \frac{P}{Q} = \log_b P - \log_b Q \tag{4}$$

$$\log_b P^n = n \log_b P \tag{5}$$

$$\log_b \sqrt[n]{P} = \frac{1}{n} \log_b P \tag{6}$$

$$\log_b 1 = 0 \tag{7}$$

$$\log_b \frac{1}{P} = -\log_b P \tag{8}$$

These rules will be derived in Secs. 11.3 to 11.8.

11.3 Multiplication

Let

$$P = b^c \quad \text{or} \quad \log_b P = c$$

and

$$Q = b^d \quad \text{or} \quad \log_b Q = d$$

Then

$$PQ = b^c \cdot b^d = b^{c+d}$$

In logarithmic notation, this becomes

$$\log_b PQ = c + d$$

But

$$c + d = \log_b P + \log_b Q$$

Therefore

$$\log_b PQ = \log_b P + \log_b Q \quad \text{[Eq. (3)]}$$

The logarithm of a product is equal to the sum of the logarithms of the separate factors.

11.4 Division

Let

$$P = b^c \quad \text{or} \quad \log_b P = c$$

and

$$Q = b^d \quad \text{or} \quad \log_b Q = d$$

Then

$$\frac{P}{Q} = \frac{b^c}{b^d} = b^{c-d}$$

In logarithmic notation, this becomes

$$\log_b \frac{P}{Q} = c - d$$

But

$$c - d = \log_b P - \log_b Q$$

Therefore

$$\log_b \frac{P}{Q} = \log_b P - \log_b Q \qquad \text{[Eq. (4)]}$$

The logarithm of a quotient is equal to the logarithm of the numerator minus the logarithm of the denominator.

11.5 Powers

If

$$P = b^c$$

then

$$\log_b P = c$$

Also

$$P^n = b^{cn}$$

which in logarithmic notation is

$$\log_b P^n = cn$$

But

$$c = \log_b P$$

Therefore

$$\log_b P^n = n \log_b P \qquad \text{[Eq. (5)]}$$

This is a form of the product rule. If we let $P = Q$ in Eq. (3), then $\log_b P^2 = 2 \log_b P$, or in general, $\log_b P^n = n \log_b P$.

The logarithm of a power of a number is equal to the exponent times the logarithm of the number.

11.6 Roots

This is a form of the power rule, for if we write

$$\log_b \sqrt[n]{P} \text{ as } \log_b P^{1/n}$$

we obtain

$$\log_b P^{1/n} = \frac{1}{n} \log_b P$$

Therefore

$$\log_b \sqrt[n]{P} = \frac{1}{n} \log_b P \qquad \text{[Eq. (6)]}$$

The logarithm of a root of a number is the logarithm of that number divided by the index of the root.

11.7 Logarithm of Unity

This is also a special case of the power rule in which $n = 0$.

$$\log_b 1 = \log_b P^0 = 0 \log_b P = 0 \qquad \text{[see Eq. (7)]}$$

The logarithm of unity to any base is zero.

11.8 Reciprocal

This is another variation of the power law. In this case $n = -1$.

$$\log_b \frac{1}{P} = \log_b P^{-1} = (-1) \log_b P = -\log_b P \qquad \text{[see Eq. (8)]}$$

The logarithm of the reciprocal of a number is the negative of the logarithm of that number.

11.9 Logarithm of the Base

If in Eq. (1) we let

$$x = b$$

then

$$b = b^y$$

Following Definition 1 in Chap. 4, it is evident that

$$y = 1$$

From Eq. (2) of this chapter,

$$1 = \log_b b \qquad\qquad\qquad (9)$$

Thus *the logarithm of the base in any system is 1.*

11.10 Systems of Logarithms

From the problems in Exercises 1 to 3 it is evident that any positive number, except 1, can be used as a base for a system of logarithms. The number 1 cannot be used as a base because 1, when raised to any power whatever, is still equal to 1.

There are two systems of logarithms in general use: the system of common logarithms to the base 10 and the system of natural logarithms to the base e, or $2.71828\cdots$. Natural logarithms will not be discussed in this text.

The system of common logarithms is most convenient for general computation because it is best suited to the decimal system. Hereafter, unless otherwise indicated, the word *logarithm* will be understood to mean common logarithm. When the base is not expressed, the base 10 is understood. Thus log $3 = 0.47712$ means $\log_{10} 3 = 0.47712$.

From the laws of exponents we know that $10^0 = 1$, $10^1 = 10$, $10^2 = 100$, $10^3 = 1,000$, $10^4 = 10,000$, $10^5 = 100,000$, etc.

In logarithmic notation the above equations become $\log_{10} 1 = 0$, $\log_{10} 10 = 1$, $\log_{10} 100 = 2$, $\log_{10} 1,000 = 3$, $\log_{10} 10,000 = 4$, $\log_{10} 100,000 = 5$, etc.

11.11 Change of Base

At times it may be convenient to change from one base to another, especially when dealing with problems involving heat transfer, mechanical shock, transient electric currents, and the like. The formula involved is

$$\log_a Q = \frac{1}{\log_b a} \log_b Q \tag{10}$$

Since a and b are both constants, $\log_b a$ is also a constant. Hence, if only a table to the base b is available, *the logarithm of a number to the base b can be converted to the logarithm of the same number to the base a by dividing the former by the constant* $\log_b a$.

The proof of this equation follows:

Let

$$Q = a^x$$

so that

$$\log_a Q = x \tag{11}$$

Also if

$$Q = a^x$$

then

$$\log_b Q = \log_b a^x = x \log_b a$$

or

$$x = \frac{1}{\log_b a} \log_b Q \tag{12}$$

Then, from Eqs. (11) and (12),

$$\log_a Q = \frac{1}{\log_b a} \log_b Q$$

11.12 Graphical Derivation of Table

If we use 10 for the base instead of the general value b, we obtain the graph shown in Fig. 11.1. This, then, is the graph of $x = 10^y$, or $y = \log_{10} x$. It is apparent that when $y = 0$, $x = 1$. When $y = \frac{1}{3}$, $x = 10^{1/3}$ or $\sqrt[3]{10}$, or 2.154. When $y = \frac{1}{2}$, $x = 10^{1/2}$, or $\sqrt{10}$, or 3.162. When $y = 1$, $x = 10$, etc.

For any given value of y, there is only one value of x; and conversely, for any given positive value of x, there is only one value of y (Sec. 11.1). Thus any positive number x may be expressed as a power of 10, just as we have considered 2.154 as the 0.333 power of 10 and 3.162 as the 0.500 power of 10.

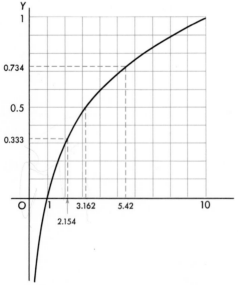

FIGURE 11.1

If this curve were drawn to a large enough scale, we should be able to read that when $x = 5.42$, $y = 0.734$ (approximately). Evidently, then, $10^{0.734} = 5.42$, or $\log_{10} 5.42 = 0.734$.

11.13 Characteristics and Mantissas

If we know that $10^{0.734} = 5.42$, then from Eqs. (1) and (2) we may write

$$\log_{10} 5.42 = 0.734$$

Now it is a simple matter to find the logarithm of 54.2, 542, 5,420, 0.00542, or any number with this particular sequence of digits, regardless of the position of the decimal point. The student should study the following examples carefully.

$\log 54.2 = \log (10 \times 5.42) = \log 10 + \log 5.42 = 1 + 0.734$
$\log 542 = \log (100 \times 5.42) = \log 100 + \log 5.42 = 2 + 0.734$
$\log 5{,}420 = \log (1{,}000 \times 5.42) = \log 1{,}000 + \log 5.42 = 3 + 0.734$
$\log 0.542 = \log (0.1 \times 5.42) = \log 0.1 + \log 5.42 = -1 + 0.734$
$\log 0.0542 = \log (0.01 \times 5.42) = \log 0.01 + \log 5.42 = -2 + 0.734$
$\log 0.00542 = \log (0.001 \times 5.42) = \log 0.001 + \log 5.42 = -3 + 0.734$

Each logarithm will be seen to consist of an integer (positive, negative, or zero) and a decimal which is a positive number or zero and less than 1.

The integral part is called the *characteristic,* and the decimal part is called the *mantissa.* The mantissa is usually an unending decimal which, by more advanced mathematics, may be computed to as many places as desired.

It will be noted that the mantissa depends only on a given sequence of digits, in this case 542. All numbers having a given sequence of digits will have the same mantissa. On the other hand, the characteristic is seen to be related to the position of the decimal point.

The relationships set forth in the beginning of this section are summarized in Table 11.1.

From Table 11.1 it will be evident that

For numbers larger than 1 the characteristic of the logarithm is zero or a positive number and one less than the number of digits at the left of the decimal point in the number. If the number is smaller than 1, the characteristic is negative and corresponds to the position of the first significant figure.

TABLE 11.1

	Characteristic		Mantissa	Common method of writing logarithm
log 5,420 =	3	+	0.734	3.734
log 542 =	2	+	0.734	2.734
log 54.2 =	1	+	0.734	1.734
log 5.42 =	0	+	0.734	0.734
log 0.542 =	−1	+	0.734	0.734 − 1
log 0.0542 =	−2	+	0.734	0.734 − 2
log 0.00542 =	−3	+	0.734	0.734 − 3

For example, in the number 0.00542 the first significant digit is in the third place to the right of the decimal point; therefore the characteristic of the logarithm is −3.

By common agreement the mantissa in a logarithm table (and elsewhere except as noted below) is always considered to be positive.

Alternative conventions of writing logarithms containing negative characteristics are illustrated below. For the most part we shall confine ourselves in this text to the form shown in the left-hand column.

0.734 − 1	9.734 − 10	$\bar{1}.734$	−0.266
0.734 − 2	8.734 − 10	$\bar{2}.734$	−1.266
0.734 − 3	7.734 − 10	$\bar{3}.734$	−2.266

The notation in the extreme right-hand column is used only under special conditions (Sec. 11.28).

If the number is expressed in scientific notation, the characteristic corresponds to the power of 10, as illustrated in Table 11.2.

A table of mantissas, or logarithms of numbers between 1 and 10, will be seen to be sufficient to determine the logarithm of any positive number, the proper positive or negative characteristic being supplied by inspection.

TABLE 11.2

Number	Scientific notation	Logarithm
5,420	5.42×10^3	3.734
542	5.42×10^2	2.734
54.2	5.42×10^1	1.734
5.42	5.42×10^0	0.734
0.542	5.42×10^{-1}	0.734 − 1
0.0542	5.42×10^{-2}	0.734 − 2
0.00542	5.42×10^{-3}	0.734 − 3

11.14 The Antilogarithm of a Number

A number N is called the *antilogarithm* (or *antilog*) of another number y if

$$y = \log N$$

For example, we see from Table 11.2 that

$$1.734 = \log 54.2$$

Therefore 54.2 is the antilogarithm of 1.734.

EXERCISE 4

1. Complete the following table:

	Number	Scientific notation	Logarithm
(a)	8.61	$8.61 \, (\times 10^0)$	0.935
(b)	861	8.61×10^2	2.935
(c)	0.0861	8.61×10^{-2}	$0.935 - 2$
(d)	8,610	———	———
(e)	———	8.61×10^5	———
(f)	———	———	4.935
(g)	0.00861	———	———
(h)	———	8.61×10^{-1}	———
(i)	———	———	$0.935 - 4$
(j)	8,610,000	———	———
(k)	———	8.61×10^8	———
(l)	———	———	$0.935 - 3$

Given $\log 3.54 = 0.549$ and $\log 7.98 = 0.902$ in Probs. 2 to 4, find the logarithms of the given numbers. In Probs. 5 to 7 find the antilogarithms of the given logarithms.

2. (a) 35.4   (b) 798   (c) 79,800
3. (a) 0.0354   (b) 0.354   (c) 7,980
4. (a) 0.00798   (b) 3,540,000   (c) 0.000354
5. (a) 2.549   (b) 1.902   (c) $8.902 - 10$
6. (a) $0.902 - 3$ (b) 4.549   (c) $0.549 - 1$
7. (a) $6.549 - 10$ (b) 3.902   (c) $0.549 - 3$

11.15 Using a Table of Logarithms

A portion of a five-place table of logarithms is reproduced in Table 11.3. (Since this table shows mantissas only, it would be more precise to refer to it as a table of mantissas. However, in ordinary usage such tables are referred to as "log tables." Therefore we shall continue to refer to them in this way.)

From this table we find that the mantissa of the logarithm of 3,124 is 0.49471. We have obtained this value by reading in from the number representing the first three digits, 312, and under the fourth digit, 4. Two space-saving conventions will be noted: the decimal point is not shown; neither are the first two digits, 49, of the mantissa shown.

Another space-saving device is revealed if it is required to look up the mantissa of the logarithm of 3,168. Reading opposite 316 and under 8, we find *079. The asterisk tells us that we have moved from mantissas 0.49··· to mantissas 0.50····. Therefore this particular mantissa is 0.50079.

Since $3,168 = 3.168 \times 10^3$, the characteristic of the logarithm is seen to be 3, and the complete logarithm of 3,168 is 3.50079.

EXERCISE 5

Referring to Table 11.3, determine the complete logarithms (characteristic and mantissa) of the following numbers:

1. 3,103   2. 313.6   3. 310   4. 3.190
5. 31,160   6. 318,200   7. 3.152   8. 3,167
9. 3,100,000   10. 31.63   11. 31   12. 3.1

TABLE 11.3

N	0	1	2	3	4	5	6	7	8	9
310	49136	150	164	178	192	206	220	234	248	262
311	276	290	304	318	332	346	360	374	388	402
312	415	429	443	457	471	485	499	513	527	541
313	554	568	582	596	610	624	638	651	665	679
314	693	707	721	734	748	762	776	790	803	817
315	831	845	859	872	886	900	914	927	941	955
316	969	982	996	*010	*024	*037	*051	*065	*079	*092
317	50106	120	133	147	161	174	188	202	215	229
318	243	256	270	284	297	311	325	338	352	365
319	379	393	406	420	433	447	461	474	488	501
320	515	529	542	556	569	583	596	610	623	637

In looking up the logarithm of a number smaller than unity, we may first express the number in scientific notation and then determine the mantissa and characteristic from the respective first and second parts of that number.

Example 1. Use Table 11.3 to read the logarithm of 0.003148. In scientific notation this becomes 3.148×10^{-3}. Therefore by inspection the characteristic is -3, and from the table, the mantissa is 0.49803. Hence the logarithm of 0.003148 is 0.49803 $-$ 3.

EXERCISE 6

Referring to Table 11.3, determine the logarithms of the numbers in Probs. 1 to 12.

1. 0.003122	2. 0.03196	3. 0.3144	4. 0.00031
5. 0.03167	6. 0.0000315	7. 0.3171	8. 0.003185
9. 0.00000319	10. 0.3111	11. 0.03105	12. 0.003162

Referring to Table 11.3, determine the numbers corresponding to the logarithms (i.e., find the antilogarithms) in Probs. 13 to 24.

13. 1.49178	14. 2.49374	15. 3.50065
16. 0.50092 $-$ 2	17. 0.50515 $-$ 3	18. 0.50284 $-$ 1
19. 4.50569	20. 0.50243	21. 5.49290
22. 0.50325 $-$ 4	23. 1.50010	24. 0.49996 $-$ 1

EXERCISE 7

Referring to a five-place log table, find the logarithms of the numbers in Probs. 1 to 3.

1. (a) 1,234	(b) 100.3	(c) 0.01234	(d) 37.05	(e) 398.4
2. (a) 79,020	(b) 0.3076	(c) 3,000	(d) 0.005006	(e) 1,003,000
3. (a) 179.7	(b) 0.1042	(c) 0.01090	(d) 709.3	(e) 0.0007065

Find the numbers corresponding to the logarithms in Probs. 4 to 6.

4. (a) 3.75051	(b) 0.56038 $-$ 2	(c) 1.55096
(d) 0.00043	(e) 0.93997 $-$ 1	
5. (a) 7.93475 $-$ 10	(b) 0.06333	(c) 5.60206
(d) 2.96440	(e) 4.76005	
6. (a) 1.21537	(b) 3.61794	(c) 0.27161 $-$ 4
(d) 5.99502	(e) 2.47012	

11.16 Interpolation

It is often necessary to find the logarithm of a number which does not exactly appear in the table. If we are satisfied to use the logarithm of the nearest number, we reduce the accuracy of a five-place table to that of a four-place table. It is not necessary to sacrifice this extra degree of accuracy if we resort to interpolation. In a five-place table interpolation will determine for us the fifth digit in the mantissa, but not a sixth.

Example 2. Find by interpolation log 25.813.

The characteristic of the required logarithm is found by inspection. In this case the characteristic is 1. For the present, therefore, we shall give our entire attention to the problem of finding the required mantissa.

From Sec. 11.12 we recall that the mantissa of the logarithm of a number depends only on the sequence of digits in that number, and not on the position of the decimal point in it. For example, the mantissa of the logarithm of 25.813 is exactly the same as the mantissa of the logarithm of 25,813, and we shall proceed to use this equivalence in the discussion to follow, not from necessity but as a matter of convenience.

First find the mantissas of 25,810 and 25,820. These are 41179 and 41196, respectively (decimal points are disregarded for the time being). The *tabular difference* between the mantissas is 17, whereas the difference between 25,810 and 25,820 is 10. It is evident that the given number 25,813 is 0.3 of the way from 25,810 to 25,820. Therefore the desired mantissa must be about 0.3 of the way from 41179 to 41196. 0.3 of 17 is 5.1, which is rounded off to 5. Adding 5 to 41179 we get 41184. Therefore the mantissa is 0.41184, and by inspection the characteristic is 1. Hence log 25.813 = 1.41184.

The interpolation may be arranged as follows:

Write down the bracketing numbers and their mantissas, leaving space for the intermediate number and its mantissa. Annex a cipher as a fifth digit to each bracketing number and write in the given number. Indicate the differences or proportional parts.

Number	Mantissa

By proportion, $3/10 = x/17$, or $x = (\frac{3}{10})(17) = 5.1$, by mental calculation or slide rule. Observe that since the mantissas are unending decimals and are rounded off to five digits, there would be no point in adding 41179 and 5.1 and calling the result 41184.1. Hence we calculate x only to the nearest whole number.

Most tables include tables of proportional parts in which x in the above proportion may be found. This is simply a table of tenths of the tabular difference, which is 17 in this example. Usually, mental computation is the quickest way to determine the proportional part x.

	17
1	1.7
2	3.4
3	5.1
⋮	⋮

Example 3. Find log 232.464973.

If, as in this example, the given number contains more than five significant figures and the tabular difference is 15 or more, round off to six significant figures and proceed as in the following example.

Rounding off to six digits, we get 232.465.

By proportion, $65/100 = x/18$, or $x = (^{65}/_{100})(18) = 11.7$, which we call 12. Therefore the mantissa is $36624 + 12$, or 0.36636. The complete logarithm is 2.36636.

The process of interpolation as applied to antilogarithms is simply the reverse of that used in looking up logarithms.

Example 4. Find antilog 3.54859.

We do not find 54859 in the table of mantissas, but we do find the bracketing mantissas 54851 and 54864. These correspond to the numbers 3,536 and 3,537, respectively. The operation is illustrated as follows:

By proportion, $u/10 = 8/13$, or $u = (^{8}/_{13})(10) = 6.2$, which we call 6. Therefore the digits of the number are $35,360 + 6$, or 35,366. The characteristic 3 determines the number to be 3,536.6.

If we choose to use the proportional-parts table, we scan the values listed under 13 until we come to the value nearest 8. This is 7.8, which corresponds to 6 (0.6 of 13).

	13
—	—
5	6.5
6	7.8
7	9.1
—	—

Note that if the difference x is less than 1 part out of 20, interpolation is omitted, since if $x < 1$, $(x/20)(10) < 0.5$, or less than 5 in the sixth place.

In this case, simply the nearest mantissa is taken.

Example 5. Find the antilogarithm of 1.21565.

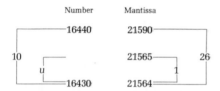

$u/10 = 1/26$, or $u = (\frac{1}{26})(10) = 0.4$, which is less than 0.5; $21564 + 0.4 = 21564$. Therefore we ignore interpolation and take the nearest mantissa, 21564. Thus we get antilog $1.21565 = 16.430$.

Interpolation may sometimes be facilitated by approaching our answer from the upper value instead of the lower.

Example 6. Find log 0.012148.

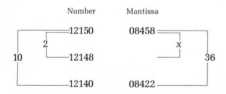

Here we write $x/36 = 2/10$, or $x = (\frac{2}{10})(36) = 7.2$, or 7; $08458 - 7 = 08451$.

This is somewhat easier than the equivalent operation:

$$y/36 = 8/10; \quad y = 28.8 \text{ or } 29; \quad 08422 + 29 = 08451.$$

In either case the complete logarithm is $0.08451 - 2$.

EXERCISE 8

Using a five-place log table, look up the logarithms of the numbers in Probs. 1 to 3, obtaining the last digit by interpolation.

1. (a) 268.18   (b) 0.0043356   (c) 21.355
   (d) 9.2354   (e) 4.1497
2. (a) 68.173   (b) 0.030689   (c) 55,071
   (d) 0.13309   (e) 8,585.4
3. (a) 11,958   (b) 37.596   (c) 0.00079364
   (d) 2,090.2   (e) 118,210,000

Find to five significant figures the numbers whose logarithms are given in Probs. 4 to 6. Obtain the fifth significant figure by interpolation.

4. (a) 3.20756   (b) $0.29577 - 1$ (c) 1.79652
   (d) 4.38896   (e) 2.59460
5. (a) $0.00309 - 3$ (b) 0.47562   (c) $0.89964 - 4$
   (d) $0.69250 - 2$ (e) 5.14521
6. (a) $0.01997 - 4$ (b) 1.27461   (c) 5.93376
   (d) 2.00215   (e) $0.37965 - 1$

11.17 Computation Using Logarithms

In carrying out computations involving logarithms, we shall make use of the equations set forth in Sec. 11.2.

11.18 Multiplication by the Use of Logarithms

According to Eq. (3), the logarithm of a product equals the sum of the logarithms of the separate factors.

Example 7. Find the product $(238.24)(0.072495)(9.5668) = N$.

From Eq. (3), $\log 238.24 + \log 0.072495 + \log 9.5668 = \log N$.

From the tables,

$$
\begin{aligned}
\log 238.24 \ \ &= 2.37701 \\
\log 0.072495 &= 0.86031 - 2 \\
\log 9.5668 \ \ &= \underline{0.98077} \\
\log N =\ &4.21809 - 2 \\
=\ &2.21809 \\
N =\ &165.23
\end{aligned}
$$

If a computation involves negative quantities, they should be treated as though they were positive, since a negative number does not have a logarithm in the set of real numbers.

The sign of the answer can be determined by inspection after completion of the computation.

EXERCISE 9

Using logarithms, perform the following computations to five significant figures:

1. $47.690 \times 32.410 \times 76.480 \times 1.9320$
2. 3.7596×159.06
3. 10.365×0.42659
4. $(2,586.0)(-169.20)(1.4230)(-0.96000)$
5. $73.650 \times 9.9000 \times 0.99990 \times 1.0010 \times 0.013090$
6. $(70.603)(0.55146)(0.00021300)(1.4020)(40.326)(-743.01)$
7. $25.726 \times 5.0789 \times 2,188.5 \times 5.2525 \times 4,660.1$
8. $2,208.6 \times 0.043680 \times 2,781,200,000 \times 0.000016976$
9. $0.0018746 \times 15.856 \times 2,650.0 \times 2.0460 \times 50,620,000$
10. $3.6492 \times 1,006.4 \times 0.00037964 \times 0.0017960 \times 36.592$

11.19 Division by the Use of Logarithms

From Eq. (4) we find that the logarithm of a quotient equals the logarithm of the numerator minus the logarithm of the denominator.

Example 8. Evaluate $845.67/0.68332 = N$.

log 845.67 = 2.92720
log 0.68332 = 0.83462 − 1
$\qquad$ log N = 2.09258 + 1
$\qquad\qquad$ = 3.09258
$\qquad\quad$ N = 1237.6

Example 9. Evaluate 37.844/510.26 = N.
$\qquad$ If we attempt to perform this subtraction directly, we write

log 37.844 = 1.57800
log 510.36 = 2.70788
$\qquad$ log N = −1.12988

$\qquad$ Although the logarithm of the product is algebraically correct, we are unable to look up the antilogarithm because the ordinary tables do not contain negative mantissas. To avoid this impasse we may write the logarithm of the numerator in the equivalent form

log 37.844 = 1.57800 = 1.57800 + 2 − 2

or

log 37.844 = 3.57800 − 2

Then the logarithmic subtraction would appear as

log 37.844 = 3.57800 − 2
log 510.36 = 2.70788
$\qquad$ log N = 0.87012 − 2

and N = 0.074152.
$\qquad$ This device can be used whenever it is necessary to subtract a larger logarithm from a smaller one in the process of calculating the logarithm of a quotient. In general, we add and subtract an integer which is large enough to yield a positive mantissa in the difference.

EXERCISE 10

Use logarithms to compute the following quotients to five significant figures:

1. $(3,796.0) \div (-472.00)$ $\qquad$ 2. $7,943.2 \div 25.793$

3. $7.5437 \div 97.604$

4. $0.047920 \div 6.3951$

5. $4,605.2 \div 56,517$

6. $99.976 \div 10.532$

7. $1.000.1 \div 3,576.5$

8. $0.37960 \div 0.020710$

9. $12.147 \div 0.00049630$

10. $-5,076.0 \div 0.043297$

11. $1 \div 53.827$

12. $1 \div 0.082941$

11.20 Combined Multiplication and Division

This is simply a combination of Eqs. (3) and (4) for multiplication and division.

Example 10. Evaluate $75.282\pi/[(6.754)(0.38949)(0.01217)] = N$.

$$\log 75.282 \ = 1.87669$$
$$\log 3.1416 \ = 0.49715$$
$$\overline{\log \text{num.} \quad = 2.37384 \qquad \longrightarrow \quad 2.37384}$$

$$\log 6.754 \quad = 0.82956$$
$$\log 0.38949 = 0.59050 - 1$$
$$\log 0.01217 = 0.08529 - 2$$
$$\overline{\log \text{den.} \quad = 1.50535 - 3 \longrightarrow \quad 1.50535 - 3}$$
$$\overline{\qquad\qquad\qquad\qquad\qquad\qquad 0.86849 + 3}$$

$$\log N = 3.86849$$
$$N = 7387.3 \quad \text{or better} \quad N = 7387$$

11.21 Cologarithms

The *cologarithm* of a number is the logarithm of the reciprocal of the number. It is therefore equal to the negative of the logarithm of the number.

$$\text{colog } N = \log \frac{1}{N} = -\log N \tag{13}$$

Since division by N is equivalent to multiplication by $1/N$, it follows that division by a given number may be accomplished by adding the cologarithm instead of subtracting the logarithm. A combined multiplication and division problem may therefore be evaluated by adding up only one column of logarithms.

For convenience, we wish to avoid becoming involved with negative mantissas when using cologs in calculations. To accomplish this we often resort to the same sort of a device as we used in Example 9. For example, following Eq. (13), we may write

$$\text{colog } N = -\log N$$
$$= -\log N + 10 - 10 \tag{14}$$

(It is usually more convenient to add and subtract 10 or some integral multiple of 10.)
We may write Eq. (14) as

$$\text{colog } N = (10 - \log N) - 10 \tag{15}$$

Example 11. Find the cologarithm of N if $N = 66.32$.
First let us evaluate the quantity $(10 - \log 66.32)$ in Eq. (15). From a five-place log table we find that

$$\log 66.32 = 1.82164$$

Then

$$\begin{aligned}
10 - \log N &= 10.00000 - 1.82164 \\
&= 8.17836
\end{aligned}$$

and from Eq. (15)

$$\text{colog } 66.32 = 8.17836 - 10 = 0.17836 - 2$$

As a practical matter, we usually evaluate the quantity $(10 - \log N)$ mentally as we read the value of $\log N$ directly from the table. Beginning at the left, we subtract each digit in the logarithm of N from 9 until we reach the last *significant* digit, which we subtract from 10. The result is the cologarithm of N.

Example 12. Find the colog of N if $N = 0.7042$.

$$\begin{aligned}
\text{colog } 0.7042 &= (10 - \log 0.7042) - 10 \\
&= [10 - (0.84770 - 1)] - 10 \\
&= 10 - 0.84770 + 1 - 10 \\
&= 9.15230 + 1 - 10 \\
&= 9.15230 - 9 = 0.15230
\end{aligned}$$

Example 13. Evaluate N in Example 10 by using cologs.

$$\begin{aligned}
\log 75.282 &= 1.87669 \\
\log \ \ 3.1416 &= 0.49715 \\
\text{colog } \ 6.754 &= 0.17044 - 1 \\
\text{colog } \ 0.38949 &= 0.40950 \\
\text{colog } \ 0.01217 &= \underline{1.91471} \\
& \qquad \quad 4.86849 - 1 \\
\log N &= 3.86849 \\
N &= 7387.3
\end{aligned}$$

EXERCISE 11

Evaluate the following to five significant figures.

1. $\dfrac{576.43 \times 976.52 \times 1.4962}{3.7425 \times 0.0096520 \times 0.017360}$

2. $\dfrac{57.040 \times 25.936 \times 0.48352}{764.32 \times 97.630 \times 0.0079860}$

3. $\dfrac{601.47 \times 93.276 \times 2.5037 \times 79.631}{927.43 \times 26.485 \times 0.0017930 \times 62.000}$

4. $\dfrac{10.243 \times 1.5006 \times 0.96016 \times 9.1207}{59.329 \times 3.8421 \times 0.027532 \times 0.50008}$

11.22 Raising to a Power

According to Eq. (5), the logarithm of the power of a number is found by multiplying the logarithm of the number by the exponent.

Example 14. Evaluate $(1.8653)^8 = N$.
    From Eq. (5) we may write

$\log N = 8 \log 1.8653$

From a table of five-place logarithms we find that

$\log 1.8653 = 0.27075$

Then

$\log N = 8 \times 0.27075 = 2.16600$

and

$N = 146.56$

Example 15. Evaluate $(5.167)^{-2.8} = N$.
    From Eq. (5) we may write

$\log N = -2.8 \log 5.167$

From a table of five-place logarithms we find

$\log 5.167 = 0.71324$

Then

$\log N = -2.8 \times 0.71324 = -1.99707$

Observe that the negative sign in the foregoing equation affects the entire logarithm, both characteristic and mantissa. To obtain a logarithm of equal value but with a positive mantissa, we again resort to the device used in Example 9 and we write

$$\log N = -1.99707 = (10 - 1.99707) - 10$$
$$= 8.00293 - 10$$
$$= 0.00293 - 2$$

Then we find

$N = 0.010068$

Alternative solution:

$$(5.167)^{-2.8} = \frac{1}{(5.167)^{2.8}} \qquad \text{(law of exponents)}$$

$$\log 5.167 = \ \ 0.71324$$
$$\underline{\hspace{1.5cm} 2.8}$$
$$570592$$
$$\underline{142648}$$
$$\log (5.167)^{2.8} = \overline{1.997072}$$

$$\log 1 = 2.00000 - 2$$
$$\log (5.167)^{2.8} = \underline{1.99707}$$
$$\log N = \overline{0.00293 - 2}$$
$$N = 0.010068$$

Example 16. Evaluate $(0.17528)^{1.75} = N$.

$$\log 0.17528 = 0.24373 - 1$$
$$\log N = 1.75 \log 0.17528$$
$$= 1.75(0.24373 - 1)$$
$$= 0.42653 - 1.75$$

Now add zero in a form to make the characteristic a whole number. In this case we add zero in the form of $(0.25 - 0.25)$ as follows:

$$0.42653 - 1.75$$
$$\underline{+0.25 \quad\quad - 0.25 \quad (=0)}$$
$$\log N = 0.67653 - 2$$
$$N = 0.047482$$

Example 17. Evaluate $(0.4627)^{-3.2} = N$.

$$\log N = -3.2 \log 0.4627$$
$$= -3.2(0.66530 - 1)$$

where $\log 0.4627 = 0.66530 - 1$

$$\log N = -2.12896 + 3.20000 = +1.07104$$
$$N = 11.777$$

Alternative solution:

$$(0.4627)^{-3.2} = \frac{1}{(0.4627)^{3.2}} = \left(\frac{1}{0.4627}\right)^{3.2}$$

$$\log 1 = 1.00000 - 1$$
$$\underline{\log 0.4627 = 0.66530 - 1}$$
$$0.33470$$
$$\underline{\quad\quad 3.2}$$
$$66940$$
$$\underline{100410}$$
$$\log N = 1.07104$$
$$N = 11.777$$

EXERCISE 12

Use logarithms to evaluate the following to five significant figures. Consider all values good to five significant figures.

1. (a) $(4.875)^5$ (b) $(-11.83)^3$ (c) $(0.6432)^4$
2. (a) $(0.09458)^3$ (b) $(0.14732)^2$ (c) $(-0.25713)^4$
3. (a) $(23.805)^{-1}$ (b) $(7.432)^{-2}$ (c) $(16.031)^{-3}$
4. (a) $(17.584)^{3/2}$ (b) $(405.76)^{2/3}$ (c) $(1807.4)^{2/5}$
5. (a) $(0.6432)^{2.2}$ (b) $(0.09473)^{-1.8}$ (c) $(2.75)^{2.75}$
6. (a) $(0.16223)^{3/4}$ (b) $(0.074962)^{4/5}$ (c) $(0.25814)^{5/2}$
7. (a) $(70.58)^{0.2}$ (b) $(13.625)^{1.3}$ (c) $(48.461)^{-0.3}$
8. (a) $(14.83)^{-1.62}$ (b) $(0.3447)^{-2.22}$ (c) $(0.0962)^{3.15}$

11.23 Extracting Roots

Referring to Eq. (6), Sec. 11.2, we read that the logarithm of the root of a number equals the logarithm of that number divided by the index of the root.

Example 18. Evaluate $\sqrt[4]{737.12} = N$.

$$\log 737.12 = 2.86754$$

$$\log N = \frac{2.86754}{4}$$

$$= 0.71688$$

$$N = 5.2106$$

Example 19. Evaluate $\sqrt[3]{0.00028864} = N$.

$$\log 0.00028864 = 0.46036 - 4$$

If we were to divide the entire logarithm by 3 as it stands, the characteristic -4 would give us the awkward -1.33333. To avoid this, we add at the left, and subtract at the right, a quantity which will make the characteristic divisible by 3. The simplest such number is 2.

$$
\begin{array}{r}
\log 0.00028864 = \quad 0.46036 - 4 \\
+2 \qquad\quad - 2 \\
\hline
3\,/\,\overline{2.46036 - 6} \\
\log N = \quad 0.82012 - 2 \\
N = 0.066087
\end{array}
$$

EXERCISE 13

Evaluate the following expressions, assuming all values accurate to five significant figures:

1. (a) $\sqrt[3]{972}$ (b) $\sqrt[3]{97.2}$ (c) $\sqrt[3]{9.72}$
2. (a) $\sqrt[3]{0.972}$ (b) $\sqrt[3]{0.0972}$ (c) $\sqrt[3]{0.00972}$
3. (a) $\sqrt{73.464}$ (b) $\sqrt[3]{168.15}$ (c) $\sqrt{3051.8}$
4. (a) $\sqrt{0.082167}$ (b) $\sqrt[3]{0.22469}$ (c) $\sqrt{0.000058274}$
5. $(8.527)^3 \times (0.7161)^2$ 6. $\sqrt{58.43} \times (6.710)^2$
7. $(185.2 \times 0.071828)^3$ 8. $\sqrt{97.422 \times 4.7881}$
9. $\sqrt{\dfrac{219.6}{0.005733}}$ 10. $\sqrt[3]{\dfrac{52.482}{371.05 \times 0.18061}}$

11. $\sqrt[3]{\dfrac{3 \times 0.28617}{4\pi}}$

12. $[\pi(0.07112)(8.859)]^2$

13. $\sqrt[3]{\dfrac{(427.5)^2}{(13.482)^5}}$

14. $\sqrt[6]{\left(\dfrac{27.42}{0.03966}\right)^5}$

15. 15. $\sqrt[5]{-381.72}$

16. $\sqrt[3]{\dfrac{-17.477}{(-0.84991)^2}}$

17. $\left(\dfrac{64.66}{-1009}\right)^{2/3}$

18. $\sqrt[4]{\dfrac{1}{8.3567}}$

11.24 Logarithmic Computation of Expressions Involving Addition and Subtraction

Since numbers cannot be added or subtracted by logarithms, we must convert to antilogarithms before performing such an operation.

Example 20. Evaluate $\sqrt{(a^5 + 1)/(a^5 - 1)} = N$, where $a = 1.0037$.

$$\log 1.0037 = 0.0016039 \qquad \text{(seven-place table)}$$

$$\dfrac{5}{}$$

$$\log (1.0037)^5 = 0.0080195$$
$$(1.0037)^5 = 1.01864 \qquad \text{(seven-place table)}$$

$$\sqrt{\dfrac{1.01864 + 1}{1.01864 - 1}} = \sqrt{\dfrac{2.01864}{0.01864}}$$

$$\log 2.0186 = 0.30505$$
Subtract: $\log 0.01864 = 0.27045 - 2$
$$\overline{0.03460 + 2}$$
$$2/2.03460$$
$$\log N = 1.01730$$
$$N = 10.41$$

Note that although we have been able to find $(1.0037)^5$ to six figures using a seven-place table, after we subtract 1 we get a denominator 0.01864 having only four significant figures. Therefore we can retain only four significant figures in our answer. This situation points up the need of extra accuracy in reading logarithms of numbers near unity. The mantissas of such numbers may have only two or three significant figures when taken from a five-place table.

EXERCISE 14

Use logarithms to compute the value of the following to five significant figures.

1. $\sqrt{(8.3150)^2 - (5.0240)^2}$ (see Sec. 11.26)
2. $\sqrt{(11.964)^2 + (4.4857)^2}$ 3. $\sqrt[3]{(4.1150)^3 + (6.3820)^3}$

4. $\sqrt[5]{(1.4860)^7 - 20.000}$ 5. $\sqrt{\dfrac{(1.3890)^3 + 1}{(1.3890)^3 - 1}}$

The short problems in Exercises 15 and 16 may be used to test the student's understanding of the properties of logarithms without spending the time usually required in consulting tables.

EXERCISE 15

Transform the equations in Probs. 1 to 4 into exponential form.

1. $\log_b (a - x) = c$ 2. $\log_b (x^2 + 3) - \log_b (2x + 1) = k$
3. $2 \log_b (3x - 2) = a + c$ 4. $\log_b y = \sqrt{2x + 5}$
5. If $\log \sqrt{x} = 0.35$, find $\log x^2$.   6. If $\log_b bx = 5$, find $\log_b x$.
7. If $\log 1/x = \frac{1}{4}$, find $\log x$.   8. If $\log x^2 = 16$, find $\log x$.
9. If $\log (x^2 - 9) - \log (x + 3) = 0$, find x.

Rewrite the expressions in Probs. 10 to 12 without exponents or radicals.

10. (a) $\log x^5$ (b) $\log 3x^2$ (c) $\log \dfrac{5}{x^2}$

11. (a) $\log \sqrt[3]{x}$ (b) $\log \dfrac{1}{x^3}$ (c) $\log \dfrac{1}{2\sqrt{x}}$

12. (a) $\log \dfrac{x^5}{8}$ (b) $\log_b bx^2$ (c) $\log_b \dfrac{2x^3}{b^2}$

Convert the expressions in Probs. 13 and 14 to single positive logarithms whose coefficient is 1.

13. (a) $6 \log x$ (b) $2 \log x - \log 3$ (c) $\dfrac{\log_b x}{2} + 1$

14. (a) $3 - \log_b x$ (b) $\frac{1}{2} (\log x + \log 2)$ (c) $-10 \log x$

15. A student wrote $\log (N + 2) + \log N = \log (2N + 2)$. This is true only for what positive value of N?
16. What is the value of $\log_5 0.04$?
17. If $\log (p/q) + \log (q/r) - \log (r/p) - 2 \log (p/r) = \log x$, what is x?
18. If $S = P(1 + r)^{-n}$, solve for n.
19. If $\log_b y = 2 - 3 \log_b x$, solve for y.

EXERCISE 16

Given only that log 2 = 0.301 and log 3 = 0.477, find the following without using tables:

1. (a) log 2,000   (b) log 0.03   (c) log $\sqrt{30}$
2. (a) log $\frac{1}{16}$ (b) log $(3)^7$ (c) log 6
3. (a) log 1   (b) log 20   (c) log 5
4. (a) log 1.5   (b) log 8   (c) log 0.0003
5. (a) log $\sqrt{15}$ (b) log $\sqrt[3]{60}$ (c) log $3\frac{1}{3}$
6. (a) log 39 − log 13   (b) log $(\frac{1}{3})^4$ (c) log $\sqrt[5]{1\frac{1}{2}}$
7. (a) log $\sqrt[4]{1\frac{1}{3}}$ (b) log $\sqrt[3]{0.0002}$ (c) log $(24)^{1.65}$
8. (a) log 46 − log 23   (b) antilog 3.477   (c) antilog 0.301 − 3
9. (a) antilog 2.301   (b) antilog 0.176 − 2   (c) antilog 3.01
10. (a) antilog 1.778   (b) antilog 7.778 − 10   (c) antilog 0.0477

11.25 Accuracy in Logarithmic Computations

The student should bear in mind that in multiplication and division the accuracy of the result is determined by the least accurate individual factor. If several six-digit numbers are to be multiplied by a number of four significant figures, all the individual factors should be rounded off to four significant figures, since the answer will be limited to this degree of accuracy. Consequently, a four-place table of logarithms will suffice. By the same token, if the utmost accuracy is to be secured from a combination of numbers of seven significant figures, a table to at least seven places must be used.

11.26 Aids to Logarithmic Computation

Computation involving logarithms may be expedited by observing a few details of technique, of which the following are representative:

1. Follow some orderly plan of development, such as that set forth in Example 10, Sec. 11.20.
2. Cancel or combine simple numbers where possible to reduce the number of logarithms to be handled.
3. Improve the arrangement for logarithmic computation, e.g.,
   (a) $2\pi r^2 + 2\pi rh = 2\pi r(r + h)$. Once r and h are added, there will be no further obstacle to logarithmic computation.
   (b) $\sqrt{c^2 - a^2} = \sqrt{(c + a)(c - a)}$. Here we enter a "straightaway" as soon as $c + a$ and $c - a$ are evaluated.
4. Negative characteristics and mantissas can often be avoided in a division problem by multiplying numerator and denominator by some integral power of 10, for example, $0.03152/0.00567 = 31.52/5.67$.

5. Another device to circumvent negative logarithms is multiplication by a^n. For example, the equation $47.59 = (1.52)^{-3}N$ may be multiplied throughout by $(1.52)^3$; thus we get $(47.59)(1.52)^3 = N$.

11.27 Power Function vs. Exponential Function

Thus far our discussion of calculation of powers and roots by means of logarithms has been limited to cases of the type x^c, where x is a variable and c is a constant. (See particularly Probs. 1 and 2, Exercise 13, page 303.)

An expression such as x^3, in which a variable is raised to a fixed power, is called a *power function*. We shall now consider *exponential functions,* that is, expressions in which the variable occurs in the exponent. Table 11.4 contrasts these functions.

In this table we note that values of x, which are evenly spaced, are in arithmetic progression because there is a constant difference between successive values. If we examine the values of 3^x, we see that they are in geometric progression because there is a constant ratio between two successive values. This is a most important property of exponential functions.

11.28 Exponential Equations

An equation in which the unknown appears in the exponent is called an *exponential equation.* An example of the simplest of this type is $3^x = 9$. Here x is obviously equal to 2, since $3^2 = 9$.

In equations in which the result cannot be determined by inspection, we must take logarithms of both sides of the equation and obtain the final solution by ordinary algebraic methods.

Example 21. Solve the following equation for x:

$$32^x = 512$$

Since we might have written $(2^5)^x = 2^9$, it is quite obvious from inspection that $5x = 9$ and that $x = 1.8$. However, we wish to assure ourselves that the logarithmic method is valid. Therefore we shall take the logarithm of each member, obtaining

TABLE 11.4

	x	0	1	2	3	4	5	6	7
Power function	x^3	0	1	8	27	64	125	216	343
Exponential function	3^x	1	3	9	27	81	243	729	2,187

$\log 32^x = \log 512$

Then

$x \log 32 = \log 512$
$1.50515x = 2.70927$

  To avoid confusion, it is suggested that the student consider this equation as an ordinary linear equation, which it is. Had this equation appeared in the section of elementary algebra devoted to linear equations in one unknown, there would be no doubt as to the method of solving for x. The possible origin of the equation did not affect the procedure there; neither does it in this case. Accordingly,

$$x = \frac{2.70927}{1.50515} = 1.8000$$

  If only two- or three-place accuracy is desired, this division may be performed on the slide rule.

Example 22. Solve the equation $9^{2x-7} = 27^{x-3}$ for x.
  Since in this example both 9 and 27 are powers of the same number, 3, we might proceed by inspection without tables:

$9^{2x-7} = 27^{x-3}$

But

$9 = 3^2$ and $27 = 3^3$

Therefore

$(3^2)^{2x-7} = (3^3)^{x-3}$
$3^{4x-14} = 3^{3x-9}$

It is evident that $4x - 14$ must equal $3x - 9$; therefore $x = 5$.

  Alternative solution: Taking logarithms of each member, we obtain

$$\log 9^{2x-7} = \log 27^{x-3}$$
$$(2x - 7) \log 9 = (x - 3) \log 27$$
$$(2x - 7)(0.95424) = (x - 3)(1.43136)$$

Here again we have an ordinary linear equation and proceed accordingly. Carrying out the indicated multiplication,

$$1.90848x - 6.67968 = 1.43136x - 4.29408$$

$$0.47712x = 2.38560$$

$$x = \frac{2.38560}{0.47712} = 5.0000$$

In some cases a few preliminary steps are necessary to obtain one term on each side of the equation before taking logarithms.

In Examples 23 and 25 below we cannot solve by inspection, and we must use the general logarithmic method.

Example 23. Solve for x the equation $\dfrac{(1.05)^x - 1}{0.05} = 18.$

$$(1.05)^x - 1 = (0.05)(18) = 0.9$$
$$(1.05)^x = 1.9$$

Proceed as before.

Example 24. Given the equation $y = 10 \, (1.065)^t$, evaluate y to 0.01 when $t = 0$, 11, 22, 33, 44. Using the principles of Sec. 11.22, we can relate y and t in the following table:

t	0	11	22	33	44
y	10.00	19.99	39.97	79.90	159.73

This is actually a table of the growth of $10 at $6\frac{1}{2}$ percent compound interest. The compound amount y is seen to approximately double in the first 11 years and predictably doubles again in each successive 11 years. We shall deal with this property at some length in Sec. 11.29.

Some equations in which the value of the exponent ultimately proves to be negative depend for their solution upon the deliberate conversion of the mantissa of a logarithm to the negative form (Sec. 11.13).

Example 25. Solve for x the equation $6^x = 0.0208.$
Taking logarithms of both sides,

$$x \log 6 = \log 0.0208$$

Then

x(0.77815) = 0.31806 − 2

Treating this as an ordinary linear equation, we obtain

0.77815x = 0.31806 − 2.00000

0.77815x = −1.68194

$$x = \frac{-1.68194}{0.77815}$$

$$= -2.1615$$

Example 26. Solve the equation $0.05 = (\frac{2}{3})^x$ for x.

In equations of this type we can avoid negative characteristics, thereby simplifying our computation by taking reciprocals of both sides. Thus we obtain $1/0.05 = (1/\frac{2}{3})^x$, or $20 = (\frac{3}{2})^x$. Then, as in Example 25, log 20 = x log 1.5, or 1.30103 = 0.17609x and x = 7.3884.

EXERCISE 17

Solve for x in the following exponential equations:

1. $2^x = 75$ 2. $(1.07)^x = 3$ 3. $125^x = 48(5)^x$

4. $5^x = 1,000$ 5. $(1.05)^{-x} = 0.36$ 6. $3^{x+1} = 5^{x-1}$

7. $4^{x+2} = 8^{2x-1}$ 8. $6^{-x} = 0.02$ 9. $(1.04)^{-x} = 0.75$

10. $\dfrac{(1.06)^x - 1}{0.06} = 12$ 11. $8^x = \dfrac{8}{2^{x-3}}$ 12. $5^x = 20^{1/x}$

11.29 Compound Interest Law (CIL), or Exponential Law of Growth and Decay

Many quantities in nature grow in much the same way as a sum of money at compound interest; i.e., the rate of growth is a fixed percentage of the amount on hand at the beginning of the interest period in question. Many chemical reactions conform to this type.

The growth of money occurs stepwise at the end of each interest-conversion period, whereas growth (and decay) in nature usually appears to be a smooth, continuous process. This simply means that the "interest" is converted exceedingly often and the analogy to compound interest remains close.

If a chemical substance is reacting, it is reacting one molecule at a time. An electrical condenser discharges one electron at a time, and a hot body is radiating its energy one quantum at a time.

As indicated above, one of the best-known examples of exponential equations is found in the growth of money at compound interest. The equation is

$$S = P(1 + r)^n \tag{16}$$

where P = original principal invested
n = number of interest periods
r = rate of interest (expressed as a decimal) applied to one interest period
S = compound amount of original principal plus accrued interest

Example 27. Make a table of compound amounts for the first 5 years if $100 is invested at 6 percent compounded annually.

Substituting in Eq. (16), we have $S = 100(1 + 0.06)^t$. Since we have one period per year, we may replace n by t.

Year, t	Principal at beginning of year	Common multiplier	Compound amount at end of year, S
1	$100.00	1.06	$100(1.06) = \$106.00$
2	106.00	1.06	$100(1.06)^2 = 112.36$
3	112.36	1.06	$100(1.06)^3 = 119.10$
4	119.10	1.06	$100(1.06)^4 = 126.25$
5	126.25	1.06	$100(1.06)^5 = 133.82$
⋮	⋮	⋮	⋮

Example 28. Repeat Example 27 except that interest is to be compounded (or converted) semiannually. Here we have two periods a year at 3 percent per period, which requires that we replace n by $2t$ and set $r = 0.03$ (3 percent per half year).

Thus we have $S = 100(1 + 0.03)^{2t}$, and the principal will grow according to the following table.

Year, t	Periods, $n = 2t$	Principal at beginning of year	Common multiplier	Compound amount at end of year, S
1	2	$100.00	$(1.03)^2$	$100(1.03)^2 = \$106.09$
2	4	106.09	$(1.03)^2$	$100(1.03)^4 = 112.55$
3	6	112.55	$(1.03)^2$	$100(1.03)^6 = 119.41$
4	8	119.41	$(1.03)^2$	$100(1.03)^8 = 126.68$
5	10	126.68	$(1.03)^2$	$100(1.03)^{10} = 134.39$
⋮	⋮	⋮	⋮	⋮

Example 29. A machine worth $1,280 new depreciates in value by a constant 25 percent annually of its value at the beginning of any given year. In this example $r = -0.25$

and the equation might be expressed

$$V = V_0(1 - 0.25)^t = V_0(0.75)^t$$

where t = time in years dating from time of purchase (= number of periods)

V_0 = original purchase price (= \$1,280)

V = book value after t years

A table of book values would appear as follows:

Year, t	Book value at beginning of year	Common multiplier	Book value at end of year, V
1	\$1,280	0.75	$1,280(0.75) = \$960$
2	960	0.75	$1,280(0.75)^2 = 720$
3	720	0.75	$1,280(0.75)^3 = 540$
4	540	0.75	$1,280(0.75)^4 = 405$
$\vdots$	$\vdots$	$\vdots$	$\vdots$

There are many physical laws which obey the CIL concept. A number of these laws are concerned with a depreciating value, as in Example 29. A few typical examples are found in radioactive decay, cooling of a warm object, extinction of light in an absorbing medium, discharge of an electrical condenser, etc. In all these cases r is negative, and the common multiplier, $1 + r$, while positive, is smaller than unity.

Example 30. The temperature of a warm body surrounded by a cooler medium is given by the equation

$$T = T_0(1 + r)^n \qquad \text{Newton's law of cooling}$$

where T_0 = initial temperature difference between temperatures of body and its surroundings

T = temperature difference after any number of periods (n)

r = rate of decay per period

Assume that an object initially 50° warmer than its surroundings cools by 20 percent per period of the temperature difference at the beginning of the period. Referring to the general equation above, we may write

$$T = 50(1 - 0.20)^n = 50(0.80)^n$$

If the period is 10 min, then

$$T = 50(0.80)^{t/10}$$

where t is the elapsed time in minutes. A table would appear as follows;

Minutes, t	Periods, $n = t/10$	Temperature difference at beginning of period	Common multiplier	Temperature difference at end of period $= T$
10	1	50	0.80	$50(0.80) = 40$
20	2	40	0.80	$50(0.80)^2 = 32$
30	3	32	0.80	$50(0.80)^3 = 25.6$
40	4	25.6	0.80	$50(0.80)^4 = 20.48$
⋮	⋮	⋮	⋮	⋮

Example 31. A radioactive sample has a half-life of 18 h; i.e., after any 18-h period the radioactive weight is one-half its value at the beginning of that period. The appropriate equation could be written $W = W_0(0.50)^{t/18}$. This is equivalent to writing $W = W_0(2)^{-t/18}$. Why? If the initial weight $W_0 = 12$ mg, the following table would apply:

Hours, t	Periods, $n = t/18$	Radioactive weight at beginning of period, mg	Common multiplier	Radioactive weight at end of period, mg
18	1	12	0.5	$12(0.5) = 6$
36	2	6	0.5	$12(0.5)^2 = 3$
54	3	3	0.5	$12(0.5)^3 = 1.5$
72	4	1.5	0.5	$12(0.5)^4 = 0.75$
⋮	⋮	⋮	⋮	⋮

A most important conclusion can be drawn from the above CIL examples: When the values of the independent variable (in these cases, time) are evenly spaced, the successive values of the dependent variable are obtained by multiplying by a constant ratio.

EXERCISE 18

Apply the compound interest law to the solution of the following problems:

1. To what amount (nearest dollar) will $5,000 grow in 22 years at 7 percent compounded semiannually?
2. If $900 grows to $3,200 in 26 years with interest compounded annually, find the rate.
3. How long will it take for a principal to double at $6\frac{1}{2}$ percent compounded annually?
4. If $100 grows to $200 in 12 years with interest compounded annually, can you determine the amount (*a*) after 6 years, (*b*) after 4 years, without using logs?
5. A chemical decomposition follows the CIL. The weights of undecomposed chemical remaining at various times are given by the table

t, hr	0	1	2	3
W, g	54	36	24	—

(a) Predict at sight W at $t = 3$ h.

(b) If the equation is $W = W_0 C^t$, find the values of W_0 and C.

(c) Find W at $t = 2.3$ h.

(d) What is the period of half-life?

6. The population of a small city in 1960 was 27,000; in 1970 it was 38,500. Find the annual percentage rate of increase, assuming it to be uniform. What was the probable population in 1967?

7. The speed of a chemical reaction doubles for each 18°F rise in temperature. What temperature rise will be needed to produce a 20-fold increase in the velocity of the reaction? (*Hint:* If $1.00 at compound interest grows to $2.00 in 18 years, what will it amount to after the first year? When will it amount to $20?)

8. When light passes through a transparent medium, its intensity is reduced according to the equation

$$I = I_0 c^{d/k}$$

where I_0 = initial light intensity
I = intensity after passing through a medium of thickness d
c, k = constants depending upon nature of light and of medium

If the intensity of sunlight is reduced to half its original value after penetrating water to a depth of 4 ft: (a) Evaluate c and k. (b) At what depth will the light intensity be 10 percent of that at the surface?

9. The temperature of a body surrounded by a medium at a different temperature is given by the equation

$$T = T_0 c^n = T_0 c^{t/3}$$

where T_0 = initial difference between temperatures of body and surroundings
T = temperature difference at any time t
c = a constant multiplier depending upon various physical conditions
n = number of periods

(See Example 30.)

If, during a power failure, a home freezer warms up from -10 to $+2$°F after standing for 3 h in a room at 50°F, how much longer will it take to warm up to 32°F?

10. If a machine costs $8,700 new and depreciates 17 percent a year, find its value after 9 years.

11. A pump cost $740 new and after 6 years is worth $110 as scrap. Assuming constant percent rate of depreciation, find the annual rate of depreciation and the value at the end of 1 year.

12. If $1 is invested at 8 percent compounded quarterly, determine (to the nearest quarter year) how long it will take to grow to (a) $2, (b) $4, (c) $8.

13. Show that a valid formula is $S = (1)(2)^{t/k}$, where t is the time in years and k is the answer obtained in part a of the preceding problem.

14. In radioactive-carbon dating, the half-life of C^{14} is estimated to be 5,500 years. If so, the equation $W_t = W_0(2)^{-t/5,500}$ would apply, where W_0 is the original amount of C^{14} and W_t is the amount of C^{14} remaining after t years. Estimate the age of a relic containing 60 percent of its original C^{14}.

15. A machine worth $10,000 new depreciates at a constant 6.7 percent per year. It can be calculated that its value drops to half, or $5,000, at the end of 10 years, approximately.

    Show that equivalent formulas for the book value V at the end of t years are (a) $V = 10,000(0.933)^t$, (b) $V = 10,000(0.5)^{t/10}$, (c) $V = 10,000(2)^{-t/10}$.

16. The magnitude, or brightness, of a star is given by the equation

$$M = 2.5 \log \frac{I_0}{I}$$

where M = magnitude of star whose light intensity is I
$\quad\quad I_0$ = light intensity of star of zero magnitude
Determine the ratio of the light intensities I/I_0 for a star of the fourth magnitude ($M = 4$). (Note that the brighter the star, the lower the magnitude; in fact, for the brighter planets M is negative.)

It can be shown that each time we multiply I by 0.1, M will increase by 2.5. Since M is the observed value of sensation and I is the stimulus producing that sensation, Fechner's law applies (see Prob. 14, Exercise 19).

As a matter of interest, in connection with the foregoing, the magnitudes of the sun, full moon, and Venus (at its brightest) are -26.5, -12.5, and -4.0, respectively. The naked-eye limit is about $+6.5$, and the 6-in telescope is $+13$.

17. The Richter scale of representing the intensity of an earthquake is exponential. An earthquake of magnitude 7 is 10 times as strong as one of magnitude 6. A quake of magnitude 8 is 10 times as strong as one of magnitude 7, etc. Thus we may state that $I \propto 10^R$, where I is the relative intensity of the shock and R is the Richter scale number.

An equivalent relation would be $I_1/I_2 = 10^{R_1 - R_2}$.

How did the San Francisco quake in 1906 (Richter scale 8.4) compare with the Los Angeles quake in 1971 (Richter scale 6.7)?

11.30 Applications from Technology

EXERCISE 19

The following problems are drawn from various fields of technology. In each case compute the answer to the degree of accuracy you think is warranted by the data.

1. A solid cast-iron sphere $4\frac{3}{4}$ in in diameter weighs 14.6 lb. Find to the nearest 0.01 in the diameter of a cast-iron sphere weighing 32 lb.
2. By the use of logarithms find the radius r of a circle inscribed in a triangle whose sides are a, b, and c if

$$r = \sqrt{\frac{(s-a)(s-b)(s-c)}{s}}$$

   where s is half of the perimeter and the sides are 53.60, 41.90, and 38.40.
3. Find the diameter of a circle inscribed in a triangle whose sides are 287.6, 303.1, and 365.9.
4. In the formula $P = 29.92e^{-h/5}$, P is the barometer reading in inches of mercury, h is the altitude in miles, and e = 2.718 (base of natural logarithms).
   (a) Show that equivalent formulas are $P = 29.92(0.8187)^h$ and $P = 29.92(2)^{-h/3.47}$.
   (b) What will the pressure be at 28,000 ft?
   (c) What altitude in feet corresponds to a pressure of 19.50 in?
   (d) What is the sea-level pressure?
5. In Fig. 11.2 a weight W is suspended by a rope over a round beam. The rope is prevented from slipping by a force F at the other end.

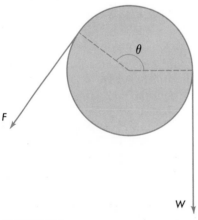

FIGURE 11.2

(a) Using the formula $F = We^{-a\theta}$, where $e = 2.718$, a is a constant, and θ is the angle of contact of the rope on the beam (expressed in turns), find a to 0.01 if 170 lb supports 200 lb when $\theta = 150°$.

(b) What force is needed to support 200 lb, using two complete turns of rope?

(c) How many turns would be required for 50 lb to support 200 lb?

6. The electrical resistance of a wire doubles (approximately) as the B & S gauge number increases 3 units. If the resistance of No. 10 copper wire is 1 Ω/1,000 ft, find the resistances of No. 13 and No. 4 wire.

7. What is the resistance ratio between two successive B & S wire sizes?

8. Find the resistance of No. 5 copper wire.

9. The diameters of successive wire gauges are in geometric progression. If the diameter of No. 16 is 0.051 in, that of No. 10 wire is 0.102 in, and that of No. 4 is 0.204 in, find the diameter of No. 6 wire.

10. What is the diameter ratio between two successive B & S wire sizes?

11. What is the resistance ratio between two wires five sizes apart?

12. What is the diameter ratio between two wires five sizes apart?

13. Find the diameter of a hemispherical cup holding exactly $\frac{1}{2}$ pt.

14. In public address systems the decibel gain $D = 10 \log (P_A/P_B)$, where P_A and P_B are the power levels of the two sounds A and B, and D is the difference in loudness (in decibels) between the sounds A and B. Show that $10^{D/10} = P_A/P_B$; also show that the following table is valid:

P_A/P_B	1	10	100	1,000
D	0	10	20	30

Thus, while D, the intensity of the sensation, is in arithmetic progression, P_A/P_B, the stimulus producing the sensation, is in geometric progression. This is known as Fechner's law. It applied also in Prob. 16, Exercise 18.

Restated, the response to any stimulus is proportional to the logarithm of that stimulus; that is, $\log (P_A/P_B) = D/10$.

15. Find the decibel gain in an electric circuit if the input is 2 W and the output is 5 W.

16. What is the decibel loss in a circuit if the output is 0.07 W and the input is 0.10 W?

17. If the power ratio between input and output of an amplifier is $\frac{1}{3}$, what is the decibel gain?

18. A loudspeaker requires 1.2 W to produce the proper volume of sound. If we assign to this sound level an arbitrary value of 1, calculate the volume range within which the speaker operates if the power varies between 1.0 and 1.5 W.

19. If the sound output of an amplifier varies from 4 dB above to 3 dB below the normal value, between what power levels does the amplifier operate?

20. The ratio of sound volume in speech may range as high as 250:1. What is the range in decibels?
21. The normal power rating of an amplifier is 1.6 W. To what should this be increased to produce a perceptible (i.e., 1-dB) increase in sound level?

     For radio and public address work, etc., 0.006 W is a common reference level from which to figure sound-energy levels. If the output of an amplifier is 0.24 W, the decibel difference between this level and 0.006 W is

$$L_{dB} = 10 \log \frac{0.24}{0.006} = 10 \log 40 = 10 \times 1.60206 = 16 \text{ dB}$$

where L_{dB} is the so-called power level at the amplifier. For convenience, the power level of this amplifier is said to be 16 dB. We understand, however, that we really should say that the power level of the amplifier output is 16 dB above 0.006 W.

22. What is the decibel level equivalent to 0.0005 W?
23. What is the decibel level equivalent to 0.46 W?
24. What is the decibel gain in an amplifier with 0.005-W input and 0.46-W output?
25. What is the difference between the decibel levels in Probs. 22 and 23?
26. What power level is equivalent to 59 W?

     Problems having to do with air conditioning, refrigeration, internal-combustion engines, air compressors, etc., often use formulas based on the following fundamental types of equations:

$$V = \frac{KT}{P} \tag{17}$$

     (By Charles' and Boyle's laws, the volume V of a given weight of gas varies directly as the absolute temperature T and inversely as the absolute pressure P.)

$$\frac{PV}{T} = K \tag{18}$$

     Under two separate sets of conditions of pressure, volume, and temperature for the *same weight* of gas we may write

$$\frac{P_1 V_1}{T_1} = K \tag{19}$$

for the first set of conditions and

$$\frac{P_2 V_2}{T_2} = K \tag{20}$$

for the second set of conditions. Any consistent set of units may be used as long as P and T are expressed on an absolute scale. Usually, P is expressed in pounds per square inch absolute (equals pounds per square inch gauge $+14.7$), and T is expressed in Rankine or Fahrenheit absolute (equals Fahrenheit $+460$).

From Eqs. (19) and (20) we obtain

$$\frac{P_1 V_1}{T_1} = \frac{P_2 V_2}{T_2} \tag{21}$$

If we wish to express the pressure-volume relationship of a given weight of gas undergoing adiabatic (with no gain or loss of heat from the system) compression or expansion without being concerned about any accompanying temperature change, we could write, for thermodynamic considerations,

$$PV^n = C \tag{22}$$

where n and C are constants.

Then for two distinct sets of conditions for the same weight of gas we may write

$$P_1 V_1{}^n = C \tag{23}$$

and

$$P_2 V_2{}^n = C \tag{24}$$

From Eqs. (23) and (24) we obtain

$$P_1 V_1{}^n = P_2 V_2{}^n \tag{25}$$

In Probs. 27 and 28 we develop formulas to be used when we are not interested in pressure changes involved or when the pressure is unknown.

27. From Eqs. (21) and (25) show that

$$\frac{T_1}{T_2} = \left(\frac{V_2}{V_1}\right)^{n-1}$$

28. From Prob. 27 show that

$$\frac{V_2}{V_1} = \left(\frac{T_1}{T_2}\right)^{1/(n-1)}$$

Formulas in Probs. 29 and 30 are used when we are not concerned with changes in volume or when data are unavailable.

29. From Eq. (25) and Prob. 28 show that

$$\frac{T_1}{T_2} = \left(\frac{P_1}{P_2}\right)^{(n-1)/n}$$

30. From Prob. 29 show that

$$\frac{P_1}{P_2} = \left(\frac{T_1}{T_2}\right)^{n/(n-1)}$$

31. Find the final gauge pressure of a quantity of gas if the volume is compressed from 220 to 150 ft^3. P_1 is atmospheric pressure and $n = 1.39$.
32. Show that if $n = 0$, the pressure is constant, that if $n = 1$, the temperature is constant, and that if $n = \infty$, the volume is constant (see formulas in Probs. 27 to 30).
33. If $n = 1.64$, find the final volume of 68 ft^3 of gas when, under compression, the temperature rises from 135 to 200°F.
34. A gas is compressed from 4.32 to 2.09 ft^3. Find the final temperature if the original temperature was 30°F and $n = 1.4$.
35. When 10 ft^3 of a vapor expanded to 32 ft^3, the temperature fell from 191 to 0°F. Find n to 0.01.
36. The pressure on 220 ft^3 of a gas is increased from 1 to 2 atm, while the volume decreases to 127 ft^3. Find n.
37. Following is a formula for the velocity of air being discharged through an orifice to the atmosphere:

$$V = 108.9\sqrt{T\left[1 - \left(\frac{14.7}{P}\right)^n\right]}$$

Find n if the air is being discharged at the rate of 323 ft^3/min through a 1⅝-in-diameter orifice. The pressure in the tank is 1.2 (lb/in^2 gauge), and the temperature in the tank is 65°F.

V = linear airflow, ft/s
T = air temperature in tank, °R (Fahrenheit absolute)
P = absolute air pressure in tank, lb/in^2

38. 1,000 ml of a salt solution contains a suspended sludge. The sludge is washed by allowing it to settle; part of the clear, overlying salt solution is poured off, replaced by an equal volume of pure water, stirred, and allowed to settle. This cycle is to

be repeated until the necessary amount of salt is washed out. The required equation is

$$C_n = \left(\frac{1,000 - w}{1,000}\right)^n C_0$$

where C_0 = original concentration of salt = 15.8 g/l
$\quad\quad C_n$ = final concentration of salt
$\quad\quad w$ = milliliters of solution removed and replaced by water, each
$\quad\quad\quad$ cycle = 400 ml
$\quad\quad n$ = number of washings
(a) If $n = 8$, find C_n.
(b) Find n to reduce C_n to less than 0.085 g/l.

39. In the measurement of heat flow through pipe insulation the log mean diameter of the insulation is computed. This is

$$D_{av} = \frac{D_2 - D_1}{2.3 \log (D_2/D_1)}$$

where D_1 = outer diameter of pipe
$\quad\quad D_2$ = outer diameter of insulation
Find D_{av} when the outer pipe diameter is $2\frac{5}{8}$ in and the insulation is $2\frac{5}{16}$ in thick. Compare with the arithmetic mean.

40. A snow-making apparatus is being used to condition a ski slope. If the initial air pressure $P_1 = 50$ lb/in² gauge, P_2 = atmospheric pressure = 14.7 lb/in² abs, and the initial air temperature $T_1 = 30°$F, find T_2 (before the air has been warmed by contact with the water spray). Assume $n = 1.4$. (See Prob. 30.)

41. Find to the nearest $\frac{1}{4}$ in the pipe diameter required to deliver natural gas. Use the formula

$$D = 0.56 \left(\frac{Q^2 G L}{H}\right)^{0.225}$$

where D = pipe diameter, in
$\quad\quad Q$ = rate of gas flow = 280 ft³/h
$\quad\quad G$ = specific gravity of the gas relative to air = 0.433
$\quad\quad L$ = pipe length, yd = $5\frac{1}{2}$ mi
$\quad\quad H$ = pressure drop = $8\frac{1}{4}$ in water

42. The equation $H = 60{,}470 \log (B_2/B_1)$ is used in the determination of altitudes by the use of barometer readings, where H is the difference in elevation (in feet) between two points at which the respective barometer readings are B_1 and B_2.

If $B_2 = 29.33$ in at the base station and $B_1 = 26.14$ in on the mountainside, find (to the nearest foot) the difference in altitude between the two stations.

43. The rate of heat transfer (Btu/h) by radiation is given by the equation

$$Q = K\left[\left(\frac{T_1}{100}\right)^4 - \left(\frac{T_2}{100}\right)^4\right]$$

where K = constant depending upon physical properties of objects involved
T_1 = absolute temperature of hot body, °F
T_2 = absolute temperature of cooler body, °F

If the rate of heat transfer is 7500 Btu/h when the hot and cold objects are at 1050 and 70°F, respectively, find the necessary hot-body temperature (in degrees Fahrenheit) to double the rate of radiant heat transfer, keeping the cooler body at 70°F.

44. Under certain conditions the drying of sheets of wallboard may be expressed by the equation

$$\log \frac{T_0 - E}{T - E} = K\theta$$

where T_0 = pounds of moisture per pound of bone-dry stock before drying
T = amount of moisture remaining (same basis) after θ h
E = pounds of moisture per pound of bone-dry stock remaining in stock after coming to equilibrium with air used for drying; that is, E represents limit to drying possible under given conditions
K = constant

If the moisture content drops from 1.2 to 0.8 lb/lb in 3 h, with a further drop to 0.64 lb/lb in another 6 h, find the amount of moisture remaining in the stock after a drying period of indefinitely great length.

The two equations are

$$\log \frac{1.2 - E}{0.8 - E} = 3K \tag{26}$$

and

$$\log \frac{0.8 - E}{0.64 - E} = 6K = 2(3K) \tag{27}$$

Replacing $3K$ in Eq. (27) by its equivalent, we obtain

$$\log \frac{0.8 - E}{0.64 - E} = 2 \log \frac{1.2 - E}{0.8 - E}$$

or

$$\frac{0.8 - E}{0.64 - E} = \left(\frac{1.2 - E}{0.8 - E}\right)^2 \tag{28}$$

Since E^3 terms cancel, Eq. (28) is actually a quadratic which is readily solved.

45. The resistance of a tungsten-lamp filament is given by the relationship

$$\frac{R_1}{R_2} = \left(\frac{T_1}{T_2}\right)^{1.2}$$

where R_1 = resistance at room temperature T_1
$ R_2$ = resistance at operating temperature T_2
Temperatures are both Kelvin (Celsius absolute) or both Rankine (Fahrenheit absolute). If the resistance at 20°C is 16 Ω and the operating resistance is 232 Ω, find the operating temperature in degrees Celsius.

46. A tungsten lamp is rated at 60 W at 115 V. If the resistance measured at 75°F is 15.2 Ω, find the operating temperature in degrees Fahrenheit. Ohms = (volts)²/watt (see Prob. 45).

47. The horsepower necessary to compress a gas in a single-stage compressor is given by the formula

$$hp = \frac{144n P_1 V_1}{33,000(n-1)}\left[\left(\frac{P_2}{P_1}\right)^{(n-1)/n} - 1\right]$$

Find the horsepower necessary to compress, in 1 min, 240 ft³ of helium from atmospheric pressure (14.7 lb/in² abs) to 250 lb/in² gauge. For helium, $n = 1.66$. Other symbols have the same meanings as in Probs. 27 to 30, pages 319, 320.

The following space problems 48 to 54 require the use of data in Appendix D.

48. Referring to Eq. (3), Appendix D, determine the period of a satellite traveling in a circular orbit 870 mi above the earth's surface. (Note: $M_b = M_e$ and $R = 3,960 + 870 = 4,830$ mi.)

49. Calculate the period for a satellite orbiting 22,300 mi above the earth's surface. Has this answer any special significance?

50. Repeat Prob. 49 for the moon if the average altitude is 237,000 mi above the earth's surface.

51. Use Eq. (3), Appendix D, to confirm the period of the earth's revolution about the sun.

52. Apollo 8 orbited around the moon at a constant altitude of 160 mi. Find its velocity and period.

53. Calculate the period of an AES traveling at an altitude of 1,200 mi above the earth's surface.
54. Confirm the velocity of escape from the moon's surface using Eq. (4) and data from the table in Appendix D.

11.31 Relation of Logarithms to the Slide Rule

An arrangement of uniform scales similar to those illustrated in Fig. 11.3 could be used for mechanically adding or subtracting numbers.

This figure illustrates the addition $0.22 + 0.26 = 0.48$. It also illustrates the subtraction $0.48 - 0.26 = 0.22$.

As an example of multiplication, multiply 1.66 by 1.82. Looking up three-place logarithms, we find 0.220 and 0.260 to be the respective mantissas. We may use the scales to add 0.220 and 0.260, obtaining 0.480, as in Fig. 11.3. Looking up the antilogarithm of 0.480, we obtain 3.02.

The slide rule of Fig. 11.3 would not be of much help to us. However, if we replace the uniform scales which we have been using to represent mantissas by scales bearing the numbers to which these mantissas correspond, we will have eliminated the necessity of looking up logarithms and antilogarithms (Fig. 11.4). Thus the scheme for mechanically adding 0.22 and 0.26 becomes a means of multiplying the antilogarithms 1.66 and 1.82 to get 3.02. Evidently, by reversal of the procedure, $0.48 - 0.26 = 0.22$ corresponds to $3.02 \div 1.82 = 1.66$.

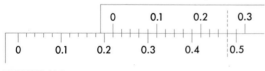

FIGURE 11.3

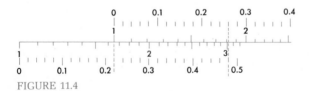

FIGURE 11.4

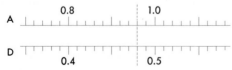

FIGURE 11.5

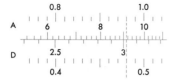

FIGURE 11.6

In Fig. 11.5 it will be observed that opposite any given number n on the D scale we read $2n$ on the A scale.

If we wished to square 3.02, we could look up the (three-place) mantissa, 0.480, double by the arrangement in Fig. 11.5, and look up the antilogarithm of 0.96, obtaining 9.12. Obviously, we stand to gain here also by placing the antilogarithms in the place of the corresponding mantissas. Thus the process of mechanically multiplying 0.48 by 2 to get 0.96 becomes the operation of squaring 3.02 to obtain 9.12. Likewise, the operation $0.96/2 = 0.48$ is replaced by $\sqrt{9.12} = 3.02$ (Fig. 11.6).

These illustrations will serve to indicate that the slide rule is basically a device to add or subtract quantities (mantissas) mechanically. A knowledge of the laws of logarithms (Sec. 11.2) will be naturally helpful in developing facility in the use of the slide rule.

11.32 Logarithmic Computations on the Slide Rule

It will be noticed that the uniform scale we discussed in Sec. 11.3 is engraved on the slide rule as the L scale. If we set the hairline over a given number on the D scale,

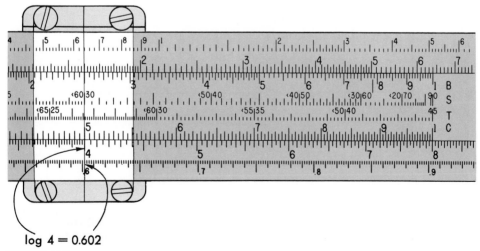

$\log 4 = 0.602$

FIGURE 11.7

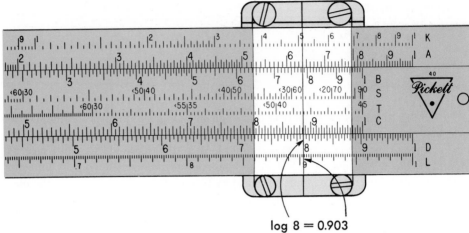

log 8 = 0.903

FIGURE 11.8

we shall at the same time find under the hairline on the L scale the mantissa of the logarithm of that number. Thus in Fig. 11.7 we find illustrated log 4 = 0.602; in Fig. 11.8, log 8 = 0.903; in Fig. 11.9, the mantissa of log 16 = 0.204, or log 16 = 1.204.

Let us illustrate the use of the L scale in certain logarithmic computations.

Example 32. Solve the equation $4^{3/2} = x$.

Taking logarithms, we have $1.5 \log 4 = \log x$ (Sec. 11.22). In Fig. 11.7 we have

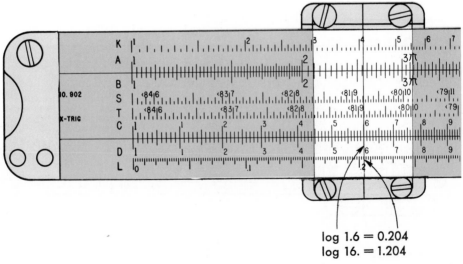

log 1.6 = 0.204
log 16. = 1.204

FIGURE 11.9

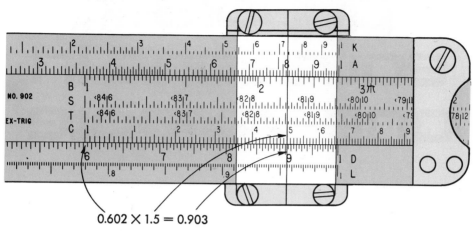

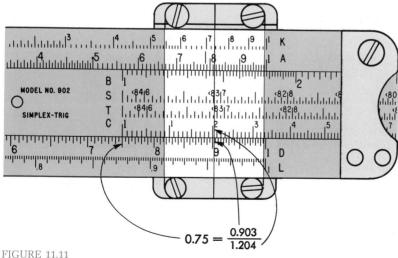

FIGURE 11.10

shown that log 4 = 0.602. Figure 11.10 shows that 1.5 × 0.602 = 0.903. If log x = 0.903, then antilog 0.903 = x = 8. This step is illustrated in Fig. 11.8.

Example 33. Solve the equation $16^x = 8$.

Taking logs of both sides, we have x log 16 = log 8. Since the settings in Figs. 11.9 and 11.8 show that log 16 = 1.204 and log 8 = 0.903, we may write 1.204x = 0.903. Finally, Fig. 11.11 illustrates the division 0.903 ÷ 1.204 = 0.75 = x.

FIGURE 11.11

quadratic equations in one variable

In Chap. 5 we discussed linear equations (or first-degree equations) in one variable.

In this chapter we shall be particularly concerned with second-degree equations in one variable in the form

$$ax^2 + bx + c = 0 \qquad a \neq 0 \tag{1}$$

where a, b and c are constants and a is not equal to zero.

12.1 Quadratic Equations

An equation equivalent to Eq. (1) is called a *quadratic equation.*

If either b or $c = 0$ (note that a cannot be zero), the equations

$$ax^2 + c = 0$$

and

$$ax^2 + bx = 0$$

are known as *incomplete quadratic equations.*

12.2 Solution of Incomplete Quadratics

Both types of incomplete quadratic equations are easily solved.

Example 1. Solve the equation $4x^2 - 9 = 0$.

Adding 9 to both sides and dividing by 4,

$$x^2 = \tfrac{9}{4}$$

Extracting the square root,

$$x = \pm \tfrac{3}{2}$$

That is,

$$x = +\tfrac{3}{2} \qquad \text{and} \qquad x = -\tfrac{3}{2}$$

Note that both $+\tfrac{3}{2}$ and $-\tfrac{3}{2}$ are actually roots of the given equation, since by substitution in the given equation we find that

$$4\left(+\frac{3}{2}\right)^2 = 9$$

and

$$4\left(-\frac{3}{2}\right)^2 = 9$$

The same example might have been solved by the factoring method as follows:

$$4x^2 - 9 = 0$$
$$(2x + 3)(2x - 3) = 0$$

We recall from Sec. 3.11 that if the product of two or more factors is zero, then at least one of these factors must be zero.

$$2x + 3 = 0$$

or

$$x = -\tfrac{3}{2}$$
$$2x - 3 = 0$$

or

$x = \frac{3}{2}$

as above.

Example 2. Solve the equation $2x^2 + 6x = 0$.
    Factoring,

$2x(x + 3) = 0$

Setting each factor equal to zero,

$\quad 2x = 0 \qquad$ or $\qquad x = 0$
$x + 3 = 0 \qquad$ or $\qquad x = -3$

EXERCISE 1

Solve the following incomplete quadratic equations, regarding x, y, z, and w as unknown quantities.

1. $y^2 - 9 = 0$
2. $z^2 - 16z = 0$
3. $x^2 = 49$
4. $144 - w^2 = 0$
5. $y^2 = 121$
6. $x^2 - 81 = 0$
7. $z^2 = 8$
8. $12 - w^2 = 0$
9. $x^2 - 30 = 0$
10. $z^2 = c^2$
11. $y^2 = 9a^2$
12. $x^2 - 16b^2 = 0$
13. $w^2 = aw$
14. $y^2 = a^2 + b^2$
15. $z^2 = 1/c^2$
16. $a^2/b^2 - y^2 = 0$
17. $16x^2 = 1$
18. $9y^2 - 25 = 0$
19. $12w^2 = 1$
20. $9z^2 = 8$
21. $w^2 + 121 = 5w^2$
22. $3y^2 - 17 = 8 - y^2$
23. $x^2/3 = 48$
24. $z^2/b = 4b$
25. $4/y^2 = 49$
26. $1/y^2 = 36$
27. $m^2/x^2 = 25$
28. $\dfrac{1}{20y^2} = 5$
29. $\dfrac{1}{z^2} = \dfrac{1}{16}$
30. $\dfrac{a^2}{c^2} = \dfrac{1}{w^2}$
31. $\dfrac{4}{x} = \dfrac{x}{9}$
32. $\dfrac{1}{2x} = \dfrac{8x}{9}$
33. $\dfrac{3w}{4} - \dfrac{9}{2w} = \dfrac{5w}{8}$
34. $\dfrac{x^2 - 1}{6} - \dfrac{x^2 + 2}{9} = 1$

12.3 Solution of the Quadratic $ax^2 + bx + c = 0$

The solution of the equation $ax^2 + bx + c = 0$ may be accomplished by the following means:

1. Plotting (approximate values—limited by accuracy of reading)
2. Factoring

3. Completing the square
4. Quadratic formula
5. Short graphical method (approximate values—limited by accuracy of reading)

12.4 Solution by Plotting

While exact answers can be obtained by plotting if they are integers, fractional values are apt to be approximated, and irrational answers will always be approximations. Imaginary roots cannot normally be found by plotting. However, by a suitable shift in the relative positions of the curve on the X axis, such roots may be approximated.

Example 3. Solve the equation $x^2 - 2x - 11 = 0$ by plotting.

Set $x^2 - 2x - 11 = y$. Substitute various values of x, and determine the corresponding values of y. For instance, if $x = 7$, then

$$y = (7)^2 - 2(7) - 11 = 49 - 14 - 11 = 24$$

In like manner make out a table, continuing to substitute various values of x until it becomes evident that y is receding from zero at an increasing rate.

x	6	5	4	3	2	1	0	−1	−2	−3	−4	...
y	13	4	−3	−8	−11	−12	−11	−8	−3	4	13	...

Plot these values on coordinate paper. Answers correspond to the intersection of the curve and the X axis. This is reasonable because, in effect, we are solving for the simultaneous equations $y = x^2 - 2x - 11$ and $y = 0$.

Figure 12.1 shows the result obtained by plotting these data. The indicated answers are roughly $x = 4.5$ and $x = -2.5$.

Tabulation may be facilitated if we take uniformly spaced values of x and list increments in y (represented by Δy) and also increments in Δy [that is, $\Delta(\Delta y)$], which we abbreviate to $\Delta^2 y$.

Δx	−1	−1	−1	−1	−1	−1	−1	−1	−1	−1
x	6	5	4	3	2	1	0	−1	−2	−3	−4
y	13	4	−3	−8	−11	−12	−11	−8	−3	4	13
Δy	−9	−7	−5	−3	−1	1	3	5	7	9
Δ²y	2	2	2	2	2	2	2	2	2

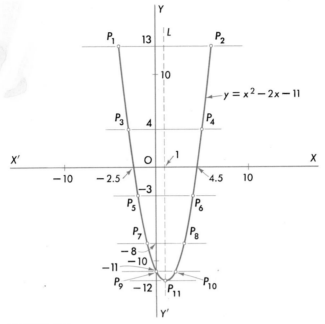

FIGURE 12.1

The table may be extended quite rapidly by taking advantage of the fact that $\Delta^2 y$ constantly equals 2.

By using the word "symmetry" in the ordinary meaning it seems quite appropriate to say that the graph in Fig. 12.1 is symmetrical with respect to the dashed line L. In this context L is said to be the *axis of symmetry*. If a series of points has been plotted on one side of the axis of symmetry, corresponding points may be located on the opposite side, as in a mirror image. The edge of the mirror rests on the axis of symmetry.

From another view, if we fold the paper on which the graph is plotted along the axis of symmetry and press it flat, we find that one branch of the curve lies exactly over the other branch.

In general, two points P_1 and P_2 (Fig. 12.1) are said to be symmetric with respect to a line L if the line L is the perpendicular bisector of the line segment P_1P_2.

Observe the lines whose equations are $y = 13, 4, -3, -8, -11$, and -12.

What are the coordinates of the midpoints on the line segments P_1P_2, P_3P_4, etc.?

If we rewrite the equation $x^2 - 2x - 11 = y$ to read $x^2 - 2x + 1 - 12 = y$ or $(x - 1)^2 - 12 = y$, it is apparent that the smallest possible value of $(x - 1)^2$ is zero. This occurs at $x = 1$. Thus y has a minimum value of -12 at $x = 1$. Note that for $x = 2$ or 0, $(x - 1)^2 = 1$ and $y = -11$. For $x = 3$ or -1, $(x - 1)^2 = 4$ and $y = -8$, etc., thus indicating a symmetry about the line $x = 1$.

If accurate plotting is not required, the parabola may be sketched rapidly by means of the intercepts and the axis of symmetry as indicated above.

EXERCISE 2

Plot as directed, labeling all curves.

1. Plot on a common area.
   (a) $y = x^2$ (b) $y = -x^2$ (c) $y = \frac{1}{4}x^2$ (d) $y = \frac{1}{2}x^2$
2. Plot on a common area.
   (a) $y = 2x^2$ (b) $x = y^2$ (c) $x = -y^2$ (d) $y = x^2 - 4$
3. Plot on a common area. Find x when $y = 0$.
   (a) $y = (x - 3)^2$ (b) $y = (x - 3)^2 - 4$
   (c) $y = (x + 3)^2$ (d) $y = x^2 - 6x + 5$
4. Plot on a common area. Find x when $y = 0$.
   (a) $y = 2(x - 4)^2$ (b) $y = 2x^2 - 16x + 14$
   (c) $y = 2(x - 4)^2 - 18$
5. Plot on a common area.
   (a) $y = -3x^2 - 12x + 15$ (b) $y = -3(x + 2)^2 + 27$
6. Plot $y = x^2 - 4x - 5$. Determine its intersection with the lines
   (a) $y = 16$ (b) $y = 7$ (c) $y = 0$ (d) $y = -5$
   (e) $y = -8$ (f) $y = -9$
   Confirm the answers in Prob. 6 by factoring.

EXERCISE 3

Solve the following quadratic equations by plotting.

1. $x^2 - 2x - 3 = 0$ 2. $x^2 + x - 6 = 0$
3. $x^2 + x = 56$ *4. $x^2 + 4 = -4x$
5. $x(x + 5) = 24$ 6. $2x^2 - 3x + 1 = 0$
7. $3x^2 - x - 2 = 0$ 8. $5x^2 - 11x + 2 = 0$
9. $(x - 2)(x - 5) = x - 5$ 10. $(x - 5)(x + 8) = x - 16$

* Note that Prob. 4 has two identical solutions.

12.5 Solution by Factoring

This method is useful when factors of an expression are fairly obvious. See Example 1. However, the student should not spend too much time looking for factors that may not exist.

Example 4. Solve the equation $x(2x - 1) = 15$ by factoring.

Carrying out the indicated multiplication,

$$2x^2 - x = 15$$

Subtracting 15 from both members of the equation,

$$2x^2 - x - 15 = 0$$

Inspection will show that the factored form is

$$(2x + 5)(x - 3) = 0$$

Therefore

$$2x + 5 = 0 \quad \text{and} \quad x = -\tfrac{5}{2}$$
$$x - 3 = 0 \quad \text{and} \quad x = 3$$

Note that, if the equation had been written $12x^2 - 6x - 90 = 0$, our first step would be the division by a common monomial factor—in this case, by 6.

EXERCISE 4

Solve the following quadratic equations by the method of factoring:

1. $x^2 - 4x + 4 = 0$
2. $x^2 - x - 30 = 0$
3. $x^2 + x = 72$
4. $x^2 - 2x = 15$
5. $x^2 + 7x = -12$
6. $x^2 - 33 = 8x$
7. $x^2 - 48 = 2x$
8. $x(x - 5) = 36$
9. $x(x + 3) = 40$
10. $x^2 + 3(x - 18) = 0$
11. $2x^2 + 3x + 1 = 0$
12. $5x^2 - 2x - 3 = 0$
13. $8x^2 + 10x - 12 = 0$
14. $21x^2 - 12x - 9 = 0$
15. $(x - 4)(x - 7) = x - 4$
16. $(x + 6)(x - 6) = 5x$
17. $\dfrac{1}{y + 4} = \dfrac{3}{y^2 + 12}$
18. $\dfrac{2}{w - 3} = \dfrac{w + 3}{8}$

12.6 Solution by Completing the Square

Under certain special conditions this method offers a ready solution. It will be used in Sec. 12.7 to derive the quadratic formula.

Example 5. Solve the equation $x^2 + 8x - 48 = 0$ by completing the square.

Rearranging so that all terms containing the unknown are on one side of the equation and all constant terms are on the other, we have

$$x^2 + 8x = +48$$

Making the left side a perfect square by adding 16 to both sides (see Prob. 10, Exercise 14, Chap. 3),

$$x^2 + 8x + 16 = 48 + 16 = 64$$

from which

$$(x + 4)^2 = 64$$

Then

$$x + 4 = \pm\sqrt{64} = \pm 8$$

Therefore

$$x + 4 = +8 \qquad x = +4$$

and

$$x + 4 = -8 \qquad x = -12$$

The selection of the quantity to be added to both sides before extracting the square root is governed by considering the identity

$$(x + k)^2 = x^2 + (2k)x + k^2$$

This identity can be verified directly by multiplication.

$$
\begin{array}{l}
x + k \\
\underline{x + k} \\
x^2 + kx \\
\underline{\qquad kx + k^2} \\
x^2 + (2k)x + k^2
\end{array}
$$

where the coefficient of x is (2k). Observe that the constant term k^2 is the square of one-half the coefficient of x.

Thus, in our example, we square one-half the coefficient of x to obtain $(\%_2)^2 = 16$.

Example 6. Solve the equation $x^2 - 6x + 45 = 0$ by completing the square.

Rearranging so that all terms containing the unknown are on one side of the equation and all constant terms are on the other, we have

$$x^2 - 6x = -45$$

Now we make the left member of the equation a perfect square by adding the square of one-half the coefficient of x to both members of the equation.

In this case the coefficient of x is −6 and

$$(-\%_2)^2 = (-3)^2 = 9$$

Thus we add 9 to both members of the equation

$$x^2 - 6x = -45$$

and obtain

$$x^2 - 6x + 9 = -36$$

Extracting the square root,

$$x - 3 = \pm 6\sqrt{-1} = \pm j6$$
$$x = +3 + j6 \quad \text{and} \quad x = +3 - j6$$

The roots of the given equation are therefore complex numbers.

Example 7. Solve, by completing the square, the equation $3x(x - 3) = 2(1 - 2x)$.

Carrying out indicated multiplications,

$$3x^2 - 9x = 2 - 4x \tag{2}$$

Adding 4x to both members, leaving the constant term on the right,

$$3x^2 - 5x = 2 \tag{3}$$

Note. *In Example 5 the coefficient of the x^2 term in the given equation is 1. Before proceeding with this example we shall have to write an equation which is equivalent to Eq. (3), in which the coefficient of the x^2 is also 1.*

We obtain such an equation by dividing both members of Eq. (3) by 3. Thus

$$x^2 - \frac{5x}{3} = \frac{2}{3} \qquad (4)$$

The term to be added to both sides will be the square of half the coefficient of the x term,

$$\left(\frac{-\frac{5}{3}}{2}\right)^2 = \frac{25}{36}$$

Adding $\frac{25}{36}$ to both sides of Eq. (4),

$$x^2 - \frac{5x}{3} + \frac{25}{36} = \frac{2}{3} + \frac{25}{36} = \frac{49}{36} \qquad (5)$$

Extracting the square root,

$$x - \frac{5}{6} = \pm\frac{7}{6} \qquad (6)$$
$$x = \frac{5}{6} + \frac{7}{6} \qquad \text{or} \qquad x = 2$$
$$x = \frac{5}{6} - \frac{7}{6} \qquad \text{or} \qquad x = -\frac{1}{3}$$

Example 8. Solve, by the method of completing the square, the equation

$$x^2 - 6cx = 4a^2 - 12ac \qquad (7)$$

In this example the terms not containing x have already been separated from those containing x. The missing quantity which will make the left side a perfect square is $(-6c/2)^2 = 9c^2$. Adding $9c^2$ to both sides of Eq. (7), we obtain

$$x^2 - 6cx + 9c^2 = 4a^2 - 12ac + 9c^2 \qquad (8)$$

or

$$(x - 3c)^2 = (2a - 3c)^2$$

Extracting the square root of both members,

$$x - 3c = \pm(2a - 3c) \qquad (9)$$

Hence

$x = 3c + (2a - 3c) = 2a$

Also

$x = 3c - (2a - 3c) = 6c - 2a$

Each of these answers will be found to check the original equation.

It is evident that the method of completing the square can be used to the greatest advantage when the coefficient of the second-degree term is unity and that of the first-degree term contains 2 as a factor.

To summarize, the solution of a quadratic by the method of completing the square consists in the following steps:

1. Write an equivalent equation if necessary so that the constant term is on one side of the equation, and terms containing the unknown are on the other side.
2. Make the coefficient of x^2 equal to unity by dividing by the coefficient of x^2 unless it is already equal to 1. The equation now has the form $x^2 + px + q$.
3. Add to both sides of the equation the square of half the coefficient of x. This transforms the left side of the equation to $x^2 + px + p^2/4$, a perfect square.
4. Extract the square roots of both sides, remembering to place a $\pm$ sign before the square root of the right-hand number.
5. Solve the two equations so formed for the unknown.
6. Check the results in the original equation.

EXERCISE 5

Solve the following equations by means of completing the square, leaving any irrational answers in radical form.

1. $x^2 + 2x - 3 = 0$

2. $x^2 - 4x - 21 = 0$

3. $x^2 + 6x - 7 = 0$

4. $x^2 - 8x - 20 = 0$

5. $x^2 + 10x - 24 = 0$

6. $y^2 - 12y + 27 = 0$

7. $x(x - 10) = 39$

8. $z(z + 8) = 48$

9. $x^2 + 2bx = a^2 - b^2$

10. $x^2 - 2dx = c^2 + 2cd$

11. $x^2 + 2ax = -6ab + 9b^2$

12. $x^2 - 4cx = 9b^2 + 12bc$

13. $x^2 + \dfrac{2x}{3} = \dfrac{15}{9}$

14. $y^2 - \dfrac{4y}{5} = \dfrac{9}{5}$

15. $y^2 - 20y + 105 = 0$

16. $x^2 - \dfrac{6x}{5} = \dfrac{8}{5}$

17. $w^2 - 4w = 8$

18. $x^2 + 10x = 55$

19. $y^2 - 14y = 11$

20. $x^2 - 8x + 27 = 0$

21. $x^2 + 8x = 32$

22. $w^2 - 12w = -16$

23. $x^2 - 14x = -59$

24. $z^2 - 12z = -43$

12.7 Solution by the Quadratic Formula

The most generally applicable method of solving a quadratic equation is that employing the *quadratic formula*. The derivation of this formula begins with the general quadratic

$$ax^2 + bx + c = 0$$

Subtracting c from both members of the above equation,

$$ax^2 + bx = -c \tag{10}$$

Dividing by the coefficient of x^2,

$$x^2 + \frac{bx}{a} = -\frac{c}{a} \tag{11}$$

Adding the square of half the coefficient of x to both members,

$$x^2 + \frac{bx}{a} + \frac{b^2}{4a^2} = \frac{b^2}{4a^2} - \frac{c}{a} \tag{12}$$

Combining fractions on the right,

$$x^2 + \frac{bx}{a} + \frac{b^2}{4a^2} = \frac{b^2 - 4ac}{4a^2} \tag{13}$$

Taking square roots of both members,

$$x + \frac{b}{2a} = \pm \frac{\sqrt{b^2 - 4ac}}{2a} \tag{14}$$

Subtracting $b/2a$ from both members of Eq. (14),

$$x = -\frac{b}{2a} \pm \frac{\sqrt{b^2 - 4ac}}{2a} \tag{15}$$

Combining fractions,

$$x = \frac{-b \pm \sqrt{b^2 - 4ac}}{2a} \qquad\qquad (16)$$

Equation (16) is commonly known as the *quadratic formula*. This formula has the advantage of being applicable to any quadratic equation. Under special conditions, the methods of factoring or completing the square may be more convenient. With practice, there should be no difficulty in selecting the shortest method for the equation at hand.

It should be remembered that a is the coefficient of the second-degree term, b is the coefficient of the first-degree term, and c includes all terms not containing the unknown.

Example 9. Use the quadratic formula to solve the quadratic equation

$$2x(4x - 1) = 15$$

Performing the indicated multiplication,

$$8x^2 - 2x = 15$$

Subtracting 15 from both members,

$$8x^2 - 2x - 15 = 0$$

According to Eq. (1), $a = 8$, $b = -2$, and $c = -15$. Then

$$x = \frac{-(-2) \pm \sqrt{(-2)^2 - 4(8)(-15)}}{(2)(8)}$$

$$= \frac{2 \pm \sqrt{4 + 480}}{16} = \frac{2 \pm 22}{16}$$

$$= \frac{2 + 22}{16} = \frac{3}{2}$$

$$= \frac{2 - 22}{16} = -\frac{5}{4}$$

EXERCISE 6

Solve the following equations by the quadratic formula, leaving any irrational answers in the radical form:

1. $x^2 - 17x + 60 = 0$ 2. $x^2 + x - 156 = 0$
3. $x^2 - 4x = 165$ 4. $3x^2 + 7x = 6$

5. $5y - 6y^2 + 1 = 0$ 6. $12z^2 + 24z = -9$

7. $x^2 + x + 1 = 0$ 8. $1/x = x/(x + 1)$

9. $3w^2 - 6w + 5 = 0$ 10. $5x^2 - 55x - 5 = 0$

11. $3x^2 - 1 = \dfrac{11x}{12}$ 12. $30y^2 + 76y + 48 = 0$

13. $w^2 + 5w + 7 = 0$ 14. $7y^2 + 12y + 4 = 0$

15. $5p^2 - 3p + 1 = 0$ 16. $\dfrac{1}{y - 1} + \dfrac{2}{y + 2} = \dfrac{1}{y + 1}$

12.8 Short Graphical Method

This method is designed for rapid, routine approximation. It depends upon separating the second-degree term from the rest of the expression and dividing by the coefficient of the second-degree term so that the equation is of the type $x^2 = mx + c$. From this equation we form the two simultaneous equations $y = x^2$ and $y = mx + c$. If we were to plot the graphs of these simultaneous equations on a common coordinate system, the roots would be indicated by the abscissas of the intersections (see Fig. 12.2 and Example 10 below).

It will be observed that the plot of $y = x^2$ is a parabola, while that of $y = mx + c$ is a straight line. The advantage of this method lies in the fact that the graph of $y = x^2$ may be drawn once and for all. A straight line scratched on the

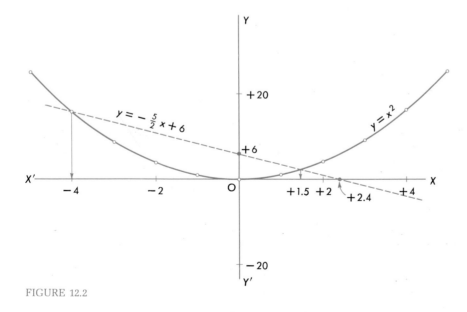

FIGURE 12.2

underside of a transparent strip will, upon proper positioning, represent the graph of $y = mx + c$. The abscissas of the intersections will represent the roots of the equation $x^2 = mx + c$.

Example 10. Solve by the short graphical method

$$2x^2 + 5x - 12 = 0$$

By subtracting $5x - 12$ from both members of the above equation,

$$2x^2 = -5x + 12$$

Dividing by the coefficient of x^2,

$$x^2 = \frac{-5x}{2} + 6$$

It is assumed that the graph of $y = x^2$, having been used in other problems, has already been drawn. The plastic strip is then laid on the graph paper so that the straight-line scratch passes through the points $(0,6)$ and $(2.4,0)$ (the y and x intercepts of the graph of $y = -5x/2 + 6$). The intersections of the two curves have abscissas of $\frac{3}{2}$ and -4, and the roots are $x = \frac{3}{2}$ and $x = -4$.

Note that, for convenience in reading, the horizontal scale is 10 times as open as the vertical scale.

12.9 Literal Quadratic Equations

In a *literal* quadratic equation some or all of the constants are literal numbers. These constants are usually represented by the first several letters of the alphabet or may involve π. In some cases it is preferable not to substitute the numerical value of π until a later stage of the solution.

The formula usually offers the easiest means of solution in the long run.

Example 11. Solve the equation $2(x^2 + p^2) + q(x + p - q) = 5px$.

Carrying out the indicated multiplication,

$$2x^2 + 2p^2 + qx + qp - q^2 = 5px$$

By subtracting $5px$ from both members of this equation and arranging in descending powers of x, we obtain

$$2x^2 + qx - 5px + 2p^2 + pq - q^2 = 0$$

Grouping and factoring x^2 and x terms,

$$2x^2 + (q - 5p)x + 2p^2 + qp - q^2 = 0$$

Tabulate the values a, b, and c of the quadratic formula [Eq. (16)], remembering that all terms not containing the unknown, no matter how many there may be, are grouped under c. Thus $a = 2$, $b = q - 5p$, $c = 2p^2 + pq - q^2$. Then

$$x = \frac{-(q - 5p) \pm \sqrt{(q - 5p)^2 - 8(2p^2 + pq - q^2)}}{2(2)}$$

$$= \frac{-q + 5p \pm \sqrt{q^2 - 10qp + 25p^2 - 16p^2 - 8pq + 8q^2}}{4}$$

$$= \frac{-q + 5p \pm \sqrt{9q^2 - 18pq + 9p^2}}{4}$$

$$= \frac{-q + 5p \pm (3q - 3p)}{4}$$

$$= \frac{-q + 5p + (3q - 3p)}{4} = \frac{2q + 2p}{4} = \frac{q + p}{2}$$

and

$$x = \frac{-q + 5p - (3q - 3p)}{4} = \frac{-4q + 8p}{4} = 2p - q$$

In cases in which the equation or the roots are more involved, checking by direct substitution may entail as much time and effort as the original solution. The process may be facilitated by assigning numerical values to the literal constants. Avoid assigning numerical values which will make a denominator equal to zero. See Example 5, page 111.

Example 12. Check the equation $(n - k)x^2 + (k - m)x + m - n = 0$, whose roots are $x = (m - n)/(n - k)$ and $x = 1$. (Note: Only the first root requires discussion.) Assigning arbitrary simple numerical values (other than 0 and 1) to m, n, and k such that x will also be a simple integer, we might select $m = 10$, $n = 4$, and $k = 2$; therefore

$$x = \frac{10 - 4}{4 - 2} = 3$$

Substituting in the original equation,

$$(4 - 2)(3)^2 + (2 - 10)(3) + 10 - 4 = 0$$

or

$$(2)(9) + (-8)(3) + 6 = 0$$

or

$$18 - 24 + 6 = 0$$
$$0 = 0$$

EXERCISE 7

Solve the following literal quadratic equations for w, x, y, or z as the case may be:

1. $15x^2 - mx - 2m^2 = 0$

2. $4x^2 + (x + 3a)^2 = 17a^2$

3. $w^2 + aw + bw - cw = c(a + b)$

4. $abcx^2 + b^2cx + ac^2x + bc^2 = 0$

5. $(b - c)y^2 + (c - a)y + a - b = 0$

6. $w^2 - \left(\dfrac{a}{b} + \dfrac{b}{a}\right) w + 1 = 0$

7. $3z^2 - hz - 5 = 0$

8. $\dfrac{c^2}{x^2} = \dfrac{c + 1}{x + 1}$

9. $(x - k)^2 + (x - h)^2 = k^2 + h^2$

10. $a = \pi x(x + 2h)$

11. Given the equation $\dfrac{x - 2}{m - 1} = \dfrac{2m}{2x - 3m}$, which is to be solved for x, write the expressions for a, b, and c of the quadratic formula [Eq. (16)], but do not solve.

12.10 Precautions Regarding Roots

At this point it is important to guard against two common errors resulting from apparently legitimate operations. These pitfalls are vanishing roots and extraneous roots. *Extraneous roots* are values which are created in the process of solving an equation but which do not check the original equation.

12.11 Vanishing Roots

It is not advisable to divide an equation through by an expression containing the unknown. By such division a root may be lost.

Example 13. Solve the equation $x^2 - 5x = 0$.
      Dividing through by x,

$$x - 5 = 0$$

or

$$x = 5$$

It is evident that another root, $x = 0$, has been lost in this process. Clearly, we should have factored and set each factor equal to zero. Thus we obtain the equation $x(x - 5) = 0$ and the roots $x = 0$ and $x = 5$.

Example 14. Solve the equation $2x^2 + 3x - 2 = x^2 - 4$.
    Dividing through by $x + 2$,

$$2x - 1 = x - 2$$

Adding $1 - x$ to both members,

$$x = -1$$

In this case, by dividing by $x + 2$, we have lost the root $x = -2$.

12.12 Extraneous Roots

If both members of a given equation are multiplied by an expression involving the unknown, the resulting equation may have solutions (or roots) which are not solutions (or roots) of the given equation. These are called *extraneous solutions,* or *extraneous roots.*
    Because of the possibility of obtaining such extraneous values it is absolutely necessary that all roots or solutions be checked by substitution in the given equation.

Example 15. Solve the equation

$$3 + \frac{4}{x + 2} = \frac{x^2}{x + 2} - 1$$

    Multiplying by $x + 2$,

$$3(x + 2) + 4 = x^2 - (x + 2)$$

Clearing of parentheses,

$$3x + 6 + 4 = x^2 - x - 2$$

By collecting terms after subtracting $3x + 10$ from both members of the above equation, we obtain

$0 = x^2 - 4x - 12$

Factoring;

$0 = (x - 6)(x + 2)$

$x - 6 = 0$ or $x = 6$

$x + 2 = 0$ or $x = -2$

The root $x = 6$ satisfies the equation, but the root $x = -2$ does not. Note that this situation could have been avoided by writing

$$3 + 1 = \frac{x^2}{x + 2} - \frac{4}{x + 2}$$

Combining fractions and reducing,

$$4 = \frac{x^2 - 4}{x + 2} = x - 2$$

or

$x = 6$

12.13 Checking

Checking is usually accomplished by direct substitution of the answers, or roots, in the original equation. However, if the original equation is in the type form for the general quadratic, a shorter method is available, especially if one or both roots are fractions. It depends upon the fact that if m and n are roots of a given equation, we may write $(x - m)(x - n) = 0$. [The *factor theorem* states that if r is a root of the equation $f(x) = 0$, then $x - r$ is a factor of the polynomial $f(x)$.] Referring back to Example 7 and the equation $3x^2 - 5x = 2$ whose roots are 2 and $-\frac{1}{3}$, we may check by writing

$(x - 2)[x - (-\frac{1}{3})] = 0$

Multiplying by 3,

$(x - 2)(3x + 1) = 0$

$3x^2 - 5x - 2 = 0$

which checks all but the preliminary multiplication and transposition.

Note. *The alert student will appreciate that we have here an aid to factoring expressions of the type $acx^2 + (ad + bc)x + bd$ (expansion 6, Sec. 3.21). If necessary, first factor the expression to free it of monomial factors. Set the resulting expression, such as $3x^2 - 5x - 2$, equal to zero, determine the roots by Eq. (16), the quadratic formula, and derive the factored form of the expression $(x - 2)(3x + 1)$ as above.*

12.14 Summary of Procedure for Solving Quadratic Equations

1. *Simplify any reducible fractions and combine any fractions having a common denominator.*
2. *Clear of fractions.*
3. *If there is a factor (not containing the unknown) common to all terms, divide through by that factor.*
4. *The choice of the method of solution might well be based upon consideration of the following suggestions in the order given:*
   a. *Use the factoring method if the factors are readily discernible.*
   b. *Use the method of completing the square if the coefficient of the second-degree term is unity and that of the first-degree term is divisible by 2.*
   c. *If neither (a) nor (b) is applicable, use the quadratic formula. This is somewhat longer but has general application to all cases.*
5. *Check all answers, discarding any extraneous roots.*

EXERCISE 8

Solve the following equations by the most convenient method, rejecting any extraneous roots and leaving any irrational answers in the radical form.

1. $\dfrac{y}{y + 1} = \dfrac{y + 2}{3y}$

2. $(w + 2)^3 - w^3 = 56$

3. $\dfrac{1}{z - 3} + \dfrac{1}{z + 4} = \dfrac{1}{12}$

4. $x + \dfrac{1}{5} = 5 + \dfrac{1}{x}$

5. $y + \dfrac{mn}{y} = m + n$

6. $\dfrac{1}{r} + r = 3 + \dfrac{3}{r}$

7. $11d^2 + 7d + 1 = 0$

8. $\dfrac{10}{x} - \dfrac{9}{x + 1} - \dfrac{8}{x + 2} = 0$

9. $\dfrac{h}{5} - \dfrac{5}{6} = \dfrac{6}{5} - \dfrac{5}{h}$

10. $\dfrac{1}{x} - \dfrac{1}{d} = \dfrac{1}{x + d}$

11. $\dfrac{1}{c} + \dfrac{1}{8 - c} = \dfrac{1}{8}$

12. $x(2x - c) + x(x - c) = bx$

13. $2y(7y - a) = (a + y)(a - y)$

14. $\dfrac{1}{x + 3} + \dfrac{1}{x + 2} - \dfrac{1}{x + 1} = 0$

15. $\pi x^2 + 2\pi n x - A = 0$

16. $8w = -3(1 + 4w^2)$

17. $\dfrac{2}{m + 5} - \dfrac{m + 3}{(m + 4)(m + 5)} = \dfrac{1}{4(m - 8)}$

18. $\dfrac{a}{x - b} + \dfrac{b}{x - a} = 2$

19. $c(x^2 - 1) = (ax + b)(x - 1)$

20. $\dfrac{1}{w - 2} + 1 = \dfrac{6 - w}{w^2 - 4} + \dfrac{1}{w + 2}$

21. $\dfrac{a}{x} + \dfrac{x}{a} = \dfrac{33a^2 - x^2}{ax}$

22. $\dfrac{1}{c + d + x} = \dfrac{1}{c} + \dfrac{1}{d} + \dfrac{1}{x}$

23. $9x^2 - hx + h^2 = 0$

24. The roots of a quadratic equation are $\frac{2}{3}$ and $-\frac{5}{2}$. Write the original equation cleared of fractions.

12.15 Discriminant

In the quadratic formula

$$x = \dfrac{-b \pm \sqrt{b^2 - 4ac}}{2a}$$

the expression $b^2 - 4ac$ is called the *discriminant*. The value of the discriminant may usually be determined by inspection and indicates the nature of the roots of the equation $ax^2 + bx + c = 0$. The relationships are summarized in Table 12.1.

Example 16. Determine the nature of the roots of the equation $2x^2 - 7x - 5 = 0$.
   Here $b^2 - 4ac = (-7)^2 - 4(2)(-5) = 49 + 40 = 89$. Since 89 is positive but not a perfect square, the roots are real, irrational, and unequal.

Example 17. Determine the nature of the roots of the equation $25x^2 - 30x + 9 = 0$.

TABLE 12.1

Discriminant	Positive and a perfect square	Positive but not a perfect square	Zero	Negative
Character of the roots	Real, rational, and unequal	Real, irrational, and unequal	Real, rational, and equal	Complex and unequal if $b \neq 0$; imaginary if $b = 0$
Graph of $ax^2 + bx + c = y$	Cuts X axis in two points		Tangent to X axis	Does not intersect X axis

In this example, $b^2 - 4ac = (-30)^2 - 4(25)(9) = 0$. Hence the roots are real, rational, and equal.

EXERCISE 9

Calculate the discriminant and describe the character of the roots without solving:

1. $x^2 + 4x - 12 = 0$
2. $x^2 - x - 20 = 0$
3. $x^2 - 6x + 9 = 0$
4. $x^2 - 3x - 5 = 0$
5. $2x^2 + 5x - 3 = 0$
6. $4x^2 + 20x + 25 = 0$
7. $6x^2 - x - 2 = 0$
8. $x^2 + 7x + 18 = 0$
9. $3x^2 - 18x + 27 = 0$
10. $2x^2 + 7x + 9 = 0$
11. $5x^2 - x - 10 = 0$
12. $28x^2 + 84x + 63 = 0$
13. $20x^2 - 7x - 3 = 0$
14. $14x^2 + 13x - 12 = 0$

12.16 Axis of Symmetry—Extreme Value

An equation in the form

$$ax^2 + bx + c = y$$

where a is a positive number, is plotted in Fig. 12.3a. Observe that this graph intersects

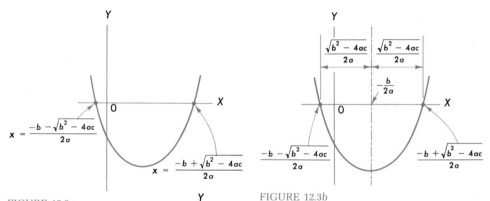

FIGURE 12.3a

FIGURE 12.3b

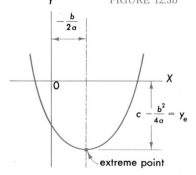

FIGURE 12.3c

the x axis at the two points

$$x = \frac{-b + \sqrt{b^2 - 4ac}}{2a}$$

and

$$x = \frac{-b - \sqrt{b^2 - 4ac}}{2a}$$

From Eq. (16) it is evident that these two values of x are the roots of the equation

$$ax^2 + bx + c = 0$$

That is, they are the values of x for which $y = 0$.

We recall from Sec. 12.4 that the graph of the equation $ax^2 + bx + c = y$ has a vertical axis of symmetry,† and that the extreme value of y, or the turning point of the curve, lies on this axis. The axis of symmetry intersects the X axis at a point midway between the x intercepts of the graph.

From the midpoint formula (Sec. 8.15) the axis of symmetry intersects the X axis at the point whose x coordinate is given by

$$\frac{\dfrac{-b + \sqrt{b^2 - 4ac}}{2a} + \dfrac{-b - \sqrt{b^2 - 4ac}}{2a}}{2}$$

or at the point whose x coordinate is $-b/2a$. See Fig. 12.3b; also Fig. 12.1. The equation of the axis of symmetry is therefore $x = -b/2a$.

By substituting $-b/2a$ for x in the given equation, we obtain

$$y_e = a\frac{b^2}{4a^2} - b\frac{b}{2a} + c = \frac{b^2}{4a} - \frac{b^2}{2a} + c$$

$$= c - \frac{b^2}{4a}$$

where y_e is the ordinate of the graph at the extreme point, or the turning point (Fig. 12.3c). The coordinates of the extreme point are therefore

$$\left(-\frac{b}{2a}, \; c - \frac{b^2}{4a}\right) \tag{17}$$

†In this instance a was equal to unity, but the symmetry still exists for any $a \neq 0$.

A brief check will show that if a is positive, the curve opens upward, $\smile$, and the extreme is a minimum. On the other hand, if a is negative, the curve opens downward, $\frown$, and the extreme is a maximum.

Example 18. Divide 6 into two parts such that their product is a maximum.

If x and $6 - x$ represent the two parts, we have

$$P = x(6 - x) = -x^2 + 6x$$

where P is the product.

By Eq. (17), x for a maximum P is

$$x = -\frac{6}{(2)(-1)} = 3$$

and the maximum product is also given by Eq. (17):

$$P_{max} = 0 - \frac{(6)^2}{(4)(-1)} = 9$$

Actually, no formal rule is necessary. If the graph of the given equation is sketched as in Fig. 12.4, the x intercepts appear at $x = 0$ and $x = 6$. The axis of symmetry intersects the X axis at

$$x = \frac{0 + 6}{2} = +3$$

As we found above, this leads to a maximum product of 9.

EXERCISE 10

In Probs. 1 to 10, determine the nature and coordinates of the extreme point of the graphs of the following equations and the equation of the axis of symmetry.

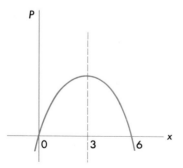

FIGURE 12.4

1. $x^2 - 2x - 15 = y$
2. $-x^2 - 4x + 12 = y$
3. $x^2 + 7x + 6 = y$
4. $2x^2 + 5x - 3 = y$
5. $-x^2 + 6x = y$
6. $5x^2 - 6x - 8 = y$
7. $-6x^2 + 7x - 2 = y$
8. $3x^2 - x - 10 = y$
9. $-4x^2 - 4x + 15 = y$
10. $-2x^2 + x + 28 = y$

11. The equation of a certain parabola is of the type $y + b = (x + a)^2$. Evaluate a and b by inspection. Find the x and y intercepts. See Fig. 12.5.

12. Divide 17 into two parts such that their product will be a maximum.

13. Find the dimensions of the largest possible rectangular area that may be enclosed by 140 ft of fencing.

14. The work done by exploding a mixture of 1 ft³ of water gas and v ft³ of air is $w = 84v - 3.15v^2$. What value of v will lead to the maximum value of w? What is the maximum value of w?

15. A ball is thrown upward with a speed of 144 ft/s from a building 340 ft high. If the height of the ball in feet h at any time t s after throwing is given by the equation $h = 340 + 144t - 16t^2$, find the maximum height reached.

16. The power delivered to an external circuit by a 32-V generator whose internal resistance is 2 Ω is $32a - 2a^2$ W, where a is the current in amperes. At what current will this generator deliver the maximum power?

17. A long sheet of copper 22 in wide is to be made into a gutter by turning strips up vertically along the two sides. How many inches should be turned up at each side to obtain the greatest carrying capacity?

18. A transit authority charges 35 cents per ride and carries an average of 2,100 passengers daily. It proposes to reduce an operating deficit by raising the fares. However, it has good reason to believe that for each 5-cent increase in the fare, 200 riders will seek other means of transportation. What fare would result in the maximum return?

    (It follows that if n is the number of 5-cent increases, there will be $2,100 - 200n$ riders, each paying a fare of $35 + 5n$ cents.)

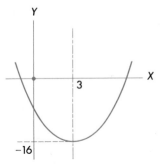

FIGURE 12.5

12.17 Applications Leading to Quadratic Equations in One Variable

In Sec. 12.12 we learned that an *extraneous* solution to a quadratic equation in one variable is a solution which is introduced in the process of solving the problem but which is not a solution to the given equation.

Many of the problems in Exercise 11 are problems in which a quadratic equation is used as a mathematical model of a certain physical situation. Very often in such cases a pair of solutions are found which do check when substituted in the basic equation but which are inconsistent with the physical restrictions imposed by the statement of the problem. Since such solutions do in fact check out in the original equation, they cannot be considered mathematically extraneous. Nevertheless, they may at times violate some of the physical conditions of the problem and are appropriately called *secondary solutions*.

Example 19. Figure 12.6 illustrates a block of concrete resting on the floor of a room and in contact with a vertical wall of the room. The concrete block is 2 ft wide and 1 ft high. The problem is to find the radius of a wheel whose circumference rests on the floor, in contact with the wall and the edge of the concrete block at point A.

We construct the right triangle ABC in which the hypotenuse is R, the vertical side is $R - 1$, and the horizontal side is $R - 2$. Then, by the Pythagorean theorem,

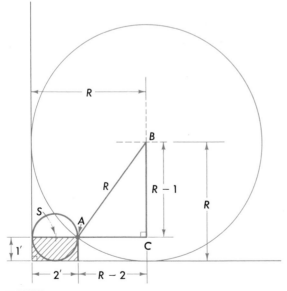

FIGURE 12.6

$$R^2 = (R - 1)^2 + (R - 2)^2$$

or

$$R^2 = R^2 - 2R + 1 + R^2 - 4R + 4$$

or

$$0 = R^2 - 6R + 5$$

Then

$$0 = (R - 5)(R - 1)$$

from which

$$R = 5 \text{ ft} \quad \text{and} \quad R = 1 \text{ ft}$$

Both of these roots are found to check in the original equation. Hence neither of them is extraneous. However, the 5-ft radius is the only one which is consistent with the physical conditions of the problem.

The circle with a 1-ft radius whose center is at point S would be an absurdity since it would require a wheel to occupy space already occupied by a block of cement.

Therefore the solution $R = 1$ ft is called a *secondary* solution.

It is frequently interesting and instructive to determine the significance of a secondary solution. (See Probs. 13, 14, and 35; also Example 20.) It may suggest that the scope of the problem is broader than the original question indicates.

Remember that a "check," unless made on the original stated problem, is no check at all.

EXERCISE 11

1. Find two numbers whose sum is 23 and whose product is 126.
2. Separate 19 into two parts whose product is 84.
3. The difference between two positive numbers is 7 and their product is 78. Find the numbers.
4. Separate 132 into two positive parts such that one part is the square of the other.
5. The sum of a positive number and its square is 72. Find the number.
6. Find two consecutive positive integers whose product is 156.
7. Find two consecutive positive even integers whose product is 224.
8. Find two consecutive positive odd integers whose product is 195.
9. Find a positive number whose square exceeds 36 by as much as 36 exceeds the number.

10. What number added to its reciprocal equals 2.9?

11. A ball is thrown downward from a building 380 ft high with a speed of 112 ft/s. Under these conditions the height h of the ball in feet at any time t s after throwing the ball is given by the equation $h = -16t^2 - 112t + 380$. When will the ball strike the ground?

12. Repeat Prob. 11 for the time at which the ball is 252 ft above the ground. Can you explain the secondary (negative) answer?

13. A model rocket is fired upward from a building 60 ft high with a speed of 112 ft/s. The height h (in feet) above the ground level at any time t (in seconds) after firing is given by the equation

$$h = -16t^2 + 112t + 60$$

   (a) Calculate h at $t = -1$, 0, 1, 2, 3, 4, 5, 6, 7, and 8 s.
   (b) Plot the graph of this equation (h vertically and t horizontally).
   (c) When did the maximum height occur? What was the height?
   (d) When did $h = 0$? What is the significance of each of the two answers?

14. A strip of metal 8 in wide is to be bent into a trough of rectangular cross section (open top) whose cross-sectional area is to be $7\frac{1}{2}$ in². Find the depth and width of the trough. How many answers?

15. The hypotenuse of a right triangle is 10 ft longer than the shorter side and 5 ft longer than the longer side. Find the sides of the triangle.

16. How high is a tree if it takes 3 s for a stone thrown over it to return to the ground? Use the formula $s = 16t^2$, where s is the distance in feet covered by a freely falling body in t s (starting from rest).

17. A bomber traveling at 375 mi/h releases a bomb at 8,200 ft. How long after its release will the bomb land? How far ahead of the target is the bomb released? Neglect air resistance.

18. If the cross-sectional area of the angle beam in Fig. 12.7 is $6\frac{1}{4}$ in², find the thickness x.

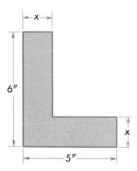

FIGURE 12.7

19. A group of boys bought a canoe for $70, planning to divide the expense equally. However, two boys dropped out, increasing the share of each boy by $1.75. How many boys were there in the original group?

20. Find the radius R in Fig. 12.8. (*Hint:* Draw the construction lines AB, BC, and CA as in Fig. 12.9. Then form the right triangle ABC. Find the missing dimensions in Fig. 12.9 in terms of R and the given constants. By means of the Pythagorean theorem relate the sides of the right triangle ABC and solve for R.)

21. Find the radius R in Fig. 12.10.

22. Find the diameter of the circle in Fig. 12.11.

23. Two resistances in parallel have a joint resistance of 4.2 Ω. The same two resistances in series have a resistance of 20 Ω. Find the value of each resistance.

24. A park is 480 yd long by 320 yd wide. It is decided to double its area, retaining the rectangular shape, by adding strips of equal width to one end and one side. Find the width of the strips.

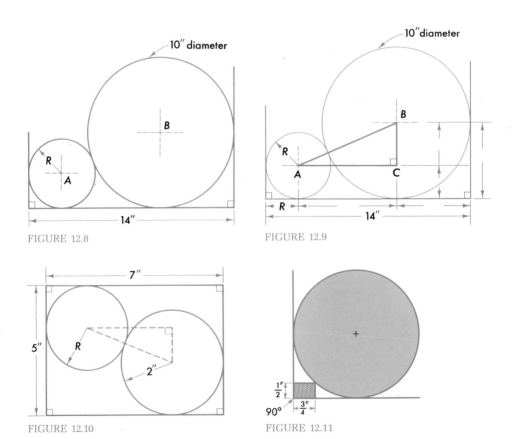

FIGURE 12.8

FIGURE 12.9

FIGURE 12.10

FIGURE 12.11

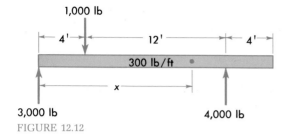

FIGURE 12.12

25. Figure 12.12 shows a beam that supports its own weight of 300 lb/ft and a concentrated load of 1,000 lb at a point 4 ft from the left end. There is a point x ft from the left end where there is no compression or tension (push or pull) in the beam. x must be larger than 4 and less than 16 and is a root of the equation

$$3,000x - 1,000(x - 4) - 150x^2 = 0$$

Find x.

26. The perimeter of a rectangular field is 274 yd, and the diagonal is 97 yd. Find its dimensions.

27. The sum of the areas of the two inner circles in Fig. 12.13 is three-quarters the area of the outer circle. Find the diameter of the smallest circle.

28. The outer portion of a garden 22 by 30 ft is to be occupied by a walk of uniform width. If the garden is to be reduced to three-quarters of its original area, find the width of the walk.

29. When a flexible rope or wire t ft long is strung between two poles l ft apart (points of attachment at same elevation), the formula $t = \sqrt{l^2 + 5.3s^2}$ applies, where s is the sag in feet. A copper telegraph wire is being installed at 60°F, the poles being 140 ft apart. Find the necessary sag (in inches) in order to provide against a temper-

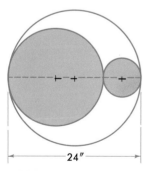

FIGURE 12.13

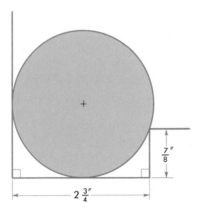

FIGURE 12.14

ature of $-20°F$. The linear coefficient of expansion of copper is 0.00001 per Fahrenheit degree.

30. An express train makes the run between two cities 250 mi apart in $1\frac{1}{4}$ h less time than a local whose speed is 10 mi/h less. Find the average speed of the express.

31. A motorboat takes 2 h 8 min longer to make a trip of 48 mi up a stream than it takes on the return trip downstream. If the average rate of the current is 4 mi/h, find the rate of the boat in still water.

32. A bomb was released from a plane in a glide at an altitude of 20,000 ft. If the vertical component of the speed of the plane was 150 mi/h, find the time required for the bomb to strike the target. $s = vt + \frac{1}{2}gt^2$, where s is the vertical distance in feet, v is the initial vertical component of the speed in feet per second, t is the time in seconds, and g is the gravitational constant (32 ft/s²). Air resistance is neglected.

33. Find the diameter of the circle in Fig. 12.14.

34. Find the radius R of the arc in Fig. 12.15.

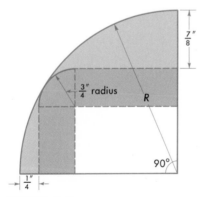

FIGURE 12.15

35. Find the radius R of the cylindrical gauge in Fig. 12.16*a*. (*Note:* In one respect an equation resembles a slide rule or computer—it will deliver only what you feed into it. In effect we "told" the equation we wanted a circle tangent to the sides of the right angle and also to the semicircle. We had no way of "programming" the restriction excluding the larger circle; so we find in our "output" both circles meeting the tangency requirement. See Fig. 12.16*b*.)

36. A 2-ft walk surrounds a circular flower bed. If the area of the walk is one-tenth the area of the flower bed, find the diameter of the bed.

37. Ice on a power-transmission line may be melted off by increasing the voltage. If $E^2/a(1 + bt) = K(t - t_0)$, find the temperature $t°$F of the wire if the voltage E is 580; a is the resistance of the line at 0°F and is equal to 52 Ω; b is the temperature coefficient of resistance of the wire, or 0.003; K is the coefficient of heat transfer by convection and conduction, or 350; t_0 is the air temperature, or 24°F.

38. A triangle has a 12-in altitude and a 20-in base. How wide a strip should be cut off by a line parallel to the base to leave 55 in² at the top?

Example 20. A 50-cd lamp and a 120-cd lamp are 30 ft apart. Find the point on a line between them which is equally illuminated by both lights. When an object is equally illuminated by two light sources, the light intensities vary directly as the squares of their respective distances from the object. What significance attaches to the secondary solution of this problem? What can you say about the locus of all such equally illuminated points (a) in the same plane, (b) in three-dimensional space? (The relationship in this problem might have been restated to read that "the intensity of illumination varies inversely as the square of the distance from the light source." A similar rela-

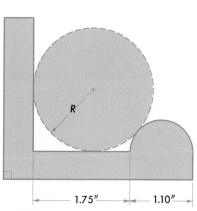

FIGURE 12.16*a*

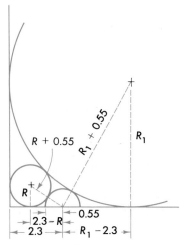

FIGURE 12.16*b*

tionship holds for other forms of energy or force, e.g., heat, electricity, magnetism, and gravity. These are examples of the *inverse square law.*)

An advantageous method of approaching the solution consists in setting up the proportion

$$\frac{(30 - x)^2}{x^2} = \frac{120}{50}$$

followed by taking the square root of both sides to obtain

$$\frac{30 - x}{x} = \pm\sqrt{\frac{120}{50}} = \pm 1.549$$

from which $x = 11.8$ and $x = -54.6$. It is true that $18.2/11.8 = \sqrt{120/50}$, but it is also true that

$$\frac{|-54.6| + |30|}{|-54.6|} = \sqrt{\frac{120}{50}}$$

(Fig. 12.17). In fact, $PB/PA = \sqrt{120/50}$, where P is any point on the dotted curve which can be shown to be a circle by analytic geometry. The circle may also represent a spherical surface of the same radius and center.

The moral of this example is: Don't reject a secondary solution out of hand. Perhaps it is trying to tell you something!

39. The solution of the problem of determining the location of the point of *minimum total* illumination between two light sources required that the real root of the equation

$$\frac{(30 - x)^3}{x^3} = \frac{120}{50} = 2.4$$

be found. Calculate x to the nearest 0.1. (*Hint:* Do not expand, but take the cube

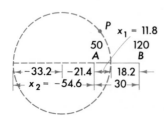

FIGURE 12.17

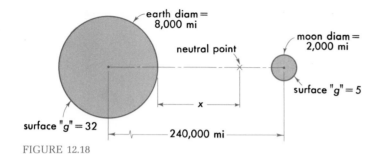

FIGURE 12.18

root of both sides.) It can be shown that this solution will yield the only real root of the equation. Compare your answer with that obtained in Example 20 above.

40. The law of gravitational attraction states that the gravitational force exerted on an object varies inversely as the square of its distance from the attracting body. At what distance x from the surface of the earth (on the earth-moon line) will the gravitational attraction of the earth just balance that of the moon? This is known as the *neutral point* (Fig. 12.18).

41. Mars will be most favorably situated for exploration when it is at its minimum distance, or about 35,000,000 mi from the earth. Locate the neutral point (where Earth's g and Mars' g just balance) under these conditions.

Example 21. An object is weighed on a platform balance and is found to balance 236.0 g. When the positions of object and weights are reversed, the indicated weight is 249.0 g because of unequal balance arms. Find the true weight correct to 0.1 g. See Fig. 12.19.

Here we find

$$236L = SW$$

and

$$LW = 249S$$

By dividing we find

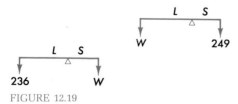

FIGURE 12.19

$$\frac{236L}{LW} = \frac{SW}{249S}$$

or

$$\frac{236}{W} = \frac{W}{249}$$

from which

$$W^2 = 236 \times 249$$

or

$$W = \sqrt{236 \times 249} = 242.4 \text{ g}$$

Note that this answer closely approximates

$$\frac{236 + 249}{2} = 242.5$$

The student may wish to know when the arithmetic mean leads to a close enough answer. The smaller the relative difference between the numbers to be averaged, the better the approximation. This is the basis of a rapid and accurate method of determining square root on a desk calculator or even by longhand.

42. Find a such that the area of the trapezoid shown in Fig. 12.20 is 300 square units if

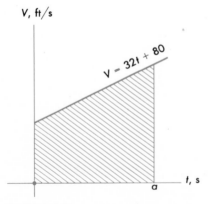

FIGURE 12.20

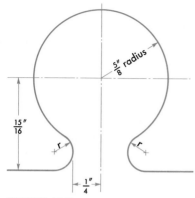

FIGURE 12.21

V = velocity, ft/s
t = time, s
Area = (ft/s)(s) = feet traveled in time t

43. Find the radius of curvature in Fig. 12.21.
44. Two steamers ply between two ports 450 mi apart. One steamer runs $2\frac{1}{2}$ mi/h faster and takes $2\frac{1}{2}$ h less for the trip than the other. Find their rates in miles per hour.
45. A man travels 30 mi by bus and returns by a train which runs 15 mi/h faster. If the total running time is 1 h 57 min, find the rates of the bus and the train.
46. Find the length of the bar in Fig. 12.22 to balance a steam pressure of 18 lb/in². An 8-lb weight is attached at the free end of the bar. The bar weighs 1.2 lb/linear ft.
47. A right triangle containing 210 ft² is roped off by a line 70 ft long. Find the three sides of the triangle.
48. Six seconds after a stone is dropped into a mine shaft the sound of the impact at the bottom reaches the top. If the velocity of sound is 1,120 ft/s and the usual

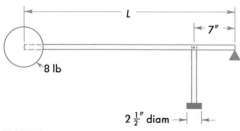

FIGURE 12.22

formula for a freely falling body applies to the falling stone, find the depth of the shaft.

49. A flat disk $1\frac{3}{8}$ in in diameter is to be pressed into an open-top cylinder $\frac{5}{8}$ in deep. If the total surface of the metal is unchanged during the operation, find the diameter of the cylinder.

Example 22. A rail-diesel coach has an acceleration of 0.8 ft/s² and a braking deceleration of 1.2 ft/s². If two stations are $1\frac{1}{4}$ mi apart, what is the maximum speed reached when the power is shut off and the brakes applied? What time is required between the two stops?

A graph of the speed-time relation is shown in Fig. 12.23. From the graph we observe that the maximum speed is given by

$$V_{\text{max}} = 0.8t_1 \tag{18}$$

and

$$V_{\text{max}} = 1.2t_2$$

Therefore

$$1.2t_2 = 0.8t_1$$

Consequently

$$t_2 = \frac{2}{3}t_1 \tag{19}$$

The graph of the speed V versus time t is a straight line. Therefore, in this case, the

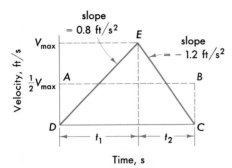

Area of triangle *DEC* = area of rectangle *ABCD* = 6600 ft

FIGURE 12.23

average speed is given by

$$V_{av} = \tfrac{1}{2}V_{max}$$

Then, from Eq. (18), we may write

$$V_{av} = \tfrac{1}{2}(0.8t_1) = 0.4t_1$$

The distance traveled is given to be $1\tfrac{1}{4}$ mi. The distance in feet d is therefore

$$d = 5{,}280 \times 1.25 = 6{,}600 \text{ ft} \tag{20}$$

Since in general

$$d = V_{av}(t)$$

we may write

$$d = (0.4t_1)(t_1 + t_2) \tag{21}$$

and by recalling Eq. (21), we may write

$$
\begin{aligned}
6{,}600 &= (0.4t_1)(t_1 + \tfrac{2}{3}t_1) \\
&= (0.4t_1)(\tfrac{5}{3}t_1) \\
&= \tfrac{2}{3}t_1{}^2
\end{aligned}
$$

Then

$$t_1{}^2 = \frac{3 \times 6{,}600}{2} = 9{,}900$$

and

$$t_1 = \sqrt{9{,}900} = 99.5 \text{ s}$$

From Eq. (19)

$$t_2 = \tfrac{2}{3}t_1 = \tfrac{2}{3} \times 99.5 = 66.3 \text{ s}$$

The total time required is

$$t_t = 99.5 + 66.3 = 165.8 \text{ s}$$

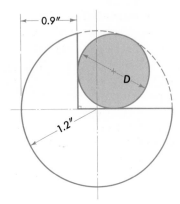

FIGURE 12.24

From Eq. (18) $V_{max} = (0.8)(99.5) = 79.6$ ft/s $= 54.3$ mi/h.

50. Find the diameter of the circle in Fig. 12.24.
51. Two boats leave simultaneously from the opposite shores of a bay which is $2\frac{1}{4}$ mi wide and pass each other in 6 min. The faster boat completes the trip $4\frac{1}{2}$ min before the other boat docks. Find the rates of the boats in miles per hour.
52. An open-top box is to be made from a 15-in square sheet of brass by cutting out squares of a certain size from each corner and turning up the resulting tabs. It was found that the capacity of the box would remain unchanged whether the squares to be cut out were of a given size or $\frac{1}{2}$ in larger. What was the smaller size of square cut out? What was the capacity of the box?
53. In the trapezoid shown in Fig. 12.25 find x such that a line drawn parallel to the base shall divide the area in half.
54. A parabola $y = ax^2 + bx + c$ passes through the points $(10, -13)$, $(30,1)$, and $(40,23)$.

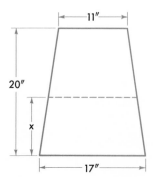

FIGURE 12.25

Substitute the x and y values of the given coordinates successively in the given equation and solve the three simultaneous equations thus formed for a, b, and c.

55. The equation of a circle is $(x - a)^2 + (y - b)^2 = R^2$, where R is the radius. The coordinates of the center of the circle are given by (a,b). The graph passes through the points (3,18), (18,13), and (7,2). Substitute the x and y coordinates of these points successively in the given equation and solve the simultaneous equations thus formed for a, b, and R.

56. The length L of a parabolic bridge cable of span a and sag h is given approximately by the formula $L = a[1 + \frac{8}{3}(h/a)^2 - \frac{32}{5}(h/a)^4]$. Determine the sag if a 200-ft cable is suspended from two points 198 ft apart and in the same horizontal plane ($a = 198$). Solve first for h/a, then for h: (a) using the entire formula, (b) by neglecting the term containing $(h/a)^4$.

57. If the radius of the earth is 3,960 mi, h is the elevation of the observer (feet above sea level), and d is the distance in miles of the horizon at sea, derive an equation for d in terms of h (Fig. 12.26). In this approximation a term is dropped, since its importance is considered to be less than that of light refraction. Show that this assumption is probably justified for such values of h as 100 ft, 1,000 ft, and even 5 mi.

58. Referring to Prob. 31, Exercise 2, Chap. 1, show that the book values v in the sum-of-years-digits schedule are related to the years t by a quadratic relationship, that is, $v = at^2 + bt + c$, where a, b, and c are constants. Select any three pairs of values and solve for the constants a, b, and c as in Prob. 54 above. Use your equation to predict the book value at $t = 1\frac{1}{2}$ years.

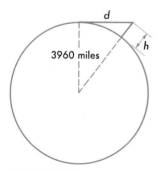

FIGURE 12.26

the right triangle

This section of the work will be concerned almost entirely with numerical trigonometry. This branch of trigonometry deals with methods by which certain sides and angles of a triangle may be computed when certain other sides and angles are known. Emphasis in this chapter will be on computation rather than analysis.

13.1 Angles

For the present purpose an angle will be described in terms of Fig. 13.1. The side r was originally coincident with the side OX. The side OX is called the initial position of r, or the *initial side* of the angle. However, r as shown in the diagram has been revolved about O in the direction of the arrow and has become the *terminal side* of the angle.

Two systems of units are ordinarily used to define the magnitude of an angle, the *degree system* and the *radian system*. The radian system will be discussed in Chap. 17. The degree system is the only one of importance in numerical trigonometry.

If r had revolved about O for one complete revolution from the initial position, an angle of 360 degrees would have been generated. By definition, 1 degree is equivalent to $\frac{1}{360}$ revolution. An angle of 90 degrees (written 90°) is equivalent to $\frac{90}{360}$, or $\frac{1}{4}$ revolution. An angle of 180° would be equivalent to $\frac{180}{360}$, or $\frac{1}{2}$ revolution.

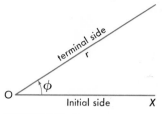

FIGURE 13.1

Often angles are measured in degrees and a decimal fraction of a degree, as for example 37.594°. Also, they are measured in degrees and subunits of a degree, called minutes and seconds. There are, by definition, 60 minutes in 1 degree and 60 seconds in 1 minute. The angular magnitude 37 degrees 42 minutes and 36 seconds would be written 37°42'36''. Since there are 60 seconds in a minute, it follows that 6'' is equivalent to $\frac{1}{10}$ minute. The angle 37°42'36'' could also have been written 37°42.6', and indeed angles are frequently measured in degrees, minutes, and tenths of minutes.

13.2 Arithmetical Operations with Degrees, Minutes, and Seconds

Addition, subtraction, multiplication, and division of angles in degrees, minutes, and seconds require somewhat different procedures from those used for similar operations in the decimal system.

Example 1. Add 27°42'38'' and 45°32'25''.

27°42'38''
45°32'25''
———————
72°74'63''

Since 63'' = 1'3'', the above answer could be written

72°75'3''

Since 75' = 1°15', the equivalent form would be

73°15'3''

Example 2. Subtract 15°47'18'' from 25°30'10''.
Notice that although 47' cannot be subtracted from 30', and 18'' cannot be sub-

tracted from 10″, the minuend can be altered to an equivalent form,

24°89′70″

and the subtraction can be carried out as shown below:

$$24°89′70″$$
$$15°47′18″$$
$$\overline{9°42′52″}$$

Example 3. Divide 25°17′27″ by 2.

$$\frac{25°}{2} = 12°$$

with a remainder of 1°, or 60′, which we add to 17′ to make 77′.

$$\frac{77′}{2} = 38′$$

with a remainder of 1′, or 60″, which we add to 27″ to make 87″.

$$\frac{87″}{2} = 43\frac{1}{2}″$$

The answer is then

12°38′43½″

Example 4. Divide 33°16′46″ by 7.

$$\frac{33°}{7} = 4°$$

with a remainder of 5°, or 300′, which we add to 16′ to make 316′.

$$\frac{316′}{7} = 45′$$

with a remainder of 1′, or 60″, which we add to 46″ to make 106″.

$$\frac{106″}{7} = 15″ \qquad \text{to the nearest second}$$

The answer is then

4°45′15″

Example 5. Multiply 35°31′42″ by 2.

$$35°31′42″$$
$$\underline{\hspace{1.2cm}2}$$
70°62′84″ or 71°03′24″

Example 6. Change 52.379° to degrees, minutes, and seconds.
   Since there are 60 minutes in 1 degree, in 0.379° there are

$$0.379° \times 60 = 22.740′$$

Since there are 60 seconds in 1 minute, in 0.74′ there are

$$0.74′ \times 60 = 44.40″$$

Therefore

$$52.379° = 52°22′44.4″$$

Example 7. Change 18°26′35″ to degrees and a decimal fraction of a degree.

$$26′ = \frac{26}{60} \text{ degrees} = 0.4333°$$

$$35″ = \frac{35}{3,600} \text{ degrees} = 0.0097°$$

$$18°26′35″ = 18.0000° + 0.4333° + 0.0097° = 18.443°$$

EXERCISE 1

Perform the operations indicated below:

1. 14°23′17″ + 28°19′5″ + 75°41′50″ + 21°13′19″
2. 53°42′39″ + 15°32′45″ + 13°15′28″ + 15°2′25″
3. 25°16′24″ + 27°10′29″ + 26°30′19″ + 10°29′37″
4. 62°3′15″ + 79°1′28″ + 16°59′28″ + 47°17′28″
5. 15°15′30″ + 26°48′13″ + 17°29′15″ + 28°15′42″
6. 15°28′17″ − 10°46′27″   7. 89°59′60″ − 25°36′10″
8. 90° − (27°42′51″)   9. 180° − (132°46′17″)

10. $360° - (42°36'25'')$
11. Multiply $26°45'48''$ by 2
12. Multiply $16°32'25''$ by 5
13. Multiply $3°00'12''$ by 6
14. Multiply $19°47'15''$ by 8
15. Multiply $22°11'38''$ by 4
16. Divide $37°42'31''$ by 5
17. Divide $10°13'27''$ by 7
18. Divide $72°19'48''$ by 9
19. Divide $52°26'11''$ by 10
20. Divide $182°15'27''$ by 4
21. Change to degrees, minutes, and seconds:
    (a) $52.763°$ (b) $0.527°$ (c) $146.895°$
22. Change to decimals:
    (a) $37°42'17''$ (b) $136°29'52''$ (c) $00°20'13''$

13.3 Unique Determination of Triangles

A triangle is *determined* when enough information is at hand so that it can be drawn to scale. (If it can be drawn to scale, it can be solved numerically.) A triangle can be drawn to scale:

1. If three sides are known
2. If one side and two angles are known
3. If two sides and one angle are known

In the latter case, if the given angle is less than 90° and one of the two given sides is opposite this angle, there may possibly be two solutions. More will be said of this later in the work.

If a triangle is known to be a right triangle, then obviously, one angle is known. Thus a right triangle is determined if in addition we know two sides or one side and an acute angle.

13.4 Solution of Triangles

A triangle is *solved* when the unknown sides and angles have been found. Triangles may be solved either by drawing them to scale and measuring the unknown sides and angles or by using suitable formulas to calculate the unknown dimensions.

Graphical solutions are common in surveying and mapping. They are also common in certain solutions to bridge- and roof-truss problems.

While computational methods will be the main concern of this chapter, the student should invariably make a scale drawing of the triangle he is solving. In this way he gets a far more vivid idea of the problem and the successive steps in the solution. Even more important, reference to the scale drawing will call his attention to gross errors in calculation.

Figure 13.2 illustrates a standard method of lettering a triangle and one which we shall use in this text.

Note that capital letters are used to denote the angles and that small letters are used to denote the sides. Furthermore, side a is opposite angle A, side b is opposite angle B, and side c is opposite angle C.

EXERCISE 2

(Even though the principal concern of this chapter is with right triangles, we have, in preparation for Chap. 15, included some oblique triangles in this exercise.)

In the following problems, the data are supposed to determine certain triangles. In some cases, these data are incomplete, inconsistent, or ambiguous. In other problems, data are correctly given. Draw these triangles full scale if correct data are given, and measure the unknown sides and angles. If the data are improperly given, tell in what respect they are improper.

1. $a = 3$ in $b = 4$ in $C = 90°$
2. $a = 4$ in $b = 3$ in $c = 5$ in $C = 90°$
3. $a = 10$ in $b = 3$ in $c = 5$ in
4. $A = 30°$ $C = 90°$ $c = 5$ in
5. $A = 30°$ $c = 3$ in $a = 1$ in
6. $A = 30°$ $c = 3$ in $a = 1.5$ in
7. $A = 30°$ $c = 3$ in $a = 2$ in
8. $A = 130°$ $B = 120°$ $c = 5$ in
9. $a = 4$ in $c = 6$ in $B = 30°$
10. $A = 30°$ $B = 120°$ $C = 30°$
11. $a = 4$ in $b = 3$ in $c = 3\frac{3}{4}$ in $C = 90°$
12. $A = 30°$ $B = 45°$ $c = 5$ in

13.5 The Trigonometric Functions

Figure 13.3 shows a right triangle whose sides are designated by a, b, and c. The side c represents the hypotenuse, and the angle C represents the right angle. This is a widely

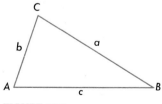

FIGURE 13.2

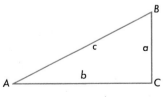

FIGURE 13.3

used method of lettering a right triangle and will be used in this text.

If these sides are used two at a time, six ratios may be formed: a/c, b/c, a/b, b/a, c/b, and c/a.

Referred to angle A these ratios have been given names as follows:

a/c is called the sine of angle A, or $sin\ A$
b/c is called the cosine of angle A, or $cos\ A$
a/b is called the tangent of angle A, or $tan\ A$
b/a is called the cotangent of angle A, or $cot\ A$
c/b is called the secant of angle A, or $sec\ A$
c/a is called the cosecant of angle A, or $csc\ A$

To emphasize the relative position of the sides of the triangle with respect to the angle A, we note that side a is the *side opposite* the angle A, the side b is the *side adjacent* to the angle A, and the side c is the hypotenuse of the triangle. Then from Fig. 13.4 we may form Table 13.1.

The size of the right triangle involved is not important. It is only the relative lengths of the sides which determine the values of the trigonometric functions. For example, consider the right triangles ABC and $AB'C'$ in Fig. 13.5. These right triangles are by no means equal in size. However, they are similar, and hence

$$\frac{a}{c} = \frac{a'}{c'} \qquad \frac{b}{c} = \frac{b'}{c'} \qquad \frac{a}{b} = \frac{a'}{b'}$$

$$\frac{b}{a} = \frac{b'}{a'} \qquad \frac{c}{b} = \frac{c'}{b'} \qquad \frac{c}{a} = \frac{c'}{a'}$$

Therefore the trigonometric functions of the angle A are the same whether the angle A is a part of the triangle ABC or the triangle $AB'C'$. In general, the values of

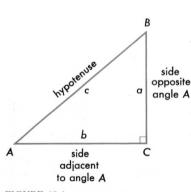

FIGURE 13.4

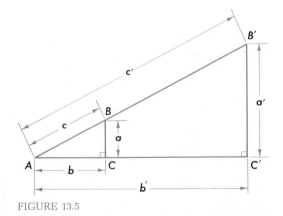

FIGURE 13.5

TABLE 13.1

$$\sin A = \frac{\text{side opposite angle } A}{\text{hypotenuse}} = \frac{a}{c}$$

$$\cos A = \frac{\text{side adjacent to angle } A}{\text{hypotenuse}} = \frac{b}{c}$$

$$\tan A = \frac{\text{side opposite angle } A}{\text{side adjacent to angle } A} = \frac{a}{b}$$

$$\cot A = \frac{\text{side adjacent to angle } A}{\text{side opposite angle } A} = \frac{b}{a}$$

$$\sec A = \frac{\text{hypotenuse}}{\text{side adjacent to angle } A} = \frac{c}{b}$$

$$\csc A = \frac{\text{hypotenuse}}{\text{side opposite angle } A} = \frac{c}{a}$$

the trigonometric functions of an angle depend entirely on the value of the angle, and not at all on the size of the triangle of which that angle is a part.

13.6 Trigonometric Tables

All six ratios associated with any angle have been calculated. The theory by which these calculations were made is beyond the scope of the present treatment, and the arithmetical work involved is too cumbersome for everyday use.

A brief table of the functions is given in Table 13.2.

To find functions of angles between 0° and 45°, look for the angle in the *left-hand*

TABLE 13.2

Angle A	a/c sin	b/c cos	a/b tan	b/a cot	c/b sec	c/a csc	
0	0.000	1.000	0.000	· · ·	1.000	· · ·	90
5	0.087	0.996	0.087	11.430	1.004	11.474	85
10	0.174	0.985	0.176	5.671	1.015	5.759	80
15	0.259	0.966	0.268	3.732	1.035	3.864	75
20	0.342	0.940	0.364	2.747	1.064	2.924	70
25	0.423	0.906	0.466	2.145	1.103	2.366	65
30	0.500	0.866	0.577	1.732	1.155	2.000	60
35	0.574	0.819	0.700	1.428	1.221	1.743	55
40	0.643	0.766	0.839	1.192	1.305	1.556	50
45	0.707	0.707	1.000	1.000	1.414	1.414	45
	cos b/c	sin a/c	cot b/a	tan a/b	csc c/a	sec c/b	Angle A

column and look for the name of the desired function at the *top* of the table. The desired function will then be at the intersection of the row containing the angle and the column containing the name of the function.

To find functions of angles between 45° and 90°, look for the angle in the *right-hand* column and the name of the desired function at the *bottom* of the table. The desired function will then be at the intersection of the row containing the angle and the column containing the name of the function.

EXERCISE 3

Verify the following:

1. $\sin 20° = 0.342$ 2. $\cos 25° = 0.906$ 3. $\tan 10° = 0.176$
4. $\cot 15° = 3.732$ 5. $\sec 40° = 1.305$ 6. $\csc 5° = 11.474$

EXERCISE 4

Verify the following:

1. $\csc 45° = 1.414$ 2. $\cos 60° = 0.500$ 3. $\sin 60° = 0.866$
4. $\cot 70° = 0.364$ 5. $\tan 75° = 3.732$ 6. $\sec 80° = 5.759$

13.7 The Solution of Right Triangles

To solve a right triangle, two parts other than the right angle must be known. At least one of these parts must be a side.

The first step in solving the problems that follow is to find an applicable formula from Table 13.1. An applicable formula is one which contains the unknown quantity we wish to find and the two quantities which are explicitly given.

Example 8. If in the right triangle shown in Fig. 13.3 angle A is 40° and side b is 5 in, find side a.

By consulting Table 13.1 we find two formulas containing the known quantities A and b as well as the unknown quantity a. These formulas are $\tan A = a/b$ and $\cot A = b/a$. By substituting the given values in these formulas and solving for a, we have

$$a = 5 \tan 40°$$

and

$$a = \frac{5}{\cot 40°}$$

Since multiplication is usually a less laborious arithmetic operation than division, we shall choose to use the first equation. Then, by consulting Table 13.2 to find the value of tan 40°, we may write

$a = 5 \times 0.839 = 4.195$ in

Example 9. If in the right triangle shown in Fig. 13.3 $A = 75°$ and $c = 25$ in, find b.
By consulting Table 13.1 we find two formulas containing the known quantities A and c as well as the unknown quantity b. These formulas are $\cos A = b/c$ and $\sec A = c/b$. By solving each formula for b, we have

$$b = \frac{25}{\sec 75°}$$

and

$$b = 25 \times \cos 75°$$

We shall choose to use the latter. By consulting Table 13.2 to find the value of $\cos 75°$, we may write

$b = 25 \times 0.259 = 6.475$ in

Example 10. If in the right triangle shown in Fig. 13.3 $A = 25°$ and $c = 15$ in, find a.
Again, by referring to Table 13.1, we find

$a = 15 \times \sin 25°$

and

$$a = \frac{15}{\csc 25°}$$

We shall choose to use the first equation. From Table 13.2, we find $\sin 25°$ to be 0.423, and we may write

$a = 15 \times 0.423 = 6.345$ in

Example 11. If in the right triangle shown in Fig. 13.3 $b = 13.4$ in and $a = 50$ in, find the angle A.
By consulting Table 13.1, we find two formulas containing the known quantities

b and a as well as the unknown quantity A. These formulas are

$$\cot A = \frac{b}{a} = \frac{13.4}{50}$$

and

$$\tan A = \frac{a}{b} = \frac{50}{13.4}$$

The division indicated in the first equation is obviously the easier. Therefore we write

$$\cot A = \frac{13.4}{50} = 0.268$$

Then from Table 13.2 we find that 0.268 is the cotangent of 75° and

$$A = 75°$$

Example 12. If in the right triangle shown in Fig. 13.3 $B = 20°$ and $c = 40$ in, find a.

Here we have given the angle B rather than the angle A. At this point in the work it is probably better to find the angle A before proceeding further.

$$A = 90° - B = 90° - 20° = 70°$$

We now proceed as in the previous examples in which the angle A is one of the known quantities.

We find by consulting Table 13.1 that there are two formulas, each containing the known quantities A and c as well as the unknown quantity a. These formulas are $\sin A = a/c$ and $\csc A = c/a$. By solving each for a we have

$$a = 40 \sin 70°$$

and

$$a = \frac{40}{\csc 70°}$$

We shall choose to use the first equation. Then, from Table 13.2, we find the value of $\sin 70°$, and we may write

$$a = 40 \times 0.940 = 37.60 \text{ in}$$

The basic formulas needed for solving right triangles were given in Table 13.1. For convenience, this information is repeated in Eqs. (1) to (12):

$\sin A = a/c$	(1)	$a = c \sin A$	(7)
$\cos A = b/c$	(2)	$b = c \cos A$	(8)
$\tan A = a/b$	(3)	$a = b \tan A$	(9)
$\cot A = b/a$	(4)	$b = a \cot A$	(10)
$\sec A = c/b$	(5)	$c = b \sec A$	(11)
$\csc A = c/a$	(6)	$c = a \csc A$	(12)

EXERCISE 5

Solve the following right triangles. The lettering for each triangle corresponds with Fig. 13.3. Use Table 13.2 for the values of the trigonometric functions.

1. $c = 7$ in and $A = 25°$ 2. $a = 38.3$ in and $A = 50°$
3. $b = 25$ in and $B = 50°$ 4. $a = 23.3$ in and $b = 50$ in
5. $c = 250$ in and $a = 125$ in   6. $c = 20$ in and $b = 10$ in

13.8 The Cofunctions

In Sec. 13.5 we expressed the trigonometric functions of angle A with respect to the right triangle shown in Fig. 13.3.

We can also express the functions of angle B with respect to the same triangle. See Figs. 13.3 and 13.6. From these figures we follow the same basic definitions given in Table 13.1 and obtain Table 13.3.

The word *cosine* means the sine of the complementary angle. In general, a cofunction is the function of the complementary angle. One angle is said to be the complement of another if their sum is 90°. In a right triangle the two acute angles are always complementary.

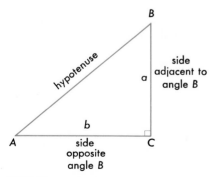

FIGURE 13.6

TABLE 13.3

$\sin B =$	$\dfrac{\text{side opposite angle } B}{\text{hypotenuse}}$	$= \dfrac{b}{c}$
$\cos B =$	$\dfrac{\text{side adjacent to angle } B}{\text{hypotenuse}}$	$= \dfrac{a}{c}$
$\tan B =$	$\dfrac{\text{side opposite angle } B}{\text{side adjacent to angle } B}$	$= \dfrac{b}{a}$
$\cot B =$	$\dfrac{\text{side adjacent to angle } B}{\text{side opposite angle } B}$	$= \dfrac{a}{b}$
$\sec B =$	$\dfrac{\text{hypotenuse}}{\text{side adjacent to angle } B}$	$= \dfrac{c}{a}$
$\csc B =$	$\dfrac{\text{hypotenuse}}{\text{side opposite angle } B}$	$= \dfrac{c}{b}$

From Tables 13.1 and 13.3, the student should verify the following relations:

$$\sin A = \cos B \qquad \cot A = \tan B$$
$$\cos A = \sin B \qquad \sec A = \csc B$$
$$\tan A = \cot B \qquad \csc A = \sec B$$

For convenience, the information contained in Table 13.3 is repeated in Eqs. (13) to (24):

$\sin B = b/c$	(13)	$b = c \sin B$	(19)
$\cos B = a/c$	(14)	$a = c \cos B$	(20)
$\tan B = b/a$	(15)	$b = a \tan B$	(21)
$\cot B = a/b$	(16)	$a = b \cot B$	(22)
$\sec B = c/a$	(17)	$c = a \sec B$	(23)
$\csc B = c/b$	(18)	$c = b \csc B$	(24)

EXERCISE 6

Express as functions of the complementary angle the following:

1. $\sin 20°$
2. $\cos 45°$
3. $\cos 60°$
4. $\tan 25°$
5. $\cot 17°$
6. $\sec 84°$
7. $\cos 38.5°$
8. $\csc 80°$
9. $\sec 25°$

13.9 Reciprocal Functions

Referring to Table 13.1, note that

$$\sin A = \frac{a}{c} \qquad \text{and} \qquad \csc A = \frac{c}{a}$$

Therefore

$$\sin A \times \csc A = \frac{a}{c} \times \frac{c}{a} = 1$$

or

$$\sin A = \frac{1}{\csc A} \quad \text{and} \quad \csc A = \frac{1}{\sin A} \tag{25}$$

Also

$$\tan A = \frac{a}{b} \quad \text{and} \quad \cot A = \frac{b}{a}$$

Therefore

$$\tan A \times \cot A = \frac{a}{b} \times \frac{b}{a} = 1$$

or

$$\tan A = \frac{1}{\cot A} \quad \text{and} \quad \cot A = \frac{1}{\tan A} \tag{26}$$

Also

$$\cos A = \frac{b}{c} \quad \text{and} \quad \sec A = \frac{c}{b}$$

Therefore

$$\cos A \times \sec A = \frac{b}{c} \times \frac{c}{b} = 1$$

or

$$\cos A = \frac{1}{\sec A} \quad \text{and} \quad \sec A = \frac{1}{\cos A} \tag{27}$$

It is apparent, then, that:

1. The sine of an angle is the reciprocal of the cosecant of the same angle.
2. The cosine of an angle is the reciprocal of the secant of the same angle.
3. The tangent of an angle is the reciprocal of the cotangent of the same angle.

13.10 Use of the Five-place Trigonometric Tables

As a practical compromise between the accuracy of larger tables and the convenience of smaller tables, the five-place table is commonly used. The values given in it for the various functions are good to about as many places as the data usually encountered in civil, mechanical, and electrical engineering. Yet there are times when a seven-place table is required. Some engineering organizations use them exclusively. On the other hand, slide-rule accuracy is sufficient in many cases. For the present we shall use the five-place table.

In the five-place table, one page is usually devoted to the functions of each degree. When we are concerned with angles less than 45°, we use the column headings at the top of the page and the minutes column on the left-hand margin of the page. When we are concerned with angles between 45° and 90°, we use the column headings at the bottom of the page and the minutes column on the right-hand margin of the page. Tabular entries are made for each minute.

EXERCISE 7

Find the values of the six functions of the following angles, using the five-place tables:

1. 20°30'
2. 40°27'
3. 44°59'
4. 45°01'
5. 25°45'
6. 60°42'
7. 80°28'
8. 42°38'
9. 00°10'
10. 89°30'
11. 13°25'
12. 48°05'
13. 55°55'
14. 81°36'
15. 01°45'

Find the angle when the following functions are given:

16. $\sin A = 0.00582$
17. $\sin A = 0.18052$
18. $\sin A = 0.68179$
19. $\sin A = 0.75414$
20. $\sin A = 0.99986$
21. $\cos A = 0.71873$
22. $\cos A = 0.54708$
23. $\cos A = 0.93190$
24. $\cos A = 0.16906$
25. $\cos A = 0.99854$
26. $\tan A = 0.06993$
27. $\tan A = 1.2124$
28. $\tan A = 0.46312$
29. $\tan A = 1.4596$
30. $\tan A = 0.90251$
31. $\cot A = 0.89883$
32. $\cot A = 1.7917$
33. $\cot A = 2.1364$
34. $\cot A = 0.62487$
35. $\cot A = 5.0658$
36. $\sec A = 1.0193$
37. $\sec A = 2.6040$
38. $\sec A = 1.2265$
39. $\sec A = 1.5601$
40. $\sec A = 1.4755$
41. $\csc A = 1.5212$
42. $\csc A = 1.2215$
43. $\csc A = 2.0466$
44. $\csc A = 6.1880$
45. $\csc A = 11.105$

13.11 Suggestions for Solving Right Triangles

In the examples given below the student should understand that all given linear dimensions are known to five significant digits and that all given angles are known to the nearest second.

The labor of solving triangles can be reduced considerably by intelligently choosing a method of attack. The student will be well advised to consider the suggestions given.

As a rule, it is better to use the formula in which the unknown side appears in the numerator. For example, suppose we are to solve the triangle in which $c = 15$ and $A = 42°$. First let us find side a. We write the fraction a/c and then check Eqs. (1) to (6), looking for the fraction a/c. We find that $a/c = \sin A$, and we shall use this formula. There is another formula which would, theoretically, be just as good: $c/a = \csc A$. However, in solving for a, the first equation leads to a multiplication, while the second leads to a division. Obviously, the one involving a multiplication is easier to use than the one involving a division.

When we have two sides given and wish to find a trigonometric function by division, it is best to choose the formula which places the number with the fewest significant digits in the denominator. This, of course, leads to an easier division.

It is always well to examine the data to see if a given decimal fraction can profitably be converted to a common fraction.

Suppose that in Fig. 13.7 the data are given as shown and the angle ϕ is to be calculated. There are two choices: either the tangent or the cotangent of ϕ may be calculated. If the tangent of ϕ is calculated, the operation involves the division of a three-figure number (0.625) by a five-figure number (1.3795); if the cotangent is calculated, the operation involves the division of a five-figure number by a three-figure number. Obviously, the arithmetic will be easier in calculating the cotangent. In this particular problem there is a still easier method if the student recognizes that 0.625 is the decimal equivalent of $\frac{5}{8}$. The arithmetic can now be reduced to a multiplication by 8 and a division by 5, as shown:

$$\cot \phi = \frac{1.3795}{\frac{5}{8}} = \frac{1.3795 \times 8}{5} = \frac{11.036}{5} = 2.2072$$

Example 13. Referring to Fig. 13.3, if $c = 22.000''$ and $A = 28°32'$, find B, a, and b.

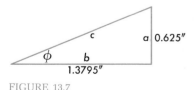

FIGURE 13.7

$B = 90° - 28°32' = 61°28'$

According to Eq. (7),

$a = c \sin A = 22 \sin 28°32'$
$= 22 \times 0.47767 = 10.509$ in

Following Eq. (8),

$b = c \cos A = 22 \cos 28°32'$
$= 22 \times 0.87854 = 19.328$ in

Example 14. Referring to Fig. 13.3, if $a = 19.000''$ and $A = 62°27'$, find B, b, and c.

$B = 90° - 62°27' = 27°33'$

Following Eq. (12),

$c = a \csc A = 19 \csc 62°27'$
$= 19 \times 1.1279 = 21.430$ in

Following Eq. (10),

$b = a \cot A = 19 \cot 62°27'$
$= 19 \times 0.52168 = 9.9119$ in

Example 15. Referring to Fig. 13.3, if $b = 26.000''$ and $A = 15°48'$, find B, a, and c.

$B = 90° - 15°48' = 74°12'$

Following Eq. (9),

$a = b \tan A = 26 \tan 15°48'$
$= 26 \times 0.28297 = 7.3572$ in

Following Eq. (11),

$c = b \sec A = 26 \sec 15°48'$
$= 26 \times 1.0393 = 27.022$ in

Example 16. Referring to Fig. 13.3, if $a = 45.000''$ and $B = 24°00'$, find A, b, and c.

$A = 90° - 24° = 66°$

Following Eq. (21),

$b = a \tan B = 45 \tan 24°$
$\quad = 45 \times 0.44523 = 20.035$ in

Following Eq. (23),

$c = a \sec B = 45 \sec 24°$
$\quad = 45 \times 1.0946 = 49.257$ in

Example 17. Referring to Fig. 13.3, if $b = 75.000''$ and $B = 85°00'$, find A, a, and c.

$A = 90° - 85° = 5°$

Following Eq. (22),

$a = b \cot B = 75 \cot 85°$
$\quad = 75 \times 0.08749 = 6.5618$ in

Following Eq. (24),

$c = b \csc B = 75 \csc 85°$
$\quad = 75 \times 1.0038 = 75.285$ in

Example 18. Referring to Fig. 13.3, if $c = 65.000''$ and $B = 8°00'$, find A, a, and b.

$A = 90° - 8° = 82°$

Following Eq. (20),

$a = c \cos B = 65 \cos 8°$
$\quad = 65 \times 0.99027 = 64.368$ in

Following Eq. (19),

$b = c \sin B = 65 \sin 8°$
$\quad = 65 \times 0.13917 = 9.0460$ in

Example 19. Referring to Fig. 13.3, if $b = 578.00''$ and $a = 483.00''$, find A, B, and c.
Following Eq. (3),

$\tan A = \dfrac{a}{b} = \dfrac{483}{578} = 0.83564$ (to five figures)

and

$$A = 39°53'$$

Following Eq. (4),

$$\cot A = \frac{b}{a} = \frac{578}{483} = 1.1967 \qquad \text{(to five figures)}$$

and

$$A = 39°53'$$
$$B = 90° - 39°53' = 50°07'$$

Compare the following solutions for side c. Following Eq. (11),

$$c = b \sec A = 578 \sec 39°53' = 578 \times 1.3032 = 753.25 \text{ in}$$

or following Eq. (12),

$$c = a \csc A = 483 \csc 39°53' = 483 \times 1.5595 = 753.24 \text{ in}$$

or following Eq. (1),

$$c = \frac{a}{\sin A} = \frac{483}{\sin 39°53'} = \frac{483}{0.64123} = 753.24 \text{ in}$$

or following Eq. (2),

$$c = \frac{b}{\cos A} = \frac{578}{\cos 39°53'} = \frac{578}{0.76735} = 753.24 \text{ in}$$

or using the Pythagorean theorem,

$$c = \sqrt{578^2 + 483^2} = \sqrt{567373} = 753.24 \text{ in}$$

Example 20. Referring to Fig. 13.3, if $c = 1{,}237.0''$ and $a = 333.00''$, find A, B, and b.
Following Eq. (1),

$$\sin A = \frac{a}{c} = \frac{333}{1{,}237} = 0.26920 \qquad \text{(to five figures)}$$

and

$A = 15°37'$

or following Eq. (6),

$$\csc A = \frac{c}{a} = \frac{1{,}237}{333} = 3.7147 \qquad \text{(to five figures)}$$

and

$A = 15°37'$
$B = 90° - 15°37' = 74°23'$
$b = c \cos A = 1{,}237 \times \cos 15°37'$
$\qquad\quad = 1{,}237 \times 0.96308 = 1{,}191.3 \text{ in}$

or

$$b = \sqrt{c^2 - a^2} = \sqrt{(1237)^2 - (333)^2} = \sqrt{1419280} = 1{,}191.3 \text{ in}$$

Example 21. The following example referring to Fig. 13.3 (when $c = 20.000''$ and $A = 6°00'$) is given without further comment to illustrate the inherent accuracy of various methods.

Using the sine,

$a = 20 \sin 6° = 20 \times 0.10453 = 2.0906 \text{ in}$

Using the cosecant,

$$a = \frac{20}{\csc 6°} = \frac{20}{9.5668} = 2.0906 \text{ in} \qquad \text{(to five figures)}$$

Using the cosine,

$b = 20 \cos 6° = 20 \times 0.99452 = 19.890 \text{ in} \qquad \text{(to five figures)}$

Using the secant,

$$b = \frac{20}{\sec 6°} = \frac{20}{1.0055} = 19.891 \text{ in} \qquad \text{(to five figures)}$$

Using seven-place tables of sines and cosines,

$a = 20 \times \sin 6° = 20 \times 0.1045285 = 2.090570$ in
$b = 20 \times \cos 6° = 20 \times 0.9945219 = 19.89044$ in

Using a and b as calculated from the five-place tables of sines and cosines,

$$\tan A = \frac{2.0906}{19.890} = 0.105108, \text{ or } 0.10511$$

By interpolation (Sec. 13.12)

$$A = 6°00'2''$$

Checking side c from the dimensions a and b calculated from five-place tables of sine and cosine,

$$c = \sqrt{(2.0906)^2 + (19.890)^2} = \sqrt{4.3706 + 395.61}$$
$$= \sqrt{399.98} = 19.9997 \text{ (approx)}$$

Actually,

$$(19.9997)^2 = 399.988$$

Checking the side c from the dimensions of a and b calculated from the seven-place tables of sine and cosine,

$$c = \sqrt{(2.090570)^2 + (19.89044)^2} = \sqrt{4.370482 + 395.6296}$$
$$= \sqrt{400.0001}$$

Actually,

$$(20.000003)^2 = 400.0001$$

EXERCISE 8

In the following right triangles find the unknown dimensions, using a five-place table. Calculate linear dimensions accurate to five figures and angles accurate to the nearest minute. The class may of course be instructed to round off both the data and the corresponding answers to whatever number of significant figures seems appropriate. In these problems all given linear dimensions are known to five significant digits and all given angles are known to the nearest second.

1. $A = 36°14'$ $c = 94.309$ ft   2. $A = 38°19'$ $c = 8.125$ in
3. $B = 62°52'$ $c = 132.00$ ft   4. $B = 11°10'$ $c = 89.048$ in

5. $A = 8°8'$	$c = 2.1919$ in	6. $B = 49°44'$	$c = 355.06$ in
7. $A = 68°22'$	$c = 250.00$ in	8. $A = 8°25'$	$c = 12.500$ ft
9. $B = 37°47'$	$c = 64.290$ ft	10. $B = 39°30'$	$c = 9.6354$ in
11. $A = 82°16'$	$a = 40,625$ ft	12. $A = 16°4'$	$a = 0.93750$ ft
13. $A = 19°21'$	$a = 47.395$ ft	14. $A = 19°31'$	$a = 31.250$ in
15. $A = 16°18'$	$a = 14.0625$ ft	16. $B = 74°24'$	$b = 93.750$ in
17. $B = 47°35'$	$a = 312.50$ in	18. $B = 50°26'$	$b = 15.625$ in
19. $A = 48°40'$	$b = 8,125.0$ in	20. $A = 72°48'$	$b = 718.75$ ft
21. $a = 347$ ft	$b = 167$ ft	22. $a = 199$ ft	$b = 160$ ft
23. $a = 67,130$ in	$b = 25,210$ in	24. $a = 46.370$ ft	$b = 94.720$ ft
25. $a = 141$ in	$b = 203$ in	26. $c = 3,477$ ft	$a = 2,638$ ft
27. $c = 2,691$ ft	$a = 839.0$ ft	28. $c = 505.0$ ft	$a = 457.0$ ft
29. $c = 43,649$ ft	$a = 17,962$ ft	30. $c = 248.09$ ft	$a = 218.54$ ft
31. $c = 11,223$ ft	$b = 10,454$ ft	32. $c = 87.02$ ft	$b = 55.43$ ft
33. $c = 2,338$ ft	$b = 1,877$ ft	34. $c = 455$ in	$b = 241$ in
35. $c = 1,029$ ft	$b = 985.0$ ft		

13.12 Interpolation in the Five-place Tables

It is often necessary to find a function of an angle intermediate between values given in a table of natural trigonometric functions. In such cases we resort to interpolation. In Sec. 11.16 we discussed the principles of interpolation in logarithmic tables. In this section we shall therefore be somewhat brief in our discussion of interpolation applied to tables of natural trigonometric functions.

Example 22. Find the sine of 26°45′12″.

From the tables we find that sin 26°45′ = 0.45010 and sin 26°46′ = 0.45036. In the process of becoming familiar with this interpolation technique, it may be convenient to use the following pattern. See Sec. 11.16.

By proportion

$$\frac{x}{0.00026} = \frac{12}{60}$$

or

$x = {}^{12}\!/_{60} \times 0.00026 = 0.00005$
$\sin 26°45'12'' = 0.45010 + 0.00005 = 0.45015$

In the following examples we shall dispense with the above diagram.

Example 23. Find the sine of 26°15.6′.

$\sin 26°16' = 0.44255$
$\sin 26°15' = \underline{0.44229}$
$\qquad\qquad 0.00026 \qquad {}^{6}\!/_{10} \times 0.00026 = 0.000156,\ \text{or}\ 0.00016$
$\sin 26°15.6' = 0.44229 + 0.00016 = 0.44245$

Example 24. Find the cosine of 61°13′52″.

$\cos 61°13' = 0.48150$
$\cos 61°14' = \underline{0.48124}$
$\qquad\qquad 0.00026 \qquad {}^{52}\!/_{60} \times 0.00026 = 0.00023$
$\cos 61°13'52'' = 0.48150 - 0.00023 = 0.48127$

Observe in the above examples that, as the angle increases, its cosine decreases.

The student should confirm from the tables that in the first quadrant, as the angle *increases,* its sine, tangent, and secant all *increase,* while its cosine, cotangent, and cosecant all *decrease.*

Example 25. Find the tangent of 53°27′19″.

$\tan 53°28' = 1.3498$
$\tan 53°27' = \underline{1.3490}$
$\qquad\qquad 0.0008 \qquad {}^{19}\!/_{60} \times 0.0008 = 0.0003$
$\tan 53°27'19'' = 1.3490 + 0.0003 = 1.3493$

Example 26. Find the cotangent of 38°41′25″.

$\cot 38°41' = 1.2489$
$\cot 38°42' = \underline{1.2482}$
$\qquad\qquad 0.0007 \qquad {}^{25}\!/_{60} \times 0.0007 = 0.0003$
$\cot 38°41'25'' = 1.2489 - 0.0003 = 1.2486$

Example 27. Find ϕ if $\sin \phi = 0.56295$.

$\sin 34°16' = 0.56305$ $\sin \phi = 0.56295$

$\sin 34°15' = \underline{0.56280}$ $\sin 34°15' = \underline{0.56280}$

         0.00025 0.00015

$\phi = 34°15' + {}^{15}\!/_{25} \times 60'' = 34°15'36''$

Example 28. Find ϕ if $\cos \phi = 0.81555$.

$\cos 35°21' = 0.81563$ $\cos 35°21' = 0.81563$

$\cos 35°22' = \underline{0.81546}$ $\cos \phi = \underline{0.81555}$

         0.00017 0.00008

$\phi = 35°21' + {}^{8}\!/_{17} \times 60'' = 35°21'28''$

If we are dealing in angular measure to the nearest second, we are in effect recognizing 60 angles between adjacent tabular entries in the angle column. However, there may be less or more than 60 five-digit numbers between corresponding tabular entries in the function column.

Therefore, if the tabular difference in the function column is small, there may be several closely grouped angles measured to the nearest second which will have the same function to five digits. If the tabular difference in the function column is large, a single angle measured to the nearest second may correspond to a range of five-figure numbers in the function column.

Refer to Example 25 above. Here, the difference between tabular entries in the tangent column is 8 in the last decimal place used. The difference in the angle is $60''$. Therefore in this range of this table a difference of one unit in the last place in the tangent column corresponds to a difference of ${}^{60}\!/_{8} = 7.5''$ in the angle column. Consequently, in this portion of the five-place tables, the tangent may not be sensitive to angular changes of less than $7.5''$.

If we were looking up the angle whose tangent is 1.3493, we probably should write the angle as $53°27'22'' \pm 4''$.

While we ordinarily do not go to this extreme in indicating the precision of interpolated angles, the student should certainly be aware of the limitations of the tables he uses.

In Example 25 above we calculated ${}^{19}\!/_{60}$ of 0.0008 and rounded off before adding on to 1.3490. We could, with equal reason, have found ${}^{19}\!/_{60}$ of 0.0008 to five decimal places and rounded off the tangent after addition. Usually it makes no difference. In the problems to follow it has been the policy to round off after addition.

In the absence of good reason to the contrary, all data in the following problems are assumed to have an accuracy consistent with the use of five-place tables with interpolation.

EXERCISE 9

Find the sine, cosine, tangent, cotangent, secant, and cosecant to five significant figures.

1. 2°28′15″	2. 6°42′29″	3. 14°17′13″	4. 23°19′52″
5. 40°12′48″	6. 49°36′27″	7. 52°47′35″	8. 71°28′18″
9. 80°59′40″	10. 85°43.4′	11. 27°15.3′	12. 42°27.6′
13. 72°19.8′	14. 85°26.2′	15. 17°15.7′	

EXERCISE 10

Find the angle to degrees, minutes, and seconds when the following functions are given:

1. $\sin \phi = 0.10572$	2. $\sin \phi = 0.32650$	3. $\sin \phi = 0.57461$
4. $\sin \phi = 0.81385$	5. $\sin \phi = 0.72645$	6. $\cos \phi = 0.49695$
7. $\cos \phi = 0.97808$	8. $\cos \phi = 0.99981$	9. $\cos \phi = 0.02391$
10. $\cos \phi = 0.54980$	11. $\tan \phi = 0.86901$	12. $\tan \phi = 1.1109$
13. $\tan \phi = 0.46430$	14. $\tan \phi = 12.271$	15. $\tan \phi = 1.0455$
16. $\cot \phi = 249.88$	17. $\cot \phi = 0.06315$	18. $\cot \phi = 2.7592$
19. $\cot \phi = 1.5476$	20. $\cot \phi = 1.3825$	21. $\sec \phi = 23.042$
22. $\sec \phi = 1.0670$	23. $\sec \phi = 1.6032$	24. $\sec \phi = 1.2130$
25. $\sec \phi = 2.3440$	26. $\csc \phi = 2.7290$	27. $\csc \phi = 9.3381$
28. $\csc \phi = 1.2880$	29. $\csc \phi = 1.3834$	30. $\csc \phi = 1.3119$

EXERCISE 11

Solve the following right triangles, using five-place functions.

The class may, of course, be instructed to round off both the data and the corresponding answers to whatever number of significant figures seems appropriate.

1. $A = 38°50′45″$	$c = 0.87500$ in	2. $A = 27°12′32″$ $c = 7.9143$ in
3. $A = 53°35.5′$	$c = 15.453$ in	4. $A = 37°50′10″$ $c = 98.268$ ft
5. $B = 22°32.1′$	$c = 2726.0$ ft	6. $A = 10°42′47″$ $c = 5.3805$ in
7. $B = 53°24′34″$	$c = 10.625$ in	8. $A = 38°14.9′$ $c = 8.1250$ ft
9. $B = 60°34′43″$	$c = 14.000$ ft	10. $A = 32°50′47″$ $c = 15.000$ ft
11. $A = 5°50′31″$	$a = 1250.0$ ft	12. $A = 11°16′44″$ $a = 457.31$ ft
13. $A = 45°15.8′$	$a = 986.91$ ft	14. $B = 36°44′2″$ $a = 12.500$ in
15. $A = 17°48.3′$	$a = 713.85$ ft	16. $B = 38°56′46″$ $b = 63.275$ ft
17. $B = 86°11.4′$	$b = 756.23$ ft	18. $B = 89°30.6′$ $b = 1.8750$ in
19. $A = 21°42′58″$	$b = 51.387$ ft	20. $A = 10°38′13″$ $b = 46.500$ ft
21. $a = 20.000$ in	$b = 37.998$ in	22. $a = 12.000$ in $b = 23.828$ in
23. $a = 4.6397$ ft	$b = 17.927$ ft	24. $a = 389.72$ in $b = 1303.1$ in
25. $a = 83.695$ ft	$b = 177.70$ ft	26. $a = 1.1250$ in $b = 1.4462$ in
27. $a = 437.92$ ft	$b = 1284.9$ ft	28. $a = 76.392$ ft $b = 191.67$ ft

29. $a = 31.250$ in $b = 44.679$ in   30. $a = 2.1875$ ft $b = 3.3878$ ft
31. $a = 147.00$ in $c = 418.58$ in   32. $a = 23.458$ in $c = 236.64$ in
33. $a = 39.738$ ft $c = 42.973$ ft   34. $a = 358.03$ ft $c = 369.16$ ft
35. $a = 939.52$ in $c = 1,136.6$ in   36. $b = 0.34297$ ft $c = 0.40083$ ft
37. $b = 125.00$ in $c = 299.80$ in   38. $b = 1.6250$ in $c = 2.5898$ in
39. $b = 13.794$ ft $c = 18.667$ ft   40. $b = 53.187$ in $c = 65.142$ in

The following problems are simplifications of various machine- and tool-design problems. Find the value of x to five significant figures or to the nearest second.

41. Solve for x in Fig. 13.8.   42. Solve for x in Fig. 13.9.
43. Solve for x in Fig. 13.10.   44. Solve for x in Fig. 13.11.
45. Solve for x in Fig. 13.12.   46. Solve for x in Fig. 13.13.
47. Solve for x in Fig. 13.14.   48. Solve for x in Fig. 13.15.
49. Solve for x in Fig. 13.16.   50. Solve for x in Fig. 13.17.

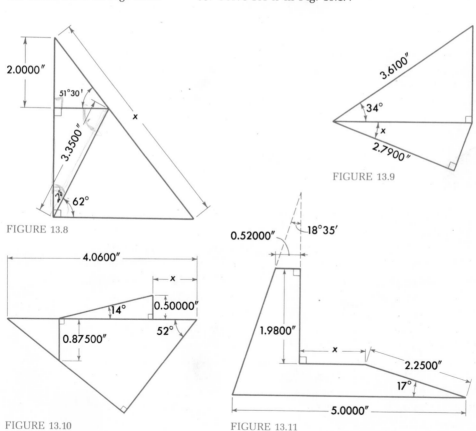

FIGURE 13.8

FIGURE 13.9

FIGURE 13.10

FIGURE 13.11

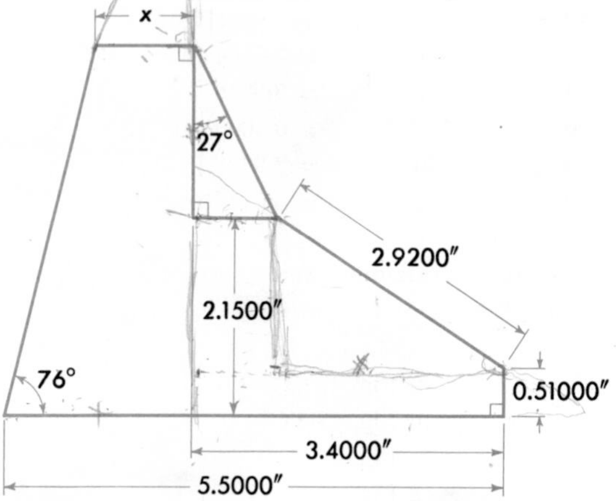

FIGURE 13.12

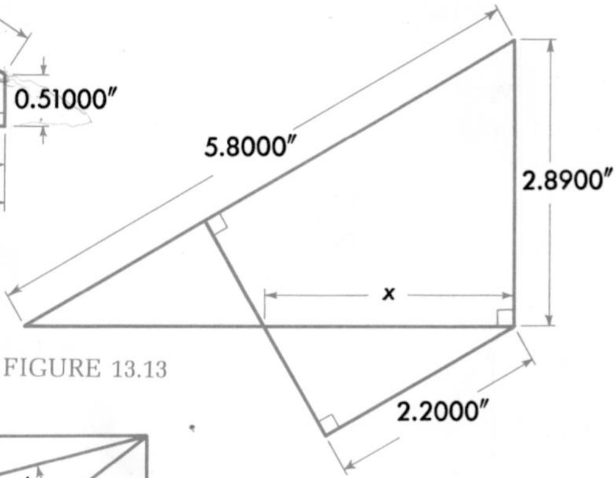

FIGURE 13.13

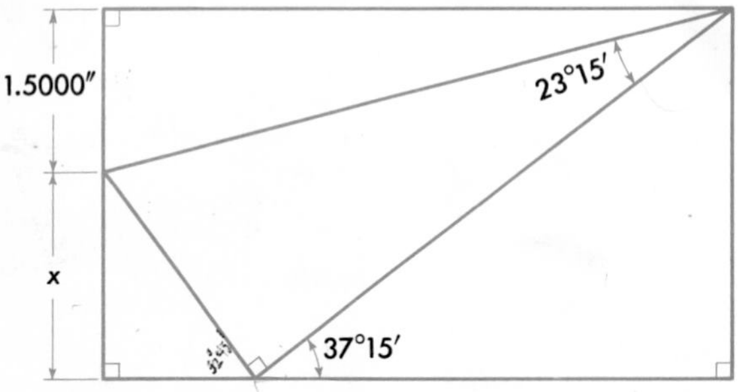

FIGURE 13.14

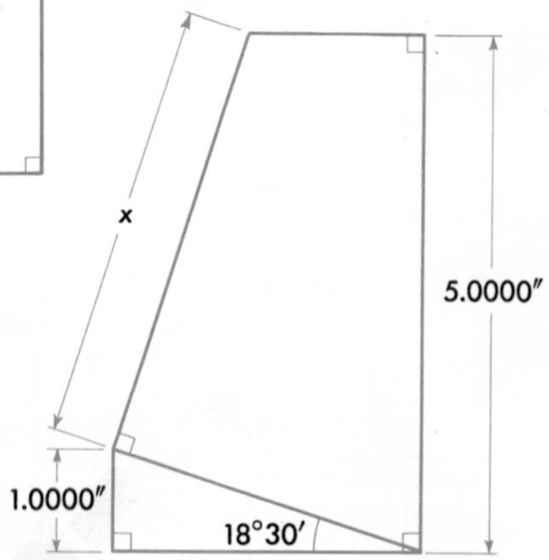

FIGURE 13.15

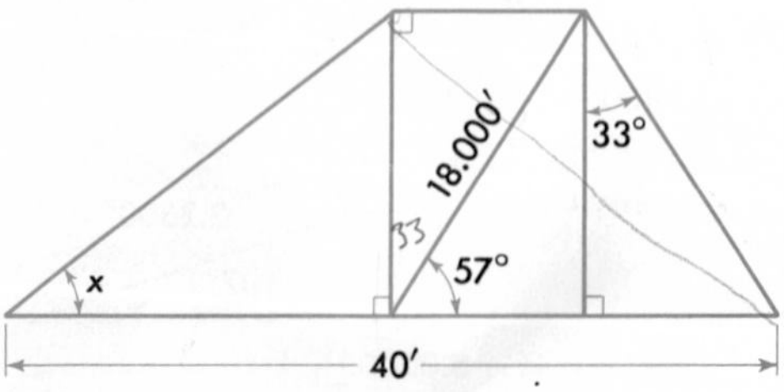

FIGURE 13.16

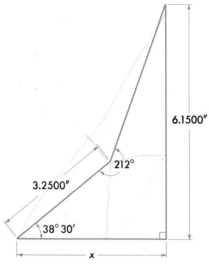

FIGURE 13.17

13.13 Isosceles Triangles

By definition, an isosceles triangle is a triangle in which two sides are equal. From this it follows that two angles must be equal. Thus in Fig. 13.18 if $a = b$, the triangle is isosceles and $A = B$. The altitude drawn to the base bisects the base and the angle C.

$$h^2 = b^2 - (\tfrac{1}{2}c)^2 \tag{28}$$
$$h = \sqrt{(b - \tfrac{1}{2}c)(b + \tfrac{1}{2}c)} \tag{29}$$

In general, for any triangle,

Area $= \tfrac{1}{2} \times$ base $\times$ altitude

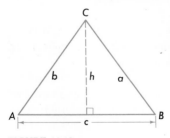

FIGURE 13.18

In the case of the triangle in Fig. 13.18,

$$h = b \sin A$$

or

$$h = b \cos \tfrac{1}{2}C$$

Therefore

$$\text{Area} = \tfrac{1}{2}cb \sin A \tag{30}$$

or

$$\text{Area} = \tfrac{1}{2}cb \cos \tfrac{1}{2}C \tag{31}$$

If angle C and either a or b are known, then

$$c = 2b \sin \tfrac{1}{2}C \tag{32}$$

or

$$c = 2a \sin \tfrac{1}{2}C \tag{33}$$

In short, by drawing the altitude h, we have divided the isosceles triangle ABC into two right triangles, the treatment of which has already been discussed.

EXERCISE 12

1. The equal sides of an isosceles triangle are each 5.86 in long, and each base angle is 23°51′. Find the length of the base and the altitude of the triangle.
2. A sheet of metal 15 in wide is bent along its centerline to form a V-shaped gutter. Will the gutter have a greater capacity when it is 6 in wide at the top, or when 6 in deep? What angle of the V will result in maximum capacity?
3. Find the angle of bend for each case in Prob. 2.
4. A right prism has for its base an equilateral triangle 7.3 in on each side. It is cut by a plane which makes an angle of 25° with the base; one side of the section includes one side of the base. Find the sides, angles, and area of the section.
5. A sphere $5\frac{1}{2}$ in in diameter is dropped into a tin cone $8\frac{1}{2}$ in in diameter and $7\frac{1}{2}$ in deep. Is the top of the sphere above or below the rim of the cone, and how far?
6. The legs of a tripod are each 4 ft $2\frac{3}{4}$ in long, and their feet form an equilateral triangle 2 ft $1\frac{1}{2}$ in on a side. Find the angle between one leg and a plum bob hung from its top.

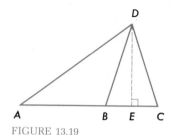

FIGURE 13.19

7. Find the area of a segment cut off from a circle 16.82 in in diameter by a line 2.73 in from the center.
8. Each leaf of a swinging double door 3 ft 6 in wide has been opened through an angle of 64°. How far apart are their edges? How far is each edge from the line of closure? Each section of the door is opened inward.
9. Figure 13.19 shows the isosceles triangle ADC in which $AD = AC$. The side AC is divided into two line segments AB and BC. We shall let the length of $BC = 1$ in. The line segment AB is constructed equal to the side DC. The line segments AB, BC, and AC are related by the proportion

$$\frac{BC}{AB} = \frac{AB}{AC} \qquad \text{(also see Prob. 8, Exercise 6, Chap. 12)}$$

Find the cosine of the angle DAC and the sine of the angle CDE to five significant digits. Then find the angles DAC, CDA, and CDE.

13.14 The Functions of 30°, 45°, and 60°

In plane geometry it can be proved that if both acute angles of a right triangle are 45° and the hypotenuse is $\sqrt{2}$ units long, then each of the other sides is 1 unit long. See Fig. 13.20a.

Also, in plane geometry it can be proved that if an acute angle of a right triangle is 30° and the hypotenuse is 2 units long, then the side opposite the 30° angle is 1

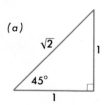
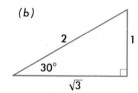

FIGURE 13.20

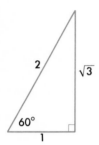

FIGURE 13.21

unit long and the adjacent side is $\sqrt{3}$ units long. Obviously, the second acute angle is 60°. See Fig. 13.20b.

   Figure 13.21 shows the same triangle as in Fig. 13.20b but with the 60° angle emphasized. The functions of these special angles appear frequently in the analysis of engineering problems. Consequently, the student will be well advised to remember them, or at least to be able to derive them at a moment's notice.

   From Fig. 13.20a and consistent with Table 13.1, it is evident that

$$\sin 45° = \frac{1}{\sqrt{2}} = \frac{\sqrt{2}}{2} = 0.707 \tag{34}$$

$$\cos 45° = \frac{1}{\sqrt{2}} = \frac{\sqrt{2}}{2} = 0.707 \tag{35}$$

$$\tan 45° = \frac{1}{1} = 1 \tag{36}$$

$$\cot 45° = \frac{1}{1} = 1 \tag{37}$$

$$\sec 45° = \frac{\sqrt{2}}{1} = \sqrt{2} = 1.414 \tag{38}$$

$$\csc 45° = \frac{\sqrt{2}}{1} = \sqrt{2} = 1.414 \tag{39}$$

Referring to Fig. 13.20b and Table 13.1,

$$\sin 30° = \frac{1}{2} = 0.500 \tag{40}$$

$$\cos 30° = \frac{\sqrt{3}}{2} = 0.866 \tag{41}$$

$$\tan 30° = \frac{1}{\sqrt{3}} = \frac{\sqrt{3}}{3} = 0.577 \tag{42}$$

$$\cot 30° = \frac{\sqrt{3}}{1} = \sqrt{3} = 1.732 \tag{43}$$

$$\sec 30° = \frac{2}{\sqrt{3}} = \frac{2\sqrt{3}}{3} = 1.155 \tag{44}$$

$$\csc 30° = \frac{2}{1} = 2 \tag{45}$$

Referring to Fig. 13.21 and again using Table 13.1,

$$\sin 60° = \frac{\sqrt{3}}{2} = 0.866 \tag{46}$$

$$\cos 60° = \frac{1}{2} = 0.500 \tag{47}$$

$$\tan 60° = \frac{\sqrt{3}}{1} = \sqrt{3} = 1.732 \tag{48}$$

$$\cot 60° = \frac{1}{\sqrt{3}} = \frac{\sqrt{3}}{3} = 0.577 \tag{49}$$

$$\sec 60° = \frac{2}{1} = 2 \tag{50}$$

$$\csc 60° = \frac{2}{\sqrt{3}} = \frac{2\sqrt{3}}{3} = 1.155 \tag{51}$$

EXERCISE 13
It is intended that the following problems will be done on a slide rule without the use of trigonometric tables.

1. Find x in Fig. 13.22a.   2. Find x in Fig. 13.22b.
3. Find x in Fig. 13.22c.   4. Find x in Fig. 13.22d.
5. Find x in Fig. 13.22e.   6. Find x in Fig. 13.22f.
7. Find x in Fig. 13.22g.

13.15 Regular Polygons

A regular polygon (Fig. 13.23) is inscribed in a circle of radius R and circumscribed about a circle of radius r. Let

s = length of one side = $\overline{ab}$
n = number of sides
R = radius of circumscribed circle
r = radius of inscribed circle, sometimes called the *apothem*

p = perimeter of polygon = ns
A_t = area of a single triangle, as for example triangle abc
A_p = area of entire polygon

It can be proved that

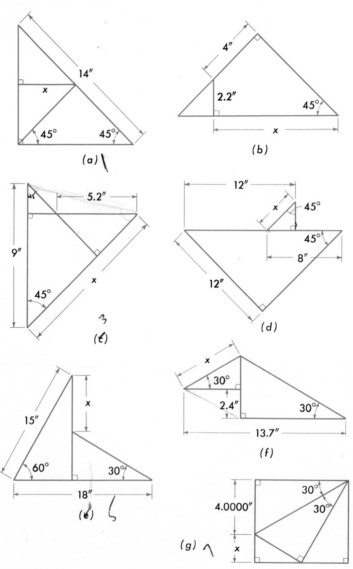

(a)

(b)

(c)

(d)

(e)

(f)

(g)

FIGURE 13.22

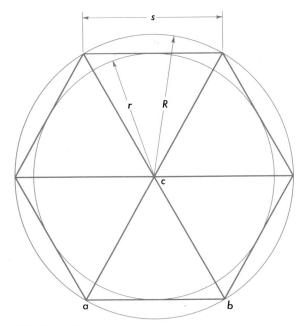

FIGURE 13.23

$$A_t = \frac{s^2}{4} \cot \frac{180°}{n} \tag{52}$$

$$A_t = \frac{R^2}{2} \sin \frac{360°}{n} \tag{53}$$

$$A_t = r^2 \tan \frac{180°}{n} \tag{54}$$

$$r = \frac{s}{2} \cot \frac{180°}{n} \tag{55}$$

$$R = \frac{s}{2} \csc \frac{180°}{n} \tag{56}$$

$$p = 2nR \sin \frac{180°}{n} \tag{57}$$

$$p = 2nr \tan \frac{180°}{n} \tag{58}$$

The derivations of Eqs. (52) to (58) are quite simple and should be performed by the student (see Prob. 1, Exercise 14).

Observe that in the special cases in which the central angle of the regular polygon is a multiple of 30°, 45°, or 60°, the quantities A_t, R, r, etc., can be calculated without the use of trigonometric tables from Eqs. (34) to (51).

TABLE 13.4

Polygon	Number of sides	Area	Radius of inscribed circle	Radius of circumscribed circle
Equilateral triangle	3	$0.433\ s^2$	$0.289\ s$	$0.577\ s$
Square	4			
Hexagon	6			
Octagon	8			

EXERCISE 14

1. Verify Eqs. (52) to (58).
2. Complete Table 13.4.
3. Prove that in a regular polygon

(a) $A_p = \dfrac{1}{4}\,ns^2 \cot \dfrac{180°}{n} = \dfrac{1}{2}\,nR^2 \sin \dfrac{360°}{n} = nr^2 \tan \dfrac{180°}{n}$

(b) $r = \dfrac{s}{2} \cot \dfrac{180°}{n}$

(c) $R = \dfrac{1}{2}\,s \csc \dfrac{180°}{n}$

(d) $p = 2nR \sin \dfrac{180°}{n} = 2nr \tan \dfrac{180°}{n}$

4. One side of a regular decagon inscribed in a circle is 3.27 in. Find the radius of the circle.
5. Find the area of the decagon of Prob. 4.
6. A ring 15 in in diameter is suspended from a point by 10 cords, each 9 in long and equally spaced. Find the angle between two adjacent cords.
7. The length of one side of a regular polygon of 6 sides is 1 in. Find the area of a regular polygon of 12 sides inscribed in the same circle.
8. Find the difference in area between a regular polygon of eight sides and a regular polygon of nine sides if the perimeter of each is 72 in.
9. The length of a side in a regular polygon of five sides is 2 in. Find the area of the ring bounded by the circumferences of the inscribed and circumscribed circles.

13.16 Logarithmic Solution of Right Triangles

The multiplication and division necessary in solving right triangles can, of course, be done by logarithms. For this purpose, tables have been published in which the logarithms of the functions are given directly. These tables are organized in exactly the same way as the corresponding tables of natural functions.

The sine or cosine of an angle can never exceed 1.00000. Therefore the logarithm of a sine or cosine always has a zero or negative characteristic.

The logarithm of the sine of 17° (log sin 17°) is listed in the table of log functions as 9.46594. It is understood that this logarithm is followed by -10, and the complete logarithm would be written

$$\log \sin 17° = 9.46594 - 10$$

or

$$\log \sin 17° = \overline{1}.46594$$

or

$$\log \sin 17° = 0.46594 - 1$$

The log tan 11° is listed as 9.28865. In this case it is understood that the logarithm is followed by -10, and the complete logarithm would be written

$$\log \tan 11° = 9.28865 - 10$$

or

$$\log \tan 11° = \overline{1}.28865$$

or

$$\log \tan 11° = 0.28865 - 1$$

Remembering that the tangent of an angle greater than 45° is itself greater than 1, it should be obvious that the log tangent of an angle greater than 45° has a positive or zero characteristic. The log tangent of 76°10′ is listed as 0.60864 and is, of course, not followed by a -10.

$$\log \tan 76°10' = 0.60864$$

Example 29. Find log sin 15°26′15″.

$$\log \sin 15°27' = 9.42553 - 10 = 0.42553 - 1$$
$$\log \sin 15°26' = \underline{9.42507 - 10} = 0.42507 - 1$$
$$0.00046 \qquad\qquad 0.00046$$

$$^{15}\!/_{60} \times 0.00046 = 0.000115$$
$$\log \sin 15°26'15'' = 9.42507 - 10 + 0.000115$$
$$ = 9.425185 - 10, \text{ or } 0.425185 - 1$$

In rounding off to five places, since the number ends in exactly 5, the last digit in the rounded-off number is left the nearest even digit, or

$$\log \sin 15°26'15'' = 9.42518 - 10 = 0.42518 - 1$$

Notice that the difference between adjacent tabular entries is published in the table under the d column. This avoids the need for an actual subtraction such as was done in the above example.

Example 30. Find ϕ if $\log \tan \phi = 0.42765$.

This number appears directly in the table and is the log tangent of 69°31'. Notice that $\log \tan 14°59' = 9.42755 - 10$, and the $\log \tan 1°32' = 8.42762 - 10$.

Therefore the student should be particularly careful to make sure that he is using the correct characteristic as well as the correct mantissa.

In the solution of right triangles, there is some question as to whether it is more efficient to use natural functions or log functions. With oblique triangles it is usually better to use log functions.

Example 31. In a certain right triangle, $a = 1.7320$ and $A = 26°30'$. Find b and c, using five-place log functions.

$$c = \frac{1.7320}{\sin 26°30'}$$

$$\log c = \log 1.7320 - \log \sin 26°30'$$
$$\log 1.7320 = 1.23855 - 1$$
$$\log \sin 26°30' = \underline{0.64953 - 1}$$
$$\log c = 0.58902 + 0$$
$$c = 3.8817$$
$$b = 1.7320 \cot 26°30'$$
$$\log b = \log 1.7320 + \log \cot 26°30'$$
$$\log 1.7320 = 0.23855$$
$$\log \cot 26°30' = \underline{0.30226}$$
$$\log b = 0.54081$$
$$b = 3.4738$$

EXERCISE 15

Solve the following right triangles, using five-place log functions:

1.	$A = 37°42'16''$	$c = 146.32$ in	2.	$A = 2°26'05''$	$c = 0.43792$ in
3.	$B = 72°19'28''$	$c = 157.65$ in	4.	$B = 36°28'45''$	$c = 29.463$ in
5.	$A = 28°36'20''$	$c = 1.3752$ in	6.	$A = 04°13'30''$	$b = 136.48$ in
7.	$A = 88°25'14''$	$a = 15.358$ in	8.	$B = 47°28'10''$	$a = 0.037940$ in
9.	$A = 89°21'38''$	$b = 15279$ in	10.	$A = 45°13'05''$	$a = 28.365$ in
11.	$a = 13.625$ in	$c = 142.98$ in	12.	$a = 76.500$ ft	$c = 92.800$ ft

13. $a = 5.4360$ in $c = 10.830$ in   14. $a = 26.9320$ in $c = 41.8670$ in

15. $a = 0.36520$ in $c = 0.58470$ in   16. $a = 10.932$ in $b = 110.36$ in

17. $a = 95.632$ in $b = 8.7305$ in   18. $a = 52.693$ in $b = 27.956$ in

19. $a = 3.6571$ in $b = 7.3058$ in   20. $a = 26.328$ in $b = 21.497$ in

13.17 The Solution of Oblique Triangles without Special Formulas

So far we have discussed only the solution of right triangles. However, the student has at this point all the information he needs to solve oblique triangles as well.

Data determining oblique triangles can be given in any of three ways. Three sides may be given, two angles and one side may be given, or two sides and one angle may be given.

In Secs. 13.18 to 13.20 all given dimensions are known to five significant digits or to seconds.

13.18 Solution of an Oblique Triangle When Three Sides Are Given

Example 32. Solve the oblique triangle shown in Fig. 13.24. The sides are known to five significant digits.

Draw h perpendicular to b. It is recommended that you draw h perpendicular to the longest side as was done in this example.

Then

$$h^2 = 17^2 - x^2 \quad \text{and} \quad h^2 = 25^2 - y^2$$

Therefore

$$17^2 - x^2 = 25^2 - y^2$$
$$y^2 - x^2 = 25^2 - 17^2 = 625 - 289 = 336$$
$$(y - x)(y + x) = 336$$

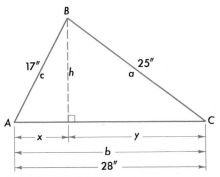

FIGURE 13.24

but

$$y + x = b = 28$$
$$(y - x)28 = 336$$
$$y - x = 336/28 = 12 \text{ in}$$

$y - x = 12$	$y - x = 12$
$y + x = 28$	$y + x = 28$
$2y \quad = 40$ (by addition)	$2x = 16$ (by subtraction)
$y = 20$ in	$x = 8$ in

$$\cos A = \frac{x}{c} = \frac{8}{17} = 0.47059$$

$$A = 61°55'39''$$

$$\cos C = \frac{y}{a} = \frac{20}{25} = 0.80000$$

$$C = 36°52'11''$$

$$B = 180° - (61°55'39'' + 36°52'11'') = 81°12'10''$$

13.19 Solution of an Oblique Triangle When Two Angles and a Side Are Given

Example 33. Solve the oblique triangle shown in Fig. 13.25. Side a is known to five significant digits, and the angles are known to the nearest second.

  Drop the perpendicular h from C to the side c.

$$h = 10 \times \sin 25° = 10 \times 0.42262 = 4.2262 \text{ in}$$
$$x = 10 \times \cos 25° = 10 \times 0.90631 = 9.0631 \text{ in}$$
$$A = 180° - (25° + 115°) = 40°$$
$$b = h \csc 40° = 4.2262 \times 1.5557 = 6.5747 \text{ in}$$
$$y = h \cot 40° = 4.2262 \times 1.1918 = 5.0368 \text{ in}$$
$$c = x + y = 9.0631 \text{ in} + 5.0368 \text{ in} = 14.0999 \text{ in, or } 14.100 \text{ in}$$

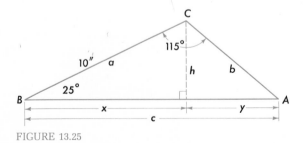

FIGURE 13.25

13.20 Solution of an Oblique Triangle When Two Sides and an Angle Are Given

Example 34. Solve the oblique triangle shown in Fig. 13.26. The sides are known to five significant digits.

In this example, *two sides and the included angle* are given.

$$C' = 180° - 136°23'51'' = 43°36'9''$$
$$h = 87 \sin 43°36'9'' = 87 \times 0.68965 = 60 \text{ in}$$
$$x = 87 \cos 43°36'9'' = 87 \times 0.72414 = 63 \text{ in}$$
$$b + x = 17 + 63 = 80 \text{ in}$$

$$\cot A = \frac{b + x}{h} = \frac{80}{60} = 1.3333$$

$$A = 36°52'15''$$
$$B = 180° - (136°23'51'' + 36°52'15'') = 6°43'54''$$
$$c = \frac{h}{\sin 36°52'15''} = \frac{60}{\sin 36°52'15''}$$
$$= \frac{60}{0.60001} = 100 \text{ in}$$

Example 35. Solve the oblique triangle in which one side is 60.000 in and the opposite angle is 41°06′44″. Another side is 73.000 in.

The data here are ambiguous. Observe that Figs. 13.27 and 13.28 are both consistent with the stated data, while these triangles are by no means equal.

Such ambiguity is possible only when two sides and the angle opposite one of them are given and when the given angle is less than 90°.

We shall solve the triangle in Fig. 13.27 first.

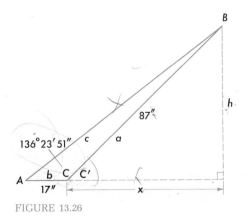

FIGURE 13.26

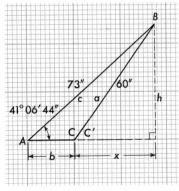

FIGURE 13.27

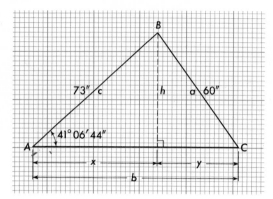

FIGURE 13.28

$$h = 73 \times \sin 41°06'44'' = 73 \times 0.65753 = 48 \text{ in}$$
$$x + b = 73 \times \cos 41°06'44'' = 73 \times 0.75342 = 55 \text{ in}$$
$$x = \sqrt{60^2 - 48^2} = 36 \text{ in}$$
$$b = 55 - 36 = 19 \text{ in}$$
$$\sin C' = \frac{h}{60} = \frac{48}{60} = 0.80000$$
$$= 53°07'49''$$
$$C = 180° - 53°07'49'' = 126°52'11''$$
$$B = 180° - (41°06'44'' + 126°52'11'') = 12°01'05''$$

The solution for Fig. 13.28 follows:

$$h = 73 \sin 41°06'44'' = 73 \times 0.65753 = 48 \text{ in}$$
$$x = 73 \cos 41°06'44'' = 73 \times 0.75342 = 55 \text{ in}$$
$$\sin C = \frac{48}{60} = 0.80000$$
$$C = 53°07'49''$$
$$y = \sqrt{60^2 - 48^2} = 36 \text{ in}$$
$$b = x + y = 55 + 36 = 91 \text{ in}$$
$$B = 180° - (53°07'49'' + 41°06'44'') = 85°45'27''$$

EXERCISE 16

Solve the following triangles, using the methods given above. If the data are inconsistent, state why. All given linear dimensions are known to five significant digits.

1. $a = 37$ in $b = 91$ in $c = 72$ in
2. $a = 52$ in $b = 25$ in $c = 63$ in

3. $a = 21$ in $\qquad b = 31$ in $\qquad c = 52$ in
4. $a = 65$ in $\qquad b = 89$ in $\qquad c = 132$ in
5. $A = 36°52'12''$ $\qquad b = 12$ in $\qquad C = 118°04'21''$
6. $A = 73°44'23''$ $\qquad b = 75$ in $\qquad C = 58°20'07''$
7. $a = 9$ in $\qquad b = 80$ in $\qquad C = 36°52'12''$
8. $b = 53$ in $\qquad c = 88$ in $\qquad A = 58°06'34''$
9. $A = 53°07'49''$ $\qquad a = 68$ in $\qquad b = 13$ in
10. $a = 20$ in $\qquad b = 5$ in $\qquad A = 22°37'12''$
11. $b = 86$ in $\qquad c = 97$ in $\qquad C = 104°15'$

13.21 The Slide-rule Solution of Right Triangles

Within its inherent limit of accuracy, the slide rule may be used to solve problems in numerical trigonometry.

On the slide rule, linear dimensions can be used and computed to about one part in a thousand. Angles can be used and computed to about the nearest 0.05°. This is not strictly true over the entire scale, but it is a good working average.

On the particular rule used here for illustration, the S and T scales (Fig. 13.29) are calibrated in degrees and decimal parts of a degree. The long markers are double-numbered in pairs of complementary angles, for example, 82°/8°, 70°/20°, 60°/30°.

When the *right-hand* numerals are used, the hairline simultaneously indicates an angle on the S and T scales and the *sine* and *tangent,* respectively, of that angle on the C scale.

For example, with the indicator in the position shown in Fig. 13.29, the hairline indicates an angle of 9.21° on the S scale and the *sine* of that angle (0.1600) on the C scale. It also indicates an angle of 9.09° on the T scale and the *tangent* of that angle (0.1600) on the C scale.

By reading the *left-hand* numerals on the S and T scales with the same setting, we observe that the cosine of 80.79° is 0.1600 and the cotangent of 80.91° is of course also 0.1600.

Other positions of the hairline are also indicated in Fig. 13.29, showing certain angles and their sines, cosines, tangents, and cotangents.

The values of the sine, cosine, tangent, and cotangent are, as we have mentioned, found on the C scale. The left-hand index of the C scale is used as 0.1, and the right-hand index is used as 1.0. Therefore on this particular rule we are limited to angles whose sine is between 0.1 and 1.0, angles whose cosine is between 0.1 and 1.0, and angles whose tangent or cotangent is between 0.1 and 1.0. The smallest angle we can use directly is about 5.7°. When using sines, we can process angles up to 90°, although the scale is crowded near 90°. When using cosines, we can use angles up to about 84.3°. The upper limit of the tangent scale and the lower limit of the cotangent scale are 45°. However, as will be illustrated in subsequent examples, this does not put any additional limitations on the usefulness of the slide rule.

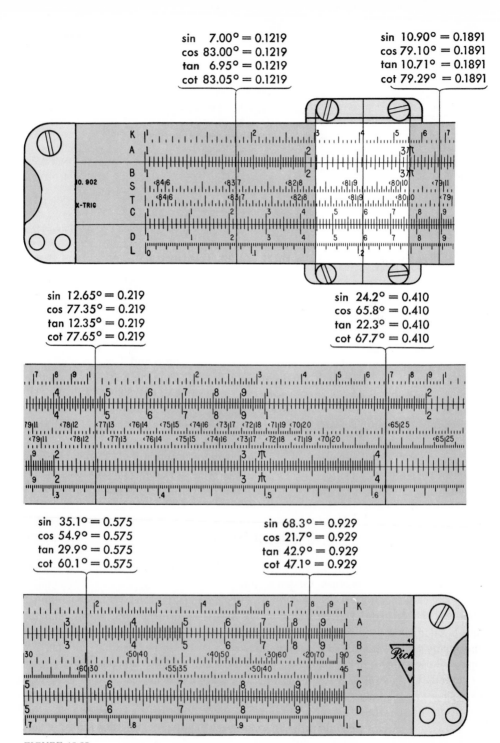

sin 7.00° = 0.1219
cos 83.00° = 0.1219
tan 6.95° = 0.1219
cot 83.05° = 0.1219

sin 10.90° = 0.1891
cos 79.10° = 0.1891
tan 10.71° = 0.1891
cot 79.29° = 0.1891

sin 12.65° = 0.219
cos 77.35° = 0.219
tan 12.35° = 0.219
cot 77.65° = 0.219

sin 24.2° = 0.410
cos 65.8° = 0.410
tan 22.3° = 0.410
cot 67.7° = 0.410

sin 35.1° = 0.575
cos 54.9° = 0.575
tan 29.9° = 0.575
cot 60.1° = 0.575

sin 68.3° = 0.929
cos 21.7° = 0.929
tan 42.9° = 0.929
cot 47.1° = 0.929

FIGURE 13.29

EXERCISE 17

1. Using a slide rule, find the sine and cosine of the following angles:

(a) 28°   (b) 8°   (c) 17°   (d) 13°   (e) 20°   (f) 30°
(g) 7.2°   (h) 12.2° (i) 26.2°   (j) 7.63°   (k) 16.75°   (l) 41.4°

The slide-rule setting illustrated in Fig. 13.30 is adequate for solving Examples 36 and 37.

Example 36. The hypotenuse of a right triangle is 8 in, and one angle is 30°. Find the side opposite the 30° angle.

Here the slide rule is set to multiply 8 in by the sine of 30°, giving 4 in as the length of the opposite side (Fig. 13.30).

Example 37. The hypotenuse of a right triangle is 8 in, and one angle is 60°. Find the side adjacent to the 60° angle.

In Fig. 13.30 the slide rule is set to multiply 8 in by the cosine of 60°.

By reference to Examples 36 and 37 and Fig. 13.30 the student should justify for himself that when solving a right triangle where the hypotenuse appears as an unknown or a known side, (1) the 90° marker on the S scale matches the length of the hypotenuse on the D scale; (2) when using the right-hand numbers on the S scale, an acute angle on this scale matches the length of its opposite side on the D scale.

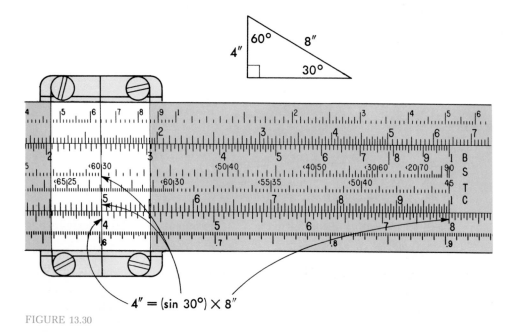

FIGURE 13.30

By using the above relations we can solve any right triangle (except those containing an angle less than 5.7°) if we know the hypotenuse and either acute angle or either leg. Similarly, we can solve right triangles if we know either leg and either acute angle.

EXERCISE 18

1. Verify the statement that the slide-rule setting shown in Fig. 2.5 is consistent with the following relations:

   (a) $\dfrac{2.53}{14.28} = \sin 10.21°$ (b) $\dfrac{2.53}{14.28} = \cos 79.79°$

   (c) $14.28 \cos 79.79° = 2.53$ (d) $14.28 \sin 10.21° = 2.53$

2. Verify the statement that the slide-rule setting in Fig. 2.7 is consistent with the following relations:

   (a) $\dfrac{8}{9.03} = \sin 62.4°$ (b) $\dfrac{8}{9.03} = \cos 27.6°$

   (c) $9.03 \sin 62.4° = 8$ (d) $9.03 \cos 27.6° = 8$

3. Write a series of four equations similar to those given in Probs. 1 and 2, but consistent with Figs. 2.4 and 2.11.

4. Using a slide rule, solve Probs. 1 to 20 and 26 to 34, Exercise 8.

The T scale is calibrated in degrees and decimal parts of a degree. The long markers are double-numbered in pairs of complementary angles in a way similar to the S scale.

When the *right-hand* numerals are used, the hairline simultaneously indicates an angle on the T scale and the *tangent* of that angle on the C scale. When the *left-hand* numerals are used, the hairline simultaneously indicates an angle on the T scale and the *cotangent* of that angle on the C scale.

EXERCISE 19

1. Using a slide rule, find
   - (a) tan 20°
   - (b) tan 35.76°
   - (c) tan 44.2°
   - (d) tan 7.9°
   - (e) tan 15.82°
   - (f) tan 29.62°
   - (g) cot 48.2°
   - (h) cot 82.1°
   - (i) cot 52.7°

The slide-rule setting illustrated in Fig. 13.31 is adequate for solving the following examples.

Example 38. One side of a right triangle is 6.93 in, and the adjacent angle is 30°. Find the side opposite the 30° angle.

Use the equation

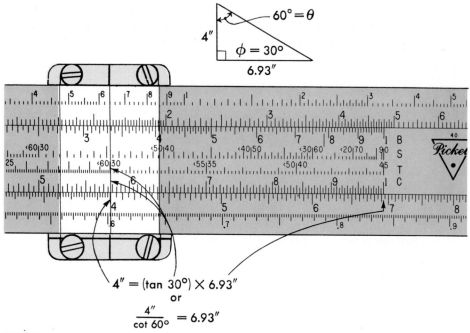

FIGURE 13.31

Opposite side = adjacent side × tangent ϕ

The slide-rule setting in Fig. 13.31 accomplishes this multiplication.

Example 39. One side of a right triangle is 4 in, and the adjacent angle is 60°. Find the side opposite the 60° angle.

Use the equation

$$\text{Opposite side} = \frac{\text{adjacent side}}{\cot \theta}$$

The slide-rule setting shown in Fig. 13.31 accomplishes this division. Thus we avoid using tangents of angles greater than 45°.

By reference to Examples 38 and 39 and Fig. 13.31 the student should justify for himself that (1) if the 45° marker on the T scale matches the longer leg on the D scale, (2) then, when using the right-hand numbers on the T scale, an acute angle on this scale matches the length of the shorter leg on the D scale.

It is assumed here, of course, that neither acute angle is smaller than about 5.7°.

By the rule above, if we are given two legs or a leg and an acute angle, we can solve the right triangle except for the hypotenuse. We have already discussed the situation in which the hypotenuse is involved.

EXERCISE 20

1. Verify the statement that the slide-rule setting shown in Fig. 2.5 is consistent with the following relations:

(a) $\dfrac{2.53}{14.28} = \tan 10.05°$ (b) $\dfrac{2.53}{14.28} = \cot 79.95°$

(c) $2.53 = 14.28 \tan 10.05°$ (d) $2.53 = 14.28 \cot 79.95°$

2. Verify the statement that the slide-rule setting shown in Fig. 2.7 is consistent with the following relations:

(a) $\dfrac{8}{9.03} = \tan 41.55°$ (b) $\dfrac{8}{9.03} = \cot 48.45°$

(c) $8 = 9.03 \tan 41.55°$ (d) $8 = 9.03 \cot 48.45°$

3. Write a series of four equations similar to those given in Probs. 1 and 2 but consistent with Figs. 2.9 and 2.11.
4. Using a slide rule, solve Probs. 21 to 25, Exercise 8.

Section 2.12 describes a method of solving Pythagorean theorem problems on the slide rule. The method illustrated in the following example is somewhat more convenient for those with a knowledge of trigonometry.

Example 40. The hypotenuse of a right triangle is 8 in, and one side is 4 in. Find the other side.

First find an acute angle, as in Fig. 13.30. Then, having found the smaller angle (in this case 30°), divide the opposite side by the tangent of the angle, as in Fig. 13.31.

Example 41. The two sides of a right triangle are 4 and 6.93 in. Find the hypotenuse.

First find an acute angle, as in Fig. 13.31. Then, having found the angle, divide the sine of this angle into the opposite side to find the hypotenuse, as in Fig. 13.30.

EXERCISE 21

1. Using a slide rule, solve Probs. 21 to 30 in Exercise 11, page 392.
2. Using a slide rule, solve Probs. 31 to 40 in Exercise 11, page 392.

13.22 The Slide-rule Solution of Oblique Triangles

The slide rule can be used to solve oblique triangles, using methods similar to those just described.

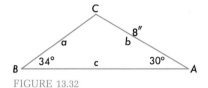

FIGURE 13.32

Example 42. Given the triangle shown in Fig. 13.32, where $b = 8$ in, $A = 30°$, and $B = 34°$, find a, c, and C.

1. Set the rule to indicate h, as in Fig. 13.33.
2. Set the rule to indicate a, as in Fig. 13.34a.
3. Set the rule to indicate x, as in Fig. 13.34b.
4. Set the rule to indicate y, as in Fig. 13.35.

  side $c = x + y$ angle $C = 180° - (A + B)$

Example 43. Given the triangle shown in Fig. 13.36, if $b = 8$ in, $c = 12.87$ in, and $A = 30°$, find a, B, and C.

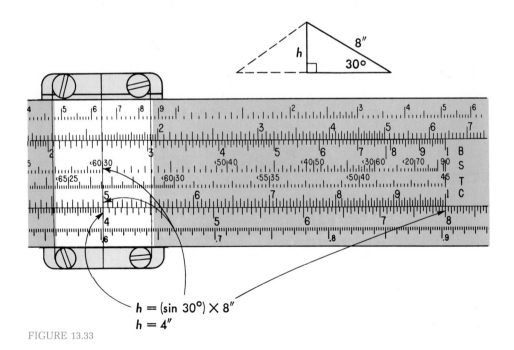

$$h = (\sin 30°) \times 8''$$
$$h = 4''$$

FIGURE 13.33

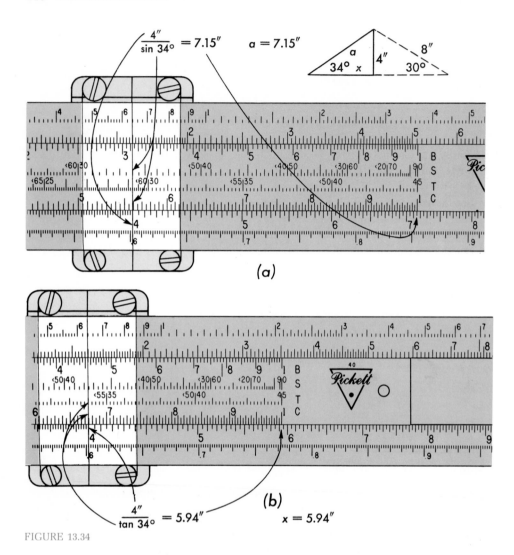

FIGURE 13.34

1. Find h and y, as in Figs. 13.33 and 13.35, respectively.
2. Since $x = 12.87 - y$, the length x may be found by subtraction.
3. Knowing h and x, find B, as in Fig. 13.34b.
4. Knowing h and B, find a, as in Fig. 13.34a.

EXERCISE 22

Solve Probs. 1 to 11 in Exercise 16 on a slide rule.

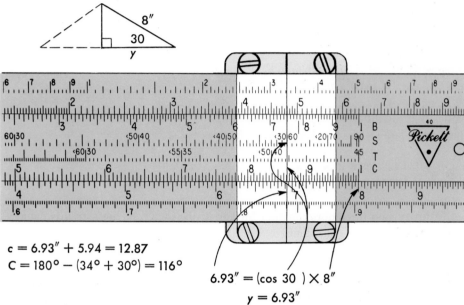

$$c = 6.93'' + 5.94 = 12.87$$
$$C = 180° - (34° + 30°) = 116°$$

$$6.93'' = (\cos 30°) \times 8''$$
$$y = 6.93''$$

FIGURE 13.35

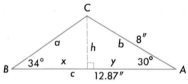

FIGURE 13.36

13.23 The Solution of Right Triangles Involving Small Angles

The slide rule illustrated in this book does not indicate angles less than about 5.7°. However, the slide rule can be used to solve triangles involving smaller angles.

Some slide rules have an "SRT" scale engraved on them. This scale is often a convenience when dealing with small angles. The left-hand C index is used as 0.01, and the right-hand index is used as 0.1. The smallest angle we can read directly on the SRT scale is about 0.57°, and the largest angle is about 5.7°.

Over this range we cannot distinguish on the scale between the tangent of an angle and the sine of the same angle. Thus the SRT scale serves both when dealing with sines or tangents of these small angles.

Other aspects of the solution of right triangles involving small angles are discussed below.

Example 44. Find the short side of a right triangle in which the hypotenuse is 134.4 in and the long side is 133.6 in.

Let the hypotenuse be c, the long side be b, and the short side be a.

$$a = \sqrt{c^2 - b^2} = \sqrt{(c - b)(c + b)} \tag{59}$$
$$c - b = 134.4 - 133.6 = 0.8 \text{ in}$$

Notice that the difference (0.8 in) is expressed to only one significant figure, whereas the c and b are given to four significant figures.

$$c + b = 134.4 + 133.6 = 268.0 \text{ in}$$
$$a = \sqrt{0.8 \times 268} = \sqrt{214.4} = 14.64 \text{ in}$$

Let us investigate this approximation further. If the length of the side b is changed by 0.1 in to make $b = 133.7$, then

$$c = \sqrt{(134.4 - 133.7)(134.4 + 133.7)} = 13.70 \text{ in}$$

Thus a change of 0.1 in in b makes a change of 0.94 in in a.

Example 45. Solve the right triangle shown in Fig. 13.37.

The side a in Fig. 13.37 is nearly equal to the arc length BX in Fig. 13.38. Of course this approximation will be close only when A is small. The arc length BX is a fraction of the circumference of a circle of radius 134 in whose center is at A. This fraction is $2.06°/360°$.

In other words,

$$\frac{2.06°}{360°} \times 2\pi \times 134 = \text{arc length } BX = \frac{\pi}{180°} \times 2.06° \times 134 \tag{60}$$

Or the side a is given by the approximate formula

$$a \approx \frac{\pi}{180°} \times 2.06° \times 134$$

$$\approx 0.0359 \times 134 = 4.82 \text{ in} \tag{61}$$

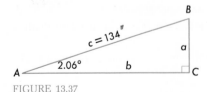

FIGURE 13.37

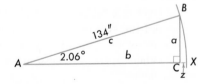

FIGURE 13.38

See Fig. 13.39 for the appropriate settings to solve Eq. (61).

The general formula for a is

$$a \approx \frac{\pi}{180} \times A \times c \qquad (62)$$

To find b, we first find the length z (Fig. 13.38). In Fig. 13.38

$$c = b + z \qquad (63)$$

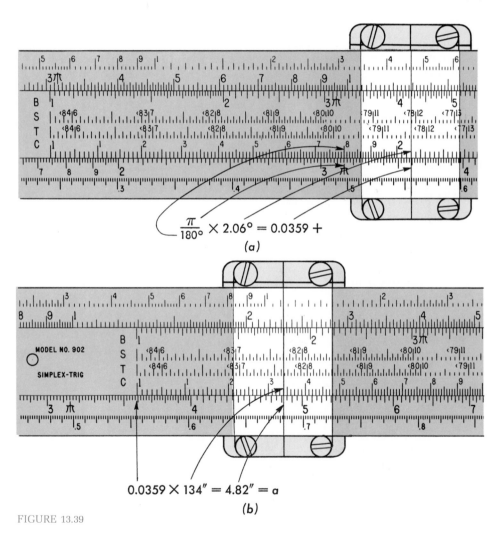

$$\frac{\pi}{180°} \times 2.06° = 0.0359 +$$

(a)

$$0.0359 \times 134'' = 4.82'' = a$$

(b)

FIGURE 13.39

and

$$c - b = z \tag{64}$$

since both c and $b + z$ are radii of the same circle. But by the Pythagorean theorem,

$$a^2 = c^2 - b^2 = (c - b)(c + b) \tag{65}$$
$$a^2 = z(c + b) \tag{66}$$

or

$$z = \frac{a^2}{c + b} \tag{67}$$

If c and b are nearly equal, we may write

$$z \approx \frac{a^2}{2c} \tag{68}$$

In the present example,

$$a = 4.82 \text{ in}$$
$$c = 134 \text{ in}$$

Therefore

$$z \approx \frac{4.82^2}{2 \times 134} = 0.0867 \text{ in} \tag{69}$$

and

$$b \approx 134.0 \text{ in} - 0.0867 \text{ in} \tag{70}$$

or within slide-rule precision,

$$b \approx 134.0 \text{ in} - 0.1 \text{ in} = 133.9 \text{ in}$$

The slide-rule settings used to solve Eq. (69) are illustrated in Fig. 13.40. Note the discrepancy of about 0.2 percent between 0.0867 obtained in Eq. (69) by using a desk calculator and 0.0865 obtained in Fig. 13.40 by slide rule.

Example 46. Solve the right triangle shown in Fig. 13.41.

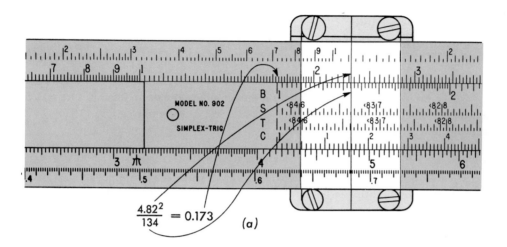

$$\frac{4.82^2}{134} = 0.173$$ *(a)*

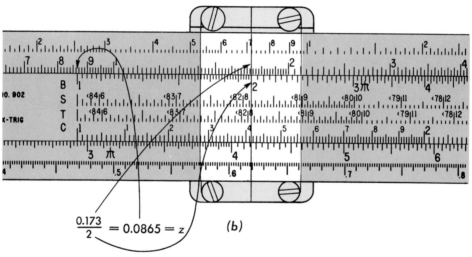

$$\frac{0.173}{2} = 0.0865 = z$$ *(b)*

FIGURE 13.40

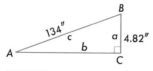

FIGURE 13.41

Following Eq. (62),

$$\frac{a}{c} \approx \frac{\pi}{180} \times A \qquad (71)$$

The ratio a/c is set on the slide rule shown in Fig. 13.39b. To this ratio we match the ratio $\pi/180$ multiplied by the value of A (as yet unknown) (see Fig. 13.39a). This requires that A be 2.06°.

The side b is calculated as in Example 45.

Example 47. Solve the triangle shown in Fig. 13.42. (Also see Example 45.)

Since b and c are nearly equal, Eq. (62) may be altered to read

$$\frac{a}{b} \approx \frac{\pi}{180°} \times A \qquad (72)$$

To solve for c, we alter Eq. (67) as follows:

$$z \approx \frac{a^2}{2b} \qquad (73)$$

Therefore

$$c \approx b + \frac{a^2}{2b} \qquad (74)$$

EXERCISE 23

1. The length of c in Fig. 13.43 is 95.000 in. If the side b is 94.000 in, find the length of side a to five significant figures using a table of squares and square roots.
2. If the side b in Prob. 1 were decreased to 93.000 in, with c the same as before (95.000 in), find side a to five significant figures using a table of squares and square roots.

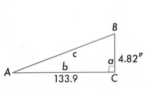

FIGURE 13.42

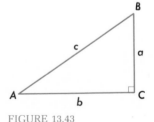

FIGURE 13.43

3. As between Probs. 1 and 2, what was the percentage decrease in the length of b? What was the percentage increase in a?

4. (a) In Fig. 13.43 the side c is 125.00 in and $A = 1.95°$. Calculate a and b on a slide rule using the approximations given above.

   (b) From a seven-place table of natural functions we find that sin $1.95° = 0.0340274$ and cos $1.95° = 0.9994209$. Using these values, calculate a and b to five significant digits and compare with the answers obtained in part a.

5. In Fig. 13.43, $a = 2.0000$ in and $b = 60.000$ in. Calculate c to five significant digits using a table of square roots. Also calculate c using the approximations above.

6. Using the approximations discussed above, find A and b in Fig. 13.43 to slide-rule accuracy if $c = 125$ in and $a = 5.00$ in.

functions of any angle

In the preceding chapter we discussed the trigonometric functions restricted to angles between 0° and 90°.

Now we shall begin to consider the trigonometric functions of angles of any magnitude.

First, however, we shall have to investigate some of the elementary properties of *vectors*.

14.1 Vector Representation

For our purpose we shall define a vector as follows: A vector is a line segment whose distinctive properties are *length, direction,* and *sense.* The *length* of a vector is the linear distance between the extremities. *Direction* is measured in terms of the angular orientation of the vector with respect to some reference line. If we consider the vector to have been generated by a moving point, the *sense* of the vector is determined by the order in which the point coincides with the extremities.

The line segment drawn from A to B in Fig. 14.1 is a vector, since it is distinguished by a *length* (7 units), a *direction* (30° referred to xy), and a *sense* as indicated by the arrow.

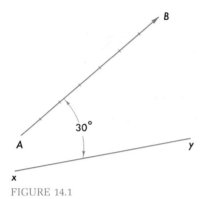

FIGURE 14.1

The vector would be referred to in written symbols as $\overrightarrow{AB}$. The arrow above the letters indicates that it is a vector quantity which is under consideration and not some other sort of quantity. As we mentioned above, it may be convenient to think of the vector $\overrightarrow{AB}$ as the path generated by a point which moved from position A to position B. Consequently, the sense is from A toward B, and is so indicated by the sequence of the letters AB.

If we happen to be interested only in the length, or magnitude, of $\overrightarrow{AB}$, we use the symbol $|\overrightarrow{AB}|$.

Two vectors are said to be equal if their lengths are equal and if their direction and sense are identical (Fig. 14.2). Here $\overrightarrow{AB} = \overrightarrow{CD}$ because of three facts: (1) their lengths are equal; (2) their directions are the same; (3) their senses are identical.

It is true that

$$|\overrightarrow{AB}| = |\overrightarrow{CD}| = |\overrightarrow{EF}| \tag{1}$$

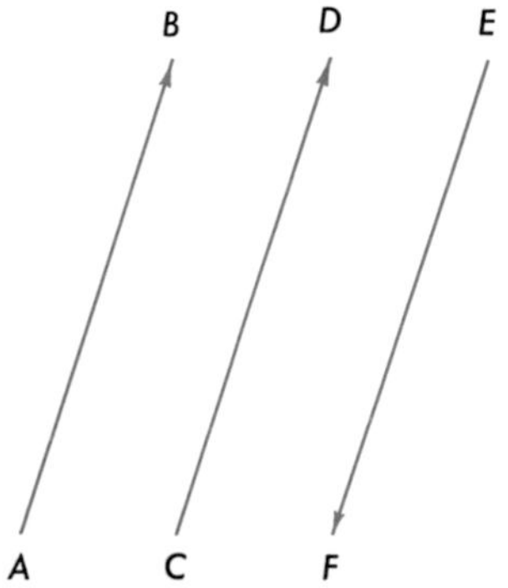

FIGURE 14.2

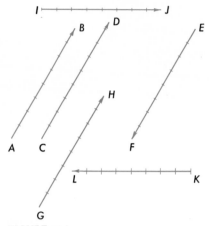

FIGURE 14.3

It is also true that the direction of all three vectors is the same, since they are parallel. However, the sense of $\overrightarrow{EF}$ is not the same as the sense of $\overrightarrow{AB}$ and $\overrightarrow{CD}$. Therefore

$$\overrightarrow{EF} \neq \overrightarrow{AB} \tag{2}$$
$$\overrightarrow{EF} \neq \overrightarrow{CD} \tag{3}$$

The sense of $\overrightarrow{EF}$ is exactly opposite to the sense of $\overrightarrow{AB}$ and $\overrightarrow{CD}$. By definition, one vector is said to be the negative of another if their senses are exactly opposite while their magnitudes and directions are equal. Therefore

$$\overrightarrow{EF} = -\overrightarrow{AB} \tag{4}$$
$$\overrightarrow{EF} = -\overrightarrow{CD} \tag{5}$$

EXERCISE 1

1. Which of the vectors in Fig. 14.3 are equal?
2. Which of these vectors are the negative of $\overrightarrow{AB}$?
3. Which of these vectors have equal magnitudes?

14.2 Position Vectors

The type of vector to which we shall now direct our attention is known as a *position vector*. A position vector is a vector drawn from the origin of a coordinate system to some other point in the coordinate plane, usually for the purpose of locating this point.

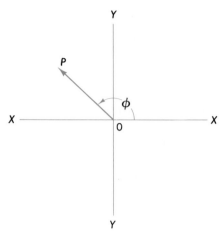

FIGURE 14.4

For example, the position vector $\overrightarrow{OP}$ in Fig. 14.4 is drawn from the origin of the coordinate system to the point P in the coordinate plane.

The direction of $\overrightarrow{OP}$ is given by the angle ϕ with respect to the positive side of the X axis.

It is sometimes convenient to think of the angle ϕ as having been generated by a rotating position vector such as $\overrightarrow{OP}$ in Fig. 14.5. Initially $\overrightarrow{OP}$ lies along the positive side of the X axis, which is said to be the *initial* position of $\overrightarrow{OP}$, or the *initial* side of angle ϕ.

Now we let $\overrightarrow{OP}$ rotate about point O to the position shown in Fig. 14.5. This

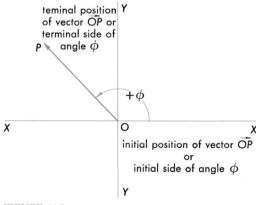

FIGURE 14.5

position is said to be the terminal position of $\overrightarrow{OP}$, or the terminal side of the angle ϕ as indicated in the drawing.

14.3 Angles in Standard Position

An angle is said to be drawn in standard position if, as in Fig. 14.4, it has its vertex at the origin of a system of rectangular coordinates and its initial side coincident with the positive side of the X axis. The terminal side of the angle may occupy any position in the coordinate plane, depending on the size of the angle.

14.4 Positive and Negative Angles

A positive angle is an angle generated by a counterclockwise rotation of a vector as in Fig. 14.5.

A negative angle is an angle generated by a clockwise rotation of a vector.

14.5 Angles of Any Magnitude

In the process of making one complete rotation, the rotating position vector generates all angles in standard position between 0° and 360°. However, the vector can be made to continue its rotation through the 360° position and go around again, generating angles between 360° and 720°. During another revolution it would generate angles between 720° and 1,080°, and so on indefinitely.

Figure 14.6 illustrates angles of several magnitudes. Note particularly that the angles +30°, +390°, −330°, and −690° all define the same direction and sense of the corresponding position vector when drawn in standard position.

EXERCISE 2

Using a protractor, draw the following angles in standard position on a system of rectangular coordinates. Dimension the angles with a curved arrow, as in Fig. 14.6. Let the arrow indicate the direction of rotation.

1.	35°	110°	185°	275°	472°
2.	70°	138°	200°	300°	1,596°
3.	10°	175°	260°	315°	720°
4.	28°	135°	210°	350°	975°
5.	30°	110°	225°	320°	643°
6.	−25°	−150°	−290°	−250°	−720°
7.	−82°	−120°	−250°	−300°	−1,256°

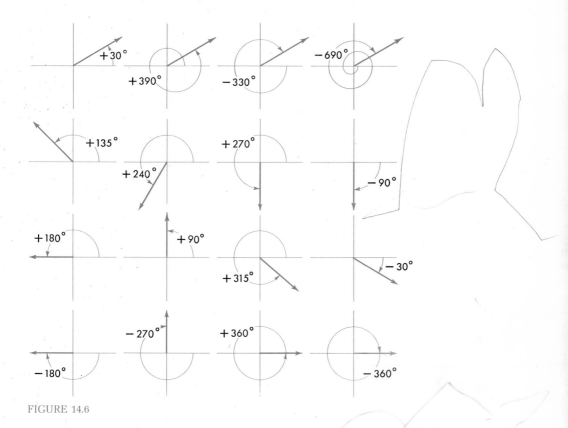

FIGURE 14.6

14.6 The Quadrants

The coordinate plane is divided into four parts as shown in Fig. 14.7. Also see Sec. 8.4.

An angle drawn in standard position is said to be a first-, second-, third-, or

second quadrant	first quadrant
third quadrant	fourth quadrant

FIGURE 14.7

fourth-quadrant angle according as its terminal side lies in the first, second, third, or fourth quadrant, respectively. See Fig. 14.8.

An angle drawn in standard position whose terminal side lies on a coordinate axis is not named in terms of any particular quadrant. Such angles are all integral multiples of 90°, and they are called *boundary angles,* or *quadrantal angles.* We shall defer a discussion of these angles until we come to Sec. 14.11.

14.7 The Definitions of the Functions in Any Quadrant

In Chap. 13 we discussed the definitions of the trigonometric functions as applied to an acute angle in a right triangle. Now we shall define the functions of any nonquadrantal angle, whether acute or not.

In Fig. 14.8 we let (x,y) be the coordinates of the point P. This point is on the terminal side of the angle at a distance $r = \sqrt{x^2 + y^2}$ from the origin. Thus, while x and y may be either positive or negative numbers depending on the quadrant in which point P lies, the distance r is always given by a positive number.

We now define the functions of angle ϕ in terms of r and the coordinates of P (Table 14.1).

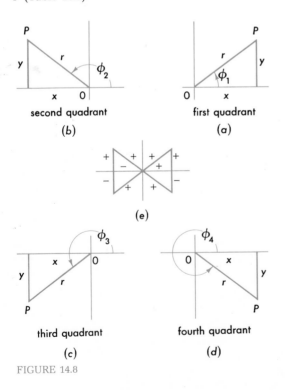

second quadrant

(b)

first quadrant

(a)

(e)

third quadrant

(c)

fourth quadrant

(d)

FIGURE 14.8

TABLE 14.1

$$\sin \phi = \frac{y}{r} \qquad \cos \phi = \frac{x}{r} \qquad \tan \phi = \frac{y}{x}$$

$$\cot \phi = \frac{x}{y} \qquad \sec \phi = \frac{r}{x} \qquad \csc \phi = \frac{r}{y}$$

Following the conventions of rectangular coordinates discussed in Sec. 8.4, when P is in the first or second quadrant, its ordinate is a positive number. When it is in the third or fourth quadrant, its ordinate is a negative number. When P is in the first or fourth quadrant, its abscissa is a positive number. When it is in the second or third quadrant, its abscissa is a negative number. These relations are indicated in Fig. 14.8e, where the $+$ or $-$ sign indicates whether the corresponding coordinate is a positive or a negative number, respectively. The length r is always a positive number and is so designated in the drawing.

By recalling the definitions of the functions from Table 14.1 and the sign conventions from Fig. 14.8, we may construct Table 14.2. In Table 14.2 the $+$ and $-$ signs indicate positive and negative numbers, respectively.

14.8 The Values of the Functions in Any Quadrant

In Chap. 13 we saw how to use the ordinary tables of the trigonometric functions directly for finding the values of the functions of angles between $0°$ and $90°$. However,

TABLE 14.2

Function	First quadrant	Second quadrant	Third quadrant	Fourth quadrant
$\sin = \dfrac{y}{r}$	$\dfrac{+}{+} = +$	$\dfrac{+}{+} = +$	$\dfrac{-}{+} = -$	$\dfrac{-}{+} = -$
$\cos = \dfrac{x}{r}$	$\dfrac{+}{+} = +$	$\dfrac{-}{+} = -$	$\dfrac{-}{+} = -$	$\dfrac{+}{+} = +$
$\tan = \dfrac{y}{x}$	$\dfrac{+}{+} = +$	$\dfrac{+}{-} = -$	$\dfrac{-}{-} = +$	$\dfrac{-}{+} = -$
$\cot = \dfrac{x}{y}$	$\dfrac{+}{+} = +$	$\dfrac{-}{+} = -$	$\dfrac{-}{-} = +$	$\dfrac{+}{-} = -$
$\sec = \dfrac{r}{x}$	$\dfrac{+}{+} = +$	$\dfrac{+}{-} = -$	$\dfrac{+}{-} = -$	$\dfrac{+}{+} = +$
$\csc = \dfrac{r}{y}$	$\dfrac{+}{+} = +$	$\dfrac{+}{+} = +$	$\dfrac{+}{-} = -$	$\dfrac{+}{-} = -$

the more general angle ϕ in Table 14.1 is not so limited. Therefore, to find the values of the functions of any angle, we have to adopt a more indirect approach.

In Fig. 14.9, r and the coordinates of P have the specific values shown. By inspection, then, the triangles OPQ are congruent for all four quadrants. Thus the acute angle A is the same for all quadrants.

The acute angle A is called the *associated acute angle* and is associated with the angles ϕ_1, ϕ_2, ϕ_3, and ϕ_4 in the following ways.

In the first quadrant,

$$A = \phi_1 \tag{6}$$

Thus, in the first quadrant, ϕ_1 is its own associated acute angle.

In the second quadrant,

$$A = 180° - \phi_2 \tag{7}$$

In the third quadrant,

$$A = \phi_3 - 180° \tag{8}$$

In the fourth quadrant,

$$A = 360° - \phi_4 \tag{9}$$

In Fig. 14.9, the six functions of angles ϕ_1, ϕ_2, ϕ_3, and ϕ_4 have been calculated and tabulated from the data given in the drawing. Now let us compare these tabulations. First, we note that the numerical values of the same-named functions of ϕ_1, ϕ_2, ϕ_3, and ϕ_4 are identical with the respective functions of angle A. Second, the *signs* of these functions follow the pattern of Table 14.2. Therefore

To find a function of any nonquadrantal angle, we proceed as follows:

1. *Find the associated acute angle.*
2. *Find the value of the required function of the associated acute angle from a table of the trigonometric functions.*
3. *Attach the proper algebraic sign.*

Example 1. Find the cosine of 115°.
1. The angle 115° is a second-quadrant angle. From Eq. (7) we find that the associated acute angle A is given by

$$A = 180° - 115° = 65°$$

$$\sin \phi_1 = \frac{+3}{+5} = +0.60$$

$$\cos \phi_1 = \frac{+4}{+5} = +0.80$$

$$\tan \phi_1 = \frac{+3}{+4} = +0.75$$

$$\cot \phi_1 = \frac{+4}{+3} = +1.33$$

$$\sec \phi_1 = \frac{+5}{+4} = +1.25$$

$$\operatorname{cosec} \phi_1 = \frac{+5}{+3} = +1.67$$

first quadrant

$$\sin \phi_2 = \frac{+3}{+5} = +0.60$$

$$\cos \phi_2 = \frac{-4}{+5} = -0.80$$

$$\tan \phi_2 = \frac{+3}{-4} = -0.75$$

$$\cot \phi_2 = \frac{-4}{+3} = -1.33$$

$$\sec \phi_2 = \frac{+5}{-4} = -1.25$$

$$\operatorname{cosec} \phi_2 = \frac{+5}{+3} = +1.67$$

second quadrant

$$\sin \phi_3 = \frac{-3}{+5} = -0.60$$

$$\cos \phi_3 = \frac{-4}{+5} = -0.80$$

$$\tan \phi_3 = \frac{-3}{-4} = +0.75$$

$$\cot \phi_3 = \frac{-4}{-3} = +1.33$$

$$\sec \phi_3 = \frac{+5}{-4} = -1.25$$

$$\operatorname{cosec} \phi_3 = \frac{+5}{-3} = -1.67$$

third quadrant

$$\sin \phi_4 = \frac{-3}{+5} = -0.60$$

$$\cos \phi_4 = \frac{+4}{+5} = +0.80$$

$$\tan \phi_4 = \frac{-3}{+4} = -0.75$$

$$\cot \phi_4 = \frac{+4}{-3} = -1.33$$

$$\sec \phi_4 = \frac{+5}{+4} = +1.25$$

$$\operatorname{cosec} \phi_4 = \frac{+5}{-3} = -1.67$$

fourth quadrant

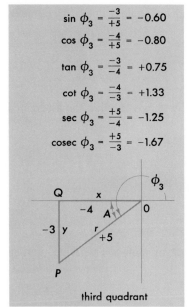

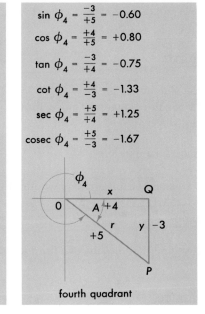

FIGURE 14.9

2. From a table of trigonometric functions we find that

$$\cos 65° = 0.42262$$

3. From Table 14.2 we find that the cosine of a second-quadrant angle is a negative number. Thus

$$\cos 115° = -0.42262$$

Example 2. Find the sine of 295°45′.
1. The angle 295°45′ is a fourth-quadrant angle. From Eq. (9) we find that the associated acute angle A is given by

$$A = 360° - 295°45' = 64°15'$$

2. From a table of trigonometric functions we find that

$$\sin 64°15' = 0.90070$$

3. From Table 14.2 we find that the sign of a fourth-quadrant angle is a negative number. Thus

$$\sin 295°45' = -0.90070$$

Example 3. Find the sine of 244°15′.
1. The angle 244°15′ is a third-quadrant angle. From Eq. (8) we find that the associated acute angle A is given by

$$A = 244°15' - 180° = 64°15'$$

2. From a table of trigonometric functions we find that

$$\sin 64°15' = 0.90070$$

3. From Table 14.2 we find that the sine of a third-quadrant angle is a negative number. Thus

$$\sin 244°15' = -0.90070$$

Example 4. Find the tangent of 236°28′.
1. The angle 236°28′ is a third-quadrant angle. From Eq. (8) we find that the associated acute angle A is given by

$$A = 236°28' - 180° = 56°28'$$

2. From a table of trigonometric functions we find that

$$\tan 56°28' = 1.5089$$

3. From Table 14.2 we find that the tangent of a third-quadrant angle is a positive number. Thus

$$\tan 236°28' = 1.5089$$

EXERCISE 3

Find the six trigonometric functions of the following angles to five significant digits.

1. 140°	2. 170°	3. 95°30'
4. 234°13'43''	5. 246°30'10''	6. 194°32'10''
7. 355°09'55''	8. 280°10'20''	9. 359°40'30''

14.9 Finding Angles When One of Their Functions Is Given

In general, there are two positive angles less than 360°, each having the same value of a given trigonometric function. For instance, in Example 2 we found that the sine of 295°45' is -0.90070. In Example 3 we found that the sine of 244°15' is also -0.90070. Thus, if $\sin \phi = -0.90070$, then $\phi = 295°45'$, or $\phi = 244°15'$. It is important to note that the same acute angle is associated with both values of ϕ.

     If a function of an unknown angle (or angles) is given, we know from Sec. 14.8 that the numerical (absolute) value of that function is equal to the value of the same-named function of the associated acute angle. Thus,

To find the angles having a given function, we proceed as follows:

1. Find the numerical (absolute) value of the given function.
2. From a table of the trigonometric functions find the associated acute angle A.
3. From the sign of the given function determine by Table 14.2 the quadrants in which the unknown angles lie.

Example 5. If $\sin \phi = 0.86310$, find all positive values of ϕ less than 360°.
1. The numerical value of the given function is 0.86310. Thus

$$\sin A = 0.86310$$

2. From a table of trigonometric functions

$$A = 59°40'$$

3. Since the sine of the unknown angles ϕ is positive, one angle must be a first-quadrant angle and the other must be a second-quadrant angle.
4. From Eqs. (6) and (7)

$$\phi = 59°40'$$

and

$$\phi = 180° - 59°40' = 120°20'$$

Example 6. If $\sin \phi = -0.58519$, find all positive values of ϕ less than 360°.
1. The numerical value of the given function is 0.58519. Thus

$$\sin A = 0.58519$$

2. From a table of trigonometric functions

$$A = 35°49'$$

3. Since the sine of the unknown angles ϕ is negative, one angle must be a third-quadrant angle and the other must be a fourth-quadrant angle.
4. From Eqs. (8) and (9)

$$\phi = 180° + 35°49' = 215°49'$$

and

$$\phi = 360° - 35°49' = 324°11'$$

Example 7. If $\cos \phi = 0.94495$, find all positive values of ϕ less than 360°.
1. The numerical value of the given function is 0.94495. Thus

$$\cos A = 0.94495$$

2. From a table of trigonometric functions

$$A = 19°06'$$

3. Since the cosine of the unknown angles ϕ is positive, one angle must be a first-quadrant and the other must be a fourth-quadrant angle.
4. From Eqs. (6) and (9)

$$\phi = 19°06'$$

and

$$\phi = 360° - 19°06' = 340°54'$$

Example 8. If $\cos \phi = -0.38462$, find all positive values of ϕ less than 360°.
1. The numerical value of the given function is 0.38462. Thus

$$\cos A = 0.38462$$

2. From a table of trigonometric functions

$$A = 67°22'47''$$

3. Since the cosine of the unknown angles ϕ is negative, one angle must be a second-quadrant angle and the other must be a third-quadrant angle.
4. From Eqs. (7) and (8)

$$\phi = 180° - 67°22'47'' = 112°37'13''$$

and

$$\phi = 180° + 67°22'47'' = 247°22'47''$$

Example 9. If $\tan \phi = 0.58513$, find all positive values of ϕ less than 360°.
1. The numerical value of the given function is 0.58513. Thus

$$\tan A = 0.58513$$

2. From a table of trigonometric functions

$$\tan A = 30°20'$$

3. Since the tangent of the unknown angles ϕ is positive, one angle must be a first-quadrant and the other must be a third-quadrant angle.

4. From Eqs. (6) and (8)

$$\phi = 30°20'$$

and

$$\phi = 180° + 30°20' = 210°20'$$

EXERCISE 4

Find two positive angles ϕ less than 360° which have each of the functions listed below.

1. $\sin \phi = 0.10572$ | 2. $\sin \phi = -0.32650$ | 3. $\sin \phi = 0.57461$

4. $\sin \phi = -0.81385$ | 5. $\sin \phi = 0.72645$ | 6. $\cos \phi = -0.49695$

7. $\cos \phi = 0.97808$ | 8. $\cos \phi = -0.99981$ | 9. $\cos \phi = 0.02391$

10. $\cos \phi = 0.54980$ | 11. $\tan \phi = -0.86901$ | 12. $\tan \phi = 1.1109$

13. $\tan \phi = -0.46430$ | 14. $\tan \phi = 12.271$ | 15. $\tan \phi = 1.0455$

16. $\cot \phi = 249.88$ | 17. $\cot \phi = -0.06315$ | 18. $\cot \phi = 2.7592$

19. $\cot \phi = -1.5476$ | 20. $\cot \phi = 1.3825$ | 21. $\sec \phi = 23.042$

22. $\sec \phi = -1.0670$ | 23. $\sec \phi = 1.6032$ | 24. $\sec \phi = -1.2130$

25. $\sec \phi = 2.3440$ | 26. $\csc \phi = -2.7290$ | 27. $\csc \phi = 9.3381$

28. $\csc \phi = 1.2880$ | 29. $\csc \phi = -1.3834$ | 30. $\csc \phi = 1.3119$

14.10 Relations among ϕ, r, and the Coordinates of P

In Fig. 14.10 we show an angle ϕ in the second quadrant. However, an angle drawn in any other quadrant would illustrate the present discussion just as well.

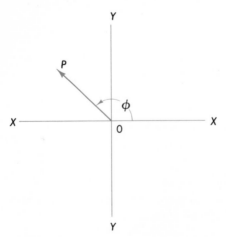

FIGURE 14.10

The coordinates of P are related to the length r of the position vector $\overrightarrow{OP}$ by the equation

$$r = |\overrightarrow{OP}| = \sqrt{x^2 + y^2}$$

From Table 14.1 we see that the coordinates of point P are related to the angle ϕ by the equation

$$\tan \phi = \frac{y}{x}$$

If both coordinates of P are given, the angle ϕ is uniquely determined. (See Example 10.)

If r and only one coordinate of P are given, there may be two angles less than 360° which are consistent with the data. (See Example 11.)

Example 10. Calculate r and ϕ if the coordinates of P are $(-4,3)$. (See Fig. 14.11.)

$$r = \sqrt{(-4)^2 + (3)^2} = 5$$

From the signs of the coordinates we see that the angle ϕ is a second-quadrant angle. From Table 14.1

$$\tan \phi = \frac{3}{-4} = -0.75000$$

The associated acute angle is the acute angle whose tangent is 0.75000. From a table of the trigonometric functions the angle A is found to be 36°52′12″. From Eq. (7)

$$\phi = 180° - 36°52′12″ = 143°07′48″$$

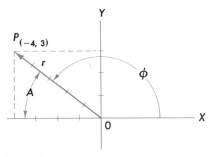

FIGURE 14.11

Example 11. Calculate the ϕ if $r = 5$ and the abscissa of P is -3. See Fig. 14.12.

From Fig. 14.12 we see that each of the two angles ϕ and ϕ' is consistent with the given data. Also observe that the associated acute angle A for both ϕ and ϕ' is given by

$$\cos A = \tfrac{3}{5} = 0.60000$$

Then

$$A = 53°07'48''$$

Since ϕ is a second-quadrant angle, we find from Eq. (7) that

$$\phi = 180° - A = 180° - 53°07'48''$$
$$= 126°52'12''$$

Since ϕ' is a third-quadrant angle, we find from Eq. (8) that

$$\phi' = 180° + 53°07'48'' = 233°07'48''$$

EXERCISE 5

Calculate r and ϕ when the rectangular coordinates of P are as given below. It is suggested that the student use a slide rule.

1. (7, 4)	2. (4, 7)	3. (−3.7, 10)
4. (−7, 7.2)	5. (7.8, −1.4)	6. (−8.6, −4.8)
7. (−3, 5.2)	8. (−9.2, −7.8)	9. (1.4, 7.9)
10. (6.2, −3)	11. (3, −8)	12. (−2, −3.5)
13. (8.5, 3.1)	14. (−6.8, −8)	15. (−10, 8.2)

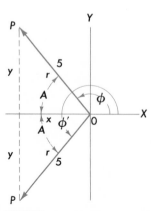

FIGURE 14.12

Calculate two angles less than 360° and the corresponding unknown projections consistent with the data given in each problem below.

16. $r = 8$, $x = +7$ 17. $r = 5$, $y = -3.2$
18. $r = 11$, $x = +8$ 19. $r = 9$, $x = -7.9$
20. $r = 12$, $x = -8$ 21. $r = 11$, $y = -3.8$
22. $r = 8$, $y = +4.5$ 23. $r = 10$, $y = +3.3$

14.11 The Functions of Zero Degrees

When $\phi = 0°$ as in Fig. 14.13a, $y = 0$ and $x = r$; therefore

$$\sin 0° = \frac{y}{r} = \frac{0}{r} = 0$$

$$\cos 0° = \frac{x}{r} = \frac{r}{r} = +1$$

$$\tan 0° = \frac{y}{x} = \frac{0}{x} = 0$$

$$\sec 0° = \frac{r}{x} = \frac{r}{r} = +1$$

The cotangent and the cosecant of zero degrees do not exist, since a calculation of them involves a division by zero. However, if ϕ is a very small positive angle, then

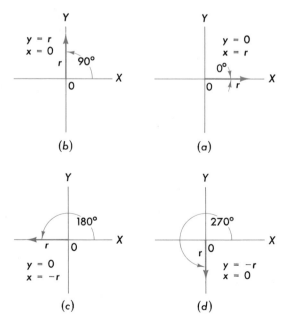

(b)   (a)

(c)   (d)

FIGURE 14.13

y will be a very small linear dimension, and both the cotangent and cosecant will be large numbers. The smaller ϕ is made while still positive, the larger the cotangent and cosecant become.

14.12 The Functions of 90°

When $\phi = 90°$ as in Fig. 14.13b, $x = 0$ and $y = r$. Thus

$$\sin 90° = \frac{y}{r} = \frac{r}{r} = +1$$

$$\cos 90° = \frac{x}{r} = \frac{0}{r} = 0$$

$$\cot 90° = \frac{x}{y} = \frac{0}{y} = 0$$

$$\csc 90° = \frac{r}{y} = \frac{r}{r} = +1$$

Here the tangent and the secant do not exist. But if ϕ is slightly less than 90°, the tangent and secant are enormously large numbers. The closer ϕ approaches 90° while remaining in the first quadrant, the larger the tangent and secant become.

It is important to observe that if ϕ is made ever so slightly greater than 90°, the signs of the tangent and secant change and become negative. The tangent and secant would then be extremely large in absolute value but negative in sign.

If the angle ϕ approaches 90° from the second quadrant, the tangent decreases without limit. If the angle ϕ approaches 90° from the first quadrant, the tangent increases without limit.

14.13 The Functions of 180°

When $\phi = 180°$ as in Fig. 14.13c, $y = 0$ and $x = -r$. Thus

$$\sin 180° = \frac{y}{r} = \frac{0}{r} = 0$$

$$\cos 180° = \frac{x}{r} = \frac{-r}{r} = -1$$

$$\tan 180° = \frac{y}{x} = \frac{0}{x} = 0$$

$$\sec 180° = \frac{r}{x} = \frac{r}{-r} = -1$$

The cotangent and cosecant of 180° do not exist since each involves a division by zero in this case.

14.14 The Functions of 270° and 360°

It is left to the student to verify the following equations by consulting Figs. 14.13*d* and 14.13*a*.

$\sin 270° = -1$	$\sin 360° = 0$
$\cos 270° = 0$	$\cos 360° = 1$
$\tan 270°$ does not exist	$\tan 360° = 0$
$\cot 270° = 0$	$\cot 360°$ does not exist
$\sec 270°$ does not exist	$\sec 360° = 1$
$\csc 270° = -1$	$\csc 360°$ does not exist

14.15 The Functions of Angles Greater Than 360° and Less Than 0°

Suppose the vector $\overrightarrow{OP}$ were initially at any given position in the coordinate plane. Then let it be rotated an integral multiple of 360° in either direction. It would then come to rest exactly coincident with its original position.

Therefore the values of the functions of the reference angle after this rotation would be exactly the same as the values of the functions of the initial reference angle.

Each 360° rotation in a counterclockwise direction is equivalent to adding 360° to the original reference angle. Each 360° rotation in a clockwise direction is equivalent to subtracting 360° from the initial reference angle.

Now suppose we wish to find the cosine of 2,635°. This angle can be expressed not only in degrees, but also in revolutions and a fraction thereof. Thus

$$2{,}635° = \frac{2{,}635}{360} = 7 + \frac{115}{360} \text{ revolutions}$$

That is,

$$2{,}635° = 7 \text{ revolutions} + 115°$$

and all the functions of 2,635° are identical with the same-named functions of 115°. The cosine of 2,635° is therefore equal to the cosine of 115°. In Example 1 we found that

$$\cos 115° = -0.42262$$

Therefore

$$\cos 2{,}635° = -0.42262$$

Similarly, the angle $-605°$ can be expressed as

$$-605° = \frac{-605}{360} = -1 \text{ revolution} - 245°$$

but an angle of $-245°$ has the same functions as an angle of $360° - 245° = 115°$.
From Example 1 we find that

$$\cos 115° = -0.42262$$

Therefore

$$\cos - 605° = -0.42262$$

14.16 Variations in the Functions

As the position vector $\overrightarrow{OP}$ rotates from the initial position through the first quadrant to the 90° position, the sine of the angle ϕ increases from 0 to 1. In the second quadrant the sine decreases from 1 at the 90° position to 0 at the 180° position. In the third quadrant, the sine decreases from 0 to -1, and in the fourth quadrant it increases from -1 to 0.

In a like manner the variations in the other five functions can be traced.

EXERCISE 6

1. Write the sine of 0°, 30°, 45°, 60°, 90°, 120°, 135°, 150°, 180°, 210°, 225°, 240°, 270°, 300°, 315°, 330°, and 360° in decimals, using proper signs.
2. In which quadrants does the sine increase as the radius vector rotates counterclockwise? In which quadrant does the cosine increase? The tangent? The cotangent? The secant? The cosecant?
3. Within what numerical limits can the sine, the cosine, the tangent, the cotangent, the secant, and the cosecant exist?
4. What are the maximum and minimum values of each of the six functions?
5. Sketch a graph of each function with the value of the function plotted vertically against the angle plotted horizontally. Plot the graph between $-360°$ and $+360°$.
6. Find the sine of 15°, 375°, 735°, 1,095°, 1,455°.
7. Find the cosine of 135°, $-225°$, $-585°$, $-945°$.

oblique triangles

For efficiency and convenience, the solution of oblique triangles is often accomplished by use of special formulas. We shall emphasize two of these formulas in this chapter. Both of these formulas will refer to the general triangle shown in Fig. 15.1.

15.1 The Sine Law

This formula is useful when we know two angles and a side or when we know two sides and the angle opposite one of them. The formula is usually stated in the form

$$\frac{a}{\sin A} = \frac{b}{\sin B} = \frac{c}{\sin C} \tag{1}$$

Equation (1) is equivalent to Eqs. (2) to (4):

$$\frac{a}{\sin A} = \frac{b}{\sin B} \tag{2}$$

$$\frac{a}{\sin A} = \frac{c}{\sin C} \tag{3}$$

$$\frac{b}{\sin B} = \frac{c}{\sin C} \tag{4}$$

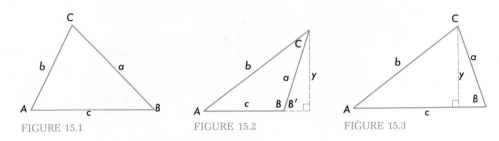

FIGURE 15.1   FIGURE 15.2   FIGURE 15.3

If one of the angles in the triangle is obtuse, its sine may be found by the following rule:

To find the sine of an obtuse angle, subtract the obtuse angle from 180°, and use the sine of the acute angle so found. See Eq. (7) in Chap. 14.

In general, when the angle A is an obtuse angle of a triangle,

$$\sin A = \sin (180° - A) \tag{5}$$

Example 1. Find the sine of 126°.

$$\sin 126° = \sin (180° - 126°) = \sin 54°$$

From the five-place table,

$$\sin 54° = 0.80902$$

Therefore

$$\sin 126° = 0.80902$$

EXERCISE 1

Find the sine of each of the following angles:

1. 140°
2. 170°
3. 95°30′
4. 125°46′17″
5. 113°29′50″
6. 165°27′50″
7. 175°09′55″
8. 100°10′20″
9. 179°40′30″

Find each of two angles less than 180° which have their sine listed below.

10. 0.05814
11. 0.33244
12. 0.57715
13. 0.72837
14. 0.90766
15. 0.99824

15.2 Derivation of the Sine Law

Referring to Fig. 15.2,

$$y = b \sin A \tag{6}$$

and

$$y = a \sin B' \tag{7}$$

But from the geometry of the figure the angles B and B' are supplementary angles; therefore, from Eq. (5), their sines are equal, and Eq. (7) may be written

$$y = a \sin B \tag{8}$$

Combining Eqs. (6) and (8),

$$a \sin B = b \sin A \tag{9}$$

Similarly, in Fig. 15.3,

$$y = a \sin B$$

and

$$y = b \sin A$$

Therefore

$$a \sin B = b \sin A \tag{10}$$

Equation (9) or (10) may be written

$$\frac{a}{\sin A} = \frac{b}{\sin B} \tag{11}$$

The same equation applies whether B is obtuse or acute. We may extend the above derivation to obtain

$$\frac{a}{\sin A} = \frac{b}{\sin B} = \frac{c}{\sin C} \tag{12}$$

Example 2. In the oblique triangle ABC, $A = 31°30'$, $B = 28°04'$, and side b equals 94.1

in. Find side a. From Eq. (2)

$$\frac{a}{\sin A} = \frac{b}{\sin B}$$

$$\frac{a}{\sin 31°30'} = \frac{94.1}{\sin 28°04'}$$

$$\frac{a}{0.52250} = \frac{94.1}{0.47050}$$

$$a = \frac{94.1 \times \overset{0.1045}{\cancel{0.52250}}}{\underset{0.0941}{\cancel{0.47050}}} = \frac{94.1 \times 0.1045}{0.0941}$$

$$= 1,000 \times 0.1045 = 104.5 \text{ in}$$

Example 3. In the oblique triangle ABC, $A = 21°06'$, $C = 35°21'$, and side b equals 46.3 in. Find side a. See Eq. (11)

$$\frac{a}{\sin A} = \frac{b}{\sin B}$$

$$B = 180° - (21°06' + 35°21')$$
$$= 180° - 56°27' = 123°33'$$

$$\frac{a}{\sin 21°06'} = \frac{b}{\sin 123°33'}$$

$$\frac{a}{0.36000} = \frac{46.3}{0.83340}$$

$$a = \frac{46.3 \times \overset{0.04}{\cancel{0.36000}}}{\underset{0.0926}{\cancel{0.83340}}}$$

$$= \frac{46.3 \times \overset{0.02}{\cancel{0.04}}}{\underset{0.0463}{\cancel{0.0926}}} = 1,000 \times 0.02 = 20 \text{ in}$$

15.3 The Ambiguous Case

If an oblique triangle exists for which two sides and the angle opposite one of them are given, there may be either one or two triangles whose dimensions are consistent with the given data.

In the following discussion (see Fig. 15.4a) we shall let

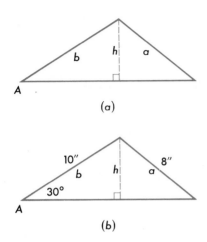

(a)

(b)

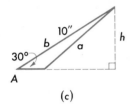

(c)

FIGURE 15.4

h = altitude drawn to unknown side
A = given angle
a = side opposite given angle
b = a side adjacent to given angle

Thus in Fig. 15.4a

$$h = b \sin A \tag{13}$$

Now suppose it is given that $A = 30°$, $b = 10$ in, and $a = 8$ in. Then either Fig. 15.4b or c applies. Both triangles are consistent with the given data. Observe that in this case

$$h < a < b$$

or from Eq. (13)

$$b \sin A < a < b$$

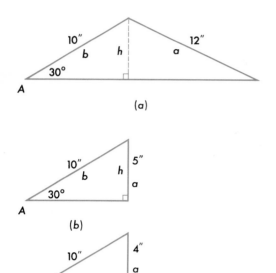

FIGURE 15.5

If, as in Fig. 15.5a, $A = 30°$, $b = 10$ in, and $a = 12$ in, there is evidently only one triangle consistent with the given data. In this case

$$a > b$$

If, as in Fig. 15.5b, $A = 30°$, $b = 10$ in, and $a = 5$ in, there is only one triangle consistent with the given data, and this is a right triangle in which $a = h$, or

$$a = b \sin A = 10 \sin 30° = 10 \cdot 0.5 = 5 \text{ in}$$

Finally, suppose the data assert that $A = 30°$, $b = 10$ in, and $a = 4$ in. In this case the side a is asserted to be *less* than the altitude h. From Fig. 15.5c it is obviously impossible to construct a triangle consistent with these data.

This information is summarized in Fig. 15.6.

Example 4. In the oblique triangle ABC, $B = 103°47'$, $c = 215$ in, and $b = 242.8$ in. Find angle C. See Fig. 15.7.

From Eq. (4)

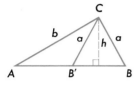

$h < a < b$
or
$b \sin A < a < b$
There are two triangles
ABC and $AB'C$

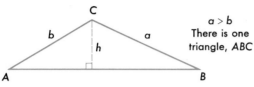

$a > b$
There is one
triangle, ABC

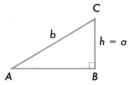

$a = h$
or
$a = b \sin A$
There is one triangle ABC
and this is a right triangle

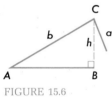

$a < h$
or
$a < b \sin A$
There is no triangle

FIGURE 15.6

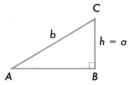

FIGURE 15.7

$$\frac{c}{\sin C} = \frac{b}{\sin B}$$

$$\frac{215}{\sin C} = \frac{242.8}{\sin 103°47'}$$

$$\frac{215}{\sin C} = \frac{242.8}{0.97120}$$

$$\sin C = \frac{215 \times \overset{0.2428}{\cancel{0.97120}}}{\underset{60.7}{\cancel{242.8}}} = \frac{215 \times 0.2428}{60.7}$$

$$= \frac{215 \times \overset{0.004}{\cancel{0.2428}}}{\underset{1}{\cancel{60.7}}} = 215 \times 0.004$$

$$= 0.86000$$

$$C = 59°19'$$

Of the two given sides in this problem, the one opposite the given angle is the larger. Therefore there is only one solution.

Example 5. In the oblique triangle ABC, the angle $C = 19°38'$, $b = 123.3$ in, and $c = 48$ in. Find angle B.

From an inspection of Figs. 15.8 and 15.9, we note that the known side opposite the known angle is larger than the altitude h and is less than the other known side b.

Thus we may expect to find two solutions.

From Eq. (4)

$$\frac{b}{\sin B} = \frac{c}{\sin C}$$

$$\frac{123.3}{\sin B} = \frac{48}{0.33600}$$

$$\sin B = \frac{123.3 \times \overset{0.056}{\cancel{0.33600}}}{\underset{8}{\cancel{48}}}$$

$$= \frac{123.3 \times 0.056}{8}$$

$$= 123.3 \times 0.007 = 0.86310$$

From Example 5, Chap. 14, we recall that 0.86310 is the sine of either of two angles. One of them is the first quadrant angle 59°40′, and the other is the second quadrant angle 120°20′. Thus the angle B in Fig. 15.8 is 59°40′ and the angle B' in Fig. 15.9 is 120°20′.

EXERCISE 2

The data for the problems given below refer to the general oblique triangle of which Fig. 15.1 is typical. These data have been chosen to minimize the arithmetical work. The student should reduce the fractions he encounters as far as he can.

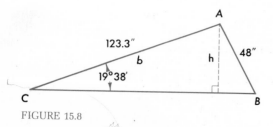

FIGURE 15.8

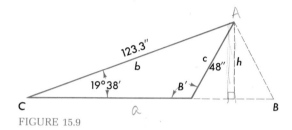

FIGURE 15.9

Where ambiguous data are given, both answers should be calculated.

1. Find side a if $C = 99°20'$, $B = 41°29'$, and $b = 7.36$ in.
2. Find side c if $A = 119°42'$, $C = 48°50'$, and $b = 4.97$ in.
3. Find side b if $A = 95°02'$, $B = 37°09'$, and $c = 2.47$ in.
4. Find side a if $A = 114°13'$, $B = 10°18'$, and $b = 1.49$ in.
5. Find side c if $B = 138°42'$, $C = 24°01'$, and $b = 60$ in.
6. Find angle B if $A = 39°11'$, $a = 54$ in, and $b = 48$ in.
7. Find angle A if $B = 34°10'$, $a = 54$ in, and $b = 48$ in.

15.4 Logarithmic Solutions

Ordinarily, the most efficient way to perform the calculations required in the problems to follow would be with a digital computer, or even with a desk calculator.

If one of these machines is not available, and often they are not, then logarithms can be used to good advantage. Also see Sec. 13.16.

Example 6. Solve the triangle $A = 42°10'00''$, $B = 78°40'00''$, $c = 150.00$ in.

$$C = 180° - (42°10' + 78°40') = 59°10'$$

$$\frac{a}{\sin A} = \frac{c}{\sin C} \qquad\qquad \frac{b}{\sin B} = \frac{c}{\sin C}$$

$$\frac{a}{\sin 42°10'} = \frac{150}{\sin 59°10'} \qquad\qquad \frac{b}{\sin 78°40'} = \frac{150}{\sin 59°10'}$$

$$a = \frac{150 \sin 42°10'}{\sin 59°10'} \qquad\qquad b = \frac{150 \sin 78°40'}{\sin 59°10'}$$

log 150 in $=$	2.17609	log 150 in $=$	2.17609
log sin 42°10' $=$	9.82691 − 10	log sin 78°40' $=$	9.99145 − 10
	12.00300 − 10		12.16754 − 10
log sin 59°10' $=$	9.93382 − 10	log sin 59°10' $=$	9.93382 − 10
log $a =$	2.06918	log $b =$	2.23372
$a =$	117.27 in	$b =$	171.28 in

The labor of calculation can be reduced somewhat by the use of cologarithms (see Sec. 11.21).

$$
\begin{aligned}
\log 150 \text{ in} &= 2.17609 \\
\log \sin 42°10' &= 9.82691 - 10 \\
\text{colog} \sin 59°10' &= 0.06618 \\
\hline
\log a &= 12.06918 - 10 \\
a &= 117.27 \text{ in}
\end{aligned}
\qquad
\begin{aligned}
\log 150 \text{ in} &= 2.17609 \\
\log \sin 78°40' &= 9.99145 - 10 \\
\text{colog} \sin 59°10' &= 0.06618 \\
\hline
\log b &= 12.23372 - 10 \\
b &= 171.28 \text{ in}
\end{aligned}
$$

Example 7. Solve the oblique triangle $B = 40°00'00''$, $C = 25°00'00''$, and $a = 23.529$ in.

$$A = 180° - (40° + 25°) = 180° - 65° = 115°$$

$$\frac{b}{\sin B} = \frac{a}{\sin A} \qquad \frac{c}{\sin C} = \frac{a}{\sin A}$$

$$\log \sin 115° = \log \sin (180° - 115°) = \log \sin 65°$$
$$\log \sin 65 = 9.95728 - 10$$
$$\log \sin 115° = 9.95728 - 10$$

$$
\begin{aligned}
\log 23.529 \text{ in} &= 1.37160 \\
\log \sin 40° &= 9.80807 - 10 \\
\text{colog} \sin 115° &= 0.04272 \\
\hline
\log b &= 11.22239 - 10 \\
b &= 16.687 \text{ in}
\end{aligned}
\qquad
\begin{aligned}
\log 23.529 \text{ in} &= 1.37160 \\
\log \sin 25° &= 9.62595 - 10 \\
\text{colog} \sin 115° &= 0.04272 \\
\hline
\log c &= 11.04027 - 10 \\
c &= 10.972 \text{ in}
\end{aligned}
$$

Example 8. Solve the oblique triangle $A = 25°00'00''$, $b = 125.00$ in, and $a = 80.000$ in.

Here, two sides and the angle opposite one of them are given. In Sec. 15.3 it was indicated that under these conditions there is a possibility of two solutions consistent with the given data.

Figure 15.10a and b is drawn approximately to scale. We shall solve both triangles.

$$\frac{b}{\sin B} = \frac{a}{\sin A} \quad \text{or} \quad \frac{\sin B}{b} = \frac{\sin A}{a} \quad \text{or} \quad \sin B = \frac{b \sin A}{a}$$

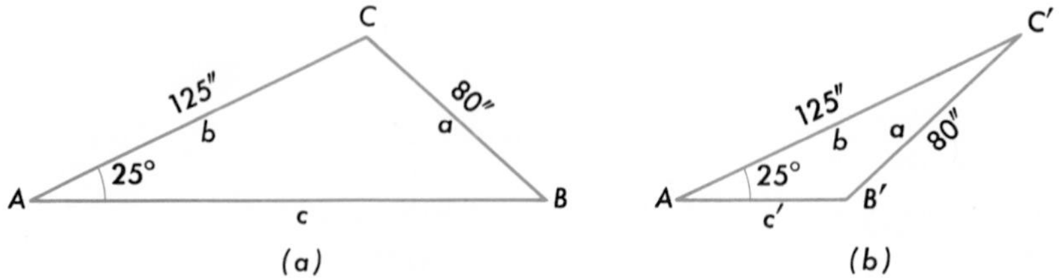

(a)

(b)

FIGURE 15.10

$$\sin B = \frac{125 \sin 25°}{80}$$

$$
\begin{aligned}
\log 125 &= 2.09691 \\
\log \sin 25° &= 9.62595 - 10 \\
\text{colog } 80 &= \underline{8.09691 - 10} \\
\log \sin B &= 19.81977 - 20 \\
B &= 41°19'34''
\end{aligned}
$$

$$C = 180° - (25° + 41°19'34'') = 180° - 66°19'34'' = 113°40'26''$$

$$\frac{c}{\sin C} = \frac{a}{\sin A}$$

$$\frac{c}{\sin 113°40'26''} = \frac{80}{\sin 25°}$$

$$c = \frac{80 \sin 113°40'26''}{\sin 25°}$$

$$\log \sin 113°40'26'' = \log \sin (180° - 113°40'26'') = \log \sin 66°19'34''$$

$$
\begin{aligned}
\log \sin 66°19'34'' &= 9.96182 - 10 \\
\log \sin 113°40'26'' &= 9.96182 - 10
\end{aligned}
$$

$$
\begin{aligned}
\log 80 \text{ in} &= 1.90309 \\
\log \sin 113°40'26'' &= 9.96182 - 10 \\
\text{colog } \sin 25° &= \underline{0.37405} \\
\log c &= 12.23896 - 10 \\
c &= 173.36 \text{ in}
\end{aligned}
$$

$$
\begin{aligned}
B' &= 180° - B = 180° - 41°19'34'' = 138°40'26'' \\
C' &= 180° - (138°40'26'' + 25°) = 16°19'34''
\end{aligned}
$$

$$c' = \frac{a \sin C'}{\sin A} = \frac{80 \sin 16°19'34''}{\sin 25°}$$

$$
\begin{aligned}
\log 80 \text{ in} &= 1.90309 \\
\log \sin 16°19'34'' &= 9.44886 - 10 \\
\text{colog } \sin 25° &= \underline{0.37405} \\
\log c' &= 11.72600 - 10 \\
c' &= 53.211 \text{ in}
\end{aligned}
$$

15.5 Sine-law Solution by the Slide Rule

The slide rule is also a convenient device for solving a triangle by use of the sine law provided slide-rule accuracy is adequate.

For example, suppose we wish to find the side a by means of a slide rule when $A = 39.5°$, $B = 57°$, and $b = 6.00$ in. The angles are expressed in degrees and decimal parts of a degree to conform with the usual slide-rule calibration. See Fig. 15.11.

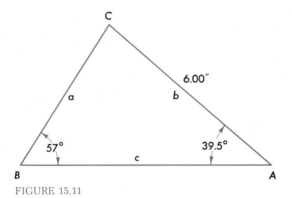

FIGURE 15.11

The general equation

$$\frac{a}{\sin A} = \frac{b}{\sin B}$$

may be applied to this problem by writing

$$\frac{a}{\sin 39.5°} = \frac{6.00}{\sin 57°} \tag{14}$$

Figure 15.12 shows the appropriate slide-rule setting for solving Eq. (14) for side a.

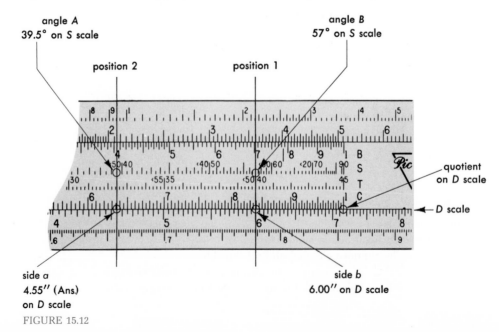

FIGURE 15.12

Observe that with the hairline in "position 1," it simultaneously indicates side b on the D scale and angle B on the S scale. With this setting the quotient of b divided by sin B may be read on the D scale as shown in Fig. 15.12. (In this problem there is no occasion actually to read this number.)

Without moving the slide, if we move the hairline to "position 2," it will then simultaneously indicate angle A on the S scale and side a on the D scale. This setting indicates the quotient of a divided by sin A. From Eq. (14) it is quite evident that this quotient is equal to the one obtained before.

By direct reading of the slide rule (Fig. 15.12)

$a = 4.55$ in

Now, the triangle shown in Fig. 15.13 has a different appearance from the one shown in Fig. 15.11. However, we observe that

$\sin 123° = \sin 57°$

since

$123° = 180° - 57°$

(See Sec. 14.8.)

Thus the slide-rule setting shown in Fig. 15.12 applies to Fig. 15.13 as well as to Fig. 15.11.

To find the side c we first find angle C, where

$C = 180° - (57° + 39.5°) = 83.5°$ (for Fig. 15.11)
$C = 180° - (123° + 39.5°) = 17.5°$ (for Fig. 15.13)

Then reset the indicator to find the side c in the corresponding drawing.

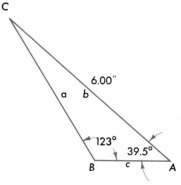

FIGURE 15.13

EXERCISE 3

1. Redraw Figs. 15.2 and 15.3, and by the use of suitable construction lines prove that

$$\frac{b}{\sin B} = \frac{c}{\sin C}$$

2. In Fig. 15.14 prove that

$$2R = \frac{a}{\sin A} = \frac{b}{\sin B} = \frac{c}{\sin C}$$

3. Rewrite Eq. (2) for the case in which $B = 90°$.
4. In a certain triangle, the angles A, B, and C are, respectively, as $3:4:5$. Side a is 10.000 in. Find sides b and c.
5. The sides of a triangle a, b, and c are, respectively, 25, 38, and 51 in. The angles (not listed in the same order as the sides) are approximately $28°05'$, $106°15'$, and $45°40'$. Identify the angles A, B, and C.
6. Given $a = 10\sqrt{2}$, $A = 30°$, and $C = 105°$, find b and c.
7. If $A = 45°$ and $B = 30°$, find the sides a and b when $c = 10$ in.
8. In Fig. 15.15 prove by the law of sines that

$$\frac{ab}{bc} = \frac{oa}{oc}$$

9. If $A = 30°$, $B = 120°$, and $c = 1,000$, find a, b, and the altitude drawn to c.
10. Referring to Fig. 15.16:

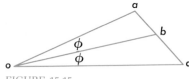
FIGURE 15.15

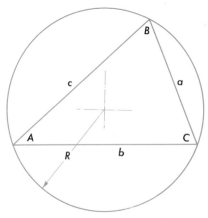
FIGURE 15.14

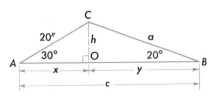
FIGURE 15.16

(a) Find h, x, and $\angle ACO$.

(b) Knowing h, find a, y, and $\angle BCO$.

(c) Knowing x, y, $\angle ACO$, and $\angle BCO$, find C and side c.

11. (a) Angle A of a certain triangle is $17°27.7'$, side a is 30 in, and angle B is $42°$. Draw the triangle to scale and calculate side b.

(b) If angle A is $162°32.3'$, $a = 30$ in, and $B = 10°$, draw the triangle to scale and solve for side b.

12. The base of a triangle is 4,500 ft, and the angles at the base are $10°20.4'$ and $15°34'$. Find the unknown sides.

In the problems below solve for unknown sides to five figures and unknown angles to the nearest minute.

13. $b = 1.5570$ ft	$A = 38°19'$	$C = 88°10'$
14. $b = 3.3492$ ft	$B = 144°10'$	$C = 13°02'$
15. $b = 4.1759$ in	$A = 31°20'$	$C = 18°27'$
16. $b = 4.2997$ in	$A = 7°36'$	$C = 10°29'$
17. $b = 6.8926$ ft	$A = 9°36'$	$C = 47°38'$
18. $a = 1.1831$ in	$B = 8°41'$	$C = 29°54'$
19. $c = 11.863$ in	$A = 5°24'$	$B = 83°50'$
20. $b = 3.8372$ in	$A = 25°26'$	$B = 124°36'$
21. $a = 6.3397$ in	$C = 26°23'$	$B = 49°07'$
22. $c = 10.165$ in	$B = 20°12'$	$C = 75°50'$
23. $c = 39.862$ in	$B = 39°11'$	$A = 12°37'$
24. $a = 2.7570$ ft	$B = 43°7'$	$C = 35°14'$
25. $b = 1.7089$ in	$c = 1.2788$ in	$C = 48°21'$
26. $b = 4.8600$ ft	$a = 3.5683$ ft	$B = 68°01'$
27. $b = 7.2199$ ft	$A = 57°37'$	$a = 6.5867$ ft
28. $a = 1.1690$ in	$c = .63966$ in	$C = 28°04'$
29. $b = 4.4682$ ft	$A = 61°22'$	$a = 4.1244$ ft
30. $b = 1.0226$ ft	$c = 1.3378$ ft	$C = 70°14'$
31. $c = 3.7198$ in	$A = 62°31'$	$a = 3.3145$ in
32. $b = 5.2979$ ft	$A = 74°3'$	$C = 11°17'$
33. $a = 3.7956$ ft	$B = 38°7'$	$C = 82°41'$
34. $b = 0.43972$ in	$A = 51°8'$	$C = 36°28'$

Find the unknown sides to five figures and the unknown angles to the nearest second, using the sine law, in the following problems.

35. $a = 17.230$ in	$A = 56°22'13''$	$C = 35°53'16''$
36. $c = 11.855$ in	$A = 7°30'47''$	$C = 47°36'12''$
37. $c = 133.70$ in	$A = 4°17'22''$	$B = 165°29'23''$

38. $b = 105.46$ in $B = 85°10'31''$ $c = 105.09$ in
39. $b = 145.70$ in $a = 145.10$ in $B = 85°09'55''$
40. $b = 16.683$ in $A = 79°50'30''$ $a = 17.938$ in
41. $c = 16.481$ in $b = 32.675$ in $B = 109°13'16''$
42. $c = 12.781$ in $a = 12.412$ in $C = 46°28'50''$
43. $a = 17.219$ in $c = 19.751$ in $C = 88°56'3''$
44. $a = 6.6435$ in $B = 53°21'9''$ $C = 48°48'37''$
45. $a = 10.959$ in $A = 80°43'53''$ $B = 33°24'58''$
46. $c = 30.361$ in $A = 21°14'28''$ $B = 146°40'26''$
47. $a = 23.293$ in $A = 24°15'35''$ $C = 25°25'25''$
48. $b = 10.878$ in $A = 44°40'38''$ $B = 49°41'43''$
49. $b = 36.234$ in $A = 20°6'20''$ $C = 48°16'42''$
50. $a = 11.306$ in $A = 25°24'44''$ $B = 92°24'3''$
51. $a = 6527.6$ in $B = 70°55'29''$ $C = 52°9'43''$
52. $c = 1004.0$ in $A = 79°19'25''$ $B = 53°27'10''$
53. $b = 14.752$ in $B = 13°19.7'$ $C = 59°13.6'$
54. $b = 999.90$ in $A = 37°58.7'$ $C = 65°2.9'$
55. $a = 497.32$ in $A = 10°36.4'$ $B = 46°37.9'$
56. $a = 832.76$ in $A = 82°36'42''$ $B = 45°32'10''$
57. $a = 796.38$ in $A = 99°36'24''$ $C = 49°37'45''$
58. $a = 827.56$ in $C = 12°48.3'$ $B = 140°59.7'$
59. $a = 143.62$ in $B = 37°42.7'$ $C = 28°26.5'$

60. A destroyer A (Fig. 15.17) is to intercept a ship B by traveling in a straight line. Ship B is just detected 5 nautical mi due north of A. The speed of B is determined to be 24 knots on a course N70°E and is assumed to continue this course until interception. Determine the course of the destroyer (the angle ϕ) and the time required for interception if the destroyer maintains a speed of 30 knots. One knot is a speed of one nautical mile per hour.

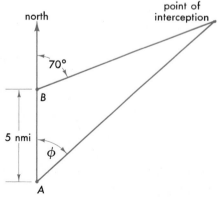

FIGURE 15.17

15.6 The Cosine Law

This law is useful when two sides and the included angle are known or when three sides are known.

Referring to Fig. 15.1,

$$a^2 = b^2 + c^2 - 2 \times b \times c \times \cos A \tag{15}$$
$$b^2 = a^2 + c^2 - 2 \times a \times c \times \cos B \tag{16}$$
$$c^2 = a^2 + b^2 - 2 \times a \times b \times \cos C \tag{17}$$

If one of the angles in the triangle is obtuse, its cosine may be found by the rule below:

To find the cosine of an obtuse angle, subtract the obtuse angle from 180° and use the negative of the cosine of the acute angle so found. (See Sec. 14.8.)

In general, when the angle A is obtuse,

$$\cos A = -\cos (180° - A) \tag{18}$$

Example 9. Find the cosine of 126°.

$$\cos 126° = -\cos (180° - 126°) = -\cos 54°$$

From the five-place tables,

$$\cos 54° = 0.58779$$

Therefore

$$\cos 126° = -0.58779$$

If the angle A, B, or C is obtuse, then Eqs. (15), (16), and (17) may be written

$$a^2 = b^2 + c^2 + 2 \times b \times c \times \cos (180° - A) \tag{19}$$
$$b^2 = a^2 + c^2 + 2 \times a \times c \times \cos (180° - B) \tag{20}$$
$$c^2 = a^2 + b^2 + 2 \times a \times b \times \cos (180° - C) \tag{21}$$

15.7 Derivation of the Cosine Law

Given the oblique triangle ABC in Fig. 15.18, where A, b, and c are known and where A is an acute angle, find a formula for a in terms of A, b, and c.

Drop the perpendicular h from B to side b.

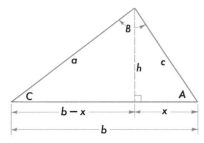

FIGURE 15.18

$$x = c \times \cos A \qquad (22)$$
$$h = c \times \sin A \qquad (23)$$
$$a^2 = h^2 + (b - x)^2 \qquad (24)$$

But

$$h^2 = c^2 - x^2 \qquad (25)$$

Substituting the above value of h^2 in Eq. (24), we obtain

$$a^2 = c^2 - x^2 + (b - x)^2 \qquad (26)$$

or

$$a^2 = c^2 - x^2 + b^2 - 2bx + x^2 \qquad (27)$$

or

$$a^2 = c^2 + b^2 - 2bx \qquad (28)$$

But

$$x = c \cos A \qquad (29)$$

Substituting the above value of x in Eq. (28),

$$a^2 = c^2 + b^2 - 2bc \cos A \qquad (30)$$

Equations (16) and (17) can be derived in a similar way.

If the known angle is obtuse, the cosine law can be derived from Fig. 15.19. Here the angle A and the sides b and c are given. Angle A is obtuse.

Drop the perpendicular h from B to the side b extended.

$h = c \sin A'$
$x = c \cos A'$
$a^2 = h^2 + (b + x)^2 \hfill (31)$

But

$$h^2 = c^2 - x^2 \hfill (32)$$

Substituting the above value of h^2 in Eq. (31),

$a^2 = c^2 - x^2 + (b + x)^2$
$a^2 = c^2 - x^2 + b^2 + 2bx + x^2$
$a^2 = c^2 + b^2 + 2bx$

But

$x = c \cos A'$

Therefore

$a^2 = b^2 + c^2 + 2bc \cos A'$
$A' = 180° - A \hfill (33)$

and

$$a^2 = b^2 + c^2 - 2bc \cos A \hfill (34)$$

Therefore the cosine law as stated in Eqs. (15) to (17) is valid for both acute and obtuse angles.

It is not at all unusual to encounter situations in which it is more convenient to use the acute exterior angle A' than to use the obtuse interior angle A (Fig. 15.19).

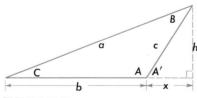

FIGURE 15.19

Such cases frequently occur in alternating-current and in concurrent-force problems. In these problems the side b, the side c, and the angle A' would naturally appear in the data.

If the angles A', B', and C' are exterior angles corresponding to the interior angles A, B, and C,

$$a^2 = b^2 + c^2 + 2bc \cos A' \tag{35}$$
$$b^2 = a^2 + c^2 + 2ac \cos B' \tag{36}$$
$$c^2 = a^2 + b^2 + 2ab \cos C' \tag{37}$$

As a special case, if $A' = 90°$, Eq. (35) becomes

$$a^2 = b^2 + c^2 + 2bc \cos 90°$$
$$= b^2 + c^2 + 2bc(0)$$
$$= b^2 + c^2$$

which is the Pythagorean theorem. The cosine law is sometimes called the *generalized Pythagorean theorem* because, with the $(\pm 2bc \cos A)$ term included, the traditional Pythagorean theorem applies to all triangles.

Example 10. Solve the triangle $a = 25$ in, $b = 56$ in, $C = 36°52'12''$.

$$c^2 = a^2 + b^2 - 2ab \cos 36°52'12''$$
$$= 25^2 + 56^2 - 2 \times 25 \times 56 \times 0.8$$
$$= 625 + 3,136 - 2,240$$
$$= 1,521$$
$$c = \sqrt{1,521} = 39$$

One of the unknown angles can now be computed by use of the sine law.

$$\frac{c}{\sin C} = \frac{b}{\sin B} \qquad \frac{39}{\sin 36°52'12''} = \frac{56}{\sin B}$$

$$\sin B = \frac{56 \sin 36°52'12''}{39}$$

$$\begin{aligned} \log 56 &= 1.74819 \\ \log \sin 36°52'12'' &= 0.77815 - 1 \\ \text{colog } 39 &= \underline{0.40894 - 2} \\ \log \sin B &= 0.93528 - 1 \end{aligned}$$

B would equal $59°29'26''$ *if B were acute*. However, from the scale drawing of the problem, B is obviously obtuse. Therefore

$$B = 180° - 59°29'26'' = 120°30'34''$$

Example 11. Solve the triangle $a = 25.000$ in, $b = 16.000$ in, $C = 143°7'48''$.

Here the known angle is obtuse, and there are two ways of thinking when attacking the problem. Equation (37) can be used, and

$$\begin{aligned}
c^2 &= a^2 + b^2 + 2ab \cos(180° - 143°7'48'') \\
&= a^2 + b^2 + 2ab \cos 36°52'12'' \\
&= 25^2 + 16^2 + 2 \times 25 \times 16 \times 0.8 = 1,521 \\
c &= \sqrt{1,521} = 39.000 \text{ in}
\end{aligned}$$

Or the fact stated in Eq. (18) may be used, and therefore

$$\begin{aligned}
\cos 143°7'48'' &= -\cos(180° - 143°7'48'') = -\cos 36°52'12'' = -0.8 \\
c^2 &= a^2 + b^2 - 2ab \cos C \\
&= 25^2 + 16^2 - 2 \times 25 \times 16(-0.8) \\
&= 25^2 + 16^2 + 2 \times 25 \times 16 \times 0.8 = 1,521 \\
c &= \sqrt{1,521} = 39
\end{aligned}$$

Angle B can be calculated by the sine law, which is here illustrated by the use of natural functions rather than log functions.

$$\frac{\sin B}{b} = \frac{\sin C}{c}$$

$$\frac{\sin B}{16} = \frac{\sin 143°7'48''}{39}$$

Referring to Eq. (5),

$$\sin 143°7'48'' = \sin(180° - 143°7'48'') = \sin 36°52'12'' = 0.60000$$

$$\sin B = \frac{16 \times 0.6}{39} = 0.24615$$

$$B = 14°15'$$

$$A = 180° - (14°15' + 143°7'48'') = 22°37'12''$$

The cosine law is also useful when three sides are known.

Example 12. Solve for all the angles in the triangle $a = 13$ in, $b = 21$ in, $c = 20$ in. (Data are known to five significant digits.)

Use Eqs. (15) to (17).

$$a^2 = b^2 + c^2 - 2bc \cos A$$
$$13^2 = 21^2 + 20^2 - 2 \times 21 \times 20 \cos A$$
$$169 = 441 + 400 - 840 \cos A$$
$$= 841 - 840 \cos A$$
$$\cos A = {}^{672}\!/_{840} = 0.80000 \qquad\qquad \text{ANS.: } A = 36°52'11''$$

$$b^2 = a^2 + c^2 - 2ac \cos B$$
$$21^2 = 13^2 + 20^2 - 2 \times 13 \times 20 \cos B$$
$$441 = 169 + 400 - 520 \cos B$$
$$= 569 - 520 \cos B$$
$$\cos B = {}^{128}\!/_{520} = 0.24615 \qquad\qquad \text{ANS.: } B = 75°45'00''$$

$$c^2 = a^2 + b^2 - 2ab \cos C$$
$$20^2 = 13^2 + 21^2 - 2 \times 13 \times 21 \cos C$$
$$400 = 169 + 441 - 546 \cos C$$
$$= 610 - 546 \cos C$$
$$\cos C = {}^{210}\!/_{546} = 0.38462 \qquad\qquad \text{ANS.: } C = \underline{\quad 67°22'47''}$$
$$\text{CHECK: } = 179°59'58''$$

Example 13. Find the angle C in the triangle $a = 78$ in, $b = 35$ in, $c = 97$ in. (Data are known to five significant digits.)

$$c^2 = a^2 + b^2 - 2ab \cos C$$
$$97^2 = 78^2 + 35^2 - 2 \times 78 \times 35 \cos C$$
$$9,409 = 6,084 + 1,225 - 5,460 \cos C$$
$$= 7,309 - 5,460 \cos C$$

Here C is obviously obtuse, since c^2 actually is larger than $a^2 + b^2$; in other words, c^2 actually is larger than it would be if C were 90°. A scale drawing would indicate the same fact. Also,

$$-\cos C = \frac{2,100}{5,460}$$

Therefore, from Eq. (18),

$$\cos C = -2,100/5,460 = -0.38462$$
$$C = 180° - 67°22'47'' = 112°37'13''$$

EXERCISE 4

1. Verify the following:

(a) $\cos 42° = 0.74314$ (b) $\cos 138° = -0.74314$
(c) $\cos 130° = -0.64279$ (d) $\cos 179° = -0.99985$

2. Given the cosine of the angle, verify the following:
   (a) -0.89101 is the cosine of $153°$.   (b) 0.89101 is the cosine of $27°$.
   (c) -0.96126 is the cosine of $164°$.   (d) 0.96126 is the cosine of $16°$.

3. Show that with proper regard to signs

$$a = b \cos C + c \cos B$$

whether B is acute or obtuse.

4. Prove that in a triangle with sides a, b, and c,

$$a^2 + b^2 + c^2 = 2(ab \cos C + bc \cos A + ca \cos B)$$

5. Show that if

$$\frac{\cos A}{b} = \frac{\cos B}{a}$$

the triangle is either an isosceles triangle or a right triangle.

6. Prove that

$$\frac{\cos A}{a} + \frac{\cos B}{b} + \frac{\cos C}{c} = \frac{a^2 + b^2 + c^2}{2abc}$$

7. Prove that

$$\frac{c^2}{b} \cos B + \frac{b^2}{a} \cos A + \frac{a^2}{c} \cos C = \frac{a^4 + b^4 + c^4}{2abc}$$

Also show that the equation is dimensionally correct.

Find the side opposite the given angle, using the cosine law.

8. $a = 4.0000$ in $b = 5.0000$ in $C = 66°25'18''$
9. $b = 7.0000$ in $c = 10.000$ in $A = 45°34'23''$
10. $a = 20.000$ in $b = 35.000$ in $C = 60°$
11. $a = 7.0000$ in $b = 12.000$ in $C = 123°22'02''$
12. $c = 11.000$ in $b = 9.0000$ in $A = 31°47'19''$
13. $a = 13.000$ in $b = 18.000$ in $C = 36°52'11''$
14. $a = 5.0000$ in $b = 10.000$ in $C = 126°52'13''$
15. $a = 15.000$ in $b = 22.000$ in $C = 130°32'30''$
16. $a = 17.000$ in $b = 15.000$ in $C = 154°9'28''$
17. $b = 10.000$ in $c = 20.000$ in $A = 36°52.2'$

Find all the angles. Make three separate calculations, one for each angle. Find angles to seconds.

18. $a = 61.000$ in $\qquad$ $b = 87.000$ in $\qquad$ $c = 74.000$ in
19. $a = 65.000$ in $\qquad$ $b = 87.000$ in $\qquad$ $c = 44.000$ in
20. $a = 78.000$ in $\qquad$ $b = 95.000$ in $\qquad$ $c = 97.000$ in
21. $a = 25.000$ in $\qquad$ $b = 28.000$ in $\qquad$ $c = 17.000$ in
22. $a = 3.0000$ in $\qquad$ $b = 4.0000$ in $\qquad$ $c = 6.0000$ in
23. $a = 4.0000$ in $\qquad$ $b = 5.0000$ in $\qquad$ $c = 7.0000$ in
24. $a = 5.0000$ in $\qquad$ $b = 7.0000$ in $\qquad$ $c = 9.0000$ in
25. $a = 5.0000$ in $\qquad$ $b = 6.0000$ in $\qquad$ $c = 7.0000$ in
26. $a = 6.0000$ in $\qquad$ $b = 9.0000$ in $\qquad$ $c = 11.000$ in
27. $a = 13.000$ in $\qquad$ $b = 16.000$ in $\qquad$ $c = 19.000$ in
28. $a = 5.0000$ in $\qquad$ $b = 8.0000$ in $\qquad$ $c = 12.000$ in
29. $a = 5.8750$ in $\qquad$ $b = 3.2500$ in $\qquad$ $c = 8.5000$ in
30. Given an oblique triangle in which $A = 60°$, $a = 7$ in, and $c = 8$ in, calculate the side b. Instead of using the sine law, apply the cosine law and solve for b in the quadratic equation

$$a^2 = b^2 + c^2 - 2bc \cos A$$

Discuss the geometric significance of the answer you obtain for b.

EXERCISE 5

1. Solve for x in Fig 15.20. $\qquad$ 2. Solve for x in Fig. 15.21.
3. Solve for x in Fig. 15.22. $\qquad$ 4. Solve for x in Fig. 15.23.
5. Solve for x in Fig. 15.24. $\qquad$ 6. Solve for x in Fig. 15.25.
7. Solve for x in Fig. 15.26.

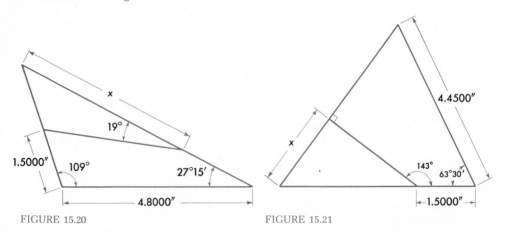

FIGURE 15.20 $\qquad\qquad\qquad\qquad$ FIGURE 15.21

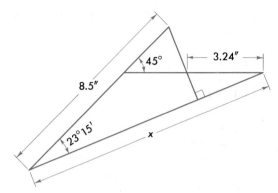

FIGURE 15.22

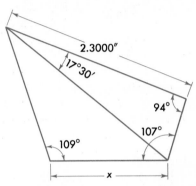

FIGURE 15.23

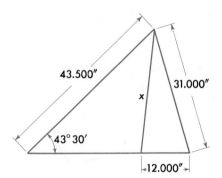

FIGURE 15.24

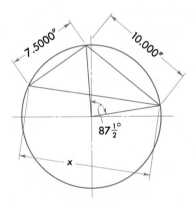

FIGURE 15.25

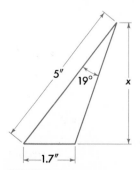

FIGURE 15.26

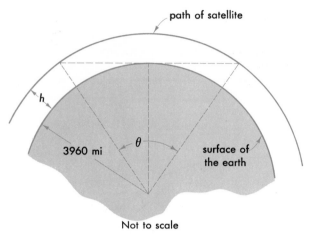

FIGURE 15.27

EXERCISE 6

1. An artificial earth satellite orbits the earth in the same plane as the earth's equator. It passes over a given point every 104 min. It is in sight for 17 min, the horizon being unrestricted. If the radius of the earth is 3,960 mi, what is the altitude of the satellite? (See Fig. 15.27.)

2. An earth satellite follows a circular orbit such that both the sun and the center of the earth lie in the plane of this orbit. The period of the satellite (time for one orbit) is 155 min. How many minutes of each period is the satellite in the earth's shadow? The diameter of the earth is 7,920 mi, and the altitude of the satellite is 1,980 mi. See Fig. 15.28.

3. The Telstar is an earth satellite in equatorial, synchronous orbit. It therefore appears to hover over one spot (on the equator) and has a period of 24 h. The required altitude for a 24-h period is 22,300 mi.

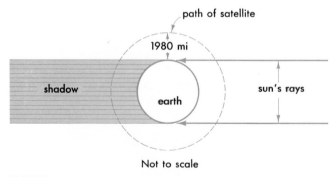

FIGURE 15.28

(a) Find the range of longitude (θ) over which a Telstar is visible at the equator. See Fig. 15.29.

(b) If there are three Telstars equally spaced 120° apart, over how many degrees of longitude at the equator are two Telstars in view at the same time?

(c) If a Telstar is due south of a ship at an angle of elevation of 77.5°, find the latitude of the ship.

The problems to follow refer to Fig. 15.30, in which

S = satellite
O = observing station on earth's surface
C = earth's center
P = subsatellite point on surface of earth
e = angle of elevation of satellite from observing station at O
h = height of satellite
slant range = range from O to S, measured by Doppler methods

Note. *Points S, O, C, and P in Fig. 15.30 all lie in the same plane.*

Taking the radius of the earth to be 3,960 statute mi, the great circle circumference is given by

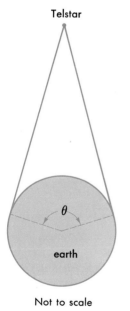

Telstar

earth

Not to scale

FIGURE 15.29

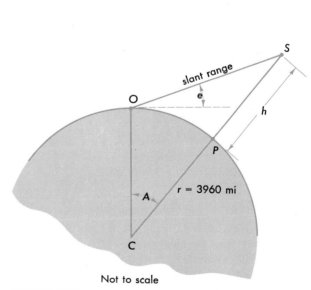

Not to scale

FIGURE 15.30

Circumference $= 2 \times \pi \times 3{,}960$

$\qquad\qquad\quad = 24{,}881$ statute mi

The arc length $\overset{\frown}{OP}$ is then given by

$$\overset{\frown}{OP} = \frac{A^\circ}{360^\circ} \times 24{,}881 \tag{38}$$

Example 14. An earth satellite has an angle of elevation of 57° as observed from point O in Fig. 15.30. The slant range OS at that time was 907 mi. Find the altitude h above the subsatellite point P and the great circle distance $\overset{\frown}{OP}$.

From Fig. 15.30 the angle SOC is $57^\circ + 90^\circ = 147^\circ$. Then, by the cosine law,

$(h + 3{,}960)^2 = (907)^2 + (3{,}960)^2 - 2 \times 907 \times 3{,}960 \times \cos 147^\circ$

$(h + 3{,}960)^2 = 822{,}649 + 15{,}681{,}600 + 2 \times 907 \times 3{,}960 \times 0.83867$

$\qquad\qquad\quad = 16{,}504{,}249 + 6{,}024{,}536$

$(h + 3{,}960)^2 = 22{,}528{,}785$

$\quad h + 3{,}960 = \sqrt{22{,}528{,}785} = 4{,}746$ mi $\qquad$ (to the nearest mile)

Then

$h = 786$ mi $\qquad$ (to the nearest mile)

By the sine law

$$\frac{\sin A}{907} = \frac{\sin 147^\circ}{4{,}746} = \frac{0.54464}{4{,}746}$$

$$\sin A = \frac{907 \times 0.54464}{4{,}746} = 0.10408$$

$$A = 5.97^\circ$$

Following Eq. (38),

$$\overset{\frown}{OP} = \frac{5.97}{360} \times 24{,}881 = 413 \text{ statute mi}$$

4. If $h = 2{,}970$ mi and $\overset{\frown}{OP} = 1{,}980$ statute mi, find angle e and the slant range.
5. If $\overset{\frown}{OP} = 530$ mi and $e = 32^\circ$, find h and the slant range.
6. If $e = 24^\circ$ and the slant range is 720 mi, find h and the great circle distance $\overset{\frown}{OP}$.
7. If h is 940 mi and $e = 28.3^\circ$, find $\overset{\frown}{OP}$ and the slant range.
8. If $h = 660$ mi and the slant range is 1,210 mi, find angles e and A.
9. In Prob. 8 find the maximum slant range (limit of visibility) for an altitude of 660 mi.

10. If $\overset{\frown}{OP} = 425$ statute mi and the slant range is 710 mi, find angles e and A.

15.8 Areas of Triangles (Fig. 15.31)

Given the Base and Altitude

From geometry,

$$A_t = \tfrac{1}{2}bh \tag{39}$$

where A_t = area of triangle
b = base of triangle
h = altitude of triangle

Given Two Sides and the Included Angle

Let the two known sides be c and b. Let the known angle be A.

$$h = c \sin A$$

Substituting the above value of h in Eq. (39),

$$A_t = \tfrac{1}{2}b \times c \sin A \tag{40}$$

Similarly,

$$A_t = \tfrac{1}{2}ab \sin C \tag{41}$$
$$= \tfrac{1}{2}ac \sin B \tag{42}$$

Given Two Angles and Any Side

$$\frac{\sin A}{\sin B} = \frac{a}{b}$$

[See Eq. (2), page 445.] Then we may write

$$b = \frac{a \sin B}{\sin A} \tag{43}$$

Substituting the value of b above in Eq. (41),

$$A_t = \frac{a^2 \sin B \times \sin C}{2 \sin A} \tag{44}$$

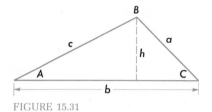

FIGURE 15.31

Similarly,

$$A_t = \frac{b^2 \sin A \times \sin C}{2 \sin B}$$

$$= \frac{c^2 \sin A \times \sin B}{2 \sin C}$$

Given Three Sides

If the sides of a triangle are a, b, and c, it can be proved that the area of the triangle is given by

$$\text{Area} = \sqrt{s(s-a)(s-b)(s-c)} \qquad\qquad \textbf{(45)}$$

where

$$s = \tfrac{1}{2}(a + b + c)$$

EXERCISE 7

Find the areas of the following triangles:

	a	b	c	A	B	C
1.	38.000		56		49°	
2.	15.000	12.000	13.000			
3.		65.000		36°	75°	
4.		48.000	84.000	73°		
5.	124.00				115°	26°
6.	34.000	36.000	38.000			
7.			682.00	56°32′	65°10′	
8.	26.800	9.7000				102°18′
9.	364.00	812.00	526.00			

10.	3.1420				31°9.7′	62°54.5′
11.	42.700	5.6000	38.500			
12.		103.00	264.00	18°42′16″		
13.			4,122.0		154°18′27″	12°12′12″
14.	56.240	43.290	27.570			
15.	3,721.0		4,609.0		27°8.7′	

applications of numerical trigonometry

In this chapter we shall illustrate some of the applications of numerical trigonometry to civil- and mechanical-engineering problems.

16.1 Plane Surveying

We shall limit any discussion of surveying to *plane surveying*, i.e., to surveys conducted on the assumption that the earth is flat over the area under consideration. The errors involved in this approximation are small, provided the area covered by the survey is small.

16.2 Measurement of Angles

Angles are measured directly in the field with either of two instruments: the engineer's transit or the surveyor's compass. For the construction and operation of these instruments the student is referred to any standard surveying text.

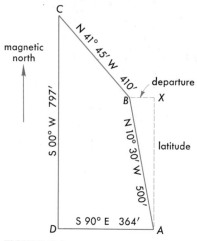

FIGURE 16.1

16.3 Mapping

A typical problem of the land surveyor is to measure the length and direction of the boundary lines of a parcel of land in order to draw a scale map of the parcel and calculate the area.

Figure 16.1 shows the map of a small parcel of land plotted from the field notes shown in Table 16.1.

The surveyor started from point A and found by observation that the line AB makes an angle of $10°30'$ with the magnetic north. Furthermore, in moving along the line AB, he moves north and also west. Hence the bearing is recorded N10°30'W. A property line is not completely described unless both *direction* and *length* are given. This particular property line was measured (probably with a tape) and found to be 500 ft long.

TABLE   16.1

Course		
AB	N10°30'W	500.00 ft
BC	N41°45'W	410.00 ft
CD	S00°00'W	797.00 ft
DA	S90°E	364.00 ft

All bearings from magnetic north.

As the notes read, we may assume that the surveyor traversed this survey in a counterclockwise direction. Therefore he moved west and north along *AB* and west and north also along *BC*. Along *CD* he moved directly south, and along *DA* he moved directly east, returning to the starting point.

We would be justified in assuming that he measured linear dimensions to the nearest 0.01 ft. For our purpose we shall assume that bearings were measured to the nearest minute.

16.4 Latitudes and Departures

In Fig. 16.1 the line *XB* is called the *departure* of the course *AB*. Since the point *B* is west of the point X, the line *XB* is a westerly departure.

Departure is defined as the distance by which the second extremity of a course is east or west of the first extremity.

The departure of the course *AB* is

$$500 \sin 10°30' = 500 \times 0.18224 = 91.12 \text{ ft west}$$

The departure of the course *BC* is

$$410 \sin 41°45' = 410 \times 0.66588 = 273.01 \text{ ft west}$$

The departure of the course *CD* is zero, since in moving along this line the surveyor moves neither east nor west.

The departure of the course *DA* is 364 ft, since in moving along this line he moves directly east.

Adding the departures in a westerly direction, we find that the total westerly departure is

$$91.12 + 273.01 = 364.13 \text{ ft west}$$

In other words, if we take the survey notes at their face value, they indicate that the surveyor moved 364.13 ft westward and 364.00 ft eastward and arrived back at the starting point. This discrepancy may be explained by the fact that it is impossible to measure with absolute precision. Yet the departures check reasonably well.

If the survey had proceeded in a clockwise direction, the departure of the course *CB* would have been an easterly departure, and the departure of the course *BA* would have been an easterly departure also.

In Fig. 16.1 the line *AX* is called the *latitude* of the course *AB*. Since the point X is north of the point *A*, the line *AX* is a northerly latitude.

Latitude is defined as the distance by which the second extremity of a course is north or south of the first extremity.

The latitude of the course AB is

500 ft cos $10°30'$ = 500 ft $\times$ 0.98325 = 491.62 ft north

The latitude of the course BC is

410 ft cos $41°45'$ = 410 ft $\times$ 0.74606 = 305.88 ft north

The latitude of the course CD is 797 ft south, since in moving along this line the surveyor moves directly south.

The latitude of the course DA is zero, since in moving along this line the surveyor moves neither north nor south.

Adding the northerly latitudes, we find that the total latitude is

491.62 ft + 305.88 ft = 797.50 ft north

Again, if we take the survey notes at their face value, they indicate that the surveyor moved 797.5 ft north and 797.0 ft south. As before, we assume that this discrepancy occurs because we cannot measure bearings and lengths with absolute precision.

This suggests a method of checking the accuracy of the field work. We know that if the surveyor ended the survey at the starting point, he must have moved south exactly the same distance that he moved north. Also he must have moved east exactly the same distance as he moved west. Stated more formally, *if northerly latitudes and easterly departures are considered positive, and southerly latitudes and westerly departures are considered negative, the algebraic sum of the departures must be zero and the algebraic sum of the latitudes must be zero if the traverse is to close on itself.*

The linear error of closure is, by definition,

$$e = \sqrt{(\text{error in latitude})^2 + (\text{error in departure})^2} \tag{1}$$

The calculations for latitudes and departures are repeated below in tabular form. The linear error is also calculated. While the latitudes and departures are calculated here to the nearest $\frac{1}{100}$ ft, a greater or fewer number of places may be retained in practical problems, depending on circumstances.

There are statistical methods of distributing the error of closure among the several courses. For our purpose such procedures would be of little importance.

Course	Bearing	Distance, ft	N lat, ft	S lat, ft	E dep, ft	W dep, ft
AB	N10°30'W	500	491.62			91.12
BC	N41°45'W	410	305.88			273.01
CD	S00°	797		797.00		
DA	S90°00'E	364			364.00	
			797.50	797.00	364.00	364.13
			797.00			364.00
		Error in lat	0.50		Error in dep	0.13

$$\text{Linear error} = \sqrt{(0.50)^2 + (0.13)^2} = 0.52'$$

16.5 Azimuth Angles

The bearings of courses can be recorded in terms of *azimuth angles*. Azimuth angles are measured from some line of reference to the line in question. Often zero degrees azimuth is taken to be true south. Azimuth angles are measured in a clockwise direction from true south. Angles are also measured from north in aviation, astronomy, and meteorology.

On maps and in field notes the line of reference should be identified in some such fashion as "00°00' azimuth true south." In terms of azimuth angles, the notes in Table 16.1 would be as follows:

Course	Azimuth 0° true south	Distance, ft
AB	169°30'	500
BC	138°15'	410
CD	00°00'	797
DA	270°00'	364

16.6 Interior Angles

The sum of the interior angles of any polygon is $180° (n - 2)$. Therefore, if we subtract 2 from the number of courses and multiply by 180°, we obtain precisely the sum of the interior angles. This total can be compared with a similar total obtained from field notes.

In Fig. 16.1 the interior angles are checked as follows:

$$\angle DAB = 90° - 10°30' \qquad\qquad = 79°30'$$
$$\angle CBA = 180° - 41°45' + 10°30' = 148°45'$$
$$\angle DCB = \qquad\qquad\qquad\qquad 41°45'$$
$$\angle CDA = \qquad\qquad\qquad\qquad 90°$$
$$\overline{\qquad\qquad\qquad\qquad\quad 358°120' = 360°}$$

The figure is a four-sided polygon, the sum of whose interior angles is $(4 - 2)\ 180° = 360°$. Thus the check is perfect.

EXERCISE 1

1. Plot the following traverses to a scale. Compute the error in latitudes and the error in departures, using five-place tables.

Course	Length, ft	Bearing
AB	683.57	N11°49′E
BC	221.63	N53°40′W
CD	412.90	S88°36′W
DE	513.27	S 1°00′E
EF	225.41	S33°17′E
FA	330.79	S74°30′E

2. Course	Length, ft	Bearing
AB	4,181.5	N10°08′E
BC	3,872.1	N58°28′W
CD	4,747.0	S41°15′W
DE	3,751.7	S16°10′E
EA	4,762.8	N77°30′E

3. Find the bearing and the length of the line DA in Prob. 1.
4. Plot the following traverse to scale. Calculate the error in the latitudes and departures.

Course	Length, ft	Bearing
AB	885.84	N10°18′W
BC	542.67	S69°45′W
CD	875.04	S12°15′E
DE	230.77	S84°00′E
EF	210.03	N14°00′E
FG	162.70	N86°20′E
GH	249.95	S14°00′W
HA	245.00	N24°00′E

5. Plot the following traverse to scale, and calculate the error in latitude and departure.

		Azimuth
Course	Length, ft	0° true south
AB	578.63	265°45′
BC	206.30	174°36′
CD	250.22	106°20′
DE	452.51	226°30′
EF	137.00	180°00′
FG	524.86	91°45′
GA	792.18	8°45′

6. The owner of the tract of land described in Prob. 5 wishes to buy the triangular piece of land CDE. What is the length of the line CE, and what is its bearing? How much area would this purchase add to his land?

7. The owner of the tract of land described in Prob. 4 wishes to sell his holdings south of the line FG. Two new bounds J and K will have to be established. J will be at the intersection of the line CD and the westward extension of the line FG. K will be at the intersection of the line BA and the eastward extension of the line FG. Find the distances CJ and BK.

16.7 Calculation of Areas

A useful method of finding the area enclosed by a traverse consisting of straight lines is illustrated below.

Ordinarily, it is more convenient to establish reference coordinates so that the entire survey is placed in one quadrant, say the northeast quadrant, as shown in Fig. 16.2. The corner points are numbered consecutively around the traverse in a counter-

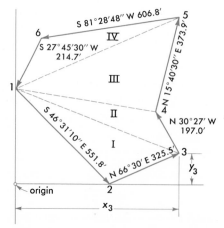

FIGURE 16.2

clockwise direction. The area is given by the equation (see Sec. 8.17)

$$A_t = \tfrac{1}{2}[x_1(y_2 - y_6) + x_2(y_3 - y_1) + x_3(y_4 - y_2) + x_4(y_5 - y_3)$$
$$+ x_5(y_6 - y_4) + x_6(y_1 - y_5)] \qquad (2)$$

To find the area enclosed by a traverse, multiply the x distance of each corner by the difference between the y distance of the following corner and the y distance of the preceding corner. The sum of these products is twice the area.

The following convention is sometimes used:

$$\text{Area} = \tfrac{1}{2}\left(\frac{x_1}{y_1} \times \frac{x_2}{y_2} \times \frac{x_3}{y_3} \times \frac{x_4}{y_4} \times \frac{x_5}{y_5} \times \frac{x_6}{y_6} \times \frac{x_1}{y_1}\right) \qquad (3)$$

Here we multiply the coordinates connected by solid lines and separately multiply the coordinates connected by dotted lines. Then we find the sum of the solid-line products and the sum of the dotted-line products. The difference between these two sums is twice the area and is a positive number.

Example 1. Find the area enclosed by the traverse shown in Fig. 16.2.

First we shall find the latitude and departure of each course by the methods already discussed in Sec. 16.4. A tabulation of the results is shown below:

Course	Bearing	Distance, ft	N lat, ft	S lat, ft	E dep, ft	W dep, ft
1-2	S46°31'10"E	551.8		379.7	400.4	
2-3	N66°30'E	325.5	129.8		298.5	
3-4	N30°27'W	197.0	169.8			99.8
4-5	N15°40'30"E	373.9	360.0		101.0	
5-6	S81°28'48"W	606.8		89.9		600.1
6-1	S27°45'30"W	214.7		190.0		100.0
			659.6	659.6	799.9	799.9

The latitudes check perfectly within the limits of the precision to which the data were obtained. The departures check within 0.1 ft. The coordinates of the several corners are found by reference to Fig. 16.2 and the tabulation above.

$x_1 = 000.0 \qquad y_1 = 379.7$

$x_2 = 400.4 \qquad y_2 = 000.0$

$x_3 = 698.9 \qquad y_3 = 129.8$

$x_4 = 599.1 \qquad y_4 = 299.6$

$x_5 = 700.1 \qquad y_5 = 659.6$

$x_6 = 100.0 \qquad y_6 = 569.7$

Substituting the x and y values just found in Eq. (2),

$$A_t = \tfrac{1}{2}[0(0 - 569.7) + 400.4(129.8 - 379.7) + 698.9(299.6 - 0)$$
$$+ 599.1(659.6 - 129.8) + 700.1(569.7 - 299.6) + 100(379.7 - 659.6)]$$
$$= \tfrac{1}{2}[(-400.4 \times 249.9) + (698.9 \times 299.6) + (599.1 \times 529.8)$$
$$+ (700.1 \times 270.1) - (100 \times 279.9)]$$
$$A_t = \tfrac{1}{2}(-100{,}060 + 209{,}390 + 317{,}403 + 189{,}097 - 27{,}990)$$
$$= \tfrac{1}{2}(587{,}840) = 293{,}900 \text{ ft}^2 \qquad \text{(to four significant figures)}$$

Using Eq. (3),

$$A_t = \frac{1}{2}\left(\frac{0}{379.7} \diagdown \frac{400.4}{0} \diagdown \frac{698.9}{129.8} \diagdown \frac{599.1}{299.6} \diagdown \frac{700.1}{659.6} \diagdown \frac{100.0}{569.7} \diagdown \frac{0}{379.7}\right)$$

Carrying out the calculations indicated above,

$0 \times 569.7 =$	00,000	$100 \times 379.7 =$	37,970
$659.6 \times 100.0 =$	65,960	$700.1 \times 569.7 =$	398,847
$700.1 \times 299.6 =$	209,750	$599.1 \times 659.6 =$	395,166
$599.1 \times 129.8 =$	77,763	$698.9 \times 299.6 =$	209,390
$698.9 \times 0 =$	00,000	$129.8 \times 400.4 =$	51,972
$400.4 \times 379.7 =$	152,032	$0 \times 0 =$	00,000
	505,505		1,093,345

$$A_t = \frac{1{,}093{,}345 - 505{,}505}{2}$$

$$= \frac{587{,}840}{2} = 293{,}900 \text{ ft}^2 \qquad \text{(to four significant figures)}$$

EXERCISE 2

1. Find the area enclosed by the traverse shown in Fig. 16.1.
2. Find the area enclosed by the traverse described in Prob. 1, Exercise 1.
3. Find the area enclosed by the traverse described in Prob. 2, Exercise 1.

16.8 Measurement of Distances in a Vertical Plane

Practically all the distances and angles used in surveying are measured in either a vertical or a horizontal plane.

The property line ab in Fig. 16.3 slopes uphill. However, the distance recorded for the length of this property line is not the actual length ab but the horizontal projection of the line ab, which is ax. Here, ab and ax designate the lengths of lines ab and ax and not the product of two numbers.

FIGURE 16.3

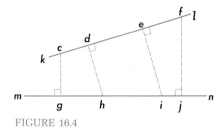

FIGURE 16.4

In general, the projection of one line on another is defined in Fig. 16.4. Here gj is the projection of the line cf on the line mn, and de is the projection of the line hi on the line kl. Notice the positions of the right angles.

Ordinarily in the field it is not difficult to measure the distance ab in Fig. 16.3; neither is it difficult to measure the difference in elevation between a and b. As a rule, it is a relatively simple matter to measure the vertical angle ϕ. It may, however, be somewhat more difficult to arrive at the distance ax by direct measurement in the field.

The distance ax in Fig. 16.3 can be calculated from field notes, provided that the length ab and the difference in elevation between a and b are known.

Following the Pythagorean theorem,

$$ax = \sqrt{(ab)^2 - (bx)^2} \tag{4}$$

If the angle ϕ is small, the approximation developed below may be perfectly satisfactory and will effect a considerable saving of time, particularly when there are many data to be reduced. Referring to Fig. 16.3,

$$xy = ab - ax \tag{5}$$

$$(bx)^2 = (ab)^2 - (ax)^2 = (ab - ax)(ab + ax)$$

$$ab - ax = \frac{(bx)^2}{ab + ax} \tag{6}$$

Substituting Eq. (6) in Eq. (5),

$$xy = \frac{(bx)^2}{ab + ax}$$

or if ab and ax are nearly equal, as they will be if ϕ is small,

$$xy \approx \frac{(bx)^2}{2(ab)} \qquad \text{[See Eq. (68) in Sec. 13.23]}$$

The distance xy is the correction which, when subtracted from ab, gives the horizontal projection ax. Therefore

$$ax \approx ab - \frac{(bx)^2}{2(ab)} \tag{8}$$

If the field notes give the angle ϕ instead of the difference in elevation between a and b and if ϕ is small, the correction bx can be found by the approximation developed below.

For small angles, the sine of the angle is very nearly proportional to the angle. That is,

$$\sin \phi \approx k\phi \tag{9}$$

where k is the proportionality constant.

Up to almost 8° in the five-place tables, each increase of 1′ increases the sine of the angle 0.00029. That is, within the limits of the five-place table and up to 7°48′,

$$\sin \phi \approx 0.00029\phi \tag{10}$$

where ϕ is measured in minutes. Or

$$\sin \phi \approx 0.00029 \times 60 \times \phi = 0.0174\phi \tag{11}$$

where ϕ is measured in degrees.

In Fig. 16.3,

$$bx = (ab) \sin \phi \tag{12}$$

or substituting Eq. (11) in Eq. (12),

$$bx \approx 0.0174 \times ab \times \phi \tag{13}$$

where ϕ is measured in degrees.

Following Eq. (7),

$$xy \approx \frac{(bx)^2}{2(ab)} \approx \frac{0.0174^2 \times (ab)^2 \times \phi^2}{2(ab)} \tag{14}$$

$$xy \approx \frac{0.00030275 \times (ab) \times \phi^2}{2} \tag{15}$$

Within about 1 percent, Eq. (16) is valid.

$$xy \approx \frac{3}{2} \times \phi^2 \times 10^{-4}(ab) \tag{16}$$

Another method may be used if a table of versines is available. By definition,

$$\text{vers } \phi = 1 - \cos \phi \tag{17}$$

In Fig. 16.3

$$xy = ab - (ab) \cos \phi = (ab)(1 - \cos \phi) \tag{18}$$
$$xy = (ab) \text{ vers } \phi \tag{19}$$

Example 2. In Fig. 16.3, if $ab = 375.00$ ft and $bx = 30.00$ ft, calculate ax, using both formulas

$$ax = \sqrt{(ab)^2 - (bx)^2}$$

and

$$ax \approx ab - \frac{(bx)^2}{2ab}$$

Using the Pythagorean theorem,

$$ax = \sqrt{(375)^2 - (30)^2} = 373.80 \text{ ft} \qquad \text{(to five figures)}$$

Using the approximate formula,

$$ax \approx 375 - \frac{(30)^2}{2 \times 375} = 375 - 1.20 = 373.80 \text{ ft} \qquad \text{(to five figures)}$$

Note that the accuracy required of this correction factor 1.20 is not beyond the capabilities of the slide rule.

16.9 Surveying around Obstructions

Equation (11) is convenient in another connection (Fig. 16.5). If ϕ is small and is

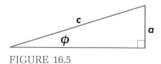

FIGURE 16.5

measured in degrees,

$$a \approx 0.0174 \times c \times \phi \tag{20}$$

to a close approximation.

A problem sometimes arises in which it is desired to locate a point x (Fig. 16.6) on an extension of the line ab. The exact distance ax is not important, but the point x must be located so that the points a, b, and x are on the same straight line. Between b and x there are obstructions such as houses, trees, and boulders, so that a direct line of sight cannot be used. The distances bc, cd, de, and ex can be measured. Also the angles u, v, w, and y can be measured. The surveyor runs a traverse bc, cd, and de, and at the point e he sets the angle y at some convenient value. He wishes to find the distance ex such that x will fall on the continuation of the line ab. The offsets fc, gd, he, and ix can be calculated, to a close approximation, as follows:

$$fc \approx 0.0174(bc)u \tag{21}$$
$$gd \approx 0.0174(cd)v \tag{22}$$
$$he \approx 0.0174(de)w \tag{23}$$
$$ix \approx 0.0174(ex)y \tag{24}$$

If the offsets downward are considered to be negative and the offsets upward are considered to be positive, then from the geometry of the figure

$$fc + he = gd + ix \tag{25}$$

Substituting Eqs. (21) to (24) in Eq. (25) and dividing out the 0.0174,

$$(bc)u + (de)w = (cd)v + (ex)y \tag{26}$$

In Eq. (26) the angles u, w, v, and y may be in any units, but we shall use minutes. To find ex,

$$ex = \frac{(bc)u + (de)w - (cd)v}{y} \tag{27}$$

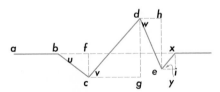

FIGURE 16.6

Example 3. Find ex in Fig. 16.6 if $bc = 120$ ft, $cd = 375$ ft, $de = 500$ ft, $\angle u = 50'$, $\angle v = 200'$, $\angle w = 240'$, and $\angle y = 80'$.

Substituting in Eq. (27),

$$ex = \frac{(50 \times 120) + (240 \times 500) - (200 \times 375)}{80}$$

$$= 638 \text{ ft}$$

EXERCISE 3

1. Plot a graph of $\sin \phi$ vertically and ϕ in minutes horizontally for angles between $0'$ and $500'$.
   (a) What is the shape of this graph?
   (b) What is the slope?
   (c) Over this range about how much does $\sin \phi$ change for a change of each $60'$, in ϕ?
2. In Fig. 16.3, if $ab = 300.00$ ft and the difference in elevation between points a and b is 15 ft, find the distance ax, using Eq. (8). Find the percentage of error that results from using this approximation.
3. In Fig. 16.3 if $ab = 650.00$ ft and $\phi = 5°$, find ax, using Eq. (16).
4. Solve Prob. 3 using a table of versines (if possible) and Eq. (19).
5. In Fig. 16.6 find the distance ex if $bc = 100$ ft, $cd = 200$ ft, $de = 150$ ft, $\angle u = 1°$, $\angle v = 1°30'$, $\angle w = 2°$, and $\angle y = 45'$.

EXERCISE 4

The problems in this exercise have their application in measuring inaccessible distances, highway curves, etc. It is beyond the scope of this book to dwell in detail on the engineering principles involved, but the student should have no particular difficulty in sensing the general nature of the applications.

1. Referring to Fig. 16.7, prove that

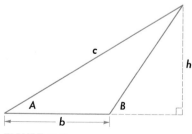

FIGURE 16.7

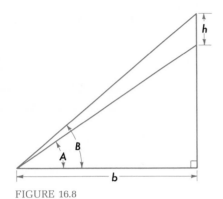

FIGURE 16.8

$$h = \frac{b}{\cot A - \cot B}$$

2. If b in Prob. 1 is 362 ft, $A = 26°48'$, and $B = 48°32'$, find h.
3. In Fig. 16.8 show that

$$h = b(\tan B - \tan A)$$

4. At point e on the surface of the earth a satellite is observed directly overhead at point B (Fig. 16.9). Find the altitude h of the satellite to the nearest mile. Use slide rule for multiplication and division.
5. If in Prob. 3 the side $b = 986$ ft, angle $A = 15°46'$, and angle $B = 23°18'$, find h.
6. Find ϕ and x in Fig. 16.10.

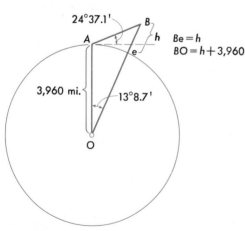

FIGURE 16.9

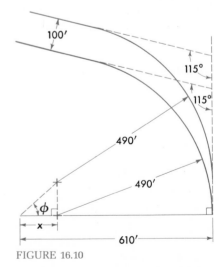

FIGURE 16.10

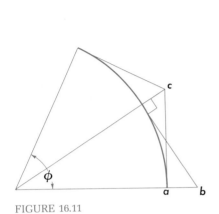

FIGURE 16.11

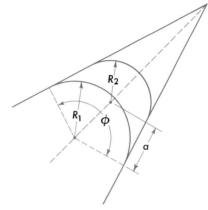

FIGURE 16.12

7. In Fig. 16.11 find an expression for the length of the line ab in terms of the line ac and the angle ϕ.

8. Referring to Fig. 16.12, show that $R_1 - R_2 = a \cot (\phi/2)$.

The following problems may be done more conveniently by the use of the oblique-triangle formulas:

9. The inaccessible distance AB across the pond is to be found from the survey data shown in Fig. 16.13. The given distances were measured from point O by means of stadia.

10. The points A and B in Fig. 16.14 are on opposite sides of the river and inaccessible from points x and y. Find the distance AB from the survey notes below:

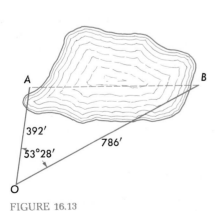

FIGURE 16.13

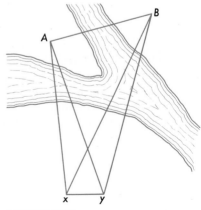

FIGURE 16.14

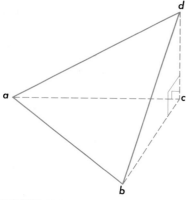

FIGURE 16.15

$$xy = 450 \text{ ft}$$
$$\angle Ayx = 32°$$
$$\angle Axy = 129°$$
$$\angle Bxy = 43°$$
$$\angle Byx = 113°$$

11. An aircraft at d (Fig. 16.15) is located from two ground stations a and b by the following data:

$$\angle dac = 25°37'$$
$$\angle dbc = 30°40'$$
$$\angle cab = 40°15'$$
$$ab = 5{,}676 \text{ yd}$$

Find da, ac, and bc.

12. Three equally spaced Telstar satellites are in equatorial, synchronous orbit. This means that the plane of the satellite orbit is the same as the plane of the equator and that the satellites orbit once in 24 h; thus each appears to hover over one spot.
    (a) Find the degrees of longitude overlap in coverage at the equator. (This is the arc over which two satellites are in sight at the same time.)
    (b) Find the northern limit (°N lat) of coverage.
    (c) How many minutes out of the 24 h is a satellite in the earth's shadow?
    The earth's radius is 3,960 mi, and the satellites orbit at an altitude of 22,300 mi.

16.10 Applications to Machine- and Tool-design Problems

The problems in the following sections will be of particular interest to students of machine and tool design. Unfortunately, no definite rules can be given for solving this type of problem. The student will have to draw on his knowledge of basic mathematics and his own native ingenuity, building up experience as he goes along. However, the few general suggestions given below may be of some help.

In practically all the problems in Exercise 5, the student will have to draw certain construction lines. In general, it is suggested that he consider drawing these lines perpendicular or parallel to known dimensions. If there are arcs of circles involved, it may be helpful to connect the centers. Often it is helpful to draw a construction line from the center of curvature of one arc and tangent to another. Figure 16.16 is an idealized drawing of a situation frequently encountered, in which we are given oa, ab, and bc. It is required to find ϕ. The procedure in solving this problem is first to find the angle boa, using the sides ab and oa. Second, the side ob is calculated and used in connection with the side bc to find the angle boc. The angle ϕ will then be the sum of aob and boc. Illustrated below is a somewhat more direct method, in which the hypotenuse ob is not actually calculated.

$$\tan boa = \frac{ba}{oa}$$

$$ob = \frac{oa}{\cos boa}$$

$$\tan cob = \frac{cb}{oa/\cos boa} = \frac{(cb)\cos boa}{oa}$$

$$\angle boa + \angle cob = \angle coa = \phi$$

This method is particularly useful when cb and oa have common factors.

In Fig. 16.17 it is desired to calculate the angle by use of the sine. Decimal equivalents could be used, but common fractions make the problem much easier.

$$\sin \phi = \frac{7/8}{25/8} = \frac{7}{8} \times \frac{8}{21} = \frac{1}{3} = 0.33333$$

If the side ob is to be calculated by means of the Pythagorean theorem, again decimal equivalents could be used, but it is much better to reason as follows: Use $\frac{1}{8}$ in as the

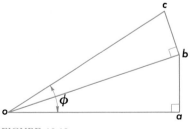

FIGURE 16.16

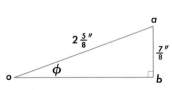

FIGURE 16.17

basic unit in the problem. See Sec. B.13. Hence the hypotenuse will be 21 units long, and the side ab will be 7 units long; therefore

$$ob = \tfrac{1}{8}\sqrt{21^2 - 7^2} = \tfrac{1}{8}\sqrt{441 - 49} = \frac{19.799}{8} = 2.475 \text{ in}$$

Figure 16.18 illustrates a type of problem which is very frequently done incorrectly. A solution is often attempted assuming that the figure obc is a right triangle. As a matter of fact, it is not even any kind of a triangle. A triangle is a figure formed by three intersecting straight lines. The line oc is not a straight line. It is a combination of the straight line oa and the arc of a circle ac; $\sin \phi$ equals ab/ob and not bc/oc.

As far as tolerances and limits are concerned, no attempt is made to follow commercial shop practices in Exercise 5. Nominal dimensions are given on the drawings, and it will be assumed that their accuracy is consistent with the use of five-place tables. Angles will be computed either to tenths of minutes or to seconds at the discretion of the instructor. A deviation from the answer of three or four seconds is not usually significant.

EXERCISE 5

1. Find the angle ϕ in Fig. 16.19.
2. Find the angle ϕ in Fig. 16.20.

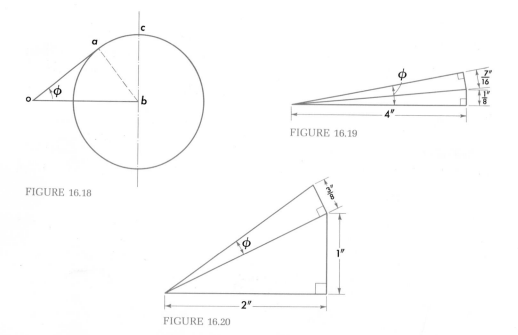

FIGURE 16.18

FIGURE 16.19

FIGURE 16.20

3. Find the diameter d in Fig. 16.21.
4. Find ϕ in Fig. 16.22.
5. Find ϕ in Fig. 16.23.
6. Find ϕ in Fig. 16.24.
7. Find ϕ in Fig. 16.25.
8. Find ϕ in Fig. 16.26.

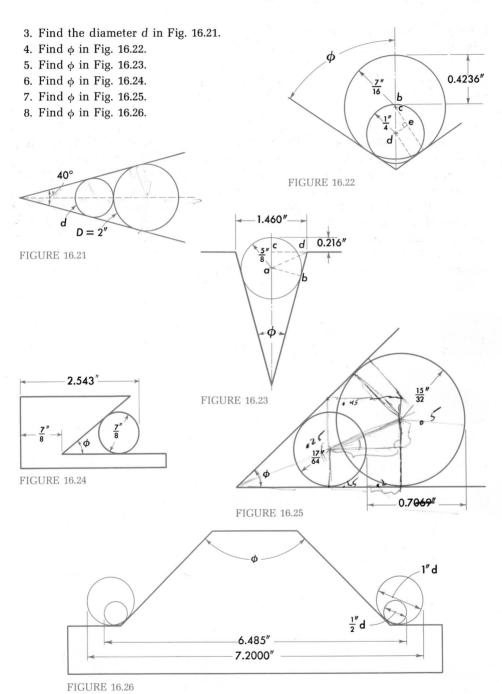

FIGURE 16.22

FIGURE 16.21

FIGURE 16.23

FIGURE 16.24

FIGURE 16.25

FIGURE 16.26

9. Find the distance $(D - d)$ in Fig. 16.27.
10. Find ϕ in Fig. 16.28.
11. Find ϕ in Fig. 16.29.
12. Find R in Fig. 16.30.
13. Find R in Fig. 16.31.

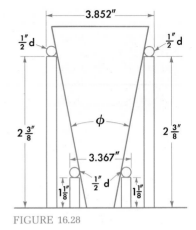

FIGURE 16.28

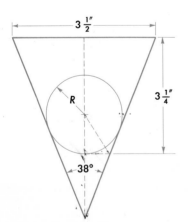

FIGURE 16.27

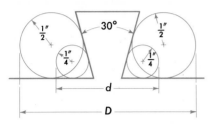

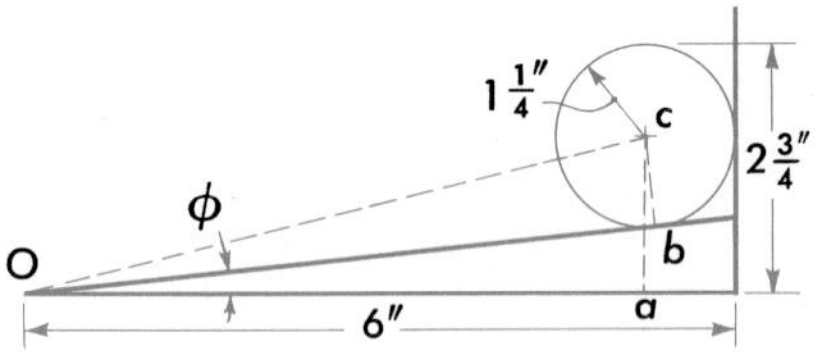

FIGURE 16.29

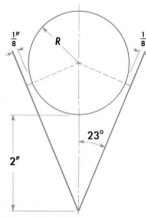

FIGURE 16.30

FIGURE 16.31

14. Find R in Fig. 16.32.
15. Find ϕ in Fig. 16.33.
16. Find ϕ in Fig. 16.34.
17. Find x in Fig. 16.35.
18. Find ϕ in Fig. 16.36.
19. Find ϕ in Fig. 16.37.

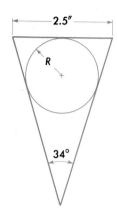

FIGURE 16.32

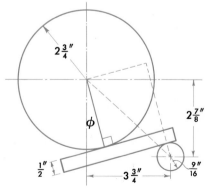

FIGURE 16.33

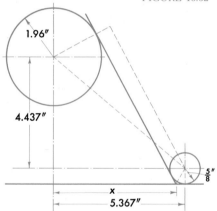

FIGURE 16.35

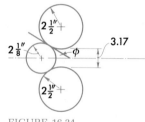

FIGURE 16.34

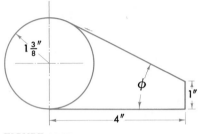

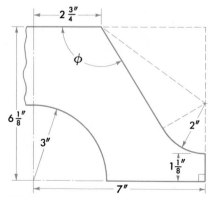

FIGURE 16.36

FIGURE 16.37

20. Find the angle ϕ and the lengths x and y in Fig. 16.38.
21. Find x in Fig. 16.39. The five holes in the $2\frac{9}{16}$-in circle are evenly spaced.
22. Find R in Fig. 16.40.
23. Find R in Fig. 16.41 if $oa = 2.400$ in, $ob = 4.000$ in, $oc = 5.000$ in, and $\angle aob = \angle boc$.
24. Find x and y in Fig. 16.42.

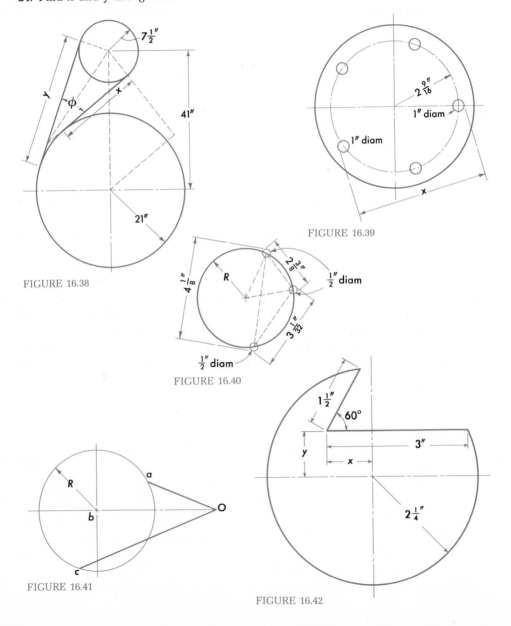

FIGURE 16.38

FIGURE 16.39

FIGURE 16.40

FIGURE 16.41

FIGURE 16.42

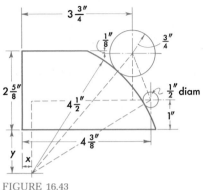

FIGURE 16.43

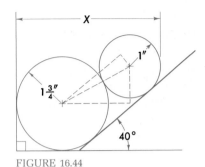

FIGURE 16.44

25. Find x and y in Fig. 16.43.
26. Find x in Fig. 16.44.

16.11 Solid Figures

Many of the problems in this chapter have involved more than one triangle, but these triangles have usually been in the same plane.

There is a wide range of practical applications in which related triangles lie in different planes. A thorough treatment of this type of problem would consume an unwarranted amount of textbook space and classroom time. Therefore the few problems which are given in this section are intended only to introduce the student to this highly specialized and restricted field of applications. The student who has reason to be interested in this particular type of problem is referred to Wolfe and Phelp's "Practical Shop Mathematics," vol. 2, 4th ed., McGraw-Hill Book Company, New York, 1960.

Figures 16.45 and 16.46 illustrate two of several figures which, for this purpose, might be called basic. A typical problem in this type of work requires that one of the

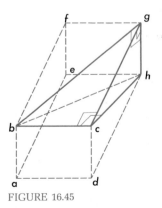

FIGURE 16.45

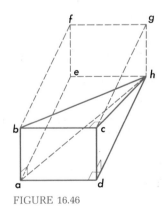

FIGURE 16.46

angles be calculated when certain other angles are given. Figures 16.45 and 16.46 both represent rectangular parallelepipeds.

Example 4. In Fig. 16.45 $\angle gch = 20°$ and $\angle bgc = 30°$ (note the positions of the 90° angles); find the angle *gbh*.

The data as given determine the *shape* of the figure, but since no linear dimensions are given, the *size* of the figure is not determined. However, this is beside the point, since the required answer is an angle, and not a side.

The general procedure in the solution is as follows:

1. Locate the known triangles. In this problem they are the triangles *gch* and *bgc*.
2. Locate the side common to both these known triangles. In this problem this is the side *cg*. Let the length of this side be 1 unit.
3. Locate the unknown triangle. In this problem it is the triangle *bgh*.
4. Locate the common sides between the unknown triangle and each of the known triangles.
5. Solve for the common sides found in step 4 in terms of functions of the known angles.
6. Solve for a function of the unknown angle in terms of the sides found in step 5.
7. Find the angle whose function was found in step 6. The arithmetical work involved in steps 5 to 7 follows in steps 8 to 10.
8. In triangle *gch*,

$$gh = 1 \times \tan gch = \tan gch = \tan 20° = 0.36397$$

9. In triangle *bgc*,

$$bg = 1 \times \sec bgc = \sec bgc = \sec 30° = 1.1547$$

10. In triangle *gbh*,

$$\tan gbh = \frac{gh}{bg} = \frac{\tan 20°}{\sec 30°} = \frac{0.36397}{1.1547} = 0.31521$$
$$\angle gbh = 17°29'43''$$

Since in general it is easier to multiply than it is to divide, the last equation probably should be altered to its equivalent:

$$\tan gbh = \tan 20° \times \cos 30° = 0.36397 \times 0.86603 = 0.31521$$
$$\angle gbh = 17°29'43''$$

EXERCISE 6

1. Draw Figs. 16.45 and 16.46 in orthographic projection. Draw enough views, including auxiliaries, so that the triangles *bcg*, *bch*, *cgh*, and *bgh* may be seen in their true dimensions. Let $ad = 1$, $ab = 1\frac{1}{2}$, and $dh = 2$ in.

2. Referring to Fig. 16.45, write equations, involving multiplication only, which can be used as follows:
   (a) Solve for $\angle gbh$ in terms of $\angle bgc$ and $\angle gch$.
   (b) Solve for $\angle gbh$ in terms of $\angle bgc$ and $\angle hbc$.
   (c) Solve for $\angle gbh$ in terms of $\angle gch$ and $\angle hbc$.
   (d) Solve for $\angle bgc$ in terms of $\angle gch$ and $\angle hbc$.
   (e) Solve for $\angle bgc$ in terms of $\angle gch$ and $\angle gbh$.
   (f) Solve for $\angle bgc$ in terms of $\angle gbh$ and $\angle hbc$.
   (g) Solve for $\angle gch$ in terms of $\angle gbh$ and $\angle bgc$.
   (h) Solve for $\angle gch$ in terms of $\angle gbh$ and $\angle hbc$.
   (i) Solve for $\angle gch$ in terms of $\angle bgc$ and $\angle hbc$.
   (j) Solve for $\angle hbc$ in terms of $\angle gbh$ and $\angle bgc$.
   (k) Solve for $\angle hbc$ in terms of $\angle gbh$ and $\angle ghc$.
   (l) Solve for $\angle hbc$ in terms of $\angle bgc$ and $\angle ghc$.

3. Prove that in Fig. 16.46 angle *bha* is less than angle *chd*.
4. Prove that in Fig. 16.46 angle *bhc* is less than angle *ahd*.
5. If in Fig. 16.46 $ad = 2$ in, $dh = 4$ in, and $ab = 3$ in, find $\angle bha$.
6. Referring to Fig. 16.46, write equations which can be used as follows:
   (a) Solve for $\angle bhc$ in terms of $\angle chd$ and $\angle ahd$.
   (b) Solve for $\angle bha$ in terms of $\angle chd$ and $\angle ahd$.
   (c) Solve for $\angle chd$ in terms of $\angle bha$ and $\angle bhc$.
   (d) Solve for $\angle ahd$ in terms of $\angle bhc$ and $\angle bha$.
7. Find the angle ϕ in Fig. 16.47.
8. Find the angles ϕ and θ in Fig. 16.48.

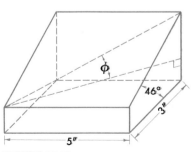

FIGURE 16.47

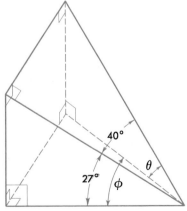

FIGURE 16.48

9. In Fig. 16.49 find the angle which the centerline ab makes with the plane $cdef$.

10. Find ϕ in Fig. 16.50.

11. Find the angle ϕ in Fig. 16.51.

12. In Fig. 16.52 $gb = 3$, $ah = 1$, $ci = 1.3$, $gi = 4$, $gh = 3$, and $hi = 2$. Find $\angle bfg$.

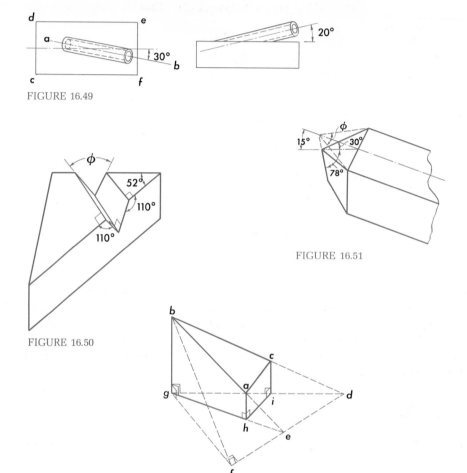

FIGURE 16.49

FIGURE 16.51

FIGURE 16.50

FIGURE 16.52

Vectors

In Sec. 14.1 we defined a vector as follows: A vector is a line segment whose distinctive properties are length, direction, and sense. The length of a vector is the linear distance between the extremities. Direction is measured in terms of the angular orientation of the vector with respect to some reference line. If we consider the vector to have been generated by a moving point, the sense of the vector is determined by the order in which the point coincides with the extremities.

17.1 Some Further Notes on Vectors and Their Projections

In Fig. 17.1 let the point P be any point in the coordinate plane. Its coordinates are x_P and y_P.

If we draw a line perpendicular to the X axis through the point P, then the point of intersection of this perpendicular with the X axis is called the X *projection of the point* P. The abscissa of this point is evidently x_P.

In a similar way, if we draw a perpendicular to the Y axis through the point P, then the point of intersection of this perpendicular with the Y axis is called the Y *projection of the point* P. The ordinate of this point is evidently y_P.

Now observe the vector $\overrightarrow{PQ}$ in Fig. 17.2. The sense of this vector is upward and to the right from point P toward point Q. Consequently, we shall name the point P

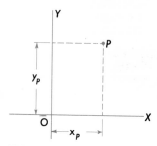

FIGURE 17.1

the *initial* point of the vector $\overrightarrow{PQ}$, and we shall name the point Q the *terminal* point of the vector $\overrightarrow{PQ}$. The X projection of the initial point of this vector is x_P. The X projection of the terminal point of this vector is x_Q.

The X projection of the *vector* $\overrightarrow{PQ}$ is defined as the *algebraic* difference obtained by subtracting the X projection of the initial point *from* the X projection of the terminal point. That is,

$$x_{PQ} = x_Q - x_P \tag{1}$$

where x_{PQ} = X projection of *vector* $\overrightarrow{PQ}$

$\quad\quad x_Q$ = X projection of terminal point of $\overrightarrow{PQ}$

$\quad\quad x_P$ = X projection of initial point of $\overrightarrow{PQ}$

In a similar manner,

$$y_{PQ} = y_Q - y_P \tag{2}$$

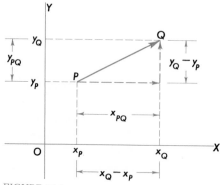

FIGURE 17.2

where y_{PQ} = Y projection of vector $\overrightarrow{PQ}$

$\qquad y_Q$ = Y projection of terminal point of $\overrightarrow{PQ}$

$\qquad y_P$ = Y projection of initial point of $\overrightarrow{PQ}$

The abscissa distances x_P and x_Q are, by the usual conventions of rectangular coordinates, signed numbers. Consequently, x_{PQ} is a signed number.

If the sign of x_{PQ} is positive, the sense of this projection is from the reader's left to his right as he observes Fig. 17.2. If the sign of x_{PQ} is negative, the sense of this projection is from the reader's right to his left as he observes Fig. 17.3. In a similar manner, if the sign of the Y projection of a vector is positive, the sense of that projection is away from the reader toward the top of the page as he observes Fig. 17.2. If the sign of the Y projection is negative, the sense of this projection is toward the reader as he observes Fig. 17.3.

The student should be careful to verify these statements by a detailed study of Examples 1 to 4.

Example 1. The coordinates of the initial point of a certain vector are $(+5,+9)$. The coordinates of the terminal point of the same vector are $(-3,+4)$. Sketch this vector, referred to a system of rectangular coordinates. Indicate the sense of the vector and its projections by arrows. Express the sense and magnitude of each projection in Fig. 17.4. Be sure to observe the proper sign conventions.

Following Eqs. (1) and (2),

X projection $= (-3) - (+5) = -8$
Y projection $= (+4) - (+9) = -5$

Example 2. The coordinates of the initial point of a certain vector are $(-3,+4)$. The

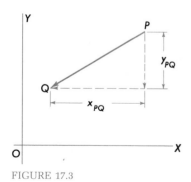

FIGURE 17.3

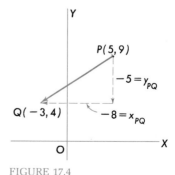

FIGURE 17.4

coordinates of the terminal point of the same vector are $(+5, +9)$. Sketch this vector, referred to a system of rectangular coordinates. Indicate the sense of the vector and its projections by arrows. Express the sense and magnitude of each projection by a signed number (Fig. 17.5).

Following Eqs. (1) and (2),

X projection $= (+5) - (-3) = +8$
Y projection $= (+9) - (+4) = +5$

Example 3. The coordinates of the initial point of a certain vector are $(-8, -5)$. The coordinates of the terminal point of the same vector are $(+3, +7)$. Sketch this vector, referred to a system of rectangular coordinates. Indicate the sense of the vector and its projections by arrows. Express the sense and magnitude of each projection by a signed number (Fig. 17.6).

Following Eqs. (1) and (2),

X projection $= (+3) - (-8) = +11$
Y projection $= (+7) - (-5) = +12$

Example 4. The coordinates of the initial point of a certain vector are $(-6, +5)$. The coordinates of the terminal point of the same vector are $(+8, -2)$. Sketch this vector, referred to a system of rectangular coordinates. Indicate the sense of the vector and its projections by arrows. Express the sense and magnitude of each projection by a signed number (Fig. 17.7).

Following Eqs. (1) and (2),

X projection $= (+8) - (-6) = +14$
Y projection $= (-2) - (+5) = -7$

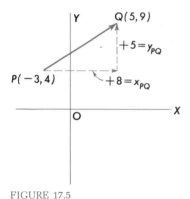

FIGURE 17.5

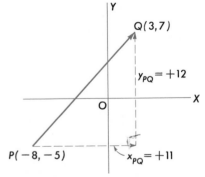

FIGURE 17.6

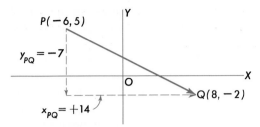

FIGURE 17.7

EXERCISE 1

Calculate the projections of the vectors described below. Draw a sketch of the vector in each case.

	Coordinates of initial point of vector	Coordinates of terminal point of vector
1.	$+8, +15$	$+5, -5$
2.	$-8, -9$	$+2, +5$
3.	$+16, +20$	$-3, +4$
4.	$+9, +4$	$+6, -7$
5.	$+15, +20$	$+25, +10$
6.	$-6, +2$	$+5, +8$
7.	$+3, -4$	$+8, -1$
8.	$+17, -12$	$-4, -9$
9.	$0, -6$	$8, 0$
10.	$-10, 0$	$0, +7$

17.2 Polar and Rectangular Coordinates

Figure 17.8 shows two equal vectors, each designated as $\overrightarrow{PQ}$. The vector $\overrightarrow{PQ}$ in Fig. 17.8a is defined in terms of its X and Y projections. The same vector $\overrightarrow{PQ}$ in Fig.

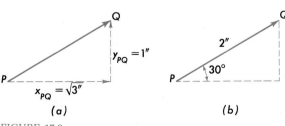

FIGURE 17.8

17.8b is defined in terms of its length, sense, and reference angle. (The student should at this point review Sec. 13.14 to assure himself of the numerical equivalence between the data in this figure.)

When, as in Fig. 17.8a, we define a vector in terms of its projections, we are using *rectangular notation*.

When, as in Fig. 17.8b, we define a vector in terms of its length, sense, and reference angle, we are using *polar notation*.

When using rectangular notation, we designate the length of the projections by x and y and the length of the vector by r with some appropriate subscript where necessary (see Fig. 17.9).

Evidently, from Fig. 17.9, we may write

$$|\overrightarrow{PQ}| = r_{PQ} = \sqrt{x_{PQ}^2 + y_{PQ}^2} \tag{3}$$

where r_{PQ} is the length of the vector $\overrightarrow{PQ}$ and is always a positive number.

In polar notation the length of the vector is designated by the symbol ρ. If ρ is a positive number, the sense of the vector is away from the vertex along the terminal side of the reference angle. If ρ is a negative number, the sense is reversed. See Fig. 17.10.

There is a compact and generally understood symbolism used to define a vector in polar notation. It is, referred to Fig. 17.8b,

$$\overrightarrow{PQ} = 2\underline{/30°}$$

In general, a position vector $\overrightarrow{PQ}$ may be designated in polar notation by the symbols

$$\overrightarrow{PQ} = \pm\rho\underline{/\phi} \tag{4}$$

where ρ (with its associated sign) defines both the length and sense of the vector.

In general, by reference to Fig. 17.9,

$$|\overrightarrow{PQ}| = \sqrt{x_{PQ}^2 + y_{PQ}^2} = r_{PQ} \tag{5}$$

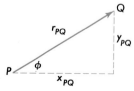

FIGURE 17.9

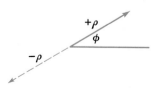

FIGURE 17.10

$$\phi = \tan^{-1}\frac{y_{PQ}}{x_{PQ}} \tag{6}$$

or in polar form,

$$\overrightarrow{PQ} = \sqrt{x_{PQ}^2 + y_{PQ}^2}\Big/\tan^{-1}\frac{y_{PQ}}{x_{PQ}} \tag{7}$$

We may define the vector $\overrightarrow{OP}$ in rectangular notation using data appropriate to polar notation as indicated:

$$x_{PQ} = |\overrightarrow{PQ}|\cos\phi \tag{8}$$
$$y_{PQ} = |\overrightarrow{PQ}|\sin\phi \tag{9}$$

Thus polar notation may be converted to rectangular notation, and vice versa, by Eqs. (5) to (9).

In Fig. 17.11, if $\overrightarrow{OP'} = -\overrightarrow{OP}$ and the rectangular coordinates of P are x_P and y_P, the rectangular coordinates of P' are $-x_{P'}$ and $-y_{P'}$. The magnitude of both $\overrightarrow{OP}$ and $\overrightarrow{OP'}$ in rectangular notation is

$$|\overrightarrow{OP}| = |\overrightarrow{OP'}| = \sqrt{x_{PQ}^2 + y_{PQ}^2} \tag{10}$$

The magnitude and sense of $\overrightarrow{OP}$ in polar notation is $+\rho$; the magnitude and sense of $\overrightarrow{OP'}$ in polar coordinates is $-\rho$. In polar notation (Fig. 17.11)

$$\overrightarrow{OP} = +\rho\underline{/\phi} \tag{11}$$

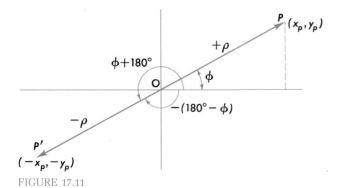

FIGURE 17.11

In polar notation (Fig. 17.11)

$$\overrightarrow{OP'} = -\rho\underline{/\phi} \tag{12}$$

or

$$\overrightarrow{OP'} = +\rho\underline{/\phi + 180°} \tag{13}$$

or

$$\overrightarrow{OP'} = +\rho\underline{/\phi - 180°} \tag{14}$$

That is, in polar notation the size of the angle shows how much the terminal side of the angle has been rotated, and the sign before ρ denotes the sense of the vector along that terminal side.

Example 5. Find the X and Y projections of the vector $6\underline{/40°}$.

X projection $= 6 \cos 40° = 6 \times 0.7660 = +4.596$
Y projection $= 6 \sin 40° = 6 \times 0.6428 = +3.857$

Example 6. Find the X and Y projections of the vector $5\underline{/120°}$.

X projection $= 5 \cos 120° = 5 \times (-0.5) = -2.5$
Y projection $= 5 \sin 120° = 5 \times (+0.8660) = +4.330$

Example 7. Find the X and Y projections of the vector $3\underline{/260°}$.

X projection $= 3 \cos 260° = 3 \times (-0.1737) = -0.5211$
Y projection $= 3 \sin 260° = 3 \times (-0.9848) = -2.9544$

Example 8. Find the X and Y projections of the vector $2\underline{/-30°}$.

X projection $= 2 \cos (-30°) = 2 \times 0.8660 = +1.732$
Y projection $= 2 \sin (-30°) = 2 \times -0.5000 = -1.000$

Example 9. Find the X and Y projections of the vector $-4\underline{/120°}$.

X projection $= -4 \cos 120° = -4 \times (-0.5) = +2.000$
Y projection $= -4 \sin 120° = -4 \times (+0.8660) = -3.464$

Example 10. If the coordinates of *P* in Fig. 17.12 are (7,11) and the coordinates of *Q* are (9,16), write the defining expression for the vector $\overrightarrow{PQ}$ in polar notation.

$$x_{PQ} = 9 - 7 = 2$$
$$y_{PQ} = 16 - 11 = 5$$

Following Eq. (7),

$$\overrightarrow{PQ} = \sqrt{2^2 + 5^2}/\tan^{-1} 5/_2$$

$$= \sqrt{29}/\tan^{-1} 5/_2$$

$$= 5.385/\tan^{-1} 5/_2$$

$$= 5.385/68°12'$$

Example 11. If the rectangular coordinates of *P* in Fig. 17.13 are (3,15) and the coordinates of *Q* are $(-2,19)$, write the defining expression for the vector $\overrightarrow{PQ}$ in polar notation.

$$x_{PQ} = -2 - 3 = -5$$
$$y_{PQ} = 19 - 15 = 4$$
$$r = \sqrt{(-5)^2 + 4^2} = \sqrt{41} = 6.403$$
$$\tan \phi = 4/-5 = -0.80000$$
$$\phi = 141°20'$$

In polar notation the vector can be expressed as

$6.403/141°20'$

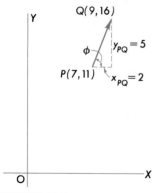

FIGURE 17.12

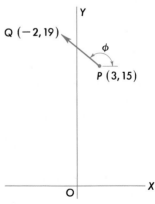

FIGURE 17.13

EXERCISE 2

1. Certain vectors are defined below in polar notation. Calculate the corresponding X and Y projections. Use a slide rule or a table of four-place natural functions.

(a) $3\underline{/27°}$ (b) $4\underline{/320°}$ (c) $-10\underline{/-30°}$

(d) $3\underline{/-270°}$ (e) $7\underline{/90°}$ (f) $5\underline{/160°}$

(g) $15\underline{/29°}$ (h) $15\underline{/-250°}$ (i) $20\underline{/-135°}$

(j) $-7\underline{/-90°}$ (k) $2\underline{/220°}$ (l) $10\underline{/-140°}$

(m) $3\underline{/270°}$ (n) $4\underline{/225°}$ (o) $5\underline{/1,470°}$

2. The coordinates of the terminal points of certain vectors are given below. The initial point is at the origin. In each case write the polar expression defining the vector. Use a slide rule or a four-place table of natural functions.

(a) $(3,6)$ (b) $(2,-3)$ (c) $(-2,+2)$

(d) $(-4,4)$ (e) $(5,-6)$ (f) $(\sqrt{3},1)$

(g) $(-3,-2)$ (h) $(-5,6)$ (i) $\left(\dfrac{\sqrt{2}}{2}, -\dfrac{\sqrt{2}}{2}\right)$

(j) $(a^2 - b^2, 2ab)$ (k) $(6,-6)$ (l) $(5,0)$

17.3 Vector Operations

Several recognized vector operations are so well defined and so frequently used that they have been given names. These names are derived, by analogy, from the names of the fundamental algebraic operations: addition, subtraction, multiplication, and division. The analogy is by no means exact, however, since the nature of a vector is quite different from the nature of an algebraic number. Therefore vector addition, subtraction, multiplication, and division are quite different processes from algebraic addition, subtraction, multiplication, and division.

In this work we shall develop the consideration of vector operations along rather restricted lines. We shall name and define operations for which we have an immediate use, but we shall have to omit consideration of those vector processes appropriate to the more advanced fields of engineering.

The vector operations of addition and subtraction will be discussed in this chapter.

17.4 Vector Addition

Since vectors in a plane are so widely used in so many fields of engineering, each with its own professional nomenclature, we shall for reference adopt the conventional system of rectangular coordinates used in Chap. 8.

A general definition of vector addition is: *The sum of several vectors is a vector, each of whose projections is the algebraic sum of the corresponding projections of the several vectors.* The term composition of vectors is sometimes used synonymously with vector addition, particularly if the vectors represent forces.

If, for example, we wish to find the sum of several vectors such as $\overrightarrow{OM}$, $\overrightarrow{ON}$, $\overrightarrow{OP}$, and $\overrightarrow{OQ}$, we first find the X projection of each vector. We then add these X projections algebraically to find the X projection of the vector sum. Next we find the Y projection of each vector. Then we add these Y projections algebraically to find the Y projection of the vector sum.

If the vectors to be added are defined in rectangular notation, the process of finding the projections of the vector sum is one of simple algebraic addition.

If the vectors to be added are defined in polar notation, we must first calculate the respective X and Y projections and then proceed as above.

17.5 Polar and Topographic Diagrams

The purely algebraic method of adding vectors described in Sec. 17.4 should be supplemented by appropriate diagrams. Such diagrams are essential for an intelligent analysis of a practical problem and will help the student to avoid gross errors.

In a polar diagram each vector is drawn with its initial point at the origin of the reference axis. This is illustrated in Fig. 17.14, in which we propose to add the vectors $\overrightarrow{OP}$ and $\overrightarrow{OQ}$.

Let us now construct the line $\overline{PS}$ equal in magnitude and direction to the vector $\overrightarrow{OQ}$ and construct the line $\overline{QS}$ equal in magnitude and direction to the vector $\overrightarrow{OP}$. We have now completed the parallelogram OPSQ.

It is quite evident from the geometry of the figure that

$$x_{PS} = x_{OQ}$$

and

$$y_{PS} = y_{OQ}$$

Also

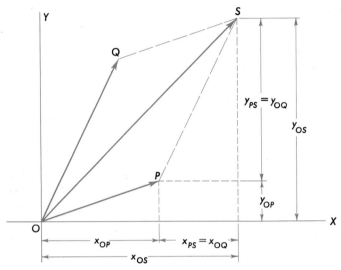

FIGURE 17.14

$$y_{OS} = y_{OP} + y_{OQ}$$

and

$$x_{OS} = x_{OP} + x_{OQ}$$

Thus the vector $\overrightarrow{OS}$ is the vector whose projections are the sum of the corresponding projections of $\overrightarrow{OP}$ and $\overrightarrow{OQ}$. By definition, therefore, $\overrightarrow{OS}$ is the vector sum of $\overrightarrow{OP}$ and $\overrightarrow{OQ}$.

When a graphical method is sufficiently accurate (and there are many such situations), two vectors may be added simply by constructing a diagram to scale. The reference angle of the sum vector can be measured with a protractor, and the length of the sum vector can be measured with a ruler.

The polar diagram is perfectly practical for adding two vectors, but it becomes rather cumbersome when more than two vectors are to be added.

A topographic diagram is more suitable if more than two vectors are to be added. To construct a topographic diagram for the graphical addition of a series of vectors, we proceed as follows: Beginning at the origin of the coordinate system, we draw the first vector chosen with its initial point on the origin. Then we draw the second vector with its initial point on the terminal point of the first vector. Then we draw the third vector with its initial point on the terminal point of the second, and so on, until the last vector in the series has been plotted. The vector whose initial point is at the origin and whose terminal point is at the terminal point of the last vector plotted is the sum of the several vectors.

The student will find that a careful study of the examples below will be more profitable than the reading of further descriptive material.

Example 12. Find the X and Y projections of the sum of the following vectors:

Vector	X projection	Y projection
$\overrightarrow{OA}$	$+320$	$+200$
$\overrightarrow{AB}$	-250	$+100$
$\overrightarrow{BC}$	-50	-500

The X projection of the vector sum is $(+320 - 250 - 50) = +20$. The Y projection of the vector sum is $(+200 + 100 - 500) = -200$. The vector sum is indicated as the vector $\overrightarrow{OC}$ in Fig. 17.15.

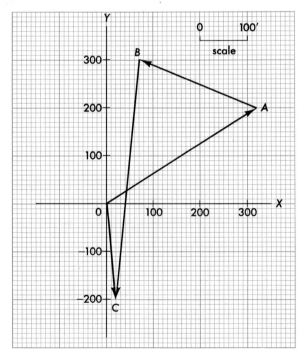

FIGURE 17.15

Example 13. Find the X and Y projections of the sum of the following vectors:

Vector	X projection	Y projection
$\overrightarrow{OA}$	-4.7	-2.1
$\overrightarrow{AB}$	$+1.0$	$+5.0$
$\overrightarrow{BC}$	$+4.5$	$+2.0$

The sum of the vectors $\overrightarrow{OA}$, $\overrightarrow{AB}$, and $\overrightarrow{BC}$ is the vector whose X projection is $(-4.7 + 1.0 + 4.5) = +0.8$ and whose Y projection is $(-2.1 + 5.0 + 2.0) = +4.9$. This vector is designated as the vector $\overrightarrow{OC}$ in Fig. 17.16.

Example 14. Find the sum of the following vectors graphically:

$$5\underline{/60°} + 2\underline{/210°} + 1\underline{/270°}$$

The three given vectors are plotted as $\overrightarrow{OA}$, $\overrightarrow{AB}$, and $\overrightarrow{BC}$ in Fig. 17.17. The vector $\overrightarrow{OC}$ is the sum of the given vectors. By scaling with a ruler, $|\overrightarrow{OC}| = 2.45$. By scaling with a protractor, $\phi = 72°$. Therefore

$$\overrightarrow{OC} = 2.45\underline{/72°}$$

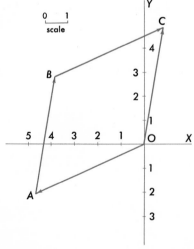

FIGURE 17.16

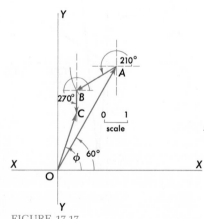

FIGURE 17.17

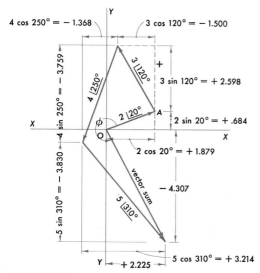

4 cos 250° = − 1.368

3 cos 120° = − 1.500

3 sin 120° = + 2.598

2 sin 20° = + .684

2 cos 20° = + 1.879

− 4.307

− 4 sin 250° = − 3.759

− 4 sin 250° = − 3.830

− 5 sin 310° = − 3.830

5 cos 310° = + 3.214

Y − + 2.225

FIGURE 17.18

Example 15. Perform the following vector addition:

$$2\underline{/20°} + 3\underline{/120°} + 4\underline{/250°} + 5\underline{/310°}$$

A topographic diagram illustrating this problem is shown in Fig. 17.18.

The work will be easier and the correct answer more certain if the solution is organized according to some orderly scheme, such as the arrangement in Table 17.1.

TABLE 17.1

Vector	cosine	X projection +	X projection −	sin	Y projection +	Y projection −
2/20°	+0.9397	1.879		+0.3420	0.684	
3/120°	−0.5000		1.500	+0.8660	2.598	
4/250°	−0.3420		1.368	−0.9397		3.759
5/310°	+0.6428	3.214		−0.7660		3.830
+X projection		5.093				
−X projection			2.868			
+Y projection					3.282	
−Y projection						7.589

Total X projection = 5.093 − 2.868 = 2.225
Total Y projection = 3.282 − 7.589 = −4.307

The magnitude of the vector sum is

$$r = \sqrt{(2.225)^2 + (-4.307)^2} = \sqrt{23.50} = 4.85$$

If ϕ is the reference angle of the vector sum,

$$\tan \phi = \frac{-4.307}{+2.225} = -1.936$$

$$\phi = 297°19'$$

In polar notation, the vector sum is $4.85\underline{/297°19'}$.

EXERCISE 3

1. The X and Y projections of each of several vectors are given below. Find the X and Y projections of the vector sums.

	Vector	X projection	Y projection
(a)	$\overrightarrow{OA}$	+ 100′	+600′
	$\overrightarrow{AB}$	+ 700′	− 200′
	$\overrightarrow{BC}$	+ 100′	− 600′
	$\overrightarrow{CD}$	− 1,000′	− 200′
(b)	$\overrightarrow{OA}$	+200′	− 200′
	$\overrightarrow{AB}$	+100′	+200′
	$\overrightarrow{BC}$	000′	+400′
	$\overrightarrow{CD}$	− 100′	− 200′
(c)	$\overrightarrow{OA}$	+100′	− 200′
	$\overrightarrow{AB}$	+100′	+400′
	$\overrightarrow{BC}$	− 200′	+300′
	$\overrightarrow{CD}$	− 100′	− 300′
	$\overrightarrow{DE}$	− 300′	+200′
	$\overrightarrow{EF}$	+100′	− 200′
	$\overrightarrow{FG}$	+500′	− 800′

2. Plot the following vectors to scale using a protractor and ruler. Find, graphically, the reference angle and length of the vector sum.

(a) 250 ft/17°

(b) 175 ft/135°

(c) 260 ft/330°

    175 ft/138°

    240 ft/60°

    210 ft/291°

    160 ft/70°

    250 ft/17°

    230 ft/315°

    400 ft/195°

    180 ft/278°

    450 ft/165°

(d) 410 ft/00°

(e) 200 ft/172°

    175 ft/270°

    240 ft/258°

    230 ft/153°

    250 ft/11°

    360 ft/205°

    150 ft/344°

    100 ft/55°

3. Find the X and Y projections of the sums of the following vectors. Also find the length and reference angle of the vector sum. Use either a slide rule or table of four-place trigonometric functions as directed by the instructor.

	Vector	X projection	Y projection
(a)	$\overrightarrow{OA}$	3.6	−4.5
	$\overrightarrow{AB}$	−2.9	−8.4
	$\overrightarrow{BC}$	5.6	10.1
(b)	$\overrightarrow{OA}$	−9.8	3.5
	$\overrightarrow{AB}$	2.6	6.8
	$\overrightarrow{BC}$	5.8	−8.1
(c)	$\overrightarrow{OA}$	5.6	0.0
	$\overrightarrow{AB}$	0.0	5.6
	$\overrightarrow{BC}$	−3.0	3.0
	$\overrightarrow{CD}$	−5.8	7.2
(d)	$\overrightarrow{OA}$	−7.0	4.0
	$\overrightarrow{AB}$	0.0	2.0
	$\overrightarrow{BC}$	5.0	−2.0

(e) $\overrightarrow{OA}$ 5.0   0.0

    $\overrightarrow{AB}$ 2.0   2.0

    $\overrightarrow{BC}$ -1.0 -5.0

4. Calculate the X and Y projections of the sums of the following vectors. Also calculate the length and reference angle of the vector sum. Use either a slide rule or a table of four-place trigonometric functions as directed by the instructor.

(a) $15\underline{/25°}$ (b) $25\underline{/120°}$ (c) $3\underline{/0°}$ (d) $2\underline{/-30°}$

    $20\underline{/140°}$ $10\underline{/-90°}$ $4\underline{/90°}$ $10\underline{/-90°}$

    $10\underline{/180°}$ $10\underline{/300°}$ $3\underline{/180°}$ $5\underline{/120°}$

(e) $10\underline{/42°} + 20\underline{/140°} + 30\underline{/240°}$

(f) $5\underline{/180°} + (-10)\underline{/30°} + 20\underline{/-30°}$

(g) $2\underline{/150°} + 6\underline{/90°} + 12\underline{/330°} + 10\underline{/225°}$

(h) $8\underline{/300°} + 5\underline{/20°} + 6\underline{/270°} + 15\underline{/220°}$

(i) $20\underline{/360°} + 20\underline{/180°} + 40\underline{/90°}$

17.6 Vector Subtraction

The process of vector subtraction may be described as follows: To subtract the vector $\overrightarrow{OQ}$ from the vector $\overrightarrow{OP}$, we subtract algebraically the projections of $\overrightarrow{OQ}$ from the corresponding projections of $\overrightarrow{OP}$ to obtain the projections of the vector difference. That is, if

$$\overrightarrow{OR} = \overrightarrow{OP} - \overrightarrow{OQ}$$

then

$$x_{OR} = x_{OP} - x_{OQ} \tag{15}$$

and

$$y_{OR} = y_{OP} - y_{OQ} \tag{16}$$

Example 16. The vector $\overrightarrow{OA}$ has an X projection of 3 in and a Y projection of 4 in.

Another vector $\overrightarrow{OB}$ has an X projection of 6 in and a Y projection of 2 in. Express in polar notation the vector $\overrightarrow{OR}$ if

$$\overrightarrow{OR} = \overrightarrow{OA} - \overrightarrow{OB}$$

Following Eq. (15),

X projection of $\overrightarrow{OR} = 3 - 6 = -3$

Following Eq. (16),

Y projection of $\overrightarrow{OR} = 4 - 2 = +2$

The magnitude of $\overrightarrow{OR}$ is

$$|OR| = \sqrt{(-3)^2 + 2^2} = \sqrt{13} = 3.61$$

$$\tan \phi_r = \frac{+2}{-3} = -0.6667$$

where

ϕ_r = reference angle of $\overrightarrow{OR}$
    $= 146°19'$

In polar notation the value of $\overrightarrow{OR}$ is $3.61\underline{/146°19'}$.

The subtractions indicated in Eqs. (15) and (16) are conventional algebraic subtractions.

Furthermore, the student will note by reference to Fig. 17.11 that a reversal of the sense of a vector automatically reverses the signs of both projections.

Therefore vector subtraction can be accomplished by reversing the sense of the vector subtrahend and proceeding as in vector addition. Thus, if

$$\overrightarrow{OR} = \overrightarrow{AB} - \overrightarrow{CD} \tag{17}$$

and if

$$\overrightarrow{CD'} = -\overrightarrow{CD} \tag{18}$$

then

$$\overrightarrow{OR} = \overrightarrow{AB} + \overrightarrow{CD'} \tag{19}$$

TABLE 17.2

Vector	cosine	X projection +	X projection −	sine	Y projection +	Y projection −
5/20°	+0.940	4.70		0.342	1.71	
8/230°	−0.643		5.144	−0.766		6.128
+X projection		4.70				
−X projection			5.144			
+Y projection					1.71	
−Y projection						6.128

Example 17. If $\overrightarrow{OR} = 5/20° − 8/50°$, express $\overrightarrow{OR}$ in polar notation.

Following Eqs. (17) to (19), we may write

$$\overrightarrow{OR} = 5/20° + 8/230°$$

Note that we have reversed the sense of the vector −8/50° and have changed the problem from a vector subtraction to a vector addition. The numerical solution is tabulated in Table 17.2.

Total X projection $= +4.70 − 5.144 = −0.444$

Total Y projection $= +1.71 − 6.128 = −4.418$

$$r_{OR} = \sqrt{(-0.444)^2 + (-4.418)^2} = 4.44$$

$$\tan \phi_{OR} = \frac{-4.418}{-0.444} = +9.95$$

$$\phi_{OR} = 264.3°$$

$$\overrightarrow{OR} = 4.44/264.3°$$

EXERCISE 4

Use a slide rule or a four-place table of natural functions as directed by the instructor.

1. Find the X and Y projections of the vector difference obtained by subtracting the vector $\overrightarrow{OA}$ from the vector $\overrightarrow{OB}$. Also find the length and the reference angle of the vector difference.

Vector	X projection	Y projection
(a) $\overrightarrow{OA}$	+ 7	+ 5
$\overrightarrow{OB}$	+ 2	+ 6

(b)	$\overrightarrow{OA}$	$-\ 3$	$-\ 5$
	$\overrightarrow{OB}$	$+\ 2$	$+\ 8$
(c)	$\overrightarrow{OA}$	$+15$	$+20$
	$\overrightarrow{OB}$	-20	$+15$
(d)	$\overrightarrow{OA}$	-20	-30
	$\overrightarrow{OB}$	-10	-10
(e)	$\overrightarrow{OA}$	-32	$+42$
	$\overrightarrow{OB}$	$+26$	-20

2. Perform the following vector operations, expressing the answer in polar notation.

(a) $30\underline{/45°} - 20\underline{/90°}$ $\qquad$ (b) $20\underline{/36°} - 5\underline{/80°}$

(c) $15\underline{/160°} - 20\underline{/330°}$ $\qquad$ (d) $90\underline{/220°} - 45\underline{/20°}$

(e) $25\underline{/110°} - 60\underline{/220°}$ $\qquad$ (f) $20\underline{/45°} - 30\underline{/30°} + 10\underline{/225°}$

(g) $100\underline{/90°} + 200\underline{/180°} - 100\underline{/0°}$ $\qquad$ (h) $-25\underline{/330°} + 40\underline{/160°} - 20\underline{/260°}$

(i) $15\underline{/60°} - 30\underline{/120°} + 50\underline{/330°}$ $\qquad$ (j) $20\underline{/115°} - 15\underline{/250°} - 30\underline{/25°}$

17.7 Units of Angular Measure

Angles are commonly measured in either of two systems of units: the degree system or the radian system.

The Degree System

The vector $\overrightarrow{OP}$ in Fig. 14.5 is shown in its initial position. If now it is rotated counterclockwise one complete revolution until it again coincides with its initial position, then, by definition, an angle of $+360$ degrees has been generated. The angle 360 degrees is usually written 360°. One degree is by definition $\frac{1}{360}$ revolution.

The Radian System

As the vector $\overrightarrow{OP}$ rotates, the point P moves in a circular path (see Fig. 17.19). If the arc length traversed by P is $\widehat{XP}$ (where $\widehat{XP}$ is measured in the same linear units as $\overrightarrow{OP}$), then ϕ measured in radians is

$$\phi_{\text{rad}} = \frac{\widehat{XP}}{|OP|} \qquad\qquad (20)$$

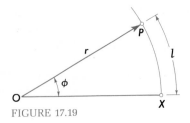

FIGURE 17.19

and

$$\widehat{XP} = |OP| \times \phi_{\text{rad}} \tag{21}$$

or

$$l = r\phi_{\text{rad}} \tag{22}$$

where l is the arc length traversed by P and the angle is measured in radians rather than in degrees.

In other words, the quotient of the arc length divided by the radius is the measure of the central angle in radians.

If the vector $\overrightarrow{OP}$ makes one complete revolution, then the arc length becomes the circumference of the circle, and

$$\phi_{\text{rad}} = \frac{2\pi|OP|}{|OP|} = 2\pi \tag{23}$$

There are 2π rad in one revolution, and 1 rad is $1/(2\pi)$ revolution.

Since by definition one revolution of the vector is equivalent to an angle of 360°, and since by definition one revolution is also equivalent to 2π rad,

$$2\pi \text{ rad} = 360°$$
$$\pi \text{ rad} = 180°$$
$$1 \text{ rad} = \frac{180°}{\pi} = 57.2957795° \tag{24}$$
$$1° = \frac{\pi}{180} = 0.01745329 \text{ rad} \tag{25}$$

Often, fractions of a degree are not written in decimals, but rather in subunits called minutes and seconds. There are no subunits attached to radian measure. For purposes of numerical calculation in solving triangles, etc., the degree system is ordinarily used. In more analytical work the radian system is common.

For simplicity, we usually choose to write π rad rather than 3.14159 rad, $\pi/2$ rad rather than 1.57080 rad, $3\pi/2$ rad rather than 4.71239 rad, etc. In such cases it is easy to change from radian measure to degree measure by substituting for π rad the degree equivalent, 180°.

$$\pi \text{ rad} = 180° \tag{26}$$

$$\frac{\pi}{2} \text{ rad} = \frac{180°}{2} = 90° \tag{27}$$

$$\frac{3\pi}{2} \text{ rad} = \frac{3 \times 180°}{2} = 270° \tag{28}$$

Example 18. Change 56.73° to radians.
Since by Eq. (25)

$$1° = \frac{\pi}{180} \text{ rad}$$

then

$$56.73° = 56.73 \times \frac{\pi}{180} = 0.99013 \text{ rad}$$

Example 19. Change 0.792 rad to degrees. Leave the answer a decimal.
Since by Eq. (24)

$$1 \text{ rad} = \frac{180°}{\pi}$$

then

$$0.792 \text{ rad} = 0.792 \times \frac{180}{\pi} = \frac{0.792 \times 180}{\pi}$$

$$= 45.38°$$

Example 20. How many degrees are there in $7\pi/15$ rad?

$$\pi \text{ rad} = 180°$$

$$\frac{7}{15} \pi \text{ rad} = \frac{7}{15} \times 180° = \frac{1,260}{15} = 84°$$

Example 21. How many revolutions are there in an angle of 5π rad?

Since 1 revolution = 2π rad, it follows that

$$5\pi \text{ rad} = \frac{5\pi}{2\pi} = 2.5 \text{ revolutions}$$

Example 22. How many revolutions are there in 2.5 rad?
Since 1 revolution = 2π or 6.2832 rad, it follows that

2.5 rad = 2.5/6.2832 = 0.40 revolution

Example 23. Refer to Fig. 17.19. If $l = 5$ in and $r = 3$ in, find the angle ϕ in radians.
Following Eq. (22),

$$l = r\phi_{rad}$$

then

$$5 = 3\phi_{rad}$$

or

$$\phi = \tfrac{5}{3} \text{ rad}$$

Example 24. What is the arc length which subtends a central angle of 2.36 rad in a circle whose radius is 4 in?
Following Eq. (22),

$$l = 4 \times 2.36 = 9.44 \text{ in}$$

Example 25. How many revolutions are equivalent to 4,320°?
Since there are 360° in one revolution, we may write

4,320/360 = 12 revolutions

EXERCISE 5

Convert into degrees, radians, and revolutions as indicated in the following tabulation:

	Radians	Degrees	Revolutions
1.	2.5		
2.			0.75
3.		1,000	
4.		2,000	

	Radians	Degrees	Revolutions
5.			400
6.	100		
7.	400π		
8.		30	
9.		315	
10.	$(3\pi)/2$		

EXERCISE 6

Referring to Fig. 17.19, find arc length, radius, or central angle as indicated in the following tabulation:

	Arc length, in	Radius, in	Central angle
1.		16	4 rad
2.	12		8 rad
3.	2.4	5	
4.		25	2 revolutions
5.	18		45°
6.	32	8	Find central angle in revolutions

17.8 Angular and Linear Velocity

In certain engineering applications we have to deal with physical phenomena which can be represented mathematically by rotating position vectors (Sec. 18.4).

In Fig. 17.19, let us consider the point P to be moving with a uniform linear velocity v along the circular arc of radius r.

The linear velocity v is expressed in some such units as feet per second, feet per minute, or inches per second.

The central angle ϕ is consequently changing with a uniform angular velocity which we shall designate as ω. The angular velocity ω is measured in units such as radians per second, degrees per minute, or revolutions per minute.

If the position vector $\overrightarrow{OP}$ in Fig. 17.19 is rotating with a uniform angular velocity ω about the point O, the angle turned in time t is

$$\phi = \omega t \tag{29}$$

If ϕ is expressed in radians, then ω is expressed as radians per second, radians per minute, etc. If ϕ is expressed in degrees, then ω is expressed as degrees per second, degrees per minute, etc.

Example 26. If the position vector $\overrightarrow{OP}$ in Fig. 17.19 is rotating with an angular velocity of 0.3 rad/s, through what angle ϕ will it rotate in 5 s?

Since

$$\phi_{\text{rad}} = \omega t$$

in this case,

$$\phi_{\text{rad}} = 0.3 \times 5 = 1.5 \text{ rad}$$

Example 27. Express the angle ϕ in Example 26 in revolutions.

$$\phi_{\text{rev}} = \frac{1.5}{2\pi} = 0.2387 \text{ revolution}$$

Since by Eq. (22)

$$l = r\phi_{\text{rad}} \tag{30}$$

by substituting in Eq. (29) we may write

$$\frac{l}{r} = \omega t \tag{31}$$

or

$$\frac{l}{t} = \omega r \tag{32}$$

But since l/t is by definition equal to linear velocity v, we may write

$$v = \omega r \tag{33}$$

Example 28. A position vector $\overrightarrow{OP}$ has a length of 1.5 ft and rotates at the rate of 400 rev/min. Find the linear velocity of point P in inches per second.

$$400 \text{ rev/min} = 2\pi \times 400 = 800\pi \text{ rad/min}$$

or

$$\frac{800\pi}{60} \text{ rad/s}$$

Furthermore,

$$1.5 \text{ ft} = 1.5 \times 12 = 18 \text{ in}$$

Now, following Eq. (33),

$$v = \frac{800\pi}{60} \times 18 = 240\pi \text{ in/s}$$

The use of a slide rule is suggested in Exercises 7 and 8.

EXERCISE 7

Find the central angle, angular velocity, or time in the units required as indicated in the following tabulation:

	Angular velocity	Time	Central angle
1.	3 rad/s	16 s	? rad
2.	? rad/s	25 s	80 rad
3.	15 rad/s	12 s	? r*
4.	? rad/s	3 min	100 r
5.	50 rad/s	2.5 min	? rad
6.	12 rad/s	? s	50 rad
7.	? rad/s	3 min	600 r
8.	6 r/s	? min	400 rad
9.	12 rad/s	? s	600
10.	4 r/s	? min	800π rad

*r is the abbreviation for revolutions.

EXERCISE 8

Find the linear velocity, angular velocity, or radius in the units required as indicated in the following tabulation:

	Linear velocity	Angular velocity	Radius
1.	? ft/s	30 rad/s	1.5 ft
2.	16 ft/s	? rad/s	2.5 ft
3.	210 in/min	35 rad/s	? ft
4.	? in/s	6 r/s*	12 in
5.	? in/s	500 r/min	15 ft
6.	4 ft/s	600 r/min	? in
7.	5 ft/s	? deg/s	16 in
8.	20 ft/s	200 r/min	? ft

*r is the abbreviation for revolutions.

graphs of the trigonometric functions

In this chapter we shall graph each of the six trigonometric functions. By observing the nature of the graphs, we can deduce many of the properties of the functions themselves.

In addition to such theoretical uses, the graphs of the functions, particularly those of the sine and cosine, are very convenient devices to use in the analysis and solution of certain engineering problems.

Figure 18.1 shows the graphs of the six trigonometric equations

$$y = \sin \phi \tag{1}$$
$$y = \cos \phi \tag{2}$$
$$y = \tan \phi \tag{3}$$
$$y = \cot \phi \tag{4}$$
$$y = \sec \phi \tag{5}$$
$$y = \csc \phi \tag{6}$$

While it is true that the choice of scale is to a certain extent arbitrary, we do obtain a somewhat more characteristic shape of the graphs if we use the same scale on both axes and plot the angles in radians rather than in degrees. Figure 18.1 is plotted in this way.

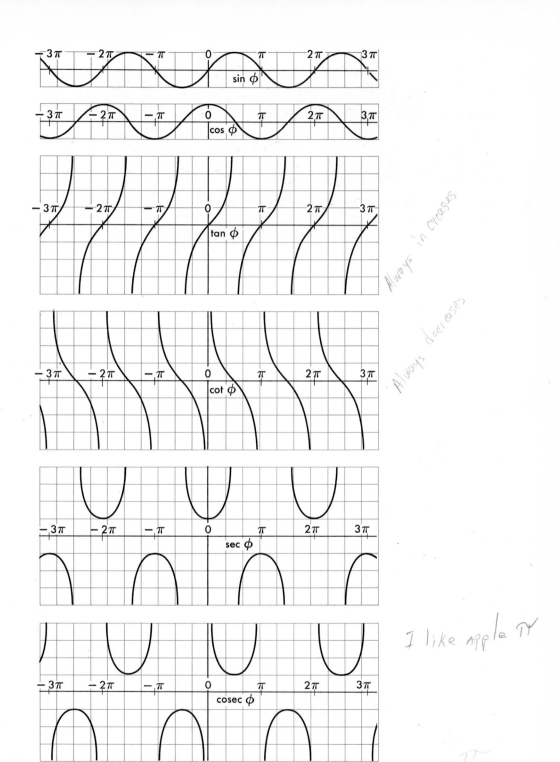

FIGURE 18.1

Always in creases

Always decreases

I like Apple TV

$3\dfrac{\pi}{2}$

However, in technology the theoretical advantage gained by using radian measure rather than degree measure does not ordinarily outweigh the disadvantage of using the irrational number π.

18.1 Some Periodic Functions

One of the most important characteristics of the trigonometric functions is their repetitive nature.

A study of Fig. 18.1 will show that as the angle increases or decreases, the six functions undergo certain characteristic changes. As the angle indefinitely increases or decreases, these changes repeat periodically. The trigonometric functions are therefore called *periodic functions*. One *cycle* has elapsed when a function has traversed once all the variations of which that particular function is capable.

The *angular period* of any of the trigonometric functions is the angular interval required for the completion of one cycle.

Beginning at $\phi = 0$, let us trace the characteristic changes of the sine function. From $\phi = 0$ to $\phi = \pi/2$ rad, the sine increases from 0 to 1; from $\pi/2$ rad to π rad, the sine decreases from 1 to 0; from π rad to $3\pi/2$ rad, the sine decreases from 0 to -1; from $3\pi/2$ rad to 2π rad, the sine increases from -1 to 0. As ϕ begins to increase beyond 2π rad, the cycle begins to repeat. Thus the angular period of the sine function is 2π rad, or $360°$.

The period of the tangent function is π rad. Over the interval $\phi = -\pi/2$ to $\phi = 0$, the tangent increases from $-\infty$ to 0. As the angle increases from 0 to $+\pi/2$ rad, the tangent increases from 0 to $+\infty$. If ϕ increases beyond $\pi/2$ rad, the tangent again traverses the complete array of values from $-\infty$ to $+\infty$, repeating the variation which took place in the previous angular period of π rad.

EXERCISE 1

1. In the equation $y = 5 \sin \beta$, find y from the tables when β is $10°$, $50°$, $160°$, $250°$, and $300°$.
2. Using the same scale, plot the six trigonometric functions between $-360°$ and $+360°$.
3. On the same axes plot the curves of $y = \sin \beta$ and $y = \tan \beta$. Use radian measure rather than degree measure, and plot over the interval $\beta = 0$ rad to $\beta = 0.02$ rad. What is the approximate slope of the graphs in this region? What would you expect the slope to be when $\beta = 0$? For any given angle in this region, which is greater, the sine or the tangent?

18.2 Sine and Cosine Graphs

For our immediate purpose, the graphs of the sine and cosine are by far the most important.

Since the period of a sine or cosine curve is 2π rad, or 360°, we usually plot only in the interval between $\phi = 0$ and $\phi = 2\pi$ rad. All the characteristic changes of these functions appear in this region.

Figure 18.2 shows a plot of the equations

$$y_1 = \sin \phi \tag{7}$$
$$y_2 = \cos \phi \tag{8}$$

The horizontal axis is calibrated in degrees rather than in radians. A comparison between the shapes of the graphs of these functions in Figs. 18.1 and 18.2 will show the practical equivalence of the two plots even when ϕ is plotted in units of degrees rather than in radians.

Notice that the sine and cosine curves in Fig. 18.2 have exactly the same shape, but the cosine curve is displaced 90° to the left with respect to the sine curve.

The *amplitude* of a sine or cosine curve is the absolute value of the maximum and minimum ordinates. In Fig. 18.2 the amplitude of both curves is 1.

Observe that all the characteristic changes in the cosine functions of ϕ occur at angles differing by 90° from the corresponding changes in the sine functions of ϕ.

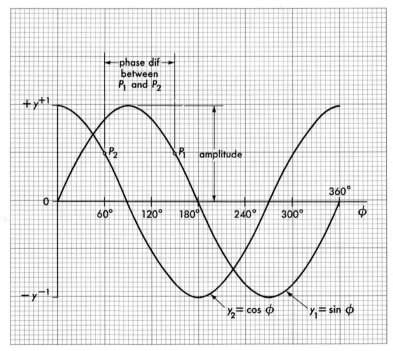

FIGURE 18.2

Therefore a cosine function of ϕ is said to have a 90° phase shift when compared with a sine function of ϕ. Since, in scanning the curve from left to right in the direction of increasing angles, we observe points on the cosine curve before we observe the corresponding points on the sine curve, we say that the cosine function *leads* the sine function.

The student should verify that by plotting the equation

$$y = \sin(\phi + 90°) \tag{9}$$

he will obtain the graph of y_2 in Fig. 18.2. Thus he may conclude that

$$\cos \phi = \sin(\phi + 90°) \tag{10}$$

The equations for the sine and cosine functions may be generalized somewhat by writing

$$y = k \sin(\phi + \psi) \tag{11}$$

and

$$y = k \cos(\phi + \psi) \tag{12}$$

where k and ψ are constants.

When $(\phi + \psi) = 90°$ in Eq. (11), the ordinate of the sine curve is a maximum and equal to k. When $(\phi + \psi) = 0°$ in Eq. (12), the ordinate of the cosine curve is a maximum and also equal to k. The factor k is called the *amplitude* of the curve and is numerically equal to the maximum value of the function.

The constant angle ψ as used in Eqs. (11) and (12) is known as the *phase shift* and has the effect of shifting the graphs of the equations

$$y = k \sin \phi$$

or

$$y = k \cos \phi$$

to the right or left along the ϕ axis. The curve is shifted to the left when ψ is positive and to the right when ψ is negative. The effect of k and ψ is illustrated in Fig. 18.3.

EXERCISE 2

1. The amplitude of a sine curve is 25. What is the ordinate of the curve at the following positions in its cycle?

(a) 20°   (b) 90°   (c) 130°   (d) 180°
(e) 210°   (f) 270°   (g) 300°   (h) 360°

2. Write an equation for a sine function in which the amplitude is 23 and the phase constant is +56°.

3. Plot the following curves:

(a) $y = \sin \phi$ (b) $y = \sin(\phi - 90°)$
(c) $y = \sin(\phi - 60°)$ (d) $y = \sin(\phi - 30°)$
(e) $y = \sin(\phi + 30°)$ (f) $y = \sin(\phi + 60°)$
(g) $y = \sin(\phi + 90°)$

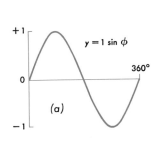

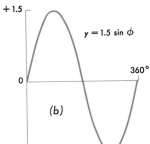

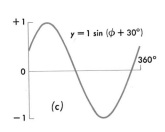

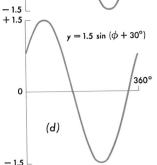

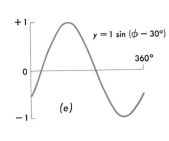

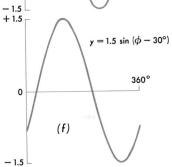

FIGURE 18.3

4. Plot the following curves:

(a) $y = \cos \phi$ (b) $y = \cos (\phi - 90°)$
(c) $y = \cos (\phi - 60°)$ (d) $y = \cos (\phi - 30°)$
(e) $y = \cos (\phi + 30°)$ (f) $y = \cos (\phi + 60°)$
(g) $y = \cos (\phi + 90°)$

Note: Be sure to compare the graphs obtained in Probs. 3g and 4a.

5. The ordinate of a sine curve is $+25$ when $100°$ of its cycle has been completed. Find the amplitude.
6. The ordinate of a sine curve is 15 when $80°$ of its cycle has been completed. What is the ordinate when $300°$ has been completed?
7. At what positive angles less than $360°$ will the absolute value of the ordinate of a cosine curve be 70 percent of its amplitude?
8. The amplitude of a sine curve is 15. Find the ordinate at a phase $20°$ after its minimum.
9. What is the phase difference between the curves of the functions $y = 3 \sin \phi$ and $y' = 7 \sin (\phi + 40°)$?

18.3 Plotting the Sine and Cosine Functions by Geometric Methods

Thus far we have plotted the sine and cosine curves by the use of tables. There is a geometric method which is extremely useful in the analysis of certain problems. This method is illustrated in Fig. 18.4a. First a base line ox of arbitrary length is established. Then a circle is drawn with its center (C) on an extension of ox. The radius of this circle is the amplitude of the sine curve to be plotted. The length ox is to scale numerically equal to the angular measure of one period.

The circle is divided into a convenient number of equal arcs, and the line ox is calibrated with the same number of equally spaced angle markers.

The points a, b, c, d, etc., are projected from the circle parallel to ox. The points a', b', c', d', etc., are projected vertically from the ox axis. Points on the sine curve are found at the intersection of corresponding projections.

Figure 18.4b is a similar construction of a cosine curve. Note that the reference point o on the circle has been advanced $90°$ counterclockwise.

18.4 Phasors

The geometric method of plotting sine functions discussed in Sec. 18.3 is highly important in explaining the procedure used in solving many important engineering problems.

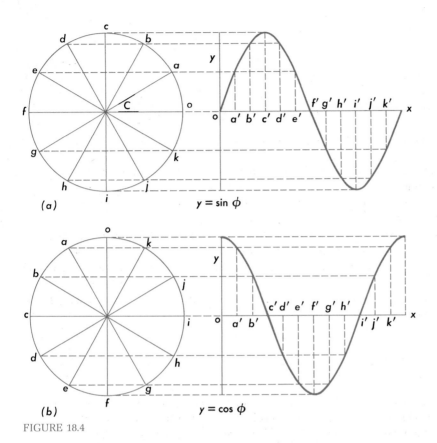

FIGURE 18.4

For these applications we shall consider that the radii in Fig. 18.5 represent successive, momentary positions of a rotating position vector.

We impose four critical conditions on such a rotating position vector:

1. *It rotates with a constant angular velocity.*
2. *It is of constant length.*
3. *It rotates in a coordinate plane about the origin of the coordinate system.*
4. *It rotates in a counterclockwise direction.*

A rotating position vector is sometimes called a *phasor*.

Referring to Fig. 18.5a, we note that this figure is similar to Fig. 18.4a. However, certain very important differences do exist. The phasors, $\overrightarrow{OR}_0$, $\overrightarrow{OR}_1$, $\overrightarrow{OR}_2$, etc., represent successive positions of the rotating phasor $\overrightarrow{OR}$ at times t_0, t_1, t_2, etc. The horizontal axis of the sine curve is calibrated with equally spaced *time* markers rather than with

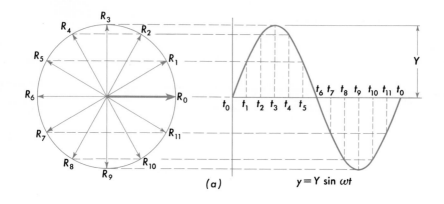

$$(a) \qquad y = Y \sin \omega t$$

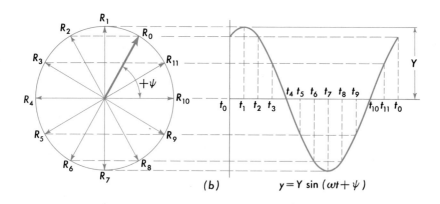

$$(b) \qquad y = Y \sin (\omega t + \psi)$$

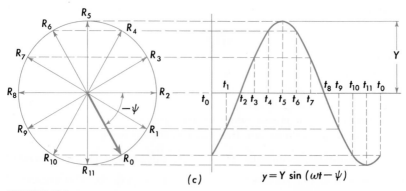

$$(c) \qquad y = Y \sin (\omega t - \psi)$$

FIGURE 18.5

equally spaced angle markers. The ordinate of the sine curve at t_0, t_1, t_2, t_3, etc., is the momentary vertical projection of the position vector $\overrightarrow{OR}$ at that time.

During each revolution the position vector $\overrightarrow{OR}$ rotates through $360°$, or 2π rad. The sequence of events which takes place during one revolution of the position vector is called a *cycle*.

The number of cycles taking place in one second is called the *frequency*. The time in seconds required for the completion of one cycle is called the *time period,* or simply the *period*.

From these definitions the following relations appear:

$$\phi = 360° \times f \times t \quad \text{(in degrees)} \tag{13}$$

$$\phi = 2\pi \times f \times t \quad \text{(in radians)} \tag{14}$$

$$T = \frac{1}{f} \tag{15}$$

where ϕ = momentary reference angle of position vector, referred to its position when we choose to begin to count time

f = frequency, Hz (or cycles per second)

t = time during which position vector has been rotating from some arbitrary zero time-reference position, s

T = period, s

The *hertz* (Hz) is now commonly used as a unit of frequency. Thus 1 Hz means 1 cycle per second, 60 Hz means 60 cycles per second, etc.

The zero reference position R_0 can be rotated either clockwise or counterclockwise, as shown in Fig. 18.5b and c. In such cases we introduce the phase shift ψ. The momentary reference angle of the position vector now becomes $(\phi - \psi)$ or $(\phi + \psi)$, according as R_0 has been rotated clockwise or counterclockwise.

The general equation for the ordinate of the sine curve as a function of time is

$$y = Y \sin (2\pi \, ft \pm \psi) \tag{16}$$

or

$$y = Y \sin (360° \, ft \pm \psi) \tag{17}$$

The term $2\pi f$ or $360° f$ is in the nature of angular velocity.

Ordinarily, we let

$$\omega = 2\pi f \tag{18}$$

(when measuring ω in radians per second), or

$$\omega = 360° f \tag{19}$$

(when measuring ω in degrees per second). Equation (16) then becomes

$$y = Y \sin (\omega t \pm \psi) \tag{20}$$

where Y = constant amplitude
ω = angular velocity, rad (or deg)/s
t = time, s
ψ = phase shift measured in same angular units as ωt
$(\omega t \pm \psi)$ = momentary reference angle of position vector referred to conventional reference position of system of rectangular coordinates

As a matter of expediency, we usually plot y against the angle ωt rather than against t itself. However, if after plotting y against ωt, it seems worthwhile to show the plot of y against t, the horizontal scale can be altered to read directly in time units. Then the phase shift can be interpreted in terms of the appropriate time units.

The terms *amplitude, frequency, angular velocity, period,* and *phase shift* have been defined. Now let us interpret these terms in relation to specific quantities appearing in the general equation for a sine function.

With reference to Eq. (20), the constant Y is the amplitude. A change in Y alters each ordinate of the sine curve, but does not destroy the characteristic shape of the sine curve. A comparison of Fig. 18.3*a, c,* and *e* with Fig. 18.3*b, d,* and *f,* respectively, should illustrate this.

The phase shift ψ has no effect on the amplitude or the shape of the curve, but does have the effect of shifting the whole curve to the right or left along the time axis. A comparison among Fig. 18.3*a, c,* and *e* should illustrate this fact.

The constant ω or $2\pi f$ has no effect on either the amplitude or the relative position of the curve along the time axis. The constant ω or $2\pi f$ determines the number of cycles occurring in a given time interval. Figure 18.6 shows four curves all having the same amplitude and phase shift. In this case the phase shift is zero. The ω in curve *a* has been replaced by 2ω, 3ω, and 4ω in curves *b, c,* and *d,* respectively. This means that, referred to curve *a*, curve *b* has twice the frequency, curve *c* has 3 times the frequency, and curve *d* has 4 times the frequency.

When the frequency of one sine function occurs as an integral multiple of the frequency of another sine function, these two functions are said to be *harmonically related*.

Referred to Fig. 18.6, the function $y = \sin \omega t$ is called the *fundamental* frequency. The function $y = \sin 2\omega t$ is called the *second harmonic* of the fundamental frequency.

The third and fourth harmonics of the function $y = \sin \omega t$ are illustrated in Fig. 18.6*c* and *d*.

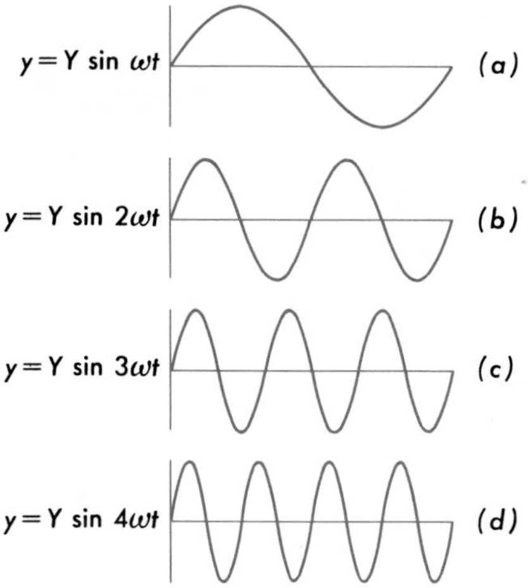

$y = Y \sin \omega t$ 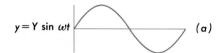 (a)

$y = Y \sin 2\omega t$  (b)

$y = Y \sin 3\omega t$  (c)

$y = Y \sin 4\omega t$ (d)

FIGURE 18.6

EXERCISE 3

1. To the same scale plot the following equations. Use y as the ordinate and ωt measured in degrees as the abscissa. Plot over an interval of 1 cycle.

(a) $y = \sin \omega t$ (b) $y = 2 \sin \omega t$
(c) $y = 3 \sin \omega t$ (d) $y = 4 \sin \omega t$

2. Plot the following equations as instructed in Prob. 1:

(a) $y = \cos \omega t$ (b) $y = 2 \cos \omega t$
(c) $y = 3 \cos \omega t$ (d) $y = 4 \cos \omega t$

3. Plot the following equations as instructed in Prob. 1:

(a) $y = \sin \omega t$ (b) $y = \sin 2\omega t$
(c) $y = \sin 3\omega t$ (d) $y = \sin 4\omega t$

4. Plot the following curves as instructed in Prob. 1:

(a) $y = \cos \omega t$ (b) $y = \cos 2\omega t$
(c) $y = \cos 3\omega t$ (d) $y = \cos 4\omega t$

5. Plot as instructed in Prob. 1:

(a) $y = \sin (\omega t + 30°)$ (b) $y = 3 \sin (\omega t + 60°)$

(c) $y = 3 \sin \omega t$ (d) $y = \sin (\omega t - 30°)$

(e) $y = \sin (\omega t - 60°)$

6. In the following equations find the amplitude, frequency, period, angular velocity, and phase shift. Express angular velocity in both radians per second and degrees per second; also express the phase shift in both radians and degrees. Note carefully when the data are given in radians and when in degrees.

(a) $y = 23 \sin (314.16t + \pi/6)$

(b) $y = 144 \sin (157.08t - \pi)$

(c) $y = 77 \sin (251.327t + 2\pi/7)$

(d) $y = 100 \sin (9 \times 10^5 t + 45°)$

(e) $y = 15 \sin (3.76992 \times 10^8 t - 3\pi/5)$

(f) $y = 147 \sin (21600t + 60°)$

(g) $y = 200 \sin (6.2832t - 2\pi/5)$

(h) $y = 300 \sin (1.6588 \times 10^{12} t + \pi/2)$

7. Using radian measure, write the equation for each of the sine curves described below:

	Amplitude	Frequency	Phase shift
(a)	200	30	$-50°$
(b)	150	3.0×10^8	$00°$
(c)	5	10^{12}	$\pi/2$
(d)	9	60	$00°$
(e)	10	1,000	$-45°$
(f)	20	500	$30°$
(g)	20	25	$00°$
(h)	115	40	-2π
(i)	100	50	$90°$

18.5 The Sum of Two Sine Functions of the Same Frequency

One of the typically characteristic problems in alternating-current electricity, vibrating mechanical bodies, certain types of wave motion, etc., involves the addition of two sine functions of the same frequency. If, for example,

$$y_1 = Y_1 \sin \omega t \tag{21}$$

and

$$y_2 = Y_2 \sin (\omega t + \psi) \tag{22}$$

let us investigate the properties of y_3, where

$$y_3 = y_1 + y_2 \tag{23}$$

That is to say, let us investigate some of the properties of the function

$$y_3 = Y_1 \sin \omega t + Y_2 \sin (\omega t + \psi) \tag{24}$$

where Y_1, Y_2, ω, and ψ are constants.

Figure 18.7 is a plot of Eqs. (21) and (22). In that figure we have arbitrarily let

$Y_1 = 2$
$Y_2 = 3$
$\psi = 60°$

The points on y_1 and y_2 were located by the geometric method discussed in Sec. 18.3. In Fig. 18.7 the phasors Y_1 and Y_2 are shown in their zero-reference position.

For any given value of ωt the ordinate of the graph of y_3 will be equal to the sum of the ordinates of y_1 and y_2. This is illustrated in Fig. 18.8. In this figure a succession of ordinates of y_1 and y_2 have been added to obtain a succession of points on the graph of y_3.

The shape of the graph of y_3 suggests that it may itself be a sine function. This fact will be demonstrated later.

Within the limits to which we can plot the graph, it would appear that the amplitude of y_3 is in the vicinity of 4.4 and that the phase of y_3 referred to y_1 is about 36°.

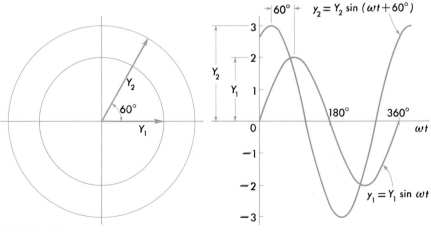

FIGURE 18.7

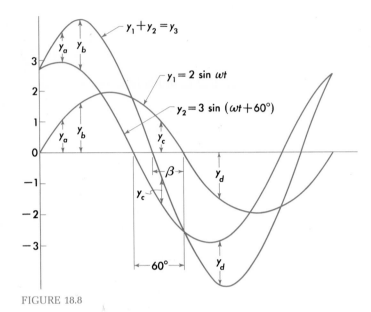

FIGURE 18.8

While this method of analysis has certain merits in describing the nature of y_3, it is quite cumbersome and its accuracy is limited by the strictly geometrical treatment.

A much more practical geometry is illustrated in Fig. 18.9. Here we have shown the three graphs plotted in Fig. 18.8, but we have also shown the phasors Y_1 and Y_2 which generated these curves, oriented in accordance with a randomly chosen value of ωt.

Now let us draw the line $\overline{P_1P_3}$ parallel to Y_2. Also let us draw the line $\overline{P_2P_3}$ parallel to Y_1, thus forming the parallelogram $OP_1P_3P_2$.

By the conditions of the problem, the lengths of Y_1 and Y_2 remain constant. Therefore the parallelogram $OP_1P_3P_2$ remains constant regardless of ωt.

Now let us draw the line $\overline{OP_3}$ and designate it as Y_3. From the geometry of the figure it is quite evident that

$$P_3a = Y_2 \sin (\omega t + \psi) \tag{25}$$

and

$$ab = Y_1 \sin \omega t \tag{26}$$

Accordingly,

$$P_3b = Y_1 \sin \omega t + Y_2 \sin (\omega t + \psi) \tag{27}$$

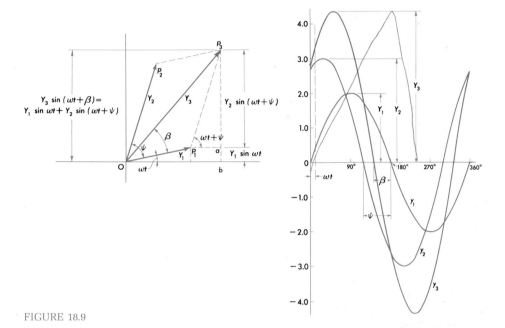

FIGURE 18.9

However,

$$P_3b = Y_3 \sin (\omega t + \beta) \tag{28}$$

where β is indicated on the diagram. The evaluation of β will be discussed presently.
Therefore

$$Y_3 \sin (\omega t + \beta) = Y_1 \sin \omega t + Y_2 \sin (\omega t + \psi) \tag{29}$$

The form of Eq. (29) shows that the sum of two sine functions of the same frequency is itself a sine function of that frequency.

A particularly important problem involving this area of mathematics is illustrated in the following example.

Example 1. If $y_1 = 2 \sin \omega t$ and $y_2 = 3 \sin (\omega t + \psi)$ and if $y_3 = y_2 + y_1$, find the amplitude of y_3 and the phase of y_3 referred to y_1, where $\psi = 60°$.

If we let Y_3 be the amplitude of y_3, and if we let β be the phase angle of y_3 referred to the phase of y_1, we can solve for Y_3 and β by reference to Fig. 18.10. We do this rather than go through the somewhat unsatisfactory procedure of plotting all three graphs.

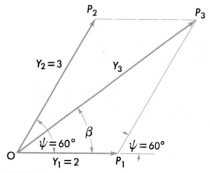

FIGURE 18.10

The student will observe that we can solve the triangle OP_3P_1 in Fig. 18.10 by the cosine law discussed in Secs. 15.6 and 15.7. That is,

$$\begin{aligned}
Y_3{}^2 &= Y_1{}^2 + Y_2{}^2 + 2Y_1Y_2 \cos \psi \\
&= 2^2 + 3^2 + 2 \times 2 \times 3 \cos 60° \\
&= 4 + 9 + 6 = 19 \\
Y_3 &= \sqrt{19} = 4.359 \qquad \text{(ANS.)}
\end{aligned} \tag{30}$$

$$\frac{\sin \beta}{3} = \frac{\sin 120°}{4.359}$$

$$\sin \beta = 3 \times \frac{0.86603}{4.359} = 0.5960$$

$$\beta = 36°35' \qquad \text{(ANS.)}$$

EXERCISE 4

Find the amplitude of the sum of the two sine functions given in Probs. 1 to 5. Also find the phase shift of the sum referred to y_1. It is suggested that the computations be done on a slide rule.

1. $y_1 = 65 \sin \omega t \qquad y_2 = 44 \sin (\omega t + 75.75°)$
2. $y_1 = 3 \sin \omega t \qquad y_2 = 4 \sin (\omega t + 62.7°)$
3. $y_1 = 5 \sin \omega t \qquad y_2 = 4 \sin (\omega t + 78.5°)$
4. $y_1 = 7 \sin \omega t \qquad y_2 = 5 \sin (\omega t - 84.2°)$
5. $y_1 = 5 \sin \omega t \qquad y_2 = 8 \sin (\omega t - 46.56°)$

appendix a
treatment of measured data

A.1 Measured data

Many of the data with which the average technical man works are obtained experimentally. There is a definite limit to their reliability. The *reliability* of a number may be expressed in terms of either precision or accuracy. *Precision* is gauged by the position of the last reliable digit relative to the decimal point, whereas *accuracy* is measured by the number of significant figures. *Significant figures* are those known to be reliable and include any zeros not merely used to locate the decimal point.

For instance, if the diameters of several wires had been measured with a micrometer and found to be 0.118, 0.056, 0.008, and 0.207 in, one might say that these diameters had been measured to a precision of 0.001 in and to accuracies of three, two, one, and three figures, respectively.

The following statements apply to significant figures:

1. *All nonzero digits are significant.*
2. *Zero digits which lie between significant digits are significant.*
3. *Zero digits which lie to the right of both the decimal point and the last nonzero digit are significant.*
4. *Zeros at the beginning of a decimal fraction are not significant.*
5. *Zeros at the end of a whole number, if used only to locate the decimal point, are not significant. When one or more such zeros are known to be significant, the tilde (~) is written over the last significant zero to indicate this fact.*

Example 1

Number	Significant figures	Number of significant figures
35.62	3,5,6,2	4
3,020	3,0,2	3
0.00046	4,6	2
0.000850	8,5,0	3
5.600	5,6,0,0	4
3.0080	3,0,0,8,0	5
12,6̄00	1,2,6,0	4
12,60̄0	1,2,6,0,0	5
40,000	4	1

Should the definition of significant figures seem somewhat arbitrary, let us consider the computation of the volume of a rectangular sheet of metal. Suppose that the measured length, width, and thickness are 165.2, 5.07, and 0.0021 in, respectively, and that these measurements are correct to the last digit given. Let us now compare the effect on the volume of changing the last digit of each measurement by one. It will be seen that such a change introduces respective errors of about $\frac{1}{16}$ of 1 percent, $\frac{1}{5}$ of 1 percent, and 5 percent. The length, then, is the most accurate, and the thickness the least accurate.

A.2 Rounding Off Numbers

Frequently, a result will be rounded off because the last several digits either are in doubt or are not required in that particular computation. The operation of rounding off is governed by the following rule:

If the figures to be rejected represent less than half a unit in the last place to be retained, they are dropped. If they represent more than half a unit in the last place to be retained, the last retained digit is increased by one. If the rejected part is known to represent just half a unit in the last place to be retained, the last retained significant digit, if even, is left even or, if odd, is raised to the nearest even number.

Example 2

Number	Four figures	Rounded off to three figures	Two figures
3.1416	3.142	3.14	3.1
14.815	14.82	14.8	15.

		Rounded off to	
Number	Four figures	three figures	Two figures
321.35	321.4	321	320
6,274.5	6,274	6,270	6,300
20,018	20,020	20,0̃00	20̃,000
71,853	71,850	71,900†	72,000

†71,853 is nearer to 71,900 than to 71,800.

In addition and subtraction the precision of the answer corresponds to the least precise of the quantities involved. *Perform the addition or subtraction and round off by eliminating any digits resulting from operations on broken columns on the right.*

Example 3. Add

$$
\begin{array}{r}
175.6 \\
2.126 \\
13.04 \\
0.0028 \\
\hline
190.7688
\end{array}
$$

Since the last unbroken column on the right is that immediately after the decimal point, we round off to 190.8.

In multiplication and division the accuracy of the answer corresponds to the least accurate of the quantities involved. *Perform the multiplication or division and round off the answer to a number of significant figures equal to that in the least accurate quantity in the computation.*

Example 4. Multiply

$$3.14159 \times 47.82 = 150.2308338$$

Although the multiplicand has six significant figures, the multiplier has only four; therefore we round off the product to four significant figures and get 150.2.

A.3 Scientific Notation

In scientific work a very large or very small number is expressed as a number between 1 and 10 times an integral power of 10. Thus 2,580,000 would be written 2.58×10^6, and 0.0000258 would be written 2.58×10^{-5}. This is called *scientific notation*. The magnitude of the number is revealed by a glance at the exponent (Table 11.2).

Several other advantages in this notation will become apparent. Space is saved, a particularly important point in tabulating data. The labor of counting figures to the right or left of the decimal point—a labor attended by risk of error—is eliminated. The accuracy with which a quantity is known is indicated by the number of figures to the right of the decimal point. For example, when we consider the number 72,000, we cannot tell whether there are two, three, four, or five significant figures. No uncertainty exists when we write 7.2×10^4, 7.20×10^4, 7.200×10^4, or 7.2000×10^4.

The ease of dealing with large and small quantities in this manner is illustrated by the following problem.

Simplify the expression

$$\frac{40\bar{0},000 \times 8,\bar{0}00,000 \times 0.0045}{60,0\bar{0}0 \times 0.025 \times 10\bar{0}} = \frac{4 \times 10^5 \times 8 \times 10^6 \times 4.5 \times 10^{-3}}{6 \times 10^4 \times 2.5 \times 10^{-2} \times 10^2}$$

$$= \frac{4 \times 8 \times 4.5}{6 \times 2.5} \times 10^{(5+6-3)-(4-2+2)} = 9.6 \times 10^4 = 96,000$$

There are two instances in which we depart from the rule of expressing a quantity as a number between 1 and 10 times a suitable power of 10. If we were to extract the square root of 2.5×10^{-7}, we should write this as 25×10^{-8} in order to make the exponent of 10 divisible by the index of the root. The square root is readily seen to be 5×10^{-4}. Also, when quantities are to be added and subtracted, they must have the same exponents. Thus $4 \times 10^{-7} + 7 \times 10^{-5} = 4 \times 10^{-7} + 700 \times 10^{-7} = 704 \times 10^{-7} = 7.04 \times 10^{-5}$.

appendix b
computational aids

B.1 Short Cuts in Computation

In this section we shall attempt to outline some of the more commonly useful short cuts in computation. With the exception of abbreviated multiplication, abbreviated division, and square root, these methods are in no sense approximations. They are intended to serve as a means of obtaining the correct answer at a saving of time and work.

Since most short cuts and approximations are soon forgotten unless constantly used, their treatment in this section is not intended to be complete, but should indicate what can be done as the need arises. The particular need of the student will depend upon his individual situation.

B.2 Aids to Addition and Subtraction

Addition may be facilitated by adding digits two or more at a time, since the order of addition has no effect on the final answer. This is known as the *associative law of addition*.

Example 1. Add

$$8 + \underbrace{2 + 5}_{10} + \underbrace{7 + 6}_{12} + \underbrace{3 + 2 + 9}_{20} = 42$$

Addition by complements may be illustrated by the addition of 237 and 98. It should be arranged as follows: $237 + 100 - 2 = 335$.

Example 2. Addition by Subtotals. If the operation of addition is interrupted for any reason, such as an insistent telephone, the whole process may need to be repeated. The use of subtotals prevents this annoyance.

1,437	
608	
2,972	
541	5,558
10,679	
825	
4,327	15,831
3,146	
655	
1,732	
47	5,580
	26,969

Example 3. Addition by Complements. This method lends itself to the addition of the following typical charges on a department-store bill:

$ 2.97 = $ 3 less $0.03
  6.95 =   7 less 0.05
  4.98 =   5 less 0.02
 11.97 = 12 less 0.03
  1.99 =   2 less 0.01
      $29 less $0.14 = $28.86

Addition may often be most conveniently checked by reversing the order of addition.

Example 4. Subtraction by Complements. If the subtraction involves "borrowing," the use of complements is often helpful.

2,114	2,114 − 738 = 2,114 − (800 − 62)
−738	= 2,114 − 800 + 62
1,376	= 1,314 + 62 = 1,376

B.3 Short Cuts in Multiplication

Multiplication tables might be extended to advantage at least to the square of 16 and preferably to the square of 20.

Some short cuts in multiplication will at once be apparent to the student. A few of these are as follows:

1. *To multiply by 25, divide by 4 and point off two places to the right, for n × 25 = n × 100/4.*
2. *To multiply by 33⅓, divide by 3 and point off two places to the right.*
3. *To multiply by 50, divide by 2 and point off two places to the right.*
4. *To multiply by 75, take ¾ of the number and point off two places to the right.*

For example, multiply 16.8 × 75 (see Example 7 below).

$$
\begin{array}{r}
4/16.8 \\
\underline{4.2} \\
12.60 = 1,260
\end{array}
$$

5. *To multiply by 125, take ⅛ of the number and point off three places to the right.*
6. *To multiply by 250, take ¼ of the number and point off three places to the right.*
7. *To multiply by 500, take ½ of the number and point off three places to the right.*
8. *Multiplication by numbers near 10, 100, etc., may be effected by the method of complements.*

For example,

$$
42 \times 98 = 42(100 - 2) = 4,200 - 84 = 4,116
$$
$$
9 \times 64 = 64(10 - 1) = 640 - 64 = 576
$$

These examples illustrate the *distributive law.*

Example 5. 25 × 5 × 24 = 25 × 120 = 3,000   (associative law)

Example 6. 75 × 17 × 12 = 75 × 12 × 17 = 900 × 17 = 15,300   (commutative law)

Simple compact arrangements for taking certain fractions of numbers are (L. Meyers, "High Speed Mathematics," D. Van Nostrand Co., Inc., Princeton, N.J., 1947):

Example 7. Multiply $\frac{2}{3} \times 147$.

$$3/\overline{147}$$
$$\underline{49} \quad \text{(subtract)}$$
$$98$$

Example 8. Multiply $\frac{7}{8} \times 344$.

$$8/\overline{344}$$
$$\underline{43} \quad \text{(subtract)}$$
$$301$$

Example 9. Multiply $1\frac{1}{4} \times 112$.

$$4/\overline{112}$$
$$\underline{28} \quad \text{(add)}$$
$$140$$

Many other combinations will occur to the student to meet his particular need.

Some multiplications are facilitated by a preliminary multiplication and division by the same quantity.

If two factors of a multiplication problem are multiplied and divided, respectively, by the same nonzero number, the product is not affected. That is,

$$ab = (na)\left(\frac{b}{n}\right)$$

When $n = 2$ as in Example 10 below, the operation is sometimes called the "double-and-halve" method.

Example 10. Multiply $21\frac{1}{2}$ by 16. If we first double $21\frac{1}{2}$ and halve 16, we get $43 \times 8 = 344$, which is easily done as an oral exercise.

Example 11. $6\frac{2}{3} \times 84 = (6\frac{2}{3})(3) \times (84)(\frac{1}{3}) = 20 \times 28 = 560$

Example 12. $175 \times 360 = (175)(4) \times (360)(\frac{1}{4}) = 700 \times 90 = 63,000$

Mixed numbers are often best multiplied by arranging as the product of one binomial by another (Sec. 3.17).

Example 13. Multiply $8\frac{1}{4} \times 6\frac{5}{16}$.
  Observe that

$$(8\tfrac{1}{4}) \times (6\tfrac{5}{16}) = (8 + \tfrac{1}{4})(6 + \tfrac{5}{16})$$

Then

$$(6)(8) + (6)(\tfrac{1}{4}) + (\tfrac{5}{16})(8) + (\tfrac{5}{16})(\tfrac{1}{4}) = 48 + 1\tfrac{1}{2} + 2\tfrac{1}{2} + \tfrac{5}{64} = 52\tfrac{5}{64}$$

This method is especially useful where large mixed numbers are involved.
  Even if the multiplier is an integer, this method is often preferable to the conventional approach of reducing to an improper fraction.

Example 14. $(112\tfrac{3}{8})(6) = (112 + \tfrac{3}{8})(6) = 672 + 2\tfrac{1}{4} = 674\tfrac{1}{4}$

B.4 Casting Out 9s

A common and quite reliable method of checking arithmetic addition and multiplication is known as *casting out 9s*. Arithmetic division can also be checked by this method.
  These checks are based on finding the remainder when a whole number is divided by 9. To find the remainder when a whole number is divided by 9, we can perform the required division as shown in the following example.

Example 15. Find the remainder when 2,895,431 is divided by 9.

```
       321714
9) 2895431
   27
   ──
    19
    18
    ──
     15
      9
     ──
     64
     63
     ──
      13
       9
      ──
      41
      36
      ──
       5      (remainder)
```

Here, the remainder is found to be 5.

Important Note: In this section, when the word *remainder* is used, it refers to the remainder when a given number is divided by 9.

While we shall offer no proof, it is true that the remainder after a given number is divided by 9 is exactly the same as the remainder obtained when the sum of the digits in the given number is divided by 9.

Example 16. Find the remainder when 2,895,431 is divided by 9. Use the *sum-of-digits method.*

The sum of the digits in the number 2,895,431 is

$$2 + 8 + 9 + 5 + 4 + 3 + 1 = 32 \tag{1}$$

The remainder after 32 is divided by 9 can also be found by the sum-of-digits method. This is given by

$$3 + 2 = 5$$

and is the same remainder as was found in Example 15.

A still more efficient way of finding the remainder is to eliminate from the digits to be added any 9s that appear and any combination of digits whose sum is 9. For example, in Eq. (1) we could have eliminated the 9 and combinations of digits whose sum is 9 as indicated:

$$2 + 8 + 9 + 5 + 4 + 3 + 1 \tag{2}$$

2   +   3   = 5   (remainder)

The student should comfirm the remainders obtained in Table B.1.

The actual operation of checking an arithmetic addition rests on the fact that the remainder in the sum is equal to the remainder in the sum of the remainders of the respective addends. In studying Example 17, refer to Table B.1, lines *a, b, c,* and *h.*

Example 17. Find and check the sum of 193,826 and 362,813.

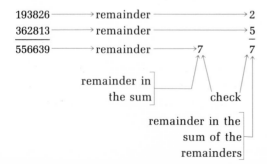

TABLE B.1

	Number	Sum of digits	Remainder†
(a)	193826	1+9+3+8+2+6	2
(b)	362813	3+6+2+8+1+3	5
(c)	556639	5+5+6+6+3+9 = 16	7
(d)	7395	7+3+9+5 = 15	6
(e)	2437	2+4+3+7	7
(f)	284	2+8+4 = 14	5
(g)	1988	1+9+8+8	8
(h)	7	7	7
(i)	3	3	3
(j)	4	4	4
(k)	852	8+5+2 = 15	6
(l)	1136	1+1+3+6	2
(m)	124108	1+2+4+1+0+8	7

† Often called "check number."

Example 18. Check the following addition.

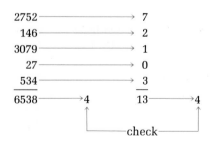

```
2752 ──────────→ 7
 146 ──────────→ 2
3079 ──────────→ 1
  27 ──────────→ 0
 534 ──────────→ 3
─────         ───
6538 ───→4    13 ───→4
      ↑            ↑
      └──── check ─┘
```

Arithmetic multiplication may be checked in a way similar to that used for checking arithmetic addition. This check rests on the fact that the remainder in a product is exactly equal to the remainder in the product of the remainders of the respective factors.

Example 19. Find and check the product of 284 and 7.

In studying this example, refer to Table B.1, lines *f*, *g*, and *h*.

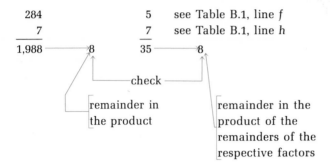

284	5	see Table B.1, line *f*
7	7	see Table B.1, line *h*
1,988 ——→8	35 ——— 8	

check

remainder in
the product

remainder in the
product of the
remainders of the
respective factors

Example 20. Find and check the product of 284 × 437.

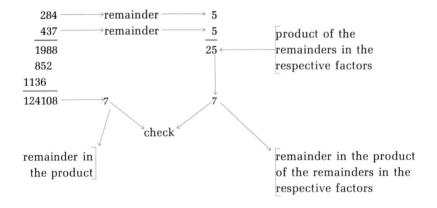

284 ——→remainder ———→ 5	
437 ——→remainder ———→ 5	product of the
1988	25 ← ———remainders in the
852	respective factors
1136	
124108 ——→7	7

check

remainder in
the product

remainder in the product
of the remainders in the
respective factors

The student is warned at this point that the checks we have been discussing are not foolproof. For example, suppose we have multiplied 327 by 485 and obtain a product of 158,559. The excess of 9s in the factors are, respectively, 3 and 8. The product of these is 24, and the excess of 9s in the answer should be 6. Upon investigation the excess of 9s in 158,559 (the product of the given factors) is indeed 6. However, as the student will discover by direct multiplication, the true product is actually 158,595. Observe that checking by casting out 9s does not detect the fact that the last two digits in the product were transposed.

In checking multiplication the overall check is much more reliable if each partial product is verified as shown in the following example.

Example 21. Find and check the product of 284 × 437.

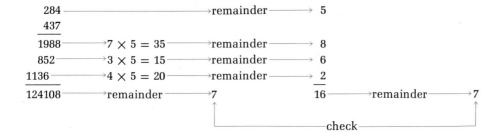

It is usually best to check subtraction by addition and to check division by multiplication. See Examples 22 and 23.

Example 22. Subtract 2,849 from 5,796 and check.

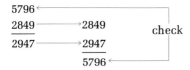

Example 23. Divide 37,592 by 28 and check.
Observe that the check consists in multiplying the integral part of the quotient by the divisor and adding to the resulting product the remainder obtained in the division being checked.

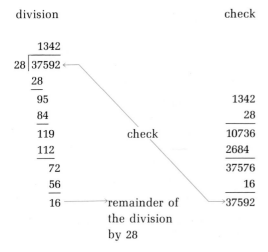

B.5 Abbreviated Multiplication

This is a rapid method of obtaining a product if it is to contain no more significant figures than the multiplier or multiplicand.

Round off both multiplier and multiplicand to one more significant figure than desired in the product. Multiply the multiplicand by the left-hand digit in the multiplier. Round off the right-hand digit of the multiplicand, and multiply by the second digit from the left in the multiplier. Continue using successive digits of the multiplier and rounding off digits in the multiplicand. Keep the right-hand digits of the product in line, and determine the position of the decimal point by inspection.

Example 24. Multiply 385.216 by 0.17278. The complete product, obtained in the usual way, is 66.55762048. Assume four significant figures are wanted in the product.

```
38522
17278
―――――
38522      (1 × 38522)
26964      (7 × 3852)
  770      (2 × 385)
  273      (7 × 39)
   32      (8 × 4)
―――――
66.561     or 66.56      (correct to four significant figures)
```

Somewhat greater accuracy can be obtained by attempting to compensate for discarded digits. For purposes of comparison we shall use the same product to illustrate this point.

```
38522
17278
―――――
38522      (1 × 38522)
26965      (7 × 3852) + 1      (adding 1 to compensate for the discarded digit 2)

  770      (2 × 385)
  270      (7 × 38) + 4        (adding 4 to compensate for the discarded digits 522)
   31      (8 × 3) + 7         (adding 7 to compensate for the discarded digits 8,522)
―――――
66.558
```

Multiplication may be checked by interchanging the multiplier and the multiplicand.

B.6 Short Division

Time and effort can be saved by extending the process of short division to two-digit divisors. For instance, the division of 2,148 by 16 can be thought out about as fast as it can be written down.

$$16\underline{/2148}$$
$$134\tfrac{1}{4}$$

In the division of one mixed number by another, the process is simplified by multiplying the numerator group and the denominator group by the least common multiple of the secondary denominators.

Example 25. $\dfrac{8\tfrac{3}{4}}{3\tfrac{1}{2}} = \dfrac{35}{14}$ $\qquad$ (multiplying top and bottom by 4)

$$= 2\tfrac{1}{2}$$

Example 26. $\dfrac{3\tfrac{1}{3}}{7\tfrac{1}{2}} \times \dfrac{6}{6} = \dfrac{20}{45} = \dfrac{4}{9}$

Where the denominator is an integer, the numerator may be divided into two parts, one of which is the largest possible integral multiple of the denominator.

Example 27. $\dfrac{13\tfrac{1}{2}}{4} = \dfrac{12 + 1\tfrac{1}{2}}{4} = 3 + \tfrac{3}{8} = 3\tfrac{3}{8}$

Or by Examples 25 and 26 we may write

$$\dfrac{13\tfrac{1}{2} \times 2}{4 \times 2} = \dfrac{27}{8} = 3\tfrac{3}{8}$$

Example 28. $\dfrac{17\tfrac{1}{4}}{3} = \dfrac{15 + 2\tfrac{1}{4}}{3} = 5 + \tfrac{3}{4} = 5\tfrac{3}{4}$

Similarly,

$$\dfrac{17\tfrac{1}{4} \times 4}{3 \times 4} = \dfrac{69}{12} = \dfrac{23}{4} = 5\tfrac{3}{4}$$

B.7 Abbreviated Division

This is a relatively fast method of obtaining a quotient containing a certain number of significant figures.

1. *The dividend and divisor are each rounded off to one more significant figure than is to be contained in the quotient. It may be necessary to retain still another digit in the dividend in order to make it larger than the divisor. Decimal points are disregarded during the operation, the position of the decimal point in the quotient being determined by inspection.*
2. *The first division is made using the entire divisor. Thereafter each successive division is made after rounding off successive digits from the right-hand side of the divisor.*

Example 29. Divide 32,586.70143 by 481.60732, carrying out the division to five significant figures in the quotient. Division carried out in the usual manner gives an answer of 67.662388 to eight figures. In the abbreviated division the dividend and divisor are rounded off to seven and six figures, respectively.

$$
\begin{array}{r}
6 \\
481607 \overline{\smash{\big)}\,3258670} \\
2889642 \\
\end{array}
$$

$$
\begin{array}{r}
48161 \overline{\smash{\big)}\,369028/7} \\
337127 \\
\end{array}
$$

$$
\begin{array}{r}
4816 \overline{\smash{\big)}\,31901/6} \\
28896 \\
\end{array}
$$

$$
\begin{array}{r}
482 \overline{\smash{\big)}\,3005/6} \\
2892 \\
\end{array}
$$

$$
\begin{array}{r}
48 \overline{\smash{\big)}\,113/2} \\
96 \\
\end{array}
$$

ANS.: 67.662

B.8 Least Common Multiple

The least common multiple (LCM) of several numbers may be determined by arranging the numbers in a line and dividing by 2, or the lowest prime number which is contained in at least two of the original numbers. When division by the first prime is exhausted, we pass to the next higher prime number which is contained in two or more of the line of quotients. In any case where division is not even, the number is repeated. The process is continued until there are no longer two or more quotients divisible by a common prime. The least common multiple is the product of all the divisors on the left and quotients remaining in the bottom line.

To illustrate, find the LCM of 75, 50, 12, 42, and 105.

```
2/75; 50; 12; 42; 105
3/75; 25;  6; 21; 105
5/25; 25;  2;  7;  35
5/ 5;  5;  2;  7;   7
7/ 1;  1;  2;  7;   7
   1;  1;  2;  1;   1
```

The LCM $= 2 \times 3 \times 5 \times 5 \times 7 \times 2 = 2{,}100$.

B.9 Greatest Common Divisor

The greatest common divisor (GCD) of several numbers is obtained by using the process for LCM, except that the entire row must be divisible. The operation ends when the entire row is no longer divisible by a given prime. The product of the primes used as divisors is the GCD.

Example 30. Find the GCD of 540, 756, and 72.

```
2/540; 756; 72
2/270; 378; 36
3/135; 189; 18
3/ 45;  63;  6
   15;  21;  2
```

The GCD $= 2 \times 2 \times 3 \times 3 = 36$.

B.10 Fractional Equivalents

Table B.2 lists the decimal and percentage equivalents of common fractions.

Knowledge of these equivalents is frequently useful, as the following common types of problems illustrate.

Example 31. Find the amount of pay for 60 h work at 1.62\frac{1}{2}$.

$60 \times 1\frac{5}{8} = 60 \times \frac{13}{8} = \$97\frac{1}{2}$

Example 32. Find the discount at $18\frac{3}{4}$ percent on $24.

$24 \times \frac{3}{16} = \frac{9}{2} = \$4\frac{1}{2}$

TABLE B.2

1/16	1/12	1/8	1/6	1/4	1/3	1/2	Decimal	Percentage
1/16							0.06250	6¼
	1/12						0.08333	8⅓
		1/8					0.12500	12½
			1/6				0.16667	16⅔
3/16							0.18750	18¾
				1/4			0.25000	25
5/16							0.31250	31¼
					1/3		0.33333	33⅓
		3/8					0.37500	37½
	5/12						0.41667	41⅔
7/16							0.43750	43¾
						1/2	0.50000	50
9/16							0.56250	56¼
	7/12						0.58333	58⅓
		5/8					0.62500	62½
					2/3		0.66667	66⅔
11/16							0.68750	68¾
				3/4			0.75000	75
13/16							0.81250	81¼
			5/6				0.83333	83⅓
		7/8					0.87500	87½
	11/12						0.91667	91⅔
15/16							0.93750	93¾

Example 33. Find the increase in cost of 150 saws if the unit price rose from $5.50 to $6.33⅓.

$$150 \times (6.33\tfrac{1}{3} - 5.50) = 150 \times 0.83\tfrac{1}{3} = 150 \times \tfrac{5}{6} = \$125$$

B.11 Percent

This section is a brief review of percentage for the student whose memory needs a refresher. It makes no attempt to treat the subject from an introductory point of view.

We recall that 7 percent (written 7%) means seven one-hundredths, or $\tfrac{7}{100}$, or 0.07.

The fact that 12 is $\tfrac{80}{100}$ of 15 may be written

$$12 = \frac{80}{100} \times 15$$

or

$12 = 0.80 \times 15$

or

$12 = 80\%$ of 15

Observe that the word "of" is sometimes used to indicate multiplication.

In the above example the number 12 is said to be the *percentage,* the number 0.80, or 80%, is said to be the *rate,* and the number 15 is said to be the *base.* The basic equation involved here is

Percentage = rate × base

or

$$p = r \times b$$

If p is the unknown, then

$$p = r \times b = (0.80)(15) = 12$$

If r is unknown, then

$$12 = (r)(15)$$

or

$$\frac{12}{15} = r = 0.80, \text{ or } 80\%$$

If b is unknown,

$$12 = 0.80b \qquad \text{or} \qquad \frac{12}{0.80} = b = 15$$

Example 34. A man receives $12 after 20% is withheld from his pay. What was his pay before taxes?

If $x = $ pay before taxes, then

$$x - 0.20x = 12$$

or

$0.80x = 12$

and

$$x = \frac{12}{0.80} = 15$$

His pay before taxes was $15.

Example 35. A dealer buys a radio for $12 and sells it for $15. What is the percent markup based on the selling price?
    Since the markup is $15 - 12 = \$3$, we may write

$3 = (r)(15)$

Then

$$\frac{3}{15} = r = 0.20, \text{ or } 20\%$$

The markup based on the selling price is 20%.
    If we wished to compute the markup based on the cost, we have

$3 = (r)(12)$

$$\frac{3}{12} = r = 0,25, \text{ or } 25\%$$

Example 36. Assuming that we know a selling price of $15 represents a 25% markup on the cost, find the cost c. Here

$c + 0.25c = 15$
$1.25c = 15$
$$c = \frac{15}{1.25} = 12$$

The cost is therefore $12.
    Observe that in Example 34 we can reduce the pay before taxes, x, to the pay after taxes by subtracting 0.20x. That is,

Pay after taxes $= x - 0.20x = 0.80x$

We can accomplish the same result by multiplying the pay before taxes by 0.80 to obtain $0.80x$.

Similarly, in Example 36, we can increase c by 25% by adding $0.25c$ to c or by multiplying c by 1.25. That is,

$$c + 0.25c = 1.25c$$

Example 37. Take-home pay is $124.10 after 16% withholding. Find the pay before taxes.

If a is the pay before taxes, then

$$a - 0.16a = \$124.10$$

or

$$0.84a = 124.10$$

and

$$a = \frac{124.10}{0.84} = \$147.74$$

Example 38. A radio sold for $16.80, which was a 25% markup on the cost. Find the cost.

If c is the cost, then

$$c + 0.25c = 16.80$$
$$1.25c = 16.80$$
$$c = \frac{16.80}{1.25} = \$13.44$$

Example 39. What total volume of unmixed cement, sand, and gravel is needed to make 80 yd^3 of concrete if there is a 12% shrinkage on mixing?

If v is the initial unmixed volume, then

$$v - 0.12v = 80$$
$$0.88v = 80$$
$$v = \frac{80}{0.88} = 90.9 \text{ yd}^3$$

Example 40. A transistor radio is listed at $9.60 less a 15% discount. A 4% sales tax is paid on the final selling price. How much must be paid for the radio?

$(9.60)(0.85) = \$8.16 =$ selling price before tax
$8.16 \times 0.04 = \$0.33 =$ tax
$8.16 + 0.33 = \$8.49 =$ selling price including tax

Example 41. A house depreciates annually 7% of its value at the beginning of each year. If the initial value is \$20,000, find its value at the end of 6 years.

$$v = (20,000)(0.93)^6 = \$12,940 \text{ approx}$$

Example 42. The basic state income tax is 14% of the taxable income, and the surtax is 20% of the basic tax. Find the total tax on a taxable income of \$15,000.

$$(0.14)(15,000) = 2,100$$
$$(0.20)(2,100) = 420$$
$$2,100 + 420 = \$2,520 \text{ total tax}$$

Example 43. What overall tax rate is represented in Example 42?

$$\frac{2,520}{15,000} = 0.168, \text{ or } 16.8\%$$

Note: A more direct solution is given by

$$(0.14)(1.20) = 0.168 = 16.8\%$$

B.12 Short Methods of Squaring

Integers and mixed numbers may be squared easily by a process based on the algebraic identity

$$a^2 - b^2 = (a + b)(a - b)$$

Adding b^2 to both sides of the equation,

$$a^2 = (a + b)(a - b) + b^2$$

Example 44. Find $(96)^2$.

Here $a = 96$, and a value of b is chosen such that either $a + b$ or $a - b$ becomes a round number, facilitating the computation. In this case we choose $b = 4$. Then

$$(96)^2 = (96 + 4)(96 - 4) + 4^2 = (100)(92) + 16 = 9,216$$

Example 45. Find $(53)^2$.

$a = 53$; take $b = 3$. Then

$(53)^2 = (53 + 3)(53 - 3) + 3^2 = (56)(50) + 9 = 2,809$

Example 46. Find $(13\frac{1}{2})^2$.

$a = 13\frac{1}{2}$; take $b = \frac{1}{2}$. Then

$(13\frac{1}{2})^2 = (13\frac{1}{2} + \frac{1}{2})(13\frac{1}{2} - \frac{1}{2}) + (\frac{1}{2})^2 = (14)(13) + \frac{1}{4} = 182\frac{1}{4}$

This may be generalized into the following rule:

To square a mixed number ending in $\frac{1}{2}$, multiply the next higher whole number by the next lower whole number and add $\frac{1}{4}$.

Example 47. Find $(135)^2$.

$a = 135$; take $b = 5$. Then

$(135)^2 = (135 + 5)(135 - 5) + (5)^2$
$\qquad = (140)(130) + 25 = 18,225$

The procedure for the above calculation may be stated as follows:

To square a number ending in 5, multiply the next lower integral multiple of 10 by the next higher integral multiple and add 25 to the result.

Example 48. Find $(7\frac{1}{4})^2$.

$a = 7\frac{1}{4}$; take $b = \frac{3}{4}$. Then

$(7\frac{1}{4})^2 = (7\frac{1}{4} + \frac{3}{4})(7\frac{1}{4} - \frac{3}{4}) + (\frac{3}{4})^2$
$\qquad = (8)(6\frac{1}{2}) + \frac{9}{16} = 52\frac{9}{16}$

Example 49. Find $(8\frac{1}{3})^2$.

$a = 8\frac{1}{3}$; take $b = \frac{2}{3}$. Then

$(8\frac{1}{3})^2 = (8\frac{1}{3} + \frac{2}{3})(8\frac{1}{3} - \frac{2}{3}) + (\frac{2}{3})^2$
$\qquad = (9)(7\frac{2}{3}) + \frac{4}{9} = 69\frac{4}{9}$

B.13 Short Cut in the Use of the Pythagorean Theorem

When the sides of a right triangle are given in the common binary system of linear measurement (for example, $\frac{1}{2}$ in, $\frac{1}{4}$ in, $\frac{1}{8}$ in, etc.), the work involved in the application of the Pythagorean theorem may be reduced by taking as a unit of length $1/n$, where n is the least common denominator.

Example 50. Find the hypotenuse of a right triangle whose sides are $1\frac{5}{8}$ and $2\frac{3}{16}$ in.

Since 16 is the LCD, we shall take $\frac{1}{16}$ in as the unit of length. Then

$1\frac{5}{8} = 26$ units

and

$2\frac{3}{16} = 35$ units

and

$$\sqrt{(26)^2 + (35)^2} = 43.60 \text{ units}$$

$$\frac{43.60}{16} = 2.725 \text{ in}$$

If the nearest $\frac{1}{64}$ in is desired,

$$\frac{43.60}{16} = \frac{x}{64}$$

$$x = 174.4 \quad \text{(call it 174)}$$

$$\frac{174}{64} = \frac{87}{32} = 2\frac{23}{32} \text{ in}$$

B.14 Suggestions for Cancellation

Cancellation involving very large or very small numbers is more easily accomplished by making use of scientific notation.

The average student recognizes numbers which contain such factors as 2, 4, 5, and 10. Further, if the digits of a number add up to a sum divisible by 3, the number itself is divisible by 3. Similarly, if the digits add up to a sum divisible by 9, the number itself is divisible by 9.

Occasionally, it will be advisable not to cancel completely. For example, 1,872/400 should be reduced only to 468/100 instead of completely to 117/25, as the answer 4.68 is at once apparent in the former expression.

Large factors may be revealed by factoring out smaller quantities, even though it can be seen that the smaller factors will not cancel out. A typical example is

$$\frac{141}{235} = \frac{3 \times \cancel{47}}{5 \times \cancel{47}} = \frac{3}{5}$$

In making computations involving both multiplication and division, cancel as much as feasible. Then carry out the necessary multiplication, followed by division.

B.15 Averaging by Averaging Departures

If a series of numbers shows but little fluctuation, the average value may be readily obtained by averaging the departures from an arbitrarily chosen reference base.

Example 51. Average the following set of numbers, assuming a reference base of 18.60.

	Departure
18.62	+2
18.65	+5
18.59	−1
18.60	0
18.64	+4
18.57	−3
18.56	−4
18.62	+2
	+5

$$\frac{0.05}{8} = +0.006$$

The average is 18.60 + 0.006, or 18.61.

B.16 Absolute and Relative Errors

The *absolute error* is the approximate value minus the true value of a number. It is positive or negative according to whether the approximate value is larger or smaller than the true value.

The *relative error* is the ratio of the absolute error to the exact value. Since the relative error is a ratio between two like quantities, it is an abstract number and is often expressed as a percentage.

Example 52. The actual length of a metal bar is 11.52 in. The length as measured with a scale with a worn end is 11.56 in. Find the actual and relative errors.

The actual error is $11.56 - 11.52 = 0.04$ in.

The relative error is $(11.56 - 11.52)/11.52 = 0.00347 = 0.3$ percent.

B.17 Iterative Process for Obtaining a Square Root

An iterative, or repetitive, process is based on assuming a value of the quantity to be found. The assumed value is tested against known information, which leads to a second and closer approximation. The second approximation is "fed back" into the process, giving us a third and still better approximation, and so on until the desired accuracy is obtained.

Specifically, in applying an iterative process to finding the square root of a number, we obtain from tables, slide rule, or mental estimate a reasonable first approximation for the square root. The number is divided by the first approximation, giving a quotient somewhat different from the divisor. Evidently the true square root must be between the values of the first trial divisor and the quotient; therefore the average of these quantities is taken for the second trial divisor. The process may be repeated (hence the term "iterative") until the desired accuracy is achieved. The process has the advantage of being self-correcting. An error, or a poor choice of the first trial divisor, will merely increase the number of iterations needed to obtain the desired answer. This method is very well adapted for use with a desk calculator.

It can be shown that if the trial divisor and quotient agree to n significant figures, their average accurately represents the square root to at least $2n$ significant figures.

Example 53. Find $\sqrt{10}$ to eight decimal places.

Assume

$$\sqrt{10} = 3 \text{ (first trial divisor)} \qquad \frac{10}{3} = 3.33$$

$$\frac{3.33 + 3}{2} = 3.16 \text{ (second trial divisor)} \qquad \frac{10}{3.16} = 3.16456$$

$$\frac{3.16456 + 3.16}{2} = 3.16228 \text{ (third trial divisor)} \quad \frac{10}{3.16228} = 3.1622753203$$

$$\frac{3.16228 + 3.1622753203}{2} = 3.1622776602$$

which we round off to 3.16227766, since only eight decimal places were required. (3.1622753203 may be rounded off to 3.16228; hence there are six significant digits of agreement and we are justified in retaining all 11 digits of 3.1622776602.)

appendix c
weights and measures

Avoirdupois Weight

1 oz = 28.35 g (gram)
1 lb = 16 oz = 453.6 g = 0.4536 kg (kilogram)
1 ton = 2,000 lb = 907.2 kg

Linear Measure

1 in = 2.54 cm (centimeter)
1 ft = 12 in = 30.48 cm
1 yd = 3 ft = 0.914 m (meter)
1 rod = 5½ yd = 16½ ft
1 statute (land) mi = 5,280 ft = 1.609 km (kilometer)
1 nautical mi = 6,070 ft = 1.151 statute mi = 1.852 km
(1 nautical mi = length of 1 minute of longitude at earth's equator)

Area Measure

1 ft^2 = 144 in^2
1 yd^2 = 9 ft^2
1 acre = 43,560 ft^2
1 mi^2 = 640 acres

Cubic Measure

1 in^3 = 16.39 cm^3 [1 cm^3 (cubic centimeter) approx = 1 ml (milliliter)]
1 ft^3 = 1,728 in^3 = 7.48 gal = 28.32 l (liters)

Liquid Measure

1 fluid oz = 29.57 cm^3 = 1.04 oz avoirdupois of freshwater
1 pt = 16 fluid oz = 0.473 l
1 qt = 2 pt = 0.946 l
1 gal = 4 qt = 231 in^3 = 3.79 l = 8.35 lb freshwater
1 barrel = 31$\frac{1}{2}$ gal

Dry Measure

2 pt = 1 qt
1 pk = 8 qt
1 bu = 4 pk = 2,150 in^3

Miscellaneous

1 ft^3 freshwater = 7.48 gal = 62.4 lb
1 knot = 1 nautical mi/h
1 light-year = distance light travels in 1 year at 186,000 mi/s = 5.86 × 10^{12} mi

Prefixes Used in Metric System

Prefix	*Meaning*	*Example*
milli	0.001 = 10^{-3}	1 millimeter = 0.001 meter
centi	0.01 = 10^{-2}	1 centimeter = 0.01 meter
deci	0.1 = 10^{-1}	1 decimeter = 0.1 meter
deka	10	1 dekameter = 10 meters
hecto	100 = 10^2	1 hectometer – 100 meters
kilo	1,000 = 10^3	1 kilometer = 1,000 meters
miga	1,000,000 = 10^6	

Other Metric Units

1 micrometer (μm) = 0.000001 meter = 10^{-6} meter
1 millimicrometer (mμm) $\quad\quad$ = 10^{-9} meter
1 micromicrometer ($\mu\mu$m) $\quad\quad$ = 10^{-12} meter
1 angstrom (Å) $\quad\quad\quad$ = 10^{-10} meter

appendix d

Formulas Used in Satellite Problems (Circular Orbits Only)

Mass: $\dfrac{M_b}{M_e} = M$

where M_b = mass of attracting body
   M_e = mass of earth
   M = mass of attracting body relative to mass of earth (earth = 1)

Acceleration of gravity: $g = \dfrac{32.2M}{(r/3{,}960)^2}$ (1)

where g = gravitational acceleration, ft/s²
   r = distance from center of attracting body, mi

Orbital velocity: $v = 4.91\sqrt{M} \cdot \sqrt{\dfrac{3{,}960}{r}}$ where v is in mi/s   (2)

Period of revolution: $T = 84.6\left(\dfrac{r}{3{,}960}\right)^{3/2} \cdot \sqrt{\dfrac{1}{M}}$ where T is in min   (3)

PLANETARY DATA
(Some of these data represent a compromise among several sources)

	Av diam, mi	Relative mass (Earth = 1)	Period of revolution	Surface gravity, ft/s²	Av distance from sun, 10⁶ mi	Surface† escape velocity, mi/s	Period of rotation on axis	Mean velocity in orbit, mi/s
Earth	7,920	1.000	365.3 days	32.2	93	6.97	23 h 56 min	18.5
Moon	2,160	0.0122	27.3 days	5.28	‡	1.49	27.3 days	0.63
Sun	864,000	330,000		900		387	25 days	
Mars	4,200	0.108	1.88 years	12.4	141.7	3.22	24 h 37 min	15
Venus	7,700	0.81	224.7 days	27.5	67.3	6.4	245 days	22
Mercury	3,100	0.041	88.0 days	8.60	36.0	2.2	88 days	30
Saturn	74,500	95	29.46 years	34.5	886	22	10 h 14 min	6
Jupiter	88,640	317	11.86 years	85.0	484	37	9 h 50 min	8
Uranus	32,000	14.7	84.0 years	29.3	1,783	13	10.8 h	4
Neptune	31,000	17.2	164.8 years	36.1	2,795	14	15.8 h	3
α Centauri					4.31 light-years§			

† Assuming no atmosphere.
‡ 239,000 mi from earth center to center (mean value).
§ 1 light-year = distance light can travel in 1 year at 186,000 mi/s = 5.86 × 10¹² mi.

$vT = 2\pi r$ (using consistent units)

Velocity of escape: $v_e = 6.97\sqrt{M} \cdot \sqrt{\dfrac{3{,}960}{r}}$ where v_e is in mi/s (4)

For an artificial earth satellite (AES) in equatorial orbit (as opposed to a polar orbit) the rotation of the earth must be allowed for in observing the period.

For W-E revolution: $\dfrac{1}{T_a} = \dfrac{1}{T} - \dfrac{1}{24}$ (5)

where T is the true period of the AES in hours and T_a is the apparent period as observed from a given point on the equator.

For E-W revolution: $\dfrac{1}{T_a} = \dfrac{1}{T} + \dfrac{1}{24}$ (6)

bibliography

General

Bakst, A.: "Mathematics: Its Magic and Mastery," D. Van Nostrand Company, Inc., Princeton, N.J., 1945.

Bell, E. T.: "Men of Mathematics," Simon & Schuster, Inc., New York, 1965.

Berkeley, E.: "A Guide to Mathematics for the Intelligent Non-mathematician," Simon & Schuster, Inc., New York, 1966.

Boehm, G.: "The New World of Mathematics," The Dial Press, Inc., New York, 1959.

Cooley, H. R., D. Gans, M. Kline, and H. E. Wahlert: "Introduction to Mathematics," Houghton Mifflin Company, Boston, 1968.

Dadourian, H. M.: "How to Study—How to Solve," Addison-Wesley Publishing Company, Inc., Reading, Mass., 1951.

Griffin, F. L.: "Introduction to Mathematical Analysis," Houghton Mifflin Company, Boston, 1936.

Kline, M.: "Mathematics: A Cultural Approach," Addison-Wesley Publishing Company, Reading, Mass., 1962.

———: "Mathematics and the Physical World," Thomas Y. Crowell Company, New York, 1959.

———: "Mathematics in Western Culture," Oxford Book Company, Inc., New York, 1953.

Kramer, E. E.: "The Main Stream of Mathematics," Fawcett World Library, New York, 1951.

"Mathematics in the Modern World," readings from *Scientific American,* W. H. Freeman and Company, San Francisco, 1968.

"Mathematics Teacher," National Council of Teachers of Mathematics, Washington. Numerous articles of interest.

Mellor, J. W.: "Higher Mathematics for Students of Chemistry and Physics," Dover Publications, Inc., New York, 1946.

Newman, J. R.: "World of Mathematics," Simon & Schuster, Inc., New York, 1956.

Polya, G.: "How to Solve It," Doubleday & Company, Inc., Garden City, N.Y., 1957.

———: "Mathematics and Plausible Reasoning," Princeton University Press, Princeton, N.J., 1954.

Struik, D. J.: "A Concise History of Mathematics," Dover Publications, Inc., New York, 1948.

Vergara, W. C.: "Mathematics in Everyday Things," Harper & Row, Publishers, Incorporated, New York, 1959.

Applications

Cell, J. W.: "Engineering Problems Illustrating Mathematics," McGraw-Hill Book Company, New York, 1943.

Cooke, N. M., and H. Adams: "Basic Mathematics for Electronics," 3d ed., McGraw-Hill Book Company, New York, 1970.

Corrington, M. S.: "Applied Mathematics for Technical Students," Harper & Row, Publishers, Incorporated, New York, 1952.

Dull, R., and R. Dull: "Mathematics for Engineers," McGraw-Hill Book Company, New York, 1951.

Mira, J. A.: "Geometry through Practical Applications," Barnes & Noble, Inc., New York, 1961.

Palmer, C. I., S. F. Bibb, J. A. Jarvis, and L. A. Mrachek: "Practical Mathematics," 5th ed., McGraw-Hill Book Company, New York, 1970.

"Source Book of Mathematical Applications: 17th Yearbook," National Council of Teachers of Mathematics, Washington, 1944.

Waldron, R. A.: "Waves and Oscillations," D. Van Nostrand Company, Inc., Princeton, N.J., 1964.

Recreational and Enrichment

Beiler, A. H.: "Recreations in the Theory of Numbers," Dover Publications, Inc., New York, 1964.

Bergamini, D.: "The Mathematics of Beauty in Nature and Art," Time-Life's Life Science Library Series, 1963, pp. 88–102.

Bowers, H., and J. E. Bowers: "Arithmetical Excursions," Dover Publications, Inc., New York, 1961.

Bryant, S. J., K. G. Wiley, and G. E. Graham: "Non-routine Problems in Algebra, Geometry, and Trigonometry," McGraw-Hill Book Company, New York, 1965.

Charosh, M. (ed.): "Mathematical Challenges," selected problems from *Mathematics Student Journal,* National Council of Teachers of Mathematics, Washington, 1965.

Courant, R., and H. Robbins: "What is Mathematics?" Oxford Book Company, Inc., New York, 1953.

Cundy, H. M., and A. P. Rollett: "Mathematical Models," Oxford University Press, London, 1953.

Dantzig, T.: "Number, the Language of Science," Doubleday & Company, Inc., Garden City, N.Y., 1954.

Gamow, G.: "One, Two, Three—Infinity," The Viking Press, Inc., New York, 1961.

Gardiner, M.: "Mathematics, Magic and Mystery," Dover Publications, Inc., New York, 1956.

Glenn, W., and D. A. Johnson: "Exploring Mathematics on Your Own," Doubleday & Company, Inc., Garden City, N.Y., 1961.

Graham, L. A.: "The Surprise Attack in Mathematical Problems," Dover Publications, Inc., New York, 1968.

————: "Ingenious Mathematical Problems and Methods," Dover Publications, Inc., New York, 1959.

Hogben, L.: "Mathematics in the Making," Doubleday & Company, Inc., Garden City, N.Y., 1960.

————: "Mathematics for the Millions," 4th ed., W. W. Norton and Company, Inc., New York, 1968.

Huntley, H. E.: "The Divine Proportion," Dover Publications, Inc., New York, 1970.

Jacobs, H. R.: "Mathematics, a Human Endeavor," W. H. Freeman and Company, San Francisco, 1970.

Kasner, E.: "Mathematics and the Imagination," Simon & Schuster, Inc., New York, 1940.

Land, F.: "The Language of Mathematics," Doubleday & Company, Inc., Garden City, N.Y., 1963.

Menninger, K. W.: "Mathematics in Your World," The Viking Press, Inc., New York, 1962.

Meyer, J. S.: "Fun with Mathematics," Fawcett World Library, New York, 1958.

Radamacher, H., and O. Toeplitz: "The Equipment of Mathematics," Princeton University Press, Princeton, N.J., 1957.

Sawyer, W. W.: "The Search for Pattern," Penguin Books, Inc., Baltimore, 1970.

————: "Prelude to Mathematics," Penguin Books, Inc., Baltimore, 1955.

————: "Mathematician's Delight," Penguin Books, Inc., Baltimore, 1952.

Schuh, F.: "The Master Book of Mathematical Recreations," Dover Publications, Inc., New York, 1968.

Singh, J.: "Great Ideas of Modern Mathematics," Dover Publications, Inc., New York, 1959.

Steinhaus, H.: "Mathematical Snapshots," Oxford Book Company, Inc., New York, 1969.

answers to odd-numbered problems

Chapter 1

Exercise 1 (*page 2*): **1.** 102,227 **3.** 300,222,296 **5.** 177 + 34 remainder
7. 1,304.8 **9.** 16.071 **11.** 66.473 **13.** 161.03 **15.** 690.40 **17.** 87.728 **19.** 156$\frac{1}{4}$
21. 13,225 **23.** 9,604 **25.** 10,609 **27.** 994,009 **29.** 125.44 **31.** 52,788 **33.** 523,476
35. 25.100 **37.** 0.86603 **39.** 4.6690 **41.** 6.9282 **43.** 35.327 **45.** 600 **47.** 3$\frac{5}{16}$
49. 52$\frac{5}{64}$ **51.** $\frac{3}{4}$ **53.** $\frac{7}{26}$ **55.** 490 **57.** 990

Exercise 2 (*page 3*): **1.** $2,143.20 **3.** 0.00488 mi **5.** 58$\frac{3}{4}$ lb **7.** $12.47 **9.** $1.61
11. 14.0 mi/h **13.** 0.0130 ft **15.** 465 **17.** $\frac{49}{64}$ in **19.** 59% **21a.** $1,458 **21b.** 24.3%
23. $135.00 **25.** $7.70 **27.** 91.3 year **29.** $3,368; $2,936; $2,504; $2,072; $1,640 **31.**

	Book value, straight line	Book value, sum-of-years digits
New	$195.00	$195.00
End year 1	171.67	155.00
End year 2	148.33	121.67
End year 3	125.00	95.00
End year 4	101.67	75.00
End year 5	78.33	61.67
End year 6	55.00	55.00

33. 6.43 h, or 6 h 26 min **35.** $152.75

Exercise 3 (*page* 5): **1.** 28 **3.** 2.4 **5.** 90 **7.** 240 **9.** 600 **11.** $\frac{1}{10}$

Chapter 2

Exercise 1 (*page* 10): **1a.** 1,270 **1c.** 1,872 **1e.** 341 **1g.** 792 **1h.** 207 **1j.** 153 **1l.** 432 **1n.** 875 **1o.** 138 **1q.** 243 **1s.** 476 **1u.** 96

Exercise 2 (*page* 12): **1.** 68.0 **3.** 29.7 **5.** 110,000 **7.** 887,000 **9.** 907,000 **11.** 415,000 **13.** 2,240 **15.** 1.16 **17.** 0.252 **19.** 2,110,000 **21.** 7,500

Exercise 3 (*page* 13): **1.** 0.360 **3.** 136 **5.** 14.6 **7.** 0.0440 **9.** 0.0267 **11.** 0.000497 **13.** 2.40

Exercise 4 (*page* 14): **1.** 1.6×10^2 **3.** 1.54×10^{-2} **5.** 1.216×10^2 **7.** 6.12×10^{-2} **9.** 3.1×10^{-3} **11.** 6.56×10^{-5}

Exercise 5 (*page* 15): **1.** 9.41 **3.** 5.44 **5.** 8.34 **7.** 11.9

Exercise 6 (*page* 20): **1.** 324 **3.** 36.6 **5.** 28,200 **7.** 0.01111 **9.** 268,000 **11.** 720,000,000 **13.** 0.136 **15.** 6.27 **17.** 1.677 **19.** 4.66 **21.** 0.0838 **23.** 0.915 **25.** 0.233 **27.** 214 **29.** 2,660 **31.** 343 **33.** 1,685 **35.** 0.00387 **37.** 12,470 **39.** 109,200,000 **41.** 480,000,000 **43.** 8.48 **45.** 0.191 **47.** 2.38 **49.** 42.4 **51.** 211 **53.** 7.79 **55.** 34.55

Exercise 7 (*page* 21): **1.** $d = 12.6$ mi; $C = 39.6$ mi; $A = 124.7$ mi^2 **3.** $d = 156$ ft; $C = 490$ ft; $A = 19,100$ ft^2 **5.** $r = 5.6$ yd; $C = 35.2$ yd; $A = 98.5$ yd^2 **7.** $r = \frac{7}{16}$ in; $C = 2.75$ in; $A = 0.601$ in^2 **9.** $r = 1.342$ in; $d = 2.683$ in; $A = 5.66$ in^2 **11.** $r = 1.785$ in; $d = 3.57$ in; $C = 11.22$ in

Exercise 8 (*page* 24): **1.** 60.4 **3.** 663 **5.** 27.4 **7.** 16.65 **9.** 4.156 **11.** 2.55

Exercise 9 (*page* 25): **1.** 52.0865 **3.** 15.0047 **5.** 1.75183

Chapter 3

Exercise 1 (*page* 39): **1a.** 13 **1c.** 14 **1e.** -14 **1g.** 4 **2a.** 3 **2c.** -6 **2e.** 8 **2g.** -22 **3a.** 4 **3c.** 5 **4a.** 40 **4c.** 48 **4e.** -24 **4g.** -36 **5a.** 4 **5c.** -5 **5e.** $\frac{4}{3}$ **5g.** $-\frac{5}{3}$

Exercise 2 (*page* 41): **1.** $3a$ **3.** y^4 **5.** $4xy^2z^3$ **7.** w^4x^2 **9.** 7 **11.** 77 **13.** 20 **15.** 15 **17.** 49 **19.** 25 **21.** 9 **23.** 5 **25.** -8 **27.** 0 **29.** 900 **31.** $11x$ **33.** $13mn$

35. $6x + 9y$ **37.** $12m + 8p$ **39.** $4k^2 - 2km + 5m^2$ **41.** $9x - 2y - 3z$ **43.** $4x - y$
45. $12p$ **47.** $2k^2 - 2km - 3m^2$ **49.** $-x - 2y + 3z$ **51.** $-a^2 + a + 1$
53. $-6x^2 - 3xy + y^2$

Exercise 3 (*page 44*): **1.** $5a + 3b$ **3.** $2x$ **5.** $4m - n$ **7.** $-6b - c$
9. $16ab - a^2b - 2ab^2$ **11a.** $5a + 2b - (4c - m + x)$
11b. $5a + 2b + (-4c + m - x)$ **13a.** $7mn + 3m^2 - 4n^2 - (8m - 5n - 2mn)$
13b. $7mn + 3m^2 - 4n^2 + (-8m + 5n + 2mn)$
15a. $9w^2 + 5y^3 - (2w^2y + 3wy^2 - w^3)$ **15b.** $9w^2 + 5y^3 + (-2w^2y - 3wy^2 + w^3)$

Exercise 4 (*page 45*): **1.** x^6 **3.** y^7 **5.** a^{10} **7.** m^{15} **9.** $25x^6$ **11.** $64b^{24}$ **13.** x^3
15. y^5 **17.** $-b$

Exercise 5 (*page 46*): **1.** $32m^2nx$ **3.** $-36ab^2c^2d$ **5.** $a^3m^3x^3$ **7.** $-72a^5b^{11}c^9m^{11}$
9. $-18my + 15ty$ **11.** $30a^4b - 42a^3b^2 - 54a^2b^3$ **13.** $24a^3h^4k^4 - 66bh^2k^5$

Exercise 6 (*page 47*): **1.** $ac + bc + ad + bd$ **3.** $h^2 - k^2$ **5.** $6m^2 + mw - 35w^2$
7. $24x^2 - 78x - 39y - 6y^2$ **9.** $b^2 - x^2 - 2xy - y^2$ **11.** $n^3 - 8$
13. $x^5 + 3x^3 + 3x^2 + 1$

Exercise 7 (*page 48*): **1.** $17 - 120x$ **3.** $120x$

Exercise 8 (*page 48*): **1.** $2x + 2y$ **3.** $2c + 2y + 6x$ **5.** $xy - ab$ **7.** $cy - 12x^2$
9. $20x^2$ **11.** Surface $= 2a^2 + ab - 2b^2$; volume $= ab(a - b)$ **13.** $(D - d)/2$

Exercise 9 (*page 50*): **1.** $3c$ **3.** $-5c^2$ **5.** $4x^9$ **7.** $8xy^3/wz^2$ **9.** $3b - 4c$
11. $-2m + 3k$ **13.** $r + 2$ **15.** $5kr/3h + 5/2r - 7h/4r$

Exercise 10 (*page 53*): **1.** $a + 8$ **3.** $4m - 5w$ **5.** $2k^4 + 4k^2 + 8$
7. $h^4 - h^3 - h + 1$ **9.** $32x^5 + 16x^4 + 8x^3 + 4x^2 + 2x + 1$
11. $1/x - 1/x^2 + 1/x^3 - 1/x^4 + \cdots$

Exercise 11 (*page 55*): **1.** $20axy - 28bxy$ **3.** $6a^2b^2cx - 8ab^3cy + 14ab^2c^2z$
5. $x^2 - 9$ **7.** $64k^2 - 9m^2$ **9.** $81p^2q^2 - 49r^2$ **11.** $64 + 16n + n^2$ **13.** $16b^2 + 24b + 9$
15. $144a^2m^2 + 168amy + 49y^2$ **17.** $36 - 12m + m^2$ **19.** $25x^2 - 10x + 1$
21. $49t^2 - 126t + 81$ **23.** $25h^2 - 120hn + 144n^2$ **25.** $m^2 + 4m - 21$
27. $a^2 - 15a + 44$ **29.** $z^2 - 16z + 48$ **31.** $8c^2 + 18c - 5$ **33.** $30x^2 - 17x + 2$

Exercise 12 (*page 56*): **1.** $3(5x - 6y)$ **3.** $6(5ab - 7km)$ **5.** $a(a - c)$
7. $c(ab + bm + mx)$ **9.** $6x^2y(7a - 4bxy + 3cx^2)$ **11.** $5(2a^2x^2 - 3abxy + 4b^2y^2)$
13. No common factors

Exercise 13 (*page 56*): **1.** $(x + 2)(x - 2)$ **3.** $(6 + y)(6 - y)$ **5.** $(7m + 1)(7m - 1)$ **7.** $(5d + 8m)(5d - 8m)$ **9.** $(10yz + 7cd)(10yz - 7cd)$ **11.** $(4n^4 + 1)(2n^2 + 1)(2n^2 - 1)$ **13.** 49

Exercise 14 (*page 57*): **1.** $10x$ **3.** $14z$ **5.** $4a$ **7.** $30n$ **9.** $80bh$ **11.** 25 **13.** $49q^2$ **15.** $25x^2$

Exercise 15 (*page 57*): **1.** $(x - 2)^2$ **3.** $(6 - m)^2$ **5.** $(3 + 7z)^2$ **7.** $(8a - 5n)^2$ **9.** $(9m^2 - 4p)^2$

Exercise 16 (*page 57*): **1.** $(y + 3)(y + 4)$ **3.** $(m - 5)(m - 7)$ **5.** $(x + 3)(x + 16)$ **7.** $(h - 5)(h + 6)$ **9.** $(k^2 + 9)(k^2 - 8)$ **11.** $(x + 3y)(x + 18y)$ **13.** $(a + 1)(2a + 1)$ **15.** $(4m - 3)(m - 2)$ **17.** $(4b + 3)(2b + 3)$ **19.** $(9h - 10)(6h + 5)$

Exercise 17 (*page 58*): **1.** $3c(3a + 5b)$ **3.** $4(a + 5)(a - 5)$ **5.** $6(b - 2)(b - 1)$ **7.** $4(4k^2 + 9)$ **9.** $(9h - 25)(h - 1)$ **11.** $(9x^2 + 4)(3x + 2)(3x - 2)$ **13.** $2(2y - 3)(y - 6)$ **15.** $4(3t + 5)^2$ **17.** No factors **19.** $(ay + 2)(a^2y^2 - 2ay + 4)$ **21.** $(x + 2y + z)^2$ **23.** $(x + 1)^2(x - 1)$ **25.** $(3a - b)(x + y)$ **27.** $b(m - n)(m^2 + mn + n^2)$ **29.** $(5c + 3d - f)^2$

Exercise 18 (*page 61*): **1.** $-\dfrac{y}{5}$ **3.** $-\dfrac{x}{2}$ **5.** $\dfrac{3 - c}{4}$ **7.** $\dfrac{1 - a}{a + 1}$ **9.** $\dfrac{b + c}{a - c}$ **11.** $\dfrac{x - y}{(z - x)(y + z)}$ or $\dfrac{y - x}{(x - z)(y + z)}$ **13.** $\dfrac{(d - c)^3}{p + r}$

Exercise 19 (*page 62*): **1.** $\dfrac{y}{w + z}$ **3.** $\dfrac{m + n}{b - c}$ **5.** $\dfrac{1}{a + b}$ **7.** $\dfrac{2}{a - 2b}$ **9.** $\dfrac{3}{2a - 3b}$ **11.** $\dfrac{a + 3}{3a}$ **13.** Not reducible **15.** $\dfrac{2(b - a)}{b + a}$ **17.** Not reducible **19.** $-\dfrac{(b - 2c)^2}{2b + c}$ **21.** Not reducible **23.** Not reducible **25.** 1

Exercise 20 (*page 63*): **1.** $\dfrac{3(a + b)}{4}$ **3.** $\dfrac{d(x^2 - x + 12)}{a(x^2 + x - 6)}$ **5.** 1 **7.** $\dfrac{3m}{2(n - m)}$ **9.** $\dfrac{2a}{3(a - 2)}$ **11.** $-\dfrac{8y}{9a}$ **13.** $\dfrac{21yz}{5x(2x - 3)}$ **15.** $\dfrac{3(a + 1)(5 - a)}{2}$

Exercise 21 (*page 67*): **1.** $\dfrac{8x}{15}$ **3.** $\dfrac{x^2 + y^2 + z^2}{xyz}$ **5.** $\dfrac{109a - 39b}{72}$ **7.** $\dfrac{24bc - 6c^2 - 4a^2 - ab - 3b^2}{24abc}$ **9.** $\dfrac{m}{30}$ **11.** $\dfrac{a^2 - c^2}{a}$ **13.** $\dfrac{b^2 + 2b + 1}{b}$

15. $\dfrac{3a}{a^2 - b^2}$ **17.** $\dfrac{x - 3}{x - 6}$ **19.** $\dfrac{11}{c^2 - 9}$ **21.** $\dfrac{3c^2 - 2cd - 6d^2}{(c - d)^3}$

23. $\dfrac{x^3}{2x + 1}$ **25.** $\dfrac{x}{(a + bx)^3}$ **27.** $-\dfrac{1}{x(a + bx)^2}$

Exercise 22 (*page 70*): **1.** $\dfrac{y - x}{y + x}$ **3.** $\dfrac{c + d}{cd}$ **5.** -1 **7.** $q - 1$ **9.** $\dfrac{1}{r - 1}$

11. $2m^2 - 1$ **13.** x

Chapter 4

Exercise 1 (*page 75*): **1.** x^6 **3.** b^{x+3} **5.** $3^5 = 243$ **7.** a^{5y} **9.** -16 **11.** $-8x^3$
13. $8x^3$ **15.** y^{12} **17.** a^{2n} **19.** a^{n^2} **21.** $25/9$ **23.** $32x^5/243$ **25.** $0.008x^3$ **27.** $a^{3n}b^6$
29. $4^n x^{2n} y^{n^2}$ **31.** $x^5 y^8$ **33.** a^2 **35.** $1/y^4$ **37.** $1/a^4$ **39.** x/y **41.** $9/16$ **43.** y^{2n-2}/x^n
45. $2/3a^2$ **47.** c^n/a^n **49.** 64

Exercise 2 (*page 77*): **1.** $\sqrt[3]{-8} = -2$ **3.** $\sqrt[4]{81} = 3$ **5.** $\sqrt{1/25} = 1/5$ **7.** $\sqrt{49} = 7$
9. 5 **11.** 8 **13.** 3 **15.** 12 **17.** 2 **19.** -1 **21.** 0.2 **23.** 40 **25.** $-3/5$ **27.** $2ab^2c^3$
29. ax^2/c^3 **31.** $0.13x^8$

Exercise 3 (*page 82*): **1.** 2 **3.** $1/25$ **5.** $-1/27$ **7.** $27/8$ **9.** -2 **11.** 2.5 **13.** 81 **15.** $1/81$
17. 0.001 **19.** $1/2$ **21.** -32 **23.** $-1/2$ **25.** $x^{7/6}$ **27.** $4x^8$ **29.** $25x^3y^4$ **31.** $243a^2b^3$
33. $18y^2z^2/x^3$ **35.** w^4/x^2y **37.** $x - 2 + (1/x)$ **39.** $a - (9/a)$ **41.** $3/2$ **43.** 2
45. 2^{n+1} **47.** ab^{-3} **49.** $4^{-2}z^2$ **51.** $x^{1/6}y^{-1/2}$ **53.** $8a^{-9}b^3x^3y^{-6}$ **55.** $\sqrt{x}$ **57.** $1/\sqrt{a}$
59. $\sqrt[4]{x^3}\sqrt[4]{y}$ **61.** $1/\sqrt[3]{5x^2}$ **63.** $a^{3/2}$ **65.** $(x + y)^{1/4}$ **67.** $a^{1/2}bc^{3/2}$ **69.** $a^{2/3}/2x^2$

Exercise 4 (*page 84*): **1.** $2\sqrt{2}$ **3.** $2\sqrt{10}$ **5.** $15\sqrt{2}$ **7.** $12\sqrt{2}$ **9.** $2\sqrt[3]{3}$ **11.** $15\sqrt[3]{2}$
13. $-5\sqrt[3]{2}$ **15.** $10\sqrt[4]{6}$ **17.** $x^3\sqrt{x}$ **19.** $xy\sqrt{xy}$ **21.** $2xy^2\sqrt{3xy}$ **23.** $x\sqrt{a + b}$
25. $2\sqrt{m^2 - 4n^2}$ **27.** $a\sqrt[n]{a}$ **29.** $a^{1-n}\sqrt{a^2}$ **31.** $a^2\sqrt{a}$ **33.** $x^2y^2\sqrt[3]{y^a}$ **35.** $x^2y^2\sqrt[3]{x^2}$
37. $10\sqrt{53} = 72.80$ **39.** $5\sqrt{55} = 37.08$ **41.** $2\sqrt[3]{754} = 18.20$

Exercise 5 (*page 86*): **1.** $\frac{1}{2}\sqrt{2} = 0.7071$ **3.** $\frac{3}{2}\sqrt{2} = 2.121$ **5.** $\frac{2}{5}\sqrt{15} = 1.549$
7. $\frac{1}{2}\sqrt[3]{4} = 0.7937$ **9.** $2\sqrt[3]{4} = 3.175$ **11.** $2\sqrt[3]{3}$ **13.** $\frac{1}{3}\sqrt{6} = 0.8165$ **15.** $\frac{3}{5}\sqrt{15} = 2.324$

17. $\frac{1}{15}\sqrt[3]{180} = 0.3764$ **19.** $\dfrac{1}{x^2}\sqrt{x}$ **21.** $\dfrac{1}{a}\sqrt[3]{a}$ **23.** $\dfrac{1}{x}\sqrt[3]{x}$ **25.** $\dfrac{2a}{9c^2}\sqrt{6abc}$

27. $\dfrac{\sqrt{a^2 - b^2}}{a + b}$ **29.** $\dfrac{\sqrt[n]{x^{n-1}}}{x}$ **31.** $\dfrac{1}{a^{n+1}}\sqrt{a}$ **33.** $3/4$

Exercise 6 (*page 87*): **1.** $\sqrt{5}$ **3.** $\sqrt[3]{4}$ **5.** $\sqrt{xy}$ **7.** $yz\sqrt[3]{5x^2z}$ **9.** $\sqrt{x}$ **11.** $\frac{1}{2}\sqrt[3]{4}$

13. $\dfrac{1}{c}\sqrt[3]{c^2}$

Exercise 7 (*page 88*): **1.** $7\sqrt{2}$ **3.** $13\sqrt{3}$ **5.** $\frac{17}{2}\sqrt{2}$ **7.** $\frac{74}{7}\sqrt{7}$ **9.** $\sqrt{6}$ **11.** $-6\sqrt{7}$

13. $(1 + x + x^2)\sqrt{x}$ **15.** $\dfrac{2y}{x^2 - y^2}\sqrt{x^2 - y^2}$ **17.** $(a + b - 1)\sqrt{a + b}$

19. $\frac{11}{4}\sqrt{2} + \frac{5}{3}\sqrt{3}$ **21.** $\left(\dfrac{\sqrt{b}}{b} + \dfrac{2\sqrt{a}}{a} - 3b\right)\sqrt{a - 3b}$

Exercise 8 (*page 91*): **1.** $\sqrt{6} = 2.449$ **3.** $7\sqrt{2} = 9.899$ **5.** 66 **7.** $2a\sqrt{3b}$
9. $bc\sqrt{ad}$
11. $3\sqrt[3]{10} = 6.463$ **13.** $2\sqrt[6]{54}$ **15.** $2\sqrt[4]{2}$ **17.** 7 **19.** 7 **21.** $9 - 6\sqrt{2} = 0.515$
23. $5 + \sqrt{6} = 7.449$ **25.** $6\sqrt{10} + 20\sqrt{6} - 6 - 4\sqrt{15} = 46.47$ **27.** $\frac{1}{3}\sqrt{21} = 1.528$

29. $\frac{6}{5}\sqrt{5} = 2.683$ **31.** $\dfrac{1}{2z}\sqrt{6xz}$ **33.** $\sqrt[6]{24}$ **35.** $\sqrt[4]{3}$ **37.** $2(\sqrt{7} + 2) = 9.292$

39. $\dfrac{2(5 + \sqrt{7})}{3} = 5.097$ **41.** $\dfrac{3(2\sqrt{5} + \sqrt{6})}{2} = 10.38$ **43.** $\dfrac{15\sqrt{2} - 4\sqrt{5}}{37} = 0.3316$

45. $\dfrac{78 + 17\sqrt{15}}{33} = 4.359$ **47.** $\dfrac{x\sqrt{z} + z\sqrt{x}}{xz}$ **49.** $\left(\dfrac{x}{\sqrt{x^2 - a^2}}\right)^3$ **51.** $\dfrac{1}{\sqrt{x^2 + a^2}}$

53. $\dfrac{x}{\sqrt{ax + b}}$

Exercise 9 (*page 94*): **1.** -1 **3.** $-j$ **5.** $-j$

Exercise 10 (*page 97*): **1.** $2(-1 + j5)\sqrt{3}$ **3.** $(-4 + j9)\sqrt{2}$ **5.** $j7\sqrt{5}$
7. $j6 + (4 + j4)\sqrt{5}$ **9.** $-j2\sqrt{2}$ **11.** $-15xy^2$ **13.** $45ab$ **15.** $\sqrt{105}$ **17.** -40
19. -6 **21.** $-5 - 2\sqrt{6}$ **23.** $-j5\sqrt{6}$ **25.** $-\sqrt{30} + \sqrt{10} - 3\sqrt{2} + \sqrt{6}$
27. $j(6\sqrt{19} + \sqrt{627} + 6\sqrt{95} + \sqrt{3,135} - 6\sqrt{10} - \sqrt{330} - 30\sqrt{2} - 5\sqrt{66})$ **29.** $\frac{7}{3}$

31. $j2\sqrt{5}$ **33.** $\dfrac{\sqrt{14}}{2}$ **35.** $\dfrac{\sqrt{2}}{2}$ **37.** $-\dfrac{j4\sqrt{5}}{5}$

Exercise 11 (*page 101*): **1.** $-1 + j31$ **3.** $-51 + j484$ **5.** -53 **7.** $4,282 - j1,475$

9. $11.0754 - j3.465$ **11.** 1 **13.** $-\dfrac{9}{20} + j\dfrac{3}{5}$ **15.** $\dfrac{1}{2} - \dfrac{j}{2}$ **17.** $\dfrac{2}{7} - \dfrac{j3\sqrt{5}}{7}$

19. $-\dfrac{1}{3} - j\dfrac{\sqrt{2}}{3}$ **21.** $-j\dfrac{\sqrt{10}}{2}$ **23.** $-\dfrac{3}{34} + j\dfrac{5}{34}$ **25.** $\dfrac{504}{157} + j\dfrac{332}{157}$

27. $0.931 + j0.361$

Chapter 5

Exercise 1 (*page 112*): **1.** 2 **3.** -2 **5.** $\frac{3}{5}$ **7.** 4 **9.** 3 **11.** 4 **13.** $\frac{2}{3}$ **15.** 1 **17.** 6
19. 8 **21.** -3 **23.** 0.7 **25.** 7 **27.** 4 **29.** 6 **31.** 7 **33.** 1 **35.** 13 **37.** -1

39. $\dfrac{t - r}{s}$ **41.** $\dfrac{a^2}{4}$ **43.** $\dfrac{m}{k}$ **45.** 3 **47.** $\dfrac{m + n}{2}$ **49.** $\dfrac{6}{a - b}$ **51.** $ac + bc$

53. $\dfrac{a}{b + c}$ **55.** $x = -\dfrac{a^2}{2b}$ **57.** $y = \dfrac{b + c}{2}$ **59.** $y = \dfrac{2r^2}{p - q - r}$ **61.** $x = -c - m$

63. $w = \dfrac{cd}{c + d}$ **65.** $z = \dfrac{n^2 - p^2}{m}$

Exercise 2 (page 116): **1.** $\dfrac{WL}{Q}$ **3.** $E - IR$ **5.** $\dfrac{eR}{E - e}$ **7.** $\dfrac{Cb}{Kb + C}$ **9.** $\dfrac{\rho(d^2 - L^2)}{2L}$

11. $T = \dfrac{2,097Q}{EI}$ **13.** $\dfrac{9}{5}C + 32$ **15.** $\dfrac{HL}{0.4\pi N}$ **17.** $\dfrac{f(n + 1)}{n}$ **19.** $\dfrac{1,299}{T - S}$

21. $\dfrac{r_1 r_2}{r_1 + r_2}$ **23.** $\dfrac{T_1 h}{T_1 h - (T_1 - T)(h_0)}$ **25.** $\dfrac{6V - Bh - bh}{4h}$ **27.** $\dfrac{273(V_1 - V_0)}{V_0}$

Exercise 3 (page 121): **1.** 23, 30 **3.** 31, 14 **5.** 23, 24, 25 **7.** 13, 15, 17, 19, 21 **9.** 18, 19 **11.** $^{24}\!/_{36}$ **13.** 33 $\times$ 47 ft **15.** $90 **17.** 22 acres **19.** $2,200 **21.** 3.2 qt **23.** $^3\!/_4$ h **25.** 11, 13 **27.** 2,400 lb **29.** 114 adults **31.** $40, $80, $160 **33.** 4 days **35.** 30 min **37.** 100 gal **39.** $12\tfrac{1}{2}$ oz, $27\tfrac{1}{2}$ oz **41.** 1,300 lb **43.** $6\tfrac{1}{4}$ in; 7 in; $7\tfrac{3}{4}$ in; $8\tfrac{1}{2}$ in; $9\tfrac{1}{4}$ in **45.** $1,115 **47.** 8.7 h **49.** 24 ft **51.** $3.30 **53.** 2.4 gal **55.** 60 mi **57.** 120 mi **59.** 40 mi from B **61.** 11,750 ft **63.** $AB = 11.93$ ft; $BC = 16.40$ ft **65.** 104 weeks **67.** 0.852 ft/s **69.** $63\tfrac{3}{4}$ years **71.** 2.83 gal/day **73.** 75 years **75.** $a = 5\tfrac{1}{4}$ in; $b = 15\tfrac{3}{4}$ in; $c = 10\tfrac{1}{2}$ in **77.** 12,500 mi **79.** $30^{10}\!/_{13}$ mi/h

Chapter 6

Exercise 1 (page 131): **1.** 6:1 **3.** 1:9 **5.** 10:3 **7.** 10:33 **9.** 3:8 **11.** 3x:4a **13.** 4:3

Exercise 2 (page 132): **1.** 6 **3.** 21 **5.** $\dfrac{ac}{b}$ **7.** $\dfrac{bc}{a + b}$ **9.** $\dfrac{ad - bc}{a - b + c - d}$ **11.** 12 **13.** $^5\!/_4$ **15.** 4abx **17.** 2(x + y) **19.** 6ab

Exercise 3 (page 138): **1.** $W = kxy$ **3.** $V = \dfrac{kx^3}{d}$ **5.** $R = \dfrac{kw\sqrt{x}}{h^3}$ **7.** $N = \dfrac{7}{y}$

9. $V = \dfrac{4.8m}{t^2}$

Exercise 4 (page 138): **1.** $k = 2,520$ in $\cdot$ lb; $V = 40$ in^3 **3.** $k = 1.8$ ft/s^2; $V = 108$ ft/s **5.** $k = 43,560$ ft^2/acre; $N = 1,440$ plants/acre **7.** $k = 1.63$ (no units); $V = 7.3$ ft/s **9.** 56 **11.** $V = 819.2$

Exercise 5 (*page* 140): **1.** 54.5 lb **3.** 2 h 59 min **5.** 506 **7.** 174½ mi **9.** 5.13 lb
11. 18 ml **13.** 83⅓ min; 145 mi **15.** 420 sheets approx **17.** 29.8 ft/s² **19.** 284 ft²
21. Streets 109°; police, etc., 79°; schools 172° **23.** 166.7 acres **25.** Brand B
27. 482 mi **29.** 6 h **31.** 225 ft **33a.** 24½ r/min **33b.** 72 teeth **33c.** 196 r/min
33d. 84 teeth **33e.** A, 60 teeth; B, 18 teeth **33f.** A, 66 teeth; B, 42 teeth
33g. A, 96 teeth; B, 30 teeth **33h.** A, 84 teeth; B, 18 teeth **35a.** 5.3 ft/s²
35b. 27.5 ft/s² **35c.** 12.4 ft/s² **35d.** 82.0 ft/s² **35e.** 222 ft/s²
37. Earth g = 0.0104 ft/s²; moon g = 0.0155 ft/s²; net g = 0.0051 ft/s² toward moon
39. 6.22 mi/s **41.** 28.8 in **43.** 199 lb **45.** 150 ft **47.** 13.7 hp **49.** 27,800 lb **51.** 1 qt,
8.32 in; 2 qt, 10.5 in **53.** 17.1 lb **55.** 6 extra men **57.** −65°F **59.** 1 part solvent to
1.47 parts varnish

Chapter 7

Exercise 1 (*page* 147†): **1.** 57°46′14″ **3.** $a = b = 132°32′$; $c = d = 47°28′$ **5.** 34°
7. $a = 29°10′$; $b = 150°50′$ **9.** 5.59 in **11.** 30°, 60°, 90°; 2:1 **13.** Side = 14 in;
area = 84 in² **15.** Side = 13.31 in; area = 165 in² **17.** 114°7′ **19.** 330 ft **21.** 9.08 in²
23. 90° **25.** 94° **27.** 45° **29.** 20.34 in **31.** 9:16 **33.** 84.8 lb **35.** 10.39 in
37. 44.22 in² **39.** 446 ft² **41.** 2,850 lb **43.** 3,550 ft **45.** 1,196 in² **47a.** No
47b. 0.154 in approx **51.** 7,880 mi **53.** 20.16 in **55.** Surface 935 in²; volume 1,780 in³
57. Volume 4,920 in³; surface 1,127 in² **59.** 5.13 in; 30° **61.** 43,400 ft² **63.** 20 mi
65. 6.37 in **67.** 76.5+ in **69.** 12 in **71.** 2 ft 10 in **73.** 8⅝ in **75.** 13 ft 8 in
77. 39.8 ft/s **79.** $A = (3\sqrt{3})/2 \, s^2 = 2.598 \, s^2$ **81.** $A = 11.196 \, s^2$ **83.** 0.0016 in
85. 0.030 in **87.** 4.8 **89.** 57½ h **91.** 107.5 acres **93.** 2,495 ft² **95a.** 0.813 mi
95b. 0.576 mi **97.** 13°39′ **99.** 6.07 in **101.** 4 in **103.** 1.27 in **105.** 115 ft **107.** 10 in
109. 3.62 in **111.** $r = c - 3w$ **113.** 5⅓ in **115.** 7.3 in **117.** 4.10 in
119. $w = (Ha - ha + hb)/H$ **121.** 4.330 in; 1.443 in **123.** 19 min
125. $V = \frac{1}{6}hw(l + 2L)$ **133.** 761 r/min **135.** 300 r/min; 1,125 to 180 r/min
137. 49.0 in³ **139.** $12\pi = 37.7$ in **141.** 0.65% high; no **143.** 4.13 ft **145.** 2¹¹⁄₃₂ in
147. $D = 1.09$ in **149.** 3,960 mi **151.** 3,800 ft approx

Chapter 8

Exercise 1 (*page* 181): **1.** −5 **3.** −5 **5.** 24 **7.** −²⁰⁄₃ **9.** $4(3a - 4)/a(8 - 3a)$

Exercise 3 (*page* 183): **1.** Square **3.** Triangle **5.** Right triangle **7.** Right triangle

Exercise 5 (*page* 187): **1.** ⅓ **3.** −¾ **5.** ⁵⁄₂ **7a.** $\sqrt{3}/3$ **7c.** $\sqrt{3}$ **7e.** −1

Exercise 6 (*page* 191): **1a.** $f(x_1) = 94$; $f(x_1 + \Delta x) = 96$; $\Delta y/\Delta x = \frac{1}{3}$ **1c.** $f(x_1) = 100$;
$f(x_1 + \Delta x) = 105$; $\Delta y/\Delta x = \frac{1}{3}$ **2a.** $f(x_1) = 7,400$; $f(x_1 + \Delta x) = 6,950$; $\Delta y/\Delta x = -150$
2c. $f(x_1) = 5,000$; $f(x_1 + \Delta x) = 3,800$; $\Delta y/\Delta x = -150$

Exercise 7 (*page 193*): **1.** $y = \frac{5}{2}x + 20$ **3.** $y = \frac{4}{5}x$ **5.** $y = \frac{8}{3}x - \frac{92}{3}$
7. $y = -\frac{24}{5}x - 60$

Exercise 8 (*page 199*): **1.** $x - 2y = -8$ **3.** $2x - y = 11$ **5.** $x - y = -4$
7. $x\sqrt{3} - y = 5 + 2\sqrt{3}$ **9.** $y = 7$ **11.** $y = -x + 8$ **13.** $y = \frac{1}{3}x - 4$
15. $y = \frac{1}{4}x + \frac{5}{2}$ **17.** $y = \frac{7}{5}x + \frac{14}{5}$ **19.** $y = \frac{3}{2}$ **21.** $y = -\frac{2}{3}x$ **23.** $x - 3y = -3$
25. $3x + 4y = -10$ **27.** $x + 4y = 3$ **29.** $y = 5$ **31.** $x - 2y = -4$
33. $2x + 3y = -18$ **35.** $3x - 4y = -24$ **37.** $2x + 3y = -30$ **39.** $\overline{OA}$, $y = 0.20x$;
$\overline{AB}$, $y = 0.22x - 80$; $\overline{BC}$, $y = 0.26x - 400$; $\overline{CD}$, $y = 0.30x - 880$; $\overline{DE}$,
$y = 0.34x - 1,520$; $\overline{EF}$, $y = 0.38x - 2,320$; $\overline{FG}$, $y = 0.43x - 3,520$

Exercise 9 (*page 204*): **1a.** $C = 0$ **1b.** $B = 0$ **1c.** $A = 0$ **1d.** $A = 24$ **1e.** $B = 2$
3. 6 **5.** 9 **7.** 5 **9.** $\sqrt{85} = 9.22$ **11a.** (5,5) **11b.** $(1\frac{1}{2},0)$ **13.** (1.2,1.2) and $(-6,6)$
15. $2x + 3y = 14$ **17.** $y = 3x - 19$ **19.** B, $y = -3x - 20$; C, $y = 3x + 20$;
D, $y = 3x - 20$; E, $y = -\frac{1}{3}x + \frac{20}{3}$; F, $y = -\frac{1}{3}x - \frac{20}{3}$; G, $y = \frac{1}{3}x + \frac{20}{3}$;
H, $y = \frac{1}{3}x - \frac{20}{3}$

Exercise 10 (*page 208*): **1.** 48 **3.** 56 **5.** $55\frac{1}{2}$ **9.** 56 **11.** $66\frac{1}{2}$

Chapter 9

Exercise 1 (*page 211*): **1.** $x = 5$; $y = 3$ **3.** $x = -\frac{1}{2}$; $y = -4$ **5.** $x = -1$; $y = 2$
7. $x = \frac{7}{2}$; $y = 1$ **9.** Dependent **11.** $x = \frac{7}{3}$; $y = -\frac{1}{3}$ **13.** $x = \frac{1}{4}$; $y = -\frac{3}{5}$

Exercise 2 (*page 216*): **1.** $x = 4$; $y = 4$ **3.** $x = 6$; $y = 6$ **5.** $x = 5$; $y = 4$
7. $x = 6$; $y = -2$ **9.** $x = -3$; $y = 11$ **11.** $x = 9$; $y = 4$ **13.** $y = 5$; $z = \frac{7}{2}$
15. $x = 4$; $w = 3$ **17.** $x = 5$; $y = -2$ **19.** $x = -2$; $y = 4$ **21.** $x = 8$; $y = 3$
23. $x = \frac{1}{5}$; $w = \frac{2}{5}$ **25.** $x = \frac{16}{3}$; $y = \frac{20}{3}$ **27.** $w = 5$; $z = 1$ **29.** $x = 36.5$; $y = 7.06$

Exercise 3 (*page 220*): **1.** $x = c + d$; $y = c - 3d$ **3.** $x = 2a - b$; $y = a - 2b$
5. $x = \dfrac{3}{a}$; $y = \dfrac{1}{2b}$ **7.** $x = \dfrac{3c}{a}$; $y = \dfrac{4c}{b}$ **9.** $x = c$; $y = b$ **11.** $x = \dfrac{m + n}{2}$; $y = \dfrac{m - n}{2}$
13. $x = \frac{1}{5}$; $y = 1$ **15.** $x = 2$; $y = 3$ **17.** $x = \frac{1}{12}$; $y = \frac{1}{18}$

Exercise 4 (*page 222*): **1.** $x = 3$; $y = 4$; $z = 5$ **3.** $x = 7$; $y = 3$; $z = -2$
5. $x = -3$; $y = 5$; $z = 8$ **7.** $x = 6$; $y = 5$; $z = 4$ **9.** $x = \frac{2}{3}$; $y = \frac{3}{4}$; $z = -1$

Exercise 5 (*page 224*): **1.** -1 **3.** 32 **5.** $9a^2 - 20b^2$

Exercise 6 (*page 227*): **1.** Inconsistent **3.** Dependent **5.** $x = a/2$; $y = a/3$
7. $x = a - b$; $y = b - a$ **9.** $x = 4$; $y = -5$

Exercise 7 (*page 235*): **1.** -47 **3.** -48 **5.** -9 **7.** $15a$ **9.** $8a + 36b + 6c$
11. -812 **13.** -445

Exercise 8 (*page 238*): **1.** $x = 7; y = 5; z = 3$ **3.** $x = -4; y = \frac{3}{2}; z = 5$
5. $x = 3a + 2b + c; y = a + 2b + 3c; z = a + 3b + c$ **7.** $x = 5; y = 4; z = 3; w = 2$

Exercise 9 (*page 239*): **1a.** $I_1 = -0.638; I_2 = 6.68$ **1c.** $I_1 = -13.50; I_2 = 14.97$
2a. $I_1 = \dfrac{E_1 - I_2 R_3}{R_1 + R_3}$; for negative I_1, $I_2 R_3 > E_1$ **3.** $R_2 = 2$

Exercise 10 (*page 240*): **1a.** $I_1 = 1.474; I_2 = 1.068; I_3 = 0.0642$ **1c.** $I_1 = 0.165;$
$I_2 = 0.101; I_3 = 0.0306$ **1e.** $I_1 = 2.252; I_2 = 1.492; I_3 = -0.0446$

Exercise 11 (*page 240*): **1.** 8 oz A, 8 oz B, 12 oz C **3.** 12.5 lb A, 28.6 lb B, 8.9 lb C
5. $W_2 = -2{,}191.5; W_3 = -3{,}331; W_4 = -2{,}606; W_5 = -1{,}548$ **7.** First group walks
4.91 mi; second group walks 3.57 mi; time for entire transfer $= 1.87$ h

Exercise 12 (*page 245*): **1.** 30, 23 **3.** Current, $\frac{3}{4}$ mi/h; boat, $6\frac{3}{4}$ mi/h
5. 30×48 ft **7.** Man 9 days, boy 36 days **9.** $A = 4\frac{1}{2}$ in; $B = 9$ in; $C = 8$ in
11. Father \$250; son \$350 **13.** 22 acres **15.** \$2,200 **17.** 3.2 qt **19.** 114 adults
21. $x = 5\frac{1}{8}$ in; $y = 8\frac{1}{8}$ in **23.** $(\pi/4)(AB)^2$ **25.** $a = \$0.50; b = \$0.08$
27. \$12 fixed charge; \$200 first order; \$280 second order; \$0.15 unit cost **29.** $\frac{1}{2}, \frac{1}{3}$
31. 100 lb **33.** $A, 9.8$ s; $B, 10.0$ s **35a.** 22 min **35b.** 132 s **37.** Ratio bus speed to
hiker speed $= 6:1$. They passed each other 30 min after starting. **39.** $a = 3; b = 4;$
$c = 2$ **41.** 5 5-cent stamps $+$ 14 8-cent stamps, or 13 5-cent stamps $+$ 9 8-cent stamps,
or 21 5-cent stamps $+$ 4 8-cent stamps

Exercise 13 (*page 255*): **1a.** $y = -1.5x + 80$; **1c.** $x = 62$ **2a.** $y = 0.08x - 4.2;$
2c. $y = -1$ **3a.** $y = 0.628x + 13.3$; **3b.** $y = 47.8$ **5.** $y = 0.696x + 4.09$
7. $h = -926P + 27{,}700$; P (sea level) $= 29.92$ in **9.** $p = -0.93t + 93$ (approx)
11. $V = 0.0758R + 2.1$ (approx); 2.1 mi/h to overcome friction
13. $V = 0.602t + 331.4$

Chapter 10

Exercise 1 (*page 263*): **1.** 45 min **3.** 60 mi **5.** 120 mi **7.** 11,800 ft

Exercise 2 (*page 268*): **1.** 8Ω **3.** $f = 4$ **5.** $6\frac{1}{4}$ h **7.** 30 min **9.** $D_o = 7$ in;
$D_i = 17\frac{1}{2}$ in

Exercise 3 (*page 271*): **1a.** 90 min **1b.** 160 min **3.** 108 min

Exercise 4 (*page* 276): **1.** 100 lb **3.** 1,300 lb 18% tungsten steel; 1,700 lb 12% tungsten steel **5.** $12\frac{1}{2}$ oz 20% silver alloy; $27\frac{1}{2}$ oz 12% silver alloy **7.** 100 gal

Exercise 5 (*page* 277): **1.** 30, 23 **3.** Current, $\frac{3}{4}$ mi/h; boat, $6\frac{3}{4}$ mi/h **14.** 35 dimes **25.** $a = \$1.00$; $b = \$0.45$ **35a.** 22 min **35b.** 2 min 12 s

Chapter 11

Exercise 1 (*page* 279): **1a.** $\log_2 8 = 3$ **1b.** $\log_2 \frac{1}{64} = -6$ **1c.** $\log_7 \frac{1}{49} = -2$ **3a.** $\log_5 625 = 4$ **3b.** $\log_3 \frac{1}{27} = -3$ **3c.** $\log_{2/3} \frac{8}{27} = 3$ **5a.** $\log_3 \frac{1}{81} = -4$ **5b.** $\log_6 36 = 2$ **5c.** $\log_{1/5} 25 = -2$ **7a.** $\log_{16} 8 = \frac{3}{4}$ **7b.** $\log_{25} 5 = \frac{1}{2}$ **7c.** $\log_{27} 9 = \frac{2}{3}$ **9a.** $\log_{27} \frac{1}{3} = -\frac{1}{3}$ **9b.** $\log_{16} \frac{1}{32} = -\frac{5}{4}$ **9c.** $\log_{36} \frac{1}{216} = -\frac{3}{2}$

Exercise 2 (*page* 280): **1a.** 2 **1b.** 3 **1c.** 4 **3a.** $\frac{3}{2}$ **3b.** $\frac{4}{3}$ **3c.** $\frac{2}{3}$ **5a.** -3 **5b.** -3 **5c.** $-\frac{5}{2}$

Exercise 3 (*page* 280): **1a.** 8 **1b.** 25 **1c.** 81 **3a.** 100 **3b.** $\frac{1}{9}$ **3c.** 9 **5a.** 3 **5b.** 2 **5c.** 8 **7a.** 15 **7b.** Any constant other than zero **7c.** 2 **9a.** $\frac{1}{27}$ **9b.** 0.1 **9c.** 8

Exercise 4 (*page* 289): **1d.** 8.61×10^3; 3.935 **1f.** 86,100; 8.61×10^4 **1h.** 0.861; $0.935 - 1$ **1j.** 8.61×10^6; 6.935 **1l.** 0.00861; 8.61×10^{-3} **3a.** $0.549 - 2$ **3b.** $0.549 - 1$ **3c.** 3.902 **5a.** 354 **5b.** 79.8 **5c.** 0.0798 **7a.** 0.000354 **7b.** 7,980 **7c.** 0.00354

Exercise 5 (*page* 290): **1.** 3.49178 **3.** 2.49136 **5.** 4.49360 **7.** 0.49859 **9.** 6.49136 **11.** 1.49136

Exercise 6 (*page* 291): **1.** $0.49443 - 3$ **3.** $0.49748 - 1$ **5.** $0.50065 - 2$ **7.** $0.50120 - 1$ **9.** $0.50379 - 6$ **11.** $0.49206 - 2$ **13.** 31.03 **15.** 3167 **17.** 0.0032 **19.** 32,040 **21.** 311,100 **23.** 31.63

Exercise 7 (*page* 291): **1a.** 3.09132 **1c.** $0.09132 - 2$ **1e.** 2.60032 **3a.** 2.25455 **3c.** $0.03743 - 2$ **3e.** $0.84911 - 4$ **5a.** 0.008605 **5c.** 400,000 **5e.** 57,550

Exercise 8 (*page* 295): **1a.** 2.42843 **1c.** 1.32950 **1e.** 0.61802 **3a.** 4.07766 **3c.** $0.89962 - 4$ **3e.** 8.07266 **5a.** 0.0010071 **5c.** 0.00079367 **5e.** 139,700

Exercise 9 (*page* 296): **1.** 228,380 **3.** 4.4216 **5.** 9.5530 **7.** 6,999,200,000 **9.** 8,158,000,000

Exercise 10 (*page* 297): **1.** -8.0425 **3.** 0.077289 **5.** 0.081484 **7.** 0.27963 **9.** 24,476 **11.** 0.018578

Exercise 11 (*page 300*): **1.** 1.3430×10^9 **3.** 4,096.4

Exercise 12 (*page 302*): **1a.** 2,753.3 **1b.** $-1,655.5$ **1c.** 0.17116 **3a.** 0.042008
3b. 0.018105 **3c.** 0.00024273 **5a.** 0.37876 **5b.** 69.553 **5c.** 16.150 **7a.** 2.3428
7b. 29.829 **7c.** 0.31216

Exercise 13 (*page 303*): **1a.** 9.9058 **1b.** 4.5979 **1c.** 2.1341 **3a.** 8.5711 **3b.** 5.5195
3c. 7.4326 **5.** 317.94 **7.** 2,354.0 **9.** 195.72 **11.** 0.40880 **13.** 0.74308 **15.** -3.2836
17. 0.16014

Exercise 14 (*page 304*): **1.** 6.6257 **3.** 6.9078 **5.** 1.2987

Exercise 17 (*page 310*): **1.** 6.2288 **3.** 1.2027 **5.** 20.939 **7.** $\frac{7}{4}$ **9.** 7.3365 **11.** $\frac{3}{2}$

Exercise 18 (*page 313*): **1.** \$22,152 **3.** 11 years **5a.** 16 g **5b.** $W_0 = 54$ g; $c = \frac{2}{3}$
5c. 21.25 g **5d.** 1.71 h **7.** $77.8°$ **9.** 13.2 h **11.** 27.2%; \$538.59 **17.** $25:1$ approx

Exercise 19 (*page 316*): **1.** 6.17 in **3.** 177.2 **5a.** 0.39 **5b.** 91.7 lb **5c.** 3.56 turns
7. $1.260:1$ **9.** 0.162 in **11.** $3.18:1$ **13.** 3.806 in **15.** 3.98 **17.** 4.77 **19.** Between 251
and 50% of normal power rating **21.** $2.0w$ **23.** 18.85 **25.** 29.64
31. 10.3 lb/in^2 gauge **33.** 57.8 ft^3 **35.** 1.30 **37.** 0.290 **39.** log mean diam = 4.57 in;
arithmetic mean diam = $4\frac{15}{16}$ in **41.** $28\frac{3}{4}$ in **43.** $1330°$F **45.** $2450°$C **47.** 83.5 hp
49. 24 h **51.** 366.2 days **53.** 125.8 min

Chapter 12

Exercise 1 (*page 330*): **1.** $y = \pm 3$ **3.** $x = \pm 7$ **5.** $y = \pm 11$ **7.** $z = \pm 2\sqrt{2}$
9. $x = \pm\sqrt{30}$ **11.** $y = \pm 3a$ **13.** $w = a, 0$ **15.** $z = \pm(1/c)$ **17.** $x = \pm\frac{1}{4}$
19. $w = \pm(\sqrt{3}/6)$ **21.** $w = \pm 1\frac{1}{2}$ **23.** $x = \pm 12$ **25.** $y = \pm\frac{2}{7}$ **27.** $x = \pm(m/5)$
29. $z = \pm 4$ **31.** $x = \pm 6$ **33.** $w = \pm 6$

Exercise 3 (*page 333*): **1.** $x = 3, -1$ **3.** $x = 7, -8$ **5.** $x = 3, -8$ **7.** $x = 1, -\frac{2}{3}$
9. $x = 5, 3$

Exercise 4 (*page 334*): **1.** $x = 2, 2$ **3.** $x = 8, -9$ **5.** $x = -3, -4$ **7.** $x = 8, -6$
9. $x = 5, -8$ **11.** $x = -\frac{1}{2}, -1$ **13.** $x = \frac{3}{4}, -2$ **15.** $x = 8, 4$ **17.** $y = 3, 0$

Exercise 5 (*page 338*): **1.** $x = 1, -3$ **3.** $x = 1, -7$ **5.** $x = 2, -12$ **7.** $x = 13,$
-3 **9.** $x = -b \pm a$ **11.** $x = 3b - 2a, -3b$ **13.** $x = 1, -\frac{5}{3}$ **15.** $y = 10 \pm j\sqrt{5}$
17. $w = 2 \pm 2\sqrt{3}$ **19.** $y = 7 \pm 2\sqrt{15}$ **21.** $x = -4 \pm 4\sqrt{3}$ **23.** $x = 7 \pm j\sqrt{10}$

Exercise 6 (*page 340*): **1.** $x = 12, 5$ **3.** $x = 15, -11$ **5.** $y = 1, -\frac{1}{6}$
7. $x = \dfrac{-1 \pm j\sqrt{3}}{2}$ **9.** $w = \dfrac{3 \pm j\sqrt{6}}{3}$ **11.** $x = \frac{3}{4}, -\frac{4}{9}$ **13.** $w = \dfrac{-5 \pm j\sqrt{3}}{2}$

15. $p = \dfrac{3 \pm j\sqrt{11}}{10}$

Exercise 7 (*page 344*): **1.** $x = \dfrac{2m}{5}, -\dfrac{m}{3}$ **3.** $w = c, -a - b$ **5.** $y = \dfrac{a - b}{b - c}, 1$

7. $z = \dfrac{h \pm \sqrt{h^2 + 60}}{6}$ **9.** $x = h + k, 0$ **11.** $a = 2; b = -3m - 4; c = 8m - 2m^2$

Exercise 8 (*page 347*): **1.** $y = 2, -\tfrac{1}{2}$ **3.** $z = 24, -1$ **5.** $y = m, n$

7. $d = \dfrac{-7 \pm \sqrt{5}}{22}$ **9.** $h = 6, {}^{25}\!\!/_{6}$ **11.** $c = 4 \pm j4\sqrt{3}$ **13.** $y = \dfrac{a}{3}, -\dfrac{a}{5}$

15. $x = -n \pm \sqrt{n^2 + \dfrac{A}{\pi}}$ **17.** $m = 12$ **19.** $x = \dfrac{b - c}{c - a}, 1$ **21.** $x = \pm 4a$

23. $x = \dfrac{h \pm jh\sqrt{35}}{18}$

Exercise 9 (*page 349*): **1.** 64, real, rational, unequal **3.** 0, real, rational, equal
5. 49, real, rational, unequal **7.** 49, real, rational, unequal **9.** 0, real, rational,
equal **11.** 201, real, irrational, unequal **13.** 289, real, rational, unequal

Exercise 10 (*page 352*): **1.** min; 1, -16; $x = 1$ **3.** min; $-\tfrac{7}{2}$, $-{}^{25}\!\!/_{4}$; $x = -\tfrac{7}{2}$
5. max; 3, 9; $x = 3$ **7.** max; $\tfrac{7}{12}$, $\tfrac{1}{24}$; $x = \tfrac{7}{12}$ **9.** max; $-\tfrac{1}{2}$, 16; $x = -\tfrac{1}{2}$
11. $a = -3; b = 16$ **13.** 35×35 ft **15.** max $h = 664$ ft **17.** $5\tfrac{1}{2}$ in

Exercise 11 (*page 354*): **1.** 14, 9 **3.** 13, 6 **5.** 8 **7.** 14, 16 **9.** 8 **11.** $2\tfrac{1}{2}$ s
13c. max $h = 256$ ft occurring at $t = 3\tfrac{1}{2}$ s **13d.** $7\tfrac{1}{2}$ s, $-\tfrac{1}{2}$ s **15.** 15 ft, 20 ft, 25 ft
17. 22.6 s, 12,450 ft **19.** 10 **21.** 1.633 in **23.** 14, 6 **25.** 15.1 ft **27.** 3.52 in **29.** 29.3 in
31. 14 mi/h **33.** 2.86 in **35.** 1.08 in **37.** 40.4°F **39.** 12.8 **41.** 8,720,000 mi from
Mars **43.** 0.230 in **45.** Bus, 25 mi/h; train, 40 mi/h **47.** 20 ft, 21 ft, 29 ft
49. 0.608 in **51.** 13.29 mi/h; 9.21 mi/h **53.** 8.94 in **55.** $a = 9; b = 11; R = \sqrt{85}$
57. $d = \sqrt{3h/2}$

Chapter 13

Exercise 1 (*page 371*): **1.** 139°37′31″ **3.** 89°26′49″ **5.** 87°48′40″ **7.** 64°23′50″
9. 47°13′43″ **11.** 53°31′36″ **13.** 18°01′12″ **15.** 88°46′32″ **17.** 1°27′38″ **19.** 5°14′37″
21a. 52°45′47″ **21b.** 00°31′37″ **21c.** 146°53′42″

Exercise 2 (*page 373*): **1.** Possible **3.** Impossible **5.** Impossible **7.** Two possible
triangles **9.** Possible **11.** Inconsistent

Exercise 5 (*page 379*): **1.** $a = 2.96$ in; $b = 6.34$ in; $B = 65°$ **3.** $a = 21.0$ in;
$c = 32.6$ in; $A = 40°$ **5.** $b = 217$ in; $A = 30°$; $B = 60°$

Exercise 6 (*page* 380): **1.** cos 70° **3.** sin 30° **5.** tan 73° **7.** sin 51.5° **9.** csc 65°

Exercise 7 (*page* 382):

	sin	*cos*	*tan*	*cot*	*sec*	*csc*
1.	0.35021	0.93667	0.37388	2.6746	1.0676	2.8555
3.	0.70690	0.70731	0.99942	1.0006	1.4138	1.4146
5.	0.43445	0.90070	0.48234	2.0732	1.1102	2.3018
7.	0.98619	0.16562	5.9545	0.16794	6.0379	1.0140
9.	0.00291	1.0000	0.00291	343.77	1.0000	343.78
11.	0.23203	0.97271	0.23854	4.1922	1.0281	4.3098
13.	0.82822	0.56040	1.4779	0.67663	1.7844	1.2074
15.	0.03054	0.99953	0.03055	32.730	1.0005	32.746

17. 10°24′ **19.** 48°57′ **21.** 44°03′ **23.** 21°16′ **25.** 3°06′ **27.** 50°29′ **29.** 55°35′
31. 48°03′ **33.** 25°05′ **35.** 11°10′ **37.** 67°25′ **39.** 50°08′ **41.** 41°06′ **43.** 29°15′
45. 5°10′

Exercise 8 (*page* 388): **1.** $a = 55.744$; $b = 76.072$; $B = 53°46′$ **3.** $a = 60.200$; $b = 117.47$; $A = 27°08′$ **5.** $a = 0.31011$; $b = 2.1698$; $B = 81°52′$ **7.** $a = 232.39$; $b = 92.168$; $B = 21°38′$ **9.** $a = 50.810$; $b = 39.389$; $A = 52°13′$ **11.** $b = 5,516.9$; $c = 40,998$; $B = 7°44′$ **13.** $b = 134.96$; $c = 143.04$; $B = 70°39′$ **15.** $b = 48.090$; $c = 50.103$; $B = 73°42′$ **17.** $b = 342.03$; $c = 463.29$; $A = 42°25′$ **19.** $a = 9,237.3$; $c = 12.302$; $B = 41°20′$ **21.** $A = 64°18′$; $B = 25°42′$; $c = 385.09$ **23.** $A = 69°25′$; $B = 20°35′$; $c = 71,707$ **25.** $A = 34°47′$; $B = 55°13′$; $c = 247.17$ **27.** $A = 18°10′$; $B = 71°50′$; $b = 2,556.9$ **29.** $A = 24°18′$; $B = 65°42′$; $b = 39,782$ **31.** $A = 21°20′$; $B = 68°40′$; $a = 4,082.8$ **33.** $A = 36°36′$; $B = 53°24′$; $a = 1,394.0$ **35.** $A = 16°49′$; $B = 73°11′$; $a = 297.70$

Exercise 9 (*page* 392):

	sin	*cos*	*tan*	*cot*	*sec*	*csc*
1.	0.04311	0.99907	0.04315	23.175	1.0009	23.196
3.	0.24678	0.96907	0.25466	3.9269	1.0319	4.0522
5.	0.64564	0.76365	0.84546	1.1827	1.3095	1.5489
7.	0.79646	0.60470	1.3172	0.75923	1.6538	1.2555
9.	0.98767	0.15653	6.3098	0.15848	6.3886	1.0125
11.	0.45795	0.88898	0.51514	1.9412	1.1249	2.1836
13.	0.95282	0.30354	3.1391	0.31856	3.2945	1.0495
15.	0.29674	0.95496	0.31073	3.2182	1.0472	3.3700

Exercise 10 (*page* 392): **1.** 6°04′06″ **3.** 35°04′20″ **5.** 46°35′24″ **7.** 12°01′10″

9. 88°37'48" **11.** 40°59'27" **13.** 24°54'20" **15.** 46°16'30" **17.** 86°23'12"
19. 32°52'07" **21.** 87°30'45" **23.** 51°24'30" **25.** 64°44'48" **27.** 6°08'51" **29.** 46°17'30"

Exercise 11 (*page* 392): **1.** $a = 0.54882$; $b = 0.68148$; $B = 51°09'15"$ **3.** $a = 12.437$; $b = 9.1720$; $B = 36°24.5'$ **5.** $a = 2,517.9$; $b = 1,044.7$; $A = 67°27.9'$ **7.** $a = 6.3335$; $b = 8.5310$; $A = 36°35'26"$ **9.** $a = 6.8772$; $b = 12.194$; $A = 29°25'17"$ **11.** $b = 12,217$; $c = 12,281$; $B = 84°09'29"$ **13.** $b = 977.88$; $c = 1,389.3$; $B = 44°44.2'$ **15.** $b = 2,222.7$; $c = 2,334.5$; $B = 72°11.7'$ **17.** $a = 50.357$; $c = 757.90$; $A = 3°48.6'$ **19.** $a = 20.466$; $c = 55.313$; $B = 68°17'02"$ **21.** $A = 27°45'34"$; $B = 62°14'26"$; $c = 42.941$
23. $A = 14°30'38"$; $B = 75°29'22"$; $c = 18.517$ **25.** $A = 25°13'12"$; $B = 64°46'48"$; $c = 196.42$ **27.** $A = 18°49'13"$; $B = 71°10'47"$; $c = 1,357.5$ **29.** $A = 34°58'13"$; $B = 55°01'47"$; $c = 54.523$ **31.** $A = 20°33'36"$; $B = 69°26'24"$; $b = 391.92$
33. $A = 67°37'33"$; $B = 22°22'27"$; $b = 16.358$ **35.** $A = 55°45'08"$; $B = 34°14'52"$; $b = 639.64$ **37.** $A = 65°21'30"$; $B = 24°38'30"$; $a = 272.50$ **39.** $A = 42°21'28"$; $B = 47°38'32"$; $a = 12.577$ **41.** 6.3351 in **43.** 0.9347 in **45.** 1.0824 in **47.** 1.9482 in
49. 36°30'40"

Exercise 12 (*page* 396): **1.** 10.719 in; 2.3694 in **3.** 47°09'24"; 73°44'22" **5.** 0.828 in above **7.** 66.002 in^2 **9.** $\cos DAC = 0.80902$, $DAC = 36°$, $CDA = 72°$

Exercise 13 (*page* 399): **1.** 4.95 in **3.** 10.04 in **5.** 6.93 in **7.** 2.00 in

Exercise 14 (*page* 402): **5.** 82.274 in^2 **7.** 3.000 in^2 **9.** π in^2

Exercise 15 (*page* 404): **1.** $a = 89.488$; $b = 115.76$ **3.** $a = 47.868$; $b = 150.21$
5. $a = 0.65841$; $b = 1.2073$ **7.** $b = 0.42347$; $c = 15.364$ **9.** $a = 1,369,100$; $c = 1,369,200$ **11.** $A = 5°28'05"$; $b = 142.33$ **13.** $A = 30°07'41"$; $b = 9.3669$
15. $A = 38°39'08"$; $b = 0.45661$ **17.** $A = 84°47'01"$; $c = 96.028$ **19.** $A = 26°35'27"$; $c = 8.1702$

Exercise 16 (*page* 408): **1.** $A = 22°37'12"$; $B = 108°55'28"$; $C = 48°27'20"$ **3.** No solution **5.** $a = 17$; $c = 25$; $B = 25°03'27"$ **7.** $c = 73.000$; $A = 4°14'32"$; $B = 138°53'16"$ **9.** $c = 75$; $B = 8°47'50"$; $C = 118°04'21"$ **11.** $a = 28.441$; $B = 59°14'24"$; $A = 16°30'36"$

Exercise 23 (*page* 422): **1.** 13.748 in **3.** 1.06%; 41% **5.** $c = 60.033$ by both methods

Chapter 14

Exercise 1 (*page* 426): **1.** $\overrightarrow{AB}$, $\overrightarrow{GH}$ **3.** $\overrightarrow{AB}$, $\overrightarrow{GH}$, $\overrightarrow{KL}$, $\overrightarrow{EF}$, $\overrightarrow{IJ}$

Exercise 3 (*page* 435):

	sin	cos	tan	cot	sec	csc
1.	0.64279	−0.76604	−0.83910	−1.1918	−1.3054	1.5557
3.	0.99540	−0.09585	−10.385	−0.09629	−10.433	1.0046
5.	−0.91708	−0.39870	+2.3001	+0.43475	−2.5081	−1.0904
7.	−0.08428	+0.99644	−0.08458	−11.823	+1.0036	−11.865
9.	−0.00568	+0.99998	−0.00568	−176.41	+1.0000	−176.41

Exercise 4 (*page* 438): **1.** 6°04′06″; 173°55′54″ **3.** 35°04′20″; 144°55′40″
5. 46°35′24″; 133°24′36″ **7.** 12°01′10″; 347°58′50″ **9.** 88°37′48″; 271°22′12″
11. 139°00′33″; 319°00′33″ **13.** 155°05′40″; 335°05′40″ **15.** 46°16′30″; 226°16′30″
17. 93°36′48″; 273°36′48″ **19.** 147°07′53″; 327°07′53″ **21.** 87°30′45″; 272°29′15″
23. 51°24′30″; 308°35′30″ **25.** 64°44′48″; 295°15′12″ **27.** 6°08′51″; 173°51′09″
29. 226°17′30″; 313°42′30″

Exercise 5 (*page* 440): **1.** $r = 8.06$; $\phi = 29.74°$ **3.** $r = 10.66$; $\phi = 110.30°$
5. $r = 7.92$; $\phi = 349.80°$ **7.** $r = 6.00$; $\phi = 119.98°$ **9.** $r = 8.02$; $\phi = 79.95°$
11. $r = 8.54$; $\phi = 290.56°$ **13.** $r = 9.05$; $\phi = 20.04°$ **15.** $r = 12.93$; $\phi = 140.65°$
17. $x_1 = +3.84$, $\phi_1 = 320.21°$; $x_2 = -3.84$, $\phi_2 = 219.79°$ **19.** $y_1 = +4.31$, $\phi_1 = 151.38°$;
$y_2 = -4.31$, $\phi_2 = 208.62°$ **21.** $x_1 = +10.32$, $\phi_1 = 339.78°$; $x_2 = -10.32$, $\phi_2 = 200.22°$
23. $x_1 = +9.44$, $\phi_1 = 19.27°$; $x_2 = -9.44$, $\phi_2 = 160.73°$

Exercise 6 (*page* 444): **1.** 0, 0.500, 0.707, 0.866, 1.000, 0.866, 0.707, 0.500, 0, −0.500,
−0.707, −0.866, −1.000, −0.866, −0.707, −0.500, 0 **3.** +1 and −1; +1 and −1:
unlimited; unlimited; does not exist between +1 and −1, unlimited otherwise; does
not exist between +1 and −1, unlimited otherwise 7. −0.707

Chapter 15

Exercise 1 (*page* 446): **1.** 0.64279 **3.** 0.99540 **5.** 0.91708 **7.** 0.08428 **9.** 0.00568
11. 19°25′; 160°35′ **13.** 46°45′; 133°15′ **15.** 86°36′; 93°24′

Exercise 2 (*page* 452): **1.** $a = 7.02$ in **3.** $b = 2.013$ in **5.** $c = 37$ in **7.** $A = 39°11′$;
$A' = 140°49′$

Exercise 3 (*page* 458): **3.** $a/\sin A = b$ **5.** $A = 28°05′$; $C = 106°15′$; $B = 45°40′$
7. $a = 7.3205$; $b = 5.1764$ **9.** $a = 1,000$; $b = 1,732.05$; $y = 866.03$ **11a.** $b = 66.913$
11b. $b = 17.365$ **13.** $a = 1.2007$; $c = 1.9355$; $B = 53°31′$ **15.** $a = 2.8438$; $c = 1.7307$;
$B = 130°13′$ **17.** $a = 1.3670$; $c = 6.0563$; $B = 122°46′$ **19.** $a = 1.1165$; $b = 11.795$;
$C = 90°46′$ **21.** $b = 4.9509$; $c = 2.9099$; $A = 104°30′$ **23.** $a = 11.080$; $b = 32.048$;
$C = 128°12′$ **25.** $a = 1.2045$; $B = 86°55′$; $A = 44°44′$; $A' = 38°34′$; $B' = 93°05′$;

$a' = 1.0669$ **27.** $c = 6.3591$; $B = 67°46'$; $C = 54°37'$; $c' = 1.3745$; $B' = 112°14'$;
$C' = 10°09'$ **29.** $c = 3.418$; $B = 71°58'$; $C = 46°40'$; $c' = 0.8644$; $B' = 108°02'$;
$C' = 10°36'$ **31.** $b = 2.0267$; $B = 32°51'$; $C = 84°38'$; $b' = 1.4067$; $B' = 22°07'$;
$C' = 95°22'$ **33.** $b = 2.7276$; $c = 4.3829$; $A = 59°12'$ **35.** $b = 20.678$; $c = 12.131$;
$B = 87°44'31''$ **37.** $a = 56.356$; $b = 188.79$; $C = 10°13'15''$ **39.** $c = 30.238$; $A = 82°54'$;
$C = 11°56'05''$ **41.** $a = 23.305$; $A = 42°20'10''$; $C = 28°26'34''$ **43.** $b = 10.002$;
$A = 60°38'51''$; $B = 30°25'06''$ **45.** $b = 6.1150$; $c = 10.132$; $C = 65°51'09''$
47. $b = 43.227$; $c = 24.339$; $B = 130°19'$ **49.** $a = 13.398$; $c = 29.090$; $B = 111°36'58''$
51. $b = 7,363.2$; $c = 6,152.9$; $A = 56°54'48''$ **53.** $a = 61.048$; $c = 54.981$; $A = 107°26.7'$
55. $b = 1,964.2$; $c = 2,272.1$; $C = 122°45.7'$ **57.** $b = 413.14$; $c = 615.36$; $B = 30°45'51''$
59. $b = 96.050$; $c = 74.784$; $A = 113°50.8'$

Exercise 4 (*page 466*): **9.** 7.1414 **11.** 16.894 **13.** 10.890 **15.** 33.734 **17.** 13.416
19. $A = 46°23'49''$; $B = 104°15'00''$; $C = 29°21'09''$ **21.** $A = 61°55'39''$; $B = 81°12'10''$;
$C = 36°52'11''$ **23.** $A = 34°02'53''$; $B = 44°24'54''$; $C = 101°32'13''$ **25.** $A = 44°24'54''$;
$B = 57°07'18''$; $C = 78°27'47''$ **27.** $A = 42°32'13''$; $B = 56°18'45''$; $C = 81°09'00''$
29. $B = 15°35'00''$; $A = 29°03'21''$; $C = 135°21'37''$

Exercise 5 (*page 468*): **1.** 4.5827 in **3.** 10.819 in **5.** 30.201 in **7.** 4.0578 in

Exercise 6 (*page 470*): **1.** $h = 587$ mi **3a.** 162.7 **3b.** 42.65° **3c.** 10.63°
5. $h = 402.77$ mi **7.** $\overparen{OP} = 1,129$ mi; sr $= 1,566$ mi **9.** 2,379 mi
sr $= 686.5$ mi

Exercise 7 (*page 474*): **1.** 803.02 **3.** 1,200.1 **5.** 4,853.5 **7.** 206,940 **9.** 72,477
11. 75.103 **13.** 4,064,600 **15.** 3,912,400

Chapter 16

Exercise 1 (*page 481*): **1.** Error in departure $= 0.08$ ft; error in latitude $= 0.28$ ft
3. S29°44'37''E; 909.90 ft **5.** Error in departure $= 0.63$ ft; error in latitude $= 0.19$ ft
7. $CJ = 711.85$ ft; $BK = 864.59$ ft (some variation permissible because of choice of
method)

Exercise 2 (*page 484*): **1.** 198,300 ft^2 **3.** 30,640,000 ft^2

Exercise 3 (*page 489*): **1a.** A straight line **1b.** Slope $= 0.00029$ **1c.** 0.0174
3. 647.56 ft **5.** 133⅓ ft

Exercise 4 (*page 489*): **5.** 146.25 ft **9.** 636.10 ft **11.** $da = 5,038.2$ yd;
$ac = 4,543.0$ yd; $bc = 3,673.5$ yd

Exercise 5 (*page* 494): **1.** 8°03′56″ **3.** 0.9806 in **5.** 24°49′56″ **7.** 43°55′10″
9. 1.1516 in **11.** 02°59′32″ **13.** 1.0775 in **15.** 25°22′11″ **17.** 4.996 in **19.** 122°10′25″
21. 5.8742 in **23.** 2.000 in **25.** x = 0.321 in; y = 1.476 in

Exercise 6 (*page* 501): **5.** 33°51′16″ **7.** 28°02′49″ **9.** 17°29′43″ **11.** 55°28′54″

Chapter 17

Exercise 1 (*page* 507):

	X projection	Y projection
1.	− 3	−20
3.	−19	−16
5.	+10	−10
7.	+ 5	+ 3
9.	+ 8	+ 6

Exercise 2 (*page* 512):

	X projection	Y projection
1a.	+ 2.67	+ 1.362
1c.	− 8.66	+ 5.00
1e.	0.00	+ 7.00
1g.	+13.12	+ 7.27
1i.	−14.14	−14.14
1k.	− 1.532	− 1.286
1m.	0.00	− 3.00
1o.	+ 4.33	+ 2.50

	ρ	ϕ
2a.	6.71	63.43°
2c.	2.83	135.00°
2e.	7.81	309.81°
2g.	3.61	213.69°
2i.	1.00	315.00°
2k.	8.49	315°

Exercise 3 (*page* 518):

	X projection	Y projection
1a.	− 100	−400
1c.	+200	−600

	Magnitude of vector sum	Reference angle
2a.	325 ft	133°
2c.	373 ft	274°
2e.	232 ft	329°

	X projection	Y projection	r	ϕ
3a.	+6.3	− 2.8	6.89	336.04°
3c.	−3.2	+15.8	16.12	101.45°
3e.	+6.0	− 3.0	6.71	333.44°

	X projection	Y projection	ρ	ϕ
4a.	−11.73	+19.20	22.5	121.42°
4c.	0.00	+ 4.00	4.00	90.00°
4e.	−22.9	− 6.43	23.8	195.70°
4g.	+ 1.589	− 6.07	6.28	284.67°
4i.	0.00	+40.00	40.00	90.00°

Exercise 4 (page 522):

	X projection	Y projection	ρ	ϕ
1a.	− 5	+ 1	5.10	168.69°
1c.	− 35	− 5	35.4	188.13°
1e.	+ 58	− 62	84.9	313.09°
2a.	21.2	1.213	21.2	3.27°
2c.	− 31.4	15.13	34.9	154.28°
2e.	37.4	62.1	72.5	58.92°
2g.	− 300	100	316	161.57°
2i.	65.8	− 38.0	76.0	330.00°

Exercise 5 (page 526):

	Radians	Degrees	Revolutions
1.		143.24	0.39789
3.	17.453		2.7778
5.	2,513.3	144,000	
7.		72,000	200.00
9.	5.4977		0.87500

Exercise 6 (page 527): **1.** 64 in **3.** 0.48 rad **5.** 22.9 in

Exercise 7 (page 529): **1.** 48 rad **3.** 90/π revolutions **5.** 7,500 rad **7.** 3.33 rad/s **9.** 314 s

Exercise 8 (*page 529*): **1.** 45 ft/s **3.** 0.00833 ft **5.** 9420 in/s **7.** 214.9 deg/s

Chapter 18

Exercise 1 (*page 532*): **1.** +0.868; +3.830; +1.710; −4.698; −4.330
3. Approximate slope = 1; tangent greater than sine.

Exercise 2 (*page 534*): **1a.** 8.55 **1b.** 25.00 **1c.** 19.15 **1d.** 0 **1e.** −12.50
1f. −25.00 **1g.** −21.65 **1h.** 0 **5.** 25.39 **7.** 45.6°; 134.4°; 225.6°; 314.4° **9.** 40°

Exercise 3 (*page 541*):

	Amplitude	*Frequency*	*Period*	*rad/s*	*deg/s*	*rad*	*deg*
6a.	23	50	0.02	100π	18,000	$\pi/6$	30
6c.	77	40	0.025	80π	14,400	$2\pi/7$	51.43
6e.	15	6×10^7	1.67×10^{-8}	$1.2\pi \times 10^8$	216×10^8	$-3\pi/5$	-108
6g.	200	1	1.00	2π	360	$-2\pi/5$	-72

7a. $y = 200 \sin (60\pi t - 5\pi/18)$ **7c.** $y = 5 \sin (2\pi \times 10^{12}t + \pi/2)$ **7e.** $y = 10 \sin (2{,}000\pi t - \pi/4)$ **7g.** $y = 20 \sin 50\pi t$ **7i.** $y = 100 \sin (100\pi t + \pi/2)$

Exercise 4 (*page 546*): **1.** Amplitude = 87; phase angle = +29.35°
3. Amplitude = 7; phase angle = +34.06° **5.** Amplitude = 12; phase angle = −28.95°

TABLE 1. NATURAL TRIGONOMETRIC FUNCTIONS*   601

	Sin	Tan	Ctn	Cos	Sec	Csc	
0° (180°)						**(359°) 179°**	
0	.00000	.00000	—	1.0000	1.0000	—	60
1	.00029	.00029	3437.7	1.0000	1.0000	3437.7	59
2	.00058	.00058	1718.9	1.0000	1.0000	1718.9	58
3	.00087	.00087	1145.9	1.0000	1.0000	1145.9	57
4	.00116	.00116	859.44	1.0000	1.0000	859.44	56
5	.00145	.00145	687.55	1.0000	1.0000	687.55	55
6	.00175	.00175	572.96	1.0000	1.0000	572.96	54
7	.00204	.00204	491.11	1.0000	1.0000	491.11	53
8	.00233	.00233	429.72	1.0000	1.0000	429.72	52
9	.00262	.00262	381.97	1.0000	1.0000	381.97	51
10	.00291	.00291	343.77	1.0000	1.0000	343.78	50
11	.00320	.00320	312.52	.99999	1.0000	312.52	49
12	.00349	.00349	286.48	.99999	1.0000	286.48	48
13	.00378	.00378	264.44	.99999	1.0000	264.44	47
14	.00407	.00407	245.55	.99999	1.0000	245.55	46
15	.00436	.00436	229.18	.99999	1.0000	229.18	45
16	.00465	.00465	214.86	.99999	1.0000	214.86	44
17	.00495	.00495	202.22	.99999	1.0000	202.22	43
18	.00524	.00524	190.98	.99999	1.0000	190.99	42
19	.00553	.00553	180.93	.99998	1.0000	180.93	41
20	.00582	.00582	171.89	.99998	1.0000	171.89	40
21	.00611	.00611	163.70	.99998	1.0000	163.70	39
22	.00640	.00640	156.26	.99998	1.0000	156.26	38
23	.00669	.00669	149.47	.99998	1.0000	149.47	37
24	.00698	.00698	143.24	.99998	1.0000	143.24	36
25	.00727	.00727	137.51	.99997	1.0000	137.51	35
26	.00756	.00756	132.22	.99997	1.0000	132.22	34
27	.00785	.00785	127.32	.99997	1.0000	127.33	33
28	.00814	.00815	122.77	.99997	1.0000	122.78	32
29	.00844	.00844	118.54	.99996	1.0000	118.54	31
30	.00873	.00873	114.59	.99996	1.0000	114.59	30
31	.00902	.00902	110.89	.99996	1.0000	110.90	29
32	.00931	.00931	107.43	.99996	1.0000	107.43	28
33	.00960	.00960	104.17	.99995	1.0000	104.18	27
34	.00989	.00989	101.11	.99995	1.0000	101.11	26
35	.01018	.01018	98.218	.99995	1.0001	98.223	25
36	.01047	.01047	95.489	.99995	1.0001	95.495	24
37	.01076	.01076	92.908	.99994	1.0001	92.914	23
38	.01105	.01105	90.463	.99994	1.0001	90.469	22
39	.01134	.01135	88.144	.99994	1.0001	88.149	21
40	.01164	.01164	85.940	.99993	1.0001	85.946	20
41	.01193	.01193	83.844	.99993	1.0001	83.849	19
42	.01222	.01222	81.847	.99993	1.0001	81.853	18
43	.01251	.01251	79.943	.99992	1.0001	79.950	17
44	.01280	.01280	78.126	.99992	1.0001	78.133	16
45	.01309	.01309	76.390	.99991	1.0001	76.397	15
46	.01338	.01338	74.729	.99991	1.0001	74.736	14
47	.01367	.01367	73.139	.99991	1.0001	73.146	13
48	.01396	.01396	71.615	.99990	1.0001	71.622	12
49	.01425	.01425	70.153	.99990	1.0001	70.160	11
50	.01454	.01455	68.750	.99989	1.0001	68.757	10
51	.01483	.01484	67.402	.99989	1.0001	67.409	9
52	.01513	.01513	66.105	.99989	1.0001	66.113	8
53	.01542	.01542	64.858	.99988	1.0001	64.866	7
54	.01571	.01571	63.657	.99988	1.0001	63.665	6
55	.01600	.01600	62.499	.99987	1.0001	62.507	5
56	.01629	.01629	61.383	.99987	1.0001	61.391	4
57	.01658	.01658	60.306	.99986	1.0001	60.314	3
58	.01687	.01687	59.266	.99986	1.0001	59.274	2
59	.01716	.01716	58.261	.99985	1.0001	58.270	1
60	.01745	.01746	57.290	.99985	1.0002	57.299	0
	Cos	Ctn	Tan	Sin	Csc	Sec	
90° (270°)						**(269°) 89°**	

	Sin	Tan	Ctn	Cos	Sec	Csc	
1° (181°)						**(358°) 178°**	
0	.01745	.01746	57.290	.99985	1.0002	57.299	60
1	.01774	.01775	56.351	.99984	1.0002	56.359	59
2	.01803	.01804	55.442	.99984	1.0002	55.451	58
3	.01832	.01833	54.561	.99983	1.0002	54.570	57
4	.01862	.01862	53.709	.99983	1.0002	53.718	56
5	.01891	.01891	52.882	.99982	1.0002	52.892	55
6	.01920	.01920	52.081	.99982	1.0002	52.090	54
7	.01949	.01949	51.303	.99981	1.0002	51.313	53
8	.01978	.01978	50.549	.99980	1.0002	50.558	52
9	.02007	.02007	49.816	.99980	1.0002	49.826	51
10	.02036	.02036	49.104	.99979	1.0002	49.114	50
11	.02065	.02066	48.412	.99979	1.0002	48.422	49
12	.02094	.02095	47.740	.99978	1.0002	47.750	48
13	.02123	.02124	47.085	.99977	1.0002	47.096	47
14	.02152	.02153	46.449	.99977	1.0002	46.460	46
15	.02181	.02182	45.829	.99976	1.0002	45.840	45
16	.02211	.02211	45.226	.99976	1.0002	45.237	44
17	.02240	.02240	44.639	.99975	1.0003	44.650	43
18	.02269	.02269	44.066	.99974	1.0003	44.077	42
19	.02298	.02298	43.508	.99974	1.0003	43.520	41
20	.02327	.02328	42.964	.99973	1.0003	42.976	40
21	.02356	.02357	42.433	.99972	1.0003	42.445	39
22	.02385	.02386	41.916	.99972	1.0003	41.928	38
23	.02414	.02415	41.411	.99971	1.0003	41.423	37
24	.02443	.02444	40.917	.99970	1.0003	40.930	36
25	.02472	.02473	40.436	.99969	1.0003	40.448	35
26	.02501	.02502	39.965	.99969	1.0003	39.978	34
27	.02530	.02531	39.506	.99968	1.0003	39.519	33
28	.02560	.02560	39.057	.99967	1.0003	39.070	32
29	.02589	.02589	38.618	.99966	1.0003	38.631	31
30	.02618	.02619	38.188	.99966	1.0003	38.202	30
31	.02647	.02648	37.769	.99965	1.0004	37.782	29
32	.02676	.02677	37.358	.99964	1.0004	37.371	28
33	.02705	.02706	36.956	.99963	1.0004	36.970	27
34	.02734	.02735	36.563	.99963	1.0004	36.576	26
35	.02763	.02764	36.178	.99962	1.0004	36.191	25
36	.02792	.02793	35.801	.99961	1.0004	35.815	24
37	.02821	.02822	35.431	.99960	1.0004	35.445	23
38	.02850	.02851	35.070	.99959	1.0004	35.084	22
39	.02879	.02881	34.715	.99959	1.0004	34.730	21
40	.02908	.02910	34.368	.99958	1.0004	34.382	20
41	.02938	.02939	34.027	.99957	1.0004	34.042	19
42	.02967	.02968	33.694	.99956	1.0004	33.708	18
43	.02996	.02997	33.366	.99955	1.0004	33.381	17
44	.03025	.03026	33.045	.99954	1.0005	33.060	16
45	.03054	.03055	32.730	.99953	1.0005	32.746	15
46	.03083	.03084	32.421	.99952	1.0005	32.437	14
47	.03112	.03114	32.118	.99952	1.0005	32.134	13
48	.03141	.03143	31.821	.99951	1.0005	31.836	12
49	.03170	.03172	31.528	.99950	1.0005	31.544	11
50	.03199	.03201	31.242	.99949	1.0005	31.258	10
51	.03228	.03230	30.960	.99948	1.0005	30.976	9
52	.03257	.03259	30.683	.99947	1.0005	30.700	8
53	.03286	.03288	30.412	.99946	1.0005	30.428	7
54	.03316	.03317	30.145	.99945	1.0006	30.161	6
55	.03345	.03346	29.882	.99944	1.0006	29.899	5
56	.03374	.03376	29.624	.99943	1.0006	29.641	4
57	.03403	.03405	29.371	.99942	1.0006	29.388	3
58	.03432	.03434	29.122	.99941	1.0006	29.139	2
59	.03461	.03463	28.877	.99940	1.0006	28.894	1
60	.03490	.03492	28.636	.99939	1.0006	28.654	0
	Cos	Ctn	Tan	Sin	Csc	Sec	
91° (271°)						**(268°) 88°**	

*From Richard Stevens Burington, "Handbook of Mathematical Tables and Formulas," 4th ed., McGraw-Hill Book Company, New York, 1965. For degrees indicated at top (bottom) of page use column headings at top (bottom). With degrees at left (right) of each block (top or bottom), use minute column at left (right).

TABLE 1. NATURAL TRIGONOMETRIC FUNCTIONS (continued) **602**

2° (182°)   (357°) 177°

′	Sin	Tan	Ctn	Cos	Sec	Csc	′
0	.03490	.03492	28.636	.99939	1.0006	28.654	60
1	.03519	.03521	28.399	.99938	1.0006	28.417	59
2	.03548	.03550	28.166	.99937	1.0006	28.184	58
3	.03577	.03579	27.937	.99936	1.0006	27.955	57
4	.03606	.03609	27.712	.99935	1.0007	27.730	56
5	.03635	.03638	27.490	.99934	1.0007	27.508	55
6	.03664	.03667	27.271	.99933	1.0007	27.290	54
7	.03693	.03696	27.057	.99932	1.0007	27.075	53
8	.03723	.03725	26.845	.99931	1.0007	26.864	52
9	.03752	.03754	26.637	.99930	1.0007	26.655	51
10	.03781	.03783	26.432	.99929	1.0007	26.451	50
11	.03810	.03812	26.230	.99927	1.0007	26.249	49
12	.03839	.03842	26.031	.99926	1.0007	26.050	48
13	.03868	.03871	25.835	.99925	1.0007	25.854	47
14	.03897	.03900	25.642	.99924	1.0008	25.661	46
15	.03926	.03929	25.452	.99923	1.0008	25.471	45
16	.03955	.03958	25.264	.99922	1.0008	25.284	44
17	.03984	.03987	25.080	.99921	1.0008	25.100	43
18	.04013	.04016	24.898	.99919	1.0008	24.918	42
19	.04042	.04046	24.719	.99918	1.0008	24.739	41
20	.04071	.04075	24.542	.99917	1.0008	24.562	40
21	.04100	.04104	24.368	.99916	1.0008	24.388	39
22	.04129	.04133	24.196	.99915	1.0009	24.216	38
23	.04159	.04162	24.026	.99913	1.0009	24.047	37
24	.04188	.04191	23.859	.99912	1.0009	23.880	36
25	.04217	.04220	23.695	.99911	1.0009	23.716	35
26	.04246	.04250	23.532	.99910	1.0009	23.553	34
27	.04275	.04279	23.372	.99909	1.0009	23.393	33
28	.04304	.04308	23.214	.99907	1.0009	23.235	32
29	.04333	.04337	23.058	.99906	1.0009	23.079	31
30	.04362	.04366	22.904	.99905	1.0010	22.926	30
31	.04391	.04395	22.752	.99904	1.0010	22.774	29
32	.04420	.04424	22.602	.99902	1.0010	22.624	28
33	.04449	.04454	22.454	.99901	1.0010	22.476	27
34	.04478	.04483	22.308	.99900	1.0010	22.330	26
35	.04507	.04512	22.164	.99898	1.0010	22.187	25
36	.04536	.04541	22.022	.99897	1.0010	22.044	24
37	.04565	.04570	21.881	.99896	1.0010	21.904	23
38	.04594	.04599	21.743	.99894	1.0011	21.766	22
39	.04623	.04628	21.606	.99893	1.0011	21.629	21
40	.04653	.04658	21.470	.99892	1.0011	21.494	20
41	.04682	.04687	21.337	.99890	1.0011	21.360	19
42	.04711	.04716	21.205	.99889	1.0011	21.229	18
43	.04740	.04745	21.075	.99888	1.0011	21.098	17
44	.04769	.04774	20.946	.99886	1.0011	20.970	16
45	.04798	.04803	20.819	.99885	1.0012	20.843	15
46	.04827	.04833	20.693	.99883	1.0012	20.717	14
47	.04856	.04862	20.569	.99882	1.0012	20.593	13
48	.04885	.04891	20.446	.99881	1.0012	20.471	12
49	.04914	.04920	20.325	.99879	1.0012	20.350	11
50	.04943	.04949	20.206	.99878	1.0012	20.230	10
51	.04972	.04978	20.087	.99876	1.0012	20.112	9
52	.05001	.05007	19.970	.99875	1.0013	19.995	8
53	.05030	.05037	19.855	.99873	1.0013	19.880	7
54	.05059	.05066	19.740	.99872	1.0013	19.766	6
55	.05088	.05095	19.627	.99870	1.0013	19.653	5
56	.05117	.05124	19.516	.99869	1.0013	19.541	4
57	.05146	.05153	19.405	.99867	1.0013	19.431	3
58	.05175	.05182	19.296	.99866	1.0013	19.322	2
59	.05205	.05212	19.188	.99864	1.0014	19.214	1
60	.05234	.05241	19.081	.99863	1.0014	19.107	0
′	Cos	Ctn	Tan	Sin	Csc	Sec	′

92° (272°)   (267°) 87°

3° (183°)   (356°) 176°

′	Sin	Tan	Ctn	Cos	Sec	Csc	′
0	.05234	.05241	19.081	.99863	1.0014	19.107	60
1	.05263	.05270	18.976	.99861	1.0014	19.002	59
2	.05292	.05299	18.871	.99860	1.0014	18.898	58
3	.05321	.05328	18.768	.99858	1.0014	18.794	57
4	.05350	.05357	18.666	.99857	1.0014	18.692	56
5	.05379	.05387	18.564	.99855	1.0014	18.591	55
6	.05408	.05416	18.464	.99854	1.0015	18.492	54
7	.05437	.05445	18.366	.99852	1.0015	18.393	53
8	.05466	.05474	18.268	.99851	1.0015	18.295	52
9	.05495	.05503	18.171	.99849	1.0015	18.198	51
10	.05524	.05533	18.075	.99847	1.0015	18.103	50
11	.05553	.05562	17.980	.99846	1.0015	18.008	49
12	.05582	.05591	17.886	.99844	1.0016	17.914	48
13	.05611	.05620	17.793	.99842	1.0016	17.822	47
14	.05640	.05649	17.702	.99841	1.0016	17.730	46
15	.05669	.05678	17.611	.99839	1.0016	17.639	45
16	.05698	.05708	17.521	.99838	1.0016	17.549	44
17	.05727	.05737	17.431	.99836	1.0016	17.460	43
18	.05756	.05766	17.343	.99834	1.0017	17.372	42
19	.05785	.05795	17.256	.99833	1.0017	17.285	41
20	.05814	.05824	17.169	.99831	1.0017	17.198	40
21	.05844	.05854	17.084	.99829	1.0017	17.113	39
22	.05873	.05883	16.999	.99827	1.0017	17.028	38
23	.05902	.05912	16.915	.99826	1.0017	16.945	37
24	.05931	.05941	16.832	.99824	1.0018	16.862	36
25	.05960	.05970	16.750	.99822	1.0018	16.779	35
26	.05989	.05999	16.668	.99821	1.0018	16.698	34
27	.06018	.06029	16.587	.99819	1.0018	16.618	33
28	.06047	.06058	16.507	.99817	1.0018	16.538	32
29	.06076	.06087	16.428	.99815	1.0019	16.459	31
30	.06105	.06116	16.350	.99813	1.0019	16.380	30
31	.06134	.06145	16.272	.99812	1.0019	16.303	29
32	.06163	.06175	16.195	.99810	1.0019	16.226	28
33	.06192	.06204	16.119	.99808	1.0019	16.150	27
34	.06221	.06233	16.043	.99806	1.0019	16.075	26
35	.06250	.06262	15.969	.99804	1.0020	16.000	25
36	.06279	.06291	15.895	.99803	1.0020	15.926	24
37	.06308	.06321	15.821	.99801	1.0020	15.853	23
38	.06337	.06350	15.748	.99799	1.0020	15.780	22
39	.06366	.06379	15.676	.99797	1.0020	15.708	21
40	.06395	.06408	15.605	.99795	1.0021	15.637	20
41	.06424	.06438	15.534	.99793	1.0021	15.566	19
42	.06453	.06467	15.464	.99792	1.0021	15.496	18
43	.06482	.06496	15.394	.99790	1.0021	15.427	17
44	.06511	.06525	15.325	.99788	1.0021	15.358	16
45	.06540	.06554	15.257	.99786	1.0021	15.290	15
46	.06569	.06584	15.189	.99784	1.0022	15.222	14
47	.06598	.06613	15.122	.99782	1.0022	15.155	13
48	.06627	.06642	15.056	.99780	1.0022	15.089	12
49	.06656	.06671	14.990	.99778	1.0022	15.023	11
50	.06685	.06700	14.924	.99776	1.0022	14.958	10
51	.06714	.06730	14.860	.99774	1.0023	14.893	9
52	.06743	.06759	14.795	.99772	1.0023	14.829	8
53	.06773	.06788	14.732	.99770	1.0023	14.766	7
54	.06802	.06817	14.669	.99768	1.0023	14.703	6
55	.06831	.06847	14.606	.99766	1.0023	14.640	5
56	.06860	.06876	14.544	.99764	1.0024	14.578	4
57	.06889	.06905	14.482	.99762	1.0024	14.517	3
58	.06918	.06934	14.421	.99760	1.0024	14.456	2
59	.06947	.06963	14.361	.99758	1.0024	14.395	1
60	.06976	.06993	14.301	.99756	1.0024	14.336	0
′	Cos	Ctn	Tan	Sin	Csc	Sec	′

93° (273°)   (266°) 86°

TABLE 1. NATURAL TRIGONOMETRIC FUNCTIONS (continued) **603**

4° (184°) (355°) 175°

′	Sin	Tan	Ctn	Cos	Sec	Csc	′
0	.06976	.06993	14.301	.99756	1.0024	14.336	60
1	.07005	.07022	14.241	.99754	1.0025	14.276	59
2	.07034	.07051	14.182	.99752	1.0025	14.217	58
3	.07063	.07080	14.124	.99750	1.0025	14.159	57
4	.07092	.07110	14.065	.99748	1.0025	14.101	56
5	.07121	.07139	14.008	.99746	1.0025	14.044	55
6	.07150	.07168	13.951	.99744	1.0026	13.987	54
7	.07179	.07197	13.894	.99742	1.0026	13.930	53
8	.07208	.07227	13.838	.99740	1.0026	13.874	52
9	.07237	.07256	13.782	.99738	1.0026	13.818	51
10	.07266	.07285	13.727	.99736	1.0027	13.763	50
11	.07295	.07314	13.672	.99734	1.0027	13.708	49
12	.07324	.07344	13.617	.99731	1.0027	13.654	48
13	.07353	.07373	13.563	.99729	1.0027	13.600	47
14	.07382	.07402	13.510	.99727	1.0027	13.547	46
15	.07411	.07431	13.457	.99725	1.0028	13.494	45
16	.07440	.07461	13.404	.99723	1.0028	13.441	44
17	.07469	.07490	13.352	.99721	1.0028	13.389	43
18	.07498	.07519	13.300	.99719	1.0028	13.337	42
19	.07527	.07548	13.248	.99716	1.0028	13.286	41
20	.07556	.07578	13.197	.99714	1.0029	13.235	40
21	.07585	.07607	13.146	.99712	1.0029	13.184	39
22	.07614	.07636	13.096	.99710	1.0029	13.134	38
23	.07643	.07665	13.046	.99708	1.0029	13.084	37
24	.07672	.07695	12.996	.99705	1.0030	13.035	36
25	.07701	.07724	12.947	.99703	1.0030	12.985	35
26	.07730	.07753	12.898	.99701	1.0030	12.937	34
27	.07759	.07782	12.850	.99699	1.0030	12.888	33
28	.07788	.07812	12.801	.99696	1.0030	12.840	32
29	.07817	.07841	12.754	.99694	1.0031	12.793	31
30	.07846	.07870	12.706	.99692	1.0031	12.745	30
31	.07875	.07899	12.659	.99689	1.0031	12.699	29
32	.07904	.07929	12.612	.99687	1.0031	12.652	28
33	.07933	.07958	12.566	.99685	1.0032	12.606	27
34	.07962	.07987	12.520	.99683	1.0032	12.560	26
35	.07991	.08017	12.474	.99680	1.0032	12.514	25
36	.08020	.08046	12.429	.99678	1.0032	12.469	24
37	.08049	.08075	12.384	.99676	1.0033	12.424	23
38	.08078	.08104	12.339	.99673	1.0033	12.379	22
39	.08107	.08134	12.295	.99671	1.0033	12.335	21
40	.08136	.08163	12.251	.99668	1.0033	12.291	20
41	.08165	.08192	12.207	.99666	1.0034	12.248	19
42	.08194	.08221	12.163	.99664	1.0034	12.204	18
43	.08223	.08251	12.120	.99661	1.0034	12.161	17
44	.08252	.08280	12.077	.99659	1.0034	12.119	16
45	.08281	.08309	12.035	.99657	1.0035	12.076	15
46	.08310	.08339	11.992	.99654	1.0035	12.034	14
47	.08339	.08368	11.950	.99652	1.0035	11.992	13
48	.08368	.08397	11.909	.99649	1.0035	11.951	12
49	.08397	.08427	11.867	.99647	1.0035	11.909	11
50	.08426	.08456	11.826	.99644	1.0036	11.868	10
51	.08455	.08485	11.785	.99642	1.0036	11.828	9
52	.08484	.08514	11.745	.99639	1.0036	11.787	8
53	.08513	.08544	11.705	.99637	1.0036	11.747	7
54	.08542	.08573	11.664	.99635	1.0037	11.707	6
55	.08571	.08602	11.625	.99632	1.0037	11.668	5
56	.08600	.08632	11.585	.99630	1.0037	11.628	4
57	.08629	.08661	11.546	.99627	1.0037	11.589	3
58	.08658	.08690	11.507	.99625	1.0038	11.551	2
59	.08687	.08720	11.468	.99622	1.0038	11.512	1
60	.08716	.08749	11.430	.99619	1.0038	11.474	0
′	Cos	Ctn	Tan	Sin	Csc	Sec	′

94° (274°) (265°) 85°

5° (185°) (354°) 174°

′	Sin	Tan	Ctn	Cos	Sec	Csc	′
0	.08716	.08749	11.430	.99619	1.0038	11.474	60
1	.08745	.08778	11.392	.99617	1.0038	11.436	59
2	.08774	.08807	11.354	.99614	1.0039	11.398	58
3	.08803	.08837	11.316	.99612	1.0039	11.360	57
4	.08831	.08866	11.279	.99609	1.0039	11.323	56
5	.08860	.08895	11.242	.99607	1.0039	11.286	55
6	.08889	.08925	11.205	.99604	1.0040	11.249	54
7	.08918	.08954	11.168	.99602	1.0040	11.213	53
8	.08947	.08983	11.132	.99599	1.0040	11.176	52
9	.08976	.09013	11.095	.99596	1.0041	11.140	51
10	.09005	.09042	11.059	.99594	1.0041	11.105	50
11	.09034	.09071	11.024	.99591	1.0041	11.069	49
12	.09063	.09101	10.988	.99588	1.0041	11.034	48
13	.09092	.09130	10.953	.99586	1.0042	10.998	47
14	.09121	.09159	10.918	.99583	1.0042	10.963	46
15	.09150	.09189	10.883	.99580	1.0042	10.929	45
16	.09179	.09218	10.848	.99578	1.0042	10.894	44
17	.09208	.09247	10.814	.99575	1.0043	10.860	43
18	.09237	.09277	10.780	.99572	1.0043	10.826	42
19	.09266	.09306	10.746	.99570	1.0043	10.792	41
20	.09295	.09335	10.712	.99567	1.0043	10.758	40
21	.09324	.09365	10.678	.99564	1.0044	10.725	39
22	.09353	.09394	10.645	.99562	1.0044	10.692	38
23	.09382	.09423	10.612	.99559	1.0044	10.659	37
24	.09411	.09453	10.579	.99556	1.0045	10.626	36
25	.09440	.09482	10.546	.99553	1.0045	10.593	35
26	.09469	.09511	10.514	.99551	1.0045	10.561	34
27	.09498	.09541	10.481	.99548	1.0045	10.529	33
28	.09527	.09570	10.449	.99545	1.0046	10.497	32
29	.09556	.09600	10.417	.99542	1.0046	10.465	31
30	.09585	.09629	10.385	.99540	1.0046	10.433	30
31	.09614	.09658	10.354	.99537	1.0047	10.402	29
32	.09642	.09688	10.322	.99534	1.0047	10.371	28
33	.09671	.09717	10.291	.99531	1.0047	10.340	27
34	.09700	.09746	10.260	.99528	1.0047	10.309	26
35	.09729	.09776	10.229	.99526	1.0048	10.278	25
36	.09758	.09805	10.199	.99523	1.0048	10.248	24
37	.09787	.09834	10.168	.99520	1.0048	10.217	23
38	.09816	.09864	10.138	.99517	1.0049	10.187	22
39	.09845	.09893	10.108	.99514	1.0049	10.157	21
40	.09874	.09923	10.078	.99511	1.0049	10.128	20
41	.09903	.09952	10.048	.99508	1.0049	10.098	19
42	.09932	.09981	10.019	.99506	1.0050	10.068	18
43	.09961	.10011	9.9893	.99503	1.0050	10.039	17
44	.09990	.10040	9.9601	.99500	1.0050	10.010	16
45	.10019	.10069	9.9310	.99497	1.0051	9.9812	15
46	.10048	.10099	9.9021	.99494	1.0051	9.9525	14
47	.10077	.10128	9.8734	.99491	1.0051	9.9239	13
48	.10106	.10158	9.8448	.99488	1.0051	9.8955	12
49	.10135	.10187	9.8164	.99485	1.0052	9.8672	11
50	.10164	.10216	9.7882	.99482	1.0052	9.8391	10
51	.10192	.10246	9.7601	.99479	1.0052	9.8112	9
52	.10221	.10275	9.7322	.99476	1.0053	9.7834	8
53	.10250	.10305	9.7044	.99473	1.0053	9.7558	7
54	.10279	.10334	9.6768	.99470	1.0053	9.7283	6
55	.10308	.10363	9.6493	.99467	1.0054	9.7010	5
56	.10337	.10393	9.6220	.99464	1.0054	9.6739	4
57	.10366	.10422	9.5949	.99461	1.0054	9.6469	3
58	.10395	.10452	9.5679	.99458	1.0054	9.6200	2
59	.10424	.10481	9.5411	.99455	1.0055	9.5933	1
60	.10453	.10510	9.5144	.99452	1.0055	9.5668	0
′	Cos	Ctn	Tan	Sin	Csc	Sec	′

95° (275°) (264°) 84°

TABLE 1. NATURAL TRIGONOMETRIC FUNCTIONS (continued)   **604**

6° (186°)   (353°) **173°**   **7° (187°)**   (352°) **172°**

′	Sin	Tan	Ctn	Cos	Sec	Csc	′		′	Sin	Tan	Ctn	Cos	Sec	Csc	′
0	.10453	.10510	9.5144	.99452	1.0055	9.5668	60		0	.12187	.12278	8.1443	.99255	1.0075	8.2055	60
1	.10482	.10540	9.4878	.99449	1.0055	9.5404	59		1	.12216	.12308	8.1248	.99251	1.0075	8.1861	59
2	.10511	.10569	9.4614	.99446	1.0056	9.5141	58		2	.12245	.12338	8.1054	.99248	1.0076	8.1668	58
3	.10540	.10599	9.4352	.99443	1.0056	9.4880	57		3	.12274	.12367	8.0860	.99244	1.0076	8.1476	57
4	.10569	.10628	9.4090	.99440	1.0056	9.4620	56		4	.12302	.12397	8.0667	.99240	1.0077	8.1285	56
5	.10597	.10657	9.3831	.99437	1.0057	9.4362	55		5	.12331	.12426	8.0476	.99237	1.0077	8.1095	55
6	.10626	.10687	9.3572	.99434	1.0057	9.4105	54		6	.12360	.12456	8.0285	.99233	1.0077	8.0905	54
7	.10655	.10716	9.3315	.99431	1.0057	9.3850	53		7	.12389	.12485	8.0095	.99230	1.0078	8.0717	53
8	.10684	.10746	9.3060	.99428	1.0058	9.3596	52		8	.12418	.12515	7.9906	.99226	1.0078	8.0529	52
9	.10713	.10775	9.2806	.99424	1.0058	9.3343	51		9	.12447	.12544	7.9718	.99222	1.0078	8.0342	51
10	.10742	.10805	9.2553	.99421	1.0058	9.3092	50		10	.12476	.12574	7.9530	.99219	1.0079	8.0156	50
11	.10771	.10834	9.2302	.99418	1.0059	9.2842	49		11	.12504	.12603	7.9344	.99215	1.0079	7.9971	49
12	.10800	.10863	9.2052	.99415	1.0059	9.2593	48		12	.12533	.12633	7.9158	.99211	1.0079	7.9787	48
13	.10829	.10893	9.1803	.99412	1.0059	9.2346	47		13	.12562	.12662	7.8973	.99208	1.0080	7.9604	47
14	.10858	.10922	9.1555	.99409	1.0059	9.2100	46		14	.12591	.12692	7.8789	.99204	1.0080	7.9422	46
15	.10887	.10952	9.1309	.99406	1.0060	9.1855	45		15	.12620	.12722	7.8606	.99200	1.0081	7.9240	45
16	.10916	.10981	9.1065	.99402	1.0060	9.1612	44		16	.12649	.12751	7.8424	.99197	1.0081	7.9059	44
17	.10945	.11011	9.0821	.99399	1.0060	9.1370	43		17	.12678	.12781	7.8243	.99193	1.0081	7.8879	43
18	.10973	.11040	9.0579	.99396	1.0061	9.1129	42		18	.12706	.12810	7.8062	.99189	1.0082	7.8700	42
19	.11002	.11070	9.0338	.99393	1.0061	9.0890	41		19	.12735	.12840	7.7882	.99186	1.0082	7.8522	41
20	.11031	.11099	9.0098	.99390	1.0061	9.0652	40		20	.12764	.12869	7.7704	.99182	1.0082	7.8344	40
21	.11060	.11128	8.9860	.99386	1.0062	9.0415	39		21	.12793	.12899	7.7525	.99178	1.0083	7.8168	39
22	.11089	.11158	8.9623	.99383	1.0062	9.0179	38		22	.12822	.12929	7.7348	.99175	1.0083	7.7992	38
23	.11118	.11187	8.9387	.99380	1.0062	8.9944	37		23	.12851	.12958	7.7171	.99171	1.0084	7.7817	37
24	.11147	.11217	8.9152	.99377	1.0063	8.9711	36		24	.12880	.12988	7.6996	.99167	1.0084	7.7642	36
25	.11176	.11246	8.8919	.99374	1.0063	8.9479	35		25	.12908	.13017	7.6821	.99163	1.0084	7.7469	35
26	.11205	.11276	8.8686	.99370	1.0063	8.9248	34		26	.12937	.13047	7.6647	.99160	1.0085	7.7296	34
27	.11234	.11305	8.8455	.99367	1.0064	8.9019	33		27	.12966	.13076	7.6473	.99156	1.0085	7.7124	33
28	.11263	.11335	8.8225	.99364	1.0064	8.8790	32		28	.12995	.13106	7.6301	.99152	1.0086	7.6953	32
29	.11291	.11364	8.7996	.99360	1.0064	8.8563	31		29	.13024	.13136	7.6129	.99148	1.0086	7.6783	31
30	.11320	.11394	8.7769	.99357	1.0065	8.8337	30		30	.13053	.13165	7.5958	.99144	1.0086	7.6613	30
31	.11349	.11423	8.7542	.99354	1.0065	8.8112	29		31	.13081	.13195	7.5787	.99141	1.0087	7.6444	29
32	.11378	.11452	8.7317	.99351	1.0065	8.7888	28		32	.13110	.13224	7.5618	.99137	1.0087	7.6276	28
33	.11407	.11482	8.7093	.99347	1.0066	8.7665	27		33	.13139	.13254	7.5449	.99133	1.0087	7.6109	27
34	.11436	.11511	8.6870	.99344	1.0066	8.7444	26		34	.13168	.13284	7.5281	.99129	1.0088	7.5942	26
35	.11465	.11541	8.6648	.99341	1.0066	8.7223	25		35	.13197	.13313	7.5113	.99125	1.0088	7.5776	25
36	.11494	.11570	8.6427	.99337	1.0067	8.7004	24		36	.13226	.13343	7.4947	.99122	1.0089	7.5611	24
37	.11523	.11600	8.6208	.99334	1.0067	8.6786	23		37	.13254	.13372	7.4781	.99118	1.0089	7.5446	23
38	.11552	.11629	8.5989	.99331	1.0067	8.6569	22		38	.13283	.13402	7.4615	.99114	1.0089	7.5282	22
39	.11580	.11659	8.5772	.99327	1.0068	8.6353	21		39	.13312	.13432	7.4451	.99110	1.0090	7.5119	21
40	.11609	.11688	8.5555	.99324	1.0068	8.6138	20		40	.13341	.13461	7.4287	.99106	1.0090	7.4957	20
41	.11638	.11718	8.5340	.99320	1.0068	8.5924	19		41	.13370	.13491	7.4124	.99102	1.0091	7.4795	19
42	.11667	.11747	8.5126	.99317	1.0069	8.5711	18		42	.13399	.13521	7.3962	.99098	1.0091	7.4635	18
43	.11696	.11777	8.4913	.99314	1.0069	8.5500	17		43	.13427	.13550	7.3800	.99094	1.0091	7.4474	17
44	.11725	.11806	8.4701	.99310	1.0069	8.5289	16		44	.13456	.13580	7.3639	.99091	1.0092	7.4315	16
45	.11754	.11836	8.4490	.99307	1.0070	8.5079	15		45	.13485	.13609	7.3479	.99087	1.0092	7.4156	15
46	.11783	.11865	8.4280	.99303	1.0070	8.4871	14		46	.13514	.13639	7.3319	.99083	1.0093	7.3998	14
47	.11812	.11895	8.4071	.99300	1.0070	8.4663	13		47	.13543	.13669	7.3160	.99079	1.0093	7.3840	13
48	.11840	.11924	8.3863	.99297	1.0071	8.4457	12		48	.13572	.13698	7.3002	.99075	1.0093	7.3684	12
49	.11869	.11954	8.3656	.99293	1.0071	8.4251	11		49	.13600	.13728	7.2844	.99071	1.0094	7.3527	11
50	.11898	.11983	8.3450	.99290	1.0072	8.4047	10		50	.13629	.13758	7.2687	.99067	1.0094	7.3372	10
51	.11927	.12013	8.3245	.99286	1.0072	8.3843	9		51	.13658	.13787	7.2531	.99063	1.0095	7.3217	9
52	.11956	.12042	8.3041	.99283	1.0072	8.3641	8		52	.13687	.13817	7.2375	.99059	1.0095	7.3063	8
53	.11985	.12072	8.2838	.99279	1.0073	8.3439	7		53	.13716	.13846	7.2220	.99055	1.0095	7.2909	7
54	.12014	.12101	8.2636	.99276	1.0073	8.3238	6		54	.13744	.13876	7.2066	.99051	1.0096	7.2757	6
55	.12043	.12131	8.2434	.99272	1.0073	8.3039	5		55	.13773	.13906	7.1912	.99047	1.0096	7.2604	5
56	.12071	.12160	8.2234	.99269	1.0074	8.2840	4		56	.13802	.13935	7.1759	.99043	1.0097	7.2453	4
57	.12100	.12190	8.2035	.99265	1.0074	8.2642	3		57	.13831	.13965	7.1607	.99039	1.0097	7.2302	3
58	.12129	.12219	8.1837	.99262	1.0074	8.2446	2		58	.13860	.13995	7.1455	.99035	1.0097	7.2152	2
59	.12158	.12249	8.1640	.99258	1.0075	8.2250	1		59	.13889	.14024	7.1304	.99031	1.0098	7.2002	1
60	.12187	.12278	8.1443	.99255	1.0075	8.2055	0		60	.13917	.14054	7.1154	.99027	1.0098	7.1853	0
′	Cos	Ctn	Tan	Sin	Csc	Sec	′		′	Cos	Ctn	Tan	Sin	Csc	Sec	′

96° (276°)   (263°) **83°**   **97° (277°)**   (262°) **82°**

TABLE 1. NATURAL TRIGONOMETRIC FUNCTIONS (continued)   605

8° (188°)   (351°) 171°

′	Sin	Tan	Ctn	Cos	Sec	Csc	′
0	.13917	.14054	7.1154	.99027	1.0098	7.1853	60
1	.13946	.14084	7.1004	.99023	1.0099	7.1705	59
2	.13975	.14113	7.0855	.99019	1.0099	7.1557	58
3	.14004	.14143	7.0706	.99015	1.0100	7.1410	57
4	.14033	.14173	7.0558	.99011	1.0100	7.1263	56
5	.14061	.14202	7.0410	.99006	1.0100	7.1117	55
6	.14090	.14232	7.0264	.99002	1.0101	7.0972	54
7	.14119	.14262	7.0117	.98998	1.0101	7.0827	53
8	.14148	.14291	6.9972	.98994	1.0102	7.0683	52
9	.14177	.14321	6.9827	.98990	1.0102	7.0539	51
10	.14205	.14351	6.9682	.98986	1.0102	7.0396	50
11	.14234	.14381	6.9538	.98982	1.0103	7.0254	49
12	.14263	.14410	6.9395	.98978	1.0103	7.0112	48
13	.14292	.14440	6.9252	.98973	1.0104	6.9971	47
14	.14320	.14470	6.9110	.98969	1.0104	6.9830	46
15	.14349	.14499	6.8969	.98965	1.0105	6.9690	45
16	.14378	.14529	6.8828	.98961	1.0105	6.9550	44
17	.14407	.14559	6.8687	.98957	1.0105	6.9411	43
18	.14436	.14588	6.8548	.98953	1.0106	6.9273	42
19	.14464	.14618	6.8408	.98948	1.0106	6.9135	41
20	.14493	.14648	6.8269	.98944	1.0107	6.8998	40
21	.14522	.14678	6.8131	.98940	1.0107	6.8861	39
22	.14551	.14707	6.7994	.98936	1.0108	6.8725	38
23	.14580	.14737	6.7856	.98931	1.0108	6.8589	37
24	.14608	.14767	6.7720	.98927	1.0108	6.8454	36
25	.14637	.14796	6.7584	.98923	1.0109	6.8320	35
26	.14666	.14826	6.7448	.98919	1.0109	6.8186	34
27	.14695	.14856	6.7313	.98914	1.0110	6.8052	33
28	.14723	.14886	6.7179	.98910	1.0110	6.7919	32
29	.14752	.14915	6.7045	.98906	1.0111	6.7787	31
30	.14781	.14945	6.6912	98902	1.0111	6.7655	30
31	.14810	.14975	6.6779	.98897	1.0112	6.7523	29
32	.14838	.15005	6.6646	.98893	1.0112	6.7392	28
33	.14867	.15034	6.6514	.98889	1.0112	6.7262	27
34	.14896	.15064	6.6383	.98884	1.0113	6.7132	26
35	.14925	.15094	6.6252	.98880	1.0113	6.7003	25
36	.14954	.15124	6.6122	.98876	1.0114	6.6874	24
37	.14982	.15153	6.5992	.98871	1.0114	6.6745	23
38	.15011	.15183	6.5863	.98867	1.0115	6.6618	22
39	.15040	.15213	6.5734	.98863	1.0115	6.6490	21
40	.15069	.15243	6.5606	.98858	1.0116	6.6363	20
41	.15097	.15272	6.5478	.98854	1.0116	6.6237	19
42	.15126	.15302	6.5350	.98849	1.0116	6.6111	18
43	.15155	.15332	6.5223	.98845	1.0117	6.5986	17
44	.15184	.15362	6.5097	.98841	1.0117	6.5861	16
45	.15212	.15391	6.4971	.98836	1.0118	6.5736	15
46	.15241	.15421	6.4846	.98832	1.0118	6.5612	14
47	.15270	.15451	6.4721	.98827	1.0119	6.5489	13
48	.15299	.15481	6.4596	.98823	1.0119	6.5366	12
49	.15327	.15511	6.4472	.98818	1.0120	6.5243	11
50	.15356	.15540	6.4348	.98814	1.0120	6.5121	10
51	.15385	.15570	6.4225	.98809	1.0120	6.4999	9
52	.15414	.15600	6.4103	.98805	1.0121	6.4878	8
53	.15442	.15630	6.3980	.98800	1.0121	6.4757	7
54	.15471	.15660	6.3859	.98796	1.0122	6.4637	6
55	.15500	.15689	6.3737	.98791	1.0122	6.4517	5
56	.15529	.15719	6.3617	.98787	1.0123	6.4398	4
57	.15557	.15749	6.3496	.98782	1.0123	6.4279	3
58	.15586	.15779	6.3376	.98778	1.0124	6.4160	2
59	.15615	.15809	6.3257	.98773	1.0124	6.4042	1
60	.15643	.15838	6.3138	.98769	1.0125	6.3925	0
′	Cos	Ctn	Tan	Sin	Csc	Sec	′

98° (278°)   (261°) 81°

9° (189°)   (350°) 170°

′	Sin	Tan	Ctn	Cos	Sec	Csc	′
0	.15643	.15838	6.3138	.98769	1.0125	6.3925	60
1	.15672	.15868	6.3019	.98764	1.0125	6.3807	59
2	.15701	.15898	6.2901	.98760	1.0126	6.3691	58
3	.15730	.15928	6.2783	.98755	1.0126	6.3574	57
4	.15758	.15958	6.2666	.98751	1.0127	6.3458	56
5	.15787	.15988	6.2549	.98746	1.0127	6.3343	55
6	.15816	.16017	6.2432	.98741	1.0127	6.3228	54
7	.15845	.16047	6.2316	.98737	1.0128	6.3113	53
8	.15873	.16077	6.2200	.98732	1.0128	6.2999	52
9	.15902	.16107	6.2085	.98728	1.0129	6.2885	51
10	.15931	.16137	6.1970	.98723	1.0129	6.2772	50
11	.15959	.16167	6.1856	.98718	1.0130	6.2659	49
12	.15988	.16196	6.1742	.98714	1.0130	6.2546	48
13	.16017	.16226	6.1628	.98709	1.0131	6.2434	47
14	.16046	.16256	6.1515	.98704	1.0131	6.2323	46
15	.16074	.16286	6.1402	.98700	1.0132	6.2211	45
16	.16103	.16316	6.1290	.98695	1.0132	6.2100	44
17	.16132	.16346	6.1178	.98690	1.0133	6.1990	43
18	.16160	.16376	6.1066	.98686	1.0133	6.1880	42
19	.16189	.16405	6.0955	.98681	1.0134	6.1770	41
20	.16218	.16435	6.0844	.98676	1.0134	6.1661	40
21	.16246	.16465	6.0734	.98671	1.0135	6.1552	39
22	.16275	.16495	6.0624	.98667	1.0135	6.1443	38
23	.16304	.16525	6.0514	.98662	1.0136	6.1335	37
24	.16333	.16555	6.0405	.98657	1.0136	6.1227	36
25	.16361	.16585	6.0296	.98652	1.0137	6.1120	35
26	.16390	.16615	6.0188	.98648	1.0137	6.1013	34
27	.16419	.16645	6.0080	.98643	1.0138	6.0906	33
28	.16447	.16674	5.9972	.98638	1.0138	6.0800	32
29	.16476	.16704	5.9865	.98633	1.0139	6.0694	31
30	.16505	.16734	5.9758	.98629	1.0139	6.0589	30
31	.16533	.16764	5.9651	.98624	1.0140	6.0483	29
32	.16562	.16794	5.9545	.98619	1.0140	6.0379	28
33	.16591	.16824	5.9439	.98614	1.0141	6.0274	27
34	.16620	.16854	5.9333	.98609	1.0141	6.0170	26
35	.16648	.16884	5.9228	.98604	1.0142	6.0067	25
36	.16677	.16914	5.9124	.98600	1.0142	5.9963	24
37	.16706	.16944	5.9019	.98595	1.0143	5.9860	23
38	.16734	.16974	5.8915	.98590	1.0143	5.9758	22
39	.16763	.17004	5.8811	.98585	1.0144	5.9656	21
40	.16792	.17033	5.8708	.98580	1.0144	5.9554	20
41	.16820	.17063	5.8605	.98575	1.0145	5.9452	19
42	.16849	.17093	5.8502	.98570	1.0145	5.9351	18
43	.16878	.17123	5.8400	.98565	1.0146	5.9250	17
44	.16906	.17153	5.8298	.98561	1.0146	5.9150	16
45	.16935	.17183	5.8197	.98556	1.0147	5.9049	15
46	.16964	.17213	5.8095	.98551	1.0147	5.8950	14
47	.16992	.17243	5.7994	.98546	1.0148	5.8850	13
48	.17021	.17273	5.7894	.98541	1.0148	5.8751	12
49	.17050	.17303	5.7794	.98536	1.0149	5.8652	11
50	.17078	.17333	5.7694	.98531	1.0149	5.8554	10
51	.17107	.17363	5.7594	.98526	1.0150	5.8456	9
52	.17136	.17393	5.7495	.98521	1.0150	5.8358	8
53	.17164	.17423	5.7396	.98516	1.0151	5.8261	7
54	.17193	.17453	5.7297	.98511	1.0151	5.8164	6
55	.17222	.17483	5.7199	.98506	1.0152	5.8067	5
56	.17250	.17513	5.7101	.98501	1.0152	5.7970	4
57	.17279	.17543	5.7004	.98496	1.0153	5.7874	3
58	.17308	.17573	5.6906	.98491	1.0153	5.7778	2
59	.17336	.17603	5.6809	.98486	1.0153	5.7683	1
60	.17365	.17633	5.6713	.98481	1.0154	5.7588	0
′	Cos	Ctn	Tan	Sin	Csc	Sec	′

99° (279°)   (260°) 80°

TABLE 1. NATURAL TRIGONOMETRIC FUNCTIONS (continued) **606**

10° (190°)   (349°) **169°**   **11° (191°)**   (348°) **168°**

′	Sin	Tan	Ctn	Cos	Sec	Csc	′
0	.17365	.17633	5.6713	.98481	1.0154	5.7588	60
1	.17393	.17663	5.6617	.98476	1.0155	5.7493	59
2	.17422	.17693	5.6521	.98471	1.0155	5.7398	58
3	.17451	.17723	5.6425	.98466	1.0156	5.7304	57
4	.17479	.17753	5.6329	.98461	1.0156	5.7210	56
5	.17508	.17783	5.6234	.98455	1.0157	5.7117	55
6	.17537	.17813	5.6140	.98450	1.0157	5.7023	54
7	.17565	.17843	5.6045	.98445	1.0158	5.6930	53
8	.17594	.17873	5.5951	.98440	1.0158	5.6838	52
9	.17623	.17903	5.5857	.98435	1.0159	5.6745	51
10	.17651	.17933	5.5764	.98430	1.0160	5.6653	50
11	.17680	.17963	5.5671	.98425	1.0160	5.6562	49
12	.17708	.17993	5.5578	.98420	1.0161	5.6470	48
13	.17737	.18023	5.5485	.98414	1.0161	5.6379	47
14	.17766	.18053	5.5393	.98409	1.0162	5.6288	46
15	.17794	.18083	5.5301	.98404	1.0162	5.6198	45
16	.17823	.18113	5.5209	.98399	1.0163	5.6107	44
17	.17852	.18143	5.5118	.98394	1.0163	5.6017	43
18	.17880	.18173	5.5026	.98389	1.0164	5.5928	42
19	.17909	.18203	5.4936	.98383	1.0164	5.5838	41
20	.17937	.18233	5.4845	.98378	1.0165	5.5749	40
21	.17966	.18263	5.4755	.98373	1.0165	5.5660	39
22	.17995	.18293	5.4665	.98368	1.0166	5.5572	38
23	.18023	.18323	5.4575	.98362	1.0166	5.5484	37
24	.18052	.18353	5.4486	.98357	1.0167	5.5396	36
25	.18081	.18384	5.4397	.98352	1.0168	5.5308	35
26	.18109	.18414	5.4308	.98347	1.0168	5.5221	34
27	.18138	.18444	5.4219	.98341	1.0169	5.5134	33
28	.18166	.18474	5.4131	.98336	1.0169	5.5047	32
29	.18195	.18504	5.4043	.98331	1.0170	5.4960	31
30	.18224	.18534	5.3955	.98325	1.0170	5.4874	30
31	.18252	.18564	5.3868	.98320	1.0171	5.4788	29
32	.18281	.18594	5.3781	.98315	1.0171	5.4702	28
33	.18309	.18624	5.3694	.98310	1.0172	5.4617	27
34	.18338	.18654	5.3607	.98304	1.0173	5.4532	26
35	.18367	.18684	5.3521	.98299	1.0173	5.4447	25
36	.18395	.18714	5.3435	.98294	1.0174	5.4362	24
37	.18424	.18745	5.3349	.98288	1.0174	5.4278	23
38	.18452	.18775	5.3263	.98283	1.0175	5.4194	22
39	.18481	.18805	5.3178	.98277	1.0175	5.4110	21
40	.18509	.18835	5.3093	.98272	1.0176	5.4026	20
41	.18538	.18865	5.3008	.98267	1.0177	5.3943	19
42	.18567	.18895	5.2924	.98261	1.0177	5.3860	18
43	.18595	.18925	5.2839	.98256	1.0178	5.3777	17
44	.18624	.18955	5.2755	.98250	1.0178	5.3695	16
45	.18652	.18986	5.2672	.98245	1.0179	5.3612	15
46	.18681	.19016	5.2588	.98240	1.0179	5.3530	14
47	.18710	.19046	5.2505	.98234	1.0180	5.3449	13
48	.18738	.19076	5.2422	.98229	1.0180	5.3367	12
49	.18767	.19106	5.2339	.98223	1.0181	5.3286	11
50	.18795	.19136	5.2257	.98218	1.0181	5.3205	10
51	.18824	.19166	5.2174	.98212	1.0182	5.3124	9
52	.18852	.19197	5.2092	.98207	1.0183	5.3044	8
53	.18881	.19227	5.2011	.98201	1.0183	5.2963	7
54	.18910	.19257	5.1929	.98196	1.0184	5.2883	6
55	.18938	.19287	5.1848	.98190	1.0184	5.2804	5
56	.18967	.19317	5.1767	.98185	1.0185	5.2724	4
57	.18995	.19347	5.1686	.98179	1.0185	5.2645	3
58	.19024	.19378	5.1606	.98174	1.0186	5.2566	2
59	.19052	.19408	5.1526	.98168	1.0187	5.2487	1
60	.19081	.19438	5.1446	.98163	1.0187	5.2408	0
′	Cos	Ctn	Tan	Sin	Csc	Sec	′

100° (280°)   (259°) **79°**

′	Sin	Tan	Ctn	Cos	Sec	Csc	′
0	.19081	.19438	5.1446	.98163	1.0187	5.2408	60
1	.19109	.19468	5.1366	.98157	1.0188	5.2330	59
2	.19138	.19498	5.1286	.98152	1.0188	5.2252	58
3	.19167	.19529	5.1207	.98146	1.0189	5.2174	57
4	.19195	.19559	5.1128	.98140	1.0189	5.2097	56
5	.19224	.19589	5.1049	.98135	1.0190	5.2019	55
6	.19252	.19619	5.0970	.98129	1.0191	5.1942	54
7	.19281	.19649	5.0892	.98124	1.0191	5.1865	53
8	.19309	.19680	5.0814	.98118	1.0192	5.1789	52
9	.19338	.19710	5.0736	.98112	1.0192	5.1712	51
10	.19366	.19740	5.0658	.98107	1.0193	5.1636	50
11	.19395	.19770	5.0581	.98101	1.0194	5.1560	49
12	.19423	.19801	5.0504	.98096	1.0194	5.1484	48
13	.19452	.19831	5.0427	.98090	1.0195	5.1409	47
14	.19481	.19861	5.0350	.98084	1.0195	5.1333	46
15	.19509	.19891	5.0273	.98079	1.0196	5.1258	45
16	.19538	.19921	5.0197	.98073	1.0197	5.1183	44
17	.19566	.19952	5.0121	.98067	1.0197	5.1109	43
18	.19595	.19982	5.0045	.98061	1.0198	5.1034	42
19	.19623	.20012	4.9969	.98056	1.0198	5.0960	41
20	.19652	.20042	4.9894	.98050	1.0199	5.0886	40
21	.19680	.20073	4.9819	.98044	1.0199	5.0813	39
22	.19709	.20103	4.9744	.98039	1.0200	5.0739	38
23	.19737	.20133	4.9669	.98033	1.0201	5.0666	37
24	.19766	.20164	4.9594	.98027	1.0201	5.0593	36
25	.19794	.20194	4.9520	.98021	1.0202	5.0520	35
26	.19823	.20224	4.9446	.98016	1.0202	5.0447	34
27	.19851	.20254	4.9372	.98010	1.0203	5.0375	33
28	.19880	.20285	4.9298	.98004	1.0204	5.0302	32
29	.19908	.20315	4.9225	.97998	1.0204	5.0230	31
30	.19937	.20345	4.9152	.97992	1.0205	5.0159	30
31	.19965	.20376	4.9078	.97987	1.0205	5.0087	29
32	.19994	.20406	4.9006	.97981	1.0206	5.0016	28
33	.20022	.20436	4.8933	.97975	1.0207	4.9944	27
34	.20051	.20466	4.8860	.97969	1.0207	4.9873	26
35	.20079	.20497	4.8788	.97963	1.0208	4.9803	25
36	.20108	.20527	4.8716	.97958	1.0209	4.9732	24
37	.20136	.20557	4.8644	.97952	1.0209	4.9662	23
38	.20165	.20588	4.8573	.97946	1.0210	4.9591	22
39	.20193	.20618	4.8501	.97940	1.0210	4.9521	21
40	.20222	.20648	4.8430	.97934	1.0211	4.9452	20
41	.20250	.20679	4.8359	.97928	1.0212	4.9382	19
42	.20279	.20709	4.8288	.97922	1.0212	4.9313	18
43	.20307	.20739	4.8218	.97916	1.0213	4.9244	17
44	.20336	.20770	4.8147	.97910	1.0213	4.9175	16
45	.20364	.20800	4.8077	.97905	1.0214	4.9106	15
46	.20393	.20830	4.8007	.97899	1.0215	4.9037	14
47	.20421	.20861	4.7937	.97893	1.0215	4.8969	13
48	.20450	.20891	4.7867	.97887	1.0216	4.8901	12
49	.20478	.20921	4.7798	.97881	1.0217	4.8833	11
50	.20507	.20952	4.7729	.97875	1.0217	4.8765	10
51	.20535	.20982	4.7659	.97869	1.0218	4.8697	9
52	.20563	.21013	4.7591	.97863	1.0218	4.8630	8
53	.20592	.21043	4.7522	.97857	1.0219	4.8563	7
54	.20620	.21073	4.7453	.97851	1.0220	4.8496	6
55	.20649	.21104	4.7385	.97845	1.0220	4.8429	5
56	.20677	.21134	4.7317	.97839	1.0221	4.8362	4
57	.20706	.21164	4.7249	.97833	1.0222	4.8296	3
58	.20734	.21195	4.7181	.97827	1.0222	4.8229	2
59	.20763	.21225	4.7114	.97821	1.0223	4.8163	1
60	.20791	.21256	4.7046	.97815	1.0223	4.8097	0
′	Cos	Ctn	Tan	Sin	Csc	Sec	′

101° (281°)   (258°) **78°**

12° (192°)   (347°) **167°**

′	Sin	Tan	Ctn	Cos	Sec	Csc	′
0	.20791	.21256	4.7046	.97815	1.0223	4.8097	60
1	.20820	.21286	4.6979	.97809	1.0224	4.8032	59
2	.20848	.21316	4.6912	.97803	1.0225	4.7966	58
3	.20877	.21347	4.6845	.97797	1.0225	4.7901	57
4	.20905	.21377	4.6779	.97791	1.0226	4.7836	56
5	.20933	.21408	4.6712	.97784	1.0227	4.7771	55
6	.20962	.21438	4.6646	.97778	1.0227	4.7706	54
7	.20990	.21469	4.6580	.97772	1.0228	4.7641	53
8	.21019	.21499	4.6514	.97766	1.0228	4.7577	52
9	.21047	.21529	4.6448	.97760	1.0229	4.7512	51
10	.21076	.21560	4.6382	.97754	1.0230	4.7448	50
11	.21104	.21590	4.6317	.97748	1.0230	4.7384	49
12	.21132	.21621	4.6252	.97742	1.0231	4.7321	48
13	.21161	.21651	4.6187	.97735	1.0232	4.7257	47
14	.21189	.21682	4.6122	.97729	1.0232	4.7194	46
15	.21218	.21712	4.6057	.97723	1.0233	4.7130	45
16	.21246	.21743	4.5993	.97717	1.0234	4.7067	44
17	.21275	.21773	4.5928	.97711	1.0234	4.7004	43
18	.21303	.21804	4.5864	.97705	1.0235	4.6942	42
19	.21331	.21834	4.5800	.97698	1.0236	4.6879	41
20	.21360	.21864	4.5736	.97692	1.0236	4.6817	40
21	.21388	.21895	4.5673	.97686	1.0237	4.6755	39
22	.21417	.21925	4.5609	.97680	1.0238	4.6693	38
23	.21445	.21956	4.5546	.97673	1.0238	4.6631	37
24	.21474	.21986	4.5483	.97667	1.0239	4.6569	36
25	.21502	.22017	4.5420	.97661	1.0240	4.6507	35
26	.21530	.22047	4.5357	.97655	1.0240	4.6446	34
27	.21559	.22078	4.5294	.97648	1.0241	4.6385	33
28	.21587	.22108	4.5232	.97642	1.0241	4.6324	32
29	.21616	.22139	4.5169	.97636	1.0242	4.6263	31
30	.21644	.22169	4.5107	.97630	1.0243	4.6202	30
31	.21672	.22200	4.5045	.97623	1.0243	4.6142	29
32	.21701	.22231	4.4983	.97617	1.0244	4.6081	28
33	.21729	.22261	4.4922	.97611	1.0245	4.6021	27
34	.21758	.22292	4.4860	.97604	1.0245	4.5961	26
35	.21786	.22322	4.4799	.97598	1.0246	4.5901	25
36	.21814	.22353	4.4737	.97592	1.0247	4.5841	24
37	.21843	.22383	4.4676	.97585	1.0247	4.5782	23
38	.21871	.22414	4.4615	.97579	1.0248	4.5722	22
39	.21899	.22444	4.4555	.97573	1.0249	4.5663	21
40	.21928	.22475	4.4494	.97566	1.0249	4.5604	20
41	.21956	.22505	4.4434	.97560	1.0250	4.5545	19
42	.21985	.22536	4.4373	.97553	1.0251	4.5486	18
43	.22013	.22567	4.4313	.97547	1.0251	4.5428	17
44	.22041	.22597	4.4253	.97541	1.0252	4.5369	16
45	.22070	.22628	4.4194	.97534	1.0253	4.5311	15
46	.22098	.22658	4.4134	.97528	1.0253	4.5253	14
47	.22126	.22689	4.4075	.97521	1.0254	4.5195	13
48	.22155	.22719	4.4015	.97515	1.0255	4.5137	12
49	.22183	.22750	4.3956	.97508	1.0256	4.5079	11
50	.22212	.22781	4.3897	.97502	1.0256	4.5022	10
51	.22240	.22811	4.3838	.97496	1.0257	4.4964	9
52	.22268	.22842	4.3779	.97489	1.0258	4.4907	8
53	.22297	.22872	4.3721	.97483	1.0258	4.4850	7
54	.22325	.22903	4.3662	.97476	1.0259	4.4793	6
55	.22353	.22934	4.3604	.97470	1.0260	4.4736	5
56	.22382	.22964	4.3546	.97463	1.0260	4.4679	4
57	.22410	.22995	4.3488	.97457	1.0261	4.4623	3
58	.22438	.23026	4.3430	.97450	1.0262	4.4566	2
59	.22467	.23056	4.3372	.97444	1.0262	4.4510	1
60	.22495	.23087	4.3315	.97437	1.0263	4.4454	0
′	Cos	Ctn	Tan	Sin	Csc	Sec	′

102° (282°)   (257°) **77°**

13° (193°)   (346°) **166°**

′	Sin	Tan	Ctn	Cos	Sec	Csc	′
0	.22495	.23087	4.3315	.97437	1.0263	4.4454	60
1	.22523	.23117	4.3257	.97430	1.0264	4.4398	59
2	.22552	.23148	4.3200	.97424	1.0264	4.4342	58
3	.22580	.23179	4.3143	.97417	1.0265	4.4287	57
4	.22608	.23209	4.3086	.97411	1.0266	4.4231	56
5	.22637	.23240	4.3029	.97404	1.0266	4.4176	55
6	.22665	.23271	4.2972	.97398	1.0267	4.4121	54
7	.22693	.23301	4.2916	.97391	1.0268	4.4066	53
8	.22722	.23332	4.2859	.97384	1.0269	4.4011	52
9	.22750	.23363	4.2803	.97378	1.0269	4.3956	51
10	.22778	.23393	4.2747	.97371	1.0270	4.3901	50
11	.22807	.23424	4.2691	.97365	1.0271	4.3847	49
12	.22835	.23455	4.2635	.97358	1.0271	4.3792	48
13	.22863	.23485	4.2580	.97351	1.0272	4.3738	47
14	.22892	.23516	4.2524	.97345	1.0273	4.3684	46
15	.22920	.23547	4.2468	.97338	1.0273	4.3630	45
16	.22948	.23578	4.2413	.97331	1.0274	4.3576	44
17	.22977	.23608	4.2358	.97325	1.0275	4.3522	43
18	.23005	.23639	4.2303	.97318	1.0276	4.3469	42
19	.23033	.23670	4.2248	.97311	1.0276	4.3415	41
20	.23062	.23700	4.2193	.97304	1.0277	4.3362	40
21	.23090	.23731	4.2139	.97298	1.0278	4.3309	39
22	.23118	.23762	4.2084	.97291	1.0278	4.3256	38
23	.23146	.23793	4.2030	.97284	1.0279	4.3203	37
24	.23175	.23823	4.1976	.97278	1.0280	4.3150	36
25	.23203	.23854	4.1922	.97271	1.0281	4.3098	35
26	.23231	.23885	4.1868	.97264	1.0281	4.3045	34
27	.23260	.23916	4.1814	.97257	1.0282	4.2993	33
28	.23288	.23946	4.1760	.97251	1.0283	4.2941	32
29	.23316	.23977	4.1706	.97244	1.0283	4.2889	31
30	.23345	.24008	4.1653	.97237	1.0284	4.2837	30
31	.23373	.24039	4.1600	.97230	1.0285	4.2785	29
32	.23401	.24069	4.1547	.97223	1.0286	4.2733	28
33	.23429	.24100	4.1493	.97217	1.0286	4.2681	27
34	.23458	.24131	4.1441	.97210	1.0287	4.2630	26
35	.23486	.24162	4.1388	.97203	1.0288	4.2579	25
36	.23514	.24193	4.1335	.97196	1.0288	4.2527	24
37	.23542	.24223	4.1282	.97189	1.0289	4.2476	23
38	.23571	.24254	4.1230	.97182	1.0290	4.2425	22
39	.23599	.24285	4.1178	.97176	1.0291	4.2375	21
40	.23627	.24316	4.1126	.97169	1.0291	4.2324	20
41	.23656	.24347	4.1074	.97162	1.0292	4.2273	19
42	.23684	.24377	4.1022	.97155	1.0293	4.2223	18
43	.23712	.24408	4.0970	.97148	1.0294	4.2173	17
44	.23740	.24439	4.0918	.97141	1.0294	4.2122	16
45	.23769	.24470	4.0867	.97134	1.0295	4.2072	15
46	.23797	.24501	4.0815	.97127	1.0296	4.2022	14
47	.23825	.24532	4.0764	.97120	1.0297	4.1973	13
48	.23853	.24562	4.0713	.97113	1.0297	4.1923	12
49	.23882	.24593	4.0662	.97106	1.0298	4.1873	11
50	.23910	.24624	4.0611	.97100	1.0299	4.1824	10
51	.23938	.24655	4.0560	.97093	1.0299	4.1774	9
52	.23966	.24686	4.0509	.97086	1.0300	4.1725	8
53	.23995	.24717	4.0459	.97079	1.0301	4.1676	7
54	.24023	.24747	4.0408	.97072	1.0302	4.1627	6
55	.24051	.24778	4.0358	.97065	1.0302	4.1578	5
56	.24079	.24809	4.0308	.97058	1.0303	4.1529	4
57	.24108	.24840	4.0257	.97051	1.0304	4.1481	3
58	.24136	.24871	4.0207	.97044	1.0305	4.1432	2
59	.24164	.24902	4.0158	.97037	1.0305	4.1384	1
60	.24192	.24933	4.0108	.97030	1.0306	4.1336	0
′	Cos	Ctn	Tan	Sin	Csc	Sec	′

103° (283°)   (256°) **76°**

TABLE 1. NATURAL TRIGONOMETRIC FUNCTIONS (continued)   608

14° (194°)   **(345°) 165°**

′	Sin	Tan	Ctn	Cos	Sec	Csc	′
0	.24192	.24933	4.0108	.97030	1.0306	4.1336	60
1	.24220	.24964	4.0058	.97023	1.0307	4.1287	59
2	.24249	.24995	4.0009	.97015	1.0308	4.1239	58
3	.24277	.25026	3.9959	.97008	1.0308	4.1191	57
4	.24305	.25056	3.9910	.97001	1.0309	4.1144	56
5	.24333	.25087	3.9861	.96994	1.0310	4.1096	55
6	.24362	.25118	3.9812	.96987	1.0311	4.1048	54
7	.24390	.25149	3.9763	.96980	1.0311	4.1001	53
8	.24418	.25180	3.9714	.96973	1.0312	4.0954	52
9	.24446	.25211	3.9665	.96966	1.0313	4.0906	51
10	.24474	.25242	3.9617	.96959	1.0314	4.0859	50
11	.24503	.25273	3.9568	.96952	1.0314	4.0812	49
12	.24531	.25304	3.9520	.96945	1.0315	4.0765	48
13	.24559	.25335	3.9471	.96937	1.0316	4.0718	47
14	.24587	.25366	3.9423	.96930	1.0317	4.0672	46
15	.24615	.25397	3.9375	.96923	1.0317	4.0625	45
16	.24644	.25428	3.9327	.96916	1.0318	4.0579	44
17	.24672	.25459	3.9279	.96909	1.0319	4.0532	43
18	.24700	.25490	3.9232	.96902	1.0320	4.0486	42
19	.24728	.25521	3.9184	.96894	1.0321	4.0440	41
20	.24756	.25552	3.9136	.96887	1.0321	4.0394	40
21	.24784	.25583	3.9089	.96880	1.0322	4.0348	39
22	.24813	.25614	3.9042	.96873	1.0323	4.0302	38
23	.24841	.25645	3.8995	.96866	1.0324	4.0256	37
24	.24869	.25676	3.8947	.96858	1.0324	4.0211	36
25	.24897	.25707	3.8900	.96851	1.0325	4.0165	35
26	.24925	.25738	3.8854	.96844	1.0326	4.0120	34
27	.24954	.25769	3.8807	.96837	1.0327	4.0075	33
28	.24982	.25800	3.8760	.96829	1.0327	4.0029	32
29	.25010	.25831	3.8714	.96822	1.0328	3.9984	31
30	.25038	.25862	3.8667	.96815	1.0329	3.9939	30
31	.25066	.25893	3.8621	.96807	1.0330	3.9894	29
32	.25094	.25924	3.8575	.96800	1.0331	3.9850	28
33	.25122	.25955	3.8528	.96793	1.0331	3.9805	27
34	.25151	.25986	3.8482	.96786	1.0332	3.9760	26
35	.25179	.26017	3.8436	.96778	1.0333	3.9716	25
36	.25207	.26048	3.8391	.96771	1.0334	3.9672	24
37	.25235	.26079	3.8345	.96764	1.0334	3.9627	23
38	.25263	.26110	3.8299	.96756	1.0335	3.9583	22
39	.25291	.26141	3.8254	.96749	1.0336	3.9539	21
40	.25320	.26172	3.8208	.96742	1.0337	3.9495	20
41	.25348	.26203	3.8163	.96734	1.0338	3.9451	19
42	.25376	.26235	3.8118	.96727	1.0338	3.9408	18
43	.25404	.26266	3.8073	.96719	1.0339	3.9364	17
44	.25432	.26297	3.8028	.96712	1.0340	3.9320	16
45	.25460	.26328	3.7983	.96705	1.0341	3.9277	15
46	.25488	.26359	3.7938	.96697	1.0342	3.9234	14
47	.25516	.26390	3.7893	.96690	1.0342	3.9190	13
48	.25545	.26421	3.7848	.96682	1.0343	3.9147	12
49	.25573	.26452	3.7804	.96675	1.0344	3.9104	11
50	.25601	.26483	3.7760	.96667	1.0345	3.9061	10
51	.25629	.26515	3.7715	.96660	1.0346	3.9018	9
52	.25657	.26546	3.7671	.96653	1.0346	3.8976	8
53	.25685	.26577	3.7627	.96645	1.0347	3.8933	7
54	.25713	.26608	3.7583	.96638	1.0348	3.8890	6
55	.25741	.26639	3.7539	.96630	1.0349	3.8848	5
56	.25769	.26670	3.7495	.96623	1.0350	3.8806	4
57	.25798	.26701	3.7451	.96615	1.0350	3.8763	3
58	.25826	.26733	3.7408	.96608	1.0351	3.8721	2
59	.25854	.26764	3.7364	.96600	1.0352	3.8679	1
60	.25882	.26795	3.7321	.96593	1.0353	3.8637	0
′	Cos	Ctn	Tan	Sin	Csc	Sec	′

104° (284°)   **(255°) 75°**

15° (195°)   **(344°) 164°**

′	Sin	Tan	Ctn	Cos	Sec	Csc	′
0	.25882	.26795	3.7321	.96593	1.0353	3.8637	60
1	.25910	.26826	3.7277	.96585	1.0354	3.8595	59
2	.25938	.26857	3.7234	.96578	1.0354	3.8553	58
3	.25966	.26888	3.7191	.96570	1.0355	3.8512	57
4	.25994	.26920	3.7148	.96562	1.0356	3.8470	56
5	.26022	.26951	3.7105	.96555	1.0357	3.8428	55
6	.26050	.26982	3.7062	.96547	1.0358	3.8387	54
7	.26079	.27013	3.7019	.96540	1.0358	3.8346	53
8	.26107	.27044	3.6976	.96532	1.0359	3.8304	52
9	.26135	.27076	3.6933	.96524	1.0360	3.8263	51
10	.26163	.27107	3.6891	.96517	1.0361	3.8222	50
11	.26191	.27138	3.6848	.96509	1.0362	3.8181	49
12	.26219	.27169	3.6806	.96502	1.0363	3.8140	48
13	.26247	.27201	3.6764	.96494	1.0363	3.8100	47
14	.26275	.27232	3.6722	.96486	1.0364	3.8059	46
15	.26303	.27263	3.6680	.96479	1.0365	3.8018	45
16	.26331	.27294	3.6638	.96471	1.0366	3.7978	44
17	.26359	.27326	3.6596	.96463	1.0367	3.7937	43
18	.26387	.27357	3.6554	.96456	1.0367	3.7897	42
19	.26415	.27388	3.6512	.96448	1.0368	3.7857	41
20	.26443	.27419	3.6470	.96440	1.0369	3.7817	40
21	.26471	.27451	3.6429	.96433	1.0370	3.7777	39
22	.26500	.27482	3.6387	.96425	1.0371	3.7737	38
23	.26528	.27513	3.6346	.96417	1.0372	3.7697	37
24	.26556	.27545	3.6305	.96410	1.0372	3.7657	36
25	.26584	.27576	3.6264	.96402	1.0373	3.7617	35
26	.26612	.27607	3.6222	.96394	1.0374	3.7577	34
27	.26640	.27638	3.6181	.96386	1.0375	3.7538	33
28	.26668	.27670	3.6140	.96379	1.0376	3.7498	32
29	.26696	.27701	3.6100	.96371	1.0377	3.7459	31
30	.26724	.27732	3.6059	.96363	1.0377	3.7420	30
31	.26752	.27764	3.6018	.96355	1.0378	3.7381	29
32	.26780	.27795	3.5978	.96347	1.0379	3.7341	28
33	.26808	.27826	3.5937	.96340	1.0380	3.7302	27
34	.26836	.27858	3.5897	.96332	1.0381	3.7263	26
35	.26864	.27889	3.5856	.96324	1.0382	3.7225	25
36	.26892	.27921	3.5816	.96316	1.0382	3.7186	24
37	.26920	.27952	3.5776	.96308	1.0383	3.7147	23
38	.26948	.27983	3.5736	.96301	1.0384	3.7108	22
39	.26976	.28015	3.5696	.96293	1.0385	3.7070	21
40	.27004	.28046	3.5656	.96285	1.0386	3.7032	20
41	.27032	.28077	3.5616	.96277	1.0387	3.6993	19
42	.27060	.28109	3.5576	.96269	1.0388	3.6955	18
43	.27088	.28140	3.5536	.96261	1.0388	3.6917	17
44	.27116	.28172	3.5497	.96253	1.0389	3.6879	16
45	.27144	.28203	3.5457	.96246	1.0390	3.6840	15
46	.27172	.28234	3.5418	.96238	1.0391	3.6803	14
47	.27200	.28266	3.5379	.96230	1.0392	3.6765	13
48	.27228	.28297	3.5339	.96222	1.0393	3.6727	12
49	.27256	.28329	3.5300	.96214	1.0394	3.6689	11
50	.27284	.28360	3.5261	.96206	1.0394	3.6652	10
51	.27312	.28391	3.5222	.96198	1.0395	3.6614	9
52	.27340	.28423	3.5183	.96190	1.0396	3.6576	8
53	.27368	.28454	3.5144	.96182	1.0397	3.6539	7
54	.27396	.28486	3.5105	.96174	1.0398	3.6502	6
55	.27424	.28517	3.5067	.96166	1.0399	3.6465	5
56	.27452	.28549	3.5028	.96158	1.0400	3.6427	4
57	.27480	.28580	3.4989	.96150	1.0400	3.6390	3
58	.27508	.28612	3.4951	.96142	1.0401	3.6353	2
59	.27536	.28643	3.4912	.96134	1.0402	3.6316	1
60	.27564	.28675	3.4874	.96126	1.0403	3.6280	0
′	Cos	Ctn	Tan	Sin	Csc	Sec	′

105° (285°)   **(254°) 74°**

TABLE 1. NATURAL TRIGONOMETRIC FUNCTIONS (continued) **609**

16° (196°) (343°) 163°

'	Sin	Tan	Ctn	Cos	Sec	Csc	'
0	.27564	.28675	3.4874	.96126	1.0403	3.6280	60
1	.27592	.28706	3.4836	.96118	1.0404	3.6243	59
2	.27620	.28738	3.4798	.96110	1.0405	3.6206	58
3	.27648	.28769	3.4760	.96102	1.0406	3.6169	57
4	.27676	.28801	3.4722	.96094	1.0406	3.6133	56
5	.27704	.28832	3.4684	.96086	1.0407	3.6097	55
6	.27731	.28864	3.4646	.96078	1.0408	3.6060	54
7	.27759	.28895	3.4608	.96070	1.0409	3.6024	53
8	.27787	.28927	3.4570	.96062	1.0410	3.5988	52
9	.27815	.28958	3.4533	.96054	1.0411	3.5951	51
10	.27843	.28990	3.4495	.96046	1.0412	3.5915	50
11	.27871	.29021	3.4458	.96037	1.0413	3.5879	49
12	.27899	.29053	3.4420	.96029	1.0413	3.5843	48
13	.27927	.29084	3.4383	.96021	1.0414	3.5808	47
14	.27955	.29116	3.4346	.96013	1.0415	3.5772	46
15	.27983	.29147	3.4308	.96005	1.0416	3.5736	45
16	.28011	.29179	3.4271	.95997	1.0417	3.5700	44
17	.28039	.29210	3.4234	.95989	1.0418	3.5665	43
18	.28067	.29242	3.4197	.95981	1.0419	3.5629	42
19	.28095	.29274	3.4160	.95972	1.0420	3.5594	41
20	.28123	.29305	3.4124	.95964	1.0421	3.5559	40
21	.28150	.29337	3.4087	.95956	1.0421	3.5523	39
22	.28178	.29368	3.4050	.95948	1.0422	3.5488	38
23	.28206	.29400	3.4014	.95940	1.0423	3.5453	37
24	.28234	.29432	3.3977	.95931	1.0424	3.5418	36
25	.28262	.29463	3.3941	.95923	1.0425	3.5383	35
26	.28290	.29495	3.3904	.95915	1.0426	3.5348	34
27	.28318	.29526	3.3868	.95907	1.0427	3.5313	33
28	.28346	.29558	3.3832	.95898	1.0428	3.5279	32
29	.28374	.29590	3.3796	.95890	1.0429	3.5244	31
30	.28402	.29621	3.3759	.95882	1.0429	3.5209	30
31	.28429	.29653	3.3723	.95874	1.0430	3.5175	29
32	.28457	.29685	3.3687	.95865	1.0431	3.5140	28
33	.28485	.29716	3.3652	.95857	1.0432	3.5106	27
34	.28513	.29748	3.3616	.95849	1.0433	3.5072	26
35	.28541	.29780	3.3580	.95841	1.0434	3.5037	25
36	.28569	.29811	3.3544	.95832	1.0435	3.5003	24
37	.28597	.29843	3.3509	.95824	1.0436	3.4969	23
38	.28625	.29875	3.3473	.95816	1.0437	3.4935	22
39	.28652	.29906	3.3438	.95807	1.0438	3.4901	21
40	.28680	.29938	3.3402	.95799	1.0439	3.4867	20
41	.28708	.29970	3.3367	.95791	1.0439	3.4833	19
42	.28736	.30001	3.3332	.95782	1.0440	3.4799	18
43	.28764	.30033	3.3297	.95774	1.0441	3.4766	17
44	.28792	.30065	3.3261	.95766	1.0442	3.4732	16
45	.28820	.30097	3.3226	.95757	1.0443	3.4699	15
46	.28847	.30128	3.3191	.95749	1.0444	3.4665	14
47	.28875	.30160	3.3156	.95740	1.0445	3.4632	13
48	.28903	.30192	3.3122	.95732	1.0446	3.4598	12
49	.28931	.30224	3.3087	.95724	1.0447	3.4565	11
50	.28959	.30255	3.3052	.95715	1.0448	3.4532	10
51	.28987	.30287	3.3017	.95707	1.0449	3.4499	9
52	.29015	.30319	3.2983	.95698	1.0450	3.4465	8
53	.29042	.30351	3.2948	.95690	1.0450	3.4432	7
54	.29070	.30382	3.2914	.95681	1.0451	3.4399	6
55	.29098	.30414	3.2879	.95673	1.0452	3.4367	5
56	.29126	.30446	3.2845	.95664	1.0453	3.4334	4
57	.29154	.30478	3.2811	.95656	1.0454	3.4301	3
58	.29182	.30509	3.2777	.95647	1.0455	3.4268	2
59	.29209	.30541	3.2743	.95639	1.0456	3.4236	1
60	.29237	.30573	3.2709	.95630	1.0457	3.4203	0
'	Cos	Ctn	Tan	Sin	Csc	Sec	'

106° (286°) (253°) 73°

17° (197°) (342°) 162°

'	Sin	Tan	Ctn	Cos	Sec	Csc	'
0	.29237	.30573	3.2709	.95630	1.0457	3.4203	60
1	.29265	.30605	3.2675	.95622	1.0458	3.4171	59
2	.29293	.30637	3.2641	.95613	1.0459	3.4138	58
3	.29321	.30669	3.2607	.95605	1.0460	3.4106	57
4	.29348	.30700	3.2573	.95596	1.0461	3.4073	56
5	.29376	.30732	3.2539	.95588	1.0462	3.4041	55
6	.29404	.30764	3.2506	.95579	1.0463	3.4009	54
7	.29432	.30796	3.2472	.95571	1.0463	3.3977	53
8	.29460	.30828	3.2438	.95562	1.0464	3.3945	52
9	.29487	.30860	3.2405	.95554	1.0465	3.3913	51
10	.29515	.30891	3.2371	.95545	1.0466	3.3881	50
11	.29543	.30923	3.2338	.95536	1.0467	3.3849	49
12	.29571	.30955	3.2305	.95528	1.0468	3.3817	48
13	.29599	.30987	3.2272	.95519	1.0469	3.3785	47
14	.29626	.31019	3.2238	.95511	1.0470	3.3754	46
15	.29654	.31051	3.2205	.95502	1.0471	3.3722	45
16	.29682	.31083	3.2172	.95493	1.0472	3.3691	44
17	.29710	.31115	3.2139	.95485	1.0473	3.3659	43
18	.29737	.31147	3.2106	.95476	1.0474	3.3628	42
19	.29765	.31178	3.2073	.95467	1.0475	3.3596	41
20	.29793	.31210	3.2041	.95459	1.0476	3.3565	40
21	.29821	.31242	3.2008	.95450	1.0477	3.3534	39
22	.29849	.31274	3.1975	.95441	1.0478	3.3502	38
23	.29876	.31306	3.1943	.95433	1.0479	3.3471	37
24	.29904	.31338	3.1910	.95424	1.0480	3.3440	36
25	.29932	.31370	3.1878	.95415	1.0480	3.3409	35
26	.29960	.31402	3.1845	.95407	1.0481	3.3378	34
27	.29987	.31434	3.1813	.95398	1.0482	3.3347	33
28	.30015	.31466	3.1780	.95389	1.0483	3.3317	32
29	.30043	.31498	3.1748	.95380	1.0484	3.3286	31
30	.30071	.31530	3.1716	.95372	1.0485	3.3255	30
31	.30098	.31562	3.1684	.95363	1.0486	3.3224	29
32	.30126	.31594	3.1652	.95354	1.0487	3.3194	28
33	.30154	.31626	3.1620	.95345	1.0488	3.3163	27
34	.30182	.31658	3.1588	.95337	1.0489	3.3133	26
35	.30209	.31690	3.1556	.95328	1.0490	3.3102	25
36	.30237	.31722	3.1524	.95319	1.0491	3.3072	24
37	.30265	.31754	3.1492	.95310	1.0492	3.3042	23
38	.30292	.31786	3.1460	.95301	1.0493	3.3012	22
39	.30320	.31818	3.1429	.95293	1.0494	3.2981	21
40	.30348	.31850	3.1397	.95284	1.0495	3.2951	20
41	.30376	.31882	3.1366	.95275	1.0496	3.2921	19
42	.30403	.31914	3.1334	.95266	1.0497	3.2891	18
43	.30431	.31946	3.1303	.95257	1.0498	3.2861	17
44	.30459	.31978	3.1271	.95248	1.0499	3.2831	16
45	.30486	.32010	3.1240	.95240	1.0500	3.2801	15
46	.30514	.32042	3.1209	.95231	1.0501	3.2772	14
47	.30542	.32074	3.1178	.95222	1.0502	3.2742	13
48	.30570	.32106	3.1146	.95213	1.0503	3.2712	12
49	.30597	.32139	3.1115	.95204	1.0504	3.2683	11
50	.30625	.32171	3.1084	.95195	1.0505	3.2653	10
51	.30653	.32203	3.1053	.95186	1.0506	3.2624	9
52	.30680	.32235	3.1022	.95177	1.0507	3.2594	8
53	.30708	.32267	3.0991	.95168	1.0508	3.2565	7
54	.30736	.32299	3.0961	.95159	1.0509	3.2535	6
55	.30763	.32331	3.0930	.95150	1.0510	3.2506	5
56	.30791	.32363	3.0899	.95142	1.0511	3.2477	4
57	.30819	.32396	3.0868	.95133	1.0512	3.2448	3
58	.30846	.32428	3.0838	.95124	1.0513	3.2419	2
59	.30874	.32460	3.0807	.95115	1.0514	3.2390	1
60	.30902	.32492	3.0777	.95106	1.0515	3.2361	0
'	Cos	Ctn	Tan	Sin	Csc	Sec	'

107° (287°) (252°) 72°

TABLE 1. NATURAL TRIGONOMETRIC FUNCTIONS (continued)   610

18° (198°)   **(341°) 161°**

′	Sin	Tan	Ctn	Cos	Sec	Csc	′
0	.30902	.32492	3.0777	.95106	1.0515	3.2361	60
1	.30929	.32524	3.0746	.95097	1.0516	3.2332	59
2	.30957	.32556	3.0716	.95088	1.0517	3.2303	58
3	.30985	.32588	3.0686	.95079	1.0518	3.2274	57
4	.31012	.32621	3.0655	.95070	1.0519	3.2245	56
5	.31040	.32653	3.0625	.95061	1.0520	3.2217	55
6	.31068	.32685	3.0595	.95052	1.0521	3.2188	54
7	.31095	.32717	3.0565	.95043	1.0522	3.2159	53
8	.31123	.32749	3.0535	.95033	1.0523	3.2131	52
9	.31151	.32782	3.0505	.95024	1.0524	3.2102	51
10	.31178	.32814	3.0475	.95015	1.0525	3.2074	50
11	.31206	.32846	3.0445	.95006	1.0526	3.2045	49
12	.31233	.32878	3.0415	.94997	1.0527	3.2017	48
13	.31261	.32911	3.0385	.94988	1.0528	3.1989	47
14	.31289	.32943	3.0356	.94979	1.0529	3.1960	46
15	.31316	.32975	3.0326	.94970	1.0530	3.1932	45
16	.31344	.33007	3.0296	.94961	1.0531	3.1904	44
17	.31372	.33040	3.0267	.94952	1.0532	3.1876	43
18	.31399	.33072	3.0237	.94943	1.0533	3.1848	42
19	.31427	.33104	3.0208	.94933	1.0534	3.1820	41
20	.31454	.33136	3.0178	.94924	1.0535	3.1792	40
21	.31482	.33169	3.0149	.94915	1.0536	3.1764	39
22	.31510	.33201	3.0120	.94906	1.0537	3.1736	38
23	.31537	.33233	3.0090	.94897	1.0538	3.1708	37
24	.31565	.33266	3.0061	.94888	1.0539	3.1681	36
25	.31593	.33298	3.0032	.94878	1.0540	3.1653	35
26	.31620	.33330	3.0003	.94869	1.0541	3.1625	34
27	.31648	.33363	2.9974	.94860	1.0542	3.1598	33
28	.31675	.33395	2.9945	.94851	1.0543	3.1570	32
29	.31703	.33427	2.9916	.94842	1.0544	3.1543	31
30	.31730	.33460	2.9887	.94832	1.0545	3.1515	30
31	.31758	.33492	2.9858	.94823	1.0546	3.1488	29
32	.31786	.33524	2.9829	.94814	1.0547	3.1461	28
33	.31813	.33557	2.9800	.94805	1.0548	3.1433	27
34	.31841	.33589	2.9772	.94795	1.0549	3.1406	26
35	.31868	.33621	2.9743	.94786	1.0550	3.1379	25
36	.31896	.33654	2.9714	.94777	1.0551	3.1352	24
37	.31923	.33686	2.9686	.94768	1.0552	3.1325	23
38	.31951	.33718	2.9657	.94758	1.0553	3.1298	22
39	.31979	.33751	2.9629	.94749	1.0554	3.1271	21
40	.32006	.33783	2.9600	.94740	1.0555	3.1244	20
41	.32034	.33816	2.9572	.94730	1.0556	3.1217	19
42	.32061	.33848	2.9544	.94721	1.0557	3.1190	18
43	.32089	.33881	2.9515	.94712	1.0558	3.1163	17
44	.32116	.33913	2.9487	.94702	1.0559	3.1137	16
45	.32144	.33945	2.9459	.94693	1.0560	3.1110	15
46	.32171	.33978	2.9431	.94684	1.0561	3.1083	14
47	.32199	.34010	2.9403	.94674	1.0563	3.1057	13
48	.32227	.34043	2.9375	.94665	1.0564	3.1030	12
49	.32254	.34075	2.9347	.94656	1.0565	3.1004	11
50	.32282	.34108	2.9319	.94646	1.0566	3.0977	10
51	.32309	.34140	2.9291	.94637	1.0567	3.0951	9
52	.32337	.34173	2.9263	.94627	1.0568	3.0925	8
53	.32364	.34205	2.9235	.94618	1.0569	3.0898	7
54	.32392	.34238	2.9208	.94609	1.0570	3.0872	6
55	.32419	.34270	2.9180	.94599	1.0571	3.0846	5
56	.32447	.34303	2.9152	.94590	1.0572	3.0820	4
57	.32474	.34335	2.9125	.94580	1.0573	3.0794	3
58	.32502	.34368	2.9097	.94571	1.0574	3.0768	2
59	.32529	.34400	2.9070	.94561	1.0575	3.0742	1
60	.32557	.34433	2.9042	.94552	1.0576	3.0716	0
′	Cos	Ctn	Tan	Sin	Csc	Sec	′

108° (288°)   **(251°) 71°**

19° (199°)   **(340°) 160°**

′	Sin	Tan	Ctn	Cos	Sec	Csc	′
0	.32557	.34433	2.9042	.94552	1.0576	3.0716	60
1	.32584	.34465	2.9015	.94542	1.0577	3.0690	59
2	.32612	.34498	2.8987	.94533	1.0578	3.0664	58
3	.32639	.34530	2.8960	.94523	1.0579	3.0638	57
4	.32667	.34563	2.8933	.94514	1.0580	3.0612	56
5	.32694	.34596	2.8905	.94504	1.0582	3.0586	55
6	.32722	.34628	2.8878	.94495	1.0583	3.0561	54
7	.32749	.34661	2.8851	.94485	1.0584	3.0535	53
8	.32777	.34693	2.8824	.94476	1.0585	3.0509	52
9	.32804	.34726	2.8797	.94466	1.0586	3.0484	51
10	.32832	.34758	2.8770	.94457	1.0587	3.0458	50
11	.32859	.34791	2.8743	.94447	1.0588	3.0433	49
12	.32887	.34824	2.8716	.94438	1.0589	3.0407	48
13	.32914	.34856	2.8689	.94428	1.0590	3.0382	47
14	.32942	.34889	2.8662	.94418	1.0591	3.0357	46
15	.32969	.34922	2.8636	.94409	1.0592	3.0331	45
16	.32997	.34954	2.8609	.94399	1.0593	3.0306	44
17	.33024	.34987	2.8582	.94390	1.0594	3.0281	43
18	.33051	.35020	2.8556	.94380	1.0595	3.0256	42
19	.33079	.35052	2.8529	.94370	1.0597	3.0231	41
20	.33106	.35085	2.8502	.94361	1.0598	3.0206	40
21	.33134	.35118	2.8476	.94351	1.0599	3.0181	39
22	.33161	.35150	2.8449	.94342	1.0600	3.0156	38
23	.33189	.35183	2.8423	.94332	1.0601	3.0131	37
24	.33216	.35216	2.8397	.94322	1.0602	3.0106	36
25	.33244	.35248	2.8370	.94313	1.0603	3.0081	35
26	.33271	.35281	2.8344	.94303	1.0604	3.0056	34
27	.33298	.35314	2.8318	.94293	1.0605	3.0031	33
28	.33326	.35346	2.8291	.94284	1.0606	3.0007	32
29	.33353	.35379	2.8265	.94274	1.0607	2.9982	31
30	.33381	.35412	2.8239	.94264	1.0608	2.9957	30
31	.33408	.35445	2.8213	.94254	1.0610	2.9933	29
32	.33436	.35477	2.8187	.94245	1.0611	2.9908	28
33	.33463	.35510	2.8161	.94235	1.0612	2.9884	27
34	.33490	.35543	2.8135	.94225	1.0613	2.9859	26
35	.33518	.35576	2.8109	.94215	1.0614	2.9835	25
36	.33545	.35608	2.8083	.94206	1.0615	2.9811	24
37	.33573	.35641	2.8057	.94196	1.0616	2.9786	23
38	.33600	.35674	2.8032	.94186	1.0617	2.9762	22
39	.33627	.35707	2.8006	.94176	1.0618	2.9738	21
40	.33655	.35740	2.7980	.94167	1.0619	2.9713	20
41	.33682	.35772	2.7955	.94157	1.0621	2.9689	19
42	.33710	.35805	2.7929	.94147	1.0622	2.9665	18
43	.33737	.35838	2.7903	.94137	1.0623	2.9641	17
44	.33764	.35871	2.7878	.94127	1.0624	2.9617	16
45	.33792	.35904	2.7852	.94118	1.0625	2.9593	15
46	.33819	.35937	2.7827	.94108	1.0626	2.9569	14
47	.33846	.35969	2.7801	.94098	1.0627	2.9545	13
48	.33874	.36002	2.7776	.94088	1.0628	2.9521	12
49	.33901	.36035	2.7751	.94078	1.0629	2.9498	11
50	.33929	.36068	2.7725	.94068	1.0631	2.9474	10
51	.33956	.36101	2.7700	.94058	1.0632	2.9450	9
52	.33983	.36134	2.7675	.94049	1.0633	2.9426	8
53	.34011	.36167	2.7650	.94039	1.0634	2.9403	7
54	.34038	.36199	2.7625	.94029	1.0635	2.9379	6
55	.34065	.36232	2.7600	.94019	1.0636	2.9355	5
56	.34093	.36265	2.7575	.94009	1.0637	2.9332	4
57	.34120	.36298	2.7550	.93999	1.0638	2.9308	3
58	.34147	.36331	2.7525	.93989	1.0640	2.9285	2
59	.34175	.36364	2.7500	.93979	1.0641	2.9261	1
60	.34202	.36397	2.7475	.93969	1.0642	2.9238	0
′	Cos	Ctn	Tan	Sin	Csc	Sec	′

109° (289°)   **(250°) 70°**

TABLE 1. NATURAL TRIGONOMETRIC FUNCTIONS (continued)   **611**

20° (200°)   (339°) **159°**

′	Sin	Tan	Ctn	Cos	Sec	Csc	′
0	.34202	.36397	2.7475	.93969	1.0642	2.9238	60
1	.34229	.36430	2.7450	.93959	1.0643	2.9215	59
2	.34257	.36463	2.7425	.93949	1.0644	2.9191	58
3	.34284	.36496	2.7400	.93939	1.0645	2.9168	57
4	.34311	.36529	2.7376	.93929	1.0646	2.9145	56
5	.34339	.36562	2.7351	.93919	1.0647	2.9122	55
6	.34366	.36595	2.7326	.93909	1.0649	2.9099	54
7	.34393	.36628	2.7302	.93899	1.0650	2.9075	53
8	.34421	.36661	2.7277	.93889	1.0651	2.9052	52
9	.34448	.36694	2.7253	.93879	1.0652	2.9029	51
10	.34475	.36727	2.7228	.93869	1.0653	2.9006	50
11	.34503	.36760	2.7204	.93859	1.0654	2.8983	49
12	.34530	.36793	2.7179	.93849	1.0655	2.8960	48
13	.34557	.36826	2.7155	.93839	1.0657	2.8938	47
14	.34584	.36859	2.7130	.93829	1.0658	2.8915	46
15	.34612	.36892	2.7106	.93819	1.0659	2.8892	45
16	.34639	.36925	2.7082	.93809	1.0660	2.8869	44
17	.34666	.36958	2.7058	.93799	1.0661	2.8846	43
18	.34694	.36991	2.7034	.93789	1.0662	2.8824	42
19	.34721	.37024	2.7009	.93779	1.0663	2.8801	41
20	.34748	.37057	2.6985	.93769	1.0665	2.8779	40
21	.34775	.37090	2.6961	.93759	1.0666	2.8756	39
22	.34803	.37123	2.6937	.93748	1.0667	2.8733	38
23	.34830	.37157	2.6913	.93738	1.0668	2.8711	37
24	.34857	.37190	2.6889	.93728	1.0669	2.8688	36
25	.34884	.37223	2.6865	.93718	1.0670	2.8666	35
26	.34912	.37256	2.6841	.93708	1.0671	2.8644	34
27	.34939	.37289	2.6818	.93698	1.0673	2.8621	33
28	.34966	.37322	2.6794	.93688	1.0674	2.8599	32
29	.34993	.37355	2.6770	.93677	1.0675	2.8577	31
30	.35021	.37388	2.6746	.93667	1.0676	2.8555	30
31	.35048	.37422	2.6723	.93657	1.0677	2.8532	29
32	.35075	.37455	2.6699	.93647	1.0678	2.8510	28
33	.35102	.37488	2.6675	.93637	1.0680	2.8488	27
34	.35130	.37521	2.6652	.93626	1.0681	2.8466	26
35	.35157	.37554	2.6628	.93616	1.0682	2.8444	25
36	.35184	.37588	2.6605	.93606	1.0683	2.8422	24
37	.35211	.37621	2.6581	.93596	1.0684	2.8400	23
38	.35239	.37654	2.6558	.93585	1.0685	2.8378	22
39	.35266	.37687	2.6534	.93575	1.0687	2.8356	21
40	.35293	.37720	2.6511	.93565	1.0688	2.8334	20
41	.35320	.37754	2.6488	.93555	1.0689	2.8312	19
42	.35347	.37787	2.6464	.93544	1.0690	2.8291	18
43	.35375	.37820	2.6441	.93534	1.0691	2.8269	17
44	.35402	.37853	2.6418	.93524	1.0692	2.8247	16
45	.35429	.37887	2.6395	.93514	1.0694	2.8225	15
46	.35456	.37920	2.6371	.93503	1.0695	2.8204	14
47	.35484	.37953	2.6348	.93493	1.0696	2.8182	13
48	.35511	.37986	2.6325	.93483	1.0697	2.8161	12
49	.35538	.38020	2.6302	.93472	1.0698	2.8139	11
50	.35565	.38053	2.6279	.93462	1.0700	2.8117	10
51	.35592	.38086	2.6256	.93452	1.0701	2.8096	9
52	.35619	.38120	2.6233	.93441	1.0702	2.8075	8
53	.35647	.38153	2.6210	.93431	1.0703	2.8053	7
54	.35674	.38186	2.6187	.93420	1.0704	2.8032	6
55	.35701	.38220	2.6165	.93410	1.0705	2.8010	5
56	.35728	.38253	2.6142	.93400	1.0707	2.7989	4
57	.35755	.38286	2.6119	.93389	1.0708	2.7968	3
58	.35782	.38320	2.6096	.93379	1.0709	2.7947	2
59	.35810	.38353	2.6074	.93368	1.0710	2.7925	1
60	.35837	.38386	2.6051	.93358	1.0711	2.7904	0
′	Cos	Ctn	Tan	Sin	Csc	Sec	′

21° (201°)   (338°) **158°**

′	Sin	Tan	Ctn	Cos	Sec	Csc	′
0	.35837	.38386	2.6051	.93358	1.0711	2.7904	60
1	.35864	.38420	2.6028	.93348	1.0713	2.7883	59
2	.35891	.38453	2.6006	.93337	1.0714	2.7862	58
3	.35918	.38487	2.5983	.93327	1.0715	2.7841	57
4	.35945	.38520	2.5961	.93316	1.0716	2.7820	56
5	.35973	.38553	2.5938	.93306	1.0717	2.7799	55
6	.36000	.38587	2.5916	.93295	1.0719	2.7778	54
7	.36027	.38620	2.5893	.93285	1.0720	2.7757	53
8	.36054	.38654	2.5871	.93274	1.0721	2.7736	52
9	.36081	.38687	2.5848	.93264	1.0722	2.7715	51
10	.36108	.38721	2.5826	.93253	1.0723	2.7695	50
11	.36135	.38754	2.5804	.93243	1.0725	2.7674	49
12	.36162	.38787	2.5782	.93232	1.0726	2.7653	48
13	.36190	.38821	2.5759	.93222	1.0727	2.7632	47
14	.36217	.38854	2.5737	.93211	1.0728	2.7612	46
15	.36244	.38888	2.5715	.93201	1.0730	2.7591	45
16	.36271	.38921	2.5693	.93190	1.0731	2.7570	44
17	.36298	.38955	2.5671	.93180	1.0732	2.7550	43
18	.36325	.38988	2.5649	.93169	1.0733	2.7529	42
19	.36352	.39022	2.5627	.93159	1.0734	2.7509	41
20	.36379	.39055	2.5605	.93148	1.0736	2.7488	40
21	.36406	.39089	2.5583	.93137	1.0737	2.7468	39
22	.36434	.39122	2.5561	.93127	1.0738	2.7447	38
23	.36461	.39156	2.5539	.93116	1.0739	2.7427	37
24	.36488	.39190	2.5517	.93106	1.0740	2.7407	36
25	.36515	.39223	2.5495	.93095	1.0742	2.7386	35
26	.36542	.39257	2.5473	.93084	1.0743	2.7366	34
27	.36569	.39290	2.5452	.93074	1.0744	2.7346	33
28	.36596	.39324	2.5430	.93063	1.0745	2.7325	32
29	.36623	.39357	2.5408	.93052	1.0747	2.7305	31
30	.36650	.39391	2.5386	.93042	1.0748	2.7285	30
31	.36677	.39425	2.5365	.93031	1.0749	2.7265	29
32	.36704	.39458	2.5343	.93020	1.0750	2.7245	28
33	.36731	.39492	2.5322	.93010	1.0752	2.7225	27
34	.36758	.39526	2.5300	.92999	1.0753	2.7205	26
35	.36785	.39559	2.5279	.92988	1.0754	2.7185	25
36	.36812	.39593	2.5257	.92978	1.0755	2.7165	24
37	.36839	.39626	2.5236	.92967	1.0757	2.7145	23
38	.36867	.39660	2.5214	.92956	1.0758	2.7125	22
39	.36894	.39694	2.5193	.92945	1.0759	2.7105	21
40	.36921	.39727	2.5172	.92935	1.0760	2.7085	20
41	.36948	.39761	2.5150	.92924	1.0761	2.7065	19
42	.36975	.39795	2.5129	.92913	1.0763	2.7046	18
43	.37002	.39829	2.5108	.92902	1.0764	2.7026	17
44	.37029	.39862	2.5086	.92892	1.0765	2.7006	16
45	.37056	.39896	2.5065	.92881	1.0766	2.6986	15
46	.37083	.39930	2.5044	.92870	1.0768	2.6967	14
47	.37110	.39963	2.5023	.92859	1.0769	2.6947	13
48	.37137	.39997	2.5002	.92849	1.0770	2.6927	12
49	.37164	.40031	2.4981	.92838	1.0771	2.6908	11
50	.37191	.40065	2.4960	.92827	1.0773	2.6888	10
51	.37218	.40098	2.4939	.92816	1.0774	2.6869	9
52	.37245	.40132	2.4918	.92805	1.0775	2.6849	8
53	.37272	.40166	2.4897	.92794	1.0777	2.6830	7
54	.37299	.40200	2.4876	.92784	1.0778	2.6811	6
55	.37326	.40234	2.4855	.92773	1.0779	2.6791	5
56	.37353	.40267	2.4834	.92762	1.0780	2.6772	4
57	.37380	.40301	2.4813	.92751	1.0782	2.6752	3
58	.37407	.40335	2.4792	.92740	1.0783	2.6733	2
59	.37434	.40369	2.4772	.92729	1.0784	2.6714	1
60	.37461	.40403	2.4751	.92718	1.0785	2.6695	0
′	Cos	Ctn	Tan	Sin	Csc	Sec	′

TABLE 1. NATURAL TRIGONOMETRIC FUNCTIONS (continued)   612

22° (202°)   (337°) **157°**

′	Sin	Tan	Ctn	Cos	Sec	Csc	′
0	.37461	.40403	2.4751	.92718	1.0785	2.6695	60
1	.37488	.40436	2.4730	.92707	1.0787	2.6675	59
2	.37515	.40470	2.4709	.92697	1.0788	2.6656	58
3	.37542	.40504	2.4689	.92686	1.0789	2.6637	57
4	.37569	.40538	2.4668	.92675	1.0790	2.6618	56
5	.37595	.40572	2.4648	.92664	1.0792	2.6599	55
6	.37622	.40606	2.4627	.92653	1.0793	2.6580	54
7	.37649	.40640	2.4606	.92642	1.0794	2.6561	53
8	.37676	.40674	2.4586	.92631	1.0796	2.6542	52
9	.37703	.40707	2.4566	.92620	1.0797	2.6523	51
10	.37730	.40741	2.4545	.92609	1.0798	2.6504	50
11	.37757	.40775	2.4525	.92598	1.0799	2.6485	49
12	.37784	.40809	2.4504	.92587	1.0801	2.6466	48
13	.37811	.40843	2.4484	.92576	1.0802	2.6447	47
14	.37838	.40877	2.4464	.92565	1.0803	2.6429	46
15	.37865	.40911	2.4443	.92554	1.0804	2.6410	45
16	.37892	.40945	2.4423	.92543	1.0806	2.6391	44
17	.37919	.40979	2.4403	.92532	1.0807	2.6372	43
18	.37946	.41013	2.4383	.92521	1.0808	2.6354	42
19	.37973	.41047	2.4362	.92510	1.0810	2.6335	41
20	.37999	.41081	2.4342	.92499	1.0811	2.6316	40
21	.38026	.41115	2.4322	.92488	1.0812	2.6298	39
22	.38053	.41149	2.4302	.92477	1.0814	2.6279	38
23	.38080	.41183	2.4282	.92466	1.0815	2.6260	37
24	.38107	.41217	2.4262	.92455	1.0816	2.6242	36
25	.38134	.41251	2.4242	.92444	1.0817	2.6223	35
26	.38161	.41285	2.4222	.92432	1.0819	2.6205	34
27	.38188	.41319	2.4202	.92421	1.0820	2.6186	33
28	.38215	.41353	2.4182	.92410	1.0821	2.6168	32
29	.38241	.41387	2.4162	.92399	1.0823	2.6150	31
30	.38268	.41421	2.4142	.92388	1.0824	2.6131	30
31	.38295	.41455	2.4122	.92377	1.0825	2.6113	29
32	.38322	.41490	2.4102	.92366	1.0827	2.6095	28
33	.38349	.41524	2.4083	.92355	1.0828	2.6076	27
34	.38376	.41558	2.4063	.92343	1.0829	2.6058	26
35	.38403	.41592	2.4043	.92332	1.0830	2.6040	25
36	.38430	.41626	2.4023	.92321	1.0832	2.6022	24
37	.38456	.41660	2.4004	.92310	1.0833	2.6003	23
38	.38483	.41694	2.3984	.92299	1.0834	2.5985	22
39	.38510	.41728	2.3964	.92287	1.0836	2.5967	21
40	.38537	.41763	2.3945	.92276	1.0837	2.5949	20
41	.38564	.41797	2.3925	.92265	1.0838	2.5931	19
42	.38591	.41831	2.3906	.92254	1.0840	2.5913	18
43	.38617	.41865	2.3886	.92243	1.0841	2.5895	17
44	.38644	.41899	2.3867	.92231	1.0842	2.5877	16
45	.38671	.41933	2.3847	.92220	1.0844	2.5859	15
46	.38698	.41968	2.3828	.92209	1.0845	2.5841	14
47	.38725	.42002	2.3808	.92198	1.0846	2.5823	13
48	.38752	.42036	2.3789	.92186	1.0848	2.5805	12
49	.38778	.42070	2.3770	.92175	1.0849	2.5788	11
50	.38805	.42105	2.3750	.92164	1.0850	2.5770	10
51	.38832	.42139	2.3731	.92152	1.0852	2.5752	9
52	.38859	.42173	2.3712	.92141	1.0853	2.5734	8
53	.38886	.42207	2.3693	.92130	1.0854	2.5716	7
54	.38912	.42242	2.3673	.92119	1.0856	2.5699	6
55	.38939	.42276	2.3654	.92107	1.0857	2.5681	5
56	.38966	.42310	2.3635	.92096	1.0858	2.5663	4
57	.38993	.42345	2.3616	.92085	1.0860	2.5646	3
58	.39020	.42379	2.3597	.92073	1.0861	2.5628	2
59	.39046	.42413	2.3578	.92062	1.0862	2.5611	1
60	.39073	.42447	2.3559	.92050	1.0864	2.5593	0
′	Cos	Ctn	Tan	Sin	Csc	Sec	′

112° (292°)   (247°) **67°**

23° (203°)   (336°) **156°**

′	Sin	Tan	Ctn	Cos	Sec	Csc	′
0	.39073	.42447	2.3559	.92050	1.0864	2.5593	60
1	.39100	.42482	2.3539	.92039	1.0865	2.5576	59
2	.39127	.42516	2.3520	.92028	1.0866	2.5558	58
3	.39153	.42551	2.3501	.92016	1.0868	2.5541	57
4	.39180	.42585	2.3483	.92005	1.0869	2.5523	56
5	.39207	.42619	2.3464	.91994	1.0870	2.5506	55
6	.39234	.42654	2.3445	.91982	1.0872	2.5488	54
7	.39260	.42688	2.3426	.91971	1.0873	2.5471	53
8	.39287	.42722	2.3407	.91959	1.0874	2.5454	52
9	.39314	.42757	2.3388	.91948	1.0876	2.5436	51
10	.39341	.42791	2.3369	.91936	1.0877	2.5419	50
11	.39367	.42826	2.3351	.91925	1.0878	2.5402	49
12	.39394	.42860	2.3332	.91914	1.0880	2.5384	48
13	.39421	.42894	2.3313	.91902	1.0881	2.5367	47
14	.39448	.42929	2.3294	.91891	1.0883	2.5350	46
15	.39474	.42963	2.3276	.91879	1.0884	2.5333	45
16	.39501	.42998	2.3257	.91868	1.0885	2.5316	44
17	.39528	.43032	2.3238	.91856	1.0887	2.5299	43
18	.39555	.43067	2.3220	.91845	1.0888	2.5282	42
19	.39581	.43101	2.3201	.91833	1.0889	2.5264	41
20	.39608	.43136	2.3183	.91822	1.0891	2.5247	40
21	.39635	.43170	2.3164	.91810	1.0892	2.5230	39
22	.39661	.43205	2.3146	.91799	1.0893	2.5213	38
23	.39688	.43239	2.3127	.91787	1.0895	2.5196	37
24	.39715	.43274	2.3109	.91775	1.0896	2.5180	36
25	.39741	.43308	2.3090	.91764	1.0898	2.5163	35
26	.39768	.43343	2.3072	.91752	1.0899	2.5146	34
27	.39795	.43378	2.3053	.91741	1.0900	2.5129	33
28	.39822	.43412	2.3035	.91729	1.0902	2.5112	32
29	.39848	.43447	2.3017	.91718	1.0903	2.5095	31
30	.39875	.43481	2.2998	.91706	1.0904	2.5078	30
31	.39902	.43516	2.2980	.91694	1.0906	2.5062	29
32	.39928	.43550	2.2962	.91683	1.0907	2.5045	28
33	.39955	.43585	2.2944	.91671	1.0909	2.5028	27
34	.39982	.43620	2.2925	.91660	1.0910	2.5012	26
35	.40008	.43654	2.2907	.91648	1.0911	2.4995	25
36	.40035	.43689	2.2889	.91636	1.0913	2.4978	24
37	.40062	.43724	2.2871	.91625	1.0914	2.4962	23
38	.40088	.43758	2.2853	.91613	1.0915	2.4945	22
39	.40115	.43793	2.2835	.91601	1.0917	2.4928	21
40	.40141	.43828	2.2817	.91590	1.0918	2.4912	20
41	.40168	.43862	2.2799	.91578	1.0920	2.4895	19
42	.40195	.43897	2.2781	.91566	1.0921	2.4879	18
43	.40221	.43932	2.2763	.91555	1.0922	2.4862	17
44	.40248	.43966	2.2745	.91543	1.0924	2.4846	16
45	.40275	.44001	2.2727	.91531	1.0925	2.4830	15
46	.40301	.44036	2.2709	.91519	1.0927	2.4813	14
47	.40328	.44071	2.2691	.91508	1.0928	2.4797	13
48	.40355	.44105	2.2673	.91496	1.0929	2.4780	12
49	.40381	.44140	2.2655	.91484	1.0931	2.4764	11
50	.40408	.44175	2.2637	.91472	1.0932	2.4748	10
51	.40434	.44210	2.2620	.91461	1.0934	2.4731	9
52	.40461	.44244	2.2602	.91449	1.0935	2.4715	8
53	.40488	.44279	2.2584	.91437	1.0936	2.4699	7
54	.40514	.44314	2.2566	.91425	1.0938	2.4683	6
55	.40541	.44349	2.2549	.91414	1.0939	2.4667	5
56	.40567	.44384	2.2531	.91402	1.0941	2.4650	4
57	.40594	.44418	2.2513	.91390	1.0942	2.4634	3
58	.40621	.44453	2.2496	.91378	1.0944	2.4618	2
59	.40647	.44488	2.2478	.91366	1.0945	2.4602	1
60	.40674	.44523	2.2460	.91355	1.0946	2.4586	0
′	Cos	Ctn	Tan	Sin	Csc	Sec	′

113° (293°)   (246°) **66°**

24° (204°)   **(335°) 155°**   **25° (205°)**   **(334°) 154°**

′	Sin	Tan	Ctn	Cos	Sec	Csc	′	′	Sin	Tan	Ctn	Cos	Sec	Csc	′
0	.40674	.44523	2.2460	.91355	1.0946	2.4586	60	0	.42262	.46631	2.1445	.90631	1.1034	2.3662	60
1	.40700	.44558	2.2443	.91343	1.0948	2.4570	59	1	.42288	.46666	2.1429	.90618	1.1035	2.3647	59
2	.40727	.44593	2.2425	.91331	1.0949	2.4554	58	2	.42315	.46702	2.1413	.90606	1.1037	2.3633	58
3	.40753	.44627	2.2408	.91319	1.0951	2.4538	57	3	.42341	.46737	2.1396	.90594	1.1038	2.3618	57
4	.40780	.44662	2.2390	.91307	1.0952	2.4522	56	4	.42367	.46772	2.1380	.90582	1.1040	2.3603	56
5	.40806	.44697	2.2373	.91295	1.0953	2.4506	55	5	.42394	.46808	2.1364	.90569	1.1041	2.3588	55
6	.40833	.44732	2.2355	.91283	1.0955	2.4490	54	6	.42420	.46843	2.1348	.90557	1.1043	2.3574	54
7	.40860	.44767	2.2338	.91272	1.0956	2.4474	53	7	.42446	.46879	2.1332	.90545	1.1044	2.3559	53
8	.40886	.44802	2.2320	.91260	1.0958	2.4458	52	8	.42473	.46914	2.1315	.90532	1.1046	2.3545	52
9	.40913	.44837	2.2303	.91248	1.0959	2.4442	51	9	.42499	.46950	2.1299	.90520	1.1047	2.3530	51
10	.40939	.44872	2.2286	.91236	1.0961	2.4426	50	10	.42525	.46985	2.1283	.90507	1.1049	2.3515	50
11	.40966	.44907	2.2268	.91224	1.0962	2.4411	49	11	.42552	.47021	2.1267	.90495	1.1050	2.3501	49
12	.40992	.44942	2.2251	.91212	1.0963	2.4395	48	12	.42578	.47056	2.1251	.90483	1.1052	2.3486	48
13	.41019	.44977	2.2234	.91200	1.0965	2.4379	47	13	.42604	.47092	2.1235	.90470	1.1053	2.3472	47
14	.41045	.45012	2.2216	.91188	1.0966	2.4363	46	14	.42631	.47128	2.1219	.90458	1.1055	2.3457	46
15	.41072	.45047	2.2199	.91176	1.0968	2.4348	45	15	.42657	.47163	2.1203	.90446	1.1056	2.3443	45
16	.41098	.45082	2.2182	.91164	1.0969	2.4332	44	16	.42683	.47199	2.1187	.90433	1.1058	2.3428	44
17	.41125	.45117	2.2165	.91152	1.0971	2.4316	43	17	.42709	.47234	2.1171	.90421	1.1059	2.3414	43
18	.41151	.45152	2.2148	.91140	1.0972	2.4300	42	18	.42736	.47270	2.1155	.90408	1.1061	2.3400	42
19	.41178	.45187	2.2130	.91128	1.0974	2.4285	41	19	.42762	.47305	2.1139	.90396	1.1062	2.3385	41
20	.41204	.45222	2.2113	.91116	1.0975	2.4269	40	20	.42788	.47341	2.1123	.90383	1.1064	2.3371	40
21	.41231	.45257	2.2096	.91104	1.0976	2.4254	39	21	.42815	.47377	2.1107	.90371	1.1066	2.3356	39
22	.41257	.45292	2.2079	.91092	1.0978	2.4238	38	22	.42841	.47412	2.1092	.90358	1.1067	2.3342	38
23	.41284	.45327	2.2062	.91080	1.0979	2.4222	37	23	.42867	.47448	2.1076	.90346	1.1069	2.3328	37
24	.41310	.45362	2.2045	.91068	1.0981	2.4207	36	24	.42894	.47483	2.1060	.90334	1.1070	2.3314	36
25	.41337	.45397	2.2028	.91056	1.0982	2.4191	35	25	.42920	.47519	2.1044	.90321	1.1072	2.3299	35
26	.41363	.45432	2.2011	.91044	1.0984	2.4176	34	26	.42946	.47555	2.1028	.90309	1.1073	2.3285	34
27	.41390	.45467	2.1994	.91032	1.0985	2.4160	33	27	.42972	.47590	2.1013	.90296	1.1075	2.3271	33
28	.41416	.45502	2.1977	.91020	1.0987	2.4145	32	28	.42999	.47626	2.0997	.90284	1.1076	2.3257	32
29	.41443	.45538	2.1960	.91008	1.0988	2.4130	31	29	.43025	.47662	2.0981	.90271	1.1078	2.3242	31
30	.41469	.45573	2.1943	.90996	1.0989	2.4114	30	30	.43051	.47698	2.0965	.90259	1.1079	2.3228	30
31	.41496	.45608	2.1926	.90984	1.0991	2.4099	29	31	.43077	.47733	2.0950	.90246	1.1081	2.3214	29
32	.41522	.45643	2.1909	.90972	1.0992	2.4083	28	32	.43104	.47769	2.0934	.90233	1.1082	2.3200	28
33	.41549	.45678	2.1892	.90960	1.0994	2.4068	27	33	.43130	.47805	2.0918	.90221	1.1084	2.3186	27
34	.41575	.45713	2.1876	.90948	1.0995	2.4053	26	34	.43156	.47840	2.0903	.90208	1.1085	2.3172	26
35	.41602	.45748	2.1859	.90936	1.0997	2.4038	25	35	.43182	.47876	2.0887	.90196	1.1087	2.3158	25
36	.41628	.45784	2.1842	.90924	1.0998	2.4022	24	36	.43209	.47912	2.0872	.90183	1.1089	2.3144	24
37	.41655	.45819	2.1825	.90911	1.1000	2.4007	23	37	.43235	.47948	2.0856	.90171	1.1090	2.3130	23
38	.41681	.45854	2.1808	.90899	1.1001	2.3992	22	38	.43261	.47984	2.0840	.90158	1.1092	2.3115	22
39	.41707	.45889	2.1792	.90887	1.1003	2.3977	21	39	.43287	.48019	2.0825	.90146	1.1093	2.3101	21
40	.41734	.45924	2.1775	.90875	1.1004	2.3961	20	40	.43313	.48055	2.0809	.90133	1.1095	2.3088	20
41	.41760	.45960	2.1758	.90863	1.1006	2.3946	19	41	.43340	.48091	2.0794	.90120	1.1096	2.3074	19
42	.41787	.45995	2.1742	.90851	1.1007	2.3931	18	42	.43366	.48127	2.0778	.90108	1.1098	2.3060	18
43	.41813	.46030	2.1725	.90839	1.1009	2.3916	17	43	.43392	.48163	2.0763	.90095	1.1099	2.3046	17
44	.41840	.46065	2.1708	.90826	1.1010	2.3901	16	44	.43418	.48198	2.0748	.90082	1.1101	2.3032	16
45	.41866	.46101	2.1692	.90814	1.1011	2.3886	15	45	.43445	.48234	2.0732	.90070	1.1102	2.3018	15
46	.41892	.46136	2.1675	.90802	1.1013	2.3871	14	46	.43471	.48270	2.0717	.90057	1.1104	2.3004	14
47	.41919	.46171	2.1659	.90790	1.1014	2.3856	13	47	.43497	.48306	2.0701	.90045	1.1106	2.2990	13
48	.41945	.46206	2.1642	.90778	1.1016	2.3841	12	48	.43523	.48342	2.0686	.90032	1.1107	2.2976	12
49	.41972	.46242	2.1625	.90766	1.1017	2.3826	11	49	.43549	.48378	2.0671	.90019	1.1109	2.2962	11
50	.41998	.46277	2.1609	.90753	1.1019	2.3811	10	50	.43575	.48414	2.0655	.90007	1.1110	2:2949	10
51	.42024	.46312	2.1592	.90741	1.1020	2.3796	9	51	.43602	.48450	2.0640	.89994	1.1112	2.2935	9
52	.42051	.46348	2.1576	.90729	1.1022	2.3781	8	52	.43628	.48486	2.0625	.89981	1.1113	2.2921	8
53	.42077	.46383	2.1560	.90717	1.1023	2.3766	7	53	.43654	.48521	2.0609	.89968	1.1115	2.2907	7
54	.42104	.46418	2.1543	.90704	1.1025	2.3751	6	54	.43680	.48557	2.0594	.89956	1.1117	2.2894	6
55	.42130	.46454	2.1527	.90692	1.1026	2.3736	5	55	.43706	.48593	2.0579	.89943	1.1118	2.2880	5
56	.42156	.46489	2.1510	.90680	1.1028	2.3721	4	56	.43733	.48629	2.0564	.89930	1.1120	2.2866	4
57	.42183	.46525	2.1494	.90668	1.1029	2.3706	3	57	.43759	.48665	2.0549	.89918	1.1121	2.2853	3
58	.42209	.46560	2.1478	.90655	1.1031	2.3692	2	58	.43785	.48701	2.0533	.89905	1.1123	2.2839	2
59	.42235	.46595	2.1461	.90643	1.1032	2.3677	1	59	.43811	.48737	2.0518	.89892	1.1124	2.2825	1
60	.42262	.46631	2.1445	.90631	1.1034	2.3662	0	60	.43837	.48773	2.0503	.89879	1.1126	2.2812	0
′	Cos	Ctn	Tan	Sin	Csc	Sec	′	′	Cos	Ctn	Tan	Sin	Csc	Sec	′

114° (294°)   **(245°) 65°**   **115° (295°)**   **(244°) 64°**

TABLE 1. NATURAL TRIGONOMETRIC FUNCTIONS (continued) 614

26° (206°) (333°) 153°

′	Sin	Tan	Ctn	Cos	Sec	Csc	′
0	.43837	.48773	2.0503	.89879	1.1126	2.2812	60
1	.43863	.48809	2.0488	.89867	1.1128	2.2798	59
2	.43889	.48845	2.0473	.89854	1.1129	2.2785	58
3	.43916	.48881	2.0458	.89841	1.1131	2.2771	57
4	.43942	.48917	2.0443	.89828	1.1132	2.2757	56
5	.43968	.48953	2.0428	.89816	1.1134	2.2744	55
6	.43994	.48989	2.0413	.89803	1.1136	2.2730	54
7	.44020	.49026	2.0398	.89790	1.1137	2.2717	53
8	.44046	.49062	2.0383	.89777	1.1139	2.2703	52
9	.44072	.49098	2.0368	.89764	1.1140	2.2690	51
10	.44098	.49134	2.0353	.89752	1.1142	2.2677	50
11	.44124	.49170	2.0338	.89739	1.1143	2.2663	49
12	.44151	.49206	2.0323	.89726	1.1145	2.2650	48
13	.44177	.49242	2.0308	.89713	1.1147	2.2636	47
14	.44203	.49278	2.0293	.89700	1.1148	2.2623	46
15	.44229	.49315	2.0278	.89687	1.1150	2.2610	45
16	.44255	.49351	2.0263	.89674	1.1151	2.2596	44
17	.44281	.49387	2.0248	.89662	1.1153	2.2583	43
18	.44307	.49423	2.0233	.89649	1.1155	2.2570	42
19	.44333	.49459	2.0219	.89636	1.1156	2.2556	41
20	.44359	.49495	2.0204	.89623	1.1158	2.2543	40
21	.44385	.49532	2.0189	.89610	1.1159	2.2530	39
22	.44411	.49568	2.0174	.89597	1.1161	2.2517	38
23	.44437	.49604	2.0160	.89584	1.1163	2.2504	37
24	.44464	.49640	2.0145	.89571	1.1164	2.2490	36
25	.44490	.49677	2.0130	.89558	1.1166	2.2477	35
26	.44516	.49713	2.0115	.89545	1.1168	2.2464	34
27	.44542	.49749	2.0101	.89532	1.1169	2.2451	33
28	.44568	.49786	2.0086	.89519	1.1171	2.2438	32
29	.44594	.49822	2.0072	.89506	1.1172	2.2425	31
30	.44620	.49858	2.0057	.89493	1.1174	2.2412	30
31	.44646	.49894	2.0042	.89480	1.1176	2.2399	29
32	.44672	.49931	2.0028	.89467	1.1177	2.2385	28
33	.44698	.49967	2.0013	.89454	1.1179	2.2372	27
34	.44724	.50004	1.9999	.89441	1.1180	2.2359	26
35	.44750	.50040	1.9984	.89428	1.1182	2.2346	25
36	.44776	.50076	1.9970	.89415	1.1184	2.2333	24
37	.44802	.50113	1.9955	.89402	1.1185	2.2320	23
38	.44828	.50149	1.9941	.89389	1.1187	2.2308	22
39	.44854	.50185	1.9926	.89376	1.1189	2.2295	21
40	.44880	.50222	1.9912	.89363	1.1190	2.2282	20
41	.44906	.50258	1.9897	.89350	1.1192	2.2269	19
42	.44932	.50295	1.9883	.89337	1.1194	2.2256	18
43	.44958	.50331	1.9868	.89324	1.1195	2.2243	17
44	.44984	.50368	1.9854	.89311	1.1197	2.2230	16
45	.45010	.50404	1.9840	.89298	1.1198	2.2217	15
46	.45036	.50441	1.9825	.89285	1.1200	2.2205	14
47	.45062	.50477	1.9811	.89272	1.1202	2.2192	13
48	.45088	.50514	1.9797	.89259	1.1203	2.2179	12
49	.45114	.50550	1.9782	.89245	1.1205	2.2166	11
50	.45140	.50587	1.9768	.89232	1.1207	2.2153	10
51	.45166	.50623	1.9754	.89219	1.1208	2.2141	9
52	.45192	.50660	1.9740	.89206	1.1210	2.2128	8
53	.45218	.50696	1.9725	.89193	1.1212	2.2115	7
54	.45243	.50733	1.9711	.89180	1.1213	2.2103	6
55	.45269	.50769	1.9697	.89167	1.1215	2.2090	5
56	.45295	.50806	1.9683	.89153	1.1217	2.2077	4
57	.45321	.50843	1.9669	.89140	1.1218	2.2065	3
58	.45347	.50879	1.9654	.89127	1.1220	2.2052	2
59	.45373	.50916	1.9640	.89114	1.1222	2.2039	1
60	.45399	.50953	1.9626	.89101	1.1223	2.2027	0
′	Cos	Ctn	Tan	Sin	Csc	Sec	′

116° (296°) (243°) 63°

27° (207°) (332°) 152°

′	Sin	Tan	Ctn	Cos	Sec	Csc	′
0	.45399	.50953	1.9626	.89101	1.1223	2.2027	60
1	.45425	.50989	1.9612	.89087	1.1225	2.2014	59
2	.45451	.51026	1.9598	.89074	1.1227	2.2002	58
3	.45477	.51063	1.9584	.89061	1.1228	2.1989	57
4	.45503	.51099	1.9570	.89048	1.1230	2.1977	56
5	.45529	.51136	1.9556	.89035	1.1232	2.1964	55
6	.45554	.51173	1.9542	.89021	1.1233	2.1952	54
7	.45580	.51209	1.9528	.89008	1.1235	2.1939	53
8	.45606	.51246	1.9514	.88995	1.1237	2.1927	52
9	.45632	.51283	1.9500	.88981	1.1238	2.1914	51
10	.45658	.51319	1.9486	.88968	1.1240	2.1902	50
11	.45684	.51356	1.9472	.88955	1.1242	2.1890	49
12	.45710	.51393	1.9458	.88942	1.1243	2.1877	48
13	.45736	.51430	1.9444	.88928	1.1245	2.1865	47
14	.45762	.51467	1.9430	.88915	1.1247	2.1852	46
15	.45787	.51503	1.9416	.88902	1.1248	2.1840	45
16	.45813	.51540	1.9402	.88888	1.1250	2.1828	44
17	.45839	.51577	1.9388	.88875	1.1252	2.1815	43
18	.45865	.51614	1.9375	.88862	1.1253	2.1803	42
19	.45891	.51651	1.9361	.88848	1.1255	2.1791	41
20	.45917	.51688	1.9347	.88835	1.1257	2.1779	40
21	.45942	.51724	1.9333	.88822	1.1259	2.1766	39
22	.45968	.51761	1.9319	.88808	1.1260	2.1754	38
23	.45994	.51798	1.9306	.88795	1.1262	2.1742	37
24	.46020	.51835	1.9292	.88782	1.1264	2.1730	36
25	.46046	.51872	1.9278	.88768	1.1265	2.1718	35
26	.46072	.51909	1.9265	.88755	1.1267	2.1705	34
27	.46097	.51946	1.9251	.88741	1.1269	2.1693	33
28	.46123	.51983	1.9237	.88728	1.1270	2.1681	32
29	.46149	.52020	1.9223	.88715	1.1272	2.1669	31
30	.46175	.52057	1.9210	.88701	1.1274	2.1657	30
31	.46201	.52094	1.9196	.88688	1.1276	2.1645	29
32	.46226	.52131	1.9183	.88674	1.1277	2.1633	28
33	.46252	.52168	1.9169	.88661	1.1279	2.1621	27
34	.46278	.52205	1.9155	.88647	1.1281	2.1609	26
35	.46304	.52242	1.9142	.88634	1.1282	2.1596	25
36	.46330	.52279	1.9128	.88620	1.1284	2.1584	24
37	.46355	.52316	1.9115	.88607	1.1286	2.1572	23
38	.46381	.52353	1.9101	.88593	1.1288	2.1560	22
39	.46407	.52390	1.9088	.88580	1.1289	2.1549	21
40	.46433	.52427	1.9074	.88566	1.1291	2.1537	20
41	.46458	.52464	1.9061	.88553	1.1293	2.1525	19
42	.46484	.52501	1.9047	.88539	1.1294	2.1513	18
43	.46510	.52538	1.9034	.88526	1.1296	2.1501	17
44	.46536	.52575	1.9020	.88512	1.1298	2.1489	16
45	.46561	.52613	1.9007	.88499	1.1300	2.1477	15
46	.46587	.52650	1.8993	.88485	1.1301	2.1465	14
47	.46613	.52687	1.8980	.88472	1.1303	2.1453	13
48	.46639	.52724	1.8967	.88458	1.1305	2.1441	12
49	.46664	.52761	1.8953	.88445	1.1307	2.1430	11
50	.46690	.52798	1.8940	.88431	1.1308	2.1418	10
51	.46716	.52836	1.8927	.88417	1.1310	2.1406	9
52	.46742	.52873	1.8913	.88404	1.1312	2.1394	8
53	.46767	.52910	1.8900	.88390	1.1313	2.1382	7
54	.46793	.52947	1.8887	.88377	1.1315	2.1371	6
55	.46819	.52985	1.8873	.88363	1.1317	2.1359	5
56	.46844	.53022	1.8860	.88349	1.1319	2.1347	4
57	.46870	.53059	1.8847	.88336	1.1320	2.1336	3
58	.46896	.53096	1.8834	.88322	1.1322	2.1324	2
59	.46921	.53134	1.8820	.88308	1.1324	2.1312	1
60	.46947	.53171	1.8807	.88295	1.1326	2.1301	0
′	Cos	Ctn	Tan	Sin	Csc	Sec	′

117° (297°) (242°) 62°

28° (208°)   **(331°) 151°**

′	Sin	Tan	Ctn	Cos	Sec	Csc	′
0	.46947	.53171	1.8807	.88295	1.1326	2.1301	60
1	.46973	.53208	1.8794	.88281	1.1327	2.1289	59
2	.46999	.53246	1.8781	.88267	1.1329	2.1277	58
3	.47024	.53283	1.8768	.88254	1.1331	2.1266	57
4	.47050	.53320	1.8755	.88240	1.1333	2.1254	56
5	.47076	.53358	1.8741	.88226	1.1334	2.1242	55
6	.47101	.53395	1.8728	.88213	1.1336	2.1231	54
7	.47127	.53432	1.8715	.88199	1.1338	2.1219	53
8	.47153	.53470	1.8702	.88185	1.1340	2.1208	52
9	.47178	.53507	1.8689	.88172	1.1342	2.1196	51
10	.47204	.53545	1.8676	.88158	1.1343	2.1185	50
11	.47229	.53582	1.8663	.88144	1.1345	2.1173	49
12	.47255	.53620	1.8650	.88130	1.1347	2.1162	48
13	.47281	.53657	1.8637	.88117	1.1349	2.1150	47
14	.47306	.53694	1.8624	.88103	1.1350	2.1139	46
15	.47332	.53732	1.8611	.88089	1.1352	2.1127	45
16	.47358	.53769	1.8598	.88075	1.1354	2.1116	44
17	.47383	.53807	1.8585	.88062	1.1356	2.1105	43
18	.47409	.53844	1.8572	.88048	1.1357	2.1093	42
19	.47434	.53882	1.8559	.88034	1.1359	2.1082	41
20	.47460	.53920	1.8546	.88020	1.1361	2.1070	40
21	.47486	.53957	1.8533	.88006	1.1363	2.1059	39
22	.47511	.53995	1.8520	.87993	1.1365	2.1048	38
23	.47537	.54032	1.8507	.87979	1.1366	2.1036	37
24	.47562	.54070	1.8495	.87965	1.1368	2.1025	36
25	.47588	.54107	1.8482	.87951	1.1370	2.1014	35
26	.47614	.54145	1.8469	.87937	1.1372	2.1002	34
27	.47639	.54183	1.8456	.87923	1.1374	2.0991	33
28	.47665	.54220	1.8443	.87909	1.1375	2.0980	32
29	.47690	.54258	1.8430	.87896	1.1377	2.0969	31
30	.47716	.54296	1.8418	.87882	1.1379	2.0957	30
31	.47741	.54333	1.8405	.87868	1.1381	2.0946	29
32	.47767	.54371	1.8392	.87854	1.1383	2.0935	28
33	.47793	.54409	1.8379	.87840	1.1384	2.0924	27
34	.47818	.54446	1.8367	.87826	1.1386	2.0913	26
35	.47844	.54484	1.8354	.87812	1.1388	2.0901	25
36	.47869	.54522	1.8341	.87798	1.1390	2.0890	24
37	.47895	.54560	1.8329	.87784	1.1392	2.0879	23
38	.47920	.54597	1.8316	.87770	1.1393	2.0868	22
39	.47946	.54635	1.8303	.87756	1.1395	2.0857	21
40	.47971	.54673	1.8291	.87743	1.1397	2.0846	20
41	.47997	.54711	1.8278	.87729	1.1399	2.0835	19
42	.48022	.54748	1.8265	.87715	1.1401	2.0824	18
43	.48048	.54786	1.8253	.87701	1.1402	2.0813	17
44	.48073	.54824	1.8240	.87687	1.1404	2.0802	16
45	.48099	.54862	1.8228	.87673	1.1406	2.0791	15
46	.48124	.54900	1.8215	.87659	1.1408	2.0779	14
47	.48150	.54938	1.8202	.87645	1.1410	2.0768	13
48	.48175	.54975	1.8190	.87631	1.1412	2.0757	12
49	.48201	.55013	1.8177	.87617	1.1413	2.0747	11
50	.48226	.55051	1.8165	.87603	1.1415	2.0736	10
51	.48252	.55089	1.8152	.87589	1.1417	2.0725	9
52	.48277	.55127	1.8140	.87575	1.1419	2.0714	8
53	.48303	.55165	1.8127	.87561	1.1421	2.0703	7
54	.48328	.55203	1.8115	.87546	1.1423	2.0692	6
55	.48354	.55241	1.8103	.87532	1.1424	2.0681	5
56	.48379	.55279	1.8090	.87518	1.1426	2.0670	4
57	.48405	.55317	1.8078	.87504	1.1428	2.0659	3
58	.48430	.55355	1.8065	.87490	1.1430	2.0648	2
59	.48456	.55393	1.8053	.87476	1.1432	2.0637	1
60	.48481	.55431	1.8040	.87462	1.1434	2.0627	0
′	Cos	Ctn	Tan	Sin	Csc	Sec	′

118° (298°)   **(241°) 61°**

29° (209°)   **(330°) 150°**

′	Sin	Tan	Ctn	Cos	Sec	Csc	′
0	.48481	.55431	1.8040	.87462	1.1434	2.0627	60
1	.48506	.55469	1.8028	.87448	1.1435	2.0616	59
2	.48532	.55507	1.8016	.87434	1.1437	2.0605	58
3	.48557	.55545	1.8003	.87420	1.1439	2.0594	57
4	.48583	.55583	1.7991	.87406	1.1441	2.0583	56
5	.48608	.55621	1.7979	.87391	1.1443	2.0573	55
6	.48634	.55659	1.7966	.87377	1.1445	2.0562	54
7	.48659	.55697	1.7954	.87363	1.1446	2.0551	53
8	.48684	.55736	1.7942	.87349	1.1448	2.0540	52
9	.48710	.55774	1.7930	.87335	1.1450	2.0530	51
10	.48735	.55812	1.7917	.87321	1.1452	2.0519	50
11	.48761	.55850	1.7905	.87306	1.1454	2.0508	49
12	.48786	.55888	1.7893	.87292	1.1456	2.0498	48
13	.48811	.55926	1.7881	.87278	1.1458	2.0487	47
14	.48837	.55964	1.7868	.87264	1.1460	2.0476	46
15	.48862	.56003	1.7856	.87250	1.1461	2.0466	45
16	.48888	.56041	1.7844	.87235	1.1463	2.0455	44
17	.48913	.56079	1.7832	.87221	1.1465	2.0445	43
18	.48938	.56117	1.7820	.87207	1.1467	2.0434	42
19	.48964	.56156	1.7808	.87193	1.1469	2.0423	41
20	.48989	.56194	1.7796	.87178	1.1471	2.0413	40
21	.49014	.56232	1.7783	.87164	1.1473	2.0402	39
22	.49040	.56270	1.7771	.87150	1.1474	2.0392	38
23	.49065	.56309	1.7759	.87136	1.1476	2.0381	37
24	.49090	.56347	1.7747	.87121	1.1478	2.0371	36
25	.49116	.56385	1.7735	.87107	1.1480	2.0360	35
26	.49141	.56424	1.7723	.87093	1.1482	2.0350	34
27	.49166	.56462	1.7711	.87079	1.1484	2.0339	33
28	.49192	.56501	1.7699	.87064	1.1486	2.0329	32
29	.49217	.56539	1.7687	.87050	1.1488	2.0318	31
30	.49242	.56577	1.7675	.87036	1.1490	2.0308	30
31	.49268	.56616	1.7663	.87021	1.1491	2.0297	29
32	.49293	.56654	1.7651	.87007	1.1493	2.0287	28
33	.49318	.56693	1.7639	.86993	1.1495	2.0276	27
34	.49344	.56731	1.7627	.86978	1.1497	2.0266	26
35	.49369	.56769	1.7615	.86964	1.1499	2.0256	25
36	.49394	.56808	1.7603	.86949	1.1501	2.0245	24
37	.49419	.56846	1.7591	.86935	1.1503	2.0235	23
38	.49445	.56885	1.7579	.86921	1.1505	2.0225	22
39	.49470	.56923	1.7567	.86906	1.1507	2.0214	21
40	.49495	.56962	1.7556	.86892	1.1509	2.0204	20
41	.49521	.57000	1.7544	.86878	1.1510	2.0194	19
42	.49546	.57039	1.7532	.86863	1.1512	2.0183	18
43	.49571	.57078	1.7520	.86849	1.1514	2.0173	17
44	.49596	.57116	1.7508	.86834	1.1516	2.0163	16
45	.49622	.57155	1.7496	.86820	1.1518	2.0152	15
46	.49647	.57193	1.7485	.86805	1.1520	2.0142	14
47	.49672	.57232	1.7473	.86791	1.1522	2.0132	13
48	.49697	.57271	1.7461	.86777	1.1524	2.0122	12
49	.49723	.57309	1.7449	.86762	1.1526	2.0112	11
50	.49748	.57348	1.7437	.86748	1.1528	2.0101	10
51	.49773	.57386	1.7426	.86733	1.1530	2.0091	9
52	.49798	.57425	1.7414	.86719	1.1532	2.0081	8
53	.49824	.57464	1.7402	.86704	1.1533	2.0071	7
54	.49849	.57503	1.7391	.86690	1.1535	2.0061	6
55	.49874	.57541	1.7379	.86675	1.1537	2.0051	5
56	.49899	.57580	1.7367	.86661	1.1539	2.0040	4
57	.49924	.57619	1.7355	.86646	1.1541	2.0030	3
58	.49950	.57657	1.7344	.86632	1.1543	2.0020	2
59	.49975	.57696	1.7332	.86617	1.1545	2.0010	1
60	.50000	.57735	1.7321	.86603	1.1547	2.0000	0
′	Cos	Ctn	Tan	Sin	Csc	Sec	′

119° (299°)   **(240°) 60°**

TABLE 1. NATURAL TRIGONOMETRIC FUNCTIONS (continued)   616

30° (210°)   (329°) 149°   **31° (211°)**   (328°) 148°

′	Sin	Tan	Ctn	Cos	Sec	Csc	′		′	Sin	Tan	Ctn	Cos	Sec	Csc	′
0	.50000	.57735	1.7321	.86603	1.1547	2.0000	60		0	.51504	.60086	1.6643	.85717	1.1666	1.9416	60
1	.50025	.57774	1.7309	.86588	1.1549	1.9990	59		1	.51529	.60126	1.6632	.85702	1.1668	1.9407	59
2	.50050	.57813	1.7297	.86573	1.1551	1.9980	58		2	.51554	.60165	1.6621	.85687	1.1670	1.9397	58
3	.50076	.57851	1.7286	.86559	1.1553	1.9970	57		3	.51579	.60205	1.6610	.85672	1.1672	1.9388	57
4	.50101	.57890	1.7274	.86544	1.1555	1.9960	56		4	.51604	.60245	1.6599	.85657	1.1675	1.9379	56
5	.50126	.57929	1.7262	.86530	1.1557	1.9950	55		5	.51628	.60284	1.6588	.85642	1.1677	1.9369	55
6	.50151	.57968	1.7251	.86515	1.1559	1.9940	54		6	.51653	.60324	1.6577	.85627	1.1679	1.9360	54
7	.50176	.58007	1.7239	.86501	1.1561	1.9930	53		7	.51678	.60364	1.6566	.85612	1.1681	1.9351	53
8	.50201	.58046	1.7228	.86486	1.1563	1.9920	52		8	.51703	.60403	1.6555	.85597	1.1683	1.9341	52
9	.50227	.58085	1.7216	.86471	1.1565	1.9910	51		9	.51728	.60443	1.6545	.85582	1.1685	1.9332	51
10	.50252	.58124	1.7205	.86457	1.1566	1.9900	50		10	.51753	.60483	1.6534	.85567	1.1687	1.9323	50
11	.50277	.58162	1.7193	.86442	1.1568	1.9890	49		11	.51778	.60522	1.6523	.85551	1.1689	1.9313	49
12	.50302	.58201	1.7182	.86427	1.1570	1.9880	48		12	.51803	.60562	1.6512	.85536	1.1691	1.9304	48
13	.50327	.58240	1.7170	.86413	1.1572	1.9870	47		13	.51828	.60602	1.6501	.85521	1.1693	1.9295	47
14	.50352	.58279	1.7159	.86398	1.1574	1.9860	46		14	.51852	.60642	1.6490	.85506	1.1695	1.9285	46
15	.50377	.58318	1.7147	.86384	1.1576	1.9850	45		15	.51877	.60681	1.6479	.85491	1.1697	1.9276	45
16	.50403	.58357	1.7136	.86369	1.1578	1.9840	44		16	.51902	.60721	1.6469	.85476	1.1699	1.9267	44
17	.50428	.58396	1.7124	.86354	1.1580	1.9830	43		17	.51927	.60761	1.6458	.85461	1.1701	1.9258	43
18	.50453	.58435	1.7113	.86340	1.1582	1.9821	42		18	.51952	.60801	1.6447	.85446	1.1703	1.9249	42
19	.50478	.58474	1.7102	.86325	1.1584	1.9811	41		19	.51977	.60841	1.6436	.85431	1.1705	1.9239	41
20	.50503	.58513	1.7090	.86310	1.1586	1.9801	40		20	.52002	.60881	1.6426	.85416	1.1707	1.9230	40
21	.50528	.58552	1.7079	.86295	1.1588	1.9791	39		21	.52026	.60921	1.6415	.85401	1.1710	1.9221	39
22	.50553	.58591	1.7067	.86281	1.1590	1.9781	38		22	.52051	.60960	1.6404	.85385	1.1712	1.9212	38
23	.50578	.58631	1.7056	.86266	1.1592	1.9771	37		23	.52076	.61000	1.6393	.85370	1.1714	1.9203	37
24	.50603	.58670	1.7045	.86251	1.1594	1.9762	36		24	.52101	.61040	1.6383	.85355	1.1716	1.9194	36
25	.50628	.58709	1.7033	.86237	1.1596	1.9752	35		25	.52126	.61080	1.6372	.85340	1.1718	1.9184	35
26	.50654	.58748	1.7022	.86222	1.1598	1.9742	34		26	.52151	.61120	1.6361	.85325	1.1720	1.9175	34
27	.50679	.58787	1.7011	.86207	1.1600	1.9732	33		27	.52175	.61160	1.6351	.85310	1.1722	1.9166	33
28	.50704	.58826	1.6999	.86192	1.1602	1.9722	32		28	.52200	.61200	1.6340	.85294	1.1724	1.9157	32
29	.50729	.58865	1.6988	.86178	1.1604	1.9713	31		29	.52225	.61240	1.6329	.85279	1.1726	1.9148	31
30	.50754	.58905	1.6977	.86163	1.1606	1.9703	30		30	.52250	.61280	1.6319	.85264	1.1728	1.9139	30
31	.50779	.58944	1.6965	.86148	1.1608	1.9693	29		31	.52275	.61320	1.6308	.85249	1.1730	1.9130	29
32	.50804	.58983	1.6954	.86133	1.1610	1.9684	28		32	.52299	.61360	1.6297	.85234	1.1732	1.9121	28
33	.50829	.59022	1.6943	.86119	1.1612	1.9674	27		33	.52324	.61400	1.6287	.85218	1.1735	1.9112	27
34	.50854	.59061	1.6932	.86104	1.1614	1.9664	26		34	.52349	.61440	1.6276	.85203	1.1737	1.9103	26
35	.50879	.59101	1.6920	.86089	1.1616	1.9654	25		35	.52374	.61480	1.6265	.85188	1.1739	1.9094	25
36	.50904	.59140	1.6909	.86074	1.1618	1.9645	24		36	.52399	.61520	1.6255	.85173	1.1741	1.9084	24
37	.50929	.59179	1.6898	.86059	1.1620	1.9635	23		37	.52423	.61561	1.6244	.85157	1.1743	1.9075	23
38	.50954	.59218	1.6887	.86045	1.1622	1.9625	22		38	.52448	.61601	1.6234	.85142	1.1745	1.9066	22
39	.50979	.59258	1.6875	.86030	1.1624	1.9616	21		39	.52473	.61641	1.6223	.85127	1.1747	1.9057	21
40	.51004	.59297	1.6864	.86015	1.1626	1.9606	20		40	.52498	.61681	1.6212	.85112	1.1749	1.9048	20
41	.51029	.59336	1.6853	.86000	1.1628	1.9597	19		41	.52522	.61721	1.6202	.85096	1.1751	1.9039	19
42	.51054	.59376	1.6842	.85985	1.1630	1.9587	18		42	.52547	.61761	1.6191	.85081	1.1753	1.9031	18
43	.51079	.59415	1.6831	.85970	1.1632	1.9577	17		43	.52572	.61801	1.6181	.85066	1.1756	1.9022	17
44	.51104	.59454	1.6820	.85956	1.1634	1.9568	16		44	.52597	.61842	1.6170	.85051	1.1758	1.9013	16
45	.51129	.59494	1.6808	.85941	1.1636	1.9558	15		45	.52621	.61882	1.6160	.85035	1.1760	1.9004	15
46	.51154	.59533	1.6797	.85926	1.1638	1.9549	14		46	.52646	.61922	1.6149	.85020	1.1762	1.8995	14
47	.51179	.59573	1.6786	.85911	1.1640	1.9539	13		47	.52671	.61962	1.6139	.85005	1.1764	1.8986	13
48	.51204	.59612	1.6775	.85896	1.1642	1.9530	12		48	.52696	.62003	1.6128	.84989	1.1766	1.8977	12
49	.51229	.59651	1.6764	.85881	1.1644	1.9520	11		49	.52720	.62043	1.6118	.84974	1.1768	1.8968	11
50	.51254	.59691	1.6753	.85866	1.1646	1.9511	10		50	.52745	.62083	1.6107	.84959	1.1770	1.8959	10
51	.51279	.59730	1.6742	.85851	1.1648	1.9501	9		51	.52770	.62124	1.6097	.84943	1.1773	1.8950	9
52	.51304	.59770	1.6731	.85836	1.1650	1.9492	8		52	.52794	.62164	1.6087	.84928	1.1775	1.8941	8
53	.51329	.59809	1.6720	.85821	1.1652	1.9482	7		53	.52819	.62204	1.6076	.84913	1.1777	1.8933	7
54	.51354	.59849	1.6709	.85806	1.1654	1.9473	6		54	.52844	.62245	1.6066	.84897	1.1779	1.8924	6
55	.51379	.59888	1.6698	.85792	1.1656	1.9463	5		55	.52869	.62285	1.6055	.84882	1.1781	1.8915	5
56	.51404	.59928	1.6687	.85777	1.1658	1.9454	4		56	.52893	.62325	1.6045	.84866	1.1783	1.8906	4
57	.51429	.59967	1.6676	.85762	1.1660	1.9444	3		57	.52918	.62366	1.6034	.84851	1.1785	1.8897	3
58	.51454	.60007	1.6665	.85747	1.1662	1.9435	2		58	.52943	.62406	1.6024	.84836	1.1788	1.8888	2
59	.51479	.60046	1.6654	.85732	1.1664	1.9425	1		59	.52967	.62446	1.6014	.84820	1.1790	1.8880	1
60	.51504	.60086	1.6643	.85717	1.1666	1.9416	0		60	.52992	.62487	1.6003	.84805	1.1792	1.8871	0
′	Cos	Ctn	Tan	Sin	Csc	Sec	′		′	Cos	Ctn	Tan	Sin	Csc	Sec	′

120° (300°)   (239°) 59°   **121° (301°)**   (238°) 58°

32° (212°)   (327°) **147°**

'	Sin	Tan	Ctn	Cos	Sec	Csc	'
0	.52992	.62487	1.6003	.84805	1.1792	1.8871	60
1	.53017	.62527	1.5993	.84789	1.1794	1.8862	59
2	.53041	.62568	1.5983	.84774	1.1796	1.8853	58
3	.53066	.62608	1.5972	.84759	1.1798	1.8844	57
4	.53091	.62649	1.5962	.84743	1.1800	1.8836	56
5	.53115	.62689	1.5952	.84728	1.1803	1.8827	55
6	.53140	.62730	1.5941	.84712	1.1805	1.8818	54
7	.53164	.62770	1.5931	.84697	1.1807	1.8810	53
8	.53189	.62811	1.5921	.84681	1.1809	1.8801	52
9	.53214	.62852	1.5911	.84666	1.1811	1.8792	51
10	.53238	.62892	1.5900	.84650	1.1813	1.8783	50
11	.53263	.62933	1.5890	.84635	1.1815	1.8775	49
12	.53288	.62973	1.5880	.84619	1.1818	1.8766	48
13	.53312	.63014	1.5869	.84604	1.1820	1.8757	47
14	.53337	.63055	1.5859	.84588	1.1822	1.8749	46
15	.53361	.63095	1.5849	.84573	1.1824	1.8740	45
16	.53386	.63136	1.5839	.84557	1.1826	1.8731	44
17	.53411	.63177	1.5829	.84542	1.1828	1.8723	43
18	.53435	.63217	1.5818	.84526	1.1831	1.8714	42
19	.53460	.63258	1.5808	.84511	1.1833	1.8706	41
20	.53484	.63299	1.5798	.84495	1.1835	1.8697	40
21	.53509	.63340	1.5788	.84480	1.1837	1.8688	39
22	.53534	.63380	1.5778	.84464	1.1839	1.8680	38
23	.53558	.63421	1.5768	.84448	1.1842	1.8671	37
24	.53583	.63462	1.5757	.84433	1.1844	1.8663	36
25	.53607	.63503	1.5747	.84417	1.1846	1.8654	35
26	.53632	.63544	1.5737	.84402	1.1848	1.8646	34
27	.53656	.63584	1.5727	.84386	1.1850	1.8637	33
28	.53681	.63625	1.5717	.84370	1.1852	1.8629	32
29	.53705	.63666	1.5707	.84355	1.1855	1.8620	31
30	.53730	.63707	1.5697	.84339	1.1857	1.8612	30
31	.53754	.63748	1.5687	.84324	1.1859	1.8603	29
32	.53779	.63789	1.5677	.84308	1.1861	1.8595	28
33	.53804	.63830	1.5667	.84292	1.1863	1.8586	27
34	.53828	.63871	1.5657	.84277	1.1866	1.8578	26
35	.53853	.63912	1.5647	.84261	1.1868	1.8569	25
36	.53877	.63953	1.5637	.84245	1.1870	1.8561	24
37	.53902	.63994	1.5627	.84230	1.1872	1.8552	23
38	.53926	.64035	1.5617	.84214	1.1875	1.8544	22
39	.53951	.64076	1.5607	.84198	1.1877	1.8535	21
40	.53975	.64117	1.5597	.84182	1.1879	1.8527	20
41	.54000	.64158	1.5587	.84167	1.1881	1.8519	19
42	.54024	.64199	1.5577	.84151	1.1883	1.8510	18
43	.54049	.64240	1.5567	.84135	1.1886	1.8502	17
44	.54073	.64281	1.5557	.84120	1.1888	1.8494	16
45	.54097	.64322	1.5547	.84104	1.1890	1.8485	15
46	.54122	.64363	1.5537	.84088	1.1892	1.8477	14
47	.54146	.64404	1.5527	.84072	1.1895	1.8468	13
48	.54171	.64446	1.5517	.84057	1.1897	1.8460	12
49	.54195	.64487	1.5507	.84041	1.1899	1.8452	11
50	.54220	.64528	1.5497	.84025	1.1901	1.8443	10
51	.54244	.64569	1.5487	.84009	1.1903	1.8435	9
52	.54269	.64610	1.5477	.83994	1.1906	1.8427	8
53	.54293	.64652	1.5468	.83978	1.1908	1.8419	7
54	.54317	.64693	1.5458	.83962	1.1910	1.8410	6
55	.54342	.64734	1.5448	.83946	1.1912	1.8402	5
56	.54366	.64775	1.5438	.83930	1.1915	1.8394	4
57	.54391	.64817	1.5428	.83915	1.1917	1.8385	3
58	.54415	.64858	1.5418	.83899	1.1919	1.8377	2
59	.54440	.64899	1.5408	.83883	1.1921	1.8369	1
60	.54464	.64941	1.5399	.83867	1.1924	1.8361	0
'	Cos	Ctn	Tan	Sin	Csc	Sec	'

122° (302°)   (237°) **57°**

33° (213°)   (326°) **146°**

'	Sin	Tan	Ctn	Cos	Sec	Csc	'
0	.54464	.64941	1.5399	.83867	1.1924	1.8361	60
1	.54488	.64982	1.5389	.83851	1.1926	1.8353	59
2	.54513	.65024	1.5379	.83835	1.1928	1.8344	58
3	.54537	.65065	1.5369	.83819	1.1930	1.8336	57
4	.54561	.65106	1.5359	.83804	1.1933	1.8328	56
5	.54586	.65148	1.5350	.83788	1.1935	1.8320	55
6	.54610	.65189	1.5340	.83772	1.1937	1.8312	54
7	.54635	.65231	1.5330	.83756	1.1939	1.8303	53
8	.54659	.65272	1.5320	.83740	1.1942	1.8295	52
9	.54683	.65314	1.5311	.83724	1.1944	1.8287	51
10	.54708	.65355	1.5301	.83708	1.1946	1.8279	50
11	.54732	.65397	1.5291	.83692	1.1949	1.8271	49
12	.54756	.65438	1.5282	.83676	1.1951	1.8263	48
13	.54781	.65480	1.5272	.83660	1.1953	1.8255	47
14	.54805	.65521	1.5262	.83645	1.1955	1.8247	46
15	.54829	.65563	1.5253	.83629	1.1958	1.8238	45
16	.54854	.65604	1.5243	.83613	1.1960	1.8230	44
17	.54878	.65646	1.5233	.83597	1.1962	1.8222	43
18	.54902	.65688	1.5224	.83581	1.1964	1.8214	42
19	.54927	.65729	1.5214	.83565	1.1967	1.8206	41
20	.54951	.65771	1.5204	.83549	1.1969	1.8198	40
21	.54975	.65813	1.5195	.83533	1.1971	1.8190	39
22	.54999	.65854	1.5185	.83517	1.1974	1.8182	38
23	.55024	.65896	1.5175	.83501	1.1976	1.8174	37
24	.55048	.65938	1.5166	.83485	1.1978	1.8166	36
25	.55072	.65980	1.5156	.83469	1.1981	1.8158	35
26	.55097	.66021	1.5147	.83453	1.1983	1.8150	34
27	.55121	.66063	1.5137	.83437	1.1985	1.8142	33
28	.55145	.66105	1.5127	.83421	1.1987	1.8134	32
29	.55169	.66147	1.5118	.83405	1.1990	1.8126	31
30	.55194	.66189	1.5108	.83389	1.1992	1.8118	30
31	.55218	.66230	1.5099	.83373	1.1994	1.8110	29
32	.55242	.66272	1.5089	.83356	1.1997	1.8102	28
33	.55266	.66314	1.5080	.83340	1.1999	1.8094	27
34	.55291	.66356	1.5070	.83324	1.2001	1.8086	26
35	.55315	.66398	1.5061	.83308	1.2004	1.8078	25
36	.55339	.66440	1.5051	.83292	1.2006	1.8070	24
37	.55363	.66482	1.5042	.83276	1.2008	1.8062	23
38	.55388	.66524	1.5032	.83260	1.2011	1.8055	22
39	.55412	.66566	1.5023	.83244	1.2013	1.8047	21
40	.55436	.66608	1.5013	.83228	1.2015	1.8039	20
41	.55460	.66650	1.5004	.83212	1.2018	1.8031	19
42	.55484	.66692	1.4994	.83195	1.2020	1.8023	18
43	.55509	.66734	1.4985	.83179	1.2022	1.8015	17
44	.55533	.66776	1.4975	.83163	1.2025	1.8007	16
45	.55557	.66818	1.4966	.83147	1.2027	1.8000	15
46	.55581	.66860	1.4957	.83131	1.2029	1.7992	14
47	.55605	.66902	1.4947	.83115	1.2032	1.7984	13
48	.55630	.66944	1.4938	.83098	1.2034	1.7976	12
49	.55654	.66986	1.4928	.83082	1.2036	1.7968	11
50	.55678	.67028	1.4919	.83066	1.2039	1.7960	10
51	.55702	.67071	1.4910	.83050	1.2041	1.7953	9
52	.55726	.67113	1.4900	.83034	1.2043	1.7945	8
53	.55750	.67155	1.4891	.83017	1.2046	1.7937	7
54	.55775	.67197	1.4882	.83001	1.2048	1.7929	6
55	.55799	.67239	1.4872	.82985	1.2050	1.7922	5
56	.55823	.67282	1.4863	.82969	1.2053	1.7914	4
57	.55847	.67324	1.4854	.82953	1.2055	1.7906	3
58	.55871	.67366	1.4844	.82936	1.2057	1.7898	2
59	.55895	.67409	1.4835	.82920	1.2060	1.7891	1
60	.55919	.67451	1.4826	.82904	1.2062	1.7883	0
'	Cos	Ctn	Tan	Sin	Csc	Sec	'

123° (303°)   (236°) **56°**

TABLE 1. NATURAL TRIGONOMETRIC FUNCTIONS (continued)   618

34° (214°)   **(325°) 145°**

′	Sin	Tan	Ctn	Cos	Sec	Csc	′
0	.55919	.67451	1.4826	.82904	1.2062	1.7883	60
1	.55943	.67493	1.4816	.82887	1.2065	1.7875	59
2	.55968	.67536	1.4807	.82871	1.2067	1.7868	58
3	.55992	.67578	1.4798	.82855	1.2069	1.7860	57
4	.56016	.67620	1.4788	.82839	1.2072	1.7852	56
5	.56040	.67663	1.4779	.82822	1.2074	1.7844	55
6	.56064	.67705	1.4770	.82806	1.2076	1.7837	54
7	.56088	.67748	1.4761	.82790	1.2079	1.7829	53
8	.56112	.67790	1.4751	.82773	1.2081	1.7821	52
9	.56136	.67832	1.4742	.82757	1.2084	1.7814	51
10	.56160	.67875	1.4733	.82741	1.2086	1.7806	50
11	.56184	.67917	1.4724	.82724	1.2088	1.7799	49
12	.56208	.67960	1.4715	.82708	1.2091	1.7791	48
13	.56232	.68002	1.4705	.82692	1.2093	1.7783	47
14	.56256	.68045	1.4696	.82675	1.2096	1.7776	46
15	.56280	.68088	1.4687	.82659	1.2098	1.7768	45
16	.56305	.68130	1.4678	.82643	1.2100	1.7761	44
17	.56329	.68173	1.4669	.82626	1.2103	1.7753	43
18	.56353	.68215	1.4659	.82610	1.2105	1.7745	42
19	.56377	.68258	1.4650	.82593	1.2108	1.7738	41
20	.56401	.68301	1.4641	.82577	1.2110	1.7730	40
21	.56425	.68343	1.4632	.82561	1.2112	1.7723	39
22	.56449	.68386	1.4623	.82544	1.2115	1.7715	38
23	.56473	.68429	1.4614	.82528	1.2117	1.7708	37
24	.56497	.68471	1.4605	.82511	1.2120	1.7700	36
25	.56521	.68514	1.4596	.82495	1.2122	1.7693	35
26	.56545	.68557	1.4586	.82478	1.2124	1.7685	34
27	.56569	.68600	1.4577	.82462	1.2127	1.7678	33
28	.56593	.68642	1.4568	.82446	1.2129	1.7670	32
29	.56617	.68685	1.4559	.82429	1.2132	1.7663	31
30	.56641	.68728	1.4550	.82413	1.2134	1.7655	30
31	.56665	.68771	1.4541	.82396	1.2136	1.7648	29
32	.56689	.68814	1.4532	.82380	1.2139	1.7640	28
33	.56713	.68857	1.4523	.82363	1.2141	1.7633	27
34	.56736	.68900	1.4514	.82347	1.2144	1.7625	26
35	.56760	.68942	1.4505	.82330	1.2146	1.7618	25
36	.56784	.68985	1.4496	.82314	1.2149	1.7610	24
37	.56808	.69028	1.4487	.82297	1.2151	1.7603	23
38	.56832	.69071	1.4478	.82281	1.2154	1.7596	22
39	.56856	.69114	1.4469	.82264	1.2156	1.7588	21
40	.56880	.69157	1.4460	.82248	1.2158	1.7581	20
41	.56904	.69200	1.4451	.82231	1.2161	1.7573	19
42	.56928	.69243	1.4442	.82214	1.2163	1.7566	18
43	.56952	.69286	1.4433	.82198	1.2166	1.7559	17
44	.56976	.69329	1.4424	.82181	1.2168	1.7551	16
45	.57000	.69372	1.4415	.82165	1.2171	1.7544	15
46	.57024	.69416	1.4406	.82148	1.2173	1.7537	14
47	.57047	.69459	1.4397	.82132	1.2176	1.7529	13
48	.57071	.69502	1.4388	.82115	1.2178	1.7522	12
49	.57095	.69545	1.4379	.82098	1.2181	1.7515	11
50	.57119	.69588	1.4370	.82082	1.2183	1.7507	10
51	.57143	.69631	1.4361	.82065	1.2185	1.7500	9
52	.57167	.69675	1.4352	.82048	1.2188	1.7493	8
53	.57191	.69718	1.4344	.82032	1.2190	1.7485	7
54	.57215	.69761	1.4335	.82015	1.2193	1.7478	6
55	.57238	.69804	1.4326	.81999	1.2195	1.7471	5
56	.57262	.69847	1.4317	.81982	1.2198	1.7463	4
57	.57286	.69891	1.4308	.81965	1.2200	1.7456	3
58	.57310	.69934	1.4299	.81949	1.2203	1.7449	2
59	.57334	.69977	1.4290	.81932	1.2205	1.7442	1
60	.57358	.70021	1.4281	.81915	1.2208	1.7434	0
′	Cos	Ctn	Tan	Sin	Csc	Sec	′

124° (304°)   **(235°) 55°**

35° (215°)   **(324°) 144°**

′	Sin	Tan	Ctn	Cos	Sec	Csc	′
0	.57358	.70021	1.4281	.81915	1.2208	1.7434	60
1	.57381	.70064	1.4273	.81899	1.2210	1.7427	59
2	.57405	.70107	1.4264	.81882	1.2213	1.7420	58
3	.57429	.70151	1.4255	.81865	1.2215	1.7413	57
4	.57453	.70194	1.4246	.81848	1.2218	1.7406	56
5	.57477	.70238	1.4237	.81832	1.2220	1.7398	55
6	.57501	.70281	1.4229	.81815	1.2223	1.7391	54
7	.57524	.70325	1.4220	.81798	1.2225	1.7384	53
8	.57548	.70368	1.4211	.81782	1.2228	1.7377	52
9	.57572	.70412	1.4202	.81765	1.2230	1.7370	51
10	.57596	.70455	1.4193	.81748	1.2233	1.7362	50
11	.57619	.70499	1.4185	.81731	1.2235	1.7355	49
12	.57643	.70542	1.4176	.81714	1.2238	1.7348	48
13	.57667	.70586	1.4167	.81698	1.2240	1.7341	47
14	.57691	.70629	1.4158	.81681	1.2243	1.7334	46
15	.57715	.70673	1.4150	.81664	1.2245	1.7327	45
16	.57738	.70717	1.4141	.81647	1.2248	1.7320	44
17	.57762	.70760	1.4132	.81631	1.2250	1.7312	43
18	.57786	.70804	1.4124	.81614	1.2253	1.7305	42
19	.57810	.70848	1.4115	.81597	1.2255	1.7298	41
20	.57833	.70891	1.4106	.81580	1.2258	1.7291	40
21	.57857	.70935	1.4097	.81563	1.2260	1.7284	39
22	.57881	.70979	1.4089	.81546	1.2263	1.7277	38
23	.57904	.71023	1.4080	.81530	1.2265	1.7270	37
24	.57928	.71066	1.4071	.81513	1.2268	1.7263	36
25	.57952	.71110	1.4063	.81496	1.2271	1.7256	35
26	.57976	.71154	1.4054	.81479	1.2273	1.7249	34
27	.57999	.71198	1.4045	.81462	1.2276	1.7242	33
28	.58023	.71242	1.4037	.81445	1.2278	1.7235	32
29	.58047	.71285	1.4028	.81428	1.2281	1.7228	31
30	.58070	.71329	1.4019	.81412	1.2283	1.7221	30
31	.58094	.71373	1.4011	.81395	1.2286	1.7213	29
32	.58118	.71417	1.4002	.81378	1.2288	1.7206	28
33	.58141	.71461	1.3994	.81361	1.2291	1.7199	27
34	.58165	.71505	1.3985	.81344	1.2293	1.7192	26
35	.58189	.71549	1.3976	.81327	1.2296	1.7185	25
36	.58212	.71593	1.3968	.81310	1.2299	1.7179	24
37	.58236	.71637	1.3959	.81293	1.2301	1.7172	23
38	.58260	.71681	1.3951	.81276	1.2304	1.7165	22
39	.58283	.71725	1.3942	.81259	1.2306	1.7158	21
40	.58307	.71769	1.3934	.81242	1.2309	1.7151	20
41	.58330	.71813	1.3925	.81225	1.2311	1.7144	19
42	.58354	.71857	1.3916	.81208	1.2314	1.7137	18
43	.58378	.71901	1.3908	.81191	1.2317	1.7130	17
44	.58401	.71946	1.3899	.81174	1.2319	1.7123	16
45	.58425	.71990	1.3891	.81157	1.2322	1.7116	15
46	.58449	.72034	1.3882	.81140	1.2324	1.7109	14
47	.58472	.72078	1.3874	.81123	1.2327	1.7102	13
48	.58496	.72122	1.3865	.81106	1.2329	1.7095	12
49	.58519	.72167	1.3857	.81089	1.2332	1.7088	11
50	.58543	.72211	1.3848	.81072	1.2335	1.7081	10
51	.58567	.72255	1.3840	.81055	1.2337	1.7075	9
52	.58590	.72299	1.3831	.81038	1.2340	1.7068	8
53	.58614	.72344	1.3823	.81021	1.2342	1.7061	7
54	.58637	.72388	1.3814	.81004	1.2345	1.7054	6
55	.58661	.72432	1.3806	.80987	1.2348	1.7047	5
56	.58684	.72477	1.3798	.80970	1.2350	1.7040	4
57	.58708	.72521	1.3789	.80953	1.2353	1.7033	3
58	.58731	.72565	1.3781	.80936	1.2355	1.7027	2
59	.58755	.72610	1.3772	.80919	1.2358	1.7020	1
60	.58779	.72654	1.3764	.80902	1.2361	1.7013	0
′	Cos	Ctn	Tan	Sin	Csc	Sec	′

125° (305°)   **(234°) 54°**

TABLE 1. NATURAL TRIGONOMETRIC FUNCTIONS (continued)   **619**

36° (216°)   (323°) 143°   37° (217°)   (322°) 142°

'	Sin	Tan	Ctn	Cos	Sec	Csc	'	'	Sin	Tan	Ctn	Cos	Sec	Csc	'
0	.58779	.72654	1.3764	.80902	1.2361	1.7013	60	0	.60182	.75355	1.3270	.79864	1.2521	1.6616	60
1	.58802	.72699	1.3755	.80885	1.2363	1.7006	59	1	.60205	.75401	1.3262	.79846	1.2524	1.6610	59
2	.58826	.72743	1.3747	.80867	1.2366	1.6999	58	2	.60228	.75447	1.3254	.79829	1.2527	1.6604	58
3	.58849	.72788	1.3739	.80850	1.2369	1.6993	57	3	.60251	.75492	1.3246	.79811	1.2530	1.6597	57
4	.58873	.72832	1.3730	.80833	1.2371	1.6986	56	4	.60274	.75538	1.3238	.79793	1.2532	1.6591	56
5	.58896	.72877	1.3722	.80816	1.2374	1.6979	55	5	.60298	.75584	1.3230	.79776	1.2535	1.6584	55
6	.58920	.72921	1.3713	.80799	1.2376	1.6972	54	6	.60321	.75629	1.3222	.79758	1.2538	1.6578	54
7	.58943	.72966	1.3705	.80782	1.2379	1.6966	53	7	.60344	.75675	1.3214	.79741	1.2541	1.6572	53
8	.58967	.73010	1.3697	.80765	1.2382	1.6959	52	8	.60367	.75721	1.3206	.79723	1.2543	1.6565	52
9	.58990	.73055	1.3688	.80748	1.2384	1.6952	51	9	.60390	.75767	1.3198	.79706	1.2546	1.6559	51
10	.59014	.73100	1.3680	.80730	1.2387	1.6945	50	10	.60414	.75812	1.3190	.79688	1.2549	1.6553	50
11	.59037	.73144	1.3672	.80713	1.2390	1.6939	49	11	.60437	.75858	1.3182	.79671	1.2552	1.6546	49
12	.59061	.73189	1.3663	.80696	1.2392	1.6932	48	12	.60460	.75904	1.3175	.79653	1.2554	1.6540	48
13	.59084	.73234	1.3655	.80679	1.2395	1.6925	47	13	.60483	.75950	1.3167	.79635	1.2557	1.6534	47
14	.59108	.73278	1.3647	.80662	1.2397	1.6918	46	14	.60506	.75996	1.3159	.79618	1.2560	1.6527	46
15	.59131	.73323	1.3638	.80644	1.2400	1.6912	45	15	.60529	.76042	1.3151	.79600	1.2563	1.6521	45
16	.59154	.73368	1.3630	.80627	1.2403	1.6905	44	16	.60553	.76088	1.3143	.79583	1.2566	1.6515	44
17	.59178	.73413	1.3622	.80610	1.2405	1.6898	43	17	.60576	.76134	1.3135	.79565	1.2568	1.6508	43
18	.59201	.73457	1.3613	.80593	1.2408	1.6892	42	18	.60599	.76180	1.3127	.79547	1.2571	1.6502	42
19	.59225	.73502	1.3605	.80576	1.2411	1.6885	41	19	.60622	.76226	1.3119	.79530	1.2574	1.6496	41
20	.59248	.73547	1.3597	.80558	1.2413	1.6878	40	20	.60645	.76272	1.3111	.79512	1.2577	1.6489	40
21	.59272	.73592	1.3588	.80541	1.2416	1.6871	39	21	.60668	.76318	1.3103	.79494	1.2579	1.6483	39
22	.59295	.73637	1.3580	.80524	1.2419	1.6865	38	22	.60691	.76364	1.3095	.79477	1.2582	1.6477	38
23	.59318	.73681	1.3572	.80507	1.2421	1.6858	37	23	.60714	.76410	1.3087	.79459	1.2585	1.6471	37
24	.59342	.73726	1.3564	.80489	1.2424	1.6852	36	24	.60738	.76456	1.3079	.79441	1.2588	1.6464	36
25	.59365	.73771	1.3555	.80472	1.2427	1.6845	35	25	.60761	.76502	1.3072	.79424	1.2591	1.6458	35
26	.59389	.73816	1.3547	.80455	1.2429	1.6838	34	26	.60784	.76548	1.3064	.79406	1.2593	1.6452	34
27	.59412	.73861	1.3539	.80438	1.2432	1.6832	33	27	.60807	.76594	1.3056	.79388	1.2596	1.6446	33
28	.59436	.73906	1.3531	.80420	1.2435	1.6825	32	28	.60830	.76640	1.3048	.79371	1.2599	1.6439	32
29	.59459	.73951	1.3522	.80403	1.2437	1.6818	31	29	.60853	.76686	1.3040	.79353	1.2602	1.6433	31
30	.59482	.73996	1.3514	.80386	1.2440	1.6812	30	30	.60876	.76733	1.3032	.79335	1.2605	1.6427	30
31	.59506	.74041	1.3506	.80368	1.2443	1.6805	29	31	.60899	.76779	1.3024	.79318	1.2608	1.6421	29
32	.59529	.74086	1.3498	.80351	1.2445	1.6799	28	32	.60922	.76825	1.3017	.79300	1.2610	1.6414	28
33	.59552	.74131	1.3490	.80334	1.2448	1.6792	27	33	.60945	.76871	1.3009	.79282	1.2613	1.6408	27
34	.59576	.74176	1.3481	.80316	1.2451	1.6785	26	34	.60968	.76918	1.3001	.79264	1.2616	1.6402	26
35	.59599	.74221	1.3473	.80299	1.2453	1.6779	25	35	.60991	.76964	1.2993	.79247	1.2619	1.6396	25
36	.59622	.74267	1.3465	.80282	1.2456	1.6772	24	36	.61015	.77010	1.2985	.79229	1.2622	1.6390	24
37	.59646	.74312	1.3457	.80264	1.2459	1.6766	23	37	.61038	.77057	1.2977	.79211	1.2624	1.6383	23
38	.59669	.74357	1.3449	.80247	1.2462	1.6759	22	38	.61061	.77103	1.2970	.79193	1.2627	1.6377	22
39	.59693	.74402	1.3440	.80230	1.2464	1.6753	21	39	.61084	.77149	1.2962	.79176	1.2630	1.6371	21
40	.59716	.74447	1.3432	.80212	1.2467	1.6746	20	40	.61107	.77196	1.2954	.79158	1.2633	1.6365	20
41	.59739	.74492	1.3424	.80195	1.2470	1.6739	19	41	.61130	.77242	1.2946	.79140	1.2636	1.6359	19
42	.59763	.74538	1.3416	.80178	1.2472	1.6733	18	42	.61153	.77289	1.2938	.79122	1.2639	1.6353	18
43	.59786	.74583	1.3408	.80160	1.2475	1.6726	17	43	.61176	.77335	1.2931	.79105	1.2641	1.6346	17
44	.59809	.74628	1.3400	.80143	1.2478	1.6720	16	44	.61199	.77382	1.2923	.79087	1.2644	1.6340	16
45	.59832	.74674	1.3392	.80125	1.2480	1.6713	15	45	.61222	.77428	1.2915	.79069	1.2647	1.6334	15
46	.59856	.74719	1.3384	.80108	1.2483	1.6707	14	46	.61245	.77475	1.2907	.79051	1.2650	1.6328	14
47	.59879	.74764	1.3375	.80091	1.2486	1.6700	13	47	.61268	.77521	1.2900	.79033	1.2653	1.6322	13
48	.59902	.74810	1.3367	.80073	1.2489	1.6694	12	48	.61291	.77568	1.2892	.79016	1.2656	1.6316	12
49	.59926	.74855	1.3359	.80056	1.2491	1.6687	11	49	.61314	.77615	1.2884	.78998	1.2659	1.6310	11
50	.59949	.74900	1.3351	.80038	1.2494	1.6681	10	50	.61337	.77661	1.2876	.78980	1.2661	1.6303	10
51	.59972	.74946	1.3343	.80021	1.2497	1.6674	9	51	.61360	.77708	1.2869	.78962	1.2664	1.6297	9
52	.59995	.74991	1.3335	.80003	1.2499	1.6668	8	52	.61383	.77754	1.2861	.78944	1.2667	1.6291	8
53	.60019	.75037	1.3327	.79986	1.2502	1.6661	7	53	.61406	.77801	1.2853	.78926	1.2670	1.6285	7
54	.60042	.75082	1.3319	.79968	1.2505	1.6655	6	54	.61429	.77848	1.2846	.78908	1.2673	1.6279	6
55	.60065	.75128	1.3311	.79951	1.2508	1.6649	5	55	.61451	.77895	1.2838	.78891	1.2676	1.6273	5
56	.60089	.75173	1.3303	.79934	1.2510	1.6642	4	56	.61474	.77941	1.2830	.78873	1.2679	1.6267	4
57	.60112	.75219	1.3295	.79916	1.2513	1.6636	3	57	.61497	.77988	1.2822	.78855	1.2682	1.6261	3
58	.60135	.75264	1.3287	.79899	1.2516	1.6629	2	58	.61520	.78035	1.2815	.78837	1.2684	1.6255	2
59	.60158	.75310	1.3278	.79881	1.2519	1.6623	1	59	.61543	.78082	1.2807	.78819	1.2687	1.6249	1
60	.60182	.75355	1.3270	.79864	1.2521	1.6616	0	60	.61566	.78129	1.2799	.78801	1.2690	1.6243	0
'	Cos	Ctn	Tan	Sin	Csc	Sec	'	'	Cos	Ctn	Tan	Sin	Csc	Sec	'

126° (306°)   (233°) 53°   127° (307°)   (232°) 52°

TABLE 1. NATURAL TRIGONOMETRIC FUNCTIONS (continued)   **620**

38° (218°)   (321°) **141°**

′	Sin	Tan	Ctn	Cos	Sec	Csc	′
0	.61566	.78129	1.2799	.78801	1.2690	1.6243	60
1	.61589	.78175	1.2792	.78783	1.2693	1.6237	59
2	.61612	.78222	1.2784	.78765	1.2696	1.6231	58
3	.61635	.78269	1.2776	.78747	1.2699	1.6225	57
4	.61658	.78316	1.2769	.78729	1.2702	1.6219	56
5	.61681	.78363	1.2761	.78711	1.2705	1.6213	55
6	.61704	.78410	1.2753	.78694	1.2708	1.6207	54
7	.61726	.78457	1.2746	.78676	1.2710	1.6201	53
8	.61749	.78504	1.2738	.78658	1.2713	1.6195	52
9	.61772	.78551	1.2731	.78640	1.2716	1.6189	51
10	.61795	.78598	1.2723	.78622	1.2719	1.6183	50
11	.61818	.78645	1.2715	.78604	1.2722	1.6177	49
12	.61841	.78692	1.2708	.78586	1.2725	1.6171	48
13	.61864	.78739	1.2700	.78568	1.2728	1.6165	47
14	.61887	.78786	1.2693	.78550	1.2731	1.6159	46
15	.61909	.78834	1.2685	.78532	1.2734	1.6153	45
16	.61932	.78881	1.2677	.78514	1.2737	1.6147	44
17	.61955	.78928	1.2670	.78496	1.2740	1.6141	43
18	.61978	.78975	1.2662	.78478	1.2742	1.6135	42
19	.62001	.79022	1.2655	.78460	1.2745	1.6129	41
20	.62024	.79070	1.2647	.78442	1.2748	1.6123	40
21	.62046	.79117	1.2640	.78424	1.2751	1.6117	39
22	.62069	.79164	1.2632	.78405	1.2754	1.6111	38
23	.62092	.79212	1.2624	.78387	1.2757	1.6105	37
24	.62115	.79259	1.2617	.78369	1.2760	1.6099	36
25	.62138	.79306	1.2609	.78351	1.2763	1.6093	35
26	.62160	.79354	1.2602	.78333	1.2766	1.6087	34
27	.62183	.79401	1.2594	.78315	1.2769	1.6082	33
28	.62206	.79449	1.2587	.78297	1.2772	1.6076	32
29	.62229	.79496	1.2579	.78279	1.2775	1.6070	31
30	.62251	.79544	1.2572	.78261	1.2778	1.6064	30
31	.62274	.79591	1.2564	.78243	1.2781	1.6058	29
32	.62297	.79639	1.2557	.78225	1.2784	1.6052	28
33	.62320	.79686	1.2549	.78206	1.2787	1.6046	27
34	.62342	.79734	1.2542	.78188	1.2790	1.6040	26
35	.62365	.79781	1.2534	.78170	1.2793	1.6035	25
36	.62388	.79829	1.2527	.78152	1.2796	1.6029	24
37	.62411	.79877	1.2519	.78134	1.2799	1.6023	23
38	.62433	.79924	1.2512	.78116	1.2802	1.6017	22
39	.62456	.79972	1.2504	.78098	1.2804	1.6011	21
40	.62479	.80020	1.2497	.78079	1.2807	1.6005	20
41	.62502	.80067	1.2489	.78061	1.2810	1.6000	19
42	.62524	.80115	1.2482	.78043	1.2813	1.5994	18
43	.62547	.80163	1.2475	.78025	1.2816	1.5988	17
44	.62570	.80211	1.2467	.78007	1.2819	1.5982	16
45	.62592	.80258	1.2460	.77988	1.2822	1.5976	15
46	.62615	.80306	1.2452	.77970	1.2825	1.5971	14
47	.62638	.80354	1.2445	.77952	1.2828	1.5965	13
48	.62660	.80402	1.2437	.77934	1.2831	1.5959	12
49	.62683	.80450	1.2430	.77916	1.2834	1.5953	11
50	.62706	.80498	1.2423	.77897	1.2837	1.5948	10
51	.62728	.80546	1.2415	.77879	1.2840	1.5942	9
52	.62751	.80594	1.2408	.77861	1.2843	1.5936	8
53	.62774	.80642	1.2401	.77843	1.2846	1.5930	7
54	.62796	.80690	1.2393	.77824	1.2849	1.5925	6
55	.62819	.80738	1.2386	.77806	1.2852	1.5919	5
56	.62842	.80786	1.2378	.77788	1.2855	1.5913	4
57	.62864	.80834	1.2371	.77769	1.2859	1.5907	3
58	.62887	.80882	1.2364	.77751	1.2862	1.5902	2
59	.62909	.80930	1.2356	.77733	1.2865	1.5896	1
60	.62932	.80978	1.2349	.77715	1.2868	1.5890	0
′	Cos	Ctn	Tan	Sin	Csc	Sec	′

128° (308°)   (231°) **51°**

39° (219°)   (320°) **140°**

′	Sin	Tan	Ctn	Cos	Sec	Csc	′
0	.62932	.80978	1.2349	.77715	1.2868	1.5890	60
1	.62955	.81027	1.2342	.77696	1.2871	1.5884	59
2	.62977	.81075	1.2334	.77678	1.2874	1.5879	58
3	.63000	.81123	1.2327	.77660	1.2877	1.5873	57
4	.63022	.81171	1.2320	.77641	1.2880	1.5867	56
5	.63045	.81220	1.2312	.77623	1.2883	1.5862	55
6	.63068	.81268	1.2305	.77605	1.2886	1.5856	54
7	.63090	.81316	1.2298	.77586	1.2889	1.5850	53
8	.63113	.81364	1.2290	.77568	1.2892	1.5845	52
9	.63135	.81413	1.2283	.77550	1.2895	1.5839	51
10	.63158	.81461	1.2276	.77531	1.2898	1.5833	50
11	.63180	.81510	1.2268	.77513	1.2901	1.5828	49
12	.63203	.81558	1.2261	.77494	1.2904	1.5822	48
13	.63225	.81606	1.2254	.77476	1.2907	1.5816	47
14	.63248	.81655	1.2247	.77458	1.2910	1.5811	46
15	.63271	.81703	1.2239	.77439	1.2913	1.5805	45
16	.63293	.81752	1.2232	.77421	1.2916	1.5800	44
17	.63316	.81800	1.2225	.77402	1.2919	1.5794	43
18	.63338	.81849	1.2218	.77384	1.2923	1.5788	42
19	.63361	.81898	1.2210	.77366	1.2926	1.5783	41
20	.63383	.81946	1.2203	.77347	1.2929	1.5777	40
21	.63406	.81995	1.2196	.77329	1.2932	1.5771	39
22	.63428	.82044	1.2189	.77310	1.2935	1.5766	38
23	.63451	.82092	1.2181	.77292	1.2938	1.5760	37
24	.63473	.82141	1.2174	.77273	1.2941	1.5755	36
25	.63496	.82190	1.2167	.77255	1.2944	1.5749	35
26	.63518	.82238	1.2160	.77236	1.2947	1.5744	34
27	.63540	.82287	1.2153	.77218	1.2950	1.5738	33
28	.63563	.82336	1.2145	.77199	1.2953	1.5732	32
29	.63585	.82385	1.2138	.77181	1.2957	1.5727	31
30	.63608	.82434	1.2131	.77162	1.2960	1.5721	30
31	.63630	.82483	1.2124	.77144	1.2963	1.5716	29
32	.63653	.82531	1.2117	.77125	1.2966	1.5710	28
33	.63675	.82580	1.2109	.77107	1.2969	1.5705	27
34	.63698	.82629	1.2102	.77088	1.2972	1.5699	26
35	.63720	.82678	1.2095	.77070	1.2975	1.5694	25
36	.63742	.82727	1.2088	.77051	1.2978	1.5688	24
37	.63765	.82776	1.2081	.77033	1.2981	1.5683	23
38	.63787	.82825	1.2074	.77014	1.2985	1.5677	22
39	.63810	.82874	1.2066	.76996	1.2988	1.5672	21
40	.63832	.82923	1.2059	.76977	1.2991	1.5666	20
41	.63854	.82972	1.2052	.76959	1.2994	1.5661	19
42	.63877	.83022	1.2045	.76940	1.2997	1.5655	18
43	.63899	.83071	1.2038	.76921	1.3000	1.5650	17
44	.63922	.83120	1.2031	.76903	1.3003	1.5644	16
45	.63944	.83169	1.2024	.76884	1.3007	1.5639	15
46	.63966	.83218	1.2017	.76866	1.3010	1.5633	14
47	.63989	.83268	1.2009	.76847	1.3013	1.5628	13
48	.64011	.83317	1.2002	.76828	1.3016	1.5622	12
49	.64033	.83366	1.1995	.76810	1.3019	1.5617	11
50	.64056	.83415	1.1988	.76791	1.3022	1.5611	10
51	.64078	.83465	1.1981	.76772	1.3026	1.5606	9
52	.64100	.83514	1.1974	.76754	1.3029	1.5601	8
53	.64123	.83564	1.1967	.76735	1.3032	1.5595	7
54	.64145	.83613	1.1960	.76717	1.3035	1.5590	6
55	.64167	.83662	1.1953	.76698	1.3038	1.5584	5
56	.64190	.83712	1.1946	.76679	1.3041	1.5579	4
57	.64212	.83761	1.1939	.76661	1.3045	1.5573	3
58	.64234	.83811	1.1932	.76642	1.3048	1.5568	2
59	.64256	.83860	1.1925	.76623	1.3051	1.5563	1
60	.64279	.83910	1.1918	.76604	1.3054	1.5557	0
′	Cos	Ctn	Tan	Sin	Csc	Sec	′

129° (309°)   (230°) **50°**

TABLE 1. NATURAL TRIGONOMETRIC FUNCTIONS (continued) **621**

40° (220°) (319°) 139°   **41° (221°)** (318°) 138°

′	Sin	Tan	Ctn	Cos	Sec	Csc	′		′	Sin	Tan	Ctn	Cos	Sec	Csc	′
0	.64279	.83910	1.1918	.76604	1.3054	1.5557	60		0	.65606	.86929	1.1504	.75471	1.3250	1.5243	60
1	.64301	.83960	1.1910	.76586	1.3057	1.5552	59		1	.65628	.86980	1.1497	.75452	1.3253	1.5237	59
2	.64323	.84009	1.1903	.76567	1.3060	1.5546	58		2	.65650	.87031	1.1490	.75433	1.3257	1.5232	58
3	.64346	.84059	1.1896	.76548	1.3064	1.5541	57		3	.65672	.87082	1.1483	.75414	1.3260	1.5227	57
4	.64368	.84108	1.1889	.76530	1.3067	1.5536	56		4	.65694	.87133	1.1477	.75395	1.3264	1.5222	56
5	.64390	.84158	1.1882	.76511	1.3070	1.5530	55		5	.65716	.87184	1.1470	.75375	1.3267	1.5217	55
6	.64412	.84208	1.1875	.76492	1.3073	1.5525	54		6	.65738	.87236	1.1463	.75356	1.3270	1.5212	54
7	.64435	.84258	1.1868	.76473	1.3076	1.5520	53		7	.65759	.87287	1.1456	.75337	1.3274	1.5207	53
8	.64457	.84307	1.1861	.76455	1.3080	1.5514	52		8	.65781	.87338	1.1450	.75318	1.3277	1.5202	52
9	.64479	.84357	1.1854	.76436	1.3083	1.5509	51		9	.65803	.87389	1.1443	.75299	1.3280	1.5197	51
10	.64501	.84407	1.1847	.76417	1.3086	1.5504	50		10	.65825	.87441	1.1436	.75280	1.3284	1.5192	50
11	.64524	.84457	1.1840	.76398	1.3089	1.5498	49		11	.65847	.87492	1.1430	.75261	1.3287	1.5187	49
12	.64546	.84507	1.1833	.76380	1.3093	1.5493	48		12	.65869	.87543	1.1423	.75241	1.3291	1.5182	48
13	.64568	.84556	1.1826	.76361	1.3096	1.5488	47		13	.65891	.87595	1.1416	.75222	1.3294	1.5177	47
14	.64590	.84606	1.1819	.76342	1.3099	1.5482	46		14	.65913	.87646	1.1410	.75203	1.3297	1.5172	46
15	.64612	.84656	1.1812	.76323	1.3102	1.5477	45		15	.65935	.87698	1.1403	.75184	1.3301	1.5167	45
16	.64635	.84706	1.1806	.76304	1.3105	1.5472	44		16	.65956	.87749	1.1396	.75165	1.3304	1.5162	44
17	.64657	.84756	1.1799	.76286	1.3109	1.5466	43		17	.65978	.87801	1.1389	.75146	1.3307	1.5156	43
18	.64679	.84806	1.1792	.76267	1.3112	1.5461	42		18	.66000	.87852	1.1383	.75126	1.3311	1.5151	42
19	.64701	.84856	1.1785	.76248	1.3115	1.5456	41		19	.66022	.87904	1.1376	.75107	1.3314	1.5146	41
20	.64723	.84906	1.1778	.76229	1.3118	1.5450	40		20	.66044	.87955	1.1369	.75088	1.3318	1.5141	40
21	.64746	.84956	1.1771	.76210	1.3122	1.5445	39		21	.66066	.88007	1.1363	.75069	1.3321	1.5136	39
22	.64768	.85006	1.1764	.76192	1.3125	1.5440	38		22	.66088	.88059	1.1356	.75050	1.3325	1.5131	38
23	.64790	.85057	1.1757	.76173	1.3128	1.5435	37		23	.66109	.88110	1.1349	.75030	1.3328	1.5126	37
24	.64812	.85107	1.1750	.76154	1.3131	1.5429	36		24	.66131	.88162	1.1343	.75011	1.3331	1.5121	36
25	.64834	.85157	1.1743	.76135	1.3135	1.5424	35		25	.66153	.88214	1.1336	.74992	1.3335	1.5116	35
26	.64856	.85207	1.1736	.76116	1.3138	1.5419	34		26	.66175	.88265	1.1329	.74973	1.3338	1.5111	34
27	.64878	.85257	1.1729	.76097	1.3141	1.5413	33		27	.66197	.88317	1.1323	.74953	1.3342	1.5107	33
28	.64901	.85308	1.1722	.76078	1.3144	1.5408	32		28	.66218	.88369	1.1316	.74934	1.3345	1.5102	32
29	.64923	.85358	1.1715	.76059	1.3148	1.5403	31		29	.66240	.88421	1.1310	.74915	1.3348	1.5097	31
30	.64945	.85408	1.1708	.76041	1.3151	1.5398	30		30	.66262	.88473	1.1303	.74896	1.3352	1.5092	30
31	.64967	.85458	1.1702	.76022	1.3154	1.5392	29		31	.66284	.88524	1.1296	.74876	1.3355	1.5087	29
32	.64989	.85509	1.1695	.76003	1.3157	1.5387	28		32	.66306	.88576	1.1290	.74857	1.3359	1.5082	28
33	.65011	.85559	1.1688	.75984	1.3161	1.5382	27		33	.66327	.88628	1.1283	.74838	1.3362	1.5077	27
34	.65033	.85609	1.1681	.75965	1.3164	1.5377	26		34	.66349	.88680	1.1276	.74818	1.3366	1.5072	26
35	.65055	.85660	1.1674	.75946	1.3167	1.5372	25		35	.66371	.88732	1.1270	.74799	1.3369	1.5067	25
36	.65077	.85710	1.1667	.75927	1.3171	1.5366	24		36	.66393	.88784	1.1263	.74780	1.3373	1.5062	24
37	.65100	.85761	1.1660	.75908	1.3174	1.5361	23		37	.66414	.88836	1.1257	.74760	1.3376	1.5057	23
38	.65122	.85811	1.1653	.75889	1.3177	1.5356	22		38	.66436	.88888	1.1250	.74741	1.3380	1.5052	22
39	.65144	.85862	1.1647	.75870	1.3180	1.5351	21		39	.66458	.88940	1.1243	.74722	1.3383	1.5047	21
40	.65166	.85912	1.1640	.75851	1.3184	1.5345	20		40	.66480	.88992	1.1237	.74703	1.3386	1.5042	20
41	.65188	.85963	1.1633	.75832	1.3187	1.5340	19		41	.66501	.89045	1.1230	.74683	1.3390	1.5037	19
42	.65210	.86014	1.1626	.75813	1.3190	1.5335	18		42	.66523	.89097	1.1224	.74664	1.3393	1.5032	18
43	.65232	.86064	1.1619	.75794	1.3194	1.5330	17		43	.66545	.89149	1.1217	.74644	1.3397	1.5027	17
44	.65254	.86115	1.1612	.75775	1.3197	1.5325	16		44	.66566	.89201	1.1211	.74625	1.3400	1.5023	16
45	.65276	.86166	1.1606	.75756	1.3200	1.5320	15		45	.66588	.89253	1.1204	.74606	1.3404	1.5018	15
46	.65298	.86216	1.1599	.75738	1.3203	1.5314	14		46	.66610	.89306	1.1197	.74586	1.3407	1.5013	14
47	.65320	.86267	1.1592	.75719	1.3207	1.5309	13		47	.66632	.89358	1.1191	.74567	1.3411	1.5008	13
48	.65342	.86318	1.1585	.75700	1.3210	1.5304	12		48	.66653	.89410	1.1184	.74548	1.3414	1.5003	12
49	.65364	.86368	1.1578	.75680	1.3213	1.5299	11		49	.66675	.89463	1.1178	.74528	1.3418	1.4998	11
50	.65386	.86419	1.1571	.75661	1.3217	1.5294	10		50	.66697	.89515	1.1171	.74509	1.3421	1.4993	10
51	.65408	.86470	1.1565	.75642	1.3220	1.5289	9		51	.66718	.89567	1.1165	.74489	1.3425	1.4988	9
52	.65430	.86521	1.1558	.75623	1.3223	1.5283	8		52	.66740	.89620	1.1158	.74470	1.3428	1.4984	8
53	.65452	.86572	1.1551	.75604	1.3227	1.5278	7		53	.66762	.89672	1.1152	.74451	1.3432	1.4979	7
54	.65474	.86623	1.1544	.75585	1.3230	1.5273	6		54	.66783	.89725	1.1145	.74431	1.3435	1.4974	6
55	.65496	.86674	1.1538	.75566	1.3233	1.5268	5		55	.66805	.89777	1.1139	.74412	1.3439	1.4969	5
56	.65518	.86725	1.1531	.75547	1.3237	1.5263	4		56	.66827	.89830	1.1132	.74392	1.3442	1.4964	4
57	.65540	.86776	1.1524	.75528	1.3240	1.5258	3		57	.66848	.89883	1.1126	.74373	1.3446	1.4959	3
58	.65562	.86827	1.1517	.75509	1.3243	1.5253	2		58	.66870	.89935	1.1119	.74353	1.3449	1.4954	2
59	.65584	.86878	1.1510	.75490	1.3247	1.5248	1		59	.66891	.89988	1.1113	.74334	1.3453	1.4950	1
60	.65606	.86929	1.1504	.75471	1.3250	1.5243	0		60	.66913	.90040	1.1106	.74314	1.3456	1.4945	0
′	Cos	Ctn	Tan	Sin	Csc	Sec	′		′	Cos	Ctn	Tan	Sin	Csc	Sec	′

130° (310°) (229°) 49°   **131° (311°)** (228°) 48°

TABLE 1. NATURAL TRIGONOMETRIC FUNCTIONS (continued) **622**

42° (222°)   (317°) **137°**   **43° (223°)**   (316°) **136°**

′	Sin	Tan	Ctn	Cos	Sec	Csc	′	′	Sin	Tan	Ctn	Cos	Sec	Csc	′
0	.66913	.90040	1.1106	.74314	1.3456	1.4945	60	0	.68200	.93252	1.0724	.73135	1.3673	1.4663	60
1	.66935	.90093	1.1100	.74295	1.3460	1.4940	59	1	.68221	.93306	1.0717	.73116	1.3677	1.4658	59
2	.66956	.90146	1.1093	.74276	1.3463	1.4935	58	2	.68242	.93360	1.0711	.73096	1.3681	1.4654	58
3	.66978	.90199	1.1087	.74256	1.3467	1.4930	57	3	.68264	.93415	1.0705	.73076	1.3684	1.4649	57
4	.66999	.90251	1.1080	.74237	1.3470	1.4925	56	4	.68285	.93469	1.0699	.73056	1.3688	1.4645	56
5	.67021	.90304	1.1074	.74217	1.3474	1.4921	55	5	.68306	.93524	1.0692	.73036	1.3692	1.4640	55
6	.67043	.90357	1.1067	.74198	1.3478	1.4916	54	6	.68327	.93578	1.0686	.73016	1.3696	1.4635	54
7	.67064	.90410	1.1061	.74178	1.3481	1.4911	53	7	.68349	.93633	1.0680	.72996	1.3699	1.4631	53
8	.67086	.90463	1.1054	.74159	1.3485	1.4906	52	8	.68370	.93688	1.0674	.72976	1.3703	1.4626	52
9	.67107	.90516	1.1048	.74139	1.3488	1.4901	51	9	.68391	.93742	1.0668	.72957	1.3707	1.4622	51
10	.67129	.90569	1.1041	.74120	1.3492	1.4897	50	10	.68412	.93797	1.0661	.72937	1.3711	1.4617	50
11	.67151	.90621	1.1035	.74100	1.3495	1.4892	49	11	.68434	.93852	1.0655	.72917	1.3714	1.4613	49
12	.67172	.90674	1.1028	.74080	1.3499	1.4887	48	12	.68455	.93906	1.0649	.72897	1.3718	1.4608	48
13	.67194	.90727	1.1022	.74061	1.3502	1.4882	47	13	.68476	.93961	1.0643	.72877	1.3722	1.4604	47
14	.67215	.90781	1.1016	.74041	1.3506	1.4878	46	14	.68497	.94016	1.0637	.72857	1.3726	1.4599	46
15	.67237	.90834	1.1009	.74022	1.3510	1.4873	45	15	.68518	.94071	1.0630	.72837	1.3729	1.4595	45
16	.67258	.90887	1.1003	.74002	1.3513	1.4868	44	16	.68539	.94125	1.0624	.72817	1.3733	1.4590	44
17	.67280	.90940	1.0996	.73983	1.3517	1.4863	43	17	.68561	.94180	1.0618	.72797	1.3737	1.4586	43
18	.67301	.90993	1.0990	.73963	1.3520	1.4859	42	18	.68582	.94235	1.0612	.72777	1.3741	1.4581	42
19	.67323	.91046	1.0983	.73944	1.3524	1.4854	41	19	.68603	.94290	1.0606	.72757	1.3744	1.4577	41
20	.67344	.91099	1.0977	.73924	1.3527	1.4849	40	20	.68624	.94345	1.0599	.72737	1.3748	1.4572	40
21	.67366	.91153	1.0971	.73904	1.3531	1.4844	39	21	.68645	.94400	1.0593	.72717	1.3752	1.4568	39
22	.67387	.91206	1.0964	.73885	1.3535	1.4840	38	22	.68666	.94455	1.0587	.72697	1.3756	1.4563	38
23	.67409	.91259	1.0958	.73865	1.3538	1.4835	37	23	.68688	.94510	1.0581	.72677	1.3759	1.4559	37
24	.67430	.91313	1.0951	.73846	1.3542	1.4830	36	24	.68709	.94565	1.0575	.72657	1.3763	1.4554	36
25	.67452	.91366	1.0945	.73826	1.3545	1.4825	35	25	.68730	.94620	1.0569	.72637	1.3767	1.4550	35
26	.67473	.91419	1.0939	.73806	1.3549	1.4821	34	26	.68751	.94676	1.0562	.72617	1.3771	1.4545	34
27	.67495	.91473	1.0932	.73787	1.3553	1.4816	33	27	.68772	.94731	1.0556	.72597	1.3775	1.4541	33
28	.67516	.91526	1.0926	.73767	1.3556	1.4811	32	28	.68793	.94786	1.0550	.72577	1.3778	1.4536	32
29	.67538	.91580	1.0919	.73747	1.3560	1.4807	31	29	.68814	.94841	1.0544	.72557	1.3782	1.4532	31
30	.67559	.91633	1.0913	.73728	1.3563	1.4802	30	30	.68835	.94896	1.0538	.72537	1.3786	1.4527	30
31	.67580	.91687	1.0907	.73708	1.3567	1.4797	29	31	.68857	.94952	1.0532	.72517	1.3790	1.4523	29
32	.67602	.91740	1.0900	.73688	1.3571	1.4792	28	32	.68878	.95007	1.0526	.72497	1.3794	1.4518	28
33	.67623	.91794	1.0894	.73669	1.3574	1.4788	27	33	.68899	.95062	1.0519	.72477	1.3797	1.4514	27
34	.67645	.91847	1.0888	.73649	1.3578	1.4783	26	34	.68920	.95118	1.0513	.72457	1.3801	1.4510	26
35	.67666	.91901	1.0881	.73629	1.3582	1.4778	25	35	.68941	.95173	1.0507	.72437	1.3805	1.4505	25
36	.67688	.91955	1.0875	.73610	1.3585	1.4774	24	36	.68962	.95229	1.0501	.72417	1.3809	1.4501	24
37	.67709	.92008	1.0869	.73590	1.3589	1.4769	23	37	.68983	.95284	1.0495	.72397	1.3813	1.4496	23
38	.67730	.92062	1.0862	.73570	1.3592	1.4764	22	38	.69004	.95340	1.0489	.72377	1.3817	1.4492	22
39	.67752	.92116	1.0856	.73551	1.3596	1.4760	21	39	.69025	.95395	1.0483	.72357	1.3820	1.4487	21
40	.67773	.92170	1.0850	.73531	1.3600	1.4755	20	40	.69046	.95451	1.0477	.72337	1.3824	1.4483	20
41	.67795	.92224	1.0843	.73511	1.3603	1.4750	19	41	.69067	.95506	1.0470	.72317	1.3828	1.4479	19
42	.67816	.92277	1.0837	.73491	1.3607	1.4746	18	42	.69088	.95562	1.0464	.72297	1.3832	1.4474	18
43	.67837	.92331	1.0831	.73472	1.3611	1.4741	17	43	.69109	.95618	1.0458	.72277	1.3836	1.4470	17
44	.67859	.92385	1.0824	.73452	1.3614	1.4737	16	44	.69130	.95673	1.0452	.72257	1.3840	1.4465	16
45	.67880	.92439	1.0818	.73432	1.3618	1.4732	15	45	.69151	.95729	1.0446	.72236	1.3843	1.4461	15
46	.67901	.92493	1.0812	.73413	1.3622	1.4727	14	46	.69172	.95785	1.0440	.72216	1.3847	1.4457	14
47	.67923	.92547	1.0805	.73393	1.3625	1.4723	13	47	.69193	.95841	1.0434	.72196	1.3851	1.4452	13
48	.67944	.92601	1.0799	.73373	1.3629	1.4718	12	48	.69214	.95897	1.0428	.72176	1.3855	1.4448	12
49	.67965	.92655	1.0793	.73353	1.3633	1.4713	11	49	.69235	.95952	1.0422	.72156	1.3859	1.4443	11
50	.67987	.92709	1.0786	.73333	1.3636	1.4709	10	50	.69256	.96008	1.0416	.72136	1.3863	1.4439	10
51	.68008	.92763	1.0780	.73314	1.3640	1.4704	9	51	.69277	.96064	1.0410	.72116	1.3867	1.4435	9
52	.68029	.92817	1.0774	.73294	1.3644	1.4700	8	52	.69298	.96120	1.0404	.72095	1.3871	1.4430	8
53	.68051	.92872	1.0768	.73274	1.3647	1.4695	7	53	.69319	.96176	1.0398	.72075	1.3874	1.4426	7
54	.68072	.92926	1.0761	.73254	1.3651	1.4690	6	54	.69340	.96232	1.0392	.72055	1.3878	1.4422	6
55	.68093	.92980	1.0755	.73234	1.3655	1.4686	5	55	.69361	.96288	1.0385	.72035	1.3882	1.4417	5
56	.68115	.93034	1.0749	.73215	1.3658	1.4681	4	56	.69382	.96344	1.0379	.72015	1.3886	1.4413	4
57	.68136	.93088	1.0742	.73195	1.3662	1.4677	3	57	.69403	.96400	1.0373	.71995	1.3890	1.4409	3
58	.68157	.93143	1.0736	.73175	1.3666	1.4672	2	58	.69424	.96457	1.0367	.71974	1.3894	1.4404	2
59	.68179	.93197	1.0730	.73155	1.3670	1.4667	1	59	.69445	.96513	1.0361	.71954	1.3898	1.4400	1
60	.68200	.93252	1.0724	.73135	1.3673	1.4663	0	60	.69466	.96569	1.0355	.71934	1.3902	1.4396	0
′	Cos	Ctn	Tan	Sin	Csc	Sec	′	′	Cos	Ctn	Tan	Sin	Csc	Sec	′

132° (312°)   (227°) **47°**   **133° (313°)**   (226°) **46°**

TABLE 1. NATURAL TRIGONOMETRIC FUNCTIONS (continued) **623**

44° (224°)   **(315°) 135°**

′	Sin	Tan	Ctn	Cos	Sec	Csc	′
0	.69466	.96569	1.0355	.71934	1.3902	1.4396	60
1	.69487	.96625	1.0349	.71914	1.3906	1.4391	59
2	.69508	.96681	1.0343	.71894	1.3909	1.4387	58
3	.69529	.96738	1.0337	.71873	1.3913	1.4383	57
4	.69549	.96794	1.0331	.71853	1.3917	1.4378	56
5	.69570	.96850	1.0325	.71833	1.3921	1.4374	55
6	.69591	.96907	1.0319	.71813	1.3925	1.4370	54
7	.69612	.96963	1.0313	.71792	1.3929	1.4365	53
8	.69633	.97020	1.0307	.71772	1.3933	1.4361	52
9	.69654	.97076	1.0301	.71752	1.3937	1.4357	51
10	.69675	.97133	1.0295	.71732	1.3941	1.4352	50
11	.69696	.97189	1.0289	.71711	1.3945	1.4348	49
12	.69717	.97246	1.0283	.71691	1.3949	1.4344	48
13	.69737	.97302	1.0277	.71671	1.3953	1.4340	47
14	.69758	.97359	1.0271	.71650	1.3957	1.4335	46
15	.69779	.97416	1.0265	.71630	1.3961	1.4331	45
16	.69800	.97472	1.0259	.71610	1.3965	1.4327	44
17	.69821	.97529	1.0253	.71590	1.3969	1.4322	43
18	.69842	.97586	1.0247	.71569	1.3972	1.4318	42
19	.69862	.97643	1.0241	.71549	1.3976	1.4314	41
20	.69883	.97700	1.0235	.71529	1.3980	1.4310	40
21	.69904	.97756	1.0230	.71508	1.3984	1.4305	39
22	.69925	.97813	1.0224	.71488	1.3988	1.4301	38
23	.69946	.97870	1.0218	.71468	1.3992	1.4297	37
24	.69966	.97927	1.0212	.71447	1.3996	1.4293	36
25	.69987	.97984	1.0206	.71427	1.4000	1.4288	35
26	.70008	.98041	1.0200	.71407	1.4004	1.4284	34
27	.70029	.98098	1.0194	.71386	1.4008	1.4280	33
28	.70049	.98155	1.0188	.71366	1.4012	1.4276	32
29	.70070	.98213	1.0182	.71345	1.4016	1.4271	31
30	.70091	.98270	1.0176	.71325	1.4020	1.4267	30
31	.70112	.98327	1.0170	.71305	1.4024	1.4263	29
32	.70132	.98384	1.0164	.71284	1.4028	1.4259	28
33	.70153	.98441	1.0158	.71264	1.4032	1.4255	27
34	.70174	.98499	1.0152	.71243	1.4036	1.4250	26
35	.70195	.98556	1.0147	.71223	1.4040	1.4246	25
36	.70215	.98613	1.0141	.71203	1.4044	1.4242	24
37	.70236	.98671	1.0135	.71182	1.4048	1.4238	23
38	.70257	.98728	1.0129	.71162	1.4052	1.4234	22
39	.70277	.98786	1.0123	.71141	1.4057	1.4229	21
40	.70298	.98843	1.0117	.71121	1.4061	1.4225	20
41	.70319	.98901	1.0111	.71100	1.4065	1.4221	19
42	.70339	.98958	1.0105	.71080	1.4069	1.4217	18
43	.70360	.99016	1.0099	.71059	1.4073	1.4213	17
44	.70381	.99073	1.0094	.71039	1.4077	1.4208	16
45	.70401	.99131	1.0088	.71019	1.4081	1.4204	15
46	.70422	.99189	1.0082	.70998	1.4085	1.4200	14
47	.70443	.99247	1.0076	.70978	1.4089	1.4196	13
48	.70463	.99304	1.0070	.70957	1.4093	1.4192	12
49	.70484	.99362	1.0064	.70937	1.4097	1.4188	11
50	.70505	.99420	1.0058	.70916	1.4101	1.4183	10
51	.70525	.99478	1.0052	.70896	1.4105	1.4179	9
52	.70546	.99536	1.0047	.70875	1.4109	1.4175	8
53	.70567	.99594	1.0041	.70855	1.4113	1.4171	7
54	.70587	.99652	1.0035	.70834	1.4118	1.4167	6
55	.70608	.99710	1.0029	.70813	1.4122	1.4163	5
56	.70628	.99768	1.0023	.70793	1.4126	1.4159	4
57	.70649	.99826	1.0017	.70772	1.4130	1.4154	3
58	.70670	.99884	1.0012	.70752	1.4134	1.4150	2
59	.70690	.99942	1.0006	.70731	1.4138	1.4146	1
60	.70711	1.0000	1.0000	.70711	1.4142	1.4142	0
′	Cos	Ctn	Tan	Sin	Csc	Sec	′

134° (314°)   **(225°) 45°**

100-150

N.	0	1	2	3	4	5	6	7	8	9
100	00 000	043	087	130	·173	217	260	303	346	389
101	432	475	518	561	604	647	689	732	775	817
102	860	903	945	988	*030	*072	*115	*157	*199	*242
103	01 284	326	368	410	452	494	536	578	620	662
104	703	745	787	828	870	912	953	995	*036	*078
105	02 119	160	202	243	284	325	366	407	449	490
106	531	572	612	653	694	735	776	816	857	898
107	938	979	*019	*060	*100	*141	*181	*222	*262	*302
108	03 342	383	423	463	503	543	583	623	663	703
109	743	782	822	862	902	941	981	*021	*060	*100
110	04 139	179	218	258	297	336	376	415	454	493
111	532	571	610	650	689	727	766	805	844	883
112	922	961	999	*038	*077	*115	*154	*192	*231	*269
113	05 308	346	385	423	461	500	538	576	614	652
114	690	729	767	805	843	881	918	956	994	*032
115	06 070	108	145	183	221	258	296	333	371	408
116	446	483	521	558	595	633	670	707	744	781
117	819	856	893	930	967	*004	*041	*078	*115	*151
118	07 188	225	262	298	335	372	408	445	482	518
119	555	591	628	664	700	737	773	809	846	882
120	918	954	990	*027	*063	*099	*135	*171	*207	*243
121	08 279	314	350	386	422	458	493	529	565	600
122	636	672	707	743	778	814	849	884	920	955
123	991	*026	*061	*096	*132	*167	*202	*237	*272	*307
124	09 342	377	412	447	482	517	552	587	621	656
125	691	726	760	795	830	864	899	934	968	*003
126	10 037	072	106	140	175	209	243	278	312	346
127	380	415	449	483	517	551	585	619	653	687
128	721	755	789	823	857	890	924	958	992	*025
129	11 059	093	126	160	193	227	261	294	327	361
130	394	428	461	494	528	561	594	628	661	694
131	727	760	793	826	860	893	926	959	992	*024
132	12 057	090	123	156	189	222	254	287	320	352
133	385	418	450	483	516	548	581	613	646	678
134	710	743	775	808	840	872	905	937	969	*001
135	13 033	066	098	130	162	194	226	258	290	322
136	354	386	418	450	481	513	545	577	609	640
137	672	704	735	767	799	830	862	893	925	956
138	988	*019	*051	*082	*114	*145	*176	*208	*239	*270
139	14 301	333	364	395	426	457	489	520	551	582
140	613	644	675	706	737	768	799	829	860	891
141	922	953	983	*014	*045	*076	*106	*137	*168	*198
142	15 229	259	290	320	351	381	412	442	473	503
143	534	564	594	625	655	685	715	746	776	806
144	836	866	897	927	957	987	*017	*047	*077	*107
145	16 137	167	197	227	256	286	316	346	376	406
146	435	465	495	524	554	584	613	643	673	702
147	732	761	791	820	850	879	909	938	967	997
148	17 026	056	085	114	143	173	202	231	260	289
149	319	348	377	406	435	464	493	522	551	580
150	609	638	667	696	725	754	782	811	840	869
N.	0	1	2	3	4	5	6	7	8	9

Prop. Pts.

	44	43	42
1	4.4	4.3	4.2
2	8.8	8.6	8.4
3	13.2	12.9	12.6
4	17.6	17.2	16.8
5	22.0	21.5	21.0
6	26.4	25.8	25.2
7	30.8	30.1	29.4
8	35.2	34.4	33.6
9	39.6	38.7	37.8

	41	40	39
1	4.1	4.0	3.9
2	8.2	8.0	7.8
3	12.3	12.0	11.7
4	16.4	16.0	15.6
5	20.5	20.0	19.5
6	24.6	24.0	23.4
7	28.7	28.0	27.3
8	32.8	32.0	31.2
9	36.9	36.0	35.1

	38	37	36
1	3.8	3.7	3.6
2	7.6	7.4	7.2
3	11.4	11.1	10.8
4	15.2	14.8	14.4
5	19.0	18.5	18.0
6	22.8	22.2	21.6
7	26.6	25.9	25.2
8	30.4	29.6	28.8
9	34.2	33.3	32.4

	35	34	33
1	3.5	3.4	3.3
2	7.0	6.8	6.6
3	10.5	10.2	9.9
4	14.0	13.6	13.2
5	17.5	17.0	16.5
6	21.0	20.4	19.8
7	24.5	23.8	23.1
8	28.0	27.2	26.4
9	31.5	30.6	29.7

	32	31	30
1	3.2	3.1	3.0
2	6.4	6.2	6.0
3	9.6	9.3	9.0
4	12.8	12.4	12.0
5	16.0	15.5	15.0
6	19.2	18.6	18.0
7	22.4	21.7	21.0
8	25.6	24.8	24.0
9	28.8	27.9	27.0

*From E. Richard Heineman, "Plane Trigonometry with Tables," 3d ed., McGraw-Hill Book Company, New York, 1964.

TABLE 2. MANTISSAS OF COMMON LOGARITHMS (continued)   625

150–200

N.	0	1	2	3	4	5	6	7	8	9
150	17 609	638	667	696	725	754	782	811	840	869
151	898	926	955	984	*013	*041	*070	*099	*127	*156
152	18 184	213	241	270	298	327	355	384	412	441
153	469	498	526	554	583	611	639	667	696	724
154	752	780	808	837	865	893	921	949	977	*005
155	19 033	061	089	117	145	173	201	229	257	285
156	312	340	368	396	424	451	479	507	535	562
157	590	618	645	673	700	728	756	783	811	838
158	866	893	921	948	976	*003	*030	*058	*085	*112
159	20 140	167	194	222	249	276	303	330	358	385
160	412	439	466	493	520	548	575	602	629	656
161	683	710	737	763	790	817	844	871	898	925
162	952	978	*005	*032	*059	*085	*112	*139	*165	*192
163	21 219	245	272	299	325	352	378	405	431	458
164	484	511	537	564	590	617	643	669	696	722
165	748	775	801	827	854	880	906	932	958	985
166	22 011	037	063	089	115	141	167	194	220	246
167	272	298	324	350	376	401	427	453	479	505
168	531	557	583	608	634	660	686	712	737	763
169	789	814	840	866	891	917	943	968	994	*019
170	23 045	070	096	121	147	172	198	223	249	274
171	300	325	350	376	401	426	452	477	502	528
172	553	578	603	629	654	679	704	729	754	779
173	805	830	855	880	905	930	955	980	*005	*030
174	24 055	080	105	130	155	180	204	229	254	279
175	304	329	353	378	403	428	452	477	502	527
176	551	576	601	625	650	674	699	724	748	773
177	797	822	846	871	895	920	944	969	993	*018
178	25 042	066	091	115	139	164	188	212	237	261
179	285	310	334	358	382	406	431	455	479	503
180	527	551	575	600	624	648	672	696	720	744
181	768	792	816	840	864	888	912	935	959	983
182	26 007	031	055	079	102	126	150	174	198	221
183	245	269	293	316	340	364	387	411	435	458
184	482	505	529	553	576	600	623	647	670	694
185	717	741	764	788	811	834	858	881	905	928
186	951	975	998	*021	*045	*068	*091	*114	*138	*161
187	27 184	207	231	254	277	300	323	346	370	393
188	416	439	462	485	508	531	554	577	600	623
189	646	669	692	715	738	761	784	807	830	852
190	875	898	921	944	967	989	*012	*035	*058	*081
191	28 103	126	149	171	194	217	240	262	285	307
192	330	353	375	398	421	443	466	488	511	533
193	556	578	601	623	646	668	691	713	735	758
194	780	803	825	847	870	892	914	937	959	981
195	29 003	026	048	070	092	115	137	159	181	203
196	226	248	270	292	314	336	358	380	403	425
197	447	469	491	513	535	557	579	601	623	645
198	667	688	710	732	754	776	798	820	842	863
199	885	907	929	951	973	994	*016	*038	*060	*081
200	30 103	125	146	168	190	211	233	255	276	298
N.	0	1	2	3	4	5	6	7	8	9

Prop. Pts.

	29	28
1	2.9	2.8
2	5.8	5.6
3	8.7	8.4
4	11.6	11.2
5	14.5	14.0
6	17.4	16.8
7	20.3	19.6
8	23.2	22.4
9	26.1	25.2

	27	26
1	2.7	2.6
2	5.4	5.2
3	8.1	7.8
4	10.8	10.4
5	13.5	13.0
6	16.2	15.6
7	18.9	18.2
8	21.6	20.8
9	24.3	23.4

	25
1	2.5
2	5.0
3	7.5
4	10.0
5	12.5
6	15.0
7	17.5
8	20.0
9	22.5

	24	23
1	2.4	2.3
2	4.8	4.6
3	7.2	6.9
4	9.6	9.2
5	12.0	11.5
6	14.4	13.8
7	16.8	16.1
8	19.2	18.4
9	21.6	20.7

	22	21
1	2.2	2.1
2	4.4	4.2
3	6.6	6.3
4	8.8	8.4
5	11.0	10.5
6	13.2	12.6
7	15.4	14.7
8	17.6	16.8
9	19.8	18.9

TABLE 2. MANTISSAS OF COMMON LOGARITHMS (continued)   626

200-250

N.	0	1	2	3	4	5	6	7	8	9
200	30 103	125	146	168	190	211	233	255	276	298
201	320	341	363	384	406	428	449	471	492	514
202	535	557	578	600	621	643	664	685	707	728
203	750	771	792	814	835	856	878	899	920	942
204	963	984	*006	*027	*048	*069	*091	*112	*133	*154
205	31 175	197	218	239	260	281	302	323	345	366
206	387	408	429	450	471	492	513	534	555	576
207	597	618	639	660	681	702	723	744	765	785
208	806	827	848	869	890	911	931	952	973	994
209	32 015	035	056	077	098	118	139	160	181	201
210	222	243	263	284	305	325	346	366	387	408
211	428	449	469	490	510	531	552	572	593	613
212	634	654	675	695	715	736	756	777	797	818
213	838	858	879	899	919	940	960	980	*001	*021
214	33 041	062	082	102	122	143	163	183	203	224
215	244	264	284	304	325	345	365	385	405	425
216	445	465	486	506	526	546	566	586	606	626
217	646	666	686	706	726	746	766	786	806	826
218	846	866	885	905	925	945	965	985	*005	*025
219	34 044	064	084	104	124	143	163	183	203	223
220	242	262	282	301	321	341	361	380	400	420
221	439	459	479	498	518	537	557	577	596	616
222	635	655	674	694	713	733	753	772	792	811
223	830	850	869	889	908	928	947	967	986	*005
224	35 025	044	064	083	102	122	141	160	180	199
225	218	238	257	276	295	315	334	353	372	392
226	411	430	449	468	488	507	526	545	564	583
227	603	622	641	660	679	698	717	736	755	774
228	793	813	832	851	870	889	908	927	946	965
229	984	*003	*021	*040	*059	*078	*097	*116	*135	*154
230	36 173	192	211	229	248	267	286	305	324	342
231	361	380	399	418	436	455	474	493	511	530
232	549	568	586	605	624	642	661	680	698	717
233	736	754	773	791	810	829	847	866	884	903
234	922	940	959	977	996	*014	*033	*051	*070	*088
235	37 107	125	144	162	181	199	218	236	254	273
236	291	310	328	346	365	383	401	420	438	457
237	475	493	511	530	548	566	585	603	621	639
238	658	676	694	712	731	749	767	785	803	822
239	840	858	876	894	912	931	949	967	985	*003
240	38 021	039	057	075	093	112	130	148	166	184
241	202	220	238	256	274	292	310	328	346	364
242	382	399	417	435	453	471	489	507	525	543
243	561	578	596	614	632	650	668	686	703	721
244	739	757	775	792	810	828	846	863	881	899
245	917	934	952	970	987	*005	*023	*041	*058	*076
246	39 094	111	129	146	164	182	199	217	235	252
247	270	287	305	322	340	358	375	393	410	428
248	445	463	480	498	515	533	550	568	585	602
249	620	637	655	672	690	707	724	742	759	777
250	794	811	829	846	863	881	898	915	933	950
N.	0	1	2	3	4	5	6	7	8	9

Prop. Pts.

	22	21
1	2.2	2.1
2	4.4	4.2
3	6.6	6.3
4	8.8	8.4
5	11.0	10.5
6	13.2	12.6
7	15.4	14.7
8	17.6	16.8
9	19.8	18.9

	20
1	2.0
2	4.0
3	6.0
4	8.0
5	10.0
6	12.0
7	14.0
8	16.0
9	18.0

	19
1	1.9
2	3.8
3	5.7
4	7.6
5	9.5
6	11.4
7	13.3
8	15.2
9	17.1

	18
1	1.8
2	3.6
3	5.4
4	7.2
5	9.0
6	10.8
7	12.6
8	14.4
9	16.2

	17
1	1.7
2	3.4
3	5.1
4	6.8
5	8.5
6	10.2
7	11.9
8	13.6
9	15.3

TABLE 2. MANTISSAS OF COMMON LOGARITHMS (continued)   627

250-300

N.	0	1	2	3	4	5	6	7	8	9	Prop. Pts.
250	39 794	811	829	846	863	881	898	915	933	950	
251	967	985	*002	*019	*037	*054	*071	*088	*106	*123	**18**
252	40 140	157	175	192	209	226	243	261	278	295	1 1.8
253	312	329	346	364	381	398	415	432	449	466	2 3.6
											3 5.4
254	483	500	518	535	552	569	586	603	620	637	4 7.2
255	654	671	688	705	722	739	756	773	790	807	5 9.0
256	824	841	858	875	892	909	926	943	960	976	6 10.8
											7 12.6
257	993	*010	*027	*044	*061	*078	*095	*111	*128	*145	8 14.4
258	41 162	179	196	212	229	246	263	280	296	313	9 16.2
259	330	347	363	380	397	414	430	447	464	481	
260	497	514	531	547	564	581	597	614	631	647	
261	664	681	697	714	731	747	764	780	797	814	**17**
262	830	847	863	880	896	913	929	946	963	979	1 1.7
263	996	*012	*029	*045	*062	*078	*095	*111	*127	*144	2 3.4
											3 5.1
264	42 160	177	193	210	226	243	259	275	292	308	4 6.8
265	325	341	357	374	390	406	423	439	455	472	5 8.5
266	488	504	521	537	553	570	586	602	619	635	6 10.2
											7 11.9
267	651	667	684	700	716	732	749	765	781	797	8 13.6
268	813	830	846	862	878	894	911	927	943	959	9 15.3
269	975	991	*008	*024	*040	*056	*072	*088	*104	*120	
270	43 136	152	169	185	201	217	233	249	265	281	log e = 0.43429
271	297	313	329	345	361	377	393	409	425	441	
272	457	473	489	505	521	537	553	569	584	600	**16**
273	616	632	648	664	680	696	712	727	743	759	1 1.6
											2 3.2
274	775	791	807	823	838	854	870	886	902	917	3 4.8
275	933	949	965	981	996	*012	*028	*044	*059	*075	4 6.4
276	44 091	107	122	138	154	170	185	201	217	232	5 8.0
											6 9.6
277	248	264	279	295	311	326	342	358	373	389	7 11.2
278	404	420	436	451	467	483	498	514	529	545	8 12.8
279	560	576	592	607	623	638	654	669	685	700	9 14.4
280	716	731	747	762	778	793	809	824	840	855	
281	871	886	902	917	932	948	963	979	994	*010	**15**
282	45 025	040	056	071	086	102	117	133	148	163	1 1.5
283	179	194	209	225	240	255	271	286	301	317	2 3.0
											3 4.5
284	332	347	362	378	393	408	423	439	454	469	4 6.0
285	484	500	515	530	545	561	576	591	606	621	5 7.5
286	637	652	667	682	697	712	728	743	758	773	6 9.0
											7 10.5
287	788	803	818	834	849	864	879	894	909	924	8 12.0
288	939	954	969	984	*000	*015	*030	*045	*060	*075	9 13.5
289	46 090	105	120	135	150	165	180	195	210	225	
290	240	255	270	285	300	315	330	345	359	374	
291	389	404	419	434	449	464	479	494	509	523	**14**
292	538	553	568	583	598	613	627	642	657	672	1 1.4
293	687	702	716	731	746	761	776	790	805	820	2 2.8
											3 4.2
294	835	850	864	879	894	909	923	938	953	967	4 5.6
295	982	997	*012	*026	*041	*056	*070	*085	*100	*114	5 7.0
296	47 129	144	159	173	188	202	217	232	246	261	6 8.4
											7 9.8
297	276	290	305	319	334	349	363	378	392	407	8 11.2
298	422	436	451	465	480	494	509	524	538	553	9 12.6
299	567	582	596	611	625	640	654	669	683	698	
300	712	727	741	756	770	784	799	813	828	842	
N.	0	1	2	3	4	5	6	7	8	9	Prop. Pts.

TABLE 2. MANTISSAS OF COMMON LOGARITHMS (continued) 628

300-350

N.	0	1	2	3	4	5	6	7	8	9
300	47 712	727	741	756	770	784	799	813	828	842
301	857	871	885	900	914	929	943	958	972	986
302	48 001	015	029	044	058	073	087	101	116	130
303	144	159	173	187	202	216	230	244	259	273
304	287	302	316	330	344	359	373	387	401	416
305	430	444	458	473	487	501	515	530	544	558
306	572	586	601	615	629	643	657	671	686	700
307	714	728	742	756	770	785	799	813	827	841
308	855	869	883	897	911	926	940	954	968	982
309	996	*010	*024	*038	*052	*066	*080	*094	*108	*122
310	49 136	150	164	178	192	206	220	234	248	262
311	276	290	304	318	332	346	360	374	388	402
312	415	429	443	457	471	485	499	513	527	541
313	554	568	582	596	610	624	638	651	665	679
314	693	707	721	734	748	762	776	790	803	817
315	831	845	859	872	886	900	914	927	941	955
316	969	982	996	*010	*024	*037	*051	*065	*079	*092
317	50 106	120	133	147	161	174	188	202	215	229
318	243	256	270	284	297	311	325	338	352	365
319	379	393	406	420	433	447	461	474	488	501
320	515	529	542	556	569	583	596	610	623	637
321	651	664	678	691	705	718	732	745	759	772
322	786	799	813	826	840	853	866	880	893	907
323	920	934	947	961	974	987	*001	*014	*028	*041
324	51 055	068	081	095	108	121	135	148	162	175
325	188	202	215	228	242	255	268	282	295	308
326	322	335	348	362	375	388	402	415	428	441
327	455	468	481	495	508	521	534	548	561	574
328	587	601	614	627	640	654	667	680	693	706
329	720	733	746	759	772	786	799	812	825	838
330	851	865	878	891	904	917	930	943	957	970
331	983	996	*009	*022	*035	*048	*061	*075	*088	*101
332	52 114	127	140	153	166	179	192	205	218	231
333	244	257	270	284	297	310	323	336	349	362
334	375	388	401	414	427	440	453	466	479	492
335	504	517	530	543	556	569	582	595	608	621
336	634	647	660	673	686	699	711	724	737	750
337	763	776	789	802	815	827	840	853	866	879
338	892	905	917	930	943	956	969	982	994	*007
339	53 020	033	046	058	071	084	097	110	122	135
340	148	161	173	186	199	212	224	237	250	263
341	275	288	301	314	326	339	352	364	377	390
342	403	415	428	441	453	466	479	491	504	517
343	529	542	555	567	580	593	605	618	631	643
344	656	668	681	694	706	719	732	744	757	769
345	782	794	807	820	832	845	857	870	882	895
346	908	920	933	945	958	970	983	995	*008	*020
347	54 033	045	058	070	083	095	108	120	133	145
348	158	170	183	195	208	220	233	245	258	270
349	283	295	307	320	332	345	357	370	382	394
350	407	419	432	444	456	469	481	494	506	518
N.	0	1	2	3	4	5	6	7	8	9

Prop. Pts.

15		14		13		12	
1	1.5	1	1.4	1	1.3	1	1.2
2	3.0	2	2.8	2	2.6	2	2.4
3	4.5	3	4.2	3	3.9	3	3.6
4	6.0	4	5.6	4	5.2	4	4.8
5	7.5	5	7.0	5	6.5	5	6.0
6	9.0	6	8.4	6	7.8	6	7.2
7	10.5	7	9.8	7	9.1	7	8.4
8	12.0	8	11.2	8	10.4	8	9.6
9	13.5	9	12.6	9	11.7	9	10.8

$\log \pi = 0.49715$

TABLE 2. MANTISSAS OF COMMON LOGARITHMS (continued)   629

350-400

N.	0	1	2	3	4	5	6	7	8	9
350	54 407	419	432	444	456	469	481	494	506	518
351	531	543	555	568	580	593	605	617	630	642
352	654	667	679	691	704	716	728	741	753	765
353	777	790	802	814	827	839	851	864	876	888
354	900	913	925	937	949	962	974	986	998	*011
355	55 023	035	047	060	072	084	096	108	121	133
356	145	157	169	182	194	206	218	230	242	255
357	267	279	291	303	315	328	340	352	364	376
358	388	400	413	425	437	449	461	473	485	497
359	509	522	534	546	558	570	582	594	606	618
360	630	642	654	666	678	691	703	715	727	739
361	751	763	775	787	799	811	823	835	847	859
362	871	883	895	907	919	931	943	955	967	979
363	991	*003	*015	*027	*038	*050	*062	*074	*086	*098
364	56 110	122	134	146	158	170	182	194	205	217
365	229	241	253	265	277	289	301	312	324	336
366	348	360	372	384	396	407	419	431	443	455
367	467	478	490	502	514	526	538	549	561	573
368	585	597	608	620	632	644	656	667	679	691
369	703	714	726	738	750	761	773	785	797	808
370	820	832	844	855	867	879	891	902	914	926
371	937	949	961	972	984	996	*008	*019	*031	*043
372	57 054	066	078	089	101	113	124	136	148	159
373	171	183	194	206	217	229	241	252	264	276
374	287	299	310	322	334	345	357	368	380	392
375	403	415	426	438	449	461	473	484	496	507
376	519	530	542	553	565	576	588	600	611	623
377	634	646	657	669	680	692	703	715	726	738
378	749	761	772	784	795	807	818	830	841	852
379	864	875	887	898	910	921	933	944	955	967
380	978	990	*001	*013	*024	*035	*047	*058	*070	*081
381	58 092	104	115	127	138	149	161	172	184	195
382	206	218	229	240	252	263	274	286	297	309
383	320	331	343	354	365	377	388	399	410	422
384	433	444	456	467	478	490	501	512	524	535
385	546	557	569	580	591	602	614	625	636	647
386	659	670	681	692	704	715	726	737	749	760
387	771	782	794	805	816	827	838	850	861	872
388	883	894	906	917	928	939	950	961	973	984
389	995	*006	*017	*028	*040	*051	*062	*073	*084	*095
390	59 106	118	129	140	151	162	173	184	195	207
391	218	229	240	251	262	273	284	295	306	318
392	329	340	351	362	373	384	395	406	417	428
393	439	450	461	472	483	494	506	517	528	539
394	550	561	572	583	594	605	616	627	638	649
395	660	671	682	693	704	715	726	737	748	759
396	770	780	791	802	813	824	835	846	857	868
397	879	890	901	912	923	934	945	956	966	977
398	988	999	*010	*021	*032	*043	*054	*065	*076	*086
399	60 097	108	119	130	141	152	163	173	184	195
400	206	217	228	239	249	260	271	282	293	304
N.	0	1	2	3	4	5	6	7	8	9

Prop. Pts.

13		12		11		10	
1	1.3	1	1.2	1	1.1	1	1.0
2	2.6	2	2.4	2	2.2	2	2.0
3	3.9	3	3.6	3	3.3	3	3.0
4	5.2	4	4.8	4	4.4	4	4.0
5	6.5	5	6.0	5	5.5	5	5.0
6	7.8	6	7.2	6	6.6	6	6.0
7	9.1	7	8.4	7	7.7	7	7.0
8	10.4	8	9.6	8	8.8	8	8.0
9	11.7	9	10.8	9	9.9	9	9.0

TABLE 2. MANTISSAS OF COMMON LOGARITHMS (continued)   630

400–450

N.	0	1	2	3	4	5	6	7	8	9	Prop. Pts.
400	60 206	217	228	239	249	260	271	282	293	304	
401	314	325	336	347	358	369	379	390	401	412	
402	423	433	444	455	466	477	487	498	509	520	
403	531	541	552	563	574	584	595	606	617	627	
404	638	649	660	670	681	692	703	713	724	735	
405	746	756	767	778	788	799	810	821	831	842	
406	853	863	874	885	895	906	917	927	938	949	
407	959	970	981	991	*002	*013	*023	*034	*045	*055	
408	61 066	077	087	098	109	119	130	140	151	162	
409	172	183	194	204	215	225	236	247	257	268	
410	278	289	300	310	321	331	342	352	363	374	
411	384	395	405	416	426	437	448	458	469	479	
412	490	500	511	521	532	542	553	563	574	584	
413	595	606	616	627	637	648	658	669	679	690	
414	700	711	721	731	742	752	763	773	784	794	
415	805	815	826	836	847	857	868	878	888	899	
416	909	920	930	941	951	962	972	982	993	*003	
417	62 014	024	034	045	055	066	076	086	097	107	
418	118	128	138	149	159	170	180	190	201	211	
419	221	232	242	252	263	273	284	294	304	315	
420	325	335	346	356	366	377	387	397	408	418	
421	428	439	449	459	469	480	490	500	511	521	
422	531	542	552	562	572	583	593	603	613	624	
423	634	644	655	665	675	685	696	706	716	726	
424	737	747	757	767	778	788	798	808	818	829	
425	839	849	859	870	880	890	900	910	921	931	
426	941	951	961	972	982	992	*002	*012	*022	*033	
427	63 043	053	063	073	083	094	104	114	124	134	
428	144	155	165	175	185	195	205	215	225	236	
429	246	256	266	276	286	296	306	317	327	337	
430	347	357	367	377	387	397	407	417	428	438	
431	448	458	468	478	488	498	508	518	528	538	
432	548	558	568	579	589	599	609	619	629	639	
433	649	659	669	679	689	699	709	719	729	739	
434	749	759	769	779	789	799	809	819	829	839	
435	849	859	869	879	889	899	909	919	929	939	
436	949	959	969	979	988	998	*008	*018	*028	*038	
437	64 048	058	068	078	088	098	108	118	128	137	
438	147	157	167	177	187	197	207	217	227	237	
439	246	256	266	276	286	296	306	316	326	335	
440	345	355	365	375	385	395	404	414	424	434	
441	444	454	464	473	483	493	503	513	523	532	
442	542	552	562	572	582	591	601	611	621	631	
443	640	650	660	670	680	689	699	709	719	729	
444	738	748	758	768	777	787	797	807	816	826	
445	836	846	856	865	875	885	895	904	914	924	
446	933	943	953	963	972	982	992	*002	*011	*021	
447	65 031	040	050	060	070	079	089	099	108	118	
448	128	137	147	157	167	176	186	196	205	215	
449	225	234	244	254	263	273	283	292	302	312	
450	321	331	341	350	360	369	379	389	398	408	
N.	0	1	2	3	4	5	6	7	8	9	Prop. Pts.

Prop. Pts.

11

1	1.1
2	2.2
3	3.3
4	4.4
5	5.5
6	6.6
7	7.7
8	8.8
9	9.9

10

1	1.0
2	2.0
3	3.0
4	4.0
5	5.0
6	6.0
7	7.0
8	8.0
9	9.0

9

1	0.9
2	1.8
3	2.7
4	3.6
5	4.5
6	5.4
7	6.3
8	7.2
9	8.1

TABLE 2. MANTISSAS OF COMMON LOGARITHMS (continued)   631

450-500

N.	0	1	2	3	4	5	6	7	8	9	Prop. Pts.
450	65 321	331	341	350	360	369	379	389	398	408	
451	418	427	437	447	456	466	475	485	495	504	
452	514	523	533	543	552	562	571	581	591	600	
453	610	619	629	639	648	658	667	677	686	696	
454	706	715	725	734	744	753	763	772	782	792	
455	801	811	820	830	839	849	858	868	877	887	
456	896	906	916	925	935	944	954	963	973	982	
457	992	*001	*011	*020	*030	*039	*049	*058	*068	*077	
458	66 087	096	106	115	124	134	143	153	162	172	
459	181	191	200	210	219	229	238	247	257	266	
460	276	285	295	304	314	323	332	342	351	361	
461	370	380	389	398	408	417	427	436	445	455	
462	464	474	483	492	502	511	521	530	539	549	
463	558	567	577	586	596	605	614	624	633	642	
464	652	661	671	680	689	699	708	717	727	736	
465	745	755	764	773	783	792	801	811	820	829	
466	839	848	857	867	876	885	894	904	913	922	
467	932	941	950	960	969	978	987	997	*006	*015	
468	67 025	034	043	052	062	071	080	089	099	108	
469	117	127	136	145	154	164	173	182	191	201	
470	210	219	228	237	247	256	265	274	284	293	
471	302	311	321	330	339	348	357	367	376	385	
472	394	403	413	422	431	440	449	459	468	477	
473	486	495	504	514	523	532	541	550	560	569	
474	578	587	596	605	614	624	633	642	651	660	
475	669	679	688	697	706	715	724	733	742	752	
476	761	770	779	788	797	806	815	825	834	843	
477	852	861	870	879	888	897	906	916	925	934	
478	943	952	961	970	979	988	997	*006	*015	*024	
479	68 034	043	052	061	070	079	088	097	106	115	
480	124	133	142	151	160	169	178	187	196	205	
481	215	224	233	242	251	260	269	278	287	296	
482	305	314	323	332	341	350	359	368	377	386	
483	395	404	413	422	431	440	449	458	467	476	
484	485	494	502	511	520	529	538	547	556	565	
485	574	583	592	601	610	619	628	637	646	655	
486	664	673	681	690	699	708	717	726	735	744	
487	753	762	771	780	789	797	806	815	824	833	
488	842	851	860	869	878	886	895	904	913	922	
489	931	940	949	958	966	975	984	993	*002	*011	
490	69 020	028	037	046	055	064	073	082	090	099	
491	108	117	126	135	144	152	161	170	179	188	
492	197	205	214	223	232	241	249	258	267	276	
493	285	294	302	311	320	329	338	346	355	364	
494	373	381	390	399	408	417	425	434	443	452	
495	461	469	478	487	496	504	513	522	531	539	
496	548	557	566	574	583	592	601	609	618	627	
497	636	644	653	662	671	679	688	697	705	714	
498	723	732	740	749	758	767	775	784	793	801	
499	810	819	827	836	845	854	862	871	880	888	
500	897	906	914	923	932	940	949	958	966	975	
N.	0	1	2	3	4	5	6	7	8	9	Prop. Pts.

Prop. Pts.

10

1	1.0
2	2.0
3	3.0
4	4.0
5	5.0
6	6.0
7	7.0
8	8.0
9	9.0

9

1	0.9
2	1.8
3	2.7
4	3.6
5	4.5
6	5.4
7	6.3
8	7.2
9	8.1

8

1	0.8
2	1.6
3	2.4
4	3.2
5	4.0
6	4.8
7	5.6
8	6.4
9	7.2

TABLE 2. MANTISSAS OF COMMON LOGARITHMS (continued)   632

500-550

N.	0	1	2	3	4	5	6	7	8	9
500	69 897	906	914	923	932	940	949	958	966	975
501	984	992	*001	*010	*018	*027	*036	*044	*053	*062
502	70 070	079	088	096	105	114	122	131	140	148
503	157	165	174	183	191	200	209	217	226	234
504	243	252	260	269	278	286	295	303	312	321
505	329	338	346	355	364	372	381	389	398	406
506	415	424	432	441	449	458	467	475	484	492
507	501	509	518	526	535	544	552	561	569	578
508	586	595	603	612	621	629	638	646	655	663
509	672	680	689	697	706	714	723	731	740	749
510	757	766	774	783	791	800	808	817	825	834
511	842	851	859	868	876	885	893	902	910	919
512	927	935	944	952	961	969	978	986	995	*003
513	71 012	020	029	037	046	054	063	071	079	088
514	096	105	113	122	130	139	147	155	164	172
515	181	189	198	206	214	223	231	240	248	257
516	265	273	282	290	299	307	315	324	332	341
517	349	357	366	374	383	391	399	408	416	425
518	433	441	450	458	466	475	483	492	500	508
519	517	525	533	542	550	559	567	575	584	592
520	600	609	617	625	634	642	650	659	667	675
521	684	692	700	709	717	725	734	742	750	759
522	767	775	784	792	800	809	817	825	834	842
523	850	858	867	875	883	892	900	908	917	925
524	933	941	950	958	966	975	983	991	999	*008
525	72 016	024	032	041	049	057	066	074	082	090
526	099	107	115	123	132	140	148	156	165	173
527	181	189	198	206	214	222	230	239	247	255
528	263	272	280	288	296	304	313	321	329	337
529	346	354	362	370	378	387	395	403	411	419
530	428	436	444	452	460	469	477	485	493	501
531	509	518	526	534	542	550	558	567	575	583
532	591	599	607	616	624	632	640	648	656	665
533	673	681	689	697	705	713	722	730	738	746
534	754	762	770	779	787	795	803	811	819	827
535	835	843	852	860	868	876	884	892	900	908
536	916	925	933	941	949	957	965	973	981	989
537	997	*006	*014	*022	*030	*038	*046	*054	*062	*070
538	73 078	086	094	102	111	119	127	135	143	151
539	159	167	175	183	191	199	207	215	223	231
540	239	247	255	263	272	280	288	296	304	312
541	320	328	336	344	352	360	368	376	384	392
542	400	408	416	424	432	440	448	456	464	472
543	480	488	496	504	512	520	528	536	544	552
544	560	568	576	584	592	600	608	616	624	632
545	640	648	656	664	672	679	687	695	703	711
546	719	727	735	743	751	759	767	775	783	791
547	799	807	815	823	830	838	846	854	862	870
548	878	886	894	902	910	918	926	933	941	949
549	957	965	973	981	989	997	*005	*013	*020	*028
550	74 036	044	052	060	068	076	084	092	099	107

Prop. Pts.

9
1 0.9
2 1.8
3 2.7
4 3.6
5 4.5
6 5.4
7 6.3
8 7.2
9 8.1

8
1 0.8
2 1.6
3 2.4
4 3.2
5 4.0
6 4.8
7 5.6
8 6.4
9 7.2

7
1 0.7
2 1.4
3 2.1
4 2.8
5 3.5
6 4.2
7 4.9
8 5.6
9 6.3

TABLE 2. MANTISSAS OF COMMON LOGARITHMS (continued)   633

550-600

N.	0	1	2	3	4	5	6	7	8	9	Prop. Pts.
550	74 036	044	052	060	068	076	084	092	099	107	
551	115	123	131	139	147	155	162	170	178	186	
552	194	202	210	218	225	233	241	249	257	265	
553	273	280	288	296	304	312	320	327	335	343	
554	351	359	367	374	382	390	398	406	414	421	
555	429	437	445	453	461	468	476	484	492	500	
556	507	515	523	531	539	547	554	562	570	578	
557	586	593	601	609	617	624	632	640	648	656	
558	663	671	679	687	695	702	710	718	726	733	
559	741	749	757	764	772	780	788	796	803	811	
560	819	827	834	842	850	858	865	873	881	889	
561	896	904	912	920	927	935	943	950	958	966	
562	974	981	989	997	*005	*012	*020	*028	*035	*043	
563	75 051	059	066	074	082	089	097	105	113	120	
564	128	136	143	151	159	166	174	182	189	197	
565	205	213	220	228	236	243	251	259	266	274	
566	282	289	297	305	312	320	328	335	343	351	
567	358	366	374	381	389	397	404	412	420	427	
568	435	442	450	458	465	473	481	488	496	504	
569	511	519	526	534	542	549	557	565	572	580	
570	587	595	603	610	618	626	633	641	648	656	
571	664	671	679	686	694	702	709	717	724	732	
572	740	747	755	762	770	778	785	793	800	808	
573	815	823	831	838	846	853	861	868	876	884	
574	891	899	906	914	921	929	937	944	952	959	
575	967	974	982	989	997	*005	*012	*020	*027	*035	
576	76 042	050	057	065	072	080	087	095	103	110	
577	118	125	133	140	148	155	163	170	178	185	
578	193	200	208	215	223	230	238	245	253	260	
579	268	275	283	290	298	305	313	320	328	335	
580	343	350	358	365	373	380	388	395	403	410	
581	418	425	433	440	448	455	462	470	477	485	
582	492	500	507	515	522	530	537	545	552	559	
583	567	574	582	589	597	604	612	619	626	634	
584	641	649	656	664	671	678	686	693	701	708	
585	716	723	730	738	745	753	760	768	775	782	
586	790	797	805	812	819	827	834	842	849	856	
587	864	871	879	886	893	901	908	916	923	930	
588	938	945	953	960	967	975	982	989	997	*004	
589	77 012	019	026	034	041	048	056	063	070	078	
590	085	093	100	107	115	122	129	137	144	151	
591	159	166	173	181	188	195	203	210	217	225	
592	232	240	247	254	262	269	276	283	291	298	
593	305	313	320	327	335	342	349	357	364	371	
594	379	386	393	401	408	415	422	430	437	444	
595	452	459	466	474	481	488	495	503	510	517	
596	525	532	539	546	554	561	568	576	583	590	
597	597	605	612	619	627	634	641	648	656	663	
598	670	677	685	692	699	706	714	721	728	735	
599	743	750	757	764	772	779	786	793	801	808	
600	815	822	830	837	844	851	859	866	873	880	
N.	0	1	2	3	4	5	6	7	8	9	Prop. Pts.

Proportional parts:

	8
1	0.8
2	1.6
3	2.4
4	3.2
5	4.0
6	4.8
7	5.6
8	6.4
9	7.2

	7
1	0.7
2	1.4
3	2.1
4	2.8
5	3.5
6	4.2
7	4.9
8	5.6
9	6.3

TABLE 2. MANTISSAS OF COMMON LOGARITHMS (continued)   **634**

600-650

N.	0	1	2	3	4	5	6	7	8	9	Prop. Pts.
600	77 815	822	830	837	844	851	859	866	873	880	
601	887	895	902	909	916	924	931	938	945	952	
602	960	967	974	981	988	996	*003	*010	*017	*025	
603	78 032	039	046	053	061	068	075	082	089	097	
604	104	111	118	125	132	140	147	154	161	168	
605	176	183	190	197	204	211	219	226	233	240	
606	247	254	262	269	276	283	290	297	305	312	
607	319	326	333	340	347	355	362	369	376	383	
608	390	398	405	412	419	426	433	440	447	455	
609	462	469	476	483	490	497	504	512	519	526	
610	533	540	547	554	561	569	576	583	590	597	
611	604	611	618	625	633	640	647	654	661	668	
612	675	682	689	696	704	711	718	725	732	739	
613	746	753	760	767	774	781	789	796	803	810	
614	817	824	831	838	845	852	859	866	873	880	
615	888	895	902	909	916	923	930	937	944	951	
616	958	965	972	979	986	993	*000	*007	*014	*021	
617	79 029	036	043	050	057	064	071	078	085	092	
618	099	106	113	120	127	134	141	148	155	162	
619	169	176	183	190	197	204	211	218	225	232	
620	239	246	253	260	267	274	281	288	295	302	
621	309	316	323	330	337	344	351	358	365	372	
622	379	386	393	400	407	414	421	428	435	442	
623	449	456	463	470	477	484	491	498	505	511	
624	518	525	532	539	546	553	560	567	574	581	
625	588	595	602	609	616	623	630	637	644	650	
626	657	664	671	678	685	692	699	706	713	720	
627	727	734	741	748	754	761	768	775	782	789	
628	796	803	810	817	824	831	837	844	851	858	
629	865	872	879	886	893	900	906	913	920	927	
630	934	941	948	955	962	969	975	982	989	996	
631	80 003	010	017	024	030	037	044	051	058	065	
632	072	079	085	092	099	106	113	120	127	134	
633	140	147	154	161	168	175	182	188	195	202	
634	209	216	223	229	236	243	250	257	264	271	
635	277	284	291	298	305	312	318	325	332	339	
636	346	353	359	366	373	380	387	393	400	407	
637	414	421	428	434	441	448	455	462	468	475	
638	482	489	496	502	509	516	523	530	536	543	
639	550	557	564	570	577	584	591	598	604	611	
640	618	625	632	638	645	652	659	665	672	679	
641	686	693	699	706	713	720	726	733	740	747	
642	754	760	767	774	781	787	794	801	808	814	
643	821	828	835	841	848	855	862	868	875	882	
644	889	895	902	909	916	922	929	936	943	949	
645	956	963	969	976	983	990	996	*003	*010	*017	
646	81 023	030	037	043	050	057	064	070	077	084	
647	090	097	104	111	117	124	131	137	144	151	
648	158	164	171	178	184	191	198	204	211	218	
649	224	231	238	245	251	258	265	271	278	285	
650	291	298	305	311	318	325	331	338	345	351	
N.	**0**	**1**	**2**	**3**	**4**	**5**	**6**	**7**	**8**	**9**	**Prop. Pts.**

Prop. Pts.

	8
1	0.8
2	1.6
3	2.4
4	3.2
5	4.0
6	4.8
7	5.6
8	6.4
9	7.2

	7
1	0.7
2	1.4
3	2.1
4	2.8
5	3.5
6	4.2
7	4.9
8	5.6
9	6.3

	6
1	0.6
2	1.2
3	1.8
4	2.4
5	3.0
6	3.6
7	4.2
8	4.8
9	5.4

TABLE 2. MANTISSAS OF COMMON LOGARITHMS (continued) 635

650-700

N.	0	1	2	3	4	5	6	7	8	9
650	81 291	298	305	311	318	325	331	338	345	351
651	358	365	371	378	385	391	398	405	411	418
652	425	431	438	445	451	458	465	471	478	485
653	491	498	505	511	518	525	531	538	544	551
654	558	564	571	578	584	591	598	604	611	617
655	624	631	637	644	651	657	664	671	677	684
656	690	697	704	710	717	723	730	737	743	750
657	757	763	770	776	783	790	796	803	809	816
658	823	829	836	842	849	856	862	869	875	882
659	889	895	902	908	915	921	928	935	941	948
660	954	961	968	974	981	987	994	*000	*007	*014
661	82 020	027	033	040	046	053	060	066	073	079
662	086	092	099	105	112	119	125	132	138	145
663	151	158	164	171	178	184	191	197	204	210
664	217	223	230	236	243	249	256	263	269	276
665	282	289	295	302	308	315	321	328	334	341
666	347	354	360	367	373	380	387	393	400	406
667	413	419	426	432	439	445	452	458	465	471
668	478	484	491	497	504	510	517	523	530	536
669	543	549	556	562	569	575	582	588	595	601
670	607	614	620	627	633	640	646	653	659	666
671	672	679	685	692	698	705	711	718	724	730
672	737	743	750	756	763	769	776	782	789	795
673	802	808	814	821	827	834	840	847	853	860
674	866	872	879	885	892	898	905	911	918	924
675	930	937	943	950	956	963	969	975	982	988
676	995	*001	*008	*014	*020	*027	*033	*040	*046	*052
677	83 059	065	072	078	085	091	097	104	110	117
678	123	129	136	142	149	155	161	168	174	181
679	187	193	200	206	213	219	225	232	238	245
680	251	257	264	270	276	283	289	296	302	308
681	315	321	327	334	340	347	353	359	366	372
682	378	385	391	398	404	410	417	423	429	436
683	442	448	455	461	467	474	480	487	493	499
684	506	512	518	525	531	537	544	550	556	563
685	569	575	582	588	594	601	607	613	620	626
686	632	639	645	651	658	664	670	677	683	689
687	696	702	708	715	721	727	734	740	746	753
688	759	765	771	778	784	790	797	803	809	816
689	822	828	835	841	847	853	860	866	872	879
690	885	891	897	904	910	916	923	929	935	942
691	948	954	960	967	973	979	985	992	998	*004
692	84 011	017	023	029	036	042	048	055	061	067
693	073	080	086	092	098	105	111	117	123	130
694	136	142	148	155	161	167	173	180	186	192
695	198	205	211	217	223	230	236	242	248	255
696	261	267	273	280	286	292	298	305	311	317
697	323	330	336	342	348	354	361	367	373	379
698	386	392	398	404	410	417	423	429	435	442
699	448	454	460	466	473	479	485	491	497	504
700	510	516	522	528	535	541	547	553	559	566
N.	0	1	2	3	4	5	6	7	8	9

Prop. Pts.

7	
1	0.7
2	1.4
3	2.1
4	2.8
5	3.5
6	4.2
7	4.9
8	5.6
9	6.3

6	
1	0.6
2	1.2
3	1.8
4	2.4
5	3.0
6	3.6
7	4.2
8	4.8
9	5.4

TABLE 2. MANTISSAS OF COMMON LOGARITHMS (continued) 636

700-750

N.	0	1	2	3	4	5	6	7	8	9	Prop. Pts.
700	84 510	516	522	528	535	541	547	553	559	566	
701	572	578	584	590	597	603	609	615	621	628	
702	634	640	646	652	658	665	671	677	683	689	
703	696	702	708	714	720	726	733	739	745	751	
704	757	763	770	776	782	788	794	800	807	813	
705	819	825	831	837	844	850	856	862	868	874	
706	880	887	893	899	905	911	917	924	930	936	
707	942	948	954	960	967	973	979	985	991	997	
708	85 003	009	016	022	028	034	040	046	052	058	
709	065	071	077	083	089	095	101	107	114	120	
710	126	132	138	144	150	156	163	169	175	181	
711	187	193	199	205	211	217	224	230	236	242	
712	248	254	260	266	272	278	285	291	297	303	
713	309	315	321	327	333	339	345	352	358	364	
714	370	376	382	388	394	400	406	412	418	425	
715	431	437	443	449	455	461	467	473	479	485	
716	491	497	503	509	516	522	528	534	540	546	
717	552	558	564	570	576	582	588	594	600	606	
718	612	618	625	631	637	643	649	655	661	667	
719	673	679	685	691	697	703	709	715	721	727	
720	733	739	745	751	757	763	769	775	781	788	
721	794	800	806	812	818	824	830	836	842	848	
722	854	860	866	872	878	884	890	896	902	908	
723	914	920	926	932	938	944	950	956	962	968	
724	974	980	986	992	998	*004	*010	*016	*022	*028	
725	86 034	040	046	052	058	064	070	076	082	088	
726	094	100	106	112	118	124	130	136	141	147	
727	153	159	165	171	177	183	189	195	201	207	
728	213	219	225	231	237	243	249	255	261	267	
729	273	279	285	291	297	303	308	314	320	326	
730	332	338	344	350	356	362	368	374	380	386	
731	392	398	404	410	415	421	427	433	439	445	
732	451	457	463	469	475	481	487	493	499	504	
733	510	516	522	528	534	540	546	552	558	564	
734	570	576	581	587	593	599	605	611	617	623	
735	629	635	641	646	652	658	664	670	676	682	
736	688	694	700	705	711	717	723	729	735	741	
737	747	753	759	764	770	776	782	788	794	800	
738	806	812	817	823	829	835	841	847	853	859	
739	864	870	876	882	888	894	900	906	911	917	
740	923	929	935	941	947	953	958	964	970	976	
741	982	988	994	999	*005	*011	*017	*023	*029	*035	
742	87 040	046	052	058	064	070	075	081	087	093	
743	099	105	111	116	122	128	134	140	146	151	
744	157	163	169	175	181	186	192	198	204	210	
745	216	221	227	233	239	245	251	256	262	268	
746	274	280	286	291	297	303	309	315	320	326	
747	332	338	344	349	355	361	367	373	379	384	
748	390	396	402	408	413	419	425	431	437	442	
749	448	454	460	466	471	477	483	489	495	500	
750	506	512	518	523	529	535	541	547	552	558	
N.	0	1	2	3	4	5	6	7	8	9	Prop. Pts.

Prop. Pts.

7
1 | 0.7
2 | 1.4
3 | 2.1
4 | 2.8
5 | 3.5
6 | 4.2
7 | 4.9
8 | 5.6
9 | 6.3

6
1 | 0.6
2 | 1.2
3 | 1.8
4 | 2.4
5 | 3.0
6 | 3.6
7 | 4.2
8 | 4.8
9 | 5.4

5
1 | 0.5
2 | 1.0
3 | 1.5
4 | 2.0
5 | 2.5
6 | 3.0
7 | 3.5
8 | 4.0
9 | 4.5

TABLE 2. MANTISSAS OF COMMON LOGARITHMS (continued)   637

750-800

N.	0	1	2	3	4	5	6	7	8	9	Prop. Pts.
750	87 506	512	518	523	529	535	541	547	552	558	
751	564	570	576	581	587	593	599	604	610	616	
752	622	628	633	639	645	651	656	662	668	674	
753	679	685	691	697	703	708	714	720	726	731	
754	737	743	749	754	760	766	772	777	783	789	
755	795	800	806	812	818	823	829	835	841	846	
756	852	858	864	869	875	881	887	892	898	904	
757	910	915	921	927	933	938	944	950	955	961	
758	967	973	978	984	990	996	*001	*007	*013	*018	
759	88 024	030	036	041	047	053	058	064	070	076	
760	081	087	093	098	104	110	116	121	127	133	
761	138	144	150	156	161	167	173	178	184	190	
762	195	201	207	213	218	224	230	235	241	247	
763	252	258	264	270	275	281	287	292	298	304	
764	309	315	321	326	332	338	343	349	355	360	
765	366	372	377	383	389	395	400	406	412	417	
766	423	429	434	440	446	451	457	463	468	474	
767	480	485	491	497	502	508	513	519	525	530	
768	536	542	547	553	559	564	570	576	581	587	
769	593	598	604	610	615	621	627	632	638	643	
770	649	655	660	666	672	677	683	689	694	700	
771	705	711	717	722	728	734	739	745	750	756	
772	762	767	773	779	784	790	795	801	807	812	
773	818	824	829	835	840	846	852	857	863	868	
774	874	880	885	891	897	902	908	913	919	925	
775	930	936	941	947	953	958	964	969	975	981	
776	986	992	997	*003	*009	*014	*020	*025	*031	*037	
777	89 042	048	053	059	064	070	076	081	087	092	
778	098	104	109	115	120	126	131	137	143	148	
779	154	159	165	170	176	182	187	193	198	204	
780	209	215	221	226	232	237	243	248	254	260	
781	265	271	276	282	287	293	298	304	310	315	
782	321	326	332	337	343	348	354	360	365	371	
783	376	382	387	393	398	404	409	415	421	426	
784	432	437	443	448	454	459	465	470	476	481	
785	487	492	498	504	509	515	520	526	531	537	
786	542	548	553	559	564	570	575	581	586	592	
787	597	603	609	614	620	625	631	636	642	647	
788	653	658	664	669	675	680	686	691	697	702	
789	708	713	719	724	730	735	741	746	752	757	
790	763	768	774	779	785	790	796	801	807	812	
791	818	823	829	834	840	845	851	856	862	867	
792	873	878	883	889	894	900	905	911	916	922	
793	927	933	938	944	949	955	960	966	971	977	
794	982	988	993	998	*004	*009	*015	*020	*026	*031	
795	90 037	042	048	053	059	064	069	075	080	086	
796	091	097	102	108	113	119	124	129	135	140	
797	146	151	157	162	168	173	179	184	189	195	
798	200	206	211	217	222	227	233	238	244	249	
799	255	260	266	271	276	282	287	293	298	304	
800	309	314	320	325	331	336	342	347	352	358	
N.	0	1	2	3	4	5	6	7	8	9	Prop. Pts.

Prop. Pts.

	6
1	0.6
2	1.2
3	1.8
4	2.4
5	3.0
6	3.6
7	4.2
8	4.8
9	5.4

	5
1	0.5
2	1.0
3	1.5
4	2.0
5	2.5
6	3.0
7	3.5
8	4.0
9	4.5

TABLE 2. MANTISSAS OF COMMON LOGARITHMS (continued)   638

800-850

N.	0	1	2	3	4	5	6	7	8	9	Prop. Pts.
800	90 309	314	320	325	331	336	342	347	352	358	
801	363	369	374	380	385	390	396	401	407	412	
802	417	423	428	434	439	445	450	455	461	466	
803	472	477	482	488	493	499	504	509	515	520	
804	526	531	536	542	547	553	558	563	569	574	
805	580	585	590	596	601	607	612	617	623	628	
806	634	639	644	650	655	660	666	671	677	682	
807	687	693	698	703	709	714	720	725	730	736	
808	741	747	752	757	763	768	773	779	784	789	
809	795	800	806	811	816	822	827	832	838	843	
810	849	854	859	865	870	875	881	886	891	897	
811	902	907	913	918	924	929	934	940	945	950	
812	956	961	966	972	977	982	988	993	998	*004	
813	91 009	014	020	025	030	036	041	046	052	057	
814	062	068	073	078	084	089	094	100	105	110	
815	116	121	126	132	137	142	148	153	158	164	
816	169	174	180	185	190	196	201	206	212	217	
817	222	228	233	238	243	249	254	259	265	270	
818	275	281	286	291	297	302	307	312	318	323	
819	328	334	339	344	350	355	360	365	371	376	
820	381	387	392	397	403	408	413	418	424	429	
821	434	440	445	450	455	461	466	471	477	482	
822	487	492	498	503	508	514	519	524	529	535	
823	540	545	551	556	561	566	572	577	582	587	
824	593	598	603	609	614	619	624	630	635	640	
825	645	651	656	661	666	672	677	682	687	693	
826	698	703	709	714	719	724	730	735	740	745	
827	751	756	761	766	772	777	782	787	793	798	
828	803	808	814	819	824	829	834	840	845	850	
829	855	861	866	871	876	882	887	892	897	903	
830	908	913	918	924	929	934	939	944	950	955	
831	960	965	971	976	981	986	991	997	*002	*007	
832	92 012	018	023	028	033	038	044	049	054	059	
833	065	070	075	080	085	091	096	101	106	111	
834	117	122	127	132	137	143	148	153	158	163	
835	169	174	179	184	189	195	200	205	210	215	
836	221	226	231	236	241	247	252	257	262	267	
837	273	278	283	288	293	298	304	309	314	319	
838	324	330	335	340	345	350	355	361	366	371	
839	376	381	387	392	397	402	407	412	418	423	
840	428	433	438	443	449	454	459	464	469	474	
841	480	485	490	495	500	505	511	516	521	526	
842	531	536	542	547	552	557	562	567	572	578	
843	583	588	593	598	603	609	614	619	624	629	
844	634	639	645	650	655	660	665	670	675	681	
845	686	691	696	701	706	711	716	722	727	732	
846	737	742	747	752	758	763	768	773	778	783	
847	788	793	799	804	809	814	819	824	829	834	
848	840	845	850	855	860	865	870	875	881	886	
849	891	896	901	906	911	916	921	927	932	937	
850	942	947	952	957	962	967	973	978	983	988	
N.	0	1	2	3	4	5	6	7	8	9	Prop. Pts.

Prop. Pts.

	6
1	0.6
2	1.2
3	1.8
4	2.4
5	3.0
6	3.6
7	4.2
8	4.8
9	5.4

	5
1	0.5
2	1.0
3	1.5
4	2.0
5	2.5
6	3.0
7	3.5
8	4.0
9	4.5

TABLE 2. MANTISSAS OF COMMON LOGARITHMS (continued)   **639**

850-900

N.	0	1	2	3	4	5	6	7	8	9
850	92 942	947	952	957	962	967	973	978	983	988
851	993	998	*003	*008	*013	*018	*024	*029	*034	*039
852	93 044	049	054	059	064	069	075	080	085	090
853	095	100	105	110	115	120	125	131	136	141
854	146	151	156	161	166	171	176	181	186	192
855	197	202	207	212	217	222	227	232	237	242
856	247	252	258	263	268	273	278	283	288	293
857	298	303	308	313	318	323	328	334	339	344
858	349	354	359	364	369	374	379	384	389	394
859	399	404	409	414	420	425	430	435	440	445
860	450	455	460	465	470	475	480	485	490	495
861	500	505	510	515	520	526	531	536	541	546
862	551	556	561	566	571	576	581	586	591	596
863	601	606	611	616	621	626	631	636	641	646
864	651	656	661	666	671	676	682	687	692	697
865	702	707	712	717	722	727	732	737	742	747
866	752	757	762	767	772	777	782	787	792	797
867	802	807	812	817	822	827	832	837	842	847
868	852	857	862	867	872	877	882	887	892	897
869	902	907	912	917	922	927	932	937	942	947
870	952	957	962	967	972	977	982	987	992	997
871	94 002	007	012	017	022	027	032	037	042	047
872	052	057	062	067	072	077	082	086	091	096
873	101	106	111	116	121	126	131	136	141	146
874	151	156	161	166	171	176	181	186	191	196
875	201	206	211	216	221	226	231	236	240	245
876	250	255	260	265	270	275	280	285	290	295
877	300	305	310	315	320	325	330	335	340	345
878	349	354	359	364	369	374	379	384	389	394
879	399	404	409	414	419	424	429	433	438	443
880	448	453	458	463	468	473	478	483	488	493
881	498	503	507	512	517	522	527	532	537	542
882	547	552	557	562	567	571	576	581	586	591
883	596	601	606	611	616	621	626	630	635	640
884	645	650	655	660	665	670	675	680	685	689
885	694	699	704	709	714	719	724	729	734	738
886	743	748	753	758	763	768	773	778	783	787
887	792	797	802	807	812	817	822	827	832	836
888	841	846	851	856	861	866	871	876	880	885
889	890	895	900	905	910	915	919	924	929	934
890	939	944	949	954	959	963	968	973	978	983
891	988	993	998	*002	*007	*012	*017	*022	*027	*032
892	95 036	041	046	051	056	061	066	071	075	080
893	085	090	095	100	105	109	114	119	124	129
894	134	139	143	148	153	158	163	168	173	177
895	182	187	192	197	202	207	211	216	221	226
896	231	236	240	245	250	255	260	265	270	274
897	279	284	289	294	299	303	308	313	318	323
898	328	332	337	342	347	352	357	361	366	371
899	376	381	386	390	395	400	405	410	415	419
900	424	429	434	439	444	448	453	458	463	468
N.	0	1	2	3	4	5	6	7	8	9

Prop. Pts.

6		5		4	
1	0.6	1	0.5	1	0.4
2	1.2	2	1.0	2	0.8
3	1.8	3	1.5	3	1.2
4	2.4	4	2.0	4	1.6
5	3.0	5	2.5	5	2.0
6	3.6	6	3.0	6	2.4
7	4.2	7	3.5	7	2.8
8	4.8	8	4.0	8	3.2
9	5.4	9	4.5	9	3.6

TABLE 2. MANTISSAS OF COMMON LOGARITHMS (continued) **640**

900-950

N.	0	1	2	3	4	5	6	7	8	9	Prop. Pts.
900	95 424	429	434	439	444	448	453	458	463	468	
901	472	477	482	487	492	497	501	506	511	516	
902	521	525	530	535	540	545	550	554	559	564	
903	569	574	578	583	588	593	598	602	607	612	
904	617	622	626	631	636	641	646	650	655	660	
905	665	670	674	679	684	689	694	698	703	708	
906	713	718	722	727	732	737	742	746	751	756	
907	761	766	770	775	780	785	789	794	799	804	
908	809	813	818	823	828	832	837	842	847	852	
909	856	861	866	871	875	880	885	890	895	899	
910	904	909	914	918	923	928	933	938	942	947	
911	952	957	961	966	971	976	980	985	990	995	
912	999	*004	*009	*014	*019	*023	*028	*033	*038	*042	
913	96 047	052	057	061	066	071	076	080	085	090	
914	095	099	104	109	114	118	123	128	133	137	
915	142	147	152	156	161	166	171	175	180	185	
916	190	194	199	204	209	213	218	223	227	232	
917	237	242	246	251	256	261	265	270	275	280	
918	284	289	294	298	303	308	313	317	322	327	
919	332	336	341	346	350	355	360	365	369	374	
920	379	384	388	393	398	402	407	412	417	421	
921	426	431	435	440	445	450	454	459	464	468	
922	473	478	483	487	492	497	501	506	511	515	
923	520	525	530	534	539	544	548	553	558	562	
924	567	572	577	581	586	591	595	600	605	609	
925	614	619	624	628	633	638	642	647	652	656	
926	661	666	670	675	680	685	689	694	699	703	
927	708	713	717	722	727	731	736	741	745	750	
928	755	759	764	769	774	778	783	788	792	797	
929	802	806	811	816	820	825	830	834	839	844	
930	848	853	858	862	867	872	876	881	886	890	
931	895	900	904	909	914	918	923	928	932	937	
932	942	946	951	956	960	965	970	974	979	984	
933	988	993	997	*002	*007	*011	*016	*021	*025	*030	
934	97 035	039	044	049	053	058	063	067	072	077	
935	081	086	090	095	100	104	109	114	118	123	
936	128	132	137	142	146	151	155	160	165	169	
937	174	179	183	188	192	197	202	206	211	216	
938	220	225	230	234	239	243	248	253	257	262	
939	267	271	276	280	285	290	294	299	304	308	
940	313	317	322	327	331	336	340	345	350	354	
941	359	364	368	373	377	382	387	391	396	400	
942	405	410	414	419	424	428	433	437	442	447	
943	451	456	460	465	470	474	479	483	488	493	
944	497	502	506	511	516	520	525	529	534	539	
945	543	548	552	557	562	566	571	575	580	585	
946	589	594	598	603	607	612	617	621	626	630	
947	635	640	644	649	653	658	663	667	672	676	
948	681	685	690	695	699	704	708	713	717	722	
949	727	731	736	740	745	749	754	759	763	768	
950	772	777	782	786	791	795	800	804	809	813	
N.	0	1	2	3	4	5	6	7	8	9	Prop. Pts.

Prop. Pts.

5	
1	0.5
2	1.0
3	1.5
4	2.0
5	2.5
6	3.0
7	3.5
8	4.0
9	4.5

4	
1	0.4
2	0.8
3	1.2
4	1.6
5	2.0
6	2.4
7	2.8
8	3.2
9	3.6

TABLE 2. MANTISSAS OF COMMON LOGARITHMS (continued) **641**

950-1000

N.	0	1	2	3	4	5	6	7	8	9	Prop. Pts.
950	97 772	777	782	786	791	795	800	804	809	813	
951	818	823	827	832	836	841	845	850	855	859	
952	864	868	873	877	882	886	891	896	900	905	
953	909	914	918	923	928	932	937	941	946	950	
954	955	959	964	968	973	978	982	987	991	996	
955	98 000	005	009	014	019	023	028	032	037	041	
956	046	050	055	059	064	068	073	078	082	087	
957	091	096	100	105	109	114	118	123	127	132	
958	137	141	146	150	155	159	164	168	173	177	
959	182	186	191	195	200	204	209	214	218	223	
960	227	232	236	241	245	250	254	259	263	268	
961	272	277	281	286	290	295	299	304	308	313	
962	318	322	327	331	336	340	345	349	354	358	
963	363	367	372	376	381	385	390	394	399	403	
964	408	412	417	421	426	430	435	439	444	448	
965	453	457	462	466	471	475	480	484	489	493	
966	498	502	507	511	516	520	525	529	534	538	
967	543	547	552	556	561	565	570	574	579	583	
968	588	592	597	601	605	610	614	619	623	628	
969	632	637	641	646	650	655	659	664	668	673	
970	677	682	686	691	695	700	704	709	713	717	
971	722	726	731	735	740	744	749	753	758	762	
972	767	771	776	780	784	789	793	798	802	807	
973	811	816	820	825	829	834	838	843	847	851	
974	856	860	865	869	874	878	883	887	892	896	
975	900	905	909	914	918	923	927	932	936	941	
976	945	949	954	958	963	967	972	976	981	985	
977	989	994	998	*003	*007	*012	*016	*021	*025	*029	
978	99 034	038	043	047	052	056	061	065	069	074	
979	078	083	087	092	096	100	105	109	114	118	
980	123	127	131	136	140	145	149	154	158	162	
981	167	171	176	180	185	189	193	198	202	207	
982	211	216	220	224	229	233	238	242	247	251	
983	255	260	264	269	273	277	282	286	291	295	
984	300	304	308	313	317	322	326	330	335	339	
985	344	348	352	357	361	366	370	374	379	383	
986	388	392	396	401	405	410	414	419	423	427	
987	432	436	441	445	449	454	458	463	467	471	
988	476	480	484	489	493	498	502	506	511	515	
989	520	524	528	533	537	542	546	550	555	559	
990	564	568	572	577	581	585	590	594	599	603	
991	607	612	616	621	625	629	634	638	642	647	
992	651	656	660	664	669	673	677	682	686	691	
993	695	699	704	708	712	717	721	726	730	734	
994	739	743	747	752	756	760	765	769	774	778	
995	782	787	791	795	800	804	808	813	817	822	
996	826	830	835	839	843	848	852	856	861	865	
997	870	874	878	883	887	891	896	900	904	909	
998	913	917	922	926	930	935	939	944	948	952	
999	957	961	965	970	974	978	983	987	991	996	
1000	00 000	004	009	013	017	022	026	030	035	039	
N.	**0**	**1**	**2**	**3**	**4**	**5**	**6**	**7**	**8**	**9**	**Prop. Pts.**

Prop. Pts.

5
1 | 0.5
2 | 1.0
3 | 1.5
4 | 2.0
5 | 2.5
6 | 3.0
7 | 3.5
8 | 4.0
9 | 4.5

4
1 | 0.4
2 | 0.8
3 | 1.2
4 | 1.6
5 | 2.0
6 | 2.4
7 | 2.8
8 | 3.2
9 | 3.6

0°

′	L Sin	d	L Tan	c d	L Cot	L Cos	
0	———		———		———	0.00 000	**60**
1	6.46 373	30103	6.46 373	30103	13.53 627	0.00 000	59
2	6.76 476	17609	6.76 476	17609	13.23 524	0.00 000	58
3	6.94 085	12494	6.94 085	12494	13.05 915	0.00 000	57
4	7.06 579	9691	7.06 579	9691	12.93 421	0.00 000	56
5	7.16 270	7918	7.16 270	7918	12.83 730	0.00 000	55
6	7.24 188	6694	7.24 188	6694	12.75 812	0.00 000	54
7	7.30 882	5800	7.30 882	5800	12.69 118	0.00 000	53
8	7.36 682	5115	7 36 682	5115	12.63 318	0.00 000	52
9	7.41 797	4576	7.41 797	4576	12.58 203	0.00 000	51
10	7.46 373	4139	7.46 373	4139	12.53 627	0.00 000	**50**
11	7.50 512	3779	7.50 512	3779	12.49 488	0.00 000	49
12	7.54 291	3476	7.54 291	3476	12.45 709	0.00 000	48
13	7.57 767	3218	7.57 767	3219	12.42 233	0.00 000	47
14	7.60 985	2997	7.60 986	2996	12.39 014	0.00 000	46
15	7.63 982	2802	7.63 982	2803	12.36 018	0.00 000	45
16	7.66 784	2633	7.66 785	2633	12.33 215	0.00 000	44
17	7.69 417	2483	7.69 418	2482	12.30 582	9.99 999	43
18	7.71 900	2348	7.71 900	2348	12.28 100	9.99 999	42
19	7.74 248	2227	7.74 248	2228	12.25 752	9.99 999	41
20	7.76 475	2119	7.76 476	2119	12.23 524	9.99 999	**40**
21	7.78 594	2021	7.78 595	2020	12.21 405	9.99 999	39
22	7.80 615	1930	7.80 615	1931	12.19 385	9.99 999	38
23	7.82 545	1848	7.82 546	1848	12.17 454	9.99 999	37
24	7.84 393	1773	7.84 394	1773	12.15 606	9.99 999	36
25	7.86 166	1704	7.86 167	1704	12.13 833	9.99 999	35
26	7.87 870	1639	7.87 871	1639	12.12 129	9.99 999	34
27	7.89 509	1579	7.89 510	1579	12.10 490	9.99 999	33
28	7.91 088	1524	7.91 089	1524	12.08 911	9.99 999	32
29	7.92 612	1472	7.92 613	1473	12.07 387	9.99 998	31
30	7.94 084	1424	7.94 086	1424	12.05 914	9.99 998	**30**
31	7.95 508	1379	7.95 510	1379	12.04 490	9.99 998	29
32	7.96 887	1336	7.96 889	1336	12.03 111	9.99 998	28
33	7.98 223	1297	7.98 225	1297	12.01 775	9.99 998	27
34	7.99 520	1259	7.99 522	1259	12.00 478	9.99 998	26
35	8.00 779	1223	8.00 781	1223	11.99 219	9.99 998	25
36	8.02 002	1190	8.02 004	1190	11.97 996	9.99 998	24
37	8.03 192	1158	8.03 194	1159	11.96 806	9.99 997	23
38	8.04 350	1128	8.04 353	1128	11.95 647	9.99 997	22
39	8.05 478	1100	8.05 481	1100	11.94 519	9.99 997	21
40	8.06 578	1072	8.06 581	1072	11.93 419	9.99 997	**20**
41	8.07 650	1046	8.07 653	1047	11.92 347	9.99 997	19
42	8.08 696	1022	8.08 700	1022	11.91 300	9.99 997	18
43	8.09 718	999	8.09 722	998	11.90 278	9.99 997	17
44	8.10 717	976	8.10 720	976	11.89 280	9.99 996	16
45	8.11 693	954	8.11 696	955	11.88 304	9.99 996	15
46	8.12 647	934	8.12 651	934	11.87 349	9.99 996	14
47	8.13 581	914	8.13 585	915	11.86 415	9.99 996	13
48	8.14 495	896	8.14 500	895	11.85 500	9.99 996	12
49	8.15 391	877	8.15 395	878	11.84 605	9.99 996	11
50	8.16 268	860	8.16 273	860	11.83 727	9.99 995	**10**
51	8.17 128	843	8.17 133	843	11.82 867	9.99 995	9
52	8.17 971	827	8.17 976	828	11.82 024	9.99 995	8
53	8.18 798	812	8.18 804	812	11.81 196	9.99 995	7
54	8.19 610	797	8.19 616	797	11.80 384	9.99 995	6
55	8.20 407	782	8.20 413	782	11.79 587	9.99 994	5
56	8.21 189	769	8.21 195	769	11.78 805	9.99 994	4
57	8.21 958	755	8.21 964	756	11.78 036	9.99 994	3
58	8.22 713	743	8.22 720	742	11.77 280	9.99 994	2
59	8.23 456	730	8.23 462	730	11.76 538	9.99 994	1
60	8.24 186		8.24 192		11.75 808	9.99 993	**0**
	L Cos	d	L Cot	c d	L Tan	L Sin	′

Since the tabular differences in the first three columns of this page, and on each of the two pages following, are so large and change so rapidly in value that ordinary linear interpolation does not give results accurate to five places of decimals, special methods of interpolation are necessary. A brief account of these special methods is given on page 270 of the tables.

89°

*From E. Richard Heineman, "Plane Trigonometry with Tables," 3d ed., McGraw-Hill Book Company, New York, 1964. Subtract 10 from each entry.

TABLE 3. LOGARITHMS OF TRIGONOMETRIC FUNCTIONS (continued) **643**

1°

′	L Sin	d	L Tan	c d	L Cot	L Cos	
0	8.24 186	717	8.24 192	718	11.75 808	9.99 993	60
1	8.24 903	706	8.24 910	706	11.75 090	9.99 993	59
2	8.25 609	695	8.25 616	696	11.74 384	9.99 993	58
3	8.26 304	684	8.26 312	684	11.73 688	9.99 993	57
4	8.26 988	673	8.26 996	673	11.73 004	9.99 992	56
5	8.27 661	663	8.27 669	663	11.72 331	9.99 992	55
6	8.28 324	653	8.28 332	654	11.71 668	9.99 992	54
7	8.28 977	644	8.28 986	643	11.71 014	9.99 992	53
8	8.29 621	634	8.29 629	634	11.70 371	9.99 992	52
9	8.30 255	624	8.30 263	625	11.69 737	9.99 991	51
10	8.30 879	616	8.30 888	617	11.69 112	9.99 991	50
11	8.31 495	608	8.31 505	607	11.68 495	9.99 991	49
12	8.32 103	599	8.32 112	599	11.67 888	9.99 990	48
13	8.32 702	590	8.32 711	591	11.67 289	9.99 990	47
14	8.33 292	583	8.33 302	584	11.66 698	9.99 990	46
15	8.33 875	575	8.33 886	575	11.66 114	9.99 990	45
16	8.34 450	568	8.34 461	568	11.65 539	9.99 989	44
17	8.35 018	560	8.35 029	561	11.64 971	9.99 989	43
18	8.35 578	553	8.35 590	553	11.64 410	9.99 989	42
19	8.36 131	547	8.36 143	546	11.63 857	9.99 989	41
20	8.36 678	539	8.36 689	540	11.63 311	9.99 988	40
21	8.37 217	533	8.37 229	533	11.62 771	9.99 988	39
22	8.37 750	526	8.37 762	527	11.62 238	9.99 988	38
23	8.83 276	520	8.38 289	520	11.61 711	9.99 987	37
24	8.38 796	514	8.38 809	514	11.61 191	9.99 987	36
25	8.39 310	508	8.39 323	509	11.60 677	9.99 987	35
26	8.39 818	502	8.39 832	502	11.60 168	9.99 986	34
27	8.40 320	496	8.40 334	496	11.59 666	9.99 986	33
28	8.40 816	491	8.40 830	491	11.59 170	9.99 986	32
29	8.41 307	485	8.41 321	486	11.58 679	9.99 985	31
30	8.41 792	480	8.41 807	480	11.58 193	9.99 985	30
31	8.42 272	474	8.42 287	475	11.57 713	9.99 985	29
32	8.42 746	470	8.42 762	470	11.57 238	9.99 984	28
33	8.43 216	464	8.43 232	464	11.56 768	9.99 984	27
34	8.43 680	459	8.43 696	460	11.56 304	9.99 984	26
35	8.44 139	455	8.44 156	455	11.55 844	9.99 983	25
36	8.44 594	450	8.44 611	450	11.55 389	9.99 983	24
37	8.45 044	445	8.45 061	446	11.54 939	9.99 983	23
38	8.45 489	441	8.45 507	441	11.54 493	9.99 982	22
39	8.45 930	436	8.45 948	437	11.54 052	9.99 982	21
40	8.46 366	433	8.46 385	432	11.53 615	9.99 982	20
41	8.46 799	427	8.46 817	428	11.53 183	9.99 981	19
42	8.47 226	424	8.47 245	424	11.52 755	9.99 981	18
43	8.47 650	419	8.47 669	420	11.52 331	9.99 981	17
44	8.48 069	416	8.48 089	416	11.51 911	9.99 980	16
45	8.48 485	411	8.48 505	412	11.51 495	9.99 980	15
46	8.48 896	408	8.48 917	408	11.51 083	9.99 979	14
47	8.49 304	404	8.49 325	404	11.50 675	9.99 979	13
48	8.49 708	400	8.49 729	401	11.50 271	9.99 979	12
49	8.50 108	396	8.50 130	397	11.49 870	9.99 978	11
50	8.50 504	393	8.50 527	393	11.49 473	9.99 978	10
51	8.50 897	390	8.50 920	390	11.49 080	9.99 977	9
52	8.51 287	386	8.51 310	386	11.48 690	9.99 977	8
53	8.51 673	382	8.51 696	383	11.48 304	9.99 977	7
54	8.52 055	379	8.52 079	380	11.47 921	9.99 976	6
55	8.52 434	376	8.52 459	376	11.47 541	9.99 976	5
56	8.52 810	373	8.52 835	373	11.47 165	9.99 975	4
57	8.53 183	369	8.53 208	370	11.46 792	9.99 975	3
58	8.53 552	367	8.53 578	367	11.46 422	9.99 974	2
59	8.53 919	363	8.53 945	363	11.46 055	9.99 974	1
60	8.54 282		8.54 308		11.45 692	9.99 974	0
	L Cos	d	L Cot	c d	L Tan	L Sin	′

If ordinary linear interpolation is not sufficiently accurate, use the special methods described on page 270 of the tables.

88°

TABLE 3. LOGARITHMS OF TRIGONOMETRIC FUNCTIONS (continued)   644

2°

'	L Sin	d	L Tan	c d	L Cot	L Cos	
0	8.54 282		8.54 308		11.45 692	9.99 974	60
1	8.54 642	360	8.54 669	361	11.45 331	9.99 973	59
2	8.54 999	357	8.55 027	358	11.44 973	9.99 973	58
3	8.55 354	355	8.55 382	355	11.44 618	9.99 972	57
4	8.55 705	351	8.55 734	352	11.44 266	9.99 972	56
5	8.56 054	349	8.56 083	349	11.43 917	9.99 971	55
6	8.56 400	346	8.56 429	346	11.43 571	9.99 971	54
7	8.56 743	343	8.56 773	344	11.43 227	9.99 970	53
8	8.57 084	341	8.57 114	341	11.42 886	9.99 970	52
9	8.57 421	337	8.57 452	338	11.42 548	9.99 969	51
10	8.57 757	336	8.57 788	336	11.42 212	9.99 969	50
11	8.58 089	332	8.58 121	333	11.41 879	9.99 968	49
12	8.58 419	330	8.58 451	330	11.41 549	9.99 968	48
13	8.58 747	328	8.58 779	328	11.41 221	9.99 967	47
14	8.59 072	325	8.59 105	326	11.40 895	9.99 967	46
15	8.59 395	323	8.59 428	323	11.40 572	9.99 967	45
16	8.59 715	320	8.59 749	321	11.40 251	9.99 966	44
17	8.60 033	318	8.60 068	319	11.39 932	9.99 966	43
18	8.60 349	316	8.60 384	316	11.39 616	9.99 965	42
19	8.60 662	313	8.60 698	314	11.39 302	9.99 964	41
20	8.60 973	311	8.61 009	311	11.38 991	9.99 964	40
21	8.61 282	309	8.61 319	310	11.38 681	9.99 963	39
22	8.61 589	307	8.61 626	307	11.38 374	9.99 963	38
23	8.61 894	305	8.61 931	305	11.38 069	9.99 962	37
24	8.62 196	302	8.62 234	303	11.37 766	9.99 962	36
25	8.62 497	301	8.62 535	301	11.37 465	9.99 961	35
26	8.62 795	298	8.62 834	299	11.37 166	9.99 961	34
27	8.63 091	296	8.63 131	297	11.36 869	9.99 960	33
28	8.63 385	294	8.63 426	295	11.36 574	9.99 960	32
29	8.63 678	293	8.63 718	292	11.36 282	9.99 959	31
30	8.63 968	290	8.64 009	291	11.35 991	9.99 959	30
31	8.64 256	288	8.64 298	289	11.35 702	9.99 958	29
32	8.64 543	287	8.64 585	287	11.35 415	9.99 958	28
33	8.64 827	284	8.64 870	285	11.35 130	9.99 957	27
34	8.65 110	283	8.65 154	284	11.34 846	9.99 956	26
35	8.65 391	281	8.65 435	281	11.34 565	9.99 956	25
36	8.65 670	279	8.65 715	280	11.34 285	9.99 955	24
37	8.65 947	277	8.65 993	278	11.34 007	9.99 955	23
38	8 66 223	276	8.66 269	276	11.33 731	9.99 954	22
39	8.66 497	274	8.66 543	274	11.33 457	9.99 954	21
40	8.66 769	272	8.66 816	273	11.33 184	9.99 953	20
41	8.67 039	270	8.67 087	271	11.32 913	9.99 952	19
42	8.67 308	269	8.67 356	269	11.32 644	9.99 952	18
43	8.67 575	267	8.67 624	268	11.32 376	9.99 951	17
44	8.67 841	266	8.67 890	266	11.32 110	9.99 951	16
45	8.68 104	263	8.68 154	264	11.31 846	9.99 950	15
46	8.68 367	263	8.68 417	263	11.31 583	9.99 949	14
47	8.68 627	260	8.68 678	261	11.31 322	9.99 949	13
48	8.68 886	259	8.68 938	260	11.31 062	9.99 948	12
49	8.69 144	258	8.69 196	258	11.30 804	9.99 948	11
50	8.69 400	256	8.69 453	257	11.30 547	9.99 947	10
51	8.69 654	254	8.69 708	255	11.30 292	9.99 946	9
52	8.69 907	253	8.69 962	254	11.30 038	9.99 946	8
53	8.70 159	252	8.70 214	252	11.29 786	9.99 945	7
54	8.70 409	250	8.70 465	251	11.29 535	9.99 944	6
55	8.70 658	249	8.70 714	249	11.29 286	9.99 944	5
56	8.70 905	247	8.70 962	248	11.29 038	9.99 943	4
57	8.71 151	246	8.71 208	246	11.28 792	9.99 942	3
58	8.71 395	244	8.71 453	245	11.28 547	9.99 942	2
59	8.71 638	243	8.71 697	244	11.28 303	9.99 941	1
60	8.71 880	242	8.71 940	243	11.28 060	9.99 940	0
	L Cos	d	L Cot	c d	L Tan	L Sin	'

If ordinary linear interpolation is not sufficiently accurate, use the special methods described on page 270 of the tables.

87°

TABLE 3. LOGARITHMS OF TRIGONOMETRIC FUNCTIONS (continued)   **645**

3°

′	L Sin	d	L Tan	c d	L Cot	L Cos	′
0	8.71 880	240	8.71 940	241	11.28 060	9.99 940	60
1	8.72 120	239	8.72 181	239	11.27 819	9.99 940	59
2	8.72 359	238	8.72 420	239	11.27 580	9.99 939	58
3	8.72 597	237	8.72 659	237	11.27 341	9.99 938	57
4	8.72 834	235	8.72 896	236	11.27 104	9.99 938	56
5	8.73 069	234	8.73 132	234	11.26 868	9.99 937	55
6	8.73 303	232	8.73 366	234	11.26 634	9.99 936	54
7	8.73 535	232	8.73 600	232	11.26 400	9.99 936	53
8	8.73 767	230	8.73 832	231	11.26 168	9.99 935	52
9	8.73 997	229	8.74 063	229	11.25 937	9.99 934	51
10	8.74 226	228	8.74 292	229	11.25 708	9.99 934	50
11	8.74 454	226	8.74 521	227	11.25 479	9.99 933	49
12	8.74 680	226	8.74 748	226	11.25 252	9.99 932	48
13	8.74 906	224	8.74 974	225	11.25 026	9.99 932	47
14	8.75 130	223	8.75 199	224	11.24 801	9.99 931	46
15	8.75 353	222	8.75 423	222	11.24 577	9.99 930	45
16	8.75 575	220	8.75 645	222	11.24 355	9.99 929	44
17	8.75 795	220	8.75 867	220	11.24 133	9.99 929	43
18	8.76 015	219	8.76 087	219	11.23 913	9.99 928	42
19	8.76 234	217	8.76 306	219	11.23 694	9.99 927	41
20	8.76 451	216	8.76 525	217	11.23 475	9.99 926	40
21	8.76 667	216	8.76 742	216	11.23 258	9.99 926	39
22	8.76 883	214	8.76 958	215	11.23 042	9.99 925	38
23	8.77 097	213	8.77 173	214	11.22 827	9.99 924	37
24	8.77 310	212	8.77 387	213	11.22 613	9.99 923	36
25	8.77 522	211	8.77 600	211	11.22 400	9.99 923	35
26	8.77 733	210	8.77 811	211	11.22 189	9.99 922	34
27	8.77 943	209	8.78 022	210	11.21 978	9.99 921	33
28	8.78 152	208	8.78 232	209	11.21 768	9.99 920	32
29	8.78 360	208	8.78 441	208	11.21 559	9.99 920	31
30	8.78 568	206	8.78 649	206	11.21 351	9.99 919	30
31	8.78 774	205	8.78 855	206	11.21 145	9.99 918	29
32	8.78 979	204	8.79 061	205	11.20 939	9.99 917	28
33	8.79 183	203	8.79 266	204	11.20 734	9.99 917	27
34	8.79 386	202	8.79 470	203	11.20 530	9.99 916	26
35	8.79 588	201	8.79 673	202	11.20 327	9.99 915	25
36	8.79 789	201	8.79 875	201	11.20 125	9.99 914	24
37	8.79 990	199	8.80 076	201	11.19 924	9.99 913	23
38	8.80 189	199	8.80 277	199	11.19 723	9.99 913	22
39	8.80 388	197	8.80 476	198	11.19 524	9.99 912	21
40	8.80 585	197	8.80 674	198	11.19 326	9.99 911	20
41	8.80 782	196	8.80 872	196	11.19 128	9.99 910	19
42	8.80 978	195	8.81 068	196	11.18 932	9.99 909	18
43	8.81 173	194	8.81 264	195	11.18 736	9.99 909	17
44	8.81 367	193	8.81 459	194	11.18 541	9.99 908	16
45	8.81 560	192	8.81 653	193	11.18 347	9.99 907	15
46	8.81 752	192	8.81 846	192	11.18 154	9.99 906	14
47	8.81 944	190	8.82 038	192	11.17 962	9.99 905	13
48	8.82 134	190	8.82 230	190	11.17 770	9.99 904	12
49	8.82 324	189	8.82 420	190	11.17 580	9.99 904	11
50	8.82 513	188	8.82 610	189	11.17 390	9.99 903	10
51	8.82 701	187	8.82 799	188	11.17 201	9.99 902	9
52	8.82 888	187	8.82 987	188	11.17 013	9.99 901	8
53	8.83 075	186	8.83 175	186	11.16 825	9.99 900	7
54	8.83 261	185	8.83 361	186	11.16 639	9.99 899	6
55	8.83 446	184	8.83 547	185	11.16 453	9.99 898	5
56	8.83 630	183	8.83 732	184	11.16 268	9.99 898	4
57	8.83 813	183	8.83 916	184	11.16 084	9.99 897	3
58	8.83 996	181	8.84 100	182	11.15 900	9.99 896	2
59	8.84 177	181	8.84 282	182	11.15 718	9.99 895	1
60	8.84 358		8.84 464		11.15 536	9.99 894	0
	L Cos	d	L Cot	c d	L Tan	L Sin	′

Prop. Pts.

	239	237	235	234
2	47.8	47.4	47.0	46.8
3	71.7	71.1	70.5	70.2
4	95.6	94.8	94.0	93.6
5	119.5	118.5	117.5	117.0
6	143.4	142.2	141.0	140.4
7	167.3	165.9	164.5	163.8
8	191.2	189.6	188.0	187.2
9	215.1	213.3	211.5	210.6

	232	229	227	226
2	46.4	45.8	45.4	45.2
3	69.6	68.7	68.1	67.8
4	92.8	91.6	90.8	90.4
5	116.0	114.5	113.5	113.0
6	139.2	137.4	136.2	135.6
7	162.4	160.3	158.9	158.2
8	185.6	183.2	181.6	180.8
9	208.8	206.1	204.3	203.4

	224	222	220	219
2	44.8	44.4	44.0	43.8
3	67.2	66.6	66.0	65.7
4	89.6	88.8	88.0	87.6
5	112.0	111.0	110.0	109.5
6	134.4	133.2	132.0	131.4
7	156.8	155.4	154.0	153.3
8	179.2	177.6	176.0	175.2
9	201.6	199.8	198.0	197.1

	217	215	213	211
2	43.4	43.0	42.6	42.2
3	65.1	64.5	63.9	63.3
4	86.8	86.0	85.2	84.4
5	108.5	107.5	106.5	105.5
6	130.2	129.0	127.8	126.6
7	151.9	150.5	149.1	147.7
8	173.6	172.0	170.4	168.8
9	195.3	193.5	191.7	189.9

	208	206	203	201
2	41.6	41.2	40.6	40.2
3	62.4	61.8	60.9	60.3
4	83.2	82.4	81.2	80.4
5	104.0	103.0	101.5	100.5
6	124.8	123.6	121.8	120.6
7	145.6	144.2	142.1	140.7
8	166.4	164.8	162.4	160.8
9	187.2	185.4	182.7	180.9

	199	197	195	193
2	39.8	39.4	39.0	38.6
3	59.7	59.1	58.5	57.9
4	79.6	78.8	78.0	77.2
5	99.5	98.5	97.5	96.5
6	119.4	118.2	117.0	115.8
7	139.3	137.9	136.5	135.1
8	159.2	157.6	156.0	154.4
9	179.1	177.3	175.5	173.7

	192	190	188	186
2	38.4	38.0	37.6	37.2
3	57.6	57.0	56.4	55.8
4	76.8	76.0	75.2	74.4
5	96.0	95.0	94.0	93.0
6	115.2	114.0	112.8	111.6
7	134.4	133.0	131.6	130.2
8	153.6	152.0	150.4	148.8
9	172.8	171.0	169.2	167.4

	184	183	182	181
2	36.8	36.6	36.4	36.2
3	55.2	54.9	54.6	54.3
4	73.6	73.2	72.8	72.4
5	92.0	91.5	91.0	90.5
6	110.4	109.8	109.2	108.6
7	128.8	128.1	127.4	126.7
8	147.2	146.4	145.6	144.8
9	165.6	164.7	163.8	162.9

TABLE 3. LOGARITHMS OF TRIGONOMETRIC FUNCTIONS (continued)   646

4°

′	L Sin	d	L Tan	c d	L Cot	L Cos	
0	8.84 358	181	8.84 464	182	11.15 536	9.99 894	**60**
1	8.84 539	179	8.84 646	180	11.15 354	9.99 893	59
2	8.84 718	179	8.84 826	180	11.15 174	9.99 892	58
3	8.84 897	178	8.85 006	179	11.14 994	9.99 891	57
4	8.85 075	177	8.85 185	178	11.14 815	9.99 891	56
5	8.85 252	177	8.85 363	177	11.14 637	9.99 890	55
6	8.85 429	176	8.85 540	177	11.14 460	9.99 889	54
7	8.85 605	175	8.85 717	176	11.14 283	9.99 888	53
8	8.85 780	175	8.85 893	176	11.14 107	9.99 887	52
9	8.85 955	173	8.86 069	174	11.13 931	9.99 886	51
10	8.86 128	173	8.86 243	174	11.13 757	9.99 885	**50**
11	8.86 301	173	8.86 417	174	11.13 583	9.99 884	49
12	8.86 474	171	8.86 591	172	11.13 409	9.99 883	48
13	8.86 645	171	8.86 763	172	11.13 237	9.99 882	47
14	8.86 816	171	8.86 935	172	11.13 065	9.99 881	46
15	8.86 987	169	8.87 106	171	11.12 894	9.99 880	45
16	8.87 156	169	8.87 277	170	11.12 723	9.99 879	44
17	8.87 325	169	8.87 447	169	11.12 553	9.99 879	43
18	8.87 494	167	8.87 616	169	11.12 384	9.99 878	42
19	8.87 661	168	8.87 785	168	11.12 215	9.99 877	41
20	8.87 829	166	8.87 953	167	11.12 047	9.99 876	**40**
21	8.87 995	166	8.88 120	167	11.11 880	9.99 875	39
22	8.88 161	165	8.88 287	166	11.11 713	9.99 874	38
23	8.88 326	164	8.88 453	165	11.11 547	9.99 873	37
24	8.88 490	164	8.88 618	165	11.11 382	9.99 872	36
25	8.88 654	163	8.88 783	165	11.11 217	9.99 871	35
26	8.88 817	163	8.88 948	163	11.11 052	9.99 870	34
27	8.88 980	162	8.89 111	163	11.10 889	9.99 869	33
28	8.89 142	162	8.89 274	163	11.10 726	9.99 868	32
29	8.89 304	160	8.89 437	161	11.10 563	9.99 867	31
30	8.89 464	161	8.89 598	162	11.10 402	9.99 866	**30**
31	8.89 625	159	8.89 760	160	11.10 240	9.99 865	29
32	8.89 784	159	8.89 920	160	11.10 080	9.99 864	28
33	8.89 943	159	8.90 080	160	11.09 920	9.99 863	27
34	8.90 102	158	8.90 240	159	11.09 760	9.99 862	26
35	8.90 260	157	8.90 399	158	11.09 601	9.99 861	25
36	8.90 417	157	8.90 557	158	11.09 443	9.99 860	24
37	8.90 574	156	8.90 715	157	11.09 285	9.99 859	23
38	8.90 730	155	8.90 872	157	11.09 128	9.99 858	22
39	8.90 885	155	8.91 029	156	11.08 971	9.99 857	21
40	8.91 040	155	8.91 185	155	11.08 815	9.99 856	**20**
41	8.91 195	154	8.91 340	155	11.08 660	9.99 855	19
42	8.91 349	153	8.91 495	155	11.08 505	9.99 854	18
43	8.91 502	153	8.91 650	153	11.08 350	9.99 853	17
44	8.91 655	152	8.91 803	154	11.08 197	9.99 852	16
45	8.91 807	152	8.91 957	153	11.08 043	9.99 851	15
46	8.91 959	151	8.92 110	152	11.07 890	9.99 850	14
47	8.92 110	151	8.92 262	152	11.07 738	9.99 848	13
48	8.92 261	150	8.92 414	151	11.07 586	9.99 847	12
49	8.92 411	150	8.92 565	151	11.07 435	9.99 846	11
50	8.92 561	149	8.92 716	150	11.07 284	9.99 845	**10**
51	8.92 710	149	8.92 866	150	11.07 134	9.99 844	9
52	8.92 859	148	8.93 016	149	11.06 984	9.99 843	8
53	8.93 007	147	8.93 165	148	11.06 835	9.99 842	7
54	8.93 154	147	8.93 313	149	11.06 687	9.99 841	6
55	8.93 301	147	8.93 462	147	11.06 538	9.99 840	5
56	8.93 448	146	8.93 609	147	11.06 391	9.99 839	4
57	8.93 594	146	8.93 756	147	11.06 244	9.99 838	3
58	8.93 740	145	8.93 903	146	11.06 097	9.99 837	2
59	8.93 885	145	8.94 049	146	11.05 951	9.99 836	1
60	8.94 030		8.94 195		11.05 805	9.99 834	**0**
	L Cos	d	L Cot	c d	L Tan	L Sin	′

Prop. Pts.

	182	181	180	179
2	36.4	36.2	36.0	35.8
3	54.6	54.3	54.0	53.7
4	72.8	72.4	72.0	71.6
5	91.0	90.5	90.0	89.5
6	109.2	108.6	108.0	107.4
7	127.4	126.7	126.0	125.3
8	145.6	144.8	144.0	143.2
9	163.8	162.9	162.0	161.1

	178	177	176	175
2	35.6	35.4	35.2	35.0
3	53.4	53.1	52.8	52.5
4	71.2	70.8	70.4	70.0
5	89.0	88.5	88.0	87.5
6	106.8	106.2	105.6	105.0
7	124.6	123.9	123.2	122.5
8	142.4	141.6	140.8	140.0
9	160.2	159.3	158.4	157.5

	174	173	172	171
2	34.8	34.6	34.4	34.2
3	52.2	51.9	51.6	51.3
4	69.6	69.2	68.8	68.4
5	87.0	86.5	86.0	85.5
6	104.4	103.8	103.2	102.6
7	121.8	121.1	120.4	119.7
8	139.2	138.4	137.6	136.8
9	156.6	155.7	154.8	153.9

	170	169	168	167
2	34.0	33.8	33.6	33.4
3	51.0	50.7	50.4	50.1
4	68.0	67.6	67.2	66.8
5	85.0	84.5	84.0	83.5
6	102.0	101.4	100.8	100.2
7	119.0	118.3	117.6	116.9
8	136.0	135.2	134.4	133.6
9	153.0	152.1	151.2	150.3

	166	165	164	163
2	33.2	33.0	32.8	32.6
3	49.8	49.5	49.2	48.9
4	66.4	66.0	65.6	65.2
5	83.0	82.5	82.0	81.5
6	99.6	99.0	98.4	97.8
7	116.2	115.5	114.8	114.1
8	132.8	132.0	131.2	130.4
9	149.4	148.5	147.6	146.7

	162	161	160	159
2	32.4	32.2	32.0	31.8
3	48.6	48.3	48.0	47.7
4	64.8	64.4	64.0	63.6
5	81.0	80.5	80.0	79.5
6	97.2	96.6	96.0	95.4
7	113.4	112.7	112.0	111.3
8	129.6	128.8	128.0	127.2
9	145.8	144.9	144.0	143.1

	158	157	156	155
2	31.6	31.4	31.2	31.0
2	47.4	47.1	46.8	46.5
4	63.2	62.8	62.4	62.0
5	79.0	78.5	78.0	77.5
6	94.8	94.2	93.6	93.0
7	110.6	109.9	109.2	108.5
8	126.4	125.6	124.8	124.0
9	142.2	141.3	140.4	139.5

	154	153	152
2	30.8	30.6	30.4
3	46.2	45.9	45.6
4	61.6	61.2	60.8
5	77.0	76.5	76.0
6	92.4	91.8	91.2
7	107.8	107.1	106.4
8	123.2	122.4	121.6
9	138.6	137.7	136.8

TABLE 3. LOGARITHMS OF TRIGONOMETRIC FUNCTIONS (continued)   **647**

5°

′	L Sin	d	L Tan	c d	L Cot	L Cos	
0	8.94 030		8.94 195		11.05 805	9.99 834	60
1	8.94 174	144	8.94 340	145	11.05 660	9.99 833	59
2	8.94 317	143	8.94 485	145	11.05 515	9.99 832	58
3	8.94 461	144	8.94 630	145	11.05 370	9.99 831	57
4	8.94 603	142	8.94 773	143	11.05 227	9.99 830	56
5	8.94 746	143	8.94 917	144	11.05 083	9.99 829	55
6	8.94 887	141	8.95 060	143	11.04 940	9.99 828	54
7	8.95 029	142	8.95 202	142	11.04 798	9.99 827	53
8	8.95 170	141	8.95 344	142	11.04 656	9.99 825	52
9	8.95 310	140	8.95 486	142	11.04 514	9.99 824	51
10	8.95 450	140	8.95 627	141	11.04 373	9.99 823	50
11	8.95 589	139	8.95 767	140	11.04 233	9.99 822	49
12	8.95 728	139	8.95 908	141	11.04 092	9.99 821	48
13	8.95 867	139	8.96 047	139	11.03 953	9.99 820	47
14	8.96 005	138	8.96 187	140	11.03 813	9.99 819	46
15	8.96 143	138	8.96 325	138	11.03 675	9.99 817	45
16	8.96 280	137	8.96 464	139	11.03 536	9.99 816	44
17	8.96 417	137	8.96 602	138	11.03 398	9.99 815	43
18	8.96 553	136	8.96 739	137	11.03 261	9.99 814	42
19	8.96 689	136	8.96 877	138	11.03 123	9.99 813	41
20	8.96 825	136	8.97 013	136	11.02 987	9.99 812	40
21	8.96 960	135	8.97 150	137	11.02 850	9.99 810	39
22	8.97 095	135	8.97 285	135	11.02 715	9.99 809	38
23	8.97 229	134	8.97 421	136	11.02 579	9.99 808	37
24	8.97 363	134	8.97 556	135	11.02 444	9.99 807	36
25	8.97 496	133	8.97 691	135	11.02 309	9.99 806	35
26	8.97 629	133	8.97 825	134	11.02 175	9.99 804	34
27	8.97 762	133	8.97 959	134	11.02 041	9.99 803	33
28	8.97 894	132	8.98 092	133	11.01 908	9.99 802	32
29	8.98 026	132	8.98 225	133	11.01 775	9.99 801	31
30	8.98 157	131	8.98 358	133	11.01 642	9.99 800	30
31	8.98 288	131	8.98 490	132	11.01 510	9.99 798	29
32	8.98 419	131	8.98 622	132	11.01 378	9.99 797	28
33	8.98 549	130	8.98 753	131	11.01 247	9.99 796	27
34	8.98 679	130	8.98 884	131	11.01 116	9.99 795	26
35	8.98 808	129	8.99 015	131	11.00 985	9.99 793	25
36	8.98 937	129	8.99 145	130	11.00 855	9.99 792	24
37	8.99 066	129	8.99 275	130	11.00 725	9.99 791	23
38	8.99 194	128	8.99 405	130	11.00 595	9.99 790	22
39	8.99 322	128	8.99 534	129	11.00 466	9.99 788	21
40	8.99 450	128	8.99 662	128	11.00 338	9.99 787	20
41	8.99 577	127	8.99 791	129	11.00 209	9.99 786	19
42	8.99 704	127	8.99 919	128	11.00 081	9.99 785	18
43	8.99 830	126	9.00 046	127	10.99 954	9.99 783	17
44	8.99 956	126	9.00 174	128	10.99 826	9.99 782	16
45	9.00 082	126	9.00 301	127	10.99 699	9.99 781	15
46	9.00 207	125	9.00 427	126	10.99 573	9.99 780	14
47	9.00 332	125	9.00 553	126	10.99 447	9.99 778	13
48	9.00 456	124	9.00 679	126	10.99 321	9.99 777	12
49	9.00 581	125	9.00 805	125	10.99 195	9.99 776	11
50	9.00 704	123	9.00 930	125	10.99 070	9.99 775	10
51	9.00 828	124	9.01 055	124	10.98 945	9.99 773	9
52	9.00 951	123	9.01 179	124	10.98 821	9.99 772	8
53	9.01 074	123	9.01 303	124	10.98 697	9.99 771	7
54	9.01 196	122	9.01 427	123	10.98 573	9.99 769	6
55	9.01 318	122	9.01 550	123	10.98 450	9.99 768	5
56	9.01 440	122	9.01 673	123	10.98 327	9.99 767	4
57	9.01 561	121	9.01 796	122	10.98 204	9.99 765	3
58	9.01 682	121	9.01 918	122	10.98 082	9.99 764	2
59	9.01 803	121	9.02 040	122	10.97 960	9.99 763	1
60	9.01 923	120	9.02 162	122	10.97 838	9.99 761	0
	L Cos	d	L Cot	c d	L Tan	L Sin	′

Prop. Pts.

	151	150	149	148
2	30.2	30.0	29.8	29.6
3	45.3	45.0	44.7	44.4
4	60.4	60.0	59.6	59.2
5	75.5	75.0	74.5	74.0
6	90.6	90.0	89.4	88.8
7	105.7	105.0	104.3	103.6
8	120.8	120.0	119.2	118.4
9	135.9	135.0	134.1	133.2

	147	146	145	144
2	29.4	29.2	29.0	28.8
3	44.1	43.8	43.5	43.2
4	58.8	58.4	58.0	57.6
5	73.5	73.0	72.5	72.0
6	88.2	87.6	87.0	86.4
7	102.9	102.2	101.5	100.8
8	117.6	116.8	116.0	115.2
9	132.3	131.4	130.5	129.6

	143	142	141	140
2	28.6	28.4	28.2	28.0
3	42.9	42.6	42.3	42.0
4	57.2	56.8	56.4	56.0
5	71.5	71.0	70.5	70.0
6	85.8	85.2	84.6	84.0
7	100.1	99.4	98.7	98.0
8	114.4	113.6	112.8	112.0
9	128.7	127.8	126.9	126.0

	139	138	137	136
2	27.8	27.6	27.4	27.2
3	41.7	41.4	41.1	40.8
4	55.6	55.2	54.8	54.4
5	69.5	69.0	68.5	68.0
6	83.4	82.8	82.2	81.6
7	97.3	96.6	95.9	95.2
8	111.2	110.4	109.6	108.8
9	125.1	124.2	123.3	122.4

	135	134	133	132
2	27.0	26.8	26.6	26.4
3	40.5	40.2	39.9	39.6
4	54.0	53.6	53.2	52.8
5	67.5	67.0	66.5	66.0
6	81.0	80.4	79.8	79.2
7	94.5	93.8	93.1	92.4
8	108.0	107.2	106.4	105.6
9	121.5	120.6	119.7	118.8

	131	130	129	128
2	26.2	26.0	25.8	25.6
3	39.3	39.0	38.7	38.4
4	52.4	52.0	51.6	51.2
5	65.5	65.0	64.5	64.0
6	78.6	78.0	77.4	76.8
7	91.7	91.0	90.3	89.6
8	104.8	104.0	103.2	102.4
9	117.9	117.0	116.1	115.2

	127	126	125	124
2	25.4	25.2	25.0	24.8
3	38.1	37.8	37.5	37.2
4	50.8	50.4	50.0	49.6
5	63.5	63.0	62.5	62.0
6	76.2	75.6	75.0	74.4
7	88.9	88.2	87.5	86.8
8	101.6	100.8	100.0	99.2
9	114.3	113.4	112.5	111.6

	123	122	121	120
2	24.6	24.4	24.2	24.0
3	36.9	36.6	36.3	36.0
4	49.2	48.8	48.4	48.0
5	61.5	61.0	60.5	60.0
6	73.8	73.2	72.6	72.0
7	86.1	85.4	84.7	84.0
8	98.4	97.6	96.8	96.0
9	110.7	109.8	108.9	108.0

TABLE 3. LOGARITHMS OF TRIGONOMETRIC FUNCTIONS (continued)   648

6°

'	L Sin	d	L Tan	c d	L Cot	L Cos	'
0	9.01 923	120	9.02 162	121	10.97 838	9.99 761	60
1	9.02 043	120	9.02 283	121	10.97 717	9.99 760	59
2	9.02 163	120	9.02 404	121	10.97 596	9.99 759	58
3	9.02 283	119	9.02 525	120	10.97 475	9.99 757	57
4	9.02 402	118	9.02 645	121	10.97 355	9.99 756	56
5	9.02 520	119	9.02 766	119	10.97 234	9.99 755	55
6	9.02 639	118	9.02 885	120	10.97 115	9.99 753	54
7	9.02 757	117	9.03 005	119	10.96 995	9.99 752	53
8	9.02 874	118	9.03 124	118	10.96 876	9.99 751	52
9	9.02 992	117	9.03 242	119	10.96 758	9.99 749	51
10	9.03 109	117	9.03 361	118	10.96 639	9.99 748	50
11	9.03 226	116	9.03 479	118	10.96 521	9.99 747	49
12	9.03 342	116	9.03 597	117	10.96 403	9.99 745	48
13	9.03 458	116	9.03 714	118	10.96 286	9.99 744	47
14	9.03 574	116	9.03 832	116	10.96 168	9.99 742	46
15	9.03 690	115	9.03 948	117	10.96 052	9.99 741	45
16	9.03 805	115	9.04 065	116	10.95 935	9.99 740	44
17	9.03 920	114	9.04 181	116	10.95 819	9.99 738	43
18	9.04 034	115	9.04 297	116	10.95 703	9.99 737	42
19	9.04 149	113	9.04 413	115	10.95 587	9.99 736	41
20	9.04 262	114	9.04 528	115	10.95 472	9.99 734	40
21	9.04 376	114	9.04 643	115	10.95 357	9.99 733	39
22	9.04 490	113	9.04 758	115	10.95 242	9.99 731	38
23	9.04 603	112	9.04 873	114	10.95 127	9.99 730	37
24	9.04 715	113	9.04 987	114	10.95 013	9.99 728	36
25	9.04 828	112	9.05 101	113	10.94 899	9.99 727	35
26	9.04 940	112	9.05 214	114	10.94 786	9.99 726	34
27	9.05 052	112	9.05 328	113	10.94 672	9.99 724	33
28	9.05 164	111	9.05 441	112	10.94 559	9.99 723	32
29	9.05 275	111	9.05 553	113	10.94 447	9.99 721	31
30	9.05 386	111	9.05 666	112	10.94 334	9.99 720	30
31	9.05 497	110	9.05 778	112	10.94 222	9.99 718	29
32	9.05 607	110	9.05 890	112	10.94 110	9.99 717	28
33	9.05 717	110	9.06 002	111	10.93 998	9.99 716	27
34	9.05 827	110	9.06 113	111	10.93 887	9.99 714	26
35	9.05 937	109	9.06 224	111	10.93 776	9.99 713	25
36	9.06 046	109	9.06 335	110	10.93 665	9.99 711	24
37	9.06 155	109	9.06 445	111	10.93 555	9.99 710	23
38	9.06 264	108	9.06 556	110	10.93 444	9.99 708	22
39	9.06 372	109	9.06 666	109	10.93 334	9.99 707	21
40	9.06 481	108	9.06 775	110	10.93 225	9.99 705	20
41	9.06 589	107	9.06 885	109	10.93 115	9.99 704	19
42	9.06 696	108	9.06 994	109	10.93 006	9.99 702	18
43	9.06 804	107	9.07 103	108	10.92 897	9.99 701	17
44	9.06 911	107	9.07 211	109	10.92 789	9.99 699	16
45	9.07 018	106	9.07 320	108	10.92 680	9.99 698	15
46	9.07 124	107	9.07 428	108	10.92 572	9.99 696	14
47	9.07 231	106	9.07 536	107	10.92 464	9.99 695	13
48	9.07 337	105	9.07 643	108	10.92 357	9.99 693	12
49	9.07 442	106	9.07 751	107	10.92 249	9.99 692	11
50	9.07 548	105	9.07 858	106	10.92 142	9.99 690	10
51	9.07 653	105	9.07 964	107	10.92 036	9.99 689	9
52	9.07 758	105	9.08 071	106	10.91 929	9.99 687	8
53	9.07 863	105	9.08 177	106	10.91 823	9.99 686	7
54	9.07 968	104	9.08 283	106	10.91 717	9.99 684	6
55	9.08 072	104	9.08 389	106	10.91 611	9.99 683	5
56	9.08 176	104	9.08 495	105	10.91 505	9.99 681	4
57	9.08 280	103	9.08 600	105	10.91 400	9.99 680	3
58	9.08 383	103	9.08 705	105	10.91 295	9.99 678	2
59	9.08 486	103	9.08 810	104	10.91 190	9.99 677	1
60	9.08 589		9.08 914		10.91 086	9.99 675	0
	L Cos	d	L Cot	c d	L Tan	L Sin	'

Prop. Pts.

	121	120	119
1	12.1	12.0	11.9
2	24.2	24.0	23.8
3	36.3	36.0	35.7
4	48.4	48.0	47.6
5	60.5	60.0	59.5
6	72.6	72.0	71.4
7	84.7	84.0	83.3
8	96.8	96.0	95.2
9	108.9	108.0	107.1

	118	117	116
1	11.8	11.7	11.6
2	23.6	23.4	23.2
3	35.4	35.1	34.8
4	47.2	46.8	46.4
5	59.0	58.5	58.0
6	70.8	70.2	69.6
7	82.6	81.9	81.2
8	94.4	93.6	92.8
9	106.2	105.3	104.4

	115	114	113
1	11.5	11.4	11.3
2	23.0	22.8	22.6
3	34.5	34.2	33.9
4	46.0	45.6	45.2
5	57.5	57.0	56.5
6	69.0	68.4	67.8
7	80.5	79.8	79.1
8	92.0	91.2	90.4
9	103.5	102.6	101.7

	112	111	110
1	11.2	11.1	11.0
2	22.4	22.2	22.0
3	33.6	33.3	33.0
4	44.8	44.4	44.0
5	56.0	55.5	55.0
6	67.2	66.6	66.0
7	78.4	77.7	77.0
8	89.6	88.8	88.0
9	100.8	99.9	99.0

	109	108	107	106
1	10.9	10.8	10.7	10.6
2	21.8	21.6	21.4	21.2
3	32.7	32.4	32.1	31.8
4	43.6	43.2	42.8	42.4
5	54.5	54.0	53.5	53.0
6	65.4	64.8	64.2	63.6
7	76.3	75.6	74.9	74.2
8	87.2	86.4	85.6	84.8
9	98.1	97.2	96.3	95.4

83°

TABLE 3. LOGARITHMS OF TRIGONOMETRIC FUNCTIONS (continued)   649

7°

'	L Sin	d	L Tan	c d	L Cot	L Cos		Prop. Pts.
0	9.08 589	103	9.08 914	105	10.91 086	9.99 675	60	
1	9.08 692	103	9.09 019	104	10.90 981	9.99 674	59	
2	9.08 795	102	9.09 123	104	10.90 877	9.99 672	58	
3	9.08 897	102	9.09 227	103	10.90 773	9.99 670	57	**105 | 104 | 103**
4	9.08 999	102	9.09 330	104	10.90 670	9.99 669	56	1 10.5 10.4 10.3
5	9.09 101	101	9.09 434	103	10.90 566	9.99 667	55	2 21.0 20.8 20.6
6	9.09 202	102	9.09 537	103	10.90 463	9.99 666	54	3 31.5 31.2 30.9
7	9.09 304	101	9.09 640	102	10.90 360	9.99 664	53	4 42.0 41.6 41.2
8	9.09 405	101	9.09 742	103	10.90 258	9.99 663	52	5 52.5 52.0 51.5
9	9.09 506	100	9.09 845	102	10.90 155	9.99 661	51	6 63.0 62.4 61.8
10	9.09 606	101	9.09 947	102	10.90 053	9.99 659	50	7 73.5 72.8 72.1
11	9.09 707	100	9.10 049	101	10.89 951	9.99 658	49	8 84.0 83.2 82.4
12	9.09 807	100	9.10 150	102	10.89 850	9.99 656	48	9 94.5 93.6 92.7
13	9.09 907	99	9.10 252	101	10.89 748	9.99 655	47	
14	9.10 006	100	9.10 353	101	10.89 647	9.99 653	46	
15	9.10 106	99	9.10 454	101	10.89 546	9.99 651	45	**102 | 101 | 99**
16	9.10 205	99	9.10 555	101	10.89 445	9.99 650	44	1 10.2 10.1 9.9
17	9.10 304	98	9.10 656	100	10.89 344	9.99 648	43	2 20.4 20.2 19.8
18	9.10 402	99	9.10 756	100	10.89 244	9.99 647	42	3 30.6 30.3 29.7
19	9.10 501	98	9.10 856	100	10.89 144	9.99 645	41	4 40.8 40.4 39.6
20	9.10 599	98	9.10 956	100	10.89 044	9.99 643	40	5 51.0 50.5 49.5
21	9.10 697	98	9.11 056	99	10.88 944	9.99 642	39	6 61.2 60.6 59.4
22	9.10 795	98	9.11 155	99	10.88 845	9.99 640	38	7 71.4 70.7 69.3
23	9.10 893	97	9.11 254	99	10.88 746	9.99 638	37	8 81.6 80.8 79.2
24	9.10 990	97	9.11 353	99	10.88 647	9.99 637	36	9 91.8 90.9 89.1
25	9.11 087	97	9.11 452	99	10.88 548	9.99 635	35	
26	9.11 184	97	9.11 551	98	10.88 449	9.99 633	34	
27	9.11 281	96	9.11 649	98	10.88 351	9.99 632	33	**98 | 97 | 96**
28	9.11 377	97	9.11 747	98	10.88 253	9.99 630	32	1 9.8 9.7 9.6
29	9.11 474	96	9.11 845	98	10.88 155	9.99 629	31	2 19.6 19.4 19.2
30	9.11 570	96	9.11 943	97	10.88 057	9.99 627	30	3 29.4 29.1 28.8
31	9.11 666	95	9.12 040	98	10.87 960	9.99 625	29	4 39.2 38.8 38.4
32	9.11 761	96	9.12 138	97	10.87 862	9.99 624	28	5 49.0 48.5 48.0
33	9.11 857	95	9.12 235	97	10.87 765	9.99 622	27	6 58.8 58.2 57.6
34	9.11 952	95	9.12 332	96	10.87 668	9.99 620	26	7 68.6 67.9 67.2
35	9.12 047	95	9.12 428	97	10.87 572	9.99 618	25	8 78.4 77.6 76.8
36	9.12 142	94	9.12 525	96	10.87 475	9.99 617	24	9 88.2 87.3 86.4
37	9.12 236	95	9.12 621	96	10.87 379	9.99 615	23	
38	9.12 331	94	9.12 717	96	10.87 283	9.99 613	22	
39	9.12 425	94	9.12 813	96	10.87 187	9.99 612	21	**95 | 94 | 93**
40	9.12 519	93	9.12 909	95	10.87 091	9.99 610	20	1 9.5 9.4 9.3
41	9.12 612	94	9.13 004	95	10.86 996	9.99 608	19	2 19.0 18.8 18.6
42	9.12 706	93	9.13 099	95	10.86 901	9.99 607	18	3 28.5 28.2 27.9
43	9.12 799	93	9.13 194	95	10.86 806	9.99 605	17	4 38.0 37.6 37.2
44	9.12 892	93	9.13 289	95	10.86 711	9.99 603	16	5 47.5 47.0 46.5
45	9.12 985	93	9.13 384	94	10.86 616	9.99 601	15	6 57.0 56.4 55.8
46	9.13 078	93	9.13 478	95	10.86 522	9.99 600	14	7 66.5 65.8 65.1
47	9 13 171	92	9.13 573	94	10.86 427	9.99 598	13	8 76.0 75.2 74.4
48	9.13 263	92	9.13 667	94	10.86 333	9.99 596	12	9 85.5 84.6 83.7
49	9.13 355	92	9.13 761	93	10.86 239	9.99 595	11	
50	9.13 447	92	9.13 854	94	10.86 146	9.99 593	10	**92 | 91 | 90**
51	9.13 539	91	9.13 948	93	10.86 052	9.99 591	9	1 9.2 9.1 9.0
52	9.13 630	92	9.14 041	93	10.85 959	9.99 589	8	2 18.4 18.2 18.0
53	9.13 722	91	9.14 134	93	10.85 866	9.99 588	7	3 27.6 27.3 27.0
54	9.13 813	91	9.14 227	93	10.85 773	9.99 586	6	4 36.8 36.4 36.0
55	9.13 904	90	9.14 320	92	10.85 680	9.99 584	5	5 46.0 45.5 45.0
56	9.13 994	91	9.14 412	92	10.85 588	9.99 582	4	6 55.2 54.6 54.0
57	9.14 085	90	9.14 504	93	10.85 496	9.99 581	3	7 64.4 63.7 63.0
58	9.14 175	91	9.14 597	91	10.85 403	9.99 579	2	8 73.6 72.8 72.0
59	9.14 266	90	9.14 688	92	10.85 312	9.99 577	1	9 82.8 81.9 81.0
60	9.14 356		9.14 780		10.85 220	9.99 575	0	
	L Cos	d	L Cot	c d	L Tan	L Sin	'	Prop. Pts.

TABLE 3. LOGARITHMS OF TRIGONOMETRIC FUNCTIONS (continued)   650

8°

′	L Sin	d	L Tan	c d	L Cot	L Cos	
0	9.14 356	89	9.14 780	92	10.85 220	9.99 575	60
1	9.14 445	90	9.14 872	91	10.85 128	9.99 574	59
2	9.14 535	89	9.14 963	91	10.85 037	9.99 572	58
3	9.14 624	90	9.15 054	91	10.84 946	9.99 570	57
4	9.14 714	89	9.15 145	91	10.84 855	9.99 568	56
5	9.14 803	88	9.15 236	91	10.84 764	9.99 566	55
6	9.14 891	89	9.15 327	90	10.84 673	9.99 565	54
7	9.14 980	89	9.15 417	91	10.84 583	9.99 563	53
8	9.15 069	88	9.15 508	90	10.84 492	9.99 561	52
9	9.15 157	88	9.15 598	90	10.84 402	9.99 559	51
10	9.15 245	88	9.15 688	89	10.84 312	9.99 557	50
11	9.15 333	88	9.15 777	90	10.84 223	9.99 556	49
12	9.15 421	87	9.15 867	89	10.84 133	9.99 554	48
13	9.15 508	88	9.15 956	90	10.84 044	9.99 552	47
14	9.15 596	87	9.16 046	89	10.83 954	9.99 550	46
15	9.15 683	87	9.16 135	89	10.83 865	9.99 548	45
16	9.15 770	87	9.16 224	88	10.83 776	9.99 546	44
17	9.15 857	87	9.16 312	89	10.83 688	9.99 545	43
18	9.15 944	86	9.16 401	88	10.83 599	9.99 543	42
19	9.16 030	86	9.16 489	88	10.83 511	9.99 541	41
20	9.16 116	87	9.16 577	88	10.83 423	9.99 539	40
21	9.16 203	86	9.16 665	88	10.83 335	9.99 537	39
22	9.16 289	85	9.16 753	88	10.83 247	9.99 535	38
23	9.16 374	86	9.16 841	87	10.83 159	9.99 533	37
24	9.16 460	85	9.16 928	88	10.83 072	9.99 532	36
25	9.16 545	86	9.17 016	87	10.82 984	9.99 530	35
26	9.16 631	85	9.17 103	87	10.82 897	9.99 528	34
27	9.16 716	85	9.17 190	87	10.82 810	9.99 526	33
28	9.16 801	85	9.17 277	86	10.82 723	9.99 524	32
29	9.16 886	84	9.17 363	87	10.82 637	9.99 522	31
30	9.16 970	85	9.17 450	86	10.82 550	9.99 520	30
31	9.17 055	84	9.17 536	86	10.82 464	9.99 518	29
32	9.17 139	84	9.17 622	86	10.82 378	9.99 517	28
33	9.17 223	84	9.17 708	86	10.82 292	9.99 515	27
34	9.17 307	84	9.17 794	86	10.82 206	9.99 513	26
35	9.17 391	83	9.17 880	85	10.82 120	9.99 511	25
36	9.17 474	84	9.17 965	86	10.82 035	9.99 509	24
37	9.17 558	83	9.18 051	85	10.81 949	9.99 507	23
38	9.17 641	83	9.18 136	85	10.81 864	9.99 505	22
39	9.17 724	83	9.18 221	85	10.81 779	9.99 503	21
40	9.17 807	83	9.18 306	85	10.81 694	9.99 501	20
41	9.17 890	83	9.18 391	84	10.81 609	9.99 499	19
42	9.17 983	82	9.18 475	85	10.81 525	9.99 497	18
43	9.18 055	82	9.18 560	84	10.81 440	9.99 495	17
44	9.18 137	83	9.18 644	84	10.81 356	9.99 494	16
45	9.18 220	82	9.18 728	84	10.81 272	9.99 492	15
46	9.18 302	81	9.18 812	84	10.81 188	9.99 490	14
47	9.18 383	82	9.18 896	83	10.81 104	9.99 488	13
48	9.18 465	82	9.18 979	84	10.81 021	9.99 486	12
49	9.18 547	81	9.19 063	83	10.80 937	9.99 484	11
50	9.18 628	81	9.19 146	83	10.80 854	9.99 482	10
51	9.18 709	81	9.19 229	83	10.80 771	9.99 480	9
52	9.18 790	81	9.19 312	83	10.80 688	9.99 478	8
53	9.18 871	81	9.19 395	83	10.80 605	9.99 476	7
54	9.18 952	81	9.19 478	83	10.80 522	9.99 474	6
55	9.19 033	80	9.19 561	82	10.80 439	9.99 472	5
56	9.19 113	80	9.19 643	82	10.80 357	9.99 470	4
57	9.19 193	80	9.19 725	82	10.80 275	9.99 468	3
58	9.19 273	80	9.19 807	82	10.80 193	9.99 466	2
59	9.19 353	80	9.19 889	82	10.80 111	9.99 464	1
60	9.19 433		9.19 971		10.80 029	9.99 462	0
	L Cos	d	L Cot	c d	L Tan	L Sin	′

Prop. Pts.

	92	91	90
1	9.2	9.1	9.0
2	18.4	18.2	18.0
3	27.6	27.3	27.0
4	36.8	36.4	36.0
5	46.0	45.5	45.0
6	55.2	54.6	54.0
7	64.4	63.7	63.0
8	73.6	72.8	72.0
9	82.8	81.9	81.0

	89	88	87
1	8.9	8.8	8.7
2	17.8	17.6	17.4
3	26.7	26.4	26.1
4	35.6	35.2	34.8
5	44.5	44.0	43.5
6	53.4	52.8	52.2
7	62.3	61.6	60.9
8	71.2	70.4	69.6
9	80.1	79.2	78.3

	86	85	84
1	8.6	8.5	8.4
2	17.2	17.0	16.8
3	25.8	25.5	25.2
4	34.4	34.0	33.6
5	43.0	42.5	42.0
6	51.6	51.0	50.4
7	60.2	59.5	58.8
8	68.8	68.0	67.2
9	77.4	76.5	75.6

	83	82	81	80
1	8.3	8.2	8.1	8.0
2	16.6	16.4	16.2	16.0
3	24.9	24.6	24.3	24.0
4	33.2	32.8	32.4	32.0
5	41.5	41.0	40.5	40.0
6	49.8	49.2	48.6	48.0
7	58.1	57.4	56.7	56.0
8	66.4	65.6	64.8	64.0
9	74.7	73.8	72.9	72.0

TABLE 3. LOGARITHMS OF TRIGONOMETRIC FUNCTIONS (continued) 651

9°

′	L Sin	d	L Tan	c d	L Cot	L Cos	
0	9.19 433	80	9.19 971	82	10.80 029	9.99 462	60
1	9.19 513	79	9.20 053	81	10.79 947	9.99 460	59
2	9.19 592	80	9.20 134	82	10.79 866	9.99 458	58
3	9.19 672	79	9.20 216	81	10.79 784	9.99 456	57
4	9.19 751	79	9.20 297	81	10.79 703	9.99 454	56
5	9.19 830	79	9.20 378	81	10.79 622	9.99 452	55
6	9.19 909	79	9.20 459	81	10.79 541	9.99 450	54
7	9.19 988	79	9.20 540	81	10.79 460	9.99 448	53
8	9.20 067	78	9.20 621	80	10.79 379	9.99 446	52
9	9.20 145	78	9.20 701	81	10.79 299	9.99 444	51
10	9.20 223	79	9.20 782	80	10.79 218	9.99 442	50
11	9.20 302	78	9.20 862	80	10.79 138	9.99 440	49
12	9.20 380	78	9.20 942	80	10.79 058	9.99 438	48
13	9.20 458	77	9.21 022	80	10.78 978	9.99 436	47
14	9.20 535	78	9.21 102	80	10.78 898	9.99 434	46
15	9.20 613	78	9.21 182	79	10.78 818	9.99 432	45
16	9.20 691	77	9.21 261	80	10.78 739	9.99 429	44
17	9.20 768	77	9.21 341	79	10.78 659	9.99 427	43
18	9.20 845	77	9.21 420	79	10.78 580	9.99 425	42
19	9.20 922	77	9.21 499	79	10.78 501	9.99 423	41
20	9.20 999	77	9.21 578	79	10.78 422	9.99 421	40
21	9.21 076	77	9.21 657	79	10.78 343	9.99 419	39
22	9.21 153	76	9.21 736	78	10.78 264	9.99 417	38
23	9.21 229	77	9.21 814	79	10.78 186	9.99 415	37
24	9.21 306	76	9.21 893	78	10.78 107	9.99 413	36
25	9.21 382	76	9.21 971	78	10.78 029	9.99 411	35
26	9.21 458	76	9.22 049	78	10.77 951	9.99 409	34
27	9.21 534	76	9.22 127	78	10.77 873	9.99 407	33
28	9.21 610	75	9.22 205	78	10.77 795	9.99 404	32
29	9.21 685	76	9.22 283	78	10.77 717	9.99 402	31
30	9.21 761	75	9.22 361	77	10.77 639	9.99 400	30
31	9.21 836	76	9.22 438	78	10.77 562	9.99 398	29
32	9.21 912	75	9.22 516	77	10.77 484	9.99 396	28
33	9.21 987	75	9.22 593	77	10.77 407	9.99 394	27
34	9.22 062	75	9.22 670	77	10.77 330	9.99 392	26
35	9.22 137	74	9.22 747	77	10.77 253	9.99 390	25
36	9.22 211	75	9.22 824	77	10.77 176	9.99 388	24
37	9.22 286	75	9.22 901	76	10.77 099	9.99 385	23
38	9.22 361	74	9.22 977	77	10.77 023	9.99 383	22
39	9.22 435	74	9.23 054	76	10.76 946	9.99 381	21
40	9.22 509	74	9.23 130	76	10.76 870	9.99 379	20
41	9.22 583	74	9.23 206	77	10.76 794	9.99 377	19
42	9.22 657	74	9.23 283	76	10.76 717	9.99 375	18
43	9.22 731	74	9.23 359	76	10.76 641	9.99 372	17
44	9.22 805	73	9.23 435	75	10.76 565	9.99 370	16
45	9.22 878	74	9.23 510	76	10.76 490	9.99 368	15
46	9.22 952	73	9.23 586	75	10.76 414	9.99 366	14
47	9.23 025	73	9.23 661	76	10.76 339	9.99 364	13
48	9.23 098	73	9.23 737	75	10.76 263	9.99 362	12
49	9.23 171	73	9.23 812	75	10.76 188	9.99 359	11
50	9.23 244	73	9.23 887	75	10.76 113	9.99 357	10
51	9.23 317	73	9.23 962	75	10.76 038	9.99 355	9
52	9.23 390	72	9.24 037	75	10.75 963	9.99 353	8
53	9.23 462	73	9.24 112	74	10.75 888	9.99 351	7
54	9.23 535	72	9.24 186	75	10.75 814	9.99 348	6
55	9.23 607	72	9.24 261	74	10.75 739	9.99 346	5
56	9.23 679	73	9.24 335	75	10.75 665	9.99 344	4
57	9.23 752	71	9.24 410	74	10.75 590	9.99 342	3
58	9.23 823	72	9.24 484	74	10.75 516	9.99 340	2
59	9.23 895	72	9.24 558	74	10.75 442	9.99 337	1
60	9.23 967		9.24 632		10.75 368	9.99 335	0
	L Cos	d	L Cot	c d	L Tan	L Sin	′

Prop. Pts.

	82	81	80
1	8.2	8.1	8.0
2	16.4	16.2	16.0
3	24.6	24.3	24.0
4	32.8	32.4	32.0
5	41.0	40.5	40.0
6	49.2	48.6	48.0
7	57.4	56.7	56.0
8	65.6	64.8	64.0
9	73.8	72.9	72.0

	79	78	77
1	7.9	7.8	7.7
2	15.8	15.6	15.4
3	23.7	23.4	23.1
4	31.6	31.2	30.8
5	39.5	39.0	38.5
6	47.4	46.8	46.2
7	55.3	54.6	53.9
8	63.2	62.4	61.6
9	71.1	70.2	69.3

	76	75	74
1	7.6	7.5	7.4
2	15.2	15.0	14.8
3	22.8	22.5	22.2
4	30.4	30.0	29.6
5	38.0	37.5	37.0
6	45.6	45.0	44.4
7	53.2	52.5	51.8
8	60.8	60.0	59.2
9	68.4	67.5	66.6

	73	72	71
1	7.3	7.2	7.1
2	14.6	14.4	14.2
3	21.9	21.6	21.3
4	29.2	28.8	28.4
5	36.5	36.0	35.5
6	43.8	43.2	42.6
7	51.1	50.4	49.7
8	58.4	57.6	56.8
9	65.7	64.8	63.9

	3	2
1	0.3	0.2
2	0.6	0.4
3	0.9	0.6
4	1.2	0.8
5	1.5	1.0
6	1.8	1.2
7	2.1	1.4
8	2.4	1.6
9	2.7	1.8

80°

TABLE 3. LOGARITHMS OF TRIGONOMETRIC FUNCTIONS (continued)   652

10°

′	L Sin	d	L Tan	c d	L Cot	L Cos	d	
0	9.23 967	72	9.24 632	74	10.75 368	9.99 335	2	60
1	9.24 039	71	9.24 706	73	10.75 294	9.99 333	2	59
2	9.24 110	71	9.24 779	74	10.75 221	9.99 331	2	58
3	9.24 181	72	9.24 853	73	10.75 147	9.99 328	3	57
4	9.24 253	71	9.24 926	74	10.75 074	9.99 326	2	56
5	9.24 324	71	9.25 000	73	10.75 000	9.99 324	2	55
6	9.24 395	71	9.25 073	73	10.74 927	9.99 322	3	54
7	9.24 466	70	9.25 146	73	10.74 854	9.99 319	2	53
8	9.24 536	71	9.25 219	73	10.74 781	9.99 317	2	52
9	9.24 607	70	9.25 292	73	10.74 708	9.99 315	2	51
10	9.24 677	71	9.25 365	72	10.74 635	9.99 313	3	50
11	9.24 748	70	9.25 437	73	10.74 563	9.99 310	2	49
12	9.24 818	70	9.25 510	72	10.74 490	9.99 308	2	48
13	9.24 888	70	9.25 582	73	10.74 418	9.99 306	2	47
14	9.24 958	70	9.25 655	72	10.74 345	9.99 304	3	46
15	9.25 028	70	9.25 727	72	10.74 273	9.99 301	2	45
16	9.25 098	70	9.25 799	72	10.74 201	9.99 299	2	44
17	9.25 168	69	9.25 871	72	10.74 129	9.99 297	3	43
18	9.25 237	70	9.25 943	72	10.74 057	9.99 294	2	42
19	9.25 307	69	9.26 015	71	10.73 985	9.99 292	2	41
20	9.25 376	69	9.26 086	72	10.73 914	9.99 290	2	40
21	9.25 445	69	9.26 158	71	10.73 842	9.99 288	3	39
22	9.25 514	69	9.26 229	72	10.73 771	9.99 285	2	38
23	9.25 583	69	9.26 301	71	10.73 699	9.99 283	2	37
24	9.25 652	69	9.26 372	71	10.73 628	9.99 281	3	36
25	9.25 721	69	9.26 443	71	10.73 557	9.99 278	2	35
26	9.25 790	68	9.26 514	71	10.73 486	9.99 276	2	34
27	9.25 858	69	9.26 585	70	10.73 415	9.99 274	3	33
28	9.25 927	68	9.26 655	71	10.73 345	9.99 271	2	32
29	9.25 995	68	9.26 726	71	10.73 274	9.99 269	2	31
30	9.26 063	68	9.26 797	70	10.73 203	9.99 267	3	30
31	9.26 131	68	9.26 867	70	10.73 133	9.99 264	2	29
32	9.26 199	68	9.26 937	71	10.73 063	9.99 262	2	28
33	9.26 267	68	9.27 008	70	10.72 992	9.99 260	3	27
34	9.26 335	68	9.27 078	70	10.72 922	9.99 257	2	26
35	9.26 403	67	9.27 148	70	10.72 852	9.99 255	3	25
36	9.26 470	68	9.27 218	70	10.72 782	9.99 252	2	24
37	9.26 538	67	9.27 288	69	10.72 712	9.99 250	2	23
38	9.26 605	67	9.27 357	70	10.72 643	9.99 248	3	22
39	9.26 672	67	9.27 427	69	10.72 573	9.99 245	2	21
40	9.26 739	67	9.27 496	70	10.72 504	9.99 243	2	20
41	9.26 806	67	9.27 566	69	10.72 434	9.99 241	3	19
42	9.26 873	67	9.27 635	69	10.72 365	9.99 238	2	18
43	9.26 940	67	9.27 704	69	10.72 296	9.99 236	3	17
44	9.27 007	66	9.27 773	69	10.72 227	9.99 233	2	16
45	9.27 073	67	9.27 842	69	10.72 158	9.99 231	2	15
46	9.27 140	66	9.27 911	69	10.72 089	9.99 229	3	14
47	9.27 206	67	9.27 980	69	10.72 020	9.99 226	2	13
48	9.27 273	66	9.28 049	68	10.71 951	9.99 224	3	12
49	9.27 339	66	9.28 117	69	10.71 883	9.99 221	2	11
50	9.27 405	66	9.28 186	68	10.71 814	9.99 219	2	10
51	9.27 471	66	9.28 254	69	10.71 746	9.99 217	3	9
52	9.27 537	65	9.28 323	68	10.71 677	9.99 214	2	8
53	9.27 602	66	9.28 391	68	10.71 609	9.99 212	3	7
54	9.27 668	66	9.28 459	68	10.71 541	9.99 209	2	6
55	9.27 734	65	9.28 527	68	10.71 473	9.99 207	3	5
56	9.27 799	65	9.28 595	67	10.71 405	9.99 204	2	4
57	9.27 864	66	9.28 662	68	10.71 338	9.99 202	2	3
58	9.27 930	65	9.28 730	68	10.71 270	9.99 200	3	2
59	9.27 995	65	9.28 798	67	10.71 202	9.99 197	2	1
60	9.28 060		9.28 865		10.71 135	9.99 195		0
	L Cos	d	L Cot	c d	L Tan	L Sin	d	′

Prop. Pts.

	74	73	72
1	7.4	7.3	7.2
2	14.8	14.6	14.4
3	22.2	21.9	21.6
4	29.6	29.2	28.8
5	37.0	36.5	36.0
6	44.4	43.8	43.2
7	51.8	51.1	50.4
8	59.2	58.4	57.6
9	66.6	65.7	64.8

	71	70	69
1	7.1	7.0	6.9
2	14.2	14.0	13.8
3	21.3	21.0	20.7
4	28.4	28.0	27.6
5	35.5	35.0	34.5
6	42.6	42.0	41.4
7	49.7	49.0	48.3
8	56.8	56.0	55.2
9	63.9	63.0	62.1

	68	67	66
1	6.8	6.7	6.6
2	13.6	13.4	13.2
3	20.4	20.1	19.8
4	27.2	26.8	26.4
5	34.0	33.5	33.0
6	40.8	40.2	39.6
7	47.6	46.9	46.2
8	54.4	53.6	52.8
9	61.2	60.3	59.4

	65	3	2
1	6.5	0.3	0.2
2	13.0	0.6	0.4
3	19.5	0.9	0.6
4	26.0	1.2	0.8
5	32.5	1.5	1.0
6	39.0	1.8	1.2
7	45.5	2.1	1.4
8	52.0	2.4	1.6
9	58.5	2.7	1.8

79°

11°

′	L Sin	d	L Tan	c d	L Cot	L Cos	d		Prop. Pts.
0	9.28 060	65	9.28 865	68	10.71 135	9.99 195	3	60	
1	9.28 125	65	9.28 933	67	10.71 067	9.99 192	2	59	
2	9.28 190	64	9.29 000	67	10.71 000	9.99 190	3	58	
3	9.28 254	65	9.29 067	67	10.70 933	9.99 187	2	57	
4	9.28 319	65	9.29 134	67	10.70 866	9.99 185	3	56	
5	9.28 384	64	9.29 201	67	10.70 799	9.99 182	2	55	
6	9.28 448	64	9.29 268	67	10.70 732	9.99 180	3	54	
7	9.28 512	65	9.29 335	67	10.70 665	9.99 177	3	53	
8	9.28 577	64	9.29 402	66	10.70 598	9.99 175	3	52	
9	9.28 641	64	9.29 468	67	10.70 532	9.99 172	2	51	
10	9.28 705	64	9.29 535	66	10.70 465	9.99 170	3	50	
11	9.28 769	64	9.29 601	67	10.70 399	9.99 167	2	49	
12	9.28 833	63	9.29 668	66	10.70 332	9.99 165	3	48	
13	9.28 896	64	9.29 734	66	10.70 266	9.99 162	2	47	
14	9.28 960	64	9.29 800	66	10.70 200	9.99 160	3	46	
15	9.29 024	63	9.29 866	66	10.70 134	9.99 157	2	45	
16	9.29 087	63	9.29 932	66	10.70 068	9.99 155	3	44	
17	9.29 150	64	9.29 998	66	10.70 002	9.99 152	2	43	
18	9.29 214	63	9.30 064	66	10.69 936	9.99 150	3	42	
19	9.29 277	63	9.30 130	65	10.69 870	9.99 147	2	41	
20	9.29 340	63	9.30 195	66	10.69 805	9.99 145	3	40	
21	9.29 403	63	9.30 261	65	10.69 739	9.99 142	2	39	
22	9.29 466	63	9.30 326	65	10.69 674	9.99 140	3	38	
23	9.29 529	62	9.30 391	66	10.69 609	9.99 137	2	37	
24	9.29 591	63	9.30 457	65	10.69 543	9.99 135	3	36	
25	9.29 654	62	9.30 522	65	10.69 478	9.99 132	2	35	
26	9.29 716	63	9.30 587	65	10.69 413	9.99 130	3	34	
27	9.29 779	62	9.30 652	65	10.69 348	9.99 127	3	33	
28	9.29 841	62	9.30 717	65	10.69 283	9.99 124	2	32	
29	9.29 903	63	9.30 782	64	10.69 218	9.99 122	3	31	
30	9.29 966	62	9.30 846	65	10.69 154	9.99 119	2	30	
31	9.30 028	62	9.30 911	64	10.69 089	9.99 117	3	29	
32	9.30 090	61	9.30 975	65	10.69 025	9.99 114	2	28	
33	9.30 151	62	9.31 040	64	10.68 960	9.99 112	3	27	
34	9.30 213	62	9.31 104	64	10.68 896	9.99 109	3	26	
35	9.30 275	61	9.31 168	65	10.68 832	9.99 106	2	25	
36	9.30 336	62	9.31 233	64	10.68 767	9.99 104	3	24	
37	9.30 398	61	9.31 297	64	10.68 703	9.99 101	2	23	
38	9.30 459	62	9.31 361	64	10.68 639	9.99 099	3	22	
39	9.30 521	61	9.31 425	64	10.68 575	9.99 096	3	21	
40	9.30 582	61	9.31 489	63	10.68 511	9.99 093	2	20	
41	9.30 643	61	9.31 552	64	10.68 448	9.99 091	3	19	
42	9.30 704	61	9.31 616	63	10.68 384	9.99 088	2	18	
43	9.30 765	61	9.31 679	64	10.68 321	9.99 086	3	17	
44	9.30 826	61	9.31 743	63	10.68 257	9.99 083	3	16	
45	9.30 887	60	9.31 806	64	10.68 194	9.99 080	2	15	
46	9.30 947	61	9.31 870	63	10.68 130	9.99 078	3	14	
47	9.31 008	60	9.31 933	63	10.68 067	9.99 075	3	13	
48	9.31 068	61	9.31 996	63	10.68 004	9.99 072	2	12	
49	9.31 129	60	9.32 059	63	10.67 941	9.99 070	3	11	
50	9.31 189	61	9.32 122	63	10.67 878	9.99 067	3	10	
51	9.31 250	60	9.32 185	63	10.67 815	9.99 064	2	9	
52	9.31 310	60	9.32 248	63	10.67 752	9.99 062	3	8	
53	9.31 370	60	9.32 311	62	10.67 689	9.99 059	3	7	
54	9.31 430	60	9.32 373	63	10.67 627	9.99 056	2	6	
55	9.31 490	59	9.32 436	62	10.67 564	9.99 054	3	5	
56	9.31 549	60	9.32 498	63	10.67 502	9.99 051	3	4	
57	9.31 609	60	9.32 561	62	10.67 439	9.99 048	2	3	
58	9.31 669	59	9.32 623	62	10.67 377	9.99 046	3	2	
59	9.31 728	60	9.32 685	62	10.67 315	9.99 043	3	1	
60	9.31 788		9.32 747		10.67 253	9.99 040		0	
	L Cos	d	L Cot	c d	L Tan	L Sin	d	′	Prop. Pts.

Prop. Pts.

	68	67	66
1	6.8	6.7	6.6
2	13.6	13.4	13.2
3	20.4	20.1	19.8
4	27.2	26.8	26.4
5	34.0	33.5	33.0
6	40.8	40.2	39.6
7	47.6	46.9	46.2
8	54.4	53.6	52.8
9	61.2	60.3	59.4

	65	64	63
1	6.5	6.4	6.3
2	13.0	12.8	12.6
3	19.5	19.2	18.9
4	26.0	25.6	25.2
5	32.5	32.0	31.5
6	39.0	38.4	37.8
7	45.5	44.8	44.1
8	52.0	51.2	50.4
9	58.5	57.6	56.7

	62	61	60
1	6.2	6.1	6.0
2	12.4	12.2	12.0
3	18.6	18.3	18.0
4	24.8	24.4	24.0
5	31.0	30.5	30.0
6	37.2	36.6	36.0
7	43.4	42.7	42.0
8	49.6	48.8	48.0
9	55.8	54.9	54.0

	59	3	2
1	5.9	0.3	0.2
2	11.8	0.6	0.4
3	17.7	0.9	0.6
4	23.6	1.2	0.8
5	29.5	1.5	1.0
6	35.4	1.8	1.2
7	41.3	2.1	1.4
8	47.2	2.4	1.6
9	53.1	2.7	1.8

78°

TABLE 3. LOGARITHMS OF TRIGONOMETRIC FUNCTIONS (continued)   654

12°

′	L Sin	d	L Tan	c d	L Cot	L Cos	d		Prop. Pts.
0	9.31 788	59	9.32 747	63	10.67 253	9.99 040	2	60	
1	9.31 847	60	9.32 810	62	10.67 190	9.99 038	3	59	
2	9.31 907	59	9.32 872	61	10.67 128	9.99 035	3	58	
3	9.31 966	59	9.32 933	62	10.67 067	9.99 032	2	57	
4	9.32 025	59	9.32 995	62	10.67 005	9.99 030	3	56	
5	9.32 084	59	9.33 057	62	10.66 943	9.99 027	3	55	**63 ǀ 62 ǀ 61**
6	9.32 143	59	9.33 119	61	10.66 881	9.99 024	2	54	1 6.3 ǀ 6.2 ǀ 6.1
7	9.32 202	59	9.33 180	62	10.66 820	9.99 022	3	53	2 12.6 ǀ 12.4 ǀ 12.2
8	9.32 261	58	9.33 242	61	10.66 758	9.99 019	3	52	3 18.9 ǀ 18.6 ǀ 18.3
9	9.32 319	59	9.33 303	62	10.66 697	9.99 016	3	51	4 25.2 ǀ 24.8 ǀ 24.4
10	9.32 378	59	9.33 365	61	10.66 635	9.99 013	2	50	5 31.5 ǀ 31.0 ǀ 30.5
11	9.32 437	58	9.33 426	61	10.66 574	9.99 011	3	49	6 37.8 ǀ 37.2 ǀ 36.6
12	9.32 495	58	9.33 487	61	10.66 513	9.99 008	3	48	7 44.1 ǀ 43.4 ǀ 42.7
13	9.32 553	59	9.33 548	61	10.66 452	9.99 005	3	47	8 50.4 ǀ 49.6 ǀ 48.8
14	9.32 612	58	9.33 609	61	10.66 391	9.99 002	2	46	9 56.7 ǀ 55.8 ǀ 54.9
15	9.32 670	58	9.33 670	61	10.66 330	9.99 000	3	45	
16	9.32 728	58	9.33 731	61	10.66 269	9.98 997	3	44	
17	9.32 786	58	9.33 792	61	10.66 208	9.98 994	3	43	
18	9.32 844	58	9.33 853	60	10.66 147	9.98 991	2	42	
19	9.32 902	58	9.33 913	61	10.66 087	9.98 989	3	41	
20	9.32 960	58	9.33 974	60	10.66 026	9.98 986	3	40	**60 ǀ 59 ǀ 58**
21	9.33 018	57	9.34 034	61	10.65 966	9.98 983	3	39	1 6.0 ǀ 5.9 ǀ 5.8
22	9.33 075	58	9.34 095	60	10.65 905	9.98 980	2	38	2 12.0 ǀ 11.8 ǀ 11.6
23	9.33 133	57	9.34 155	60	10.65 845	9.98 978	3	37	3 18.0 ǀ 17.7 ǀ 17.4
24	9.33 190	58	9.34 215	61	10.65 785	9.98 975	3	36	4 24.0 ǀ 23.6 ǀ 23.2
25	9.33 248	57	9.34 276	60	10.65 724	9.98 972	3	35	5 30.0 ǀ 29.5 ǀ 29.0
26	9.33 305	57	9.34 336	60	10.65 664	9.98 969	2	34	6 36.0 ǀ 35.4 ǀ 34.8
27	9.33 362	58	9.34 396	60	10.65 604	9.98 967	3	33	7 42.0 ǀ 41.3 ǀ 40.6
28	9.33 420	57	9.34 456	60	10.65 544	9.98 964	3	32	8 48.0 ǀ 47.2 ǀ 46.4
29	9.33 477	57	9.34 516	60	10.65 484	9.98 961	3	31	9 54.0 ǀ 53.1 ǀ 52.2
30	9.33 534	57	9.34 576	59	10.65 424	9.98 958	3	30	
31	9.33 591	56	9.34 635	60	10.65 365	9.98 955	2	29	
32	9.33 647	57	9.34 695	60	10.65 305	9.98 953	3	28	
33	9.33 704	57	9.34 755	59	10.65 245	9.98 950	3	27	**57 ǀ 56 ǀ 55**
34	9.33 761	57	9.34 814	60	10.65 186	9.98 947	3	26	1 5.7 ǀ 5.6 ǀ 5.5
35	9.33 818	56	9.34 874	59	10.65 126	9.98 944	3	25	2 11.4 ǀ 11.2 ǀ 11.0
36	9.33 874	57	9.34 933	59	10.65 067	9.98 941	3	24	3 17.1 ǀ 16.8 ǀ 16.5
37	9.33 931	56	9.34 992	59	10.65 008	9.98 938	2	23	4 22.8 ǀ 22.4 ǀ 22.0
38	9.33 987	56	9.35 051	60	10.64 949	9.98 936	3	22	5 28.5 ǀ 28.0 ǀ 27.5
39	9.34 043	57	9.35 111	59	10.64 889	9.98 933	3	21	6 34.2 ǀ 33.6 ǀ 33.0
40	9.34 100	56	9.35 170	59	10.64 830	9.98 930	3	20	7 39.9 ǀ 39.2 ǀ 38.5
41	9.34 156	56	9.35 229	59	10.64 771	9.98 927	3	19	8 45.6 ǀ 44.8 ǀ 44.0
42	9.34 212	56	9.35 288	59	10.64 712	9.98 924	3	18	9 51.3 ǀ 50.4 ǀ 49.5
43	9.34 268	56	9.35 347	58	10.64 653	9.98 921	2	17	
44	9.34 324	56	9.35 405	59	10.64 595	9.98 919	3	16	
45	9.34 380	56	9.35 464	59	10.64 536	9.98 916	3	15	
46	9.34 436	55	9.35 523	58	10.64 477	9.98 913	3	14	
47	9.34 491	56	9.35 581	59	10.64 419	9.98 910	3	13	**3 ǀ 2**
48	9.34 547	55	9.35 640	58	10.64 360	9.98 907	3	12	1 0.3 ǀ 0.2
49	9.34 602	56	9.35 698	59	10.64 302	9.98 904	3	11	2 0.6 ǀ 0.4
50	9.34 658	55	9.35 757	58	10.64 243	9.98 901	3	10	3 0.9 ǀ 0.6
51	9.34 713	56	9.35 815	58	10.64 185	9.98 898	2	9	4 1.2 ǀ 0.8
52	9.34 769	55	9.35 873	58	10.64 127	9.98 896	3	8	5 1.5 ǀ 1.0
53	9.34 824	55	9.35 931	58	10.64 069	9.98 893	3	7	6 1.8 ǀ 1.2
54	9.34 879	55	9.35 989	58	10.64 011	9.98 890	3	6	7 2.1 ǀ 1.4
55	9.34 934	55	9.36 047	58	10.63 953	9.98 887	3	5	8 2.4 ǀ 1.6
56	9.34 989	55	9.36 105	58	10.63 895	9.98 884	3	4	9 2.7 ǀ 1.8
57	9.35 044	55	9.36 163	58	10.63 837	9.98 881	3	3	
58	9.35 099	55	9.36 221	58	10.63 779	9.98 878	3	2	
59	9.35 154	55	9.36 279	57	10.63 721	9.98 875	3	1	
60	9.35 209		9.36 336		10.63 664	9.98 872		0	

	L Cos	d	L Cot	c d	L Tan	L Sin	d	′	Prop. Pts.

77°

TABLE 3. LOGARITHMS OF TRIGONOMETRIC FUNCTIONS (continued)   655

13°

′	L Sin	d	L Tan	c d	L Cot	L Cos	d		Prop. Pts.
0	9.35 209	54	9.36 336	58	10.63 664	9.98 872	3	60	
1	9.35 263	55	9.36 394	58	10.63 606	9.98 869	2	59	
2	9.35 318	55	9.36 452	57	10.63 548	9.98 867	3	58	
3	9.35 373	54	9.36 509	57	10.63 491	9.98 864	3	57	
4	9.35 427	54	9.36 566	58	10.63 434	9.98 861	3	56	
5	9.35 481	55	9.36 624	57	10.63 376	9.98 858	3	55	**58 | 57 | 56**
6	9.35 536	54	9.36 681	57	10.63 319	9.98 855	3	54	1 5.8 5.7 5.6
7	9.35 590	54	9.36 738	57	10.63 262	9.98 852	3	53	2 11.6 11.4 11.2
8	9.35 644	54	9.36 795	57	10.63 205	9.98 849	3	52	3 17.4 17.1 16.8
9	9.35 698	54	9.36 852	57	10.63 148	9.98 846	3	51	4 23.2 22.8 22.4
10	9 35 752	54	9.36 909	57	10.63 091	9.98 843	3	50	5 29.0 28.5 28.0
11	9.35 806	54	9.36 966	57	10.63 034	9.98 840	3	49	6 34.8 34.2 33.6
12	9.35 860	54	9.37 023	57	10.62 977	9.98 837	3	48	7 40.6 39.9 39.2
13	9.35 914	54	9.37 080	57	10.62 920	9.98 834	3	47	8 46.4 45.6 44.8
14	9.35 968	54	9.37 137	56	10.62 863	9.98 831	3	46	9 52.2 51.3 50.4
15	9.36 022	53	9.37 193	57	10.62 807	9.98 828	3	45	
16	9.36 075	54	9.37 250	56	10.62 750	9.98 825	3	44	
17	9.36 129	53	9.37 306	57	10.62 694	9.98 822	3	43	
18	9.36 182	54	9.37 363	56	10.62 637	9.98 819	3	42	
19	9.36 236	53	9.37 419	57	10.62 581	9.98 816	3	41	**55 | 54 | 53**
20	9.36 289	53	9.37 476	56	10.62 524	9.98 813	3	40	1 5.5 5.4 5.3
21	9.36 342	53	9.37 532	56	10.62 468	9.98 810	3	39	2 11.0 10.8 10.6
22	9.36 395	54	9.37 588	56	10.62 412	9.98 807	3	38	3 16.5 16.2 15.9
23	9.36 449	53	9.37 644	56	10.62 356	9.98 804	3	37	4 22.0 21.6 21.2
24	9.36 502	53	9.37 700	56	10.62 300	9.98 801	3	36	5 27.5 27.0 26.5
25	9.36 555	53	9.37 756	56	10.62 244	9.98 798	3	35	6 33.0 32.4 31.8
26	9.36 608	52	9.37 812	56	10.62 188	9.98 795	3	34	7 38.5 37.8 37.1
27	9.36 660	53	9.37 868	56	10.62 132	9.98 792	3	33	8 44.0 43.2 42.4
28	9.36 713	53	9.37 924	56	10.62 076	9.98 789	3	32	9 49.5 48.6 47.7
29	9.36 766	53	9.37 980	55	10.62 020	9.98 786	3	31	
30	9.36 819	52	9.38 035	56	10.61 965	9.98 783	3	30	
31	9.36 871	53	9.38 091	56	10.61 909	9.98 780	3	29	
32	9.36 924	52	9.38 147	55	10.61 853	9.98 777	3	28	
33	9.36 976	52	9.38 202	55	10.61 798	9.98 774	3	27	**52 | 51**
34	9.37 028	53	9.38 257	56	10.61 743	9.98 771	3	26	1 5.2 5.1
35	9.37 081	52	9.38 313	55	10.61 687	9.98 768	3	25	2 10.4 10.2
36	9.37 133	52	9.38 368	55	10.61 632	9.98 765	3	24	3 15.6 15.3
37	9.37 185	52	9.38 423	56	10.61 577	9.98 762	3	23	4 20.8 20.4
38	9.37 237	52	9.38 479	55	10.61 521	9.98 759	3	22	5 26.0 25.5
39	9.37 289	52	9.38 534	55	10.61 466	9.98 756	3	21	6 31.2 30.6
40	9.37 341	52	9.38 589	55	10.61 411	9.98 753	3	20	7 36.4 35.7
41	9.37 393	52	9.38 644	55	10.61 356	9.98 750	4	19	8 41.6 40.8
42	9.37 445	52	9.38 699	55	10.61 301	9.98 746	3	18	9 46.8 45.9
43	9.37 497	52	9.38 754	54	10.61 246	9.98 743	3	17	
44	9.37 549	51	9.38 808	55	10.61 192	9.98 740	3	16	
45	9.37 600	52	9.38 863	55	10.61 137	9.98 737	3	15	
46	9.37 652	51	9.38 918	54	10.61 082	9.98 734	3	14	
47	9.37 703	52	9.38 972	55	10.61 028	9.98 731	3	13	**4 | 3 | 2**
48	9.37 755	51	9.39 027	55	10.60 973	9.98 728	3	12	1 0.4 0.3 0.2
49	9.37 806	52	9.39 082	54	10.60 918	9.98 725	3	11	2 0.8 0.6 0.4
50	9.37 858	51	9.39 136	54	10.60 864	9.98 722	3	10	3 1.2 0.9 0.6
51	9.37 909	51	9.39 190	55	10.60 810	9.98 719	4	9	4 1.6 1.2 0.8
52	9.37 960	51	9.39 245	54	10.60 755	9.98 715	3	8	5 2.0 1.5 1.0
53	9.38 011	51	9.39 299	54	10.60 701	9.98 712	3	7	6 2.4 1.8 1.2
54	9.38 062	51	9.39 353	54	10.60 647	9.98 709	3	6	7 2.8 2.1 1.4
55	9.38 113	51	9.39 407	54	10.60 593	9.98 706	3	5	8 3.2 2.4 1.6
56	9.38 164	51	9.39 461	54	10.60 539	9.98 703	3	4	9 3.6 2.7 1.8
57	9.38 215	51	9.39 515	54	10.60 485	9.98 700	3	3	
58	9.38 266	51	9.39 569	54	10.60 431	9.98 697	3	2	
59	9.38 317	51	9.39 623	54	10.60 377	9.98 694	4	1	
60	9.38 368		9.39 677		10.60 323	9.98 690		0	
	L Cos	d	L Cot	c d	L Tan	L Sin	d	′	Prop. Pts.

76°

TABLE 3. LOGARITHMS OF TRIGONOMETRIC FUNCTIONS (continued)   656

14°

′	L Sin	d	L Tan	c d	L Cot	L Cos	d	′
0	9.38 368		9.39 677		10.60 323	9.98 690		**60**
1	9.38 418	50	9.39 731	54	10.60 269	9.98 687	3	59
2	9.38 469	51	9.39 785	54	10.60 215	9.98 684	3	58
3	9.38 519	50	9.39 838	53	10.60 162	9.98 681	3	57
4	9.38 570	51	9.39 892	54	10.60 108	9.98 678	3	56
5	9.38 620	50	9.39 945	53	10.60 055	9.98 675	3	55
6	9.38 670	50	9.39 999	54	10.60 001	9.98 671	4	54
7	9.38 721	51	9.40 052	53	10.59 948	9.98 668	3	53
8	9.38 771	50	9.40 106	54	10.59 894	9.98 665	3	52
9	9.38 821	50	9.40 159	53	10.59 841	9.98 662	3	51
10	9.38 871	50	9.40 212	53	10.59 788	9.98 659	3	**50**
11	9.38 921	50	9.40 266	54	10.59 734	9.98 656	3	49
12	9.38 971	50	9.40 319	53	10.59 681	9.98 652	4	48
13	9.39 021	50	9.40 372	53	10.59 628	9.98 649	3	47
14	9.39 071	50	9.40 425	53	10.59 575	9.98 646	3	46
15	9.39 121	50	9.40 478	53	10.59 522	9.98 643	3	45
16	9.39 170	49	9.40 531	53	10.59 469	9.98 640	4	44
17	9.39 220	50	9.40 584	53	10.59 416	9.98 636	3	43
18	9.39 270	50	9.40 636	52	10.59 364	9.98 633	3	42
19	9.39 319	49	9.40 689	53	10.59 311	9.98 630	3	41
20	9.39 369	50	9.40 742	53	10.59 258	9.98 627	4	**40**
21	9.39 418	49	9.40 795	53	10.59 205	9.98 623	3	39
22	9.39 467	49	9.40 847	52	10.59 153	9.98 620	3	38
23	9.39 517	50	9.40 900	53	10.59 100	9.98 617	3	37
24	9.39 566	49	9.40 952	52	10.59 048	9.98 614	4	36
25	9.39 615	49	9.41 005	53	10.58 995	9.98 610	3	35
26	9.39 664	49	9.41 057	52	10.58 943	9.98 607	3	34
27	9.39 713	49	9.41 109	52	10.58 891	9.98 604	3	33
28	9.39 762	49	9.41 161	53	10.58 839	9.98 601	4	32
29	9.39 811	49	9.41 214	52	10.58 786	9.98 597	3	31
30	9.39 860	49	9.41 266	52	10.58 734	9.98 594	3	**30**
31	9.39 909	49	9.41 318	52	10.58 682	9.98 591	3	29
32	9.93 958	48	9.41 370	52	10.58 630	9.98 588	4	28
33	9.40 006	49	9.41 422	52	10.58 578	9.98 584	3	27
34	9.40 055	48	9.41 474	52	10.58 526	9.98 581	3	26
35	9.40 103	49	9.41 526	52	10.58 474	9.98 578	4	25
36	9.40 152	48	9.41 578	51	10.58 422	9.98 574	3	24
37	9.40 200	49	9.41 629	52	10.58 371	9.98 571	3	23
38	9.40 249	48	9.41 681	52	10.58 319	9.98 568	3	22
39	9.40 297	49	9.41 733	51	10.58 267	9.98 565	4	21
40	9.40 346	48	9.41 784	52	10.58 216	9.98 561	3	**20**
41	9.40 394	48	9.41 836	51	10.58 164	9.98 558	3	19
42	9.40 442	48	9.41 887	52	10.58 113	9.98 555	4	18
43	9.40 490	48	9.41 939	51	10.58 061	9.98 551	3	17
44	9.40 538	48	9.41 990	51	10.58 010	9.98 548	3	16
45	9.40 586	48	9.42 041	52	10.57 959	9.98 545	4	15
46	9.40 634	48	9.42 093	51	10.57 907	9.98 541	3	14
47	9.40 682	48	9.42 144	51	10.57 856	9.98 538	3	13
48	9.40 730	48	9.42 195	51	10.57 805	9.98 535	4	12
49	9.40 778	47	9.42 246	51	10.57 754	9.98 531	3	11
50	9.40 825	48	9.42 297	51	10.57 703	9.98 528	3	**10**
51	9.40 873	48	9.42 348	51	10.57 652	9.98 525	4	9
52	9.40 921	47	9.42 399	51	10.57 601	9.98 521	3	8
53	9.40 968	48	9.42 450	51	10.57 550	9.98 518	3	7
54	9.41 016	47	9.42 501	51	10.57 499	9.98 515	4	6
55	9.41 063	48	9.42 552	51	10.57 448	9.98 511	3	5
56	9.41 111	47	9.42 603	50	10.57 397	9.98 508	3	4
57	9.41 158	47	9.42 653	51	10.57 347	9.98 505	3	3
58	9.41 205	47	9.42 704	51	10.57 296	9.98 501	3	2
59	9.41 252	48	9.42 755	50	10.57 245	9.98 498	4	1
60	9.41 300		9.42 805		10.57 195	9.98 494		**0**
	L Cos	d	L Cot	c d	L Tan	L Sin	d	′

Prop. Pts.

	54	53	52
1	5.4	5.3	5.2
2	10.8	10.6	10.4
3	16.2	15.9	15.6
4	21.6	21.2	20.8
5	27.0	26.5	26.0
6	32.4	31.8	31.2
7	37.8	37.1	36.4
8	43.2	42.4	41.6
9	48.6	47.7	46.8

	51	50	49
1	5.1	5.0	4.9
2	10.2	10.0	9.8
3	15.3	15.0	14.7
4	20.4	20.0	19.6
5	25.5	25.0	24.5
6	30.6	30.0	29.4
7	35.7	35.0	34.3
8	40.8	40.0	39.2
9	45.9	45.0	44.1

	48	47
1	4.8	4.7
2	9.6	9.4
3	14.4	14.1
4	19.2	18.8
5	24.0	23.5
6	28.8	28.2
7	33.6	32.9
8	38.4	37.6
9	43.2	42.3

	4	3
1	0.4	0.3
2	0.8	0.6
3	1.2	0.9
4	1.6	1.2
5	2.0	1.5
6	2.4	1.8
7	2.8	2.1
8	3.2	2.4
9	3.6	2.7

TABLE 3. LOGARITHMS OF TRIGONOMETRIC FUNCTIONS (continued)   657

15°

′	L Sin	d	L Tan	c d	L Cot	L Cos	d		Prop. Pts.
0	9.41 300		9.42 805		10.57 195	9.98 494		60	
1	9.41 347	47	9.42 856	51	10.57 144	9.98 491	3	59	
2	9.41 394	47	9.42 906	50	10.57 094	9.98 488	3	58	
3	9.41 441	47	9.42 957	51	10.57 043	9.98 484	4	57	
4	9.41 488	47	9.43 007	50	10.56 993	9.98 481	3	56	
5	9.41 535	47	9.43 057	50	10.56 943	9.98 477	4	55	**51** **50** **49**
6	9.41 582	47	9.43 108	51	10.56 892	9.98 474	3	54	1 5.1 5.0 4.9
7	9.41 628	46	9.43 158	50	10.56 842	9.98 471	3	53	2 10.2 10.0 9.8
8	9.41 675	47	9.43 208	50	10.56 792	9.98 467	4	52	3 15.3 15.0 14.7
9	9.41 722	47	9.43 258	50	10.56 742	9.98 464	3	51	4 20.4 20.0 19.6
10	9.41 768	46	9.43 308	50	10.56 692	9.98 460	4	50	5 25.5 25.0 24.5
11	9.41 815	47	9.43 358	50	10.56 642	9.98 457	3	49	6 30.6 30.0 29.4
12	9.41 861	46	9.43 408	50	10.56 592	9.98 453	4	48	7 35.7 35.0 34.3
13	9.41 908	47	9.43 458	50	10.56 542	9.98 450	3	47	8 40.8 40.0 39.2
14	9.41 954	46	9.43 508	50	10.56 492	9.98 447	4	46	9 45.9 45.0 44.1
15	9.42 001	47	9.43 558	49	10.56 442	9.98 443	3	45	
16	9.42 047	46	9.43 607	50	10.56 393	9.98 440	4	44	
17	9.42 093	46	9.43 657	50	10.56 343	9.98 436	3	43	
18	9.42 140	47	9.43 707	49	10.56 293	9.98 433	4	42	
19	9.42 186	46	9.43 756	50	10.56 244	9.98 429	3	41	**48** **47** **46**
20	9.42 232	46	9.43 806	49	10.56 194	9.98 426	4	40	1 4.8 4.7 4.6
21	9.42 278	46	9.43 855	50	10.56 145	9.98 422	3	39	2 9.6 9.4 9.2
22	9.42 324	46	9.43 905	49	10.56 095	9.98 419	4	38	3 14.4 14.1 13.8
23	9.42 370	46	9.43 954	50	10.56 046	9.98 415	3	37	4 19.2 18.8 18.4
24	9.42 416	45	9.44 004	49	10.55 996	9.98 412	3	36	5 24.0 23.5 23.0
25	9.42 461	46	9.44 053	49	10.55 947	9.98 409	4	35	6 28.8 28.2 27.6
26	9.42 507	46	9.44 102	49	10.55 898	9.98 405	3	34	7 33.6 32.9 32.2
27	9.42 553	46	9.44 151	50	10.55 849	9.98 402	4	33	8 38.4 37.6 36.8
28	9.42 599	45	9.44 201	49	10.55 799	9.98 398	3	32	9 43.2 42.3 41.4
29	9.42 644	46	9.44 250	49	10.55 750	9.98 395	4	31	
30	9.42 690	45	9.44 299	49	10.55 701	9.98 391	3	30	
31	9.42 735	46	9.44 348	49	10.55 652	9.98 388	4	29	
32	9.42 781	45	9.44 397	49	10.55 603	9.98 384	3	28	
33	9.42 826	46	9.44 446	49	10.55 554	9.98 381	4	27	**45** **44**
34	9.42 872	45	9.44 495	49	10.55 505	9.98 377	4	26	1 4.5 4.4
35	9.42 917	45	9.44 544	48	10.55 456	9.98 373	3	25	2 9.0 8.8
36	9.42 962	46	9.44 592	49	10.55 408	9.98 370	4	24	3 13.5 13.2
37	9.43 008	45	9.44 641	49	10.55 359	9.98 366	3	23	4 18.0 17.6
38	9.43 053	45	9.44 690	48	10.55 310	9.98 363	4	22	5 22.5 22.0
39	9.43 098	45	9.44 738	49	10.55 262	9.98 359	3	21	6 27.0 26.4
40	9.43 143	45	9.44 787	49	10.55 213	9.98 356	4	20	7 31.5 30.8
41	9.43 188	45	9.44 836	48	10.55 164	9.98 352	3	19	8 36.0 35.2
42	9.43 233	45	9.44 884	49	10.55 116	9.98 349	4	18	9 40.5 39.6
43	9.43 278	45	9.44 933	48	10.55 067	9.98 345	3	17	
44	9.43 323	44	9.44 981	48	10.55 019	9.98 342	4	16	
45	9.43 367	45	9.45 029	49	10.54 971	9.98 338	4	15	
46	9.43 412	45	9.45 078	48	10.54 922	9.98 334	3	14	
47	9.43 457	45	9.45 126	48	10.54 874	9.98 331	4	13	**4** **3**
48	9.43 502	44	9.45 174	48	10.54 826	9.98 327	3	12	1 0.4 0.3
49	9.43 546	45	9.45 222	49	10.54 778	9.98 324	4	11	2 0.8 0.6
50	9.43 591	44	9.45 271	48	10.54 729	9.98 320	3	10	3 1.2 0.9
51	9.43 635	45	9.45 319	48	10.54 681	9.98 317	4	9	4 1.6 1.2
52	9.43 680	44	9.45 367	48	10.54 633	9.98 313	4	8	5 2.0 1.5
53	9.43 724	45	9.45 415	48	10.54 585	9.98 309	3	7	6 2.4 1.8
54	9.43 769	44	9.45 463	48	10.54 537	9.98 306	4	6	7 2.8 2.1
55	9.43 813	44	9.45 511	48	10.54 489	9.98 302	3	5	8 3.2 2.4
56	9.43 857	44	9.45 559	47	10.54 441	9.98 299	4	4	9 3.6 2.7
57	9.43 901	45	9.45 606	48	10.54 394	9.98 295	4	3	
58	9.43 946	44	9.45 654	48	10.54 346	9.98 291	3	2	
59	9.43 990	44	9.45 702	48	10.54 298	9.98 288	4	1	
60	9.44 034		9.45 750		10.54 250	9.98 284		0	
	L Cos	d	L Cot	c d	L Tan	L Sin	d	′	Prop. Pts.

TABLE 3. LOGARITHMS OF TRIGONOMETRIC FUNCTIONS (continued)   658

16°

'	L Sin	d	L Tan	c d	L Cot	L Cos	d	
0	9.44 034	44	9.45 750	47	10.54 250	9.98 284	3	60
1	9.44 078	44	9.45 797	48	10.54 203	9.98 281	4	59
2	9.44 122	44	9.45 845	47	10.54 155	9.98 277	4	58
3	9.44 166	44	9.45 892	48	10.54 108	9.98 273	3	57
4	9.44 210	43	9.45 940	47	10.54 060	9.98 270	4	56
5	9.44 253	44	9.45 987	48	10.54 013	9.98 266	4	55
6	9.44 297	44	9.46 035	47	10.53 965	9.98 262	3	54
7	9.44 341	44	9.46 082	48	10.53 918	9.98 259	4	53
8	9.44 385	43	9.46 130	47	10.53 870	9.98 255	4	52
9	9.44 428	44	9.46 177	47	10.53 823	9.98 251	3	51
10	9.44 472	44	9.46 224	47	10.53 776	9.98 248	4	50
11	9.44 516	43	9.46 271	48	10.53 729	9.98 244	4	49
12	9.44 559	43	9.46 319	47	10.53 681	9.98 240	3	48
13	9.44 602	44	9.46 366	47	10.53 634	9.98 237	4	47
14	9.44 646	43	9.46 413	47	10.53 587	9.98 233	4	46
15	9.44 689	44	9.46 460	47	10.53 540	9.98 229	3	45
16	9.44 733	43	9.46 507	47	10.53 493	9.98 226	4	44
17	9.44 776	43	9.46 554	47	10.53 446	9.98 222	4	43
18	9.44 819	43	9.46 601	47	10.53 399	9.98 218	3	42
19	9.44 862	43	9.46 648	46	10.53 352	9.98 215	4	41
20	9.44 905	43	9.46 694	47	10.53 306	9.98 211	4	40
21	9.44 948	44	9.46 741	47	10.53 259	9.98 207	3	39
22	9.44 992	43	9.46 788	47	10.53 212	9.98 204	4	38
23	9.45 035	42	9.46 835	46	10.53 165	9.98 200	4	37
24	9.45 077	43	9.46 881	47	10.53 119	9.98 196	4	36
25	9.45 120	43	9.46 928	47	10.53 072	9.98 192	3	35
26	9.45 163	43	9.46 975	46	10.53 025	9.98 189	4	34
27	9.45 206	43	9.47 021	47	10.52 979	9.98 185	4	33
28	9.45 249	43	9.47 068	46	10.52 932	9.98 181	4	32
29	9.45 292	42	9.47 114	46	10.52 886	9.98 177	3	31
30	9.45 334	43	9.47 160	47	10.52 840	9.98 174	4	30
31	9.45 377	42	9.47 207	46	10.52 793	9.98 170	4	29
32	9.45 419	43	9.47 253	46	10.52 747	9.98 166	4	28
33	9.45 462	42	9.47 299	47	10.52 701	9.98 162	3	27
34	9.45 504	43	9.47 346	46	10.52 654	9.98 159	4	26
35	9.45 547	42	9.47 392	46	10.52 608	9.98 155	4	25
36	9.45 589	43	9.47 438	46	10.52 562	9.98 151	4	24
37	9.45 632	42	9.47 484	46	10.52 516	9.98 147	3	23
38	9.45 674	42	9.47 530	46	10.52 470	9.98 144	4	22
39	9.45 716	42	9.47 576	46	10.52 424	9.98 140	4	21
40	9.45 758	43	9.47 622	46	10.52 378	9.98 136	4	20
41	9.45 801	42	9.47 668	46	10.52 332	9.98 132	3	19
42	9.45 843	42	9.47 714	46	10.52 286	9.98 129	4	18
43	9.45 885	42	9.47 760	46	10.52 240	9.98 125	4	17
44	9.45 927	42	9.47 806	46	10.52 194	9.98 121	4	16
45	9.45 969	42	9.47 852	45	10.52 148	9.98 117	4	15
46	9.46 011	42	9.47 897	46	10.52 103	9.98 113	3	14
47	9.46 053	42	9.47 943	46	10.52 057	9.98 110	4	13
48	9.46 095	41	9.47 989	46	10.52 011	9.98 106	4	12
49	9.46 136	42	9.48 035	45	10.51 965	9.98 102	4	11
50	9.46 178	42	9.48 080	46	10.51 920	9.98 098	4	10
51	9.46 220	42	9.48 126	45	10.51 874	9.98 094	4	9
52	9.46 262	41	9.48 171	46	10.51 829	9.98 090	3	8
53	9.46 303	42	9.48 217	45	10.51 783	9.98 087	4	7
54	9.46 345	41	9.48 262	45	10.51 738	9.98 083	4	6
55	9.46 386	42	9.48 307	46	10.51 693	9.98 079	4	5
56	9.46 428	41	9.48 353	45	10.51 647	9.98 075	4	4
57	9.46 469	42	9.48 398	45	10.51 602	9.98 071	4	3
58	9.46 511	41	9.48 443	46	10.51 557	9.98 067	4	2
59	9.46 552	42	9.48 489	45	10.51 511	9.98 063	3	1
60	9.46 594		9.48 534		10.51 466	9.98 060		0
	L Cos	d	L Cot	c d	L Tan	L Sin	d	'

Prop. Pts.

	48	47	46
1	4.8	4.7	4.6
2	9.6	9.4	9.2
3	14.4	14.1	13.8
4	19.2	18.8	18.4
5	24.0	23.5	23.0
6	28.8	28.2	27.6
7	33.6	32.9	32.2
8	38.4	37.6	36.8
9	43.2	42.3	41.4

	45	44	43
1	4.5	4.4	4.3
2	9.0	8.8	8.6
3	13.5	13.2	12.9
4	18.0	17.6	17.2
5	22.5	22.0	21.5
6	27.0	26.4	25.8
7	31.5	30.8	30.1
8	36.0	35.2	34.4
9	40.5	39.6	38.7

	42	41
1	4.2	4.1
2	8.4	8.2
3	12.6	12.3
4	16.8	16.4
5	21.0	20.5
6	25.2	24.6
7	29.4	28.7
8	33.6	32.8
9	37.8	36.9

	4	3
1	0.4	0.3
2	0.8	0.6
3	1.2	0.9
4	1.6	1.2
5	2.0	1.5
6	2.4	1.8
7	2.8	2.1
8	3.2	2.4
9	3.6	2.7

73°

TABLE 3. LOGARITHMS OF TRIGONOMETRIC FUNCTIONS (continued)   659

17°

′	L Sin	d	L Tan	c d	L Cot	L Cos	d	
0	9.46 594	41	9.48 534	45	10.51 466	9.98 060	4	60
1	9.46 635	41	9.48 579	45	10.51 421	9.98 056	4	59
2	9.46 676	41	9.48 624	45	10.51 376	9.98 052	4	58
3	9.46 717	41	9.48 669	45	10.51 331	9.98 048	4	57
4	9.46 758	42	9.48 714	45	10.51 286	9.98 044	4	56
5	9.46 800	41	9.48 759	45	10.51 241	9.98 040	4	55
6	9.46 841	41	9.48 804	45	10.51 196	9.98 036	4	54
7	9.46 882	41	9.48 849	45	10.51 151	9.98 032	3	53
8	9.46 923	41	9.48 894	45	10.51 106	9.98 029	4	52
9	9.46 964	41	9.48 939	45	10.51 061	9.98 025	4	51
10	9.47 005	40	9.48 984	45	10.51 016	9.98 021	4	50
11	9.47 045	41	9.49 029	44	10.50 971	9.98 017	4	49
12	9.47 086	41	9.49 073	45	10.50 927	9.98 013	4	48
13	9.47 127	41	9.49 118	45	10.50 882	9.98 009	4	47
14	9.47 168	41	9.49 163	44	10.50 837	9.98 005	4	46
15	9.47 209	40	9.49 207	45	10.50 793	9.98 001	4	45
16	9.47 249	41	9.49 252	44	10.50 748	9.97 997	4	44
17	9.47 290	40	9.49 296	45	10.50 704	9.97 993	4	43
18	9.47 330	41	9.49 341	44	10.50 659	9.97 989	3	42
19	9.47 371	40	9.49 385	45	10.50 615	9.97 986	4	41
20	9.47 411	41	9.49 430	44	10.50 570	9.97 982	4	40
21	9.47 452	40	9.49 474	45	10.50 526	9.97 978	4	39
22	9.47 492	41	9.49 519	44	10.50 481	9.97 974	4	38
23	9.47 533	40	9.49 563	44	10.50 437	9.97 970	4	37
24	9.47 573	40	9.49 607	45	10.50 393	9.97 966	4	36
25	9.47 613	41	9.49 652	44	10.50 348	9.97 962	4	35
26	9.47 654	40	9.49 696	44	10.50 304	9.97 958	4	34
27	9.47 694	40	9.49 740	44	10.50 260	9.97 954	4	33
28	9.47 734	40	9.49 784	44	10.50 216	9.97 950	4	32
29	9.47 774	40	9.49 828	44	10.50 172	9.97 946	4	31
30	9.47 814	40	9.49 872	44	10.50 128	9.97 942	4	30
31	9.47 854	40	9.49 916	44	10.50 084	9.97 938	4	29
32	9.47 894	40	9.49 960	44	10.50 040	9.97 934	4	28
33	9.47 934	40	9.50 004	44	10.49 996	9.97 930	4	27
34	9.47 974	40	9.50 048	44	10.49 952	9.97 926	4	26
35	9.48 014	40	9.50 092	44	10.49 908	9.97 922	4	25
36	9.48 054	40	9.50 136	44	10.49 864	9.97 918	4	24
37	9.48 094	39	9.50 180	43	10.49 820	9.97 914	4	23
38	9.48 133	40	9.50 223	44	10.49 777	9.97 910	4	22
39	9.48 173	40	9.50 267	44	10.49 733	9.97 906	4	21
40	9.48 213	39	9.50 311	44	10.49 689	9.97 902	4	20
41	9.48 252	40	9.50 355	43	10.49 645	9.97 898	4	19
42	9.48 292	40	9.50 398	44	10.49 602	9.97 894	4	18
43	9.48 332	39	9.50 442	43	10.49 558	9.97 890	4	17
44	9.48 371	40	9.50 485	44	10.49 515	9.97 886	4	16
45	9.48 411	39	9.50 529	43	10.49 471	9.97 882	4	15
46	9.48 450	40	9.50 572	44	10.49 428	9.97 878	4	14
47	9.48 490	39	9.50 616	43	10.49 384	9.97 874	4	13
48	9.48 529	39	9.50 659	44	10.49 341	9.97 870	4	12
49	9.48 568	39	9.50 703	43	10.49 297	9.97 866	5	11
50	9.48 607	40	9.50 746	43	10.49 254	9.97 861	4	10
51	9.48 647	39	9.50 789	44	10.49 211	9.97 857	4	9
52	9.48 686	39	9.50 833	43	10.49 167	9.97 853	4	8
53	9.48 725	39	9.50 876	43	10.49 124	9.97 849	4	7
54	9.48 764	39	9.50 919	43	10.49 081	9.97 845	4	6
55	9.48 803	39	9.50 962	43	10.49 038	9.97 841	4	5
56	9.48 842	39	9.51 005	43	10.48 995	9.97 837	4	4
57	9.48 881	39	9.51 048	44	10.48 952	9.97 833	4	3
58	9.48 920	39	9.51 092	43	10.48 908	9.97 829	4	2
59	9.48 959	39	9.51 135	43	10.48 865	9.97 825	4	1
60	9.48 998		9.51 178		10.48 822	9.97 821		0
	L Cos	d	L Cot	c d	L Tan	L Sin	d	′

Prop. Pts.

	45	44	43
1	4.5	4.4	4.3
2	9.0	8.8	8.6
3	13.5	13.2	12.9
4	18.0	17.6	17.2
5	22.5	22.0	21.5
6	27.0	26.4	25.8
7	31.5	30.8	30.1
8	36.0	35.2	34.4
9	40.5	39.6	38.7

	42	41
1	4.2	4.1
2	8.4	8.2
3	12.6	12.3
4	16.8	16.4
5	21.0	20.5
6	25.2	24.6
7	29.4	28.7
8	33.6	32.8
9	37.8	36.9

	40	39
1	4.0	3.9
2	8.0	7.8
3	12.0	11.7
4	16.0	15.6
5	20.0	19.5
6	24.0	23.4
7	28.0	27.3
8	32.0	31.2
9	36.0	35.1

	5	4	3
1	0.5	0.4	0.3
2	1.0	0.8	0.6
3	1.5	1.2	0.9
4	2.0	1.6	1.2
5	2.5	2.0	1.5
6	3.0	2.4	1.8
7	3.5	2.8	2.1
8	4.0	3.2	2.4
9	4.5	3.6	2.7

72°

TABLE 3. LOGARITHMS OF TRIGONOMETRIC FUNCTIONS (continued)   660

18°

'	L Sin	d	L Tan	c d	L Cot	L Cos	d	'
0	9.48 998	39	9.51 178	43	10.48 822	9.97 821	4	60
1	9.49 037	39	9.51 221	43	10.48 779	9.97 817	5	59
2	9.49 076	39	9.51 264	43	10.48 736	9.97 812	4	58
3	9.49 115	38	9.51 306	42	10.48 694	9.97 808	4	57
4	9.49 153	39	9.51 349	43	10.48 651	9.97 804	4	56
5	9.49 192	39	9.51 392	43	10.48 608	9.97 800	4	55
6	9.49 231	38	9.51 435	43	10.48 565	9.97 796	4	54
7	9.49 269	39	9.51 478	42	10.48 522	9.97 792	4	53
8	9.49 308	39	9.51 520	43	10.48 480	9.97 788	4	52
9	9.49 347	38	9.51 563	43	10.48 437	9.97 784	5	51
10	9.49 385	39	9.51 606	42	10.48 394	9.97 779	4	50
11	9.49 424	38	9.51 648	43	10.48 352	9.97 775	4	49
12	9.49 462	38	9.51 691	43	10.48 309	9.97 771	4	48
13	9.49 500	39	9.51 734	42	10.48 266	9.97 767	4	47
14	9.49 539	38	9.51 776	43	10.48 224	9.97 763	4	46
15	9.49 577	38	9.51 819	42	10.48 181	9.97 759	5	45
16	9.49 615	39	9.51 861	42	10.48 139	9.97 754	4	44
17	9.49 654	38	9.51 903	43	10.48 097	9.97 750	4	43
18	9.49 692	38	9.51 946	42	10.48 054	9.97 746	4	42
19	9.49 730	38	9.51 988	43	10.48 012	9.97 742	4	41
20	9.49 768	38	9.52 031	42	10.47 969	9.97 738	4	40
21	9.49 806	38	9.52 073	42	10.47 927	9.97 734	5	39
22	9.49 844	38	9.52 115	42	10.47 885	9.97 729	4	38
23	9.49 882	38	9.52 157	43	10.47 843	9.97 725	4	37
24	9.49 920	38	9.52 200	42	10.47 800	9.97 721	4	36
25	9.49 958	38	9.52 242	42	10.47 758	9.97 717	4	35
26	9.49 996	38	9.52 284	42	10.47 716	9.97 713	5	34
27	9.50 034	38	9.52 326	42	10.47 674	9.97 708	4	33
28	9.50 072	38	9.52 368	42	10.47 632	9.97 704	4	32
29	9.50 110	38	9.52 410	42	10.47 590	9.97 700	4	31
30	9.50 148	37	9.52 452	42	10.47 548	9.97 696	5	30
31	9.50 185	38	9.52 494	42	10.47 506	9.97 691	4	29
32	9.50 223	38	9.52 536	42	10.47 464	9.97 687	4	28
33	9.50 261	37	9.52 578	42	10.47 422	9.97 683	4	27
34	9.50 298	38	9.52 620	41	10.47 380	9.97 679	5	26
35	9.50 336	38	9.52 661	42	10.47 339	9.97 674	4	25
36	9.50 374	37	9.52 703	42	10.47 297	9.97 670	4	24
37	9.50 411	38	9.52 745	42	10.47 255	9.97 666	4	23
38	9.50 449	37	9.52 787	42	10.47 213	9.97 662	5	22
39	9.50 486	37	9.52 829	41	10.47 171	9.97 657	4	21
40	9.50 523	38	9.52 870	42	10.47 130	9.97 653	4	20
41	9.50 561	37	9.52 912	41	10.47 088	9.97 649	4	19
42	9.50 598	37	9.52 953	42	10.47 047	9.97 645	5	18
43	9.50 635	38	9.52 995	42	10.47 005	9.97 640	4	17
44	9.50 673	37	9.53 037	41	10.46 963	9.97 636	4	16
45	9.50 710	37	9.53 078	42	10.46 922	9.97 632	4	15
46	9.50 747	37	9.53 120	41	10.46 880	9.97 628	5	14
47	9.50 784	37	9.53 161	41	10.46 839	9.97 623	4	13
48	9.50 821	37	9.53 202	42	10.46 798	9.97 619	4	12
49	9.50 858	38	9.53 244	41	10.46 756	9.97 615	5	11
50	9.50 896	37	9.53 285	42	10.46 715	9.97 610	4	10
51	9.50 933	37	9.53 327	41	10.46 673	9.97 606	4	9
52	9.50 970	37	9.53 368	41	10.46 632	9.97 602	5	8
53	9.51 007	36	9.53 409	41	10.46 591	9.97 597	4	7
54	9.51 043	37	9.53 450	42	10.46 550	9.97 593	4	6
55	9.51 080	37	9.53 492	41	10.46 508	9.97 589	5	5
56	9.51 117	37	9.53 533	41	10.46 467	9.97 584	4	4
57	9.51 154	37	9.53 574	41	10.46 426	9.97 580	4	3
58	9.51 191	36	9.53 615	41	10.46 385	9.97 576	5	2
59	9.51 227	37	9.53 656	41	10.46 344	9.97 571	4	1
60	9.51 264		9.53 697		10.46 303	9.97 567		0
	L Cos	d	L Cot	c d	L Tan	L Sin	d	'

Prop. Pts.

	43	42	41
1	4.3	4.2	4.1
2	8.6	8.4	8.2
3	12.9	12.6	12.3
4	17.2	16.8	16.4
5	21.5	21.0	20.5
6	25.8	25.2	24.6
7	30.1	29.4	28.7
8	34.4	33.6	32.8
9	38.7	37.8	36.9

	39	38	37
1	3.9	3.8	3.7
2	7.8	7.6	7.4
3	11.7	11.4	11.1
4	15.6	15.2	14.8
5	19.5	19.0	18.5
6	23.4	22.8	22.2
7	27.3	26.6	25.9
8	31.2	30.4	29.6
9	35.1	34.2	33.3

	36	5	4
1	3.6	0.5	0.4
2	7.2	1.0	0.8
3	10.8	1.5	1.2
4	14.4	2.0	1.6
5	18.0	2.5	2.0
6	21.6	3.0	2.4
7	25.2	3.5	2.8
8	28.8	4.0	3.2
9	32.4	4.5	3.6

TABLE 3. LOGARITHMS OF TRIGONOMETRIC FUNCTIONS (continued)   661

19°

′	L Sin	d	L Tan	c d	L Cot	L Cos	d	
0	9.51 264	37	9.53 697	41	10.46 303	9.97 567	4	60
1	9.51 301	37	9.53 738	41	10.46 262	9.97 563	5	59
2	9.51 338	36	9.53 779	41	10.46 221	9.97 558	4	58
3	9.51 374	37	9.53 820	41	10.46 180	9.97 554	4	57
4	9.51 411	36	9.53 861	41	10.46 139	9.97 550	5	56
5	9.51 447	37	9.53 902	41	10.46 098	9.97 545	4	55
6	9.51 484	36	9.53 943	41	10.46 057	9.97 541	5	54
7	9.51 520	37	9.53 984	41	10.46 016	9.97 536	4	53
8	9.51 557	36	9.54 025	40	10.45 975	9.97 532	4	52
9	9.51 593	36	9.54 065	41	10.45 935	9.97 528	5	51
10	9.51 629	37	9.54 106	41	10.45 894	9.97 523	4	50
11	9.51 666	36	9.54 147	40	10.45 853	9.97 519	4	49
12	9.51 702	36	9.54 187	41	10.45 813	9.97 515	5	48
13	9.51 738	36	9.54 228	41	10.45 772	9.97 510	4	47
14	9.51 774	37	9.54 269	40	10.45 731	9.97 506	5	46
15	9.51 811	36	9.54 309	41	10.45 691	9.97 501	4	45
16	9.51 847	36	9.54 350	40	10.45 650	9.97 497	5	44
17	9.51 883	36	9.54 390	41	10.45 610	9.97 492	4	43
18	9.51 919	36	9.54 431	40	10.45 569	9.97 488	4	42
19	9.51 955	36	9.54 471	41	10.45 529	9.97 484	5	41
20	9.51 991	36	9.54 512	40	10.45 488	9.97 479	4	40
21	9.52 027	36	9.54 552	41	10.45 448	9.97 475	5	39
22	9.52 063	36	9.54 593	40	10.45 407	9.97 470	4	38
23	9.52 099	36	9.54 633	40	10.45 367	9.97 466	5	37
24	9.52 135	36	9.54 673	41	10.45 327	9.97 461	4	36
25	9.52 171	36	9.54 714	40	10.45 286	9.97 457	4	35
26	9.52 207	35	9.54 754	40	10.45 246	9.97 453	5	34
27	9.52 242	36	9.54 794	41	10.45 206	9.97 448	4	33
28	9.52 278	36	9.54 835	40	10.45 165	9.97 444	5	32
29	9.52 314	36	9.54 875	40	10.45 125	9.97 439	4	31
30	9.52 350	35	9.54 915	40	10.45 085	9.97 435	5	30
31	9.52 385	36	9.54 955	40	10.45 045	9.97 430	4	29
32	9.52 421	35	9.54 995	40	10.45 005	9.97 426	5	28
33	9.52 456	36	9.55 035	40	10.44 965	9.97 421	4	27
34	9.52 492	35	9.55 075	40	10.44 925	9.97 417	5	26
35	9.52 527	36	9.55 115	40	10.44 885	9.97 412	4	25
36	9.52 563	35	9.55 155	40	10.44 845	9.97 408	5	24
37	9.52 598	36	9.55 195	40	10.44 805	9.97 403	4	23
38	9.52 634	35	9.55 235	40	10.44 765	9.97 399	5	22
39	9.52 669	36	9.55 275	40	10.44 725	9.97 394	4	21
40	9.52 705	35	9.55 315	40	10.44 685	9.97 390	5	20
41	9.52 740	35	9.55 355	40	10.44 645	9.97 385	4	19
42	9.52 775	36	9.55 395	39	10.44 605	9.97 381	5	18
43	9.52 811	35	9.55 434	40	10.44 566	9.97 376	4	17
44	9.52 846	35	9.55 474	40	10.44 526	9.97 372	5	16
45	9.52 881	35	9.55 514	40	10.44 486	9.97 367	4	15
46	9.52 916	35	9.55 554	39	10.44 446	9.97 363	5	14
47	9.52 951	35	9.55 593	40	10.44 407	9.97 358	5	13
48	9.52 986	35	9.55 633	40	10.44 367	9.97 353	4	12
49	9.53 021	35	9.55 673	39	10.44 327	9.97 349	5	11
50	9.53 056	36	9.55 712	40	10.44 288	9.97 344	4	10
51	9.53 092	34	9.55 752	39	10.44 248	9.97 340	5	9
52	9.53 126	35	9.55 791	40	10.44 209	9.97 335	4	8
53	9.53 161	35	9.55 831	39	10.44 169	9.97 331	5	7
54	9.53 196	35	9.55 870	40	10.44 130	9.97 326	4	6
55	9.53 231	35	9.55 910	39	10.44 090	9.97 322	5	5
56	9.53 266	35	9.55 949	40	10.44 051	9.97 317	5	4
57	9.53 301	35	9.55 989	39	10.44 011	9.97 312	4	3
58	9.53 336	34	9.56 028	39	10.43 972	9.97 308	5	2
59	9.53 370	35	9.56 067	40	10.43 933	9.97 303	4	1
60	9.53 405		9.56 107		10.43 893	9.97 299		0
	L Cos	d	L Cot	c d	L Tan	L Sin	d	′

Prop. Pts.

	41	40	39
1	4.1	4.0	3.9
2	8.2	8.0	7.8
3	12.3	12.0	11.7
4	16.4	16.0	15.6
5	20.5	20.0	19.5
6	24.6	24.0	23.4
7	28.7	28.0	27.3
8	32.8	32.0	31.2
9	36.9	36.0	35.1

	37	36	35
1	3.7	3.6	3.5
2	7.4	7.2	7.0
3	11.1	10.8	10.5
4	14.8	14.4	14.0
5	18.5	18.0	17.5
6	22.2	21.6	21.0
7	25.9	25.2	24.5
8	29.6	28.8	28.0
9	33.3	32.4	31.5

	34	5	4
1	3.4	0.5	0.4
2	6.8	1.0	0.8
3	10.2	1.5	1.2
4	13.6	2.0	1.6
5	17.0	2.5	2.0
6	20.4	3.0	2.4
7	23.8	3.5	2.8
8	27.2	4.0	3.2
9	30.6	4.5	3.6

70°

TABLE 3. LOGARITHMS OF TRIGONOMETRIC FUNCTIONS (continued)   662

20°

′	L Sin	d	L Tan	c d	L Cot	L Cos	d	′
0	9.53 405	35	9.56 107	39	10.43 893	9.97 299	5	60
1	9.53 440	35	9.56 146	39	10.43 854	9.97 294	5	59
2	9.53 475	34	9.56 185	39	10.43 815	9.97 289	4	58
3	9.53 509	35	9.56 224	40	10.43 776	9.97 285	5	57
4	9.53 544	34	9.56 264	39	10.43 736	9.97 280	4	56
5	9.53 578	35	9.56 303	39	10.43 697	9.97 276	5	55
6	9.53 613	34	9.56 342	39	10.43 658	9.97 271	5	54
7	9.53 647	35	9.56 381	39	10.43 619	9.97 266	4	53
8	9.53 682	34	9.56 420	39	10.43 580	9.97 262	5	52
9	9.53 716	35	9.56 459	39	10.43 541	9.97 257	5	51
10	9.53 751	34	9.56 498	39	10.43 502	9.97 252	4	50
11	9.53 785	34	9.56 537	39	10.43 463	9.97 248	5	49
12	9.53 819	35	9.56 576	39	10.43 424	9.97 243	5	48
13	9.53 854	34	9.56 615	39	10.43 385	9.97 238	4	47
14	9.53 888	34	9.56 654	39	10.43 346	9.97 234	5	46
15	9.53 922	35	9.56 693	39	10.43 307	9.97 229	5	45
16	9.53 957	34	9.56 732	39	10.43 268	9.97 224	4	44
17	9.53 991	34	9.56 771	39	10.43 229	9.97 220	5	43
18	9.54 025	34	9.56 810	39	10.43 190	9.97 215	5	42
19	9.54 059	34	9.56 849	38	10.43 151	9.97 210	4	41
20	9.54 093	34	9.56 887	39	10.43 113	9.97 206	5	40
21	9.54 127	34	9.56 926	39	10.43 074	9.97 201	5	39
22	9.54 161	34	9.56 965	39	10.43 035	9.97 196	4	38
23	9.54 195	34	9.57 004	38	10.42 996	9.97 192	5	37
24	9.54 229	34	9.57 042	39	10.42 958	9.97 187	5	36
25	9.54 263	34	9.57 081	39	10.42 919	9.97 182	4	35
26	9.54 297	34	9.57 120	38	10.42 880	9.97 178	5	34
27	9.54 331	34	9.57 158	39	10.42 842	9.97 173	5	33
28	9.54 365	34	9.57 197	38	10.42 803	9.97 168	5	32
29	9.54 399	34	9.57 235	39	10.42 765	9.97 163	4	31
30	9.54 433	33	9.57 274	38	10.42 726	9.97 159	5	30
31	9.54 466	34	9.57 312	39	10.42 688	9.97 154	5	29
32	9.54 500	34	9.57 351	38	10.42 649	9.97 149	4	28
33	9.54 534	33	9.57 389	39	10.42 611	9.97 145	5	27
34	9.54 567	34	9.57 428	38	10.42 572	9.97 140	5	26
35	9.54 601	34	9.57 466	38	10.42 534	9.97 135	5	25
36	9.54 635	33	9.57 504	39	10.42 496	9.97 130	4	24
37	9.54 668	34	9.57 543	38	10.42 457	9.97 126	5	23
38	9.54 702	33	9.57 581	38	10.42 419	9.97 121	5	22
39	9.54 735	34	9.57 619	39	10.42 381	9.97 116	5	21
40	9.54 769	33	9.57 658	38	10.42 342	9.97 111	4	20
41	9.54 802	34	9.57 696	38	10.42 304	9.97 107	5	19
42	9.54 836	33	9.57 734	38	10.42 266	9.97 102	5	18
43	9.54 869	34	9.57 772	38	10.42 228	9.97 097	5	17
44	9.54 903	33	9.57 810	39	10.42 190	9.97 092	5	16
45	9.54 936	33	9.57 849	38	10.42 151	9.97 087	4	15
46	9.54 969	34	9.57 887	38	10.42 113	9.97 083	5	14
47	9.55 003	33	9.57 925	38	10.42 075	9.97 078	5	13
48	9.55 036	33	9.57 963	38	10.42 037	9.97 073	5	12
49	9.55 069	33	9.58 001	38	10.41 999	9.97 068	5	11
50	9.55 102	34	9.58 039	38	10.41 961	9.97 063	4	10
51	9.55 136	33	9.58 077	38	10.41 923	9.97 059	5	9
52	9.55 169	33	9.58 115	38	10.41 885	9.97 054	5	8
53	9.55 202	33	9.58 153	38	10.41 847	9.97 049	5	7
54	9.55 235	33	9.58 191	38	10.41 809	9.97 044	5	6
55	9.55 268	33	9.58 229	38	10.41 771	9.97 039	4	5
56	9.55 301	33	9.58 267	37	10.41 733	9.97 035	5	4
57	9.55 334	33	9.58 304	38	10.41 696	9.97 030	5	3
58	9.55 367	33	9.58 342	38	10.41 658	9.97 025	5	2
59	9.55 400	33	9.58 380	38	10.41 620	9.97 020	5	1
60	9.55 433		9.58 418		10.41 582	9.97 015		0
	L Cos	d	L Cot	c d	L Tan	L Sin	d	′

Prop. Pts.

	40	39	38
1	4.0	3.9	3.8
2	8.0	7.8	7.6
3	12.0	11.7	11.4
4	16.0	15.6	15.2
5	20.0	19.5	19.0
6	24.0	23.4	22.8
7	28.0	27.3	26.6
8	32.0	31.2	30.4
9	36.0	35.1	34.2

	37	35	34
1	3.7	3.5	3.4
2	7.4	7.0	6.8
3	11.1	10.5	10.2
4	14.8	14.0	13.6
5	18.5	17.5	17.0
6	22.2	21.0	20.4
7	25.9	24.5	23.8
8	29.6	28.0	27.2
9	33.3	31.5	30.6

	33	5	4
1	3.3	0.5	0.4
2	6.6	1.0	0.8
3	9.9	1.5	1.2
4	13.2	2.0	1.6
5	16.5	2.5	2.0
6	19.8	3.0	2.4
7	23.1	3.5	2.8
8	26.4	4.0	3.2
9	29.7	4.5	3.6

TABLE 3. LOGARITHMS OF TRIGONOMETRIC FUNCTIONS (continued)   **663**

21°

′	L Sin	d	L Tan	c d	L Cot	L Cos	d		Prop. Pts.
0	9.55 433	33	9.58 418	37	10.41 582	9.97 015	5	60	
1	9.55 466	33	9.58 455	38	10.41 545	9.97 010	5	59	
2	9.55 499	33	9.58 493	38	10.41 507	9.97 005	4	58	
3	9.55 532	32	9.58 531	38	10.41 469	9.97 001	5	57	
4	9.55 564	33	9.58 569	37	10.41 431	9.96 996	5	56	
5	9.55 597	33	9.58 606	38	10.41 394	9.96 991	5	55	
6	9.55 630	33	9.58 644	37	10.41 356	9.96 986	5	54	
7	9.55 663	32	9.58 681	38	10.41 319	9.96 981	5	53	
8	9.55 695	33	9.58 719	38	10.41 281	9.96 976	5	52	**38 37 36**
9	9.55 728	33	9.58 757	37	10.41 243	9.96 971	5	51	1 3.8 3.7 3.6
10	9.55 761	32	9.58 794	38	10.41 206	9.96 966	4	50	2 7.6 7.4 7.2
11	9.55 793	33	9.58 832	37	10.41 168	9.96 962	5	49	3 11.4 11.1 10.8
12	9.55 826	32	9.58 869	38	10.41 131	9.96 957	5	48	4 15.2 14.8 14.4
13	9.55 858	33	9.58 907	37	10.41 093	9.96 952	5	47	5 19.0 18.5 18.0
14	9.55 891	32	9.58 944	37	10.41 056	9.96 947	5	46	6 22.8 22.2 21.6
15	9.55 923	33	9.58 981	38	10.41 019	9.96 942	5	45	7 26.6 25.9 25.2
16	9.55 956	32	9.59 019	37	10.40 981	9.96 937	5	44	8 30.4 29.6 28.8
17	9.55 988	33	9.59 056	38	10.40 944	9.96 932	5	43	9 34.2 33.3 32.4
18	9.56 021	32	9.59 094	37	10.40 906	9.96 927	5	42	
19	9.56 053	32	9.59 131	37	10.40 869	9.96 922	5	41	
20	9.56 085	33	9.59 168	37	10.40 832	9.96 917	5	40	
21	9.56 118	32	9.59 205	38	10.40 795	9.96 912	5	39	
22	9.56 150	32	9.59 243	37	10.40 757	9.96 907	4	38	
23	9.56 182	33	9.59 280	37	10.40 720	9.96 903	5	37	
24	9.56 215	32	9.59 317	37	10.40 683	9.96 898	5	36	
25	9.56 247	32	9.59 354	37	10.40 646	9.96 893	5	35	
26	9.56 279	32	9.59 391	38	10.40 609	9.96 888	5	34	**33 32 31**
27	9.56 311	32	9.59 429	37	10.40 571	9.96 883	5	33	1 3.3 3.2 3.1
28	9.56 343	32	9.59 466	37	10.40 534	9.96 878	5	32	2 6.6 6.4 6.2
29	9.56 375	33	9.59 503	37	10.40 497	9.96 873	5	31	3 9.9 9.6 9.3
30	9.56 408	32	9.59 540	37	10.40 460	9.96 868	5	30	4 13.2 12.8 12.4
31	9.56 440	32	9.59 577	37	10.40 423	9.96 863	5	29	5 16.5 16.0 15.5
32	9.56 472	32	9.59 614	37	10.40 386	9.96 858	5	28	6 19.8 19.2 18.6
33	9.56 504	32	9.59 651	37	10.40 349	9.96 853	5	27	7 23.1 22.4 21.7
34	9.56 536	32	9.59 688	37	10.40 312	9.96 848	5	26	8 26.4 25.6 24.8
35	9.56 568	31	9.59 725	37	10.40 275	9.96 843	5	25	9 29.7 28.8 27.9
36	9.56 599	32	9.59 762	37	10.40 238	9.96 838	5	24	
37	9.56 631	32	9.59 799	36	10.40 201	9.96 833	5	23	
38	9.56 663	32	9.59 835	37	10.40 165	9.96 828	5	22	
39	9.56 695	32	9.59 872	37	10.40 128	9.96 823	5	21	
40	9.56 727	32	9.59 909	37	10.40 091	9.96 818	5	20	
41	9.56 759	31	9.59 946	37	10.40 054	9.96 813	5	19	
42	9.56 790	32	9.59 983	36	10.40 017	9.96 808	5	18	
43	9.56 822	32	9.60 019	37	10.39 981	9.96 803	5	17	
44	9.56 854	32	9.60 056	37	10.39 944	9.96 798	5	16	**6 5 4**
45	9.56 886	31	9.60 093	37	10.39 907	9.96 793	5	15	1 0.6 0.5 0.4
46	9.56 917	32	9.60 130	36	10.39 870	9.96 788	5	14	2 1.2 1.0 0.8
47	9.56 949	31	9.60 166	37	10.39 834	9.96 783	5	13	3 1.8 1.5 1.2
48	9.56 980	32	9.60 203	37	10.39 797	9.96 778	6	12	4 2.4 2.0 1.6
49	9.57 012	32	9.60 240	36	10.39 760	9.96 772	5	11	5 3.0 2.5 2.0
50	9.57 044	31	9.60 276	37	10.39 724	9.96 767	5	10	6 3.6 3.0 2.4
51	9.57 075	32	9.60 313	36	10.39 687	9.96 762	5	9	7 4.2 3.5 2.8
52	9.57 107	31	9.60 349	37	10.39 651	9.96 757	5	8	8 4.8 4.0 3.2
53	9.57 138	31	9.60 386	36	10.39 614	9.96 752	5	7	9 5.4 4.5 3.6
54	9.57 169	32	9.60 422	37	10.39 578	9.96 747	5	6	
55	9.57 201	31	9.60 459	36	10.39 541	9.96 742	5	5	
56	9.57 232	32	9.60 495	37	10.39 505	9.96 737	5	4	
57	9.57 264	31	9.60 532	36	10.39 468	9.96 732	5	3	
58	9.57 295	31	9.60 568	37	10.39 432	9.96 727	5	2	
59	9.57 326	32	9.60 605	36	10.39 395	9.96 722	5	1	
60	9.57 358		9.60 641		10.39 359	9.96 717		0	
	L Cos	d	L Cot	c d	L Tan	L Sin	d	′	Prop. Pts.

68°

TABLE 3. LOGARITHMS OF TRIGONOMETRIC FUNCTIONS (continued)   664

22°

′	L Sin	d	L Tan	c d	L Cot	L Cos	d	
0	9.57 358		9.60 641		10.39 359	9.96 717		60
1	9.57 389	31	9.60 677	36	10.39 323	9.96 711	6	59
2	9.57 420	31	9.60 714	37	10.39 286	9.96 706	5	58
3	9.57 451	31	9.60 750	36	10.39 250	9.96 701	5	57
4	9.57 482	31	9.60 786	36	10.39 214	9.96 696	5	56
5	9.57 514	32	9.60 823	37	10.39 177	9.96 691	5	55
6	9.57 545	31	9.60 859	36	10.39 141	9.96 686	5	54
7	9.57 576	31	9.60 895	36	10.39 105	9.96 681	5	53
8	9.57 607	31	9.60 931	36	10.39 069	9.96 676	5	52
9	9.57 638	31	9.60 967	36	10.39 033	9.96 670	6	51
10	9.57 669	31	9.61 004	37	10.38 996	9.96 665	5	50
11	9.57 700	31	9.61 040	36	10.38 960	9.96 660	5	49
12	9.57 731	31	9.61 076	36	10.38 924	9.96 655	5	48
13	9.57 762	31	9.61 112	36	10.38 888	9.96 650	5	47
14	9.57 793	31	9.61 148	36	10.38 852	9.96 645	5	46
15	9.57 824	31	9.61 184	36	10.38 816	9.96 640	5	45
16	9.57 855	31	9.61 220	36	10.38 780	9.96 634	6	44
17	9.57 885	30	9.61 256	36	10.38 744	9.96 629	5	43
18	9.57 916	31	9.61 292	36	10.38 708	9.96 624	5	42
19	9.57 947	31	9.61 328	36	10.38 672	9.96 619	5	41
20	9.57 978	31	9.61 364	36	10.38 636	9.96 614	5	40
21	9.58 008	30	9.61 400	36	10.38 600	9.96 608	6	39
22	9.58 039	31	9.61 436	36	10.38 564	9.96 603	5	38
23	9.58 070	31	9.61 472	36	10.38 528	9.96 598	5	37
24	9.58 101	31	9.61 508	36	10.38 492	9.96 593	5	36
25	9.58 131	30	9.61 544	36	10.38 456	9.96 588	6	35
26	9.58 162	31	9.61 579	35	10.38 421	9.96 582	5	34
27	9.58 192	30	9.61 615	36	10.38 385	9.96 577	5	33
28	9.58 223	31	9.61 651	36	10.38 349	9.96 572	5	32
29	9.58 253	30	9.61 687	36	10.38 313	9.96 567	5	31
30	9.58 284	31	9.61 722	35	10.38 278	9.96 562	6	30
31	9.58 314	30	9.61 758	36	10.38 242	9.96 556	5	29
32	9.58 345	31	9.61 794	36	10.38 206	9.96 551	5	28
33	9.58 375	30	9.61 830	36	10.38 170	9.96 546	5	27
34	9.58 406	31	9.61 865	35	10.38 135	9.96 541	6	26
35	9.58 436	30	9.61 901	36	10.38 099	9.96 535	5	25
36	9.58 467	31	9.61 936	35	10.38 064	9.96 530	5	24
37	9.58 497	30	9.61 972	36	10.38 028	9.96 525	5	23
38	9.58 527	30	9.62 008	36	10.37 992	9.96 520	6	22
39	9.58 557	30	9.62 043	35	10.37 957	9.96 514	5	21
40	9.58 588	31	9.62 079	36	10.37 921	9.96 509	5	20
41	9.58 618	30	9.62 114	35	10.37 886	9.96 504	5	19
42	9.58 648	30	9.62 150	36	10.37 850	9.96 498	6	18
43	9.58 678	30	9.62 185	35	10.37 815	9.96 493	5	17
44	9.58 709	31	9.62 221	36	10.37 779	9.96 488	5	16
45	9.58 739	30	9.62 256	35	10.37 744	9.96 483	5	15
46	9.58 769	30	9.62 292	36	10.37 708	9.96 477	6	14
47	9.58 799	30	9.62 327	35	10.37 673	9.96 472	5	13
48	9.58 829	30	9.62 362	35	10.37 638	9.96 467	6	12
49	9.58 859	30	9.62 398	36	10.37 602	9.96 461	5	11
50	9.58 889	30	9.62 433	35	10.37 567	9.96 456	5	10
51	9.58 919	30	9.62 468	35	10.37 532	9.96 451	5	9
52	9.58 949	30	9.62 504	36	10.37 496	9.96 445	6	8
53	9.58 979	30	9.62 539	35	10.37 461	9.96 440	5	7
54	9.59 009	30	9.62 574	35	10.37 426	9.96 435	5	6
55	9.59 039	30	9.62 609	35	10.37 391	9.96 429	6	5
56	9.59 069	30	9.62 645	36	10.37 355	9.96 424	5	4
57	9.59 098	29	9.62 680	35	10.37 320	9.96 419	5	3
58	9.59 128	30	9.62 715	35	10.37 285	9.96 413	6	2
59	9.59 158	30	9.62 750	35	10.37 250	9.96 408	5	1
60	9.59 188	30	9.62 785	35	10.37 215	9.96 403	5	0
	L Cos	d	L Cot	c d	L Tan	L Sin	d	′

Prop. Pts.

	37	36	35
1	3.7	3.6	3.5
2	7.4	7.2	7.0
3	11.1	10.8	10.5
4	14.8	14.4	14.0
5	18.5	18.0	17.5
6	22.2	21.6	21.0
7	25.9	25.2	24.5
8	29.6	28.8	28.0
9	33.3	32.4	31.5

	32	31	30
1	3.2	3.1	3.0
2	6.4	6.2	6.0
3	9.6	9.3	9.0
4	12.8	12.4	12.0
5	16.0	15.5	15.0
6	19.2	18.6	18.0
7	22.4	21.7	21.0
8	25.6	24.8	24.0
9	28.8	27.9	27.0

	29	6	5
1	2.9	0.6	0.5
2	5.8	1.2	1.0
3	8.7	1.8	1.5
4	11.6	2.4	2.0
5	14.5	3.0	2.5
6	17.4	3.6	3.0
7	20.3	4.2	3.5
8	23.2	4.8	4.0
9	26.1	5.4	4.5

TABLE 3. LOGARITHMS OF TRIGONOMETRIC FUNCTIONS (continued)   **665**

23°

′	L Sin	d	L Tan	c d	L Cot	L Cos	d		Prop. Pts.
0	9.59 188		9.62 785		10.37 215	9.96 403		**60**	
1	9.59 218	30	9.62 820	35	10.37 180	9.96 397	6	59	
2	9.59 247	29	9.62 855	35	10.37 145	9.96 392	5	58	
3	9.59 277	30	9.62 890	35	10.37 110	9.96 387	5	57	
4	9.59 307	30	9.62 926	36	10.37 074	9.96 381	6	56	
5	9.59 336	29	9.62 961	35	10.37 039	9.96 376	5	55	
6	9.59 366	30	9.62 996	35	10.37 004	9.96 370	6	54	
7	9.59 396	30	9.63 031	35	10.36 969	9.96 365	5	53	
8	9.59 425	29	9.63 066	35	10.36 934	9.96 360	5	52	
9	9.59 455	30	9.63 101	35	10.36 899	9.96 354	6	51	
10	9.59 484	29	9.63 135	34	10.36 865	9.96 349	5	**50**	
11	9.59 514	30	9.63 170	35	10.36 830	9.96 343	6	49	
12	9.59 543	29	9.63 205	35	10.36 795	9.96 338	5	48	
13	9.59 573	30	9.63 240	35	10.36 760	9.96 333	5	47	
14	9.59 602	29	9.63 275	35	10.36 725	9.96 327	6	46	
15	9.59 632	30	9.63 310	35	10.36 690	9.96 322	5	45	
16	9.59 661	29	9.63 345	34	10.36 655	9.96 316	6	44	
17	9.59 690	29	9.63 379	35	10.36 621	9.96 311	5	43	
18	9.59 720	30	9.63 414	35	10.36 586	9.96 305	6	42	
19	9.59 749	29	9.63 449	35	10.36 551	9.96 300	5	41	
20	9.59 778	29	9.63 484	35	10.36 516	9.96 294	6	**40**	
21	9.59 808	30	9.63 519	35	10.36 481	9.96 289	5	39	
22	9.59 839	29	9.63 553	34	10.36 447	9.96 284	5	38	
23	9.59 866	29	9.63 588	35	10.36 412	9.96 278	6	37	
24	9.59 895	29	9.63 623	35	10.36 377	9.96 273	5	36	
25	9.59 924	29	9.63 657	34	10.36 343	9.96 267	6	35	
26	9.59 954	30	9.63 692	35	10.36 308	9.96 262	5	34	
27	9.59 983	29	9.63 726	34	10.36 274	9.96 256	6	33	
28	9.60 012	29	9.63 761	35	10.36 239	9.96 251	5	32	
29	9.60 041	29	9.63 796	35	10.36 204	9.96 245	6	31	
30	9.60 070	29	9.63 830	34	10.36 170	9.96 240	5	**30**	
31	9.60 099	29	9.63 865	35	10.36 135	9.96 234	6	29	
32	9.60 128	29	9.63 899	34	10.36 101	9.96 229	5	28	
33	9.60 157	29	9.63 934	35	10.36 066	9.96 223	6	27	
34	9.60 186	29	9.63 968	34	10.36 032	9.96 218	5	26	
35	9.60 215	29	9.64 003	35	10.35 997	9.96 212	6	25	
36	9.60 244	29	9.64 037	34	10.35 963	9.96 207	5	24	
37	9.60 273	29	9.64 072	35	10.35 928	9.96 201	6	23	
38	9.60 302	29	9.64 106	34	10.35 894	9.96 196	5	22	
39	9.60 331	29	9.64 140	34	10.35 860	9.96 190	6	21	
40	9.60 359	28	9.64 175	35	10.35 825	9.96 185	5	**20**	
41	9.60 388	29	9.64 209	34	10.35 791	9.96 179	6	19	
42	9.60 417	29	9.64 243	34	10.35 757	9.96 174	5	18	
43	9.60 446	29	9.64 278	35	10.35 722	9.96 168	6	17	
44	9.60 474	28	9.64 312	34	10.35 688	9.96 162	6	16	
45	9.60 503	29	9.64 346	34	10.35 654	9.96 157	5	15	
46	9.60 532	29	9.64 381	35	10.35 619	9.96 151	6	14	
47	9.60 561	29	9.64 415	34	10.35 585	9.96 146	5	13	
48	9.60 589	28	9.64 449	34	10.35 551	9.96 140	6	12	
49	9.60 618	29	9.64 483	34	10.35 517	9.96 135	5	11	
50	9.60 646	28	9.64 517	34	10.35 483	9.96 129	6	**10**	
51	9.60 675	29	9.64 552	35	10.35 448	9.96 123	6	9	
52	9.60 704	29	9.64 586	34	10.35 414	9.96 118	5	8	
53	9.60 732	28	9.64 620	34	10.35 380	9.96 112	6	7	
54	9.60 761	29	9.64 654	34	10.35 346	9.96 107	5	6	
55	9.60 789	28	9.64 688	34	10.35 312	9.96 101	6	5	
56	9.60 818	29	9.64 722	34	10.35 278	9.96 095	6	4	
57	9.60 846	28	9.64 756	34	10.35 244	9.96 090	5	3	
58	9.60 875	29	9.64 790	34	10.35 210	9.96 084	6	2	
59	9.60 903	28	9.64 824	34	10.35 176	9.96 079	5	1	
60	9.60 931	28	9.64 858	34	10.35 142	9.96 073	6	**0**	
	L Cos	d	L Cot	c d	L Tan	L Sin	d	′	Prop. Pts.

Prop. Pts.

	36	35	34
1	3.6	3.5	3.4
2	7.2	7.0	6.8
3	10.8	10.5	10.2
4	14.4	14.0	13.6
5	18.0	17.5	17.0
6	21.6	21.0	20.4
7	25.2	24.5	23.8
8	28.8	28.0	27.2
9	32.4	31.5	30.6

	30	29	28
1	3.0	2.9	2.8
2	6.0	5.8	5.6
3	9.0	8.7	8.4
4	12.0	11.6	11.2
5	15.0	14.5	14.0
6	18.0	17.4	16.8
7	21.0	20.3	19.6
8	24.0	23.2	22.4
9	27.0	26.1	25.2

	6	5
1	0.6	0.5
2	1.2	1.0
3	1.8	1.5
4	2.4	2.0
5	3.0	2.5
6	3.6	3.0
7	4.2	3.5
8	4.8	4.0
9	5.4	4.5

TABLE 3. LOGARITHMS OF TRIGONOMETRIC FUNCTIONS (continued)   666

24°

′	L Sin	d	L Tan	c d	L Cot	L Cos	d		Prop. Pts.
0	9.60 931	29	9.64 858	34	10.35 142	9.96 073	6	60	
1	9.60 960	28	9.64 892	34	10.35 108	9.96 067	5	59	
2	9.60 988	28	9.64 926	34	10.35 074	9.96 062	6	58	
3	9.61 016	29	9.64 960	34	10.35 040	9.96 056	6	57	
4	9.61 045	28	9.64 994	34	10.35 006	9.96 050	5	56	
5	9.61 073	28	9.65 028	34	10.34 972	9.96 045	6	55	
6	9.61 101	28	9.65 062	34	10.34 938	9.96 039	5	54	
7	9.61 129	29	9.65 096	34	10.34 904	9.96 034	6	53	
8	9.61 158	28	9.65 130	34	10.34 870	9.96 028	6	52	**34** / **33**
9	9.61 186	28	9.65 164	33	10.34 836	9.96 022	5	51	1 3.4 3.3
10	9.61 214	28	9.65 197	34	10.34 803	9.96 017	6	50	2 6.8 6.6
11	9.61 242	28	9.65 231	34	10.34 769	9.96 011	6	49	3 10.2 9.9
12	9.61 270	28	9.65 265	34	10.34 735	9.96 005	5	48	4 13.6 13.2
13	9.61 298	28	9.65 299	34	10.34 701	9.96 000	6	47	5 17.0 16.5
14	9.61 326	28	9.65 333	33	10.34 667	9.95 994	6	46	6 20.4 19.8
15	9.61 354	28	9.65 366	34	10.34 634	9.95 988	6	45	7 23.8 23.1
16	9.61 382	29	9.65 400	34	10.34 600	9.95 982	5	44	8 27.2 26.4
17	9.61 411	27	9.65 434	33	10.34 566	9.95 977	6	43	9 30.6 29.7
18	9.61 438	28	9.65 467	34	10.34 533	9.95 971	6	42	
19	9.61 466	28	9.65 501	34	10.34 499	9.95 965	5	41	
20	9.61 494	28	9.65 535	33	10.34 465	9.95 960	6	40	
21	9.61 522	28	9.65 568	34	10.34 432	9.95 954	6	39	
22	9.61 550	28	9.65 602	34	10.34 398	9.95 948	6	38	
23	9.61 578	28	9.65 636	33	10.34 364	9.95 942	5	37	
24	9.61 606	28	9.65 669	34	10.34 331	9.95 937	6	36	
25	9.61 634	28	9.65 703	33	10.34 297	9.95 931	6	35	
26	9.61 662	27	9.65 736	34	10.34 264	9.95 925	5	34	**29** / **28** / **27**
27	9.61 689	28	9.65 770	33	10.34 230	9.95 920	6	33	1 2.9 2.8 2.7
28	9.61 717	28	9.65 803	34	10.34 197	9.95 914	6	32	2 5.8 5.6 5.4
29	9.61 745	28	9.65 837	33	10.34 163	9.95 908	6	31	3 8.7 8.4 8.1
30	9.61 773	27	9.65 870	34	10.34 130	9.95 902	5	30	4 11.6 11.2 10.8
31	9.61 800	28	9.65 904	33	10.34 096	9.95 897	6	29	5 14.5 14.0 13.5
32	9.61 828	28	9.65 937	34	10.34 063	9.95 891	6	28	6 17.4 16.8 16.2
33	9.61 856	27	9.65 971	33	10.34 029	9.95 885	6	27	7 20.3 19.6 18.9
34	9.61 883	28	9.66 004	34	10.33 996	9.95 879	6	26	8 23.2 22.4 21.6
35	9.61 911	28	9.66 038	33	10.33 962	9.95 873	5	25	9 26.1 25.2 24.3
36	9.61 939	27	9.66 071	33	10.33 929	9.95 868	6	24	
37	9.61 966	28	9.66 104	34	10.33 896	9.95 862	6	23	
38	9.61 994	27	9.66 138	33	10.33 862	9.95 856	6	22	
39	9.62 021	28	9.66 171	33	10.33 829	9.95 850	6	21	
40	9.62 049	27	9.66 204	34	10.33 796	9.95 844	5	20	
41	9.62 076	28	9.66 238	33	10.33 762	9.95 839	6	19	
42	9.62 104	27	9.66 271	33	10.33 729	9.95 833	6	18	
43	9.62 131	28	9.66 304	33	10.33 696	9.95 827	6	17	
44	9.62 159	27	9.66 337	34	10.33 663	9.95 821	6	16	**6** / **5**
45	9.62 186	28	9.66 371	33	10.33 629	9.95 815	5	15	1 0.6 0.5
46	9.62 214	27	9.66 404	33	10.33 596	9.95 810	6	14	2 1.2 1.0
47	9.62 241	27	9.66 437	33	10.33 563	9.95 804	6	13	3 1.8 1.5
48	9.62 268	28	9.66 470	33	10.33 530	9.95 798	6	12	4 2.4 2.0
49	9.62 296	27	9.66 503	34	10.33 497	9.95 792	6	11	5 3.0 2.5
50	9.62 323	27	9.66 537	33	10.33 463	9.95 786	6	10	6 3.6 3.0
51	9.62 350	27	9.66 570	33	10.33 430	9.95 780	5	9	7 4.2 3.5
52	9.62 377	28	9.66 603	33	10.33 397	9.95 775	6	8	8 4.8 4.0
53	9.62 405	27	9.66 636	33	10.33 364	9.95 769	6	7	9 5.4 4.5
54	9.62 432	27	9.66 669	33	10.33 331	9.95 763	6	6	
55	9.62 459	27	9.66 702	33	10.33 298	9.95 757	6	5	
56	9.62 486	27	9.66 735	33	10.33 265	9.95 751	6	4	
57	9.62 513	28	9.66 768	33	10.33 232	9.95 745	6	3	
58	9.62 541	27	9.66 801	33	10.33 199	9.95 739	6	2	
59	9.62 568	27	9.66 834	33	10.33 166	9.95 733	5	1	
60	9.62 595		9.66 867		10.33 133	9.95 728		0	
	L Cos	d	L Cot	c d	L Tan	L Sin	d	′	Prop. Pts.

TABLE 3. LOGARITHMS OF TRIGONOMETRIC FUNCTIONS (continued)　　667

25°

′	L Sin	d	L Tan	c d	L Cot	L Cos	d	
0	9.62 595	27	9.66 867	33	10.33 133	9.95 728	6	60
1	9.62 622	27	9.66 900	33	10.33 100	9.95 722	6	59
2	9.62 649	27	9.66 933	33	10.33 067	9.95 716	6	58
3	9.62 676	27	9.66 966	33	10.33 034	9.95 710	6	57
4	9.62 703	27	9.66 999	33	10.33 001	9.95 704	6	56
5	9.62 730	27	9.67 032	33	10.32 968	9.95 698	6	55
6	9.62 757	27	9.67 065	33	10.32 935	9.95 692	6	54
7	9.62 784	27	9.67 098	33	10.32 902	9.95 686	6	53
8	9.62 811	27	9.67 131	32	10.32 869	9.95 680	6	52
9	9.62 838	27	9.67 163	33	10.32 837	9.95 674	6	51
10	9.62 865	27	9.67 196	33	10.32 804	9.95 668	5	50
11	9.62 892	26	9.67 229	33	10.32 771	9.95 663	6	49
12	9.62 918	27	9.67 262	33	10.32 738	9.95 657	6	48
13	9.62 945	27	9.67 295	32	10.32 705	9.95 651	6	47
14	9.62 972	27	9.67 327	33	10.32 673	9.95 645	6	46
15	9.62 999	27	9.67 360	33	10.32 640	9.95 639	6	45
16	9.63 026	26	9.67 393	33	10.32 607	9.95 633	6	44
17	9.63 052	27	9.67 426	32	10.32 574	9.95 627	6	43
18	9.63 079	27	9.67 458	33	10.32 542	9.95 621	6	42
19	9.63 106	27	9.67 491	33	10.32 509	9.95 615	6	41
20	9.63 133	26	9.67 524	32	10.32 476	9.95 609	6	40
21	9.63 159	27	9.67 556	33	10.32 444	9.95 603	6	39
22	9.63 186	27	9.67 589	33	10.32 411	9.95 597	6	38
23	9.63 213	26	9.67 622	32	10.32 378	9.95 591	6	37
24	9.63 239	27	9.67 654	33	10.32 346	9.95 585	6	36
25	9.63 266	26	9.67 687	32	10.32 313	9.95 579	6	35
26	9.63 292	27	9.67 719	33	10.32 281	9.95 573	6	34
27	9.63 319	26	9.67 752	33	10.32 248	9.95 567	6	33
28	9.63 345	27	9.67 785	32	10.32 215	9.95 561	6	32
29	9.63 372	26	9.67 817	33	10.32 183	9.95 555	6	31
30	9.63 398	27	9.67 850	32	10.32 150	9.95 549	6	30
31	9.63 425	26	9.67 882	33	10.32 118	9.95 543	6	29
32	9.63 451	27	9.67 915	32	10.32 085	9.95 537	6	28
33	9.63 478	26	9.67 947	33	10.32 053	9.95 531	6	27
34	9.63 504	27	9.67 980	32	10.32 020	9.95 525	6	26
35	9.63 531	26	9.68 012	32	10.31 988	9.95 519	6	25
36	9.63 557	26	9.68 044	33	10.31 956	9.95 513	6	24
37	9.63 583	27	9.68 077	32	10.31 923	9.95 507	7	23
38	9.63 610	26	9.68 109	33	10.31 891	9.95 500	6	22
39	9.63 636	26	9.68 142	32	10.31 858	9.95 494	6	21
40	9.63 662	27	9.68 174	32	10.31 826	9.95 488	6	20
41	9.63 689	26	9.68 206	33	10.31 794	9.95 482	6	19
42	9.63 715	26	9.68 239	32	10.31 761	9.95 476	6	18
43	9.63 741	26	9.68 271	32	10.31 729	9.95 470	6	17
44	9.63 767	27	9.68 303	33	10.31 697	9.95 464	6	16
45	9.63 794	26	9.68 336	32	10.31 664	9.95 458	6	15
46	9.63 820	26	9.68 368	32	10.31 632	9.95 452	6	14
47	9.63 846	26	9.68 400	32	10.31 600	9.95 446	6	13
48	9.63 872	26	9.68 432	33	10.31 568	9.95 440	6	12
49	9.63 898	26	9.68 465	32	10.31 535	9.95 434	7	11
50	9.63 924	26	9.68 497	32	10.31 503	9.95 427	6	10
51	9.63 950	26	9.68 529	32	10.31 471	9.95 421	6	9
52	9.63 976	26	9.68 561	32	10.31 439	9.95 415	6	8
53	9.64 002	26	9.68 593	33	10.31 407	9.95 409	6	7
54	9.64 028	26	9.68 626	32	10.31 374	9.95 403	6	6
55	9.64 054	26	9.68 658	32	10.31 342	9.95 397	6	5
56	9.64 080	26	9.68 690	32	10.31 310	9.95 391	7	4
57	9.64 106	26	9.68 722	32	10.31 278	9.95 384	6	3
58	9.64 132	26	9.68 754	32	10.31 246	9.95 378	6	2
59	9.64 158	26	9.68 786	32	10.31 214	9.95 372	6	1
60	9.64 184		9 68 818		10.31 182	9.95 366		0
	L Cos	d	L Cot	c d	L Tan	L Sin	d	′

Prop. Pts.

	33	32
1	3.3	3.2
2	6.6	6.4
3	9.9	9.6
4	13.2	12.8
5	16.5	16.0
6	19.8	19.2
7	23.1	22.4
8	26.4	25.6
9	29.7	28.8

	27	26
1	2.7	2.6
2	5.4	5.2
3	8.1	7.8
4	10.8	10.4
5	13.5	13.0
6	16.2	15.6
7	18.9	18.2
8	21.6	20.8
9	24.3	23.4

	7	6	5
1	0.7	0.6	0.5
2	1.4	1.2	1.0
3	2.1	1.8	1.5
4	2.8	2.4	2.0
5	3.5	3.0	2.5
6	4.2	3.6	3.0
7	4.9	4.2	3.5
8	5.6	4.8	4.0
9	6.3	5.4	4.5

64°

TABLE 3. LOGARITHMS OF TRIGONOMETRIC FUNCTIONS (continued)   668

26°

′	L Sin	d	L Tan	c d	L Cot	L Cos	d	
0	9.64 184	26	9.68 818	32	10.31 182	9.95 366	6	60
1	9.64 210	26	9.68 850	32	10.31 150	9.95 360	6	59
2	9.64 236	26	9.68 882	32	10.31 118	9.95 354	6	58
3	9.64 262	26	9.68 914	32	10.31 086	9.95 348	7	57
4	9.64 288	25	9.68 946	32	10.31 054	9.95 341	6	56
5	9.64 313	26	9.68 978	32	10.31 022	9.95 335	6	55
6	9.64 339	26	9.69 010	32	10.30 990	9.95 329	6	54
7	9.64 365	26	9.69 042	32	10.30 958	9.95 323	6	53
8	9.64 391	26	9.69 074	32	10.30 926	9.95 317	7	52
9	9.64 417	25	9.69 106	32	10.30 894	9.95 310	6	51
10	9.64 442	26	9.69 138	32	10.30 862	9.95 304	6	50
11	9.64 468	26	9.69 170	32	10.30 830	9.95 298	6	49
12	9.64 494	25	9.69 202	32	10.30 798	9.95 292	6	48
13	9.64 519	26	9.69 234	32	10.30 766	9.95 286	7	47
14	9.64 545	26	9.69 266	32	10.30 734	9.95 279	6	46
15	9.64 571	25	9.69 298	31	10.30 702	9.95 273	6	45
16	9.64 596	26	9.69 329	32	10.30 671	9.95 267	6	44
17	9.64 622	25	9.69 361	32	10.30 639	9.95 261	7	43
18	9.64 647	26	9.69 393	32	10.30 607	9.95 254	6	42
19	9.64 673	25	9.69 425	32	10.30 575	9.95 248	6	41
20	9.64 698	26	9.69 457	31	10.30 543	9.95 242	6	40
21	9.64 724	25	9.69 488	32	10.30 512	9.95 236	7	39
22	9.64 749	26	9.69 520	32	10.30 480	9.95 229	6	38
23	9.64 775	25	9.69 552	32	10.30 448	9.95 223	6	37
24	9.64 800	26	9.69 584	31	10.30 416	9.95 217	6	36
25	9.64 826	25	9.69 615	32	10.30 385	9.95 211	7	35
26	9.64 851	26	9.69 647	32	10.30 353	9.95 204	6	34
27	9.64 877	25	9.69 679	31	10.30 321	9.95 198	6	33
28	9.64 902	25	9.69 710	32	10.30 290	9.95 192	7	32
29	9.64 927	26	9.69 742	32	10.30 258	9.95 185	6	31
30	9.64 953	25	9.69 774	31	10.30 226	9.95 179	6	30
31	9.64 978	25	9.69 805	32	10.30 195	9.95 173	6	29
32	9.65 003	26	9.69 837	31	10.30 163	9.95 167	7	28
33	9.65 029	25	9.69 868	32	10.30 132	9.95 160	6	27
34	9.65 054	25	9.69 900	32	10.30 100	9.95 154	6	26
35	9.65 079	25	9.69 932	31	10.30 068	9.95 148	7	25
36	9.65 104	26	9.69 963	32	10.30 037	9.95 141	6	24
37	9.65 130	25	9.69 995	31	10.30 005	9.95 135	6	23
38	9.65 155	25	9.70 026	32	10.29 974	9.95 129	7	22
39	9.65 180	25	9.70 058	31	10.29 942	9.95 122	6	21
40	9.65 205	25	9.70 089	32	10.29 911	9.95 116	6	20
41	9.65 230	25	9.70 121	31	10.29 879	9.95 110	7	19
42	9.65 255	26	9.70 152	32	10.29 848	9.95 103	6	18
43	9.65 281	25	9.70 184	31	10.29 816	9.95 097	7	17
44	9.65 306	25	9.70 215	32	10.29 785	9.95 090	6	16
45	9.65 331	25	9.70 247	31	10.29 753	9.95 084	6	15
46	9.65 356	25	9.70 278	31	10.29 722	9.95 078	7	14
47	9.65 381	25	9.70 309	32	10.29 691	9.95 071	6	13
48	9.65 406	25	9.70 341	31	10.29 659	9.95 065	6	12
49	9.65 431	25	9.70 372	32	10.29 628	9.95 059	7	11
50	9.65 456	25	9.70 404	31	10.29 596	9.95 052	6	10
51	9.65 481	25	9.70 435	31	10.29 565	9.95 046	7	9
52	9.65 506	25	9.70 466	32	10.29 534	9.95 039	6	8
53	9.65 531	25	9.70 498	31	10.29 502	9.95 033	6	7
54	9.65 556	24	9.70 529	31	10.29 471	9.95 027	7	6
55	9.65 580	25	9.70 560	32	10.29 440	9.95 020	6	5
56	9.65 605	25	9.70 592	31	10.29 408	9.95 014	7	4
57	9.65 630	25	9.70 623	31	10.29 377	9.95 007	6	3
58	9.65 655	25	9.70 654	31	10.29 346	9.95 001	6	2
59	9.65 680	25	9.70 685	32	10.29 315	9.94 995	7	1
60	9.65 705		9.70 717		10.29 283	9.94 988		0
	L Cos	d	L Cot	c d	L Tan	L Sin	d	′

Prop. Pts.

	32	31
1	3.2	3.1
2	6.4	6.2
3	9.6	9.3
4	12.8	12.4
5	16.0	15.5
6	19.2	18.6
7	22.4	21.7
8	25.6	24.8
9	28.8	27.9

	26	25	24
1	2.6	2.5	2.4
2	5.2	5.0	4.8
3	7.8	7.5	7.2
4	10.4	10.0	9.6
5	13.0	12.5	12.0
6	15.6	15.0	14.4
7	18.2	17.5	16.8
8	20.8	20.0	19.2
9	23.4	22.5	21.6

	7	6
1	0.7	0.6
2	1.4	1.2
3	2.1	1.8
4	2.8	2.4
5	3.5	3.0
6	4.2	3.6
7	4.9	4.2
8	5.6	4.8
9	6.3	5.4

TABLE 3. LOGARITHMS OF TRIGONOMETRIC FUNCTIONS (continued)   669

27°

′	L Sin	d	L Tan	c d	L Cot	L Cos	d	′
0	9.65 705	24	9.70 717	31	10.29 283	9.94 988	6	60
1	9.65 729	25	9.70 748	31	10.29 252	9.94 982	7	59
2	9.65 754	25	9.70 779	31	10.29 221	9.94 975	6	58
3	9.65 779	25	9.70 810	31	10.29 190	9.94 969	7	57
4	9.65 804	24	9.70 841	32	10.29 159	9.94 962	6	56
5	9.65 828	25	9.70 873	31	10.29 127	9.94 956	7	55
6	9.65 853	25	9.70 904	31	10.29 096	9.94 949	6	54
7	9.65 878	24	9.70 935	31	10.29 065	9.94 943	7	53
8	9.65 902	25	9.70 966	31	10.29 034	9.94 936	6	52
9	9.65 927	25	9.70 997	31	10.29 003	9.94 930	7	51
10	9.65 952	24	9.71 028	31	10.28 972	9.94 923	6	50
11	9.65 976	25	9.71 059	31	10.28 941	9.94 917	6	49
12	9.66 001	24	9.71 090	31	10.28 910	9.94 911	7	48
13	9.66 025	25	9.71 121	32	10.28 879	9.94 904	6	47
14	9.66 050	25	9.71 153	31	10.28 847	9.94 898	7	46
15	9.66 075	24	9.71 184	31	10.28 816	9.94 891	6	45
16	9.66 099	25	9.71 215	31	10.28 785	9.94 885	7	44
17	9.66 124	24	9.71 246	31	10.28 754	9.94 878	7	43
18	9.66 148	25	9.71 277	31	10.28 723	9.94 871	6	42
19	9.66 173	24	9.71 308	31	10.28 692	9.94 865	7	41
20	9.66 197	24	9.71 339	31	10.28 661	9.94 858	6	40
21	9.66 221	25	9.71 370	31	10.28 630	9.94 852	7	39
22	9.66 246	24	9.71 401	30	10.28 599	9.94 845	6	38
23	9.66 270	25	9.71 431	31	10.28 569	9.94 839	7	37
24	9.66 295	24	9.71 462	31	10.28 538	9.94 832	6	36
25	9.66 319	24	9.71 493	31	10.28 507	9.94 826	7	35
26	9.66 343	25	9.71 524	31	10.28 476	9.94 819	6	34
27	9.66 368	24	9.71 555	31	10.28 445	9.94 813	7	33
28	9.66 392	24	9.71 586	31	10.28 414	9.94 806	7	32
29	9.66 416	25	9.71 617	31	10.28 383	9.94 799	6	31
30	9.66 441	24	9.71 648	31	10.28 352	9.94 793	7	30
31	9.66 465	24	9.71 679	30	10.28 321	9.94 786	6	29
32	9.66 489	24	9.71 709	31	10.28 291	9.94 780	7	28
33	9.66 513	24	9.71 740	31	10.28 260	9.94 773	6	27
34	9.66 537	25	9.71 771	31	10.28 229	9.94 767	7	26
35	9.66 562	24	9.71 802	31	10.28 198	9.94 760	7	25
36	9.66 586	24	9.71 833	30	10.28 167	9.94 753	6	24
37	9.66 610	24	9.71 863	31	10.28 137	9.94 747	7	23
38	9.66 634	24	9.71 894	31	10.28 106	9.94 740	6	22
39	9.66 658	24	9.71 925	30	10.28 075	9.94 734	7	21
40	9.66 682	24	9.71 955	31	10.28 045	9.94 727	7	20
41	9.66 706	25	9.71 986	31	10.28 014	9.94 720	6	19
42	9.66 731	24	9.72 017	31	10.27 983	9.94 714	7	18
43	9.66 755	24	9.72 048	30	10.27 952	9.94 707	7	17
44	9.66 779	24	9.72 078	31	10.27 922	9.94 700	6	16
45	9.66 803	24	9.72 109	31	10.27 891	9.94 694	7	15
46	9.66 827	24	9.72 140	30	10.27 860	9.94 687	7	14
47	9.66 851	24	9.72 170	31	10.27 830	9.94 680	6	13
48	9.66 875	24	9.72 201	30	10.27 799	9.94 674	7	12
49	9.66 899	23	9.72 231	31	10.27 769	9.94 667	7	11
50	9.66 922	24	9.72 262	31	10.27 738	9.94 660	6	10
51	9.66 946	24	9.72 293	30	10.27 707	9.94 654	7	9
52	9.66 970	24	9.72 323	31	10.27 677	9.94 647	7	8
53	9.66 994	24	9.72 354	30	10.27 646	9.94 640	6	7
54	9.67 018	24	9.72 384	31	10.27 616	9.94 634	7	6
55	9.67 042	24	9.72 415	30	10.27 585	9.94 627	7	5
56	9.67 066	24	9.72 445	31	10.27 555	9.94 620	6	4
57	9.67 090	23	9.72 476	30	10.27 524	9.94 614	7	3
58	9.67 113	24	9.72 506	31	10.27 494	9.94 607	7	2
59	9.67 137	24	9.72 537	30	10.27 463	9.94 600	7	1
60	9.67 161		9.72 567		10.27 433	9.94 593		0
	L Cos	d	L Cot	c d	L Tan	L Sin	d	′

Prop. Pts.

	32	31	30
1	3.2	3.1	3.0
2	6.4	6.2	6.0
3	9.6	9.3	9.0
4	12.8	12.4	12.0
5	16.0	15.5	15.0
6	19.2	18.6	18.0
7	22.4	21.7	21.0
8	25.6	24.8	24.0
9	28.8	27.9	27.0

	25	24	23
1	2.5	2.4	2.3
2	5.0	4.8	4.6
3	7.5	7.2	6.9
4	10.0	9.6	9.2
5	12.5	12.0	11.5
6	15.0	14.4	13.8
7	17.5	16.8	16.1
8	20.0	19.2	18.4
9	22.5	21.6	20.7

	7	6
1	0.7	0.6
2	1.4	1.2
3	2.1	1.8
4	2.8	2.4
5	3.5	3.0
6	4.2	3.6
7	4.9	4.2
8	5.6	4.8
9	6.3	5.4

62°

TABLE 3. LOGARITHMS OF TRIGONOMETRIC FUNCTIONS (continued) 670

28°

′	L Sin	d	L Tan	c d	L Cot	L Cos	d		Prop. Pts.
0	9.67 161		9.72 567		10.27 433	9.94 593		60	
1	9.67 185	24	9.72 598	31	10.27 402	9.94 587	6	59	
2	9.67 208	23	9.72 628	30	10.27 372	9.94 580	7	58	
3	9.67 232	24	9.72 659	31	10.27 341	9.94 573	7	57	
4	9.67 256	24	9.72 689	30	10.27 311	9.94 567	6	56	
5	9.67 280	24	9.72 720	31	10.27 280	9.94 560	7	55	
6	9.67 303	23	9.72 750	30	10.27 250	9.94 553	7	54	
7	9.67 327	24	9.72 780	30	10.27 220	9.94 546	7	53	
8	9.67 350	23	9.72 811	31	10.27 189	9.94 540	6	52	
9	9.67 374	24	9.72 841	30	10.27 159	9.94 533	7	51	**31** **30** **29**
10	9.67 398	24	9.72 872	31	10.27 128	9.94 526	7	50	1 3.1 3.0 2.9
11	9.67 421	23	9.72 902	30	10.27 098	9.94 519	7	49	2 6.2 6.0 5.8
12	9.67 445	24	9.72 932	30	10.27 068	9.94 513	6	48	3 9.3 9.0 8.7
13	9.67 468	23	9.72 963	31	10.27 037	9.94 506	7	47	4 12.4 12.0 11.6
14	9.67 492	24	9.72 993	30	10.27 007	9.94 499	7	46	5 15.5 15.0 14.5
15	9.67 515	23	9.73 023	30	10.26 977	9.94 492	7	45	6 18.6 18.0 17.4
16	9.67 539	24	9.73 054	31	10.26 946	9.94 485	7	44	7 21.7 21.0 20.3
17	9.67 562	23	9.73 084	30	10.26 916	9.94 479	6	43	8 24.8 24.0 23.2
18	9.67 586	24	9.73 114	30	10.26 886	9.94 472	7	42	9 27.9 27.0 26.1
19	9.67 609	23	9.73 144	30	10.26 856	9.94 465	7	41	
20	9.67 633	24	9.73 175	31	10.26 825	9.94 458	7	40	
21	9.67 656	23	9.73 205	30	10.26 795	9.94 451	7	39	
22	9.67 680	24	9.73 235	30	10.26 765	9.94 445	6	38	
23	9.67 703	23	9.73 265	30	10.26 735	9.94 438	7	37	
24	9.67 726	23	9.73 295	30	10.26 705	9.94 431	7	36	
25	9.67 750	24	9.73 326	31	10.26 674	9.94 424	7	35	
26	9.67 773	23	9.73 356	30	10.26 644	9.94 417	7	34	
27	9.67 796	23	9.73 386	30	10.26 614	9.94 410	7	33	**24** **23** **22**
28	9.67 820	24	9.73 416	30	10.26 584	9.94 404	6	32	1 2.4 2.3 2.2
29	9.67 843	23	9.73 446	30	10.26 554	9.94 397	7	31	2 4.8 4.6 4.4
30	9.67 866	23	9.73 476	30	10.26 524	9.94 390	7	30	3 7.2 6.9 6.6
31	9.67 890	24	9.73 507	31	10.26 493	9.94 383	7	29	4 9.6 9.2 8.8
32	9.67 913	23	9.73 537	30	10.26 463	9.94 376	7	28	5 12.0 11.5 11.0
33	9.67 936	23	9.73 567	30	10.26 433	9.94 369	7	27	6 14.4 13.8 13.2
34	9.67 959	23	9.73 597	30	10.26 403	9.94 362	7	26	7 16.8 16.1 15.4
35	9.67 982	23	9.73 627	30	10.26 373	9.94 355	7	25	8 19.2 18.4 17.6
36	9.68 006	24	9.73 657	30	10.26 343	9.94 349	6	24	9 21.6 20.7 19.8
37	9.68 029	23	9.73 687	30	10.26 313	9.94 342	7	23	
38	9.68 052	23	9.73 717	30	10.26 283	9.94 335	7	22	
39	9.68 075	23	9.73 747	30	10.26 253	9.94 328	7	21	
40	9.68 098	23	9.73 777	30	10.26 223	9.94 321	7	20	
41	9.68 121	23	9.73 807	30	10.26 193	9.94 314	7	19	
42	9.68 144	23	9.73 837	30	10.26 163	9.94 307	7	18	
43	9.68 167	23	9.73 867	30	10.26 133	9.94 300	7	17	
44	9.68 190	23	9.73 897	30	10.26 103	9.94 293	7	16	**7** **6**
45	9.68 213	24	9.73 927	30	10.26 073	9.94 286	7	15	1 0.7 0.6
46	9.68 237	23	9.73 957	30	10.26 043	9.94 279	6	14	2 1.4 1.2
47	9.68 260	23	9.73 987	30	10.26 013	9.94 273	7	13	3 2.1 1.8
48	9.68 283	22	9.74 017	30	10.25 983	9.94 266	7	12	4 2.8 2.4
49	9.68 305	23	9.74 047	30	10.25 953	9.94 259	7	11	5 3.5 3.0
50	9.68 328	23	9.74 077	30	10.25 923	9.94 252	7	10	6 4.2 3.6
51	9.68 351	23	9.74 107	30	10.25 893	9.94 245	7	9	7 4.9 4.2
52	9.68 374	23	9.74 137	29	10.25 863	9.94 238	7	8	8 5.6 4.8
53	9.68 397	23	9.74 166	30	10.25 834	9.94 231	7	7	9 6.3 5.4
54	9.68 420	23	9.74 196	30	10.25 804	9.94 224	7	6	
55	9.68 443	23	9.74 226	30	10.25 774	9.94 217	7	5	
56	9.68 466	23	9.74 256	30	10.25 744	9.94 210	7	4	
57	9.68 489	23	9.74 286	30	10.25 714	9.94 203	7	3	
58	9.68 512	22	9.74 316	29	10.25 684	9.94 196	7	2	
59	9.68 534	23	9.74 345	30	10.25 655	9.94 189	7	1	
60	9.68 557		9.74 375		10.25 625	9.94 182		0	
	L Cos	d	L Cot	c d	L Tan	L Sin	d	′	Prop. Pts.

TABLE 3. LOGARITHMS OF TRIGONOMETRIC FUNCTIONS (continued)   671

29°

′	L Sin	d	L Tan	c d	L Cot	L Cos	d	′
0	9.68 557	23	9.74 375	30	10.25 625	9.94 182	7	**60**
1	9.68 580	23	9.74 405	30	10.25 595	9.94 175	7	59
2	9.68 603	22	9.74 435	30	10.25 565	9.94 168	7	58
3	9.68 625	23	9.74 465	29	10.25 535	9.94 161	7	57
4	9.68 648	23	9.74 494	30	10.25 506	9.94 154	7	56
5	9.68 671	23	9.74 524	30	10.25 476	9.94 147	7	55
6	9.68 694	22	9.74 554	29	10.25 446	9.94 140	7	54
7	9.68 716	23	9.74 583	30	10.25 417	9.94 133	7	53
8	9.68 739	23	9.74 613	30	10.25 387	9.94 126	7	52
9	9.68 762	22	9.74 643	30	10.25 357	9.94 119	7	51
10	9.68 784	23	9.74 673	29	10.25 327	9.94 112	7	**50**
11	9.68 807	22	9.74 702	30	10.25 298	9.94 105	7	49
12	9.68 829	23	9.74 732	30	10.25 268	9.94 098	8	48
13	9.68 852	23	9.74 762	29	10.25 238	9.94 090	7	47
14	9.68 875	22	9.74 791	30	10.25 209	9.94 083	7	46
15	9.68 897	23	9.74 821	30	10.25 179	9.94 076	7	45
16	9.68 920	22	9.74 851	29	10.25 149	9.94 069	7	44
17	9.68 942	23	9.74 880	30	10.25 120	9.94 062	7	43
18	9.68 965	22	9.74 910	29	10.25 090	9.94 055	7	42
19	9.68 987	23	9.74 939	30	10.25 061	9.94 048	7	41
20	9.69 010	22	9.74 969	29	10.25 031	9.94 041	7	**40**
21	9.69 032	23	9.74 998	30	10.25 002	9.94 034	7	39
22	9.69 055	22	9.75 028	30	10.24 972	9.94 027	7	38
23	9.69 077	23	9.75 058	29	10.24 942	9.94 020	8	37
24	9.69 100	22	9.75 087	30	10.24 913	9.94 012	7	36
25	9.69 122	22	9.75 117	29	10.24 883	9.94 005	7	35
26	9.69 144	23	9.75 146	30	10.24 854	9.93 998	7	34
27	9.69 167	22	9.75 176	29	10.24 824	9.93 991	7	33
28	9.69 189	23	9.75 205	30	10.24 795	9.93 984	7	32
29	9.69 212	22	9.75 235	29	10.24 765	9.93 977	7	31
30	9.69 234	22	9.75 264	30	10.24 736	9.93 970	7	**30**
31	9.69 256	23	9.75 294	29	10.24 706	9.93 963	8	29
32	9.69 279	22	9.75 323	30	10.24 677	9.93 955	7	28
33	9.69 301	22	9.75 353	29	10.24 647	9.93 948	7	27
34	9.69 323	22	9.75 382	29	10.24 618	9.93 941	7	26
35	9.69 345	23	9.75 411	30	10.24 589	9.93 934	7	25
36	9.69 368	22	9.75 441	29	10.24 559	9.93 927	7	24
37	9.69 390	22	9.75 470	30	10.24 530	9.93 920	8	23
38	9.69 412	22	9.75 500	29	10.24 500	9.93 912	7	22
39	9.69 434	22	9.75 529	29	10.24 471	9.93 905	7	21
40	9.69 456	23	9.75 558	30	10.24 442	9.93 898	7	**20**
41	9.69 479	22	9.75 588	29	10.24 412	9.93 891	7	19
42	9.69 501	22	9.75 617	30	10.24 383	9.93 884	8	18
43	9.69 523	22	9.75 647	29	10.24 353	9.93 876	7	17
44	9.69 545	22	9.75 676	29	10.24 324	9.93 869	7	16
45	9.69 567	22	9.75 705	30	10.24 295	9.93 862	7	15
46	9.69 589	22	9.75 735	29	10.24 265	9.93 855	8	14
47	9.69 611	22	9.75 764	29	10.24 236	9.93 847	7	13
48	9.69 633	22	9.75 793	29	10.24 207	9.93 840	7	12
49	9.69 655	22	9.75 822	30	10.24 178	9.93 833	7	11
50	9.69 677	22	9.75 852	29	10.24 148	9.93 826	7	**10**
51	9.69 699	22	9.75 881	29	10.24 119	9.93 819	8	9
52	9.69 721	22	9.75 910	29	10.24 090	9.93 811	7	8
53	9.69 743	22	9.75 939	30	10.24 061	9.93 804	7	7
54	9.69 765	22	9.75 969	29	10.24 031	9.93 797	8	6
55	9.69 787	22	9.75 998	29	10.24 002	9.93 789	7	5
56	9.69 809	22	9.76 027	29	10.23 973	9.93 782	7	4
57	9.69 831	22	9.76 056	30	10.23 944	9.93 775	7	3
58	9.69 853	22	9.76 086	29	10.23 914	9.93 768	8	2
59	9.69 875	22	9.76 115	29	10.23 885	9.93 760	7	1
60	9.69 897		9.76 144		10.23 856	9.93 753		**0**
	L Cos	d	L Cot	c d	L Tan	L Sin	d	′

Prop. Pts.

	30	29
1	3.0	2.9
2	6.0	5.8
3	9.0	8.7
4	12.0	11.6
5	15.0	14.5
6	18.0	17.4
7	21.0	20.3
8	24.0	23.2
9	27.0	26.1

	23	22
1	2.3	2.2
2	4.6	4.4
3	6.9	6.6
4	9.2	8.8
5	11.5	11.0
6	13.8	13.2
7	16.1	15.4
8	18.4	17.6
9	20.7	19.8

	8	7
1	0.8	0.7
2	1.6	1.4
3	2.4	2.1
4	3.2	2.8
5	4.0	3.5
6	4.8	4.2
7	5.6	4.9
8	6.4	5.6
9	7.2	6.3

60°

TABLE 3. LOGARITHMS OF TRIGONOMETRIC FUNCTIONS (continued)   **672**

30°

'	L Sin	d	L Tan	c d	L Cot	L Cos	d	
0	9.69 897	22	9.76 144	29	10.23 856	9.93 753	7	60
1	9.69 919	22	9.76 173	29	10.23 827	9.93 746	8	59
2	9.69 941	22	9.76 202	29	10.23 798	9.93 738	7	58
3	9.69 963	21	9.76 231	30	10.23 769	9.93 731	7	57
4	9.69 984	22	9.76 261	29	10.23 739	9.93 724	7	56
5	9.70 006	22	9.76 290	29	10.23 710	9.93 717	8	55
6	9.70 028	22	9.76 319	29	10.23 681	9.93 709	7	54
7	9.70 050	22	9.76 348	29	10.23 652	9.93 702	7	53
8	9.70 072	21	9.76 377	29	10.23 623	9.93 695	8	52
9	9.70 093	22	9.76 406	29	10.23 594	9.93 687	7	51
10	9.70 115	22	9.76 435	29	10.23 565	9.93 680	7	50
11	9.70 137	22	9.76 464	29	10.23 536	9.93 673	8	49
12	9.70 159	21	9.76 493	29	10.23 507	9.93 665	7	48
13	9.70 180	22	9.76 522	29	10.23 478	9.93 658	8	47
14	9.70 202	22	9.76 551	29	10.23 449	9.93 650	7	46
15	9.70 224	21	9.76 580	29	10.23 420	9.93 643	7	45
16	9.70 245	22	9.76 609	30	10.23 391	9.93 636	8	44
17	9.70 267	21	9.76 639	29	10.23 361	9.93 628	7	43
18	9.70 288	22	9.76 668	29	10.23 332	9.93 621	7	42
19	9.70 310	22	9.76 697	28	10.23 303	9.93 614	8	41
20	9.70 332	21	9.76 725	29	10.23 275	9.93 606	7	40
21	9.70 353	22	9.76 754	29	10.23 246	9.93 599	8	39
22	9.70 375	21	9.76 783	29	10.23 217	9.93 591	7	38
23	9.70 396	22	9.76 812	29	10.23 188	9.93 584	7	37
24	9.70 418	21	9.76 841	29	10.23 159	9.93 577	8	36
25	9.70 439	22	9.76 870	29	10.23 130	9.93 569	7	35
26	9.70 461	21	9.76 899	29	10.23 101	9.93 562	8	34
27	9.70 482	22	9.76 928	29	10.23 072	9.93 554	7	33
28	9.70 504	21	9.76 957	29	10.23 043	9.93 547	8	32
29	9.70 525	22	9.76 986	29	10.23 014	9.93 539	7	31
30	9.70 547	21	9.77 015	29	10.22 985	9.93 532	7	30
31	9.70 568	22	9.77 044	29	10.22 956	9.93 525	8	29
32	9.70 590	21	9.77 073	28	10.22 927	9.93 517	7	28
33	9.70 611	22	9.77 101	29	10.22 899	9.93 510	8	27
34	9.70 633	21	9.77 130	29	10.22 870	9.93 502	7	26
35	9.70 654	21	9.77 159	29	10.22 841	9.93 495	8	25
36	9.70 675	22	9.77 188	29	10.22 812	9.93 487	7	24
37	9.70 697	21	9.77 217	29	10.22 783	9.93 480	8	23
38	9.70 718	21	9.77 246	28	10.22 754	9.93 472	7	22
39	9.70 739	22	9.77 274	29	10.22 726	9.93 465	8	21
40	9.70 761	21	9.77 303	29	10.22 697	9.93 457	7	20
41	9.70 782	21	9.77 332	29	10.22 668	9.93 450	8	19
42	9.70 803	21	9.77 361	29	10.22 639	9.93 442	7	18
43	9.70 824	22	9.77 390	28	10.22 610	9.93 435	8	17
44	9.70 846	21	9.77 418	29	10.22 582	9.93 427	7	16
45	9.70 867	21	9.77 447	29	10.22 553	9.93 420	8	15
46	9.70 888	21	9.77 476	29	10.22 524	9.93 412	7	14
47	9.70 909	22	9.77 505	28	10.22 495	9.93 405	8	13
48	9.70 931	21	9.77 533	29	10.22 467	9.93 397	7	12
49	9.70 952	21	9.77 562	29	10.22 438	9.93 390	8	11
50	9.70 973	21	9.77 591	28	10.22 409	9.93 382	7	10
51	9.70 994	21	9.77 619	29	10.22 381	9.93 375	8	9
52	9.71 015	21	9.77 648	29	10.22 352	9.93 367	7	8
53	9.71 036	22	9.77 677	29	10.22 323	9.93 360	8	7
54	9.71 058	21	9.77 706	28	10.22 294	9.93 352	8	6
55	9.71 079	21	9.77 734	29	10.22 266	9.93 344	7	5
56	9.71 100	21	9.77 763	28	10.22 237	9.93 337	8	4
57	9.71 121	21	9.77 791	29	10.22 209	9.93 329	7	3
58	9.71 142	21	9.77 820	29	10.22 180	9.93 322	8	2
59	9.71 163	21	9.77 849	28	10.22 151	9.93 314	7	1
60	9.71 184		9.77 877		10.22 123	9.93 307		0
	L Cos	d	L Cot	c d	L Tan	L Sin	d	'

Prop. Pts.

	30	29	28
1	3.0	2.9	2.8
2	6.0	5.8	5.6
3	9.0	8.7	8.4
4	12.0	11.6	11.2
5	15.0	14.5	14.0
6	18.0	17.4	16.8
7	21.0	20.3	19.6
8	24.0	23.2	22.4
9	27.0	26.1	25.2

	22	21
1	2.2	2.1
2	4.4	4.2
3	6.6	6.3
4	8.8	8.4
5	11.0	10.5
6	13.2	12.6
7	15.4	14.7
8	17.6	16.8
9	19.8	18.9

	8	7
1	0.8	0.7
2	1.6	1.4
3	2.4	2.1
4	3.2	2.8
5	4.0	3.5
6	4.8	4.2
7	5.6	4.9
8	6.4	5.6
9	7.2	6.3

TABLE 3. LOGARITHMS OF TRIGONOMETRIC FUNCTIONS (continued)   **673**

31°

′	L Sin	d	L Tan	c d	L Cot	L Cos	d		Prop. Pts.
0	9.71 184	21	9.77 877	29	10.22 123	9.93 307	8	**60**	
1	9.71 205	21	9.77 906	29	10.22 094	9.93 299	8	59	
2	9.71 226	21	9.77 935	28	10.22 065	9.93 291	8	58	
3	9.71 247	21	9.77 963	29	10.22 037	9.93 284	7	57	
4	9.71 268	21	9.77 992	28	10.22 008	9.93 276	8	56	
5	9.71 289	21	9.78 020	29	10.21 980	9.93 269	7	55	
6	9.71 310	21	9.78 049	28	10.21 951	9.93 261	8	54	
7	9.71 331	21	9.78 077	29	10.21 923	9.93 253	7	53	
8	9.71 352	21	9.78 106	29	10.21 894	9.93 246	8	52	
9	9.71 373	20	9.78 135	28	10.21 865	9.93 238	8	51	
10	9.71 393	21	9.78 163	29	10.21 837	9.93 230	7	**50**	
11	9.71 414	21	9.78 192	28	10.21 808	9.93 223	8	49	
12	9.71 435	21	9.78 220	29	10.21 780	9.93 215	8	48	
13	9.71 456	21	9.78 249	28	10.21 751	9.93 207	7	47	
14	9.71 477	21	9.78 277	29	10.21 723	9.93 200	8	46	
15	9.71 498	21	9.78 306	28	10.21 694	9.93 192	8	45	
16	9.71 519	20	9.78 334	29	10.21 666	9.93 184	7	44	
17	9.71 539	21	9.78 363	28	10.21 637	9.93 177	8	43	
18	9.71 560	21	9.78 391	28	10.21 609	9.93 169	8	42	
19	9.71 581	21	9.78 419	29	10.21 581	9.93 161	7	41	
20	9.71 602	20	9.78 448	28	10.21 552	9.93 154	8	**40**	
21	9.71 622	21	9.78 476	29	10.21 524	9.93 146	8	39	
22	9.71 643	21	9.78 505	28	10.21 495	9.93 138	7	38	
23	9.71 664	21	9.78 533	29	10.21 467	9.93 131	8	37	
24	9.71 685	20	9.78 562	28	10.21 438	9.93 123	8	36	
25	9.71 705	21	9.78 590	28	10.21 410	9.93 115	7	35	
26	9.71 726	21	9.78 618	29	10.21 382	9.93 108	8	34	
27	9.71 747	20	9.78 647	28	10.21 353	9.93 100	8	33	
28	9.71 767	21	9.78 675	29	10.21 325	9.93 092	8	32	
29	9.71 788	21	9.78 704	28	10.21 296	9.93 084	7	31	
30	9.71 809	20	9.78 732	28	10.21 268	9.93 077	8	**30**	
31	9.71 829	21	9.78 760	29	10.21 240	9.93 069	8	29	
32	9.71 850	20	9.78 789	28	10.21 211	9.93 061	8	28	
33	9.71 870	21	9.78 817	28	10.21 183	9.93 053	7	27	
34	9.71 891	20	9.78 845	29	10.21 155	9.93 046	8	26	
35	9.71 911	21	9.78 874	28	10.21 126	9.93 038	8	25	
36	9.71 932	20	9.78 902	28	10.21 098	9.93 030	8	24	
37	9.71 952	21	9.78 930	29	10.21 070	9.93 022	8	23	
38	9.71 973	21	9.78 959	28	10.21 041	9.93 014	7	22	
39	9.71 994	20	9.78 987	28	10.21 013	9.93 007	8	21	
40	9.72 014	20	9.79 015	28	10.20 985	9.92 999	8	**20**	
41	9.72 034	20	9.79 043	29	10.20 957	9.92 991	8	19	
42	9.72 055	20	9.79 072	28	10.20 928	9.92 983	7	18	
43	9.72 075	21	9.79 100	28	10.20 900	9.92 976	8	17	
44	9.72 096	20	9.79 128	28	10.20 872	9.92 968	8	16	
45	9.72 116	21	9.79 156	29	10.20 844	9.92 960	8	15	
46	9.72 137	20	9.79 185	28	10.20 815	9.92 952	8	14	
47	9.72 157	20	9.79 213	28	10.20 787	9.92 944	8	13	
48	9.72 177	21	9.79 241	28	10.20 759	9.92 936	7	12	
49	9.72 198	20	9.79 269	28	10.20 731	9.92 929	8	11	
50	9.72 218	20	9.79 297	29	10.20 703	9.92 921	8	**10**	
51	9.72 238	21	9.79 326	28	10.20 674	9.92 913	8	9	
52	9.72 259	20	9.79 354	28	10.20 646	9.92 905	8	8	
53	9.72 279	20	9 79 382	28	10.20 618	9.92 897	8	7	
54	9.72 299	21	9 .79 410	28	10.20 590	9.92 889	8	6	
55	9.72 320	20	9 79 438	28	10.20 562	9.92 881	7	5	
56	9.72 340	20	9.79 466	29	10.20 534	9.92 874	8	4	
57	9.72 360	21	9.79 495	28	10.20 505	9.92 866	8	3	
58	9.72 381	20	9.79 523	28	10.20 477	9.92 858	8	2	
59	9.72 401	20	9.79 551	28	10.20 449	9.92 850	8	1	
60	9.72 421		9.79 579		10.20 421	9.92 842		**0**	
	L Cos	d	L Cot	c d	L Tan	L Sin	d	′	Prop. Pts.

Prop. Pts.

	29	28
1	2.9	2.8
2	5.8	5.6
3	8.7	8.4
4	11.6	11.2
5	14.5	14.0
6	17.4	16.8
7	20.3	19.6
8	23.2	22.4
9	26.1	25.2

	21	20
1	2.1	2.0
2	4.2	4.0
3	6.3	6.0
4	8.4	8.0
5	10.5	10.0
6	12.6	12.0
7	14.7	14.0
8	16.8	16.0
9	18.9	18.0

	8	7
1	0.8	0.7
2	1.6	1.4
3	2.4	2.1
4	3.2	2.8
5	4.0	3.5
6	4.8	4.2
7	5.6	4.9
8	6.4	5.6
9	7.2	6.3

58°

TABLE 3. LOGARITHMS OF TRIGONOMETRIC FUNCTIONS (continued) 674

32°

'	L Sin	d	L Tan	c d	L Cot	L Cos	d	
0	9.72 421	20	9.79 579	28	10.20 421	9.92 842	8	60
1	9.72 441	20	9.79 607	28	10.20 393	9.92 834	8	59
2	9.72 461	21	9.79 635	28	10.20 365	9.92 826	8	58
3	9.72 482	20	9.79 663	28	10.20 337	9.92 818	8	57
4	9.72 502	20	9.79 691	28	10.20 309	9.92 810	7	56
5	9.72 522	20	9.79 719	28	10.20 281	9.92 803	8	55
6	9.72 542	20	9.79 747	29	10.20 253	9.92 795	8	54
7	9.72 562	20	9.79 776	28	10.20 224	9.92 787	8	53
8	9.72 582	20	9.79 804	28	10.20 196	9.92 779	8	52
9	9.72 602	20	9.79 832	28	10.20 168	9.92 771	8	51
10	9.72 622	21	9.79 860	28	10.20 140	9.92 763	8	50
11	9.72 643	20	9.79 888	28	10.20 112	9.92 755	8	49
12	9.72 663	20	9.79 916	28	10.20 084	9.92 747	8	48
13	9.72 683	20	9.79 944	28	10.20 056	9.92 739	8	47
14	9.72 703	20	9.79 972	28	10.20 028	9.92 731	8	46
15	9.72 723	20	9.80 000	28	10.20 000	9.92 723	8	45
16	9.72 743	20	9.80 028	28	10.19 972	9.92 715	8	44
17	9.72 763	20	9.80 056	28	10.19 944	9.92 707	8	43
18	9.72 783	20	9.80 084	28	10.19 916	9.92 699	8	42
19	9.72 803	20	9.80 112	28	10.19 888	9.92 691	8	41
20	9.72 823	20	9.80 140	28	10.19 860	9.92 683	8	40
21	9.72 843	20	9.80 168	27	10.19 832	9.92 675	8	39
22	9.72 863	20	9.80 195	28	10.19 805	9.92 667	8	38
23	9.72 883	19	9.80 223	28	10.19 777	9.92 659	8	37
24	9.72 902	20	9.80 251	28	10.19 749	9.92 651	8	36
25	9.72 922	20	9.80 279	28	10.19 721	9.92 643	8	35
26	9.72 942	20	9.80 307	28	10.19 693	9.92 635	8	34
27	9.72 962	20	9.80 335	28	10.19 665	9.92 627	8	33
28	9.72 982	20	9.80 363	28	10.19 637	9.92 619	8	32
29	9.73 002	20	9.80 391	28	10.19 609	9.92 611	8	31
30	9.73 022	19	9.80 419	28	10.19 581	9.92 603	8	30
31	9.73 041	20	9.80 447	27	10.19 553	9.92 595	8	29
32	9.73 061	20	9.80 474	28	10.19 526	9.92 587	8	28
33	9.73 081	20	9.80 502	28	10.19 498	9.92 579	8	27
34	9.73 101	20	9.80 530	28	10.19 470	9.92 571	8	26
35	9.73 121	19	9.80 558	28	10.19 442	9.92 563	8	25
36	9.73 140	20	9.80 586	28	10.19 414	9.92 555	9	24
37	9.73 160	20	9.80 614	28	10.19 386	9.92 546	8	23
38	9.73 180	20	9.80 642	27	10.19 358	9.92 538	8	22
39	9.73 200	19	9.80 669	28	10.19 331	9.92 530	8	21
40	9.73 219	20	9.80 697	28	10.19 303	9.92 522	8	20
41	9.73 239	20	9.80 725	28	10.19 275	9.92 514	8	19
42	9.73 259	19	9.80 753	28	10.19 247	9.92 506	8	18
43	9.73 278	20	9.80 781	27	10.19 219	9.92 498	8	17
44	9.73 298	20	9.80 808	28	10.19 192	9.92 490	8	16
45	9.73 318	19	9.80 836	28	10.19 164	9.92 482	9	15
46	9.73 337	20	9.80 864	28	10.19 136	9.92 473	8	14
47	9.73 357	20	9.80 892	27	10.19 108	9.92 465	8	13
48	9.73 377	19	9.80 919	28	10.19 081	9.92 457	8	12
49	9.73 396	20	9.80 947	28	10.19 053	9.92 449	8	11
50	9.73 416	19	9.80 975	28	10.19 025	9.92 441	8	10
51	9.73 435	20	9.81 003	27	10.18 997	9.92 433	8	9
52	9.73 455	19	9.81 030	28	10.18 970	9.92 425	9	8
53	9.73 474	20	9.81 058	28	10.18 942	9.92 416	8	7
54	9.73 494	19	9.81 086	27	10.18 914	9.92 408	8	6
55	9.73 513	20	9.81 113	28	10.18 887	9.92 400	8	5
56	9.73 533	19	9.81 141	28	10.18 859	9.92 392	8	4
57	9.73 552	20	9.81 169	27	10.18 831	9.92 384	8	3
58	9.73 572	19	9.81 196	28	10.18 804	9.92 376	9	2
59	9.73 591	20	9.81 224	28	10.18 776	9.92 367	8	1
60	9.73 611		9.81 252		10.18 748	9.92 359		0
	L Cos	d	L Cot	c d	L Tan	L Sin	d	'

Prop. Pts.

	29	28	27
1	2.9	2.8	2.7
2	5.8	5.6	5.4
3	8.7	8.4	8.1
4	11.6	11.2	10.8
5	14.5	14.0	13.5
6	17.4	16.8	16.2
7	20.3	19.6	18.9
8	23.2	22.4	21.6
9	26.1	25.2	24.3

	21	20	19
1	2.1	2.0	1.9
2	4.2	4.0	3.8
3	6.3	6.0	5.7
4	8.4	8.0	7.6
5	10.5	10.0	9.5
6	12.6	12.0	11.4
7	14.7	14.0	13.3
8	16.8	16.0	15.2
9	18.9	18.0	17.1

	9	8	7
1	0.9	0.8	0.7
2	1.8	1.6	1.4
3	2.7	2.4	2.1
4	3.6	3.2	2.8
5	4.5	4.0	3.5
6	5.4	4.8	4.2
7	6.3	5.6	4.9
8	7.2	6.4	5.6
9	8.1	7.2	6.3

57°

TABLE 3. LOGARITHMS OF TRIGONOMETRIC FUNCTIONS (continued)   675

33°

′	L Sin	d	L Tan	c d	L Cot	L Cos	d		Prop. Pts.
0	9.73 611	19	9.81 252	27	10.18 748	9.92 359	8	60	
1	9.73 630	20	9.81 279	28	10.18 721	9.92 351	8	59	
2	9.73 650	19	9.81 307	28	10.18 693	9.92 343	8	58	
3	9.73 669	20	9.81 335	27	10.18 665	9.92 335	9	57	
4	9.73 689	19	9.81 362	28	10.18 638	9.92 326	8	56	
5	9.73 708	19	9.81 390	28	10.18 610	9.92 318	8	55	
6	9.73 727	20	9.81 418	27	10.18 582	9.92 310	8	54	
7	9.73 747	19	9.81 445	28	10.18 555	9.92 302	9	53	
8	9.73 766	19	9.81 473	27	10.18 527	9.92 293	8	52	
9	9.73 785	20	9.81 500	28	10.18 500	9.92 285	8	51	**28 | 27**
10	9.73 805	19	9.81 528	28	10.18 472	9.92 277	8	50	1 2.8 2.7
11	9.73 824	19	9.81 556	27	10.18 444	9.92 269	9	49	2 5.6 5.4
12	9.73 843	20	9.81 583	28	10.18 417	9.92 260	8	48	3 8.4 8.1
13	9.73 863	19	9.81 611	27	10.18 389	9.92 252	8	47	4 11.2 10.8
14	9.73 882	19	9.81 638	28	10.18 362	9.92 244	9	46	5 14.0 13.5
15	9.73 901	20	9.81 666	27	10.18 334	9.92 235	8	45	6 16.8 16.2
16	9.73 921	19	9.81 693	28	10.18 307	9.92 227	8	44	7 19.6 18.9
17	9.73 940	19	9.81 721	27	10.18 279	9.92 219	8	43	8 22.4 21.6
18	9.73 959	19	9.81 748	28	10.18 252	9.92 211	9	42	9 25.2 24.3
19	9.73 978	19	9.81 776	27	10.18 224	9.92 202	8	41	
20	9.73 997	20	9.81 803	28	10.18 197	9.92 194	8	40	
21	9.74 017	19	9.81 831	27	10.18 169	9.92 186	9	39	
22	9.74 036	19	9.81 858	28	10.18 142	9.92 177	8	38	
23	9.74 055	19	9.81 886	27	10.18 114	9.92 169	8	37	
24	9.74 074	19	9.81 913	28	10.18 087	9.92 161	9	36	
25	9.74 093	20	9.81 941	27	10.18 059	9.92 152	8	35	
26	9.74 113	19	9.81 968	28	10.18 032	9.92 144	8	34	**20 | 19 | 18**
27	9.74 132	19	9.81 996	27	10.18 004	9.92 136	9	33	1 2.0 1.9 1.8
28	9.74 151	19	9.82 023	28	10.17 977	9.92 127	8	32	2 4.0 3.8 3.6
29	9.74 170	19	9.82 051	27	10.17 949	9.92 119	8	31	3 6.0 5.7 5.4
30	9.74 189	19	9.82 078	28	10.17 922	9.92 111	9	30	4 8.0 7.6 7.2
31	9.74 208	19	9.82 106	27	10.17 894	9.92 102	8	29	5 10.0 9.5 9.0
32	9.74 227	19	9.82 133	28	10.17 867	9.92 094	8	28	6 12.0 11.4 10.8
33	9.74 246	19	9.82 161	27	10.17 839	9.92 086	9	27	7 14.0 13.3 12.6
34	9.74 265	19	9.82 188	27	10.17 812	9.92 077	8	26	8 16.0 15.2 14.4
35	9.74 284	19	9.82 215	28	10.17 785	9.92 069	9	25	9 18.0 17.1 16.2
36	9.74 303	19	9.82 243	27	10.17 757	9.92 060	8	24	
37	9.74 322	19	9.82 270	28	10.17 730	9.92 052	8	23	
38	9.74 341	19	9.82 298	27	10.17 702	9.92 044	9	22	
39	9.74 360	19	9.82 325	27	10.17 675	9.92 035	8	21	
40	9.74 379	19	9.82 352	28	10.17 648	9.92 027	9	20	
41	9.74 398	19	9.82 380	27	10.17 620	9.92 018	8	19	
42	9.74 417	19	9.82 407	28	10.17 593	9.92 010	8	18	
43	9.74 436	19	9.82 435	27	10.17 565	9.92 002	9	17	
44	9.74 455	19	9.82 462	27	10.17 538	9.91 993	8	16	**9 | 8**
45	9.74 474	19	9.82 489	28	10.17 511	9.91 985	9	15	1 0.9 0.8
46	9.74 493	19	9.82 517	27	10.17 483	9.91 976	8	14	2 1.8 1.6
47	9.74 512	19	9.82 544	27	10.17 456	9.91 968	9	13	3 2.7 2.4
48	9.74 531	18	9.82 571	28	10.17 429	9.91 959	8	12	4 3.6 3.2
49	9.74 549	19	9.82 599	27	10.17 401	9.91 951	9	11	5 4.5 4.0
50	9.74 568	19	9.82 626	27	10.17 374	9.91 942	8	10	6 5.4 4.8
51	9.74 587	19	9.82 653	28	10.17 347	9.91 934	9	9	7 6.3 5.6
52	9.74 606	19	9.82 681	27	10.17 319	9.91 925	8	8	8 7.2 6.4
53	9.74 625	19	9.82 708	27	10.17 292	9.91 917	9	7	9 8.1 7.2
54	9.74 644	18	9.82 735	27	10.17 265	9.91 908	8	6	
55	9.74 662	19	9.82 762	28	10.17 238	9.91 900	9	5	
56	9.74 681	19	9.82 790	27	10.17 210	9.91 891	8	4	
57	9.74 700	19	9.82 817	27	10.17 183	9.91 883	9	3	
58	9.74 719	18	9.82 844	27	10.17 156	9.91 874	8	2	
59	9.74 737	19	9.82 871	28	10.17 129	9.91 866	9	1	
60	9.74 756		9.82 899		10.17 101	9.91 857		0	
	L Cos	d	L Cot	c d	L Tan	L Sin	d	′	Prop. Pts.

TABLE 3. LOGARITHMS OF TRIGONOMETRIC FUNCTIONS (continued)   676

34°

′	L Sin	d	L Tan	c d	L Cot	L Cos	d		′
0	9.74 756		9.82 899		10.17 101	9.91 857			60
1	9.74 775	19	9.82 926	27	10.17 074	9.91 849	8		59
2	9.74 794	19	9.82 953	27	10.17 047	9.91 840	9		58
3	9.74 812	18	9.82 980	27	10.17 020	9.91 832	8		57
4	9.74 831	19	9.83 008	28	10.16 992	9.91 823	9		56
5	9.74 850	19	9.83 035	27	10.16 965	9.91 815	8		55
6	9.74 868	18	9.83 062	27	10.16 938	9.91 806	9		54
7	9.74 887	19	9.83 089	27	10.16 911	9.91 798	8		53
8	9.74 906	19	9.83 117	28	10.16 883	9.91 789	9		52
9	9.74 924	18	9.83 144	27	10.16 856	9.91 781	8		51
10	9.74 943	19	9.83 171	27	10.16 829	9.91 772	9		50
11	9.74 961	18	9.83 198	27	10.16 802	9.91 763	9		49
12	9.74 980	19	9.83 225	27	10.16 775	9.91 755	8		48
13	9.74 999	19	9.83 252	27	10.16 748	9.91 746	9		47
14	9.75 017	18	9.83 280	28	10.16 720	9.91 738	8		46
15	9.75 036	19	9.83 307	27	10.16 693	9.91 729	9		45
16	9.75 054	18	9.83 334	27	10.16 666	9.91 720	9		44
17	9.75 073	19	9.83 361	27	10.16 639	9.91 712	8		43
18	9.75 091	18	9.83 388	27	10.16 612	9.91 703	9		42
19	9.75 110	19	9.83 415	27	10.16 585	9.91 695	8		41
20	9.75 128	18	9.83 442	27	10.16 558	9.91 686	9		40
21	9.75 147	19	9.83 470	28	10.16 530	9.91 677	9		39
22	9.75 165	18	9.83 497	27	10.16 503	9.91 669	8		38
23	9.75 184	19	9.83 524	27	10.16 476	9.91 660	9		37
24	9.75 202	18	9.83 551	27	10.16 449	9.91 651	9		36
25	9.75 221	19	9.83 578	27	10.16 422	9.91 643	8		35
26	9.75 239	18	9.83 605	27	10.16 395	9.91 634	9		34
27	9.75 258	19	9.83 632	27	10.16 368	9.91 625	9		33
28	9.75 276	18	9.83 659	27	10.16 341	9.91 617	8		32
29	9.75 294	18	9.83 686	27	10.16 314	9.91 608	9		31
30	9.75 313	19	9.83 713	27	10.16 287	9.91 599	9		30
31	9.75 331	18	9.83 740	28	10.16 260	9.91 591	8		29
32	9.75 350	18	9.83 768	27	10.16 232	9.91 582	9		28
33	9.75 368	18	9.83 795	27	10.16 205	9.91 573	9		27
34	9.75 386	19	9.83 822	27	10.16 178	9.91 565	8		26
35	9.75 405	18	9.83 849	27	10.16 151	9.91 556	9		25
36	9.75 423	18	9.83 876	27	10.16 124	9.91 547	9		24
37	9.75 441	18	9.83 903	27	10.16 097	9.91 538	9		23
38	9.75 459	19	9.83 930	27	10.16 070	9.91 530	8		22
39	9.75 478	18	9.83 957	27	10.16 043	9.91 521	9		21
40	9.75 496	18	9.83 984	27	10.16 016	9.91 512	9		20
41	9.75 514	19	9.84 011	27	10.15 989	9.91 504	8		19
42	9.75 533	18	9.84 038	27	10.15 962	9.91 495	9		18
43	9.75 551	18	9.84 065	27	10.15 935	9.91 486	9		17
44	9.75 569	18	9.84 092	27	10.15 908	9.91 477	8		16
45	9.75 587	18	9.84 119	27	10.15 881	9.91 469	9		15
46	9.75 605	19	9.84 146	27	10.15 854	9.91 460	9		14
47	9.75 624	18	9.84 173	27	10.15 827	9.91 451	9		13
48	9.75 642	18	9.84 200	27	10.15 800	9.91 442	9		12
49	9.75 660	18	9.84 227	27	10.15 773	9.91 433	8		11
50	9.75 678	18	9.84 254	26	10.15 746	9.91 425	9		10
51	9.75 696	18	9.84 280	27	10.15 720	9.91 416	9		9
52	9.75 714	19	9.84 307	27	10.15 693	9.91 407	9		8
53	9.75 733	18	9.84 334	27	10.15 666	9.91 398	9		7
54	9.75 751	18	9.84 361	27	10.15 639	9.91 389	8		6
55	9.75 769	18	9.84 388	27	10.15 612	9.91 381	9		5
56	9.75 787	18	9.84 415	27	10.15 585	9.91 372	9		4
57	9.75 805	18	9.84 442	27	10.15 558	9.91 363	9		3
58	9.75 823	18	9.84 469	27	10.15 531	9.91 354	9		2
59	9.75 841	18	9.84 496	27	10.15 504	9.91 345	9		1
60	9.75 859		9.84 523		10.15 477	9.91 336			0
	L Cos	d	L Cot	c d	L Tan	L Sin	d	′	

Prop. Pts.

	28	27	26
1	2.8	2.7	2.6
2	5.6	5.4	5.2
3	8.4	8.1	7.8
4	11.2	10.8	10.4
5	14.0	13.5	13.0
6	16.8	16.2	15.6
7	19.6	18.9	18.2
8	22.4	21.6	20.8
9	25.2	24.3	23.4

	19	18
1	1.9	1.8
2	3.8	3.6
3	5.7	5.4
4	7.6	7.2
5	9.5	9.0
6	11.4	10.8
7	13.3	12.6
8	15.2	14.4
9	17.1	16.2

	9	8
1	0.9	0.8
2	1.8	1.6
3	2.7	2.4
4	3.6	3.2
5	4.5	4.0
6	5.4	4.8
7	6.3	5.6
8	7.2	6.4
9	8.1	7.2

55°

TABLE 3. LOGARITHMS OF TRIGONOMETRIC FUNCTIONS (continued)   677

35°

'	L Sin	d	L Tan	c d	L Cot	L Cos	d		Prop. Pts.
0	9.75 859	18	9.84 523	27	10.15 477	9.91 336	8	60	
1	9.75 877	18	9.84 550	26	10.15 450	9.91 328	9	59	
2	9.75 895	18	9.84 576	27	10.15 424	9.91 319	9	58	
3	9.75 913	18	9.84 603	27	10.15 397	9.91 310	9	57	
4	9.75 931	18	9.84 630	27	10.15 370	9.91 301	9	56	
5	9.75 949	18	9.84 657	27	10.15 343	9.91 292	9	55	
6	9.75 967	18	9.84 684	27	10.15 316	9.91 283	9	54	
7	9.75 985	18	9.84 711	27	10.15 289	9.91 274	8	53	
8	9.76 003	18	9.84 738	26	10.15 262	9.91 266	9	52	**27 26**
9	9.76 021	18	9.84 764	27	10.15 236	9.91 257	9	51	1 2.7 2.6
10	9.76 039	18	9.84 791	27	10.15 209	9.91 248	9	50	2 5.4 5.2
11	9.76 057	18	9.84 818	27	10.15 182	9.91 239	9	49	3 8.1 7.8
12	9.76 075	18	9.84 845	27	10.15 155	9.91 230	9	48	4 10.8 10.4
13	9.76 093	18	9.84 872	27	10.15 128	9.91 221	9	47	5 13.5 13.0
14	9.76 111	18	9.84 899	26	10.15 101	9.91 212	9	46	6 16.2 15.6
15	9.76 129	17	9.84 925	27	10.15 075	9.91 203	9	45	7 18.9 18.2
16	9.76 146	18	9.84 952	27	10.15 048	9.91 194	9	44	8 21.6 20.8
17	9.76 164	18	9.84 979	27	10.15 021	9.91 185	9	43	9 24.3 23.4
18	9.76 182	18	9.85 006	27	10.14 994	9.91 176	9	42	
19	9.76 200	18	9.85 033	26	10.14 967	9.91 167	9	41	
20	9.76 218	18	9.85 059	27	10.14 941	9.91 158	9	40	
21	9.76 236	17	9.85 086	27	10.14 914	9.91 149	8	39	
22	9.76 253	18	9.85 113	27	19.14 887	9.91 141	9	38	
23	9.76 271	18	9.85 140	26	10.14 860	9.91 132	9	37	
24	9.76 289	18	9.85 166	27	10.14 834	9.91 123	9	36	
25	9.76 307	17	9.85 193	27	10.14 807	9.91 114	9	35	
26	9.76 324	18	9.85 220	27	10.14 780	9.91 105	9	34	**18 17**
27	9.76 342	18	9.85 247	26	10.14 753	9.91 096	9	33	1 1.8 1.7
28	9.76 360	18	9.85 273	27	10.14 727	9.91 087	9	32	2 3.6 3.4
29	9.76 378	17	9.85 300	27	10.14 700	9.91 078	9	31	3 5.4 5.1
30	9.76 395	18	9.85 327	27	10.14 673	9.91 069	9	30	4 7.2 6.8
31	9.76 413	18	9.85 354	26	10.14 646	9.91 060	9	29	5 9.0 8.5
32	9.76 431	17	9.85 380	27	10.14 620	9.91 051	9	28	6 10.8 10.2
33	9.76 448	18	9.85 407	27	10.14 593	9.91 042	9	27	7 12.6 11.9
34	9.76 466	18	9.85 434	26	10.14 566	9.91 033	10	26	8 14.4 13.6
35	9.76 484	17	9.85 460	27	10.14 540	9.91 023	9	25	9 16.2 15.3
36	9.76 501	18	9.85 487	27	10.14 513	9.91 014	9	24	
37	9.76 519	18	9.85 514	26	10.14 486	9.91 005	9	23	
38	9.76 537	17	9.85 540	27	10.14 460	9.90 996	9	22	
39	9.76 554	18	9.85 567	27	10.14 433	9.90 987	9	21	
40	9.76 572	18	9.85 594	26	10.14 406	9.90 978	9	20	
41	9.76 590	17	9.85 620	27	10.14 380	9.90 969	9	19	
42	9.76 607	18	9.85 647	27	10.14 353	9.90 960	9	18	
43	9.76 625	17	9.85 674	26	10.14 326	9.90 951	9	17	
44	9.76 642	18	9.85 700	27	10.14 300	9.90 942	9	16	**10 9 8**
45	9.76 660	17	9.85 727	27	10.14 273	9.90 933	9	15	1 1.0 0.9 0.8
46	9.76 677	18	9.85 754	26	10.14 246	9.90 924	9	14	2 2.0 1.8 1.6
47	9.76 695	17	9.85 780	27	10.14 220	9.90 915	9	13	3 3.0 2.7 2.4
48	9.76 712	18	9.85 807	27	10.14 193	9.90 906	10	12	4 4.0 3.6 3.2
49	9.76 730	17	9.85 834	26	10.14 166	9.90 896	9	11	5 5.0 4.5 4.0
50	9.76 747	18	9.85 860	27	10.14 140	9.90 887	9	10	6 6.0 5.4 4.8
51	9.76 765	17	9.85 887	26	10.14 113	9.90 878	9	9	7 7.0 6.3 5.6
52	9.76 782	18	9.85 913	27	10.14 087	9.90 869	9	8	8 8.0 7.2 6.4
53	9.76 800	17	9.85 940	27	10.14 060	9.90 860	9	7	9 9.0 8.1 7.2
54	9.76 817	18	9.85 967	26	10.14 033	9.90 851	9	6	
55	9.76 835	17	9.85 993	27	10.14 007	9.90 842	10	5	
56	9.76 852	18	9.86 020	26	10.13 980	9.90 832	9	4	
57	9.76 870	17	9.86 046	27	10.13 954	9.90 823	9	3	
58	9.76 887	17	9.86 073	27	10.13 927	9.90 814	9	2	
59	9.76 904	18	9.86 100	26	10.13 900	9.90 805	9	1	
60	9.76 922		9.86 126		10.13 874	9.90 796		0	
	L Cos	d	L Cot	c d	L Tan	L Sin	d	'	Prop. Pts.

54°

TABLE 3. LOGARITHMS OF TRIGONOMETRIC FUNCTIONS (continued)   **678**

36°

'	L Sin	d	L Tan	c d	L Cot	L Cos	d	'
0	9.76 922	17	9.86 126	27	10.13 874	9.90 796	9	60
1	9.76 939	18	9.86 153	26	10.13 847	9.90 787	10	59
2	9.76 957	17	9.86 179	27	10.13 821	9.90 777	9	58
3	9.76 974	17	9.86 206	26	10.13 794	9.90 768	9	57
4	9.76 991	18	9.86 232	27	10.13 768	9.90 759	9	56
5	9.77 009	17	9.86 259	26	10.13 741	9.90 750	9	55
6	9.77 026	17	9.86 285	27	10.13 715	9.90 741	10	54
7	9.77 043	18	9.86 312	26	10.13 688	9.90 731	9	53
8	9.77 061	17	9.86 338	27	10.13 662	9.90 722	9	52
9	9.77 078	17	9.86 365	27	10.13 635	9.90 713	9	51
10	9.77 095	17	9.86 392	26	10.13 608	9.90 704	10	50
11	9.77 112	18	9.86 418	27	10.13 582	9.90 694	9	49
12	9.77 130	17	9.86 445	26	10.13 555	9.90 685	9	48
13	9.77 147	17	9.86 471	27	10.13 529	9.90 676	9	47
14	9.77 164	17	9.86 498	26	10.13 502	9.90 667	10	46
15	9.77 181	18	9.86 524	27	10.13 476	9.90 657	9	45
16	9.77 199	17	9.86 551	26	10.13 449	9.90 648	9	44
17	9.77 216	17	9.86 577	26	10.13 423	9.90 639	9	43
18	9.77 233	17	9.86 603	27	10.13 397	9.90 630	10	42
19	9.77 250	18	9.86 630	26	10.13 370	9.90 620	9	41
20	9.77 268	17	9.86 656	27	10.13 344	9.90 611	9	40
21	9.77 285	17	9.86 683	26	10.13 317	9.90 602	10	39
22	9.77 302	17	9.86 709	27	10.13 291	9.90 592	9	38
23	9.77 319	17	9.86 736	26	10.13 264	9.90 583	9	37
24	9.77 336	17	9.86 762	27	10.13 238	9.90 574	9	36
25	9.77 353	17	9.86 789	26	10.13 211	9.90 565	10	35
26	9.77 370	17	9.86 815	27	10.13 185	9.90 555	9	34
27	9.77 387	18	9.86 842	26	10.13 158	9.90 546	9	33
28	9.77 405	17	9.86 868	26	10.13 132	9.90 537	10	32
29	9.77 422	17	9.86 894	27	10.13 106	9.90 527	9	31
30	9.77 439	17	9.86 921	26	10.13 079	9.90 518	9	30
31	9.77 456	17	9.86 947	27	10.13 053	9.90 509	10	29
32	9.77 473	17	9.86 974	26	10.13 026	9.90 499	9	28
33	9.77 490	17	9.87 000	27	10.13 000	9.90 490	10	27
34	9.77 507	17	9.87 027	26	10.12 973	9.90 480	9	26
35	9.77 524	17	9.87 053	26	10.12 947	9.90 471	9	25
36	9.77 541	17	9.87 079	27	10.12 921	9.90 462	10	24
37	9.77 558	17	9.87 106	26	10.12 894	9.90 452	9	23
38	9.77 575	17	9.87 132	26	10.12 868	9.90 443	9	22
39	9.77 592	17	9.87 158	27	10.12 842	9.90 434	10	21
40	9.77 609	17	9.87 185	26	10.12 815	9.90 424	9	20
41	9.77 626	17	9.87 211	27	10.12 789	9.90 415	10	19
42	9.77 643	17	9.87 238	26	10.12 762	9.90 405	9	18
43	9.77 660	17	9.87 264	26	10.12 736	9.90 396	10	17
44	9.77 677	17	9.87 290	27	10.12 710	9.90 386	9	16
45	9.77 694	17	9.87 317	26	10.12 683	9.90 377	9	15
46	9.77 711	17	9.87 343	26	10.12 657	9.90 368	10	14
47	9.77 728	16	9.87 369	27	10.12 631	9.90 358	9	13
48	9.77 744	17	9.87 396	26	10.12 604	9.90 349	10	12
49	9.77 761	17	9.87 422	26	10.12 578	9.90 339	9	11
50	9.77 778	17	9.87 448	27	10.12 552	9.90 330	10	10
51	9.77 795	17	9.87 475	26	10.12 525	9.90 320	9	9
52	9.77 812	17	9.87 501	26	10.12 499	9.90 311	10	8
53	9.77 829	17	9.87 527	27	10.12 473	9.90 301	9	7
54	9.77 846	16	9.87 554	26	10.12 446	9.90 292	10	6
55	9.77 862	17	9.87 580	26	10.12 420	9.90 282	9	5
56	9.77 879	17	9.87 606	27	10.12 394	9.90 273	10	4
57	9.77 896	17	9.87 633	26	10.12 367	9.90 263	9	3
58	9.77 913	17	9.87 659	26	10.12 341	9.90 254	10	2
59	9.77 930	16	9.87 685	26	10.12 315	9.90 244	9	1
60	9.77 946		9.87 711		10.12 289	9.90 235		0
	L Cos	d	L Cot	c d	L Tan	L Sin	d	'

Prop. Pts.

	27	26
1	2.7	2.6
2	5.4	5.2
3	8.1	7.8
4	10.8	10.4
5	13.5	13.0
6	16.2	15.6
7	18.9	18.2
8	21.6	20.8
9	24.3	23.4

	18	17	16
1	1.8	1.7	1.6
2	3.6	3.4	3.2
3	5.4	5.1	4.8
4	7.2	6.8	6.4
5	9.0	8.5	8.0
6	10.8	10.2	9.6
7	12.6	11.9	11.2
8	14.4	13.6	12.8
9	16.2	15.3	14.4

	10	9
1	1.0	0.9
2	2.0	1.8
3	3.0	2.7
4	4.0	3.6
5	5.0	4.5
6	6.0	5.4
7	7.0	6.3
8	8.0	7.2
9	9.0	8.1

TABLE 3. LOGARITHMS OF TRIGONOMETRIC FUNCTIONS (continued)   **679**

37°

′	L Sin	d	L Tan	c d	L Cot	L Cos	d		Prop. Pts.
0	9.77 946	17	9.87 711	27	10.12 289	9.90 235	10	60	
1	9.77 963	17	9.87 738	26	10.12 262	9.90 225	9	59	
2	9.77 980	17	9.87 764	26	10.12 236	9.90 216	10	58	
3	9.77 997	16	9.87 790	27	10.12 210	9.90 206	9	57	
4	9.78 013	17	9.87 817	26	10.12 183	9.90 197	10	56	
5	9.78 030	17	9.87 843	26	10.12 157	9.90 187	9	55	
6	9.78 047	16	9.87 869	26	10.12 131	9.90 178	10	54	
7	9.78 063	17	9.87 895	27	10.12 105	9.90 168	9	53	
8	9.78 080	17	9.87 922	26	10.12 078	9.90 159	10	52	
9	9.78 097	16	9.87 948	26	10.12 052	9.90 149	10	51	
10	9.78 113	17	9.87 974	26	10.12 026	9.90 139	9	50	**27 26**
11	9.78 130	17	9.88 000	27	10.12 000	9.90 130	10	49	1 2.7 2.6
12	9.78 147	16	9.88 027	26	10.11 973	9.90 120	9	48	2 5.4 5.2
13	9.78 163	17	9.88 053	26	10.11 947	9.90 111	10	47	3 8.1 7.8
14	9.78 180	17	9.88 079	26	10.11 921	9.90 101	10	46	4 10.8 10.4
15	9.78 197	16	9.88 105	26	10.11 895	9.90 091	9	45	5 13.5 13.0
16	9.78 213	17	9.88 131	27	10.11 869	9.90 082	10	44	6 16.2 15.6
17	9.78 230	16	9.88 158	26	10.11 842	9.90 072	9	43	7 18.9 18.2
18	9.78 246	17	9.88 184	26	10.11 816	9.90 063	10	42	8 21.6 20.8
19	9.78 263	17	9.88 210	26	10.11 790	9.90 053	10	41	9 24.3 23.4
20	9.78 280	16	9.88 236	26	10.11 764	9.90 043	9	40	
21	9.78 296	17	9.88 262	27	10.11 738	9.90 034	10	39	
22	9.78 313	16	9.88 289	26	10.11 711	9.90 024	10	38	
23	9.78 329	17	9.88 315	26	10.11 685	9.90 014	9	37	
24	9.78 346	16	9.88 341	26	10.11 659	9.90 005	10	36	
25	9.78 362	17	9.88 367	26	10.11 633	9.89 995	10	35	
26	9.78 379	16	9.88 393	27	10.11 607	9.89 985	9	34	**17 16**
27	9.78 395	17	9.88 420	26	10.11 580	9.89 976	10	33	1 1.7 1.6
28	9.78 412	16	9.88 446	26	10.11 554	9.89 966	10	32	2 3.4 3.2
29	9.78 428	17	9.88 472	26	10.11 528	9.89 956	9	31	3 5.1 4.8
30	9.78 445	16	9.88 498	26	10.11 502	9.89 947	10	30	4 6.8 6.4
31	9.78 461	17	9.88 524	26	10.11 476	9.89 937	10	29	5 8.5 8.0
32	9.78 478	16	9.88 550	27	10.11 450	9.89 927	9	28	6 10.2 9.6
33	9.78 494	16	9.88 577	26	10.11 423	9.89 918	10	27	7 11.9 11.2
34	9.78 510	17	9.88 603	26	10.11 397	9.89 908	10	26	8 13.6 12.8
35	9.78 527	16	9.88 629	26	10.11 371	9.89 898	10	25	9 15.3 14.4
36	9.78 543	17	9.88 655	26	10.11 345	9.89 888	9	24	
37	9.78 560	16	9.88 681	26	10.11 319	9.89 879	10	23	
38	9.78 576	16	9.88 707	26	10.11 293	9.89 869	10	22	
39	9.78 592	17	9.88 733	26	10.11 267	9.89 859	10	21	
40	9.78 609	16	9.88 759	27	10.11 241	9.89 849	9	20	
41	9.78 625	17	9.88 786	26	10.11 214	9.89 840	10	19	
42	9.78 642	16	9.88 812	26	10.11 188	9.89 830	10	18	
43	9.78 658	16	9.88 838	26	10.11 162	9.89 820	10	17	**10 9**
44	9.78 674	17	9.88 864	26	10.11 136	9.89 810	9	16	1 1.0 0.9
45	9.78 691	16	9.88 890	26	10.11 110	9.89 801	10	15	2 2.0 1.8
46	9.78 707	16	9.88 916	26	10.11 084	9.89 791	10	14	3 3.0 2.7
47	9.78 723	16	9.88 942	26	10.11 058	9.89 781	10	13	4 4.0 3.6
48	9.78 739	17	9.88 968	26	10.11 032	9.89 771	10	12	5 5.0 4.5
49	9.78 756	16	9.88 994	26	10.11 006	9.89 761	9	11	6 6.0 5.4
50	9.78 772	16	9.89 020	26	10.10 980	9.89 752	10	10	7 7.0 6.3
51	9.78 788	17	9.89 046	27	10.10 954	9.89 742	10	9	8 8.0 7.2
52	9.78 805	16	9.89 073	26	10.10 927	9.89 732	10	8	9 9.0 8.1
53	9.78 821	16	9.89 099	26	10.10 901	9.89 722	10	7	
54	9.78 837	16	9.89 125	26	10.10 875	9.89 712	10	6	
55	9.78 853	16	9.89 151	26	10.10 849	9.89 702	9	5	
56	9.78 869	17	9.89 177	26	10.10 823	9.89 693	10	4	
57	9.78 886	16	9.89 203	26	10.10 797	9.89 683	10	3	
58	9.78 902	16	9.89 229	26	10.10 771	9.89 673	10	2	
59	9.78 918	16	9.89 255	26	10.10 745	9.89 663	10	1	
60	9.78 934		9.89 281		10.10 719	9.89 653		0	
	L Cos	d	L Cot	c d	L Tan	L Sin	d	′	Prop. Pts.

TABLE 3. LOGARITHMS OF TRIGONOMETRIC FUNCTIONS (continued)   680

38°

′	L Sin	d	L Tan	c d	L Cot	L Cos	d		Prop. Pts.
0	9.78 934		9.89 281		10.10 719	9.89 653		60	
1	9.78 950	16	9.89 307	26	10.10 693	9.89 643	10	59	
2	9.78 967	17	9.89 333	26	10.10 667	9.89 633	10	58	
3	9.78 983	16	9.89 359	26	10.10 641	9.89 624	9	57	
4	9.78 999	16	9.89 385	26	10.10 615	9.89 614	10	56	
5	9.79 015	16	9.89 411	26	10.10 589	9.89 604	10	55	
6	9.79 031	16	9.89 437	26	10.10 563	9.89 594	10	54	
7	9.79 047	16	9.89 463	26	10.10 537	9.89 584	10	53	
8	9.79 063	16	9.89 489	26	10.10 511	9.89 574	10	52	26 25
9	9.79 079	16	9.89 515	26	10.10 485	9.89 564	10	51	1 2.6 2.5
10	9.79 095	16	9.89 541	26	10.10 459	9.89 554	10	50	2 5.2 5.0
11	9.79 111	16	9.89 567	26	10.10 433	9.89 544	10	49	3 7.8 7.5
12	9.79 128	17	9.89 593	26	10.10 407	9.89 534	10	48	4 10.4 10.0
13	9.79 144	16	9.89 619	26	10.10 381	9.89 524	10	47	5 13.0 12.5
14	9.79 160	16	9.89 645	26	10.10 355	9.89 514	10	46	6 15.6 15.0
15	9.79 176	16	9.89 671	26	10.10 329	9.89 504	9	45	7 18.2 17.5
16	9.79 192	16	9.89 697	26	10.10 303	9.89 495	10	44	8 20.8 20.0
17	9.79 208	16	9.89 723	26	10.10 277	9.89 485	10	43	9 23.4 22.5
18	9.79 224	16	9.89 749	26	10.10 251	9.89 475	10	42	
19	9.79 240	16	9.89 775	26	10.10 225	9.89 465	10	41	
20	9.79 256	16	9.89 801	26	10.10 199	9.89 455	10	40	
21	9.79 272	16	9.89 827	26	10.10 173	9.89 445	10	39	
22	9.79 288	16	9.89 853	26	10.10 147	9.89 435	10	38	
23	9.79 304	16	9.89 879	26	10.10 121	9.89 425	10	37	
24	9.79 319	15	9.89 905	26	10.10 095	9.89 415	10	36	
25	9.79 335	16	9.89 931	26	10.10 069	9.89 405	10	35	
26	9.79 351	16	9.89 957	26	10.10 043	9.89 395	10	34	17 16 15
27	9.79 367	16	9.89 983	26	10.10 017	9.89 385	10	33	1 1.7 1.6 1.5
28	9.79 383	16	9.90 009	26	10.09 991	9.89 375	11	32	2 3.4 3.2 3.0
29	9.79 399	16	9.90 035	26	10.09 965	9.89 364	10	31	3 5.1 4.8 4.5
30	9.79 415	16	9.90 061	25	10.09 939	9.89 354	10	30	4 6.8 6.4 6.0
31	9.79 431	16	9.90 086	26	10.09 914	9.89 344	10	29	5 8.5 8.0 7.5
32	9.79 447	16	9.90 112	26	10.09 888	9.89 334	10	28	6 10.2 9.6 9.0
33	9.79 463	15	9.90 138	26	10.09 862	9.89 324	10	27	7 11.9 11.2 10.5
34	9.79 478	16	9.90 164	26	10.09 836	9.89 314	10	26	8 13.6 12.8 12.0
35	9.79 494	16	9.90 190	26	10.09 810	9.89 304	10	25	9 15.3 14.4 13.5
36	9.79 510	16	9.90 216	26	10.09 784	9.89 294	10	24	
37	9.79 526	16	9.90 242	26	10.09 758	9.89 284	10	23	
38	9.79 542	16	9.90 268	26	10.09 732	9.89 274	10	22	
39	9.79 558	15	9.90 294	26	10.09 706	9.89 264	10	21	
40	9.79 573	16	9.90 320	26	10.09 680	9.89 254	10	20	
41	9.79 589	16	9.90 346	25	10.09 654	9.89 244	11	19	
42	9.79 605	16	9.90 371	26	10.09 629	9.89 233	10	18	
43	9.79 621	15	9.90 397	26	10.09 603	9.89 223	10	17	
44	9.79 636	16	9.90 423	26	10.09 577	9.89 213	10	16	11 10 9
45	9.79 652	16	9.90 449	26	10.09 551	9.89 203	10	15	1 1.1 1.0 0.9
46	9.79 668	16	9.90 475	26	10.09 525	9.89 193	10	14	2 2.2 2.0 1.8
47	9.79 684	15	9.90 501	26	10.09 499	9.89 183	10	13	3 3.3 3.0 2.7
48	9.79 699	16	9.90 527	26	10.09 473	9.89 173	11	12	4 4.4 4.0 3.6
49	9.79 715	16	9.90 553	25	10.09 447	9.89 162	10	11	5 5.5 5.0 4.5
50	9.79 731	15	9.90 578	26	10.09 422	9.89 152	10	10	6 6.6 6.0 5.4
51	9.79 746	16	9.90 604	26	10.09 396	9.89 142	10	9	7 7.7 7.0 6.3
52	9.79 762	16	9.90 630	26	10.09 370	9.89 132	10	8	8 8.8 8.0 7.2
53	9.79 778	15	9.90 656	26	10.09 344	9.89 122	10	7	9 9.9 9.0 8.1
54	9.79 793	16	9.90 682	26	10.09 318	9.89 112	11	6	
55	9.79 809	16	9.90 708	26	10.09 292	9.89 101	10	5	
56	9.79 825	15	9.90 734	25	10.09 266	9.89 091	10	4	
57	9.79 840	16	9.90 759	26	10.09 241	9.89 081	10	3	
58	9.79 856	16	9.90 785	26	10.09 215	9.89 071	11	2	
59	9.79 872	15	9.90 811	26	10.09 189	9.89 060	10	1	
60	9.79 887		9.90 837		10.09 163	9.89 050		0	
	L Cos	d	L Cot	c d	L Tan	L Sin	d	′	Prop. Pts.

TABLE 3. LOGARITHMS OF TRIGONOMETRIC FUNCTIONS (continued)   **681**

39°

′	L Sin	d	L Tan	c d	L Cot	L Cos	d		Prop. Pts.
0	9.79 887		9.90 837		10.09 163	9.89 050		**60**	
1	9.79 903	16	9.90 863	26	10.09 137	9.89 040	10	59	
2	9.79 918	15	9.90 889	26	10.09 111	9.89 030	10	58	
3	9.79 934	16	9.90 914	25	10.09 086	9.89 020	10	57	
4	9.79 950	16	9.90 940	26	10.09 060	9.89 009	11	56	
5	9.79 965	15	9.90 966	26	10.09 034	9.88 999	10	55	
6	9.79 981	16	9.90 992	26	10.09 008	9.88 989	10	54	
7	9 79 996	15	9.91 018	26	10.08 982	9.88 978	11	53	
8	9.80 012	16	9.91 043	25	10.08 957	9.88 968	10	52	**26** | **25**
9	9.80 027	15	9.91 069	26	10.08 931	9.88 958	10	51	1| 2.6| 2.5
10	9.80 043	16	9.91 095	26	10.08 905	9.88 948	10	**50**	2| 5.2| 5.0
11	9.80 058	15	9.91 121	26	10.08 879	9.88 937	11	49	3| 7.8| 7.5
12	9.80 074	16	9.91 147	26	10.08 853	9.88 927	10	48	4|10.4|10.0
13	9.80 089	15	9.91 172	25	10.08 828	9.88 917	10	47	5|13.0|12.5
14	9.80 105	16	9.91 198	26	10.08 802	9.88 906	11	46	6|15.6|15.0
15	9.80 120	15	9.91 224	26	10.08 776	9.88 896	10	45	7|18.2|17.5
16	9.80 136	16	9.91 250	26	10.08 750	9.88 886	10	44	8|20.8|20.0
17	9.80 151	15	9.91 276	26	10.08 724	9.88 875	11	43	9|23.4|22.5
18	9.80 166	15	9.91 301	25	10.08 699	9.88 865	10	42	
19	9.80 182	16	9.91 327	26	10.08 673	9.88 855	10	41	
20	9.80 197	15	9.91 353	26	10.08 647	9.88 844	11	**40**	
21	9.80 213	16	9.91 379	26	10.08 621	9.88 834	10	39	
22	9.80 228	15	9.91 404	25	10.08 596	9.88 824	10	38	
23	9.80 244	16	9.91 430	26	10.08 570	9.88 813	11	37	
24	9.80 259	15	9.91 456	26	10.08 544	9.88 803	10	36	
25	9.80 274	15	9.91 482	26	10.08 518	9.88 793	10	35	
26	9.80 290	16	9.91 507	25	10.08 493	9.88 782	11	34	**16** | **15**
27	9.80 305	15	9.91 533	26	10.08 467	9.88 772	10	33	1| 1.6| 1.5
28	9.80 320	15	9.91 559	26	10.08 441	9.88 761	11	32	2| 3.2| 3.0
29	9.80 336	16	9.91 585	26	10.08 415	9.88 751	10	31	3| 4.8| 4.5
30	9.80 351	15	9.91 610	25	10.08 390	9.88 741	10	**30**	4| 6.4| 6.0
31	9.80 366	15	9.91 636	26	10.08 364	9.88 730	11	29	5| 8.0| 7.5
32	9.80 382	16	9.91 662	26	10.08 338	9.88 720	10	28	6| 9.6| 9.0
33	9.80 397	15	9.91 688	26	10.08 312	9.88 709	11	27	7|11.2|10.5
34	9.80 412	15	9.91 713	25	10.08 287	9.88 699	10	26	8|12.8|12.0
35	9.80 428	16	9.91 739	26	10.08 261	9.88 688	11	25	9|14.4|13.5
36	9.80 443	15	9.91 765	26	10.08 235	9.88 678	10	24	
37	9.80 458	15	9.91 791	26	10.08 209	9.88 668	10	23	
38	9.80 473	15	9.91 816	25	10.08 184	9.88 657	11	22	
39	9.80 489	16	9.91 842	26	10.08 158	9.88 647	10	21	
40	9.80 504	15	9.91 868	26	10.08 132	9.88 636	11	**20**	
41	9.80 519	15	9.91 893	25	10.08 107	9.88 626	10	19	
42	9.80 534	15	9.91 919	26	10.08 081	9.88 615	11	18	
43	9.80 550	16	9.91 945	26	10.08 055	9.88 605	10	17	
44	9.80 565	15	9.91 971	26	10.08 029	9.88 594	11	16	**11** | **10**
45	9.80 580	15	9.91 996	25	10.08 004	9.88 584	10	15	1|1.1|1.0
46	9.80 595	15	9.92 022	26	10.07 978	9.88 573	11	14	2|2.2|2.0
47	9.80 610	15	9.92 048	25	10.07 952	9.88 563	10	13	3|3.3|3.0
48	9.80 625	16	9.92 073	26	10.07 927	9.88 552	11	12	4|4.4|4.0
49	9.80 641	15	9.92 099	26	10.07 901	9.88 542	10	11	5|5.5|5.0
50	9.80 656	15	9.92 125	25	10.07 875	9.88 531	11	**10**	6|6.6|6.0
51	9.80 671	15	9.92 150	26	10.07 850	9.88 521	11	9	7|7.7|7.0
52	9.80 686	15	9.92 176	26	10.07 824	9.88 510	11	8	8|8.8|8.0
53	9.80 701	15	9.92 202	25	10.07 798	9.88 499	10	7	9|9.9|9.0
54	9.80 716	15	9.92 227	26	10.07 773	9.88 489	11	6	
55	9.80 731	15	9.92 253	26	10.07 747	9.88 478	10	5	
56	9.80 746	16	9.92 279	25	10.07 721	9.88 468	11	4	
57	9.80 762	15	9.92 304	26	10.07 696	9.88 457	10	3	
58	9.80 777	15	9.92 330	26	10.07 670	9.88 447	11	2	
59	9.80 792	15	9.92 356	25	10.07 644	9.88 436	11	1	
60	9.80 807		9.92 381		10.07 619	9.88 425		**0**	
	L Cos	d	L Cot	c d	L Tan	L Sin	d	′	Prop. Pts.

50°

TABLE 3. LOGARITHMS OF TRIGONOMETRIC FUNCTIONS (continued)   682

40°

′	L Sin	d	L Tan	c d	L Cot	L Cos	d	′
0	9.80 807	15	9.92 381	26	10.07 619	9.88 425	10	60
1	9.80 822	15	9.92 407	26	10.07 593	9.88 415	11	59
2	9.80 837	15	9.92 433	25	10.07 567	9.88 404	10	58
3	9.80 852	15	9.92 458	26	10.07 542	9.88 394	11	57
4	9.80 867	15	9.92 484	26	10.07 516	9.88 383	11	56
5	9.80 882	15	9.92 510	25	10.07 490	9.88 372	10	55
6	9.80 897	15	9.92 535	26	10.07 465	9.88 362	11	54
7	9.80 912	15	9.92 561	26	10.07 439	9.88 351	11	53
8	9.80 927	15	9.92 587	25	10.07 413	9.88 340	10	52
9	9.80 942	15	9.92 612	26	10.07 388	9.88 330	11	51
10	9.80 957	15	9.92 638	25	10.07 362	9.88 319	11	50
11	9.80 972	15	9.92 663	26	10.07 337	9.88 308	10	49
12	9.80 987	15	9.92 689	26	10.07 311	9.88 298	11	48
13	9.81 002	15	9.92 715	25	10.07 285	9.88 287	11	47
14	9.81 017	15	9.92 740	26	10.07 260	9.88 276	10	46
15	9.81 032	15	9.92 766	26	10.07 234	9.88 266	11	45
16	9.81 047	14	9.92 792	25	10.07 208	9.88 255	11	44
17	9.81 061	15	9.92 817	26	10.07 183	9.88 244	10	43
18	9.81 076	15	9.92 843	25	10.07 157	9.88 234	11	42
19	9.81 091	15	9.92 868	26	10.07 132	9.88 223	11	41
20	9.81 106	15	9.92 894	26	10.07 106	9.88 212	11	40
21	9.81 121	15	9.92 920	25	10.07 080	9.88 201	10	39
22	9.81 136	15	9.92 945	26	10.07 055	9.88 191	11	38
23	9.81 151	15	9.92 971	25	10.07 029	9.88 180	11	37
24	9.81 166	14	9.92 996	26	10.07 004	9.88 169	11	36
25	9.81 180	15	9.93 022	26	10.06 978	9.88 158	10	35
26	9.81 195	15	9.93 048	25	10.06 952	9.88 148	11	34
27	9.81 210	15	9.93 073	26	10.06 927	9.88 137	11	33
28	9.81 225	15	9.93 099	25	10.06 901	9.88 126	11	32
29	9.81 240	14	9.93 124	26	10.06 876	9.88 115	10	31
30	9.81 254	15	9.93 150	25	10.06 850	9.88 105	11	30
31	9.81 269	15	9.93 175	26	10.06 825	9.88 094	11	29
32	9.81 284	15	9.93 201	26	10.06 799	9.88 083	11	28
33	9.81 299	15	9.93 227	25	10.06 773	9.88 072	11	27
34	9.81 314	14	9.93 252	26	10.06 748	9.88 061	10	26
35	9.81 328	15	9.93 278	25	10.06 722	9.88 051	11	25
36	9.81 343	15	9.93 303	26	10.06 697	9.88 040	11	24
37	9.81 358	14	9.93 329	25	10.06 671	9.88 029	11	23
38	9.81 372	15	9.93 354	26	10.06 646	9.88 018	11	22
39	9.81 387	15	9.93 380	26	10.06 620	9.88 007	11	21
40	9.81 402	15	9.93 406	25	10.06 594	9.87 996	11	20
41	9.81 417	14	9.93 431	26	10.06 569	9.87 985	10	19
42	9.81 431	15	9.93 457	25	10.06 543	9.87 975	11	18
43	9.81 446	15	9.93 482	26	10.06 518	9.87 964	11	17
44	9.81 461	14	9.93 508	25	10.06 492	9.87 953	11	16
45	9.81 475	15	9.93 533	26	10.06 467	9.87 942	11	15
46	9.81 490	15	9.93 559	25	10.06 441	9.87 931	11	14
47	9.81 505	14	9.93 584	26	10.06 416	9.87 920	11	13
48	9.81 519	15	9.93 610	26	10.06 390	9.87 909	11	12
49	9.81 534	15	9.93 636	25	10.06 364	9.87 898	11	11
50	9.81 549	14	9.93 661	26	10.06 339	9.87 887	10	10
51	9.81 563	15	9.93 687	25	10.06 313	9.87 877	11	9
52	9.81 578	14	9.93 712	26	10.06 288	9.87 866	11	8
53	9.81 592	15	9.93 738	25	10.06 262	9.87 855	11	7
54	9.81 607	15	9.93 763	26	10.06 237	9.87 844	11	6
55	9.81 622	14	9.93 789	25	10.06 211	9.87 833	11	5
56	9.81 636	15	9.93 814	26	10.06 186	9.87 822	11	4
57	9.81 651	14	9.93 840	26	10.06 160	9.87 811	11	3
58	9.81 665	15	9.93 866	25	10.06 135	9.87 800	11	2
59	9.81 680	14	9.93 891	25	10.06 109	9.87 789	11	1
60	9.81 694		9.93 916		10.06 084	9.87 778		0
	L Cos	d	L Cot	c d	L Tan	L Sin	d	′

Prop. Pts.

	26	25
1	2.6	2.5
2	5.2	5.0
3	7.8	7.5
4	10.4	10.0
5	13.0	12.5
6	15.6	15.0
7	18.2	17.5
8	20.8	20.0
9	23.4	22.5

	15	14
1	1.5	1.4
2	3.0	2.8
3	4.5	4.2
4	6.0	5.6
5	7.5	7.0
6	9.0	8.4
7	10.5	9.8
8	12.0	11.2
9	13.5	12.6

	11	10
1	1.1	1.0
2	2.2	2.0
3	3.3	3.0
4	4.4	4.0
5	5.5	5.0
6	6.6	6.0
7	7.7	7.0
8	8.8	8.0
9	9.9	9.0

TABLE 3. LOGARITHMS OF TRIGONOMETRIC FUNCTIONS (continued)   683

41°

′	L Sin	d	L Tan	c d	L Cot	L Cos	d	′
0	9.81 694	15	9.93 916	26	10.06 084	9.87 778	11	60
1	9.81 709	14	9.93 942	25	10.06 058	9.87 767	11	59
2	9.81 723	15	9.93 967	26	10.06 033	9.87 756	11	58
3	9.81 738	14	9.93 993	25	10.06 007	9.87 745	11	57
4	9.81 752	15	9.94 018	26	10.05 982	9.87 734	11	56
5	9.81 767	14	9.94 044	25	10.05 956	9.87 723	11	55
6	9.81 781	15	9.94 069	26	10.05 931	9.87 712	11	54
7	9.81 796	14	9.94 095	25	10.05 905	9.87 701	11	53
8	9.81 810	15	9.94 120	26	10.05 880	9.87 690	11	52
9	9.81 825	14	9.94 146	25	10.05 854	9.87 679	11	51
10	9.81 839	15	9.94 171	26	10.05 829	9.87 668	11	50
11	9.81 854	14	9.94 197	25	10.05 803	9.87 657	11	49
12	9.81 868	14	9.94 222	26	10.05 778	9.87 646	11	48
13	9.81 882	15	9.94 248	25	10.05 752	9.87 635	11	47
14	9.81 897	14	9.94 273	26	10.05 727	9.87 624	11	46
15	9.81 911	15	9.94 299	25	10.05 701	9.87 613	12	45
16	9.81 926	14	9.94 324	26	10.05 676	9.87 601	11	44
17	9.81 940	15	9.94 350	25	10.05 650	9.87 590	11	43
18	9.81 955	14	9.94 375	26	10.05 625	9.87 579	11	42
19	9.81 969	14	9.94 401	25	10.05 599	9.87 568	11	41
20	9.81 983	15	9.94 426	26	10.05 574	9.87 557	11	40
21	9.81 998	14	9.94 452	25	10.05 548	9.87 546	11	39
22	9.82 012	14	9.94 477	26	10.05 523	9.87 535	11	38
23	9.82 026	15	9.94 503	25	10.05 497	9.87 524	11	37
24	9.82 041	14	9.94 528	26	10.05 472	9.87 513	12	36
25	9.82 055	14	9.94 554	25	10.05 446	9.87 501	11	35
26	9.82 069	15	9.94 579	26	10.05 421	9.87 490	11	34
27	9.82 084	14	9.94 604	26	10.05 396	9.87 479	11	33
28	9.82 098	14	9.94 630	25	10.05 370	9.87 468	11	32
29	9.82 112	14	9.94 655	26	10.05 345	9.87 457	11	31
30	9.82 126	15	9.94 681	25	10.05 319	9.87 446	12	30
31	9.82 141	14	9.94 706	26	10.05 294	9.87 434	11	29
32	9.82 155	14	9.94 732	25	10.05 268	9.87 423	11	28
33	9.82 169	15	9.94 757	26	10.05 243	9.87 412	11	27
34	9.82 184	14	9.94 783	25	10.05 217	9.87 401	11	26
35	9.82 198	14	9.94 808	26	10.05 192	9.87 390	12	25
36	9.82 212	14	9.94 834	25	10.05 166	9.87 378	11	24
37	9.82 226	14	9.94 859	25	10.05 141	9.87 367	11	23
38	9.82 240	15	9.94 884	26	10.05 116	9.87 356	11	22
39	9.82 255	14	9.94 910	25	10.05 090	9.87 345	11	21
40	9.82 269	14	9.94 935	26	10.05 065	9.87 334	12	20
41	9.82 283	14	9.94 961	25	10.05 039	9.87 322	11	19
42	9.82 297	14	9.94 986	26	10.05 014	9.87 311	11	18
43	9.82 311	15	9.95 012	25	10.04 988	9.87 300	12	17
44	9.82 326	14	9.95 037	25	10.04 963	9.87 288	11	16
45	9.82 340	14	9.95 062	26	10.04 938	9.87 277	11	15
46	9.82 354	14	9.95 088	25	10.04 912	9.87 266	11	14
47	9.82 368	14	9.95 113	26	10.04 887	9.87 255	12	13
48	9.82 382	14	9.95 139	25	10.04 861	9.87 243	11	12
49	9.82 396	14	9.95 164	26	10.04 836	9.87 232	11	11
50	9.82 410	14	9.95 190	25	10.04 810	9.87 221	12	10
51	9.82 424	15	9.95 215	25	10.04 785	9.87 209	11	9
52	9.82 439	14	9.95 240	26	10.04 760	9.87 198	11	8
53	9.82 453	14	9.95 266	25	10.04 734	9.87 187	12	7
54	9.82 467	14	9.95 291	26	10.04 709	9.87 175	11	6
55	9.82 481	14	9.95 317	25	10.04 683	9.87 164	11	5
56	9.82 495	14	9.95 342	26	10.04 658	9.87 153	12	4
57	9.82 509	14	9.95 368	25	10.04 632	9.87 141	11	3
58	9.82 523	14	9.95 393	25	10.04 607	9.87 130	11	2
59	9.82 537	14	9.95 418	26	10.04 582	9.87 119	12	1
60	9.82 551		9.95 444		10.04 556	9.87 107		0
′	L Cos	d	L Cot	c d	L Tan	L Sin	d	′

Prop. Pts.

	26	25
1	2.6	2.5
2	5.2	5.0
3	7.8	7.5
4	10.4	10.0
5	13.0	12.5
6	15.6	15.0
7	18.2	17.5
8	20.8	20.0
9	23.4	22.5

	15	14
1	1.5	1.4
2	3.0	2.8
3	4.5	4.2
4	6.0	5.6
5	7.5	7.0
6	9.0	8.4
7	10.5	9.8
8	12.0	11.2
9	13.5	12.6

	12	11
1	1.2	1.1
2	2.4	2.2
3	3.6	3.3
4	4.8	4.4
5	6.0	5.5
6	7.2	6.6
7	8.4	7.7
8	9.6	8.8
9	10.8	9.9

48°

TABLE 3. LOGARITHMS OF TRIGONOMETRIC FUNCTIONS (continued) **684**

42°

′	L Sin	d	L Tan	c d	L Cot	L Cos	d	
0	9.82 551	14	9.95 444	25	10.04 556	9.87 107	11	60
1	9.82 565	14	9.95 469	26	10.04 531	9.87 096	11	59
2	9.82 579	14	9.95 495	25	10.04 505	9.87 085	12	58
3	9.82 593	14	9.95 520	25	10.04 480	9.87 073	11	57
4	9.82 607	14	9.95 545	26	10.04 455	9.87 062	12	56
5	9.82 621	14	9.95 571	25	10.04 429	9.87 050	11	55
6	9.82 635	14	9.95 596	26	10.04 404	9.87 039	11	54
7	9.82 649	14	9.95 622	25	10.04 378	9.87 028	12	53
8	9.82 663	14	9.95 647	25	10.04 353	9.87 016	11	52
9	9.82 677	14	9.95 672	26	10.04 328	9.87 005	12	51
10	9.82 691	14	9.95 698	25	10.04 302	9.86 993	11	50
11	9.82 705	14	9.95 723	25	10.04 277	9.86 982	12	49
12	9.82 719	14	9.95 748	26	10.04 252	9.86 970	11	48
13	9.82 733	14	9.95 774	25	10.04 226	9.86 959	12	47
14	9.82 747	14	9.95 799	26	10.04 201	9.86 947	11	46
15	9.82 761	14	9.95 825	25	10.04 175	9.86 936	12	45
16	9.82 775	13	9.95 850	25	10.04 150	9.86 924	11	44
17	9.82 788	14	9.95 875	26	10.04 125	9.86 913	11	43
18	9.82 802	14	9.95 901	25	10.04 099	9.86 902	12	42
19	9.82 816	14	9.95 926	26	10.04 074	9.86 890	11	41
20	9.82 830	14	9.95 952	25	10.04 048	9.86 879	12	40
21	9.82 844	14	9.95 977	25	10.04 023	9.86 867	12	39
22	9.82 858	14	9.96 002	26	10.03 998	9.86 855	11	38
23	9.82 872	13	9.96 028	25	10.03 972	9.86 844	12	37
24	9.82 885	14	9.96 053	25	10.03 947	9.86 832	11	36
25	9.82 899	14	9.96 078	26	10.03 922	9.86 821	12	35
26	9.82 913	14	9.96 104	25	10.03 896	9.86 809	11	34
27	9.82 927	14	9.96 129	26	10.03 871	9.86 798	12	33
28	9.82 941	14	9.96 155	25	10.03 845	9.86 786	11	32
29	9.82 955	13	9.96 180	25	10.03 820	9.86 775	12	31
30	9.82 968	14	9.96 205	26	10.03 795	9.86 763	11	30
31	9.82 982	14	9.96 231	25	10.03 769	9.86 752	12	29
32	9.82 996	14	9.96 256	25	10.03 744	9.86 740	12	28
33	9.83 010	13	9.96 281	26	10.03 719	9.86 728	11	27
34	9.83 023	14	9.96 307	25	10.03 693	9.86 717	12	26
35	9.83 037	14	9.96 332	25	10.03 668	9.86 705	11	25
36	9.83 051	14	9.96 357	26	10.03 643	9.86 694	12	24
37	9.83 065	13	9.96 383	25	10.03 617	9.86 682	12	23
38	9.83 078	14	9.96 408	25	10.03 592	9.86 670	11	22
39	9.83 092	14	9.96 433	26	10.03 567	9.86 659	12	21
40	9.83 106	14	9.96 459	25	10.03 541	9.86 647	12	20
41	9.83 120	13	9.96 484	26	10.03 516	9.86 635	11	19
42	9.83 133	14	9.96 510	25	10.03 490	9.86 624	12	18
43	9.83 147	14	9.96 535	25	10.03 465	9.86 612	12	17
44	9.83 161	13	9.96 560	26	10.03 440	9.86 600	11	16
45	9.83 174	14	9.96 586	25	10.03 414	9.86 589	12	15
46	9.83 188	14	9.96 611	25	10.03 389	9.86 577	12	14
47	9.83 202	13	9.96 636	26	10.03 364	9.86 565	11	13
48	9.83 215	14	9.96 662	25	10.03 338	9.86 554	12	12
49	9.83 229	13	9.96 687	25	10.03 313	9.86 542	12	11
50	9.83 242	14	9.96 712	26	10.03 288	9.86 530	12	10
51	9.83 256	14	9.96 738	25	10.03 262	9.86 518	11	9
52	9.83 270	13	9.96 763	25	10.03 237	9.86 507	12	8
53	9.83 283	14	9.96 788	26	10.03 212	9.86 495	12	7
54	9.83 297	13	9.96 814	25	10.03 186	9.86 483	11	6
55	9.83 310	14	9.96 839	25	10.03 161	9.86 472	12	5
56	9.83 324	14	9.96 864	26	10.03 136	9.86 460	12	4
57	9.83 338	13	9.96 890	25	10.03 110	9.86 448	12	3
58	9.83 351	14	9.96 915	25	10.03 085	9.86 436	11	2
59	9.83 365	13	9.96 940	26	10.03 060	9.86 425	12	1
60	9.83 378		9.96 966		10.03 034	9.86 413		0
	L Cos	d	L Cot	c d	L Tan	L Sin	d	′

Prop. Pts.

	26	25
1	2.6	2.5
2	5.2	5.0
3	7.8	7.5
4	10.4	10.0
5	13.0	12.5
6	15.6	15.0
7	18.2	17.5
8	20.8	20.0
9	23.4	22.5

	14	13
1	1.4	1.3
2	2.8	2.6
3	4.2	3.9
4	5.6	5.2
5	7.0	6.5
6	8.4	7.8
7	9.8	9.1
8	11.2	10.4
9	12.6	11.7

	12	11
1	1.2	1.1
2	2.4	2.2
3	3.6	3.3
4	4.8	4.4
5	6.0	5.5
6	7.2	6.6
7	8.4	7.7
8	9.6	8.8
9	10.8	9.9

47°

index